FOREWORD

Over the past five years or so, there has been a renewed interest in the use and development of supercritical fluid extraction (SFE) methods for industrial applications. The early laboratory successes did not, in general, translate into successful commercial processes. This situation cast a discouraging shadow on SFE as an emerging technology. It is now clear that the applications for which SFE is considered must be carefully chosen. In general, high cost, low volume commodities that require the use of a nontoxic solvent in their processing are potential candidates. Examples include pharmaceuticals, flavors, foods and essential oils. In addition, the use of SFE as a tool for environmental control and cleanup is being investigated, as well as applications in polymer processing.

It is clear that SFE processes are often constrained to narrow regions of feasibility due to the phase behavior of the solvent-solute system. This in turn restricts a potential industrial process to narrow windows of economic feasibility. The result of these observations has been the need for much tighter controls on both the design and control of processes using SFE. If the specifications for a given process must be more exact, then the information required to generate the design must be more exact. This includes experimental measurements, theories, and correlations (the union of experimental data with an appropriate theoretically valid model).

In this volume, we have collected a series of reviews that cover both experimental and theoretical work geared toward the more exact requirements of current SFE applications. While we have artificially divided the volume into experimental and theoretical sections, natural overlaps will be apparent. Many of the papers on experimental technique contain discussions of equation of state correlations. Indeed, a good deal of the experimental work is intimately tied to a mathematical description of fluid mixtures.

The theoretical section presents reviews that cover the modern theory of critical phenomena, methods to correlate near critical experimental results and approaches to understanding the behavior of near critical fluids from microscopic theory. It is hoped that the scope of these reviews will provide the reader with the basis to further develop our understanding of the behavior of supercritical fluids.

T. J. Bruno
J. F. Ely
Boulder, Colorado

CONTRIBUTORS

Aydin Akgerman
Department of Chemical Engineering
Texas A&M University
College Station, Texas

Eric J. Beckman
Chemical Sciences Department
Battelle, Pacific Northwest Laboratories
Richland, Washington

Joan F. Brennecke
Department of Chemical Engineering
University of Notre Dame
Notre Dame, Indiana

Thomas J. Bruno
Thermophysics Division
National Institute of Standards and
 Technology
Boulder, Colorado

Carlos A. Nieto de Castro
Departmento de Quimica
Faculdade de Ciencias da Universidade
 de Lisboa
Lisboa, Portugal

Henry D. Cochran
Chemical Technology Division
Oak Ridge National Laboratory
Oak Ridge, Tennessee

Miriam L. Cygnarowicz
United States Department of Agriculture
Agricultural Research Service
Eastern Regional Research Center
Philadelphia, Pennsylvania

Pablo G. Debenedetti
Department of Chemical Engineering
Princeton University
Princeton, New Jersey

Charles A. Eckert
School of Chemical Engineering
Georgia Institute of Technology
Atlanta, Georgia

Michael P. Ekart
Department of Chemical Engineering
University of Illinois
Urbana, Illinois

James F. Ely
Thermophysics Division
National Institute of Standards and
 Technology
Boulder, Colorado

John L. Fulton
Chemical Sciences Department
Battelle, Pacific Northwest Laboratories
Richland, Washington

Richard K. Hess
Department of Chemical Engineering
Texas A&M University
College Station, Texas

Michael C. Jones
Chemical Engineering Science Division
National Institute of Standards and
 Technology
Boulder, Colorado

Michael T. Klein
Center for Catalytic Science and
 Technology
Department of Chemical Engineering
University of Delaware
Newark, Delaware

Concetta LaMarca
Center for Catalytic Science and
 Technology
Department of Chemical Engineering
University of Delaware
Newark, Delaware

Lloyd L. Lee
School of Chemical Engineering and
 Materials Science
University of Oklahoma
Norman, Oklahoma

Joe W. Magee
Thermophysics Division
National Institute of Standards and
  Technology
Boulder, Colorado

G. Ali Mansoori
Department of Chemical Engineering
University of Illinois at Chicago
Chicago, Illinois

Eloy E. Martinelli
Department of Chemical Engineering
University of Illinois at Chicago
Chicago, Illinois

Stephen C. Paspek
BP America Research
Cleveland, Ohio

James C. Rainwater
Thermophysics Division
National Institute of Standards and
  Technology
Boulder, Colorado

Robert K. Roop
Department of Chemical Engineering
Texas A&M University
College Station, Texas

Karl Schulz
Department of Chemical Engineering
University of Illinois at Chicago
Chicago, Illinois

Warren D. Seider
Department of Chemical Engineering
University of Pennsylvania
Philadelphia, Pennsylvania

J. M. H. Levelt Sengers
Thermophysics Division
National Institute of Standards and
  Technology
Gaithersburg, Maryland

K. S. Shing
Department of Chemical Engineering
University of Southern California
Los Angeles, California

Richard D. Smith
Chemical Sciences Department
Battelle, Pacific Northwest Laboratories
Richland, Washington

V. Vesovic
IUPAC Transport Properties Project
  Centre
Department of Chemical Engineering and
  Chemical Technology
Imperial College
London, England

W. A. Wakeham
IUPAC Transport Properties Project
  Centre
Department of Chemical Engineering and
  Chemical Technology
Imperial College
London, England

Benjamin C. Wu
Center for Catalytic Science and
  Technology
Department of Chemical Engineering
University of Delaware
Newark, Delaware

Sang-Do Yeo
Department of Chemical Engineering
Texas A&M University
College Station, Texas

TABLE OF CONTENTS

Part I:
Theory of Supercritical Fluids

Chapter 1

THERMODYNAMICS OF SOLUTIONS NEAR THE SOLVENT'S CRITICAL POINT

J. M. H. Levelt Sengers

TABLE OF CONTENTS

I. INTRODUCTION

This chapter is a treatment of the behavior of dilute mixtures and solutions near the solvent's critical point. Although these mixtures form only a modest subset of the supercritical mixtures that are used in the supercritical process technology, their interest ranges beyond the confines of chemical technology. Dilute aqueous solutions, for instance, are encountered in the power industry, oceanography, and geology; when these systems are taken to the vicinity of the critical point of steam (647 K and 22 MPa), as they are in practice, the framework of solution physical chemistry, carefully erected during the 20th century, shakes on its foundation; unexpected, often spectacular effects have been reported.

Dilute solutions, likewise, are only a subclass of fluid mixtures; the thermodynamic foundations and physical models for describing fluid mixtures date to the last decades of the 19th century. Nevertheless, the large effects impurities have on highly compressible mixtures have constantly deluded and eluded scientists. Although van der Waals and his school developed a basic understanding of dilute near-critical mixtures, this knowledge had a tendency to disappear until a new, striking effect was found. In the worst case, sensational and very erroneous theories would result (for a review, see Reference 1). In the best cases,[2-5] the effect was recognized and understood as a quirk of dilute mixtures.

The development of solution physical chemistry in the early part of the 20th century led to the introduction of new concepts, such as excess, apparent, and partial molar properties; infinite-dilution standard states; and electrolyte theory. All of these theories are based on comparing the properties of a mixture with those of an "ideal" or "model" mixture at the same pressure and temperature. That many of these concepts become useless near the critical point of the solvent, however, is a relatively new insight.[6-10] As a consequence, dilute-mixture effects in near-critical electrolyte solutions are sometimes ascribed to Coulombic forces, although they are common to all dilute near-critical mixtures.

Dilute near-critical mixtures are also worth studying for fundamental reasons. They display schizoid behavior, because they "have to make up their minds" on whether they are, first and foremost, dilute, or rather, critical. They form one of the very few examples where the values of apparently well-defined thermodynamic functions depend on the path. They challenge our intuition and preconceived notions. This chapter is an attempt to give attention to all aspects mentioned.

Section II deals solely with the critical behavior of pure fluids and will serve to introduce all properties, concepts, and methods that are necessary to subsequent sections. Properties include thermodynamic first and second derivatives and the correlation function. Emphasis is given to the difference between density- and field-like properties, and to critical divergences and their dependence on the path of approach to the critical point. An expansion of the classical Helmholtz free energy is worked out in some detail, while the generalization to nonclassical behavior is summarized. This section is also intended to alert the experimenter to the particular difficulties and sources of error associated with the very large compressibility and expansion coefficient of a near-critical fluid.

Section III concerns the generalization to the critical behavior of fluid mixtures. An

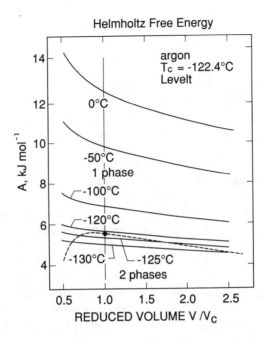

FIGURE 1.   The Helmholtz free energy of an SCF flattens out as a function of volume as the critical point is approached. Tie lines are shown in the two-phase region. The data are for argon.[12-14]

effort is made to introduce all thermodynamic properties that are encountered in supercritical fluid (SCF) technology, and to describe the character of their critical behavior in dependence on the path. Sections II and III make heavy use of the paper of Griffiths and Wheeler,[11] which laid the foundation for applying the modern theory of critical phenomena to fluid mixtures.

Section IV discusses dilute mixtures, and concentrates on what makes them different from mixtures in general. Several methods, but principally an expansion of the classical Helmholtz free energy at the critical point of the solvent, are introduced that permit the derivation of critical anomalies in the properties that are of interest to SCF technology. These properties include Henry's constants, infinite-dilution K factors, partial molar properties, osmotic susceptibilities, supercritical solubility, and the so-called total and direct correlation function integrals. The results of a generalization to nonclassical critical behavior are summarized. Aqueous systems will be frequently used as examples; there is a rich store of data, and dilute-mixture effects are very large because the components are very unlike.

Each of the sections ends with a summary. The reader might begin by reading the summary, and use it to select the subsections that draw his/her interest. The reader who is principally interested in dilute near-critical mixtures might begin by reading the last section, and draw on preceding material as required.

II. CRITICAL BEHAVIOR OF ONE-COMPONENT FLUIDS

A. CRITICALITY

For a one-component fluid to be stable, it is necessary that the isothermal Helmholtz free energy as a function of volume curve upwards (Figure 1). This means that the second derivative, $(\partial^2 A/\partial V^2)_T$, must be positive. This derivative is proportional to the inverse of the isothermal compressibility, K_T,

$$A_{VV} = (\partial^2 A/\partial V^2)_T = (VK_T)^{-1} \tag{1}$$

with

$$K_T = -(1/V)(\partial V/\partial p)_T = (VA_{VV})^{-1} \tag{2}$$

and V the molar volume, A the molar Helmholtz free energy, and p the pressure. When a critical point is approached from the one-phase region, A_{VV} approaches the value zero, and the compressibility becomes infinite. At the critical point, the fluid thus reaches a limit of stability. The third derivative of the free energy must therefore also be equal to zero; if it were not, the compressibility would be negative on the critical isotherm on one side of the critical point, which would violate stability. At criticality, the free energy is still concave upwards, because nonzero derivatives higher than the fourth make it curve upwards. It is, however, much flatter than it is at ordinary stable points (Figure 1). Below the critical point, the Helmholtz free energy develops a straight portion that is tangent to the two branches of the isothermal free energy, the liquid branch and the vapor branch. The system separates into the two phases, located on the dashed coexistence curve in Figure 1. Since $(\partial^2 A/\partial V^2)_T = 0$, the compressibility is infinite throughout the two-phase region.

The infinity in the compressibility spawns infinities in the expansion coefficient, $\alpha_p = (1/V)(\partial V/\partial T)_p$, and in the constant-pressure heat capacity, $C_p = (\partial H/\partial T)_p$, with H the molar enthalpy. This follows from the thermodynamic relations[15]

$$\alpha_p = (\partial p/\partial T)_V \cdot K_T \tag{3}$$

$$C_p = C_V + TV \, \alpha_p \cdot (\partial p/\partial T)_V \tag{4}$$

and the facts that in fluids $(\partial p/\partial T)_V$ is finite (its critical value being equal to the critical slope of the vapor pressure curve),[16,17] and C_V at most is weakly divergent (see Section II.G).

B. POWER LAWS, CRITICAL EXPONENTS, AND AMPLITUDES

The limiting behavior of thermodynamic properties near a critical point is described by means of power laws. Five parameters are required to define a power law: the property in question, Q, measured with respect to its critical value, Q_c, if it does not diverge; one independent variable, r_1, measured with respect to its critical value, r_{1c}; the path of approach, in the form of other independent variables r_2 held constant, or some other constraint; a non-negative critical exponent ϵ; and an amplitude E. The prototype power law is of the form:

$$|(Q - Q_c)/Q_c| = E \, |(r_1 - r_{1c})/r_{1c}|^{+\epsilon} \text{ on the path } r_2 = r_{2c}$$

$$(Q_c \text{ finite})$$

$$|Q^*| = E \, |(r_1 - r_{1c})/r_{1c}|^{-\epsilon} \text{ on the path } r_2 = r_{2c}$$

$$(Q \text{ divergent}) \tag{5}$$

The + sign preceding ϵ refers to properties that are finite, the − sign to properties that diverge at the critical point. The asterisk refers to a reduced property, a property that has been made dimensionless by dividing by an appropriate combination of critical parameters p_c, V_c, T_c. If variables are expressed in reduced units the critical amplitudes become dimensionless numbers. Thus, for instance, the divergence of the compressibility is expressed by the power law

$$K_T^* = p_c\, K_T \simeq \Gamma \mid (T - T_c)/T_c) \mid^{-\gamma} \text{ on the critical isochore, } \rho = \rho_c \qquad (6)$$

while the manner in which the orthobaric densities, ρ_1 of the liquid and ρ_v of the vapor, approach the critical density, ρ_c, at the critical point is described by

$$\mid (\rho_{\ell,v} - \rho_c)/\rho_c \mid \simeq B \mid (T - T_c)/T_c \mid^\beta; \text{ coex. phases} \qquad (7)$$

The coefficients Γ in Equation 6 and B in Equation 7 are dimensionless critical amplitudes that are, in general, different from substance to substance.

The critical point acts as a focal point for critical divergences. One may view it as a mountaintop rising above the plane of the independent variables. The steepness of the path of approach may vary, but the nearness of the top will be felt on any path. Thus, it is not possible for a critical divergence to appear only on an isochoric path, but not on an isothermal path to the critical point, or vice versa.

It is one of the triumphs of the theory of critical phenomena that only two universal exponents and two nonuniversal amplitudes are required to characterize the critical behavior of all thermodynamic properties along any path for all fluids and fluid mixtures. All other exponents and amplitudes on any path can be expressed in terms of these two. This principle goes by the name "two-scale-factor universality".[18-21] The so-called critical-point scaling laws are compact formulations of these power laws. Since they have been extensively reported and reviewed in the existing literature,[22,23] they are not discussed in this chapter.

The definitions of the critical exponents are given in Table 1.

C. EXPANSION OF A CLASSICAL HELMHOLTZ FREE ENERGY

At this point, some critical exponent values will be calculated for a simple model, namely that of a so-called classical Helmholtz free energy, a free energy of the van der Waals type, that can be expanded in a Taylor series in its independent variables at the critical point. Although real fluids exhibit nonclassical critical behavior and their free energies cannot be expanded, working through such a classical expansion is still a very useful exercise, because many of the important concepts, such as power laws, critical exponents and their path dependence, and the anomalous critical properties of dilute mixtures, can be obtained easily and transparently. The expansion of the molar Helmholtz free energy A(V,T) takes the following form:

$$A(V,T) = A(V_c,T_c) + A_V^c \cdot (\delta V) + A_T^c \cdot (\delta T) + A_{TT}^c \cdot (\delta T)^2/2 + \ldots$$

$$+ A_{VT}^c \cdot (\delta V)(\delta T) + A_{VVT}^c \cdot (\delta V)^2(\delta T)/2 + \ldots$$

$$+ A_{VVVV}^c \cdot (\delta V)^4/24 + \ldots \qquad (8)$$

where the subscripted indices V,T denote (repeated) partial differentiation of the Helmholtz free energy with respect to the independent variables, and the superscript c denotes critical values of these partial derivatives. Also, $\delta V = V - V_c$ and $\delta T = T - T_c$. The terms in A_{VV}^c and A_{VVV}^c are absent because these derivatives are zero at the critical point. The critical pressure equals $-A_V^c$, the critical entropy $-A_T^c$ and the critical slopes of the vapor pressure curve and critical isochore equal $-A_{VT}^c$.

It follows immediately from Equation 8 that

$$(VK_T)^{-1} = A_{VV} \approx A_{VVT}^c \cdot (\delta T) + A_{VVVV}^c \cdot (\delta V)^2/2 + \ldots \qquad (9)$$

TABLE 1
Critical Exponents and Amplitudes

Definitions

Property	Power law	Path	Exponent value	
			Classical	Nonclassical
Thermodynamic				
Isothermal compressibility	$K_T = \Gamma \, |\delta T|^{-\gamma}$	Critical isochore	$\gamma = 1$	$\gamma = 1.239 \pm 0.002$
Isochoric heat capacity	$C_V = A^{\pm} \, |\delta T|^{-\alpha}$	Critical isochore	$\alpha = 0$ nondiv.	$\alpha = 0.110 \pm 0.003$
Coexisting densities	$\rho_1 - \rho_v = 2B \, |\delta T|^{\beta}$	Two-phase	$\beta = 1/2$	$\beta = 0.326 \pm 0.002$
Pressure	$|\delta P| = D \, |\delta\rho|^{\delta}$	Critical isotherm	$\delta = 3$	$\delta = 4.80 \pm 0.02$
Fluctuation				
Correlation function	$H(r) \approx 1/r^{d-2+\eta}$	Critical point, r large	$\eta = 0$	$\eta = 0.031 \pm 0.004$
Correlation length	$H(r) \approx e^{-r/\xi}/r$	Noncritical, r large	$\nu = 0.5$	$\nu = 0.630 \pm 0.001$
	$\xi = \xi_o \, |\delta T|^{-\nu}$	Critical isochore		

Exponent Equalities

Thermodynamic	Fluctuation
$\gamma = \beta \, (\delta - 1)$	$\gamma = \nu \, (2 - \eta)$
$2 - \alpha = \beta \, (\delta + 1)$	$d\nu = (2 - \alpha)$

Two-Scale-Factor Universal Amplitude Relations

Relation	$A\Gamma/B^2$	$\Gamma D B^{\delta-1}$	$\xi_o(AP_c/k_BT_c)^{1/3}$	$\xi_o(B^2P_c/\Gamma k_BT_c)^{1/3}$
Value	0.06	1.65	0.27	0.7
	± 0.02	± 0.1	± 0.01	± 0.1

Path Dependence

Property and independent variables		Path		Exponent	Exponent value	
					Classical	Nonclassical
$K_T(\Delta T)$	(s)	Critical isochore	(w)	γ	1	1.239
$K_T(\Delta T)$	(s)	Critical isobar	(s)	$\gamma/\beta\delta$	$2/3$	0.792
$K_T(\Delta p)$	(s)	Critical isotherm	(s)	$\gamma/\beta\delta$	$2/3$	0.792
$K_T(\Delta\rho)$	(s)	Critical isotherm	(s)	γ/β	2	3.8
$C_V(\Delta T)$	(w)	Critical isochore	(w)	α	0	0.110
$C_V(\Delta T)$	(w)	Critical isobar	(s)	$\alpha/\beta\delta$	0	0.070
$C_V(\Delta p)$	(w)	Critical isotherm	(s)	$\alpha/\beta\delta$	0	0.070
$C_V(\Delta\rho)$	(w)	Critical isotherm	(s)	α/β	0	0.337

On the critical isochore, $\delta V = 0$, we find from Equation 9

$$(VK_T)^{-1} \simeq A^c_{VVT} \cdot (\delta T)$$

$$K_T^* \simeq -A^c_V (V_c \cdot A^c_{VVT})^{-1} \cdot (\delta T)^{-1}; \quad V = V_c \qquad (10)$$

so that $\gamma = 1$, and Γ, the amplitude of the power law for the compressibility, equals $(-A^c_V/T_c) (V_c \cdot A^c_{VVT})^{-1}$. Divergences with exponent values of order 1 are called strong divergences. Those are the only ones present in classical equations.

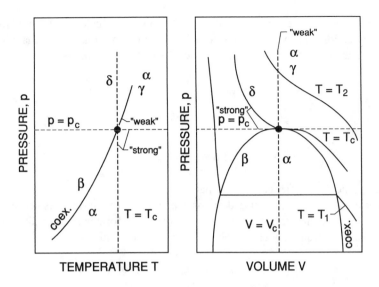

FIGURE 2. The weak and strong directions along which critical exponents are defined are indicated in p-T space (left) and in p-V space (right).

D. PATH DEPENDENCE; EXPONENT RENORMALIZATION

The exponent characterizing the divergence depends not only on the property considered but also on *the path of approach* to the critical point. The classical expansion, Equations 8 and 9, can be used to illustrate this path dependence. One might, for instance, ask for the character of the divergence of the compressibility on the critical isobar, ($p = p_c$) instead of the critical isochore, ($V = V_c$). In order to define this path it is necessary to expand the pressure, $p = -A_V$. From Equation 8, we derive:

$$p = -A_V^c - A_{VT}^c \cdot (\delta T) - A_{VVT}^c (\delta V)(\delta T) - A_{VVVV}^c (\delta V)^3/6 - \ldots \qquad (11)$$

so that the critical isobar, $p = -A_V^c = p_c$, is asymptotically defined by

$$\delta T = -(A_{VVVV}^c/6A_{VT}^c) \cdot (\delta V)^3; \quad p = p_c \qquad (12)$$

The elimination of δV in Equation 9 by means of Equation 12 then yields for the compressibility on the critical isobar:

$$K*_T = (-2A_V^c/V_c) \, A_{VVVV}^c{}^{-1/3} \, 6A_{VT}^c{}^{-2/3} \cdot \delta T^{-2/3}; \quad p = p_c \qquad (13)$$

The divergence, although still strong with $\gamma = 2/3$, is not as strong as on the critical isochore with $\gamma = 1$. Likewise, one will find that on the critical isochore, K_T behaves as $|\delta p|^{-1}$, while on the critical isotherm, $T = T_c$, K_T will behave as $|\delta p|^{-2/3}$. Thus, critical exponents have different values on the critical isochore compared to other paths, such as the critical isobar or isotherm (Figure 2). This dependence of the exponent value on the path is called *exponent renormalization*. Examples are given in Table 1.

E. OTHER CRITICAL EXPONENTS

The shape of the top of the coexistence curve (Figure 2), in molar volume or density coordinate vs. temperature, is described by means of an exponent β, defined as

$$|\delta V/V_c| \simeq |\delta\rho/\rho_c| \simeq B \, |\delta T/T_c|^\beta; \quad \text{coex. phases} \qquad (14)$$

Here, $\delta\rho = \rho - \rho_c$. The classical values of β and B are calculated from the pressure expansion, Equation 11, by setting δT to a finite, nonzero value and demanding that for two volumes, $\pm \delta V$, a liquid $(-)$ and a vapor $(+)$, the pressures be equal. This leads to

$$\delta V^2 \simeq |\ 6\ A^c_{VVT}/A^c_{VVVV}\ | \cdot |\delta T|; \text{ coex. phases;}$$

$$\beta = {}^1/_2$$

$$B = V_c^{-1}\ T_c^{1/2} \cdot |\ 6\ A^c_{VVT}/A^c_{VVVV}\ |^{1/2} \tag{15}$$

so that the classical coexistence curve is parabolic.

Another critical exponent is that defining the shapes of the critical isotherm and isobar (the strong direction) in terms of volume or density. The former follows directly from the pressure expansion, Equation 11, by setting $T = T_c$. We find

$$p - p_c = -A^c_{VVVV} \cdot (V - V_c)^3/6; \ T = T_c$$

$$\delta = 3$$

$$D = V_c^3 \cdot A^c_{VVVV}/(-6A^c_V) \tag{16}$$

The same exponent, but a different amplitude, is found for the shape of the critical isobar. In addition, the chemical potential μ is found to have the same power law behavior as the pressure, except for a factor of V_c:

$$\mu - \mu_c = -\ V_c \cdot A^c_{VVVV} \cdot (V - V_c)^3/6; \quad T = T_c$$

$$\delta = 3 \tag{17}$$

This follows from the isothermal relation $d\mu = V\,dp$, and the fact that near the critical point V is close to V_c, so that to lowest order, μ varies like $V_c p$.

Another critical exponent, α, of major theoretical significance, characterizes the behavior of the isochoric heat capacity $C_V = -T\ (\partial^2 A/\partial T^2)_V$. Like γ, it is defined along the weak direction, the critical isochore. For classical equations, as is seen from the expansion, Equation 8, the specific heat in the one-phase region is finite both above and below T_c so that $\alpha = 0$. The classical heat capacity experiences a finite jump when the phase boundary is crossed (Reference 22; Reference 24, p. 3). The classical values of the critical exponents defined thus far are summarized in Table 1.

F. ORDER PARAMETER

The concept of order parameter was first introduced when critical behavior was studied in systems other than fluids. For an early overview, see Reference 24. The spontaneous magnetization exhibited by magnetic materials diminishes with temperature and disappears rather abruptly at what is called the Curie point. Below the Curie point an ordered state exists in which the atomic spins are at least partially aligned; above this point no such order exists. The spontaneous magnetization is called the order parameter, and at the critical point it reaches the value zero; however, in uniaxial ferromagnets and in the Ising model, the spin can only assume two values, up or down, and is therefore represented as a scalar. In different ferromagnetic materials and models the spin can be multidimensional.

The behavior of a fluid in its weak direction is reminiscent of that of the uniaxial ferromagnet. Below the critical point, there are two states of different density corresponding

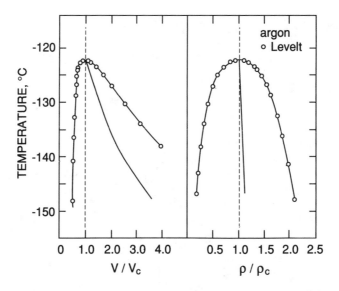

FIGURE 3. The coexistence curve is much more symmetric in T-ρ space (right) than in T-V space (left). The data are for argon.[12-14,17]

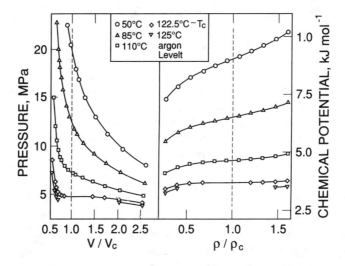

FIGURE 4. Pressure-volume isotherms show very limited antisymmetry with respect to V_c. Chemical potential-density isotherms show a large range of antisymmetry around ρ_c. The data are for argon.[12-14,17]

to the states of aligned spins pointing up or down; above it, the density difference has disappeared. Therefore, a scalar order parameter is ascribed to the critical behavior of fluids. In one-component fluids, this order parameter is assumed to be the density difference with the critical density; however, this definition is rather arbitrary. In principle, the difference noted in the critical value of any extensive property of the fluid, or combination of several of those, could be a valid order parameter. In pure fluids, the choice is dictated principally by the observed symmetry of the fluid around the critical density. The coexisting densities, and the chemical potential-density isotherms, show a near-symmetric behavior over a larger range than do other choices of variables, such as coexisting volumes and pressure-volume isotherms (Figures 3 and 4). A reasonably complete understanding of the analogy between a pure-fluid order parameter and that of the Ising model has been reached, and the best

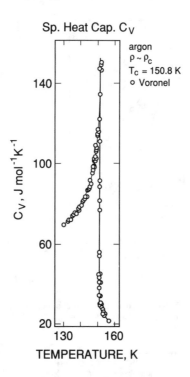

FIGURE 5.   Voronel and co-workers[31] discovered the weak divergence of the constant-volume heat capacity.

theoretical models for pure fluids[25-27] incorporate the corrections induced by the lesser symmetry typical of fluids. In fluid mixtures, especially those of the vapor-liquid type, the proper choice of order parameter is a much harder problem than in one-component fluids; a workable understanding has not been reached, and this is an impediment to the development of good models for engineering applications.

G. NONCLASSICAL CRITICAL EXPONENTS OF REAL FLUIDS; UNIVERSALITY

Since the end of the 19th century[17,24,28,29] it was noticed that critical exponents in fluids do not assume the values predicted by classical equations of the type analyzed above. Around 1900, for instance, the coexistence curve was already known to be roughly cubic, rather than parabolic. Only a few scientists realized at an early stage that this was a serious failure of classical theory.[29] The failure of classical, or mean-field theory, was brought to general consciousness by Onsager's solution of the two-dimensional Ising model in 1941. The impact of this discovery mainly bypassed the community of scientists working on fluid properties, although there were some notable exceptions.[30] A major scientific breakthrough came with the discovery by Voronel and co-workers[31] that the isochoric heat capacity C_v diverges weakly at the critical point (Figure 5). As a consequence of this experiment, fluid critical behavior reentered the mainstream of the study of critical phenomena.

The divergence of the isochoric specific heat is weak compared to that of the compressibility. The critical exponent α equals about 0.1. The critical behavior of real fluids is called nonclassical. Refined measurement of critical exponent values in pure fluids[32] and fluid mixtures (for reviews, see References 33 and 34) has revealed that the value of any individual critical exponent is independent of the fluid or fluid mixture studied. The fluid critical exponents thus have a character of universality. The theory of critical phenomena predicts that this universality encompasses a class much wider than simply fluids. It claims that all systems having the same spatial dimensionality, having short-range interactions between their constituents, and having a scalar order parameter, belong to the same universality class.[35] The critical exponents of this universality class have been calculated theoretically by means of several methods. The most accurate values result from calculations by means of the renormalization-group theory of Wilson and Fisher.[36,37] The calculated values[38] are presently much more accurate than can be hoped to be measured in fluids. In Table 1, the most likely values[23] of the critical exponents of the Ising universality class are listed.

H. WEAK AND STRONG DIRECTIONS; DENSITIES AND FIELDS

Although it might appear from Figure 2 that there is an infinite number of paths to the critical point, in one-component fluids there are only two paths that are distinct in the sense of critical behavior. These are (1) the special direction singled out by the phase transition, i.e., the direction of the vapor pressure curve and its extension, the critical isochore, and (2) any path that intersects the special direction. The special direction is called the *weak* direction. Members of this class are the critical isochore, isentrope, and isenthalpe. The directions intersecting it are called the *strong* direction. Members are the critical isotherm, isobar, and isochemical-potential curve.

The difference between these two types of isocurve classes is that the second class refers to variables (pressure, temperature, and chemical potential) that are identical in the coexisting phase. Griffiths and Wheeler[11] call such variables *fields*. The first class refers to the variables volume, density, enthalpy, and entropy that are generally not equal in coexisting phases; Griffiths and Wheeler call such variables *densities*. These definitions of fields and densities are distinct from the intensive and extensive variables encountered in standard thermodynamics. Not all properties that are dissimiliar in coexisting phases are densities. The densities defined by Griffiths and Wheeler are first derivatives of those thermodynamic potentials that are strictly defined in terms of fields. For one-component fluids, the potential $\mu(p,T)$ is an example; equally acceptable are $p(\mu,T)$ and $T(\mu,p)$. For the unit of amount of substance, the mole is used; thus, μ is the chemical potential for a mole of substance, V is the molar volume, and S the molar entropy.

The first derivatives of $\mu(p,T)$ are

$$(\partial\mu/\partial T)_p = -S \qquad (\partial\mu/\partial p)_T = V \qquad (18)$$

while the first derivatives of $p(\mu,T)$ are

$$(\partial p/\partial T)_\mu = S/V = s \qquad (\partial p/\partial\mu)_T = \rho \qquad (19)$$

with s the entropy density and ρ the molar density. All of the derivatives in Equations 18 and 19 are densities in the sense of Griffiths and Wheeler. The chemical potential isotherms, however, are approximately antisymmetric in density with respect to the critical density over a large density range, while the pressure isotherms have only a very limited range of antisymmetry in terms of volume (Figures 3 and 4). Consequently, the $\mu - \rho$ set of variables is more suitable for describing gas-liquid critical points than the $p - V$ set, although the two sets are asymptotically equivalent.

The second derivatives of the thermodynamic potentials, which are first derivatives of densities with respect to fields, yield the strongly diverging "susceptibilities", such as isothermal compressibility, isobaric heat capacity, and expansion coefficient. For instance, from Equation 18 we derive:

$$C_p \equiv T\,(\partial S/\partial T)_p \qquad = -T\,(\partial^2\mu/\partial T^2)_p$$

$$K_T \equiv -(1/V)(\partial V/\partial p)_T = -(1/V)(\partial^2\mu/\partial p^2)_T = -\rho(\partial^2\mu/\partial p^2)_T$$

$$\alpha_p \equiv (1/V)(\partial V/\partial T)_p = (1/V)(\partial^2\mu/\partial p\partial T) = \rho(\partial^2\mu/\partial p\partial T) \qquad (20)$$

and from Equation 19, we obtain the generalized susceptibility χ_T

$$\chi_T = (\partial\rho/\partial\mu)_T = (\partial^2 p/\partial\mu^2)_T = \rho^2 K_T \tag{21}$$

Along isotherms in $\chi - \rho$ space, this susceptibility is more symmetric than the compressibility in $K_T - V$ space. Also, the susceptibility χ is more naturally generalized to fluid mixtures than the compressibility K_T.

In conclusion, the structure of a strongly diverging derivative is the derivative of a density with respect to a field, while the other field is kept constant, so that the derivative is taken along a strong direction.

The structure of a weakly diverging property is clear from an example, the constant-volume heat capacity C_V, defined as

$$C_V = T (\partial S/\partial T)_V \tag{22}$$

Thus, a derivative is taken of a density with respect to a field while another *density* is kept constant. By keeping a density constant, the derivative is taken along the weak direction. A weak anomaly results.

It is also possible to take the derivative of a density with respect to another density, at constant field; or the derivative of a field with respect to a field, at constant density. Some simple examples may suffice. The first is that of $(\partial S/\partial V)_T$, and the second that of $(\partial P/\partial T)_V$. The latter derivative is finite and continuous on the vapor pressure curve below T_c and the critical isochore above T_c.[16] By a well-known Maxwell relation, however,

$$(\partial S/\partial V)_T = (\partial p/\partial T)_V \tag{23}$$

it is found that the former derivative is likewise finite and well behaved. This is generally true for any derivative of a density with respect to a density at constant field, such as $(\partial H/\partial V)_T$, or $(\partial H/\partial V)_p$, with H the enthalpy. These derivatives are simply related to the slope of the vapor pressure curve and assume values in the supercritical regime that are close to those in the two-phase region just below the critical point. Thus, all density-like variables are linearly related at the critical point.

A simple mnemonic for the distinction between strong and weak divergences is that those divergences are strong that survive in the classical theory. The counterparts of weakly diverging second derivatives remain finite in the classical theory.

I. RELATIONS BETWEEN CRITICAL EXPONENTS; TWO-SCALE-FACTOR UNIVERSALITY

The laws of thermodynamics impose relations between critical exponents. A number of inequalities between the exponents $\alpha - \delta$ were derived rigorously. On the basis of numerical evidence from model calculations it was then postulated that certain equalities were valid between the critical exponents, such as

$$2 - \alpha = \beta (\delta + 1)$$

$$\gamma = \beta (\delta - 1) \tag{24}$$

It is readily seen that classical theory obeys these equalities. All subsequent experimental and theoretical work has upheld the validity of these equalities for the nonclassical exponents of real fluids. They imply that only two independent exponent values characterize all thermodynamic critical anomalies within a universality class. For these, one could choose the strong exponent γ and the weak exponent α. Likewise, only two (nonuniversal) critical amplitudes can be chosen independently; all other amplitudes follow from these by means of relations called universal amplitude ratios. Some examples are given in Table 1.

Modern theory of critical phenomena thus allows two independent critical exponents that, however, are identical for all fluids and fluid mixtures with short-range forces between the molecules; it allows two nonuniversal amplitudes for each substance within a universality class. This principle of two-scale-factor universality[18-21] also encompasses the correlation function exponents discussed in the next section.

The principle of corresponding states is considerably more restrictive than the principle of two-scale-factor universality; for all fluids obeying this principle, the two nonuniversal critical amplitudes must be the same. It follows that in a class of systems in which departures from corresponding states can be sufficiently characterized by one additional parameter, such as the acentric factor, only one critical amplitude can be freely chosen.[39]

J. FLUCTUATIONS; CORRELATION FUNCTION; CORRELATION LENGTH

The large compressibility, or equivalently, the flatness of the Helmholtz free energy surface near a critical point, makes it possible for density fluctuations to be generated at very low cost in energy. Thus, near a critical point, large density fluctuations are thermally excited. At the turn of the century, Einstein used this insight in order to give a first explanation of the phenomenon of critical opalescence, which is the most characteristic signature of a critical-point phase transition. Ornstein and Zernike realized shortly thereafter that because of the extra cost invested in the creation of gradients, only long-wavelength density fluctuations are enhanced. They lead to enhanced scattering mostly in the forward direction.

Alternatively, one can say that local density fluctuations that arise spontaneously due to thermal agitation become correlated in space near a critical point. These correlations are expressed by means of the correlation functions $G(r)$ and $H(r) \equiv G(r) - 1$, where $\rho G(r)$ gives the molar density a distance r from a given molecule at the origin, and $\rho H(r)$ the excess molar density compared to the average, ρ.

In fluids, $H(r)$ approaches -1 for r within the repulsive core of the interaction, has a pronounced maximum roughly at the position of the minimum of the attractive well, and may have several further subsidiary maxima and minima, depending on the overall density and temperature, before decaying to zero. Near a critical point, $H(r)$ and therefore also $G(r)$ develop a subtle "tail", or a long-range character,[40] about which more is said later.

Several thermodynamic properties can be expressed in terms of the intermolecular potential $\varphi(r)$ and the pair correlation function. Of interest to us is the residual energy $U^r(V,T)$ = $U(V,T) - U$ (perfect gas, V,T):

$$U^r = \int G(r)\, \varphi(r)\, d\mathbf{r} \tag{25}$$

For intermolecular potentials that are short ranged, $\varphi(r)$ has decayed to zero in the region of r where the critical "tail" of the distribution function develops; therefore, U^r is determined by the short-range part of $G(r)$. This implies that U^r has no conspicuous critical effects, although there are some subtleties[41,42] that are discussed below.

The theory of critical phenomena is primarily concerned with the large-r behavior of $H(r)$. The fluctuation theorem connects the integral of the correlation function with the fluid compressibility:

$$RT\, \rho K_T = 1 + \rho \int H(r)\, d\mathbf{r}$$

$$\text{or } RT\, \chi_T = \rho + \rho^2 \int H(r)\, d\mathbf{r} \tag{26}$$

where $R = N_A k_B$ is the gas constant, N_A is Avogadro's number, and k_B is Boltzmann's constant. In most texts, Equation 26 is found with ρ as the number density and k_B replacing R. Following O'Connell,[43] however, ρ is used here for the molar density. As a consequence, a factor of N has implicitly been absorbed in H(r), so that the volume integral of H(r) is a molar, not a molecular, volume. For a one-component system, this volume integral is indicated by H_{11} and is called the total correlation function integral.

Since the compressibility diverges at the critical point, by virtue of Equation 26 the integral of H(r) must diverge at this point, so that H(r) becomes long ranged both in the classical and the nonclassical sense.[40] Where these cases differ is in the details of the way in which H(r) becomes long ranged. In the classical Ornstein-Zernike theory, H(r) decays as 1/r at the critical point. Away from this point, H(r) decays as

$$H(r) \approx e^{-r/\xi}/r; \text{ r large, } \xi \text{ finite} \tag{27}$$

The decay parameter ξ, which itself is a function of temperature and density, is called the correlation length. At the critical point the correlation length diverges to infinity, according to the power law

$$\xi = \xi_o \left| \delta T/T_c \right|^{-\nu}; \rho = \rho_c \tag{28}$$

In classical theory, $\nu = 1/2$, or $\gamma/2$. The parameter ξ_o is of the order of a molecular size. Typical values[22,23] for simple molecules are in the range of 0.1 to 0.3 nm. This implies that 3°C from an ambient-temperature critical point, the correlation length at ρ_c is of the order of 4 nm, or a dozen molecular diameters.

The total correlation function is to be distinguished from the direct correlation function C(r) which was introduced by Ornstein and Zernike by the definition:

$$C(r_{12}) = H(r_{12}) - \rho \int C(r_{13}) H(r_{23}) \, d\mathbf{r}_3 \tag{29}$$

In our convention, ρ is a molar density and a factor of N_A, Avogadro's number, has been absorbed in the definition of C(r) just as in the case of H(r). Ornstein and Zernike expected C(r) to be relatively short ranged. In fact, the direct correlation function integral, C_{11}, defined as

$$C_{11} = \int C(r) \, d\mathbf{r}, \tag{30}$$

must be finite, because from the definition of C(r), it can be shown that

$$1 + \rho H_{11} = 1/(1 - \rho C_{11}) = RT \rho K_T \tag{31}$$

and since K_T is always larger than 0, C_{11} cannot become infinite; at the critical point, ρC_{11} equals unity. $C_{11}(r)$ is thus definitely of shorter range than H(r). In the classical case, its second moment, $\int C(r) r^2 dr$, exists and is equal to a finite quantity often denoted as R^2. $R/N_A^{1/3}$ is a length of the order of a molecular size. From Equation 31 it is obvious that C_{11} and H_{11} contain the same information.

There are some obvious and some subtle differences between the classical and the nonclassical cases.[40] As to the obvious, γ no longer equals unity, so a first guess at the nonclassical value of the exponent ν would yield 0.62 ($= \gamma/2$). As to the subtle effects, however, in three dimensions, H(r) decays not quite as 1/r at the critical point, but as

$$H(r) \approx 1/r^{d-2+\eta}; \text{ critical point} \tag{32}$$

with d the dimensionality of the system.[40] This departure from the classical $1/r$ decay is associated with a weak divergence of the second moment of the direct correlation function integral,[41] contrary to the expectation of Ornstein and Zernike. The exponent η is very small and has been very difficult to measure or predict. The best current value is 0.03. As a consequence, the nonclassical value of the exponent ν is 0.630, subtly different from the value of 0.620, which one would naively expect on the basis of the relation $\nu = \gamma/2$. The relation between ν, γ, and η is

$$\nu = \gamma/(2-\eta) \tag{33}$$

Another exponent equality connects ν with α:

$$d\nu = 2 - \alpha \tag{34}$$

The correlation-function critical exponents and the most-used exponent equalities are summarized in Table 1.

We have seen through Equation 25 that the internal energy reflects the short-range behavior of $G(r)$. In the classical case, therefore, the internal energy has no critical anomalies; in the nonclassical case, something subtle happens with the short-range behavior of $G(r)$, since C_V, the first derivative of the internal energy with respect to temperature, has a weak anomaly. Therefore, the short-range structures of $G(r)$ and $H(r)$ must be such that this weak anomaly in the derivative property is generated, as was first noted by Stell and co-workers.[41,42] This argument can be extended to incorporate a host of fluid properties that depend solely on the short-range part of $G(r)$. Examples are the dielectric constant, Raman, NMR, and ESR line shifts, and presumably, the solvatochromic shift: those properties that depend on an electric or magnetic signal impinging on a molecule, which in turn polarizes neighboring molecules, which then interact back. These signals are attenuated with a factor that is typically proportional to $1/r^6$. It is therefore expected,[41,42,44] and has been demonstrated in some cases,[45] that such properties have a temperature dependence of $(T - T_c)^{1-\alpha}$, which means that they remain finite at the critical point, but their first temperature derivative diverges weakly. These $(1 - \alpha)$ anomalies are very hard to detect and can usually be ignored in engineering applications.

K. RANGE AND SIZE OF CRITICAL ANOMALIES

It is heard in chemical engineering circles that critical phenomena are something esoteric that happen within a fraction of a degree from a critical point and can be ignored almost everywhere else. In this section, the truth of this statement is explored. The data for some real fluids will explicate in what range the critical anomalies occur and where they can be ignored. The large compressibility confines critical anomalies to narrow pressure ranges. For the sake of easier visualization, properties are displayed here as functions of density rather than pressure.

As to the strong anomalies, Figure 6 shows data for the isothermal compressibility of near- and supercritical steam, calculated from the National Bureau of Standards/National Research Council of Canada (NBS/NRC) Steam Tables. In Figure 7, similar data are shown for the isobaric heat capacity of argon.[12-14] These properties give important information on SCF: how sensitive they are to pressure and temperature disturbances, and how anomalous the partial molar volume and enthalpy of a dilute solute are going to be. The vertical scale is logarithmic. For these strong anomalies a broad maximum occurs even at temperatures more than twice the critical; on approaching T_c, this maximum increases in height by almost two orders of magnitude in the range shown in Figures 6 and 7. It is obvious that for an SCF the strong anomalies cannot be ignored anywhere in this large range, irrespective of fine points such as the distinction between classical and nonclassical behavior.

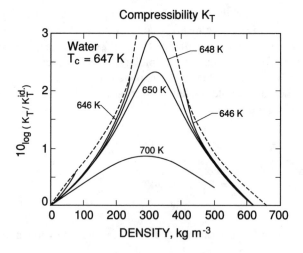

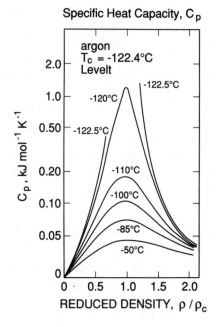

FIGURE 6.   The compressibility of supercritical steam, a strong anomaly, varies by several orders of magnitude in a range from 1 to 50 K above the critical point.

FIGURE 7.   The constant-pressure heat capacity of supercritical argon,[12-14] a strong anomaly, varies by several orders of magnitude in the range from −50 to −120°C.

A property related to the strong anomalies is the shape of the coexistence curve in $\rho - T$ space. It has long been known[29] that a simple cubic relation describes this curve over a temperature range of 90 to 100% of T_c far better than the classical parabola. This is one case where the engineer cannot ignore the results of the modern theory of critical phenomena, and chemical engineers appear to be well aware of this.

Other examples of strong critical anomalies are critical opalescence and gravitational sedimentation. Critical opalescence refers to light scattered from refractive index fluctuations

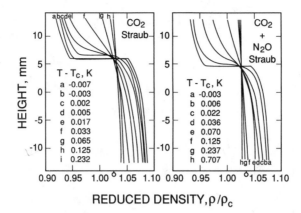

FIGURE 8. Pure fluids (left) and fluid mixtures (right) develop macroscopic density gradients, due to compression in the field of gravity, in a range of a few 0.1 K from the critical point.[50]

extending over regions comparable with the wavelength of light and due to the critical density fluctuations driven by the compressibility. Critical opalescence is visible to the naked eye within roughly 1 K from a critical point. With modern instrumentation, it can be detected over much larger ranges of temperature.[46] Critical opalescence has been known to interfere with optical studies near critical points.[47,48] Gravitational sedimentation refers to a density gradient that develops in a near-critical fluid near its critical point, due to compression under its own weight. It was first noticed and correctly explained by Gouy[49] at the end of the 19th century. Macroscopic density gradients develop within roughly 1 K from the critical point, and can be as large as 10% over the height of a few centimeters (Reference 50 and p. 13 in Reference 24) (Figure 8). They can lead to very serious error in the assessment of the density.

Another property related to the strong anomalies is the correlation length.[22,23,41] It is important to know what this length is[22,23] if one is interested in experimental design or computer simulation near a critical point. The true critical behavior is seen only if a region is probed that is much larger than the correlation length.

Turning now to the weak anomalies, which do not occur in classical theory, Figure 9 shows a plot for the isochoric heat capacity C_v of argon.[12-14] Unlike Figures 6 and 7, this plot is not on a logarithmic scale. The broad maximum develops at a temperature 35 K (20% of T_c) above the critical point, and grows to a few times its original size by the time T_c is approached more closely. A corollary is the accompanying minimum in the speed of sound[25-27] (Figure 10), which approaches zero weakly, with an exponent $\alpha/2$. The weak anomaly, therefore, cannot be ignored in a range of several degrees Kelvin (1%) from the critical temperature, at densities near the critical density.

L. EXPERIMENTAL ERROR NEAR CRITICAL POINTS

Because the response functions of the system, such as compressibility and expansion coefficient, diverge strongly at the critical point, the densities of the system become extremely sensitive to variations in the fields. As a consequence, the experimenter can be sure that ''what can go wrong will go wrong'' in experiments in near-critical fluids. To paraphrase this law: anyone who first measures a property in a near-critical fluid is apt to find a strong anomaly. An extremely cautious and critical attitude is required when experiments are performed within a few degrees Kelvin from a critical point.

Examples abound in the literature. Since Andrews' first correct experiments in carbon dioxide in the 1860s, at least three times in history was the existence of an anomalous region

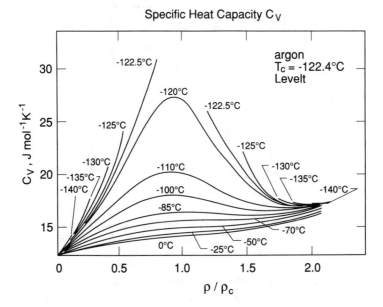

FIGURE 9.   The constant-volume heat capacity of supercritical argon,[12-14] a weak anomaly, increases by a factor of 2 in the range from -25 to $-120°C$.

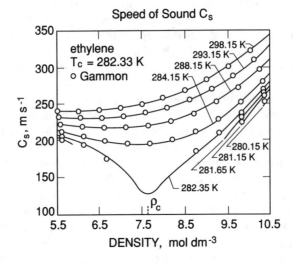

FIGURE 10.   The thermodynamic speed of sound develops a weak zero at the critical point.[26,53]

claimed in which the interface had disappeared but the isotherms were still horizontal.[1] Each time, it was newly demonstrated that the fallacious results were due to the presence of small, unequilibrated impurities.

Another example is that of the thermal conductivity of gases. Due to the divergence of the expansion coefficient, the tendency toward convection increases tremendously near a critical point. Therefore, the amount of heat transferred across a temperature differential will always be more than that due to pure conduction. The real problem was not to measure an effect, but to prove the effect was real. In the definitive experiment, it took 10 years to eliminate all spurious effects and show that an anomalous contribution to the thermal conductivity remained.[51,52]

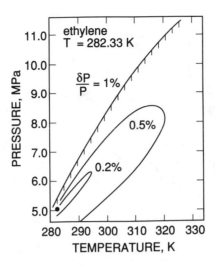

FIGURE 11.   A pressure uncertainty of 0.2% causes more than 1% error in the calculated density of ethylene[26,53] in the indicated range of 1 MPa and 10 K above T_c. The density error exceeds 1% in much larger ranges if the pressure is less well known.

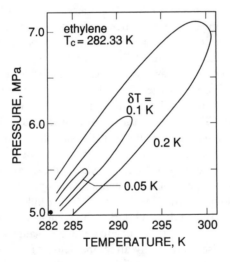

FIGURE 12.   A temperature uncertainty of 0.1 K causes errors of more than 1% in the calculated density of ethylene,[26,53] in the indicated range of 1 MPa, and 10 K around the critical point.

The most fertile source of spurious critical anomalies is the practice of obtaining the density not by direct measurement, but by calculation from the observed pressure through the use of an equation of state (EOS). Even the best EOS will fail to produce correct densities when the critical point is approached closely. Experimental error in the pressure or the temperature, impurity, gradients, and/or noxious volumes in the experiment will all lead to erroneous values of the calculated density. The divergences of the expansion coefficient and the compressibility of the fluid, and that of the partial molar volume of the impurity, cause a divergence of the density error, even if the EOS is flawless. As an example, Figures 11 and 12 illustrate in which regions of the p-T space of a supercritical solvent (ethylene) the density error will exceed 1% if the EOS is assumed to be free of error, and the pressure or temperature measurement is of the uncertainty stated. It is seen that for respectable pressure

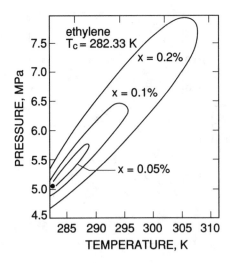

FIGURE 13.   An impurity of 0.1 mol% of methane in supercritical ethylene causes more than 1% error in the calculated
density of ethylene[25,53] in a range of 1.5 MPa and 10 K around
the critical point.

control to 0.2%, or temperature control to 0.1 K, there is a range of over 1 MPa in pressure
and over 10 K in temperature, where the density error exceeds 1%. Likewise, Figure 13
shows a similar large range in which the density is affected by more than 1% if an impurity
much more volatile than the near-critical fluid is present on the level of only 0.2 mol%.
This figure was constructed on the basis of a corresponding-states estimate for ethylene with
a methane impurity.[26,53] The very large effect of impurities on density near a critical point
was recognized by Gouy and by scientists in Leiden at the turn of the 20th century (see
Reference 1 for a review). This effect is further discussed in Section IV.

Parts of the sample system that are not at the same temperature as the sample cell, or
that are at a higher or lower level, will start "breathing" as the critical point of the solvent
is approached. In experiments in such "open" systems, the diverging expansion coefficient
drives the fluid to the colder parts of the system, setting up large density differentials. An
example is the large amount of ethylene (T_c = 9°C) that hides in the transportation lines
between the points of supply and demand during the winter, and shows up in the summer.
In a laboratory experiment, a capillary leading to a valve or gauge at a different temperature
than the sample will likewise inordinately affect the density of a near-critical sample.

The diverging compressibility causes a density gradient, resulting in denser fluid collecting at the lower level. Even accurate knowledge of the total amount of fluid in the system
is no guarantee of knowledge of the density at the point at which a property of the fluid is
measured. It was demonstrated, for instance (by repeating the experiment with the cell turned
upside down), that a noxious volume in a valve was the origin of an apparent anomaly in
the dielectric constant of near-critical helium.[54]

A reported critical anomaly cannot be taken seriously if its size depends on a density
that is not properly controlled. In experimentation near critical points, the experimenter has
the obligation to *investigate, estimate,* and *report* the effect of all sources of error on the
property of interest, and on its independent variables, for the results to be worthy of consideration.

M. SUMMARY

In this section, a framework, due principally to Griffiths and Wheeler, has been introduced for describing and classifying critical anomalies in pure fluids. The essential ingredients
are

1. The definition and characteristics of two kinds of thermodynamic variables: fields and densities
2. The existence of two different directions of approach to criticality, a strong and a weak direction
3. The characterization of critical divergences by means of critical exponents, only two of which are independent; for instance γ, describing a strong divergence, and α, describing a weak divergence
4. The definition of critical amplitudes, only two of which are independent and non-universal
5. The differences between classical or van der Waals-like, and nonclassical critical behavior
6. The universality of the nonclassical critical exponents
7. The mathematical structure of strongly, weakly, and nondiverging second derivatives

The author has introduced the power laws and critical exponents characterizing the divergence of the tail of the pair correlation function and stressed that the short-range part of G(r), just like the internal energy, is almost completely ignorant of the nearness of a critical point. By means of practical examples, the ranges have been delineated in which the critical anomalies prevail, and the implications for the design and implementation of experiments near critical points have been drawn.

III. CRITICAL BEHAVIOR OF FLUID MIXTURES

A. CRITICALITY

Before the limit of mechanical stability is reached, fluid mixtures become materially unstable — they can lower their free energy by splitting into two phases of different composition. The total Helmholtz free energy $A_t(V, T, n_1, n_2, \ldots)$ of an n-component mixture is a function of volume, temperature, and n amounts of substance n_i. For the isothermal molar Helmholtz free energy to remain concave upwards, an $n \cdot n$ determinant of second derivatives must remain positive. For a two-component mixture, this condition translates into:

$$\text{Det } A \equiv A_{VV} \cdot A_{xx} - A_{Vx}^2 \geq 0 \tag{35}$$

where x is the mole fraction $n_2/(n_1 + n_2)$ and the subscripts denote differentiation of the Helmholtz free energy with respect to the indicated variables. Condition 35 can be rewritten as[15]

$$\text{Det } A \equiv -(\partial p/\partial V)_{xT}(\partial^2 G/\partial x^2)_{pT} \geq 0 \tag{36}$$

with G, the molar Gibbs free energy. [If the symbols G and H are used for the correlation functions, they will always be in the form of G(r), H(r). The symbol H for correlation function integrals will always be subscripted, H_{ij}.]. Since classically, $(\partial p/\partial V)_{xT}$ is finite and nonzero in a mixture, the well-known criticality condition

$$(\partial^2 G/\partial x^2)_{pT} = 0 \tag{37}$$

follows from Equation 36. The step from Equation 36 to 37 needs care in the nonclassical case,[11] in the case of dilute mixtures near the critical point of the solvent, where $(\partial p/\partial V)_{xT}^{-1}$ approaches a strong divergence, and in the case of critical azeotropy,[15] where $(\partial p/\partial V)_{xT} = 0$.

Useful relations exist between the derivatives of chemical potentials μ_i of the components of a mixture and the Gibbs free energy. They are

$$(\partial x_1/\partial \mu_1)_{PT} = - (\partial x_2/\partial \mu_1)_{PT} = \{x_2 (\partial^2 G/\partial x_2^2)_{PT}\}^{-1}$$

$$(\partial x_2/\partial \mu_2)_{PT} = - (\partial x_1/\partial \mu_2)_{PT} = \{x_1 (\partial^2 G/\partial x_1^2)_{PT}\}^{-1}$$

$$= \{(1-x_2) (\partial^2 G/\partial x_2^2)_{PT}\}^{-1} \qquad (38)$$

Because $\{(\partial^2 G/\partial x_2^2)_{PT}\}^{-1}$ diverges strongly, all the above derivatives, called *osmotic susceptibilities,* diverge strongly at the critical line of the mixture as long as x_1, x_2 are unequal to zero; they are, in the mixture, the analogs of the compressibility in the pure fluid. Since the second derivative of the Gibbs free energy cannot be negative, $(\partial x_1/\partial \mu_1)_{PT}$ and $(\partial x_2/\partial \mu_2)_{PT}$ are positive, while the derivatives with mixed indices are negative.

B. THE CLASSICAL AND THE NONCLASSICAL PICTURE

In the case of the one-component fluids, Section II discussed how the critical anomalies of thermodynamic derivatives depend on the variables that are kept constant in the differentiation and on the path of approach to the critical point. Generalization to fluid mixtures is the topic of this chapter, and for the case of classical critical behavior, this generalization was carried out by van der Waals in 1891 (English translation in Reference 55). From his work, thermodynamic properties can be found that are, in general, strongly divergent. Also, many of the exceptional cases, such as critical azeotropy and the occurrences of maxima and minima in critical lines, were studied in detail by van der Waals and his followers. In 1970, Griffiths and Wheeler[11] provided a systematic generalization for the case of nonclassical critical behavior of fluid mixtures. Their work was inspired by the thermodynamics of superfluid helium, which had been well established by Buckingham, Fairbanks, Rice, and others long before.

First to be discussed are some of the results of the classical or mean-field theory of van der Waals, which incorporates only the strong anomalies. Van der Waals assumed that the Helmholtz free energy of a binary mixture of fixed composition obeys a two-parameter EOS corresponding to that of the pure components. The two parameters, a, denoting an energy of attraction, and b, an excluded volume, are both functions of the composition x. It follows that the "reduced" pressure-volume-temperature diagram of the constant-composition mixture corresponds with those of the single individual constituents (Figure 14). The mixture, however, separates into two phases because of material instability, $(\partial^2 G/\partial x^2)_{pT} = 0$, while the one-component fluid separates because of mechanical instability, $(\partial^2 A/\partial V^2)_T = 0$. In general, the critical point, or plait point, of the mixture occurs in what would be the one-phase region of the pure component (Figure 14). The plait point temperature depends on the composition so that the binary mixture develops a critical line, the ternary a critical surface, etc.

Figure 14 shows the dew-bubble curve, but it cannot show the phases the mixture splits into, because those phases have compositions different from those for which the plot was made. Figure 14 does reveal that the compressibility of the mixture, $K_{Tx} = - (1/V)(\partial V/\partial p)_{Tx}$, is finite at the critical point, from which it follows that the expansion coefficient α_{px} and heat capacity C_{px} do not diverge for a classical mixture. Strong divergences do occur, as seen in Section III.A, in the osmotic susceptibilities $(\partial x/\partial \mu_i)_{pT}$ (see Equation 38). This suggests that the strong derivatives are associated with the fact that only fields, not densities, are kept constant in the differentiation leading to the osmotic susceptibilities.

Griffiths and Wheeler[11] developed the nonclassical description of fluid mixtures on the basis of a geometric picture that they obtained for the n-component mixture in terms of a thermodynamic potential, which is itself a field, and a function of n + 1 independent field

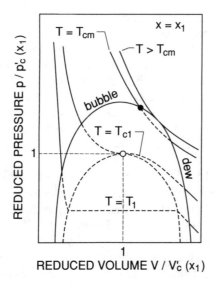

FIGURE 14. The law of corresponding states applied to a mixture of constant composition x_1. The mixture's critical point (●) and dew-bubble curve are in the one-phase region of the pure fluid, and the critical isotherm of the mixture (T_{cm}) has a finite slope.

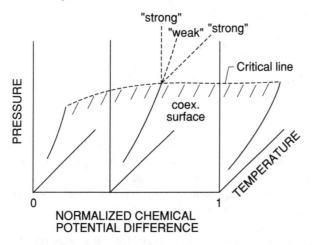

FIGURE 15. Coexistence surface, critical line, and directions of approach to criticality of a binary mixture in the space of three independent field variables.

variables. For a two-component fluid, $\mu_1(p,T,\mu_2)$, $\mu_2(\Delta,p,T)$, $p(\mu_1,\mu_2,T)$, with $\Delta = \mu_2 - \mu_1$, are all valid examples. Although asymptotically equivalent, some of these potentials are more convenient for describing incompressible binary liquid mixtures near consolute points and others for binaries near gas-liquid critical points; emphasis is placed here on the latter case. If a potential with pressure, temperature, and a chemical potential as independent variables is taken as an example, then the primary concern is with the geometry in the space of three independent variables. The chemical potentials are exponentiated so as to avoid their divergence at $x = 0$ or 1. The geometry is shown in Figure 15, which indicates the coexistence curves of the pure components, and the coexistence surface of the mixture terminated by the critical line. Note that in this field space the coexistence surface is single valued, since field variables assume identical values in coexisting phases.

In a one-component fluid, the critical point can be approached along the special direction determined by the coexistence curve (the weak direction), or at an angle to it (the strong direction). It is obvious from Figure 15 that in the binary mixture the critical point can be approached in three different ways: intersecting the coexistence surface, in the coexistence surface but not parallel to the critical line, and parallel to the critical line (or critical surface in the case of more than two components). Just as in the case of the one-component fluid, the connection is to be made between the character of a thermodynamic derivative, the variables held constant during differentiation, and the path of approach to the critical point, on the one hand, and the resulting critical exponent value, on the other hand.[11,56]

C. STRONGLY, WEAKLY, AND NONDIVERGING DERIVATIVES

The first hypothesis of Griffiths and Wheeler,[11] which is in essence the principle of critical-point universality, is that if one field variable is fixed, the binary mixture will generally behave as a one-component fluid. For the latter, if a derivative of a density is taken with respect to a field, with the addition of yet another field constant, a strongly diverging property will result; the same is therefore true for a similar derivative in the binary mixture, with two fields constant. A plane with two fields constant will generally intersect the coexistence surface and define a strong direction (Figure 15). The potential $\mu_2(\Delta, p, T)$, with $\Delta = \mu_2 - \mu_1$, leads to the densities

$$1 - x = (\partial \mu_2/\partial \Delta)_{pT}; \quad V = (\partial \mu_2/\partial p)_{\Delta T}; \quad S = - (\partial \mu_2/\partial T)_{p\Delta} \tag{39}$$

with strongly diverging second derivatives

$$(\partial^2 \mu_2/\partial \Delta^2)_{pT} = - (\partial x/\partial \Delta)_{pT};$$

$$(\partial^2 \mu_2/\partial p^2)_{\Delta T} = - VK_{T\Delta}$$

$$- (\partial^2 \mu_2/\partial T^2)_{p\Delta} = (\partial S/\partial T)_{p\Delta} = C_{p\Delta}/T \tag{40}$$

All these strong derivatives diverge with an exponent γ along the weak path or coexistence surface and its extension (see Figures 2 and 15). Along a strong path intersecting the coexistence surface, they diverge with the renormalized exponent $\gamma/\beta\delta$, just as in the one-component fluid. As $x \to 0$, the potential $\mu_1(\Delta, p, T)$ goes over smoothly into the potential $\mu(p, T)$ (Equation 18) for the one-component fluid.

It is usually not possible to devise experiments in which Δ is kept constant. Only the first of the above derivatives, an osmotic susceptibility, is readily accessible to experiment; it is the cause of critical opalescence in fluid mixtures. Another strong derivative, however, is readily observed and is the cause of the phenomenon of enhanced solubility in SCFs. It is defined here as $(\partial x/\partial P)_{T,\sigma}$; this is the isothermal increase of the mole fraction of solute with pressure in the presence of an inert phase σ of pure solute that keeps the chemical potential μ_2 of the solute effectively constant. This derivative is therefore expected to diverge strongly at the point on the critical line where the chemical potential of the solute equals μ_2.

Another choice of thermodynamic potential that is very useful is that of $p(\mu_1, \mu_2, T)$. This choice leads to the densities

$$\rho_1 = (\partial p/\partial \mu_1)_{\mu_2 T}; \quad \rho_2 = (\partial p/\partial \mu_2)_{\mu_1 T}; \quad s = (\partial p/\partial T)_{\mu_1 \mu_2} \tag{41}$$

where ρ_1, ρ_2 are the molar densities of components 1 and 2, respectively, and s is the entropy per unit volume. Some of the second derivatives associated with this potential are

$$(\partial^2 p/\partial\mu_1{}^2)_{\mu_2 T} = (\partial\rho_1/\partial\mu_1)_{\mu_2 T}$$

$$(\partial^2 p/\partial\mu_2{}^2)_{\mu_1 T} = (\partial\rho_2/\partial\mu_2)_{\mu_1 T}$$

$$(\partial^2 p/\partial\mu_1\partial\mu_2)_T = (\partial\rho_1/\partial\mu_2)_{\mu_1 T} = (\partial\rho_2/\partial\mu_1)_{\mu_2 T} \qquad (42)$$

Although none of these second derivatives appear to be very practical, they play a major role in the Kirkwood-Buff theory of solutions,[57] a theory that is being applied widely to model supercritical solubility.[72-74]

As $x \to 0$, the potential $p(\mu_1,\mu_2,T)$ goes over smoothly into the potential $p(\mu,T)$ (Equation 19), for the one-component fluid. Symmetry arguments have led to preference for that potential for gas-liquid critical points. This potential is therefore also believed to be the preferable choice for fluid mixtures near plait points (however, see Section III.D).

The discussion now turns to the weak anomalies that are not present in classical theory. If one field is kept constant, the binary mixture is brought into analogy with the pure fluid, in which differentiation of a density with respect to a field, while another density is constant, leads to a weak anomaly (Section II.H). Consequently, the following derivatives are, in general, examples of weak divergences in binary mixtures:

$$(\partial^2\mu_2/\partial p^2)_{Tx} = (\partial V/\partial p)_{Tx} = -VK_{Tx}$$

$$-(\partial^2\mu_2/\partial T^2)_{px} = (\partial S/\partial T)_{px} = C_{px}/T \qquad (43)$$

or, in alternative variables,

$$(\partial^2 p/\partial\mu_1{}^2)_{T\rho_2} = (\partial\rho_1/\partial\mu_1)_{T\rho_2}$$

$$(\partial^2 p/\partial\mu_2\partial T)_{\rho_1} = (\partial\rho_2/\partial T)_{\rho_1} \qquad (44)$$

These derivatives diverge with exponent α along the weak path, a path in the coexistence surface intersecting the critical line; they remain finite in the classical case, as discussed in Section III.B. The theory of nonclassical critical behavior therefore predicts that in general, properties that diverge strongly in pure fluids will diverge only weakly in binary mixtures of constant composition or constant molar density of one of the components. In studying the behavior of dilute mixtures, it is important to question the manner in which the crossover occurs from pure-fluid behavior to mixture behavior.

It is perhaps less obvious that a path in which two densities are held constant is asymptotically parallel to the critical line (or critical surface, for more than two components).[58] Consider a path in the coexistence surface, obtained by keeping only one density and $c - 1$ fields constant, with c the number of components. All other densities must vary relatively rapidly along this path when the critical point is approached, because their first derivatives diverge weakly. For a second density to remain constant, the corresponding path must bend over and become tangential to the critical line. Since this is an effect induced by a weak anomaly, it is not present in the classical case.

Since density-like properties vary smoothly along the critical line, second derivatives taken along this direction do not diverge. Examples are

$$(\partial^2\mu_2/\partial p^2)_{Sx} = (\partial V/\partial p)_{Sx} = -VK_{Sx}$$

$$-(\partial^2\mu_2/\partial T^2)_{Vx} = (\partial S/\partial T)_{Vx} = C_{Vx}/T \qquad (45)$$

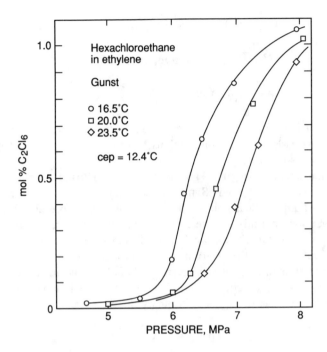

FIGURE 16.   The mole fraction of hexachloroethane in supercritical ethylene[59] along isotherms has an infinite pressure derivative at the critical endpoint ($\simeq 12.4°C$). The critical point of pure ethylene is $\simeq 9.2°C$.

the adiabatic compressibility and the constant-volume heat capacity. Nonclassical theory therefore predicts that while the adiabatic compressibility and isochoric heat capacity of a pure fluid diverge weakly, those of a mixture at constant composition in general do not diverge.

In the case of pure fluids, the question was raised: what happens to derivatives of densities with respect to densities, at constant field (Section II.A)? By an example, it was shown that those derivatives remain finite. An analogous situation exists in systems with more components.[11] An example of such a finite derivative is $(\partial V/\partial x)_{pT}$, a quantity that defines the partial molar volumes (Section IV.D). We have

$$(\partial V/\partial x)_{pT} = - (\partial V/\partial p)_{xT}/(\partial x/\partial p)_{VT} \qquad (46)$$

Both derivatives on the right side diverge weakly at the critical line, but their ratio remains finite. Exceptions occur for $x = 0, 1$, where the compressibility of the solvent becomes strongly divergent. Section IV.E is a detailed discussion of how this crossover occurs.

Another very useful derivative is $(\partial x/\partial V)_{T\mu_2}$. This derivative describes the *density* dependence of the supercritical solubility in the presence of an inert phase of the pure solute. Applying the Griffiths-Wheeler rules, one may note by inspection that this derivative must be finite at the critical line. Here, there is no exception, since the state $x = 0$ cannot be reached as long as μ_2 is finite.

This conclusion can be generalized to imply that near a critical line densities vary linearly with densities on paths for which two fields are kept constant. Thus, as a simple consequence of critical mixture thermodynamics, the sharp increase of supercritical solubility, disappears if the solubility is plotted vs. density instead of pressure[59] (Figures 16 and 17).

D. ORDER PARAMETER IN FLUID MIXTURES

Symmetry arguments demonstrate that in one-component fluids the density is a good,

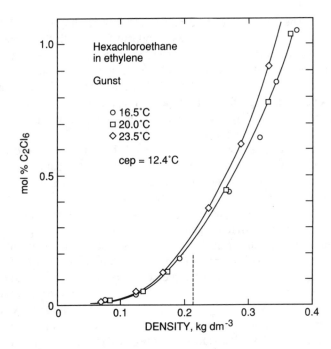

FIGURE 17. The mole fraction of hexachloroethane in supercritical ethylene along isotherms varies roughly linearly with the density of the solution.[59]

albeit not perfect, choice of an order parameter. In fluid mixtures, it is in general an open question as to what is the order parameter. Only in a few limiting cases is this choice unequivocal. For consolute points of virtually incompressible binary liquids, the potential $\mu_1(p,\Delta,T)$, with densities x and S, is often a good choice.[60] Better than the mole fraction, however, is the volume fraction,[60] especially if the two components have molecules of very different sizes. In binary mixtures near plait points, it is natural to assume that $p(\mu_1,\mu_2,T)$ is a good potential since it approaches the "good" potential $p(\mu,T)$, with density as the order parameter, for the pure components as x → 0, 1. The nonclassical model for binary gas-liquid mixtures, by Leung and Griffiths,[61] uses a potential closely related to $p(\mu_1,\mu_2,T)$, with the mixture density as the order parameter. The application of this model to fluid mixtures, and especially the variant of the model extensively applied by Rainwater et al.[62-64] to binary mixture coexistence curves, has demonstrated that the model becomes less powerful as the components become more unlike. It is not clear whether it is improper choice of order parameter or some other limitation of the model that causes this problem.

Since a liquid-liquid critical line can transform itself continuously into a gas-liquid critical line, it is clear that the two extreme forms of definition of an order parameter, mole fraction and density, must somehow be mixed. Thus far, there have been few ideas on how this can be managed. In fact, the formulation of scaled equations of state for fluid mixtures has made no essential progress since the Leung-Griffiths model was formulated for helium mixtures and adapted to other mixtures by D'Arrigo et al.,[65] Moldover and Gallagher,[66] Rainwater et al.,[62-64] and Chang and Doiron.[67]

E. EXPONENT RENORMALIZATION

Depending on whether two fields, one field and one density, or two densities are held constant in the differentiation of a density with respect to a field, properties result that are strongly, weakly, or nondivergent. These divergences are characterized by exactly the same critical exponents as in the pure fluid, which is a consequence of the principle of critical-point universality. The exponent will be renormalized depending on the path of approach to the critical point. In the case of the one-component fluid, both the strongly and weakly diverging properties have their critical exponents, γ and α, respectively, reduced by a factor of $1/\beta\delta$ ($^2/_3$, classical, or ≈ 0.64, nonclassical) if measured on a strong path, rather than on the special direction in which they are defined (Section II.D and Table 1). In the mixture, the critical exponents undergo the same type of renormalization. Thus, the supercritical solubility enhancement, $(\partial x/\partial P)_{T,\sigma}$, will diverge with an exponent $\gamma/\beta\delta$ ($^2/_3$ classically) on the isotherm through the critical endpoint.

Since there are now three types of paths, another renormalization occurs that is not present in the one-component fluid, namely a path asymptotically parallel to the critical line. This renormalization is called Fisher renormalization.[58] On such a path, two densities are held constant, and the exponents γ, δ, and β are enhanced by a factor of $1/(1 - \alpha)$, (≈ 1.1). The weak anomaly of the heat capacity C_{px}, however, is transformed into a finite cusp.

In mixtures, exceptions exist to the general rule when the directions special to the mixture, the coexistence surface and the critical line, assume special orientations with respect to the physical fields. When the critical pressure goes through a maximum as a function of composition, for instance, the path of constant pressure becomes parallel to the critical line and is no longer a strong direction at that point. This has consequences for the observable critical properties. Griffiths and Wheeler[11] give many examples of such cases. Where strong anomalies are concerned, these authors have been proven correct in all cases. With weak anomalies the evidence is less clear-cut, mainly because the range in which a renormalized exponent is to be observed is often too small to be studied.

The dilute mixtures that are of special interest here can be considered to be one of the exceptional cases mentioned above, since the dominance of the term RT ln x causes a path of constant chemical potential to almost coincide with a line of constant composition.

F. RANGE AND SIZE OF CRITICAL ANOMALIES IN MIXTURES

As in the case of pure fluids, it is important to know in which range predicted critical anomalies have been experimentally observed. The most conspicuous signature of the non-classical behavior of fluid mixtures is the cubic shape of the coexistence curve, which is as ubiquitous and well recognized as in one-component fluids.[24,33,34,60-67] In fact, the principal weakness of the popular cubic equations, which are almost always (incorrectly) forced to pass through the actual critical point of the fluid, may well be their inability to properly describe the flatness of the coexistence curve.

The case of mixtures is complicated by the fact that the strong divergences themselves are generally not observable. Exceptions are critical opalescence, gravitational sedimentation, and supercritical solubility enhancement.

Critical opalescence is governed by the osmotic susceptibilities (Equation 38) and has been observed in fluid mixtures in the same range as in one-component fluids.[46,68] It may interfere with optical measurements near critical lines.[47,48]

Gravitational sedimentation refers to a density-composition gradient that develops in near-critical mixtures.[50] Near a consolute point, the gradient is preponderantly compositional, but near a plait point, the composition gradient is coupled to a large density gradient.[69] The composition gradient is defined as $(\partial x/\partial h)_{T,\Delta\mu''}$, with $\Delta\mu''$ the difference of the unit-mass chemical potentials of the two components per gram of the mixture. Since the chemical potentials vary linearly with the height h, h acts like a field and the derivative $(\partial x/\partial h)_{T,\Delta\mu''}$ is of the form of a strong divergence. Figure 8 shows the refractive index gradients observed by Straub[50] in near-critical two-component systems near gas-liquid critical points. They are indistinguishable from those for one-component fluids, but in the mixture, the gradient is a combined density-composition gradient.

For the purposes of this chapter, the most important observable divergence is *supercritical solubility enhancement,* defined here as $(\partial x/\partial p)_{T,\sigma}$, the isothermal increase with pressure of the mole fraction of the solute in the presence of an excess of this component in an inert phase (Section III.C), which keeps the chemical potential μ_2 of the solute virtually constant. From the form of the derivative, it is obvious that this is a strongly diverging property. The steep slope of the isothermal solubility vs. pressure curves near the critical point of the solvent is very well documented and visible in a range of 20 K or more from room-temperature critical endpoints.[59,70,71] (Figure 17). The divergence is centered on *a critical endpoint,* the end of the mixture critical line starting at the critical point of the solvent; it has nothing to do with the critical point of the solvent itself. In practical cases of very low solubility, the critical line may be so short that it cannot be distinguished from a point when the pressure and temperature measurements are of limited accuracy. It is, however, important to keep the principle in mind, because dilute near-critical mixtures have subtle features that are easily blurred if precision is not maintained. This topic is discussed further in Section IV.L.

The observability of the weak anomalies is of limited interest in the present context. The reader is referred to References 11 and 56.

G. CORRELATION FUNCTIONS IN MIXTURES; OSMOTIC SUSCEPTIBILITIES

Section II.J gave the relations between the compressibility and the integral of the correlation function H(r). Here, these relations are generalized to fluid mixtures, while use is made of the fundamental work of Kirkwood and Buff,[57] O'Connell,[43] Cochran and Lee,[72] and McGuigan and Monson,[73] and the discussion is limited to binary mixtures.

Section II.K, defining H(r) for a one-component fluid, is generalized to mixtures by introducing correlation functions $H_{ij}(r)$, where $\rho_i H_{ij}(r)$ measures the excess of the molar density of molecules of type i at position r, compared to the average, if a molecule of type j is at the origin. The integrals of the $H_{ij}(r)$ over all space

$$H_{ij} = \int H_{ij}(r)\, \mathbf{dr} \qquad (47)$$

relate to susceptibilities in complete analogy to the relation between the correlation function and the compressibility in the one-component fluid:

$$RT\chi_{ij} = RT\, (\partial\rho_i/\partial\mu_j)_{T\,\mu_{k\neq j}} = (\rho_i\rho_j\, H_{ij} + \delta_{ij}\rho_j) \qquad (48)$$

with δ_{ij} the Dirac delta function. For a two-component mixture,

$$RT\chi_{11} = RT\, (\partial\rho_1/\partial\mu_1)_{T\mu_2} = \rho_1{}^2 H_{11} + \rho_1$$

$$RT\chi_{12} = RT\, (\partial\rho_1/\partial\mu_2)_{T\mu_1} = RT\, (\partial\rho_1/\partial\mu_1)_{T\mu_2} = \rho_1\rho_2 H_{12}$$

$$RT\chi_{22} = RT\, (\partial\rho_2/\partial\mu_2)_{T\mu_1} = \rho_2{}^2 H_{22} + \rho_2 \qquad (49)$$

Since all osmotic susceptibilities on the left side diverge strongly at the critical line, all integrals on the right side must do likewise, and all integrands $H_{ij}(r)$ must become long ranged,[72,73] just as in the case for the one-component fluid (Section II.J). The total correlation function integral H_{12} is sometimes singled out, and its divergence described as the "clustering" or "condensation" of a near-critical solvent around a solute.[74] This description overlooks the facts that all H_{ij} diverge, and that the divergence is due to an excess of molecules far from the reference molecule. In the first few shells, no effects of criticality are expected, except that leading to the subtle and weak nonclassical $(1 - \alpha)$ behavior discussed in Section II.J. The supercritical solubility itself is due to short-range molecular effects and is therefore not expected to have a significant critical anomaly.

For a near-critical mixture at a fixed thermodynamic state there is only one decay length ξ and all $H_{ij}(r)$ decay with that same length:

$$H_{ij}(r) \propto e^{-r/\xi}/r; \text{ large } r, \xi \text{ finite} \tag{50}$$

The decay length diverges as in the pure fluid (Equation 28), but additional constraints (paths) need to be specified, namely one or more fields to be kept constant, depending on the number of components. The amplitude ξ_o in Equation 28 will vary with the composition of the mixture.

In the case of dilute binary mixtures, $x_2 \to 0$, the osmotic susceptibility $(\partial \rho_1/\partial \mu_1)_{T\mu_2}$ approaches $(\partial \rho/\partial \mu)_T = \rho^2 K_T$, with K_T the compressibility of the pure solvent, while the osmotic susceptibilities $(\partial \rho_1/\partial \mu_2)_{T\mu_1}$ and $(\partial \rho_2/\partial \mu_2)_{T\mu_1}$ approach the value 0.

By choosing the conventional system of variables p and T, the expressions for the relevant susceptibilities, which were worked out by Kirkwood and Buff,[57] become more complex. Thus, they have

$$(\partial \mu_2/\partial \rho_2)_{pT} RT = 1/\rho_2 + \frac{H_{12} - H_{22}}{1 + \rho_2(H_{22} - H_{12})} \tag{51}$$

$$(\partial \mu_1/\partial \rho_1)_{pT}/RT = 1/\rho_1 + \frac{H_{12} - H_{11}}{1 + \rho_1(H_{11} - H_{12})} \tag{52}$$

In terms of our convention: $x = (\rho_2/\rho_1 + \rho_2)$; $\rho = (\rho_1 + \rho_2) = N_A V^{-1}$, with N_A Avogadro's number and V the molar volume, the osmotic susceptibilities defined in Equation 38 (by simplifying the corresponding expression in the paper by Kirkwood and Buff) become

$$RT(\partial x_1/\partial \mu_1)_{PT} = - RT(\partial x_2/\partial \mu_1)_{PT} = x_1 (1 - \rho x_1 x_2 \Delta H)$$

$$RT(\partial x_2/\partial \mu_2)_{PT} = - RT(\partial x_1/\partial \mu_2)_{PT} = x_2 (1 - \rho x_1 x_2 \Delta H) \tag{53}$$

where $\Delta H = 2 H_{12} - H_{11} - H_{22}$, with a factor of N_A absorbed as before. These more familiar osmotic susceptibilities diverge strongly at the critical line in the same fashion as their earlier-defined counterparts (Equations 42 and 49). Therefore, all osmotic susceptibilities diverge in the same way, except for the prefactor of x_i^2 which dampens the amplitude of the susceptibilities with respect to μ_i, as x_i approaches zero. It is also seen from Equation 53 that for finite ΔH the osmotic susceptibilites with respect to μ_i approach zero when x_i goes to zero, the appropriate infinite-dilution limit. The interesting question is how osmotic susceptibilities of dilute mixtures decide whether they are near criticality or near infinite dilution (see Section IV.K).

As was pointed out by O'Connnell,[43] in fluid mixtures as well as in one-component fluids it is advantageous to define and use the direct correlation function $C_{ij}(\mathbf{r})$ and its integral C_{ij}, since this integral does not diverge. The Ornstein-Zernike equation for an n-component mixture is[43]

$$H_{ik}(\mathbf{r}) = C_{ik}(\mathbf{r}) + \sum_{j=1}^{n} \rho_j \int C_{ij}(|\mathbf{r'}|) H_{jk}(|\mathbf{r} - \mathbf{r'}|) \, d\mathbf{r'} \tag{54}$$

where the indices i, j, and k refer to the species of molecule considered. O'Connell[43] gives the expressions for a number of thermodynamic derivatives in terms of the direct correlation function integrals $C_{ij} = \int C_{ij}(r)\, \mathbf{dr}$. From these are obtained

$$A_{VV} = (VK_{Tx})^{-1} = RT\, \rho^2\, [1 - \{(1 - x)^2 C_{11} + 2x\, (1 - x)\, C_{12} + x_2 C_{22}\}]$$

$$A_{Vx} = -\, (\partial p/\partial x)_{VT} = -\, RT\, \rho\{(1 - x)\, (C_{11} - C_{12}) - x\, (C_{12} - C_{22})\} \qquad (55)$$

The equation for A_{VV} is in apparent analogy with Equation 31. In the one-component case, however, the expression for the compressibility diverges at the critical point: ρC_{11}, known to be finite, approaches the value of 1. In the classical case, where K_{Tx} is finite at the critical line, the right side of Equation 55 is not equal to zero on the critical line, except at $x = 0, 1$, where it goes to zero strongly. In nonclassical theory the expression approaches zero weakly on the critical line, except at $x = 0, 1$, where it does so strongly.[7,9] The true analog of Equation 30 for the fluid mixture relates an expression involving a matrix of direct correlation function integrals[57] to the osmotic susceptibilities that diverge strongly.

Thus, the generalization of the concepts of the total and direct correlation functions to fluid mixtures leads to a number of interesting expressions for both strongly and weakly diverging thermodynamic derivatives. As long as only correlation function *integrals* are used, however, one has done no more than give existing thermodynamic derivatives new names. This may or may not be useful, depending on whether the integral evokes a correct or an incorrect intuitive image of molecular interactions. The introduction of correlation function integrals becomes more than a game when one actually tries to calculate them from specified molecular interactions, either by approximate mathematical procedures or by computer simulation. If such calculations are extended to the nearness of a critical state, many problems are encountered. Mathematical approximations lead to internal inconsistencies in many cases.[75-78] Computer simulations of the total correlation function must be carried out for box lengths larger than the correlation length, which is appreciable in quite a large range around the critical point.[22,23] Since the critical anomaly is a subtle divergence of the tail of the correlation function, it is not clear that this anomaly has any relevance to short-range chemical effects of interest in supercritical solubility. Just like the internal energy in the one-component fluid (Section II.J), such short-range effects have an $(1 - \alpha)$ type anomaly that will not be observable except under the most carefully controlled circumstances.

H. SUMMARY

This section generalized the concepts of strong and weak critical divergences to fluid mixtures by means of the principles of Griffiths and Wheeler. Two types of path-dependent renormalization of critical exponents have been discussed. Range and size of critical anomalies in mixtures, and their possible effects on experiments, have been described. The properties that are believed to be of interest in supercritical extraction have been emphasized. A number of osmotic susceptibilities have been introduced and related to correlation function integrals. Their behavior at the critical line and in the limit of infinite dilution has been discussed. It is argued that the long-range behavior of the correlation function should be of little relevance to supercritical solubility.

IV. DILUTE NEAR-CRITICAL SOLUTIONS

A. EXPANSION OF A CLASSICAL HELMHOLTZ FREE ENERGY

As was done for the one-component fluid, the critical behavior of several much-used thermodynamic properties of mixtures for a simple model will be calculated. This behavior is the classical Helmholtz free energy $A(V,T,x)$ of a binary mixture from which all other mixture properties can be derived by means of the relations summarized in Table 2. In order

TABLE 2
Thermodynamic Properties in Terms of
Derivatives of the Helmholtz Free Energy
A(V,T,x)

Pressure p	$-A_V$
Entropy S	$-A_T$
Energy U	$A - TA_T$
Enthalpy H	$A - TA_T - VA_V$
Chemical potential Δ	A_x
Chemical potential μ_1	$A - xA_x - VA_V$
Chemical potential μ_2	$A + (1 - x)A_x - VA_V$
Heat capacity C_V	$-TA_{TT}$
Heat capacity C_p	$-TA_{TT} + T(A_{VT})^2/A_{VV}$

TABLE 3
Thermodynamic Derivatives in Terms of Derivatives of the Helmholtz
Free Energy A(V,T,x)

$(\partial p/\partial V)_{Tx} = (VK_{Tx})^{-1}$	A_{VV}
$(\partial p/\partial T)_{Vx}$	$-A_{VT}$
$(\partial V/\partial T)_{px} = (V\alpha_p)_{Tx}$	$-A_{VT}/A_{VV}$
$(\partial p/\partial x)_{VT}$	$-A_{Vx}$
$(\partial V/\partial x)_{pT}$	$-A_{Vx}/A_{VV}$
$(\partial H/\partial x)_{pT}$	$A_x - TA_{Tx} + TA_{Vx}A_{VT}/A_{VV}$
$(\partial C_{px}/\partial x)_{pT}/T$	$-A_{TTx} + (A_{VVT}A_{Vx} + 2A_{VT}S_{VTx})/A_{VV} +$
	$+ [-2A_{VT}A_{Vx}A_{VVT} - (A_{VT})^2A_{VVx}]/A_{VV}^2 +$
	$+ A_{Vx}(A_{VT})^2A_{VVV}/(A_{VV})^3$
$(\partial \mu_1/\partial x)_{pT}$	$-x \, \text{Det} \, A/A_{VV}$
$(\partial \mu_2/\partial x)_{pT}$	$(1 - x) \, \text{Det} \, A/A_{VV}$
Det A	$A_{VV}A_{xx} - A_{Vx}^2$

to specialize to dilute mixtures, the Helmholtz free energy is expanded around the critical point of the major component, the solvent, or component 1. The second, minor component is the solute; its mole fraction is denoted by x_2 or by x, if no confusion is possible. The molar Helmholtz free energy of a binary fluid mixture can be separated into the following parts[15,79]

1. A finite analytic residual part, $A^r(V,T,x)$, which represents the difference in free energy of the real mixture and a perfect gas at the same volume and temperature
2. A(T) which depends solely on the temperature and is analytic except at T = 0 K
3. A perfect-gas volume dependence $RT \cdot \ln V$ and an ideal-mixing term $RT \cdot \{x\ln x + (1 - x) \ln (1 - x)\}$

All parts of the mixture Helmholtz free energy are well behaved at the mixture critical point, except for the ideal mixing term at x = 0, 1. Therefore, before expanding the Helmholtz free energy at the critical point of the solvent the ideal-mixing term is subtracted. The part of the Helmholtz free energy that is expanded by A'' (V,T,x) is denoted with

$$A''(V,T,x) = A(V,T,x) - RT \cdot \{x\ln x + (1 - x) \ln (1 - x)\} \qquad (56)$$

Thus,

$$A''(V,T,x) = A''^c + A_V^c(\delta V) + A_T''^c(\delta T) + A_x''^c x +$$

$$+ A_{VT}^c(\delta V)(\delta T) + A_{Vx}^c(\delta V)x + \ldots$$

$$+ A_{VVT}^c(\delta V)^2(\delta T)/2 + A_{VVx}(\delta V)^2 x/2 \ldots$$

$$+ \ldots$$

$$+ A_{VVVV}^c(\delta V)^4/24 + \ldots \tag{57}$$

where the subscripts V, T, x, appended to A, denote (repeated) partial differentiation with respect to the pertaining variables volume V, temperature T, and mole fraction of solute x; they do therefore not denote variables held constant. The symbol δ denotes the difference between the value of the variable that follows it, and its value at the critical point of the solvent. The superscript c denotes that the pertaining derivative is evaluated at the critical point of the solvent. The symbol $''$ has been omitted in all cases in which the derivatives of A and of A$''$ are identical. Note that $-A_V^c$ equals the critical pressure of the solvent, $-A_{VT}^c$, the limiting slope of its vapor pressure curve and critical isochore at the critical point, and A_{Vx}^c equals $(-\partial p/\partial x)_{VT}^c$.

Asymptotic expansions for the pressure p and isothermal compressibility K_T are readily obtained from Equation 57:

$$p = -A_V = -A_V^c - A_{VT}^c(\delta T) - A_{Vx}^c x - A_{VVT}^c(\delta V)(\delta T) +$$

$$- A_{VVx}^c(\delta V)x - A_{VVVV}^c(\delta V)^3/6 - \ldots \tag{58}$$

$$(VK_T)^{-1} = A_{VV} = A_{VVT}^c(\delta T) + A_{VVx}^c x + A_{VVVV}^c(\delta V)^2/2 + \ldots \tag{59}$$

Other useful expansions are those of the chemical potentials μ_1, μ_2, and their difference. Since $A_x = \mu_2 - \mu_1$, often denoted as Δ,

$$\Delta = \mu_2 - \mu_1 = A_x''^c + A_{Vx}^c(\delta V) + A_{xx}'' x + A_{VVx}^c(\delta V)^2/2 + \ldots \tag{60}$$
$$+ RT \ln x - RT \ln (1 - x)$$

The expansions for μ_1, μ_2 individually are obtained by means of the relations given in Table 2:

$$\mu_1 = A''^c + A_T''^c(\delta T) + V_c p - A_{Vx}^c(\delta V)x - A_{VVVV}^c(\delta V)^4/8 + \ldots - RT x + \ldots \tag{61}$$

$$\mu_2 = A''^c + A_x''^c + A_{Vx}^c(\delta V) + (A_T''^c + A_{Tx}''^c)(\delta T) + A_{xx}'' x + V_c p + \ldots \tag{62}$$

$$+ RT \ln x + \ldots$$

where, in both Equations 61 and 62, the expansion for the pressure (Equation 58), is to be substituted for the symbol p. For dilute mixtures, the leading behavior of μ_1 is the same as that of the pressure; for μ_2, the RT ln x term dominates, while the term $A_{Vx}^c(\delta V)$ is of lower order than the pressure.

Dilute near-critical mixtures were extensively studied in the U.S.S.R. in the 1960s. Many experiments were performed by Krichevskii's group,[2,3] while Rozen,[5] in 1976, did a careful analysis of the classical behavior of such mixtures by means of an expansion of the type introduced here.

B. INITIAL SLOPE OF THE CRITICAL LINE

The initial slope of the critical line (CRL) is obtained from the expansion (Equation 57) by imposing the first criticality condition (Equation 35). For small x is obtained:

$$A_{VV} = 0 + A^c_{VVx}x + A_{VVT}(\delta T) + A_{VVVV}(\delta V)^2 + ..$$

$$A_{Vx} = A^c_{Vx} + ..$$

$$A_{xx} = RT_c/x + .. \tag{63}$$

so that

$$\text{Det } A = [A^c_{VVx}x + A_{VVT}(\delta T)] \cdot RT_c/x - A^{c\,2}_{Vx} = 0 \tag{64}$$

From Equation 64 it is found that the initial rise of the critical temperature due to the admixture of the second component, $dT/dx|^c_{CRL}$, is given by

$$dT/dx|^c_{CRL} = \frac{(A^c_{Vx})^2 - A^c_{VVx} RT_c}{RT_c A^c_{VVT}} \tag{65}$$

as was derived by Redlich and Kister.[80] The initial pressure increase, $dp/dx|^c_{CRL}$, is obtained from the temperature slope (Equation 65) by means of a thermodynamic identity noted by Krichevskii[2]

$$dp/dx|^c_{CRL} = (\partial p/\partial x)^c_{VT} + dp/dT|^c_{CXC}\, dT/dx|^c_{CRL} \tag{66}$$

Here, $dp/dT|^c_{CXC}$ stands for the slope of the vapor pressure curve of the pure solvent at the critical point, which is known to equal the slope of the critical isochore.[16] Therefore,

$$dp/dx|^c_{CRL} = -A^c_{Vx} - A^c_{VT} \cdot dT/dx|^c_{CRL} \tag{67}$$

with $dT/dx|^c_{CRL}$ from Equation 65. From Equations 65 and 67 it follows that

$$dp/dT/|^c_{CRL} = -A^c_{VT} - A^c_{Vx}/(dT/dx|^c_{CRL}) \tag{68}$$

The initial slope of the critical volume is obtained from the second criticality condition[15] (refer to the paper by Morrison[10] for details).

Equation 68 states that in the p-T plane the slope of the critical line differs from that of the vapor pressure curve (or of the critical isochore, $-A^c_{VT}$) by an amount of $-A^c_{Vx}/(dT/dx|^c_{CRL})$. The sign of A^c_{Vx} determines whether the critical line in the p-T plane starts off in the half-plane below the vapor pressure and its extension, or in the half-plane above it (Figure 18). If the pressure drops when some solvent molecules are replaced by solute molecules at constant volume and temperature, A^c_{Vx} is positive and the critical line takes off in the lower half-plane. It is customary to associate such a pressure drop with a solute volatility lower than that of the solvent. If the pressure rises on increasing the mole fraction of the solute, one associates this with volatility of the solute, and the critical line takes off in the half-plane above the vapor pressure curve and its extension, irrespective of the sign of $dT/dx|^c_{CRL}$. As an example, Figure 19 is a schematic of the initial part of the critical line for several volatile solutes in near-critical steam. Gases far above their critical points have critical lines in the upper half-plane. There are two categories, those with weak interactions such as helium,[81] hydrogen,[81] and, as a model, the ideal gas,[82] and those with

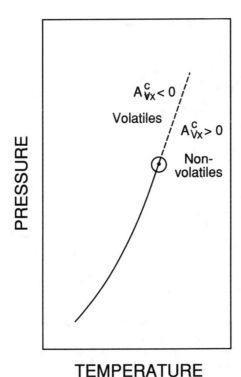

FIGURE 18. The pure solvent vapor pressure curve and its extension, the critical isochore, divide the p-T plane into two regions: where the critical lines of the nonvolatiles originate (lower, right half plane), and where those of the volatiles originate (upper, left half plane).

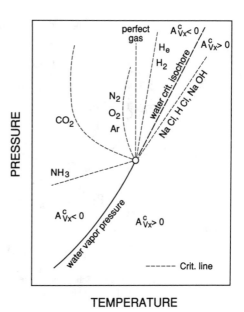

FIGURE 19. Schematic of critical lines of aqueous solvents. Nonvolatile solutes have critical lines originating in the lower, right half plane, $A^c_{Vx} > 0$. Strongly interacting nonvolatile solutes, such as HCl, NaCl, and NaOH, have critical lines very close to the extension of the water vapor pressure curve.[8] Weakly and noninteracting volatile solutes, such as H_2, He, and the ideal gas, have critical lines close to the vertical.[81,82] More strongly interacting volatile solutes, such as CO_2 and NH_3, have critical lines running to much lower temperatures.[81,83]

somewhat stronger interactions, such as nitrogen,[81] oxygen,[81] carbon dioxide,[81] and ammonia.[83] The ideal gas, with zero interaction energy and zero excluded volume, forms the dividing line between these two groups of gases and its critical line rises vertically in the p-T plane.[82] All categories, however, have a negative value of A^c_{Vx}. The gases with weak molecular attractions but non-negligible excluded volume are found in the segment between the vertical and the extended vapor pressure curve. The gases with somewhat stronger molecular attractions and larger excluded volume cause an initial decrease of the critical temperature and are found in the part of the upper half-plane left of the vertical. The transition across the dividing line, the critical isochore, leads from strongly attractive to weakly, mostly repulsively interacting solutes, and is discontinuous.

The critical lines of electrolytes in steam appear to be indistinguishable from the extension of the vapor pressure curve in the p-T plane. This is due to the fact that $dT/dx|^c_{CRL}$ is extremely large for such systems (see Equation 68). That these critical lines actually run in the lower half-plane is deduced from other pieces of evidence, namely the sign of the critical anomalies in partial molar properties (Sections IV.F to IV.H). The geometry of critical lines is one of many examples of the decisive role of the derivative $A^c_{Vx} = (\partial p/\partial x)^c_{VT}$ in determining the behavior of dilute near-critical mixtures.

C. THE DEW-BUBBLE CURVE

Figure 20 depicts two cases of the isothermal dew-bubble curve near the critical point of the solvent, that of a nonvolatile (left) and that of a volatile (right) solute. For simplicity, assume that the mixtures have a gas-liquid critical line connecting the critical points of the

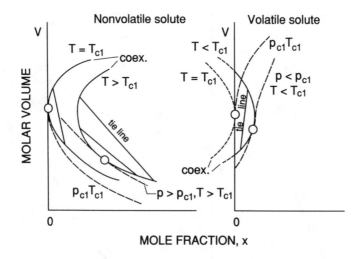

FIGURE 20. Schematic of the effect on volume of the addition of a nonvolatile (left) or volatile (right) solute to a near-critical solvent.

two components; appropriate examples are ethane as a nonvolatile solute, in methane near
its critical point, and methane as a volatile solute, in ethane near its critical point.

Isothermal addition of the nonvolatile solute to the critical solvent induces a phase
separation and a two-phase region opens up because the solute is far below its critical point
(Figure 20, left). At a temperature higher than the critical temperature of the solvent, this
two-phase region pulls away from the x = 0 axis and the critical point moves away from
the extremum (Figure 20, left). In the case of isothermal addition of a volatile (Figure 20,
right), no two-phase region appears because the solute is far above its critical point. One
would have to subtract solute, or make x negative, to induce a virtual phase separation (see
the dashed coexistence curve in Figure 20, right). By lowering the temperature, however,
the ''virtual'' phase separation slips across the x = 0 axis (see the full coexistence curve
for $T < T_{c1}$ in Figure 20, right). A tie line and the critical point of the mixture are indicated.
The critical isotherm-isobar runs in opposite directions in these two cases. In the case of
the nonvolatile solute, isothermal-isobaric addition of the solute forces the system into the
liquid state (Figure 20, left). Adding a volatile solute forces the isotherm-isobar into the
vapor phase.

In Figure 21, the two cases in the p-x plane are shown. In the simple case of a connected
critical line, the two cases are mirror images of each other, and the difference between them
is one of perspective only, because the x = 0 case is called the solvent, and is plotted on
the left.

The case of the volatile addition (Figure 21, right) leads to isothermal dew-bubble loops
that begin at the pure-solvent axis, the solvent being at, or below its critical point. For
temperatures between the critical temperatures of pure solvent and pure solute, such curves
reach a maximum on the critical line (Figure 21, right, curve a). Dew-bubble curves of type
a shrink to a point at the critical temperature of the solvent.

In the case of nonvolatile addition (Figure 21, left), both the sub- and supercritical
isothermal dew-bubble loops stretch all the way to x = 1, because the solute is always
below its critical point. Subcritical loops extend from x = 0 to x = 1; supercritical loops
do not reach the x = 0 axis but have a maximum on the critical line (Figure 21, left, curve
a). The interesting phenomenon here is the transition from the supercritical case of a smooth
curve, (a), to the subcritical case where the dew and bubble curves meet at an angle, (c).
This occurs by the formation of a ''bird's beak'', a cusp, (b). The equation for the limiting
slope of the ''birds's beak'' will be derived; to this end, the relation between the volumes
and compositions of two coexisting phases at $T = T_c$ must be determined.

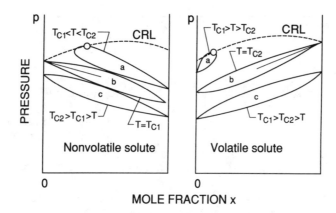

FIGURE 21. Schematic p-x dew-bubble curves for nonvolatile (left) or
volatile (right) solutes added to a near-critical solvent, in the simple case
of a connected gas-liquid critical line.

Before providing the derivation, the intuitive or pictorial argument that underlies the
derivation is given. For coexistence, the pressures and the chemical potentials of each of
the components must be equal. To lowest order, the equations for the pressure (Equation
58), and chemical potential of the solvent (Equation 61), are equivalent and uninteresting;
that for the solute (Equation 60), however, is dominated by the term RT ln x, which varies
strongly with composition when x is small. Changes in composition must be compensated
for by the term linear in δV in order to keep the chemical potential of the solute the same
in coexisting phases. Therefore, the compositions of the two coexisting phases must remain
close to each other while the molar volumes grow apart rapidly. Figure 20 represents this
argument pictorially: the volumes of coexisting phases change strongly in opposite directions
starting at V_c, while the tie lines remain almost vertical, indicating nearly identical compositions.

Equality of the pressures requires, at $T = T_c$, and by Equation 58

$$A^c_{Vx}(x_v - x_1) = - A^c_{VVVV} [(\delta V_v)^3 - (\delta V_1)^3]/6 \qquad (69)$$

and equality of the chemical potential of the second component:

$$RT \ln x_\ell/x_v = - A^c_{Vx} [(\delta V_\ell) - (\delta V_v)] \qquad (70)$$

Here, the subscripts v and 1 refer to the coexisting vapor and liquid phases, respectively.

Under the assumption that $|x_v - x_1| \ll x_1, x_v$ (as intuited above, and to be verified later),
one may write

$$RT \ln x_\ell/x_v \ (= - A^c_{Vx} [(\delta V_\ell) - (\delta V_v)]) \simeq RT (x_\ell - x_v)/x_{\ell,v} \qquad (71)$$

With the "Ansatz" $\delta V_1 \simeq - \delta V_v$, the following equation is obtained from Equations 69
and 71

$$(\delta V_v)^2 \simeq (\delta V_\ell)^2 = 6 (A^c_{Vx})^2 x_{v,1}/(A^c_{VVVV}RT_c) \qquad (72)$$

or

$$(\delta V_{v,\ell}/V_c)^2 = B_2^2 x_{v,1} \qquad (73)$$

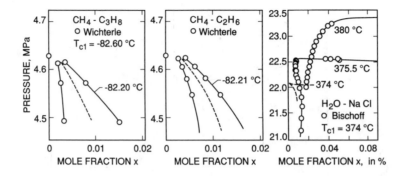

FIGURE 22.   The "bird's beak" developing in dew-bubble curves of slightly supercritical methane solutions (left two)[84] and aqueous NaCl[85,86] (right). The pure
solvent critical pressure is indicated at x = 0.

with the critical amplitude B_2 given by

$$B_2 = [6 \, (A^c_{Vx})^2/(RT_c \, A^c_{VVVV})]^{1/2}/V_c \tag{74}$$

We now develop the expression for the dew-bubble curve in p-x space. If the expressions
(Equation 73) for δV_1, δV_v are substituted in the relation for $x_1 - x_v$ (Equation 69), it is
found that $x_1 - x_v$ varies as $x^{3/2}$, and therefore this difference disappears more rapidly than
the individual x values, as assumed earlier. If these expressions for δV_1, δV_v are substituted
in Equation 58 for the pressure on the dew and bubble side at the critical temperature of
the solvent, one finds that they lead to terms of order higher than x, so that the term
$- A^c_{Vx}x$ is the leading term on both sides. The same result is obtained for nonclassical critical
behavior. Thus, the shape of the dew-bubble curve is as indicated in Figure 21 (left, curve
b), and the dew and bubble curves have a joint slope given by the now-familiar quantity
$- A^c_{Vx}$. For examples of ethane and propane in methane[84,85] and of NaCl in H_2O,[85,86] see
Figure 22.

D. PARTIAL AND APPARENT MOLAR PROPERTIES

The total volume V_t, enthalpy H_t, Gibbs free energy G_t, and heat capacity C_{pt} of a fluid
mixture are all extensive properties. At constant pressure and temperature, they are linear
in the amount of substance. The values they assume depend on the number of moles n_i of
each species i, and on temperature and pressure. If any of the above properties are denoted
by $F_t(n_1, n_2, \ldots, p, T)$, then the partial molar property $\overline{F}_i$ is defined as[15]

$$\overline{F}_i = (\partial F_t/\partial n_i)_{pTn_{j(j \neq i)}} \tag{75}$$

For a mole of a binary mixture, the total value of any extensive property, F_t, equals its
molar value F. In terms of the mole fraction x of the second component,[15]

$$\overline{F}_1 = F - x \, (\partial F/\partial x)_{pT}$$

$$\overline{F}_2 = F + (1 - x) \, (\partial F/\partial x)_{pT} \tag{76}$$

As an example, Figure 23 shows a plot of the volume of a mole of mixture as a function
of x. The partial molar volume, $\overline{V}_1$ and $\overline{V}_2$, at a mole fraction x_1 are obtained by drawing
a tangent to the volume curve at x_1 and reading off the intercepts of the tangent with the
verticals at x = 0, 1. The plot also shows the excess volume V^E, defined as the difference

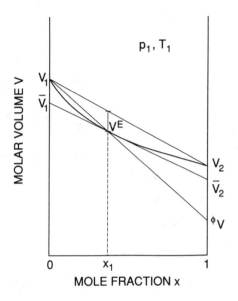

FIGURE 23. The molar volume of a mixture as function of composition at fixed pressure and temperature, and the geometrical definition of excess (V^E), partial (V_2), and apparent (ϕV) molar volumes.

of the actual volume with the mole fraction-averaged volume of the two pure components at the same pressure and temperature.

In aqueous solutions it is customary to consider the so-called *apparent* molar property ϕF. It is defined as the increase in F_t per mole of solute, as m moles of solute are added to 1 kg of water, while pressure and temperature are kept constant.

$$\phi F = [F_t (1 \text{ kg water, m moles solute}) - F_t (1 \text{ kg water})]/m \qquad (77)$$

In terms of the molecular weight of water, M_w, that of the solute, M_s, and the mole fraction, x,

$$x^\phi F = F - F_w (1 - x) \qquad (78)$$

where F is the molar property of the solution and F_w is the molar property of water. Figure 23 shows a construction of the apparent molar volume ϕV in terms of mole fraction (Equation 78), while Figure 24 shows the construction of ϕV in terms of coordinates appropriate for an aqueous solution (Equation 77). In both cases, the construction involves a chord, not a tangent. Although thermodynamically the apparent molar property is less fundamental than the partial, the apparent property is much closer to experiment because it does not require tangent construction. For low concentrations of the solute the apparent molar property approaches the more fundamental partial molar property, except at the critical point of the solvent (Section IV.E).

E. DIVERGENCE OF THE PARTIAL MOLAR PROPERTIES OF THE SOLUTE

In order to understand the behavior of the partial and apparent molar properties as the critical point of the solvent is approached, the volume is used as an example, with Figure 25 as a pictorial answer.

Figure 25 shows the volume of a mixture at fixed temperature slightly above the critical

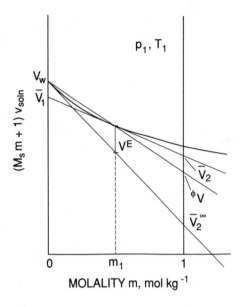

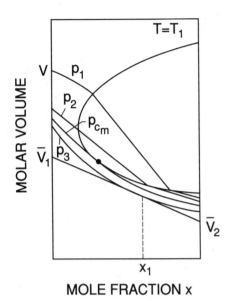

FIGURE 24. The total volume of an aqueous solution containing 1 kg of water, as a function of molality m at fixed pressure and temperature, and the geometric definition of excess, partial, apparent, and infinite-dilution ($\overline{V}_2^\infty$) volumes of the solute.

FIGURE 25. The volume vs. mole fraction of a solution of a nonvolatile solute in a supercritical solvent. The isothermal coexistence curve, mixture critical point (●), a number of isobars including the critical isobar, p_{c_m}, and some tie lines are indicated.

point of the solvent. An involatile solute is added, which induces a phase separation. The critical point of the mixture is indicated on the dew-bubble curve, and a few tie lines are shown. In order to construct the partial molar volumes, one needs to draw the tangent to the isotherm-isobar. Several isotherm-isobars, p_1, p_2, p_{cm}, at the critical pressure of the mixture, and p_3 are shown in Figure 25; p_1 passes through the two-phase region. An example of the tangent construction is shown at x_1 on the isotherm-isobar p_3. At the critical temperature of the solvent (Figure 26), the coexistence loop opens up at $x = 0$, and the critical point is on the $x = 0$ axis, so that the tie lines near $x = 0$ are almost vertical. For an involatile solute, as shown, the pressure drops along the coexistence curve as x increases, the solute being very far below its critical temperature. The critical isotherm-isobar is therefore confined to the one-phase region and to liquid-like densities. It is evident from Figure 26 that at the critical point of the solvent the tangent to the isotherm-isobar, defining the partials, will be vertical and the intercept defining $\overline{V}_2$ will tend to $-\infty$. It is not clear from Figure 26 what happens to $\overline{V}_1$, but this is the subject of a future discussion.

Several power laws can be conjectured by an inspection of Figure 26. For classical theory, the coexistence curve must be quadratic and the isotherm-isobar a cubic, $(V - V_c)^3 \simeq Bx$. It follows that $(\partial V/\partial x)_{pT}$ behaves as $x^{-2/3}$, so that $\overline{V}_2$ diverges to $-\infty$ as $(B/3)(Bx)^{-2/3}$. It is clear from Equation 76 that $\overline{V}_1$ approaches V_c, as one would expect; however, this is not so on other paths of approach to the critical point. The apparent molar volume, given by $x^\varphi V = V - V_c$ (Equation 78), will behave as $^\varphi V = (Bx)^{1/3} x^{-1}$, or $B (Bx)^{-2/3}$, which is a divergence as that of the partial molar volume, but with a three times larger amplitude (Figure 26). The discussion that follows is limited to the partial molar properties; however, the reader should not assume that in the limit of infinite dilution the results will always equal those for apparent molar properties.

A second method of appreciating the divergence of $\overline{V}_2$ is by referring to the thermodynamic identity

$$(\partial V/\partial x)_{pT} = -(\partial V/\partial p)_{Tx} \cdot (\partial p/\partial x)_{VT} = -A_{Vx}/A_{VV} \qquad (79)$$

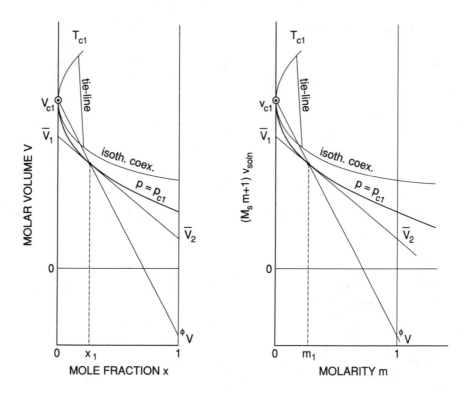

FIGURE 26. (Left) The volume vs. mole fraction of a solution of a nonvolatile solute in a critical solvent. The isothermal coexistence curve, the critical isotherm-isobar, and two tie lines are indicated. (Right) The corresponding plot for the convention used in aqueous solutions.

Since A_{Vx} is finite in classical theory, and A_{VV} is nonzero in the mixture, $(\partial V/\partial x)_{pT}$ is finite in a mixture everywhere, including at the critical line. The only exceptions are the cases $x = 0, 1$. In those cases, the derivative $(\partial V/\partial x)_{pT}$ becomes proportional to the compressibility of one of the pure components, and must therefore diverge strongly as this component approaches its critical point. As a consequence, infinite-dilution partial molar volumes diverge as the solvent's compressibility at the critical point of the solvent, irrespective of whether the solute is more or less volatile. From Equation 79 it is obvious, however, that the sign and amplitude of the divergence are determined by the sign and value of A_{Vx}^c. For volatile solutes in a near-critical solvent, this coefficient has a negative value, and therefore their infinite-dilution partial molar volume is positive. As an example, see Figure 27, which displays the measurements of Biggerstaff and Wood[87] for argon in supercritical steam. Nonvolatile solutes have negative infinite-dilution partial molar volumes (see data for naphthalene in carbon dioxide by van Wasen et al.[89] and by Eckert et al.[90]). Amplitude and sign of the critical anomaly are thus governed by the value of A_{Vx}^c.

The relation for the derivative $(\partial H/\partial x)_{pT}$ of the enthalpy in Table 2 contains the same term $-A_{Vx}/A_{VV}$ as that of the volume, but multiplied by a factor of $-TA_{VT}$. This factor equals $T(\partial p/\partial T)_{Vx}$, which at $x = 0$ represents the finite slope of the vapor pressure curve of the solvent. Thus, the partial molar enthalpy diverges in similar fashion as the partial molar volume. As an example, see Figure 28, which shows the isobaric heat of dilution data of Busey et al.[91] for NaCl in water at temperatures 20 to 50 K below the critical point of steam as a function of composition. The intercept becomes strongly negative as T_c is approached. Alternatively, one can write $(\partial H/\partial x)_{pT}$ as $TA_{Vx}^c(V\alpha_P)$, which relates the divergence of $\bar{H}_2$ to that of the expansion coefficient.[92]

For the partial molar heat capacity, however, the situation is more complex and further discussion is postponed until the case of the partial molar volume is explored in more depth.

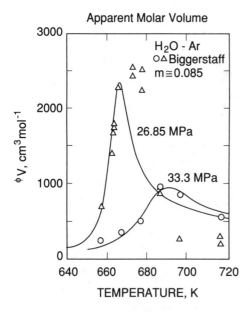

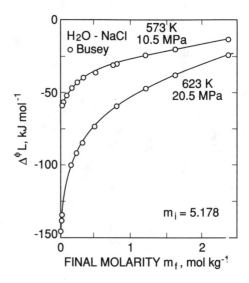

FIGURE 27.   Effectively infinite-dilution isobaric apparent molar volumes of argon as a function of temperature in supercritical steam.[87] The full curves are predictions from a corresponding-states model.[88]

FIGURE 28.   Apparent isothermal-isobaric heats of dilution of NaCl in near-critical liquid water as a function of composition.[91]

F. CLASSICAL PATH DEPENDENCE OF THE PARTIAL MOLAR VOLUMES

In the previous subsection, the partial molar properties of the solute were found to diverge at the critical point of the solvent, and along the infinite-dilution path the divergence takes a particularly simple form. In this subsection, the nature of the divergence on other paths is explored, again with the partial molar volume as an example.[5,7-10] In the process, some remarkable results are seen: a nonlinear initial composition dependence for $\overline{V}_2$ and a path-dependent limit for the (finite) value of $\overline{V}_1$.

We will work through an example for the classical case, where the expansion, Equation 57, is valid. To this end, we expand the factors A_{VV} and A_{Vx} in Equation 79. We obtain, asymptotically near the solvent's critical point

$$A_{Vx} = A_{Vx}^c + \ldots$$

$$A_{VV} = A_{VVx}^c x + A_{VVT}^c (\delta T) + A_{VVVV}^c (\delta V)^2/2 + \ldots \tag{80}$$

so that

$$(\partial V/\partial x)_{pT} = \frac{-A_{Vx}^c + \ldots}{A_{VVx}^c x + A_{VVT}^c(\delta T) + A_{VVVV}^c(\delta V)^2/2 + \ldots} \tag{81}$$

Equation 81 makes clear that the way the denominator approaches the value 0 will depend on the path of approach to the critical point.

The first path, that of infinite dilution, was discussed in Section IV.E. In that case, the denominator consisted of $A_{VVT}^c (\delta T) + A_{VVVV}^c (\delta V)^2/2 + \ldots$, which is the expansion of the inverse of the pure-solvent compressibility. The path dependence of the divergence of this property was the topic of Section II.D.

The next path that is of interest is the critical line. It has already been noted that

$(\partial V/\partial x)_{pT}$ remains finite on the critical line, except at $x = 0, 1$. On the critical line, T initially varies linearly with x (see Equation 65). The same is true for V, so that the term in $(\delta V)^2$ is of higher order. It follows from Equations 65 and 81 that for small x

$$(\partial V/\partial x)_{pT} = - (RT_c/A^c_{Vx}) \cdot x^{-1}; \text{ critical line} \tag{82}$$

Thus, by the definition of Equation 76, $\overline{V}_2$ diverges as x^{-1} while $\overline{V}_1$ approaches the value of

$$\overline{V}_1 \rightarrow V_c + RT_c/A^c_{Vx}; \text{ critical line} \tag{83}$$

which is, in general, finite but not equal to V_c.

Along the coexistence curve at $T = T_c$ (Figure 26), $(\delta V)^2$ varies linearly in x, so the two contributions in the denominator of Equation 81 are of the same order. With the expression (Equation 72) for the coexistence curve, we derive from Equation 81

$$(\partial V/\partial x)_{pT} = - [A^c_{Vx}/\{A^c_{VVx} + 3 (A^c_{Vx})^2/RT_c\}] \cdot x^{-1}; \text{ coexistence curve} \tag{84}$$

so that $\overline{V}_2$ diverges as x^{-1} and V_1 approaches the limiting value of

$$\overline{V}_1 (x = 0) = V_c - A^c_{Vx}/\{A^c_{VVx} + 3 (A^c_{Vx})^2/RT_c\} \tag{85}$$

Again, $\overline{V}_1$ does not approach V_c.

Another path of interest, for which the power law has already been conjectured, is the critical isotherm-isobar (Figure 26). This path is determined by means of the pressure expansion (Equation 58). At $T = T_c$, $p = p_c$,

$$(\delta V)^3 = - (6 A^c_{Vx}/A^c_{VVVV}) \cdot x; \text{ critical isotherm-isobar} \tag{86}$$

This equation shows why near-critical fluids are so sensitive to impurity: if the prefactor in Equation 86 is of order unity, a 0.1% addition of an impurity causes a change of 10% in density at constant T, p.

The partial molar volume of the solute is found from

$$(\partial V/\partial x)_{pT} = (-\text{sign } A^c_{Vx}) \left|2 A^c_{Vx}/9 A^c_{VVVV}\right|^{1/3} \cdot x^{-2/3} \tag{87}$$

so that $\overline{V}_2$ diverges as $x^{-2/3}$. The factor of $(-\text{sign } A^c_{Vx})$ ensures that for nonvolatile solutes, $A^c_{Vx} > 0$, $\overline{V}_2$ diverges to $-\infty$, and vice versa. The partial molar volume of the solvent, $\overline{V}_1$, is given by

$$\overline{V}_1 = V_c + (-\text{sign } A^c_{Vx}) \left|2 A^c_{Vx}/9 A^c_{VVVV}\right|^{1/3} \cdot x^{1/3} \tag{88}$$

On the critical isotherm-isobar, $\overline{V}_1$ approaches V_c, but it does not vary linearly with the composition. On paths near to, but not at the critical isotherm-isobar, the different terms in the denominator of Equation 81 compete and can even produce an extremum in the composition dependence of $\overline{V}_2$. The Russian literature[2,3,5] reports this anomalous composition dependence of $\overline{V}_2$.

G. THE PARTIAL MOLAR ENTHALPY

The partial molar enthalpies follow from the expression for $(\partial H/\partial x)_{pT}$ in terms of the derivatives of the Helmholtz free energy given in Table 2

$$(\partial H/\partial x)_{pT} = A_x - TA_{Tx} + TA_{Vx} A_{VT}/A_{VV} = A_x'' - TA_{Tx}'' - TA_{Vx} (V\alpha_p) \quad (89)$$

with α_p the thermal expansion coefficient at constant x, $\alpha_p = (1/V) (\partial V/\partial T)_{px}$.

In the limit of infinite dilution, $(\partial H/\partial x)_{pT}$ will diverge as the expansion coefficient. Since the expansion coefficient and the compressibility have similar critical behavior, the partial molar enthalpies will depend on the path of approach to the critical point of the solvent in a way similar to path dependence of the partial molar volumes. The anomalous composition dependence of the apparent molar heat of dilution was noted by Busey et al.[91] for aqueous NaCl (Figure 28).

H. PARTIAL AND APPARENT MOLAR HEAT CAPACITIES

The manner in which the partial molar volume and enthalpy of a solute, which are composition derivatives of finite and well-behaved extensive properties of the mixture, develop a strong divergence as the critical point of the solvent is approached has been worked through in this chapter. It is therefore not surprising that the partial molar heat capacity $\overline{C}_{p2}$, the composition derivative of a property that already exhibits a strong divergence, behaves more strongly anomalously than the apparent molar volume. A simple thermodynamic argument given by Wood[92] shows this in a different way. Since

$$\overline{C}_{p2} = (\partial \overline{H}_2/\partial T)_p, \quad (90)$$

and since $\overline{H}_2^\infty$ diverges as the expansion coefficient α_p (Equation 89), $\overline{C}_{p2}^\infty$ must diverge as the temperature derivative of α_p, although it must be borne in mind that the prefactor A_{Vx} and the other terms in Equation 89 lead to additional divergences on differentiation. Since α_p has an extremum close to ρ_c on near-critical isobars, the anomaly in $\overline{C}_{p2}^\infty$ must change signs near the point at which the system passes through the critical density of the solvent. This is indeed what was observed in the original experiments performed by Wood and co-workers[93,94] in somewhat supercritical volatile and nonvolatile aqueous systems (Figure 29).

For an analysis of the path dependence of $\overline{C}_{p2}^\infty$ and for a determination of the critical exponent on various paths, in principle, one can proceed as with the partial molar volume. The latter quantity was very simply written as the ratio of two derivatives of the Helmholtz free energy; however, this is not so for the partial molar heat capacity. The full expression for the partial molar heat capacity at finite x, in terms of derivatives of the Helmholtz free energy, is[8]

$$\overline{C}_{p2}/T = C_{px}/T - A_{TTx} (1-x) +$$

$$+ \frac{(1-x)}{A_{VV}} [A_{VVT} A_{Vx} + 2A_{VT} A_{VTx}] +$$

$$+ \frac{(1-x)}{A_{VV}^2} [-2A_{VT} A_{Vx} A_{VVT} - (A_{VT})^2 A_{VVx}] +$$

$$- \frac{(1-x)}{A_{VV}^3} (A_{VT})^2 A_{Vx} A_{VVV} \quad (91)$$

where $C_p(x)$, the molar heat capacity at mole fraction x, is given by

$$C_{px}/T = -A_{TT} + \frac{(A_{VT})^2}{A_{VV}} \quad (92)$$

There are, therefore, four types of terms in increasing powers of A_{VV}^{-1}. In the infinite-dilution

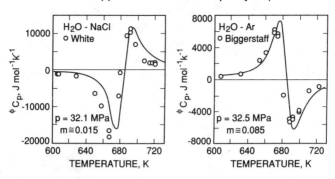

FIGURE 29. Isobaric, effectively infinite-dilution apparent molar heat capacities of a nonvolatile (NaCl, left) and a volatile (Ar, right) solute in supercritical water.[93,94] The full curves are predictions from a corresponding-states model.[88]

limit, these terms diverge with increasing strength. The last term, however, contains a factor, A_{VVV}, in the numerator, and this factor approaches zero in the limit. Very roughly speaking, the classical $\overline{C}_{p2}^{\infty}$ diverges as the square of the compressibility, which is a very strong divergence indeed. In fact, the behavior of $\overline{C}_{p2}^{\infty}$ is quite subtle,[8,10] and it is necessary to keep careful track of the path of approach to the critical point in determining the value of the critical exponent. The behavior of the terms A_{VV}, A_{VVV} near the critical point of the solvent is given by

$$A_{VV} = A_{VVx}^c x + A_{VVT}^c(\delta T) + A_{VVVV}^c(\delta V)^2/2 + ...$$

$$A_{VVV} = A_{VVVx}^c x + A_{VVVT}^c(\delta T) + A_{VVVV}^c(\delta V) \qquad (93)$$

With respect to V_c, A_{VV} is symmetric and A_{VVV} is antisymmetric in volume. Along the critical isotherm-isobar, the last term in Equation 91 dominates and leads to an $x^{-5/3}$ behavior. At infinite dilution and on the critical isochore, however, the one-but-last term dominates and leads to a divergence of $(\delta T)^{-2}$.

The behavior of $\overline{C}_{p2}$ on other paths can be sorted out just as was done for the apparent molar volume. Thus, along the critical line, $\overline{C}_{p2}$ diverges as $1/x^2$ classically. For further detail, refer to References 7 to 10. The behavior of $\overline{C}_{P2}$ is summarized in Table 4.

I. NONCLASSICAL PATH DEPENDENCE OF $\overline{V}_2$, $\overline{H}_2$, AND $\overline{C}_{p2}$

The nonclassical counterparts of the classical expressions that were obtained for the divergences of the partial molar volume and partial molar enthalpy are as follows. Since in Figure 26 the shape of the coexistence curve is characterized by the exponent β, and the shape of the critical isotherm-isobar by the exponent δ, $(\partial V/\partial x)_{pT}$ must diverge as $x^{(1/\delta - 1)}$ = $x^{-\gamma/\beta\delta}$. This implies a $\gamma/\beta\delta$ ($\simeq 0.80$) divergence for $\overline{V}_2$, $\overline{H}_2$, while $\overline{V}_1$ approaches V_c (Equation 76). It also implies that in the nonclassical case, the x dependence of the molar volume of the solution and of the partial molar volume of the solvent is as $x^{1/\delta}$, which is close to $x^{1/5}$. Thus, even a 10-ppm impurity can change the partial molar volume of the solvent by an amount of the order of 10%.

On the other paths studied previously, it is found that $(\partial V/\partial x)_{pT}$ and $(\partial H/\partial x)_{pT}$ diverge as $1/x$, but with an amplitude that depends on the path. As in the classical case, the limiting value of $\overline{V}_1$ is not V_c.

The behavior of the partial molar heat capacity is investigated in a similar fashion. The

TABLE 4
Path Dependence of the Critical
Anomalies of the Partial Molar
Properties

Property	$\overline{V}_2, \overline{H}_2$	$\overline{C}_{p2}$	$\overline{V}_1$
Path			
Classical			
p_c, T_c	$x^{-2/3}$	$x^{-5/3}$	V_c
p_c, ρ_c	x^{-1}	x^{-2}	Not V_c
T_c, ρ_c	x^{-1}	x^{-2}	Not V_c
CRL	x^{-1}	x^{-2}	Not V_c
Nonclassical			
p_c, T_c	$x^{-(1-1/\delta)}$	$x^{-(2-1/\delta)}$	V_c
p_c, ρ_c	x^{-1}	$x^{-(3-\gamma)}$	Not V_c
T_c, ρ_c	x^{-1}	$x^{-(3-\gamma)}$	Not V_c
CRL	x^{-1}	NA	Not V_c

$x^{-5/3}$ divergence noted along the critical isotherm-isobar becomes, nonclassically, an $x^{-(2-1/\delta)}$ divergence. The x^{-2} divergence on the critical isochore-isotherm becomes an $x^{3-\gamma}$ divergence. Since C_{px} is (weakly) infinite on the critical line, it is not sensible to ask for the composition dependence of $\overline{C}_{p2}$ on this path. The classical and nonclassical results for the path dependence of the partials discussed here are summarized in Table 4.

In the customary treatment of aqeuous solutions the infinite-dilution partial molar properties of the solute are used as standard states.[95] Dilute-mixture thermodynamics near critical point, and the experiments referred to, make it clear that such standard states cease to be useful in a large range around the critical point of steam.

J. HENRY'S CONSTANT AND THE K FACTOR

Henry's constant, k_H, of the volatile solute 2 in solvent 1 at infinite dilution is defined as:

$$k_H = \lim_{x \to 0} f_2/x \tag{94}$$

with f_2 the fugacity of the solute and x its mole fraction. The Henry constant will be given a superscript I or II, depending on the number of phases present. If the system is in two phases, the compositions of the two phases, x for the liquid and y for the vapor, will be different. The K factor or distribution coefficient of the solute in the solvent is defined as

$$K = y/x \tag{95}$$

and, in the limit of infinite dilution,

$$K^\infty = \lim_{x \to 0} y/x \tag{96}$$

It is well known that both Henry's constant and the distribution coefficient behave anomalously near the critical point of the solvent. As the temperature is raised above the triple point of the solvent, the volatile solute evaporates and Henry's constant increases.

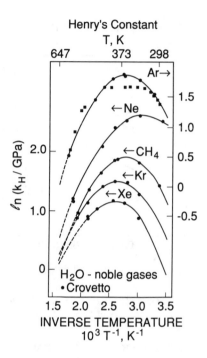

Henry's Constant

FIGURE 30. Henry's constants as functions of temperature for noble gases in water, according to Crovetto and Fernández-Prini.[97]

When the temperature is increased to the point that the liquid is quite expanded and the density of the vapor becomes appreciable, this process reverses. Thus, Henry's constant passes through a maximum and then declines, reaching an apparently finite value at the critical point of the solvent[96,97] (Figure 30). The K factor, which is very large at low temperatures, approaches the value of unity at the critical point. When plotted against temperature, both coefficients appear to approach their critical values with infinite slope.

From the classical expansion, Equation 70, arises the following asymptotic expression for the K factor[98]

$$RT \ln (y/x) = A_{Vx}^c \, [(\delta V_\ell) - (\delta V_v)] \tag{96a}$$

so that

$$\ln K^\infty = \lim_{x \to 0} \ln (y/x) = - \frac{2 \, A_{Vx}^c}{RT} \, B|\delta T|^{1/2} \tag{97}$$

with B the prefactor or amplitude of the coexistence curve of the pure solvent, given by Equation 15 for the classical case. Thus, RT ln K^∞ is linear in the volume difference of the vapor and liquid phases of the pure solvent. If RT ln K^∞ is plotted vs. δV, a straight line asymptote is expected. Density is usually found to be a better order parameter than volume, and this straight-line behavior is also expected when density is the independent variable. In the nonclassical case, the relation between density and temperature is governed by a different exponent than in Equation 97, but the predicted asymptotic linearity prevails.[98] The slope of this asymptotic line is $2A_{Vx}^c$, and thus intimately connected to the slope of the critical line (Section IV.A) and of the dew-bubble curve in p-x space (Section IV.B). Figure 31 is an example of a K plot for the system CO in C_6H_6. The linearity appears to prevail over a

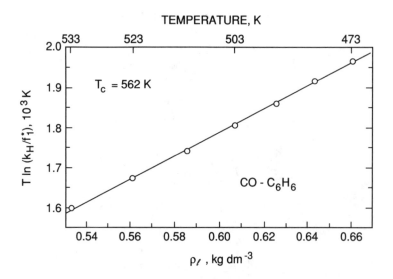

FIGURE 31.   Henry's constant of CO in benzene, according to Connolly and Kandalic,[100] and linearized by the procedure of Japas and Levelt Sengers.[98]

substantial range of temperatures, and the line extrapolates without difficulty to the value of 1 at $\delta\rho = 0$.

Both the infinite-dilution K factor and Henry's constant are simply related to composition derivatives of the residual Helmholtz free energy A^r, which is defined[79] as the difference of the Helmholtz free energy of the mixture and that of an ideal gas with the same internal degrees of freedom, at the same composition, temperature, and density. The relations are[98]

$$RT \ln K = RT \ln y/x = A_x^r - A_y^r \tag{98}$$

$$RT \ln k_H^{II}/f_1 = A_x^{r\infty} \tag{99}$$

where f_1 is the fugacity of the pure solvent.

The free energy A'' which was expanded earlier in the chapter and A^r introduced here, are different properties since A'', contrary to A^r, contains the ideal-gas part of the free energy. At saturation, however, it is the term $A_{vx}^c(\delta V) x$ (Equation 57) that dominates the behavior of A_x near the critical point because it varies with temperature more rapidly than the term $A''^c_{Tx}(\delta T) x$. The derivative A_{vx} is the same for all Helmholtz free energies considered. Since asymptotically near the critical point δV_ℓ is positive and equal to $-\delta V_v$, the interesting conclusion is that Henry's constant, divided by the fugacity of the solvent, varies half as fast as the infinite-dilution K factor, and with the same sign. Both properties vary asymptotically linearly with respect to the volume or density of the solvent (Figures 31 and 32).

Empirically, the linearity in density is found to prevail over a larger range than that in volume. Henry's constants for some nonpolar gases in water are known over most of the liquid range, from the triple point to the critical point of water. For N_2 in water, linearity of $T \ln k_H$ vs. liquid density is found from the critical point (647 K) to 400 K, close to the boiling point.[99]

Harvey et al.[101] have recently shown that the slope of the linearized Henry constant plot does not equal the value A_{vx}^c as deduced from the critical line (Equation 66). In the aqueous systems that have been checked[98-100] the Henry constant slope usually comes out about 50% larger than the critical-line slope. Although there are problems due to limited accuracy of critical line data near the critical point of steam, and large corrections to be applied in the

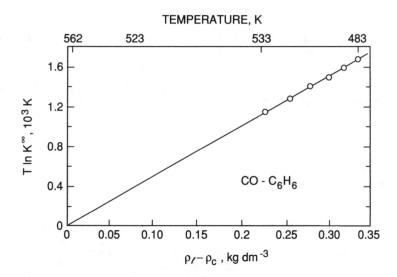

FIGURE 32.   The K factor of CO in benzene,[100] and linearized by the procedure of Japas and Levelt Sengers.[98] This plot must extrapolate to the origin.

extrapolation of solubilities to infinite solution, the systematic discrepancy with the critical-line slope suggests that it is not the asymptotic value of A_{Vx} that is observed in the linearized Henry constant plots. The slope of the K^∞ plot, however, appears to come close to twice the asymptotic value derived from the critical line slope.

K. THE OSMOTIC SUSCEPTIBILITIES

Since osmotic susceptibilities strongly diverge at the critical line, it is of some importance to determine the way in which they approach their finite or zero limiting values at infinite dilution. In order to determine this behavior, the reader should return to the expansions of the chemical potentials (Equations 61 and 62), and the definition of the susceptibilities χ_{ij} in Equation 49 and their better-known counterparts in Equation 53. It is obvious from the definition that one needs to determine only one of the derivatives $(\partial \mu_i / \partial x_j)_{pT}$ in order to know all the others.

As a specific example, consider the behavior of $(\partial \mu_2 / \partial x)_{pT}$ on the critical isotherm-isobar. If, in addition to the temperature, the pressure is to be kept constant in the differentiation, the asymptotic relation between volume and composition that prevails at constant pressure should be imposed. From Equation 58 at constant p and $\delta T = 0$:

$$(\delta V)^3 = - (6 A^c_{Vx} / A^c_{VVVV}) x \qquad (100)$$

and therefore, from Equation 39,

$$(\partial \mu_2 / \partial x)_{pT} = A''^c_{xx} + RT\, x^{-1} - |A^c_{Vx}||6\, A^c_{Vx}/A^c_{VVVV}|^{1/3}\, x^{-2/3}/3 + \dots \qquad (101)$$

Equation 101 reveals that on this path the dilute-mixture behavior, $RT\, x^{-1}$, prevails, and the osmotic susceptibility $(\partial x / \partial \mu_2)_{pT}$ approaches zero. From Equation 58, $(\partial x / \partial \mu_1)_{pT} = -\{x/(1-x)\}(\partial x / \partial \mu_2)_{pT}$, therefore it remains finite. Its leading x-dependence is as $x^{1/3}$.

For a view of the competition between infinite dilution and criticality it is enlightening to consider the exact relations

$$(\partial x / \partial \mu_2)_{pT}^{-1} = (1-x)\, [A_{xx} - A^2_{Vx}/A_{VV}] = (1-x)\, \text{Det}\, A/\, A_{VV}$$
$$(\partial x / \partial \mu_1)_{pT}^{-1} = -\, x\, [A_{xx} - A^2_{Vx}] = -\, x\, \text{Det}\, A/\, A_{VV} \qquad (102)$$

TABLE 5
Nature of Critical Behavior of Mixture Thermodynamic Derivatives Near
$$x (= x_2) = 0$$

Property	Structure	$\zeta = 0$, then CR		CRL, ζ not 0	
		Classical	Nonclassical	Classical	Nonclassical
K_{Tx}	$\dfrac{s + \zeta\, s \cdot w}{f + \zeta\, s}$	s	s	$f\,x^{-1}$	$w + x^{-1}$
$(\partial p/\partial x)_{VT}$	$\dfrac{s}{s + \zeta\, s \cdot w}$	f	f	f	$(1 + x\,w)^{-1}$
$(\partial V/\partial x)_{pT}$	$\dfrac{s}{f + \zeta\, s}$	s	s	$f\,x^{-1}$	$f\,x^{-1}$
$-RT(\partial x/\partial \mu_1)_{pT}$	$[1 + \zeta\, s]$	1	1	$x\,s$	$x\,s$
$RT(\partial x/\partial \mu_2)_{pT}$	$[1 + \zeta\, s]\,\zeta$	0	0	$x^2 s$	$x^2 s$

Abbreviations: f = finite, nonzero; s = strong divergence; w = weak divergence; w^{-1} = weak zero; ζ = generalized activity that is equal to x for small x.

that follow from Equation 36. The term A_{xx} contains the part RT/x that forces the osmotic susceptibility of the solute $(\partial x/\partial \mu_2)_{pT}$ to zero at infinite dilution. The RT/x divergence is canceled by the multiplier x in the case of the osmotic susceptibility of the solvent. For finite x, both osmotic susceptibilities diverge on the critical line where Det A equals zero. Away from the critical line, Det A is nonzero and both susceptibilities are proportional to the compressibility of the mixture. As x approaches zero away from criticality, the susceptibility of the solute $(\partial x/\partial \mu_2)_{pT}$ goes to zero while that of the solvent, $(\partial x/\partial \mu_1)_{pT}$, becomes proportional to the compressibility of the pure solvent. These results hold for both classical and nonclassical critical behavior. For any path to the critical point of the solvent, a detailed analysis must be done, as in the example of Section IV.F. Chang and Levelt Sengers[9] have worked out the classical and nonclassical path dependences for the osmotic susceptibility and other properties of interest to chemical engineers. Table 5 summarizes some of their results.

L. SUPERCRITICAL SOLUBILITY AND PARTIAL MOLAR VOLUMES

The solubility of a low-volatile solute is enhanced as the density of the volatile solvent increases. This is a general effect that is due to the short-range attractive forces between solvent and solute. Short-range effects have no particular or peculiar features near a critical point. Thus, the relation between supercritical solubility and solution density is roughly linear (Figure 17), whereas a true critical enhancement peaks at the density of the critical point (Figures 6, 7, 9, and 10).

At the critical endpoint, the solubility does change with infinite slope on the isobar as a function of temperature, or on the isotherm as a function of pressure (Figure 16). This, in a sense, is a trivial effect of exactly the same type as the infinite slopes that enthalpy, entropy, and energy of one-component fluids display when graphed against pressure on the critical isotherm. None of these properties have strictly critical anomalies. The infinite slopes are induced by the infinite compressibility of the fluid at the critical point in the one-component case and by the infinite susceptibility of the solution at the critical endpoint in the two-component case.

Strictly speaking, the theory of critical phenomena has little to contribute to the description of supercritical solubility, since it will not provide a model for short-range solute-solvent interactions. There is, however, a very simple relation between the infinite slopes $dx/dp|_{T,\sigma}$, $dx/dT|_{p,\sigma}$ and the partial molar properties that suggests that something dramatic might happen in very dilute supercritical mixtures. These relations follow from thermodynamic principles known as Gibbs-Konowalow rules, and they have been applied under the simplifying, but nonessential assumption that the solid phase is unaffected by the solvent; its partial properties are therefore replaced by its molar properties, H_s, V_s:

$$dx/dp|_{T,\sigma} = \frac{V_s - \overline{V}_2}{(1 - x)(\partial^2 G/\partial x^2)_{pT}}$$

$$dx/dT|_{p,\sigma} = \frac{\overline{H}_2 - H_s}{(1 - x)(\partial^2 G/\partial x^2)_{pT}} \tag{103}$$

These relations make it obvious that the infinity in the slope is induced by the zero in the second derivative of the Gibbs free energy at the critical endpoint. It is now tempting to conclude that in a dilute mixture, where the numerator in Equation 103 is very large because of the divergence of the partial molar volume at infinite dilution, the amplitude of the divergence that is due to the criticality of the mixture is enhanced.[102-104] As pointed out by Wheeler and Petschek[47] in a slightly different context, this idea is incorrect. Within the framework of the classical expansion, it can be demonstrated here why this is so.

For the sake of simplicity, consider only the first relation in Equation 103, and assume that the chemical potential of the solute is kept constant by the presence of an inert solid phase of the second component. Next, replace the condition indicated by σ in Equation 103 by the condition of constant μ_2. For low solubility, a path of constant μ_2 intersects the critical line at a point near the critical point of the solvent, at a composition indicated by x^{cm}. The question is whether the solute with the smaller x^{cm} will have the larger amplitude of the solubility enhancement $(\partial x/\partial p)_{T\mu_2}$.

The relation between x and P, at constant T and μ_2, is most readily obtained by considering the Gibbs free energy G as a function of P and x. A point is chosen on the critical line of the mixture, and the values of G, μ_2, P, V, T, and x are denoted at this point by a superscript cm. Expanding G around a point on the critical line at constant $T = T^{cm}$, we find

$$G = G^{cm} + G_x^{cm} (x - x^{cm}) + G_{xxxx}^{cm} (x - x^{cm})^4/24 + \\ + G_{Px}^{cm}(P - P^{cm})(x - x^{cm}) + ... \tag{104}$$

since the second and third derivatives of G with respect to x are zero because of criticality. From the relation between μ_2 and G, $\mu_2 = G + (1 - x) (\partial G/\partial x)_{pT}$ (Equation 76), it follows that, to leading order, at T^{cm}

$$\mu_2 = G_x^{cm} (1 - x^{cm}) + G_{xxxx}^{cm} (x - x^{cm})^3/6 + G_{px}^{cm} (p - p^{cm}) + ... \tag{105}$$

The relation between pressure and composition is obtained by setting μ_2 equal to a constant. For the relation between pressure and composition at constant μ_2, T, we therefore obtain

$$p - p_c = \frac{G_{xxxx}^{cm}(x - x^{cm})^3/6}{G_{px}^{cm}} \tag{106}$$

which is a cubic, as could have been anticipated.

If x^{cm} is small, G is dominated by the term $RT x^{cm}\ln x^{cm}$, which yields an amplitude of $2RT^{cm}/(x^{cm})^3$ for G_{xxxx}^{cm}. Furthermore, $G_{px}^{cm} = (\partial V/\partial x)_{pT}$, which, according to Equation 79, assumes the value of $- A_{Vx}^{cm}/A_{VV}^{cm}$ on the critical line. If x^{cm} is small, A_{Vx}^{cm} assumes the value A_{Vx}^c, while A_{VV}^{cm}, from the expansion in Equation 59 and the equation of the critical line (Equation 65), is found to assume the value of $(A_{Vx}^c)^2 x^{cm}/RT^{cm}$. Consequently,

$$(\partial x/\partial p)_{T\mu_2} = (1/A^c_{Vx}) \{(x - x^{cm})/x^{cm}\}^{-2} \tag{107}$$

is obtained. Contrary to statements in References 101 to 103, as x^{cm} becomes small (solute of low solubility), the amplitude of the slope $(\partial x/\partial P)_{T\mu_2}$ as a function of the absolute composition $(x - x^{cm})$ actually decreases, as $(x^{cm})^2$, while it is constant if the composition is measured relative to x^{cm}. Thus, there is no thermodynamic mechanism that makes the solubility enhancement $(\partial x/\partial P)_{T\mu_2}$ larger for poorly soluble (smaller x^{cm}) solutes in a supercritical solvent, even though the partial molar property values in Equation 103 diverge for dilute solutions. Note that for low-solubility solutes, the solubility enhancement is determined solely by the coefficient A^c_{Vx}.

M. SUMMARY

The anomalous properties of dilute mixtures near the critical point of the solvent have been investigated on the basis of an expansion of the classical Helmholtz free energy at that critical point. The properties studied include the initial slope of the critical line; the shape of the dew-bubble curve; osmotic susceptibilities and the related correlation function integrals; partial, apparent, and infinite-dilution molar volumes, enthalpies and heat capacities; Henry's constant; and supercritical solubility enhancement. The nonclassical counterparts of the results are stated. The derivative $(\partial p/\partial x)_{VT}$ at the critical point of the solvent governs the sign and strength of all near-critical dilute-mixture behavior, including supercritical solubility. Its sign can be used to distinguish between volatile and involatile solutes. An attempt has been made to clarify several cases about which there appears to be confusion in the literature. It has been stressed, for instance, how criticality affects the tail of the correlation function, not its short-range structure, so that properties depending on the short-range structure do not have significant critical effects. Also, contrary to statements in the literature, the divergence of the infinite-dilution partial molar volume and partial molar enthalpy does not lead to an additional enhancement of supercritical solubility in systems of low solubility.

ACKNOWLEDGMENTS

This work was supported in part by the Office of Standard Reference Data at the National Institute of Standards and Technology (NIST). Enlightening discussions with Professors G. S. Stell and R. H. Wood, and with Drs. A. H. Harvey and R. Crovetto, are gratefully acknowledged.

REFERENCES

1. **Levelt Sengers, J. M. H.,** Liquidons and gasons; controversies about the continuity of states, *Physica,* 98A, 363, 1979.
2. **Krichevskii, I. R.,** Thermodynamics of critical phenomena in infinitely dilute binary solutions, *Russ. J. Phys. Chem.,* 41, 1332, 1967.
3. **Khazanova, N. E. and Sominskaya, E. E.,** Partial molar volumes in the ethane-carbon dioxide systems near the critical points of the pure components, *Russ. J. Phys. Chem.,* 45, 1485, 1971.
4. **Wheeler, J. C.,** Behavior of a solute near the critical point of an almost pure solvent, *Ber. Bunsen Ges. Physik. Chem.,* 76, 308, 1972.
5. **Rozen, A. M.,** The unusual properties of solutions in the vicinity of the critical point of the solvent, *Russ. J. Phys. Chem.,* 50, 837, 1976.
6. **Levelt Sengers, J. M. H., Morrison, G., and Chang, R. F.,** Critical behavior in fluids and fluid mixtures, *Fluid Phase Equilibria,* 14, 19, 1983.

7. **Chang, R. F., Morrison, G., and Levelt Sengers, J. M. H.,** The critical dilemma of dilute mixtures, *J. Phys. Chem.,* 88, 3389, 1984.
8. **Levelt Sengers, J. M. H., Everhart, C. M., Morrison, G., and Pitzer, K. S.,** Thermodynamic anomalies in near-critical aqueous NaCl solutions, *Chem. Eng. Commun.,* 47, 315, 1986.
9. **Chang, R. F. and Levelt Sengers, J. M. H.,** Behavior of dilute mixtures near the solvent's critical point, *J. Phys. Chem.,* 90, 5921, 1986.
10. **Morrison, G.,** Modelling aqueous solutions near the critical point of water, *J. Solution Chem.,* 17, 887, 1988.
11. **Griffiths, R. B. and Wheeler, J. C.,** Critical points in multicomponent systems, *Phys. Rev. A,* 2, 1047, 1970.
12. **Levelt, J. M. H.,** Measurements of the Compressibility of Argon in the Gaseous and Liquid Phase: Comparison of the Results with Existing Theories, Ph.D. thesis, Municipal University, Amsterdam, 1958.
13. **Michels, A., Levelt, J. M., and Wolkers, G. J.,** Thermodynamic properties of argon at temperatures between 0°C and − 140°C and at densities up to 640 amagat (pressures up to 1050 atm), *Physica,* 24, 769, 1958.
14. **Michels, A., Levelt, J. M., and De Graaff, W.,** Compressibility isotherms of argon at temperatures between 25°C and − 155°C, and at densities up to 640 amagat (pressures up to 1050 atmospheres), *Physica,* 24, 659, 1958.
15. **Rowlinson, J. S. and Swinton, P. L.,** *Liquids and Liquid Mixtures,* 3rd ed., Butterworth, London, 1982.
16. **Van der Waals, J. D. and Kohnstamm, Ph.,** *Lehrbuch der Thermodynamik,* Vol. 1, Verlag J. A. Barth, Leipzig, 1923, 35.
17. **Levelt Sengers, J. M. H.,** From Van der Waals' equation to the scaling laws, *Physica,* 73, 73, 1974.
18. **Aharony, A. and Hohenberg, P. C.,** Universal relations among thermodynamic critical amplitudes, *Phys. Rev. B,* 13, 3081, 1976.
19. **Stauffer, D., Ferer, M., and Wortis, M.,** Universality of second-order phase transitions: the scale factor for the correlation length, *Phys. Rev. Lett.,* 29, 345, 1972.
20. **Hohenberg, P. C., Aharony, A., Halperin, B. I., and Siggia, E. D.,** Two-scale-factor universality and the renormalization group, *Phys. Rev. B,* 13, 2986, 1976.
21. **Sengers, J. V. and Moldover, M. R.,** Two-scale-factor universality near the critical point of fluids, *Phys. Lett.,* 66A, 1, 1978.
22. **Sengers, J. V. and Levelt Sengers, J. M. H.,** Critical phenomena in classical fluids, in *Progress in Liquid Physics,* Croxton, C. A., Ed., John Wiley & Sons, Chichester, U.K., 1978, 103.
23. **Sengers, J. V. and Levelt Sengers, J. M. H.,** Thermodynamic behavior of fluids near the critical point, *Ann. Rev. Phys. Chem.,* 37, 189, 1986.
24. *Critical Phenomena, Proc. of a Conference Held in Washington, D.C.,* NBS Misc. Publ. 273, Green, M. S. and Sengers, J. V., Eds., National Bureau of Standards, Washington, D.C., 1986.
25. **Kamgar-Parsi, B., Balfour, F. W., and Sengers, J. V.,** Thermodynamic properties of steam in the critical region, *J. Phys. Chem. Ref. Data,* 12, 1, 1983.
26. **Levelt Sengers, J. M. H., Olchowy, G. A., Kamgar-Parsi, G., and Sengers, J. V.,** *A Thermodynamic Surface for the Critical Region of Ethylene,* NBS Tech. Note 1189, National Bureau of Standards, Washington, D.C., 1984.
27. **Albright, P. C., Chen, Z. Y., and Sengers, J. V.,** Crossover from singular to regular thermodynamic behavior of fluids in the critical region, *Phys. Rev. B,* 36, 877, 1987.
28. **Levelt Sengers, J. M. H.,** Critical exponents at the turn of the century, *Physica,* 82A, 319, 1976.
29. **Verschaffelt, J. E.,** On the critical isothermal line and the densities of saturated vapour and liquid in isopentane and carbon dioxide, *Proc. Kon. Akad. Sci. Amsterdam,* 2, 588, 1900.
30. **Widom, B. and Rice, O. K.,** Critical isotherm and the equation of state of liquid-vapor systems, *J. Chem. Phys.,* 23, 1250, 1955.
31. **Bagatskii, M. I., Voronel, A. V., and Gusak, V. G.,** Measurement of the specific heat C_v of argon in the immediate vicinity of the critical region, *Sov. Phys. JETP,* 43, 517, 1962 (letter to the editor, English translation).
32. **Hocken, R. and Moldover, M. R.,** Ising exponents in real fluids: an experiment, *Phys. Rev. Lett.,* 37, 29, 1976.
33. **Moldover, M. R.,** Thermodynamic anomalies near the liqud-vapor critical point: a review of experiments, in *Phase Transitions: Cargese 1980,* Levy, M. and Le Guillou, J.-C., Eds., Plenum Press, New York, 1982, 63.
34. **Greer, S. C. and Moldover, M. R.,** Thermodynamic anomalies at critical points of fluids, *Ann. Rev. Phys. Chem.,* 32, 233, 1981.
35. **Kadanoff, L. P.,** Scaling laws for Ising models near T_c, *Physics,* 2, 263, 1966.
36. **Wilson, K. G. and Fisher, M. E.,** Critical exponents in 3.99 dimensions, *Phys. Rev. Lett.,* 28, 240, 1972.
37. **Wilson, K. G.,** Renormalization group and critical phenomena. I. Renormalization group and the Kadanoff scaling picture, *Phys. Rev. B,* 4, 3174, 1971.

38. **Le Guillou, J.-C. and Zinn-Justin, J.**, Accurate critical exponents from the ε-expansion, *J. Phys. Lett.*, 46, L-137, 1985.

39. **Singh, R. R . and Pitzer, K. S.**, Rectilinear diameters and extended corresponding states theory, *J. Chem. Phys.*, 92, 3096, 1990.

40. **Fisher, M. E.**, Correlation functions and the critical region of simple fluids, *J. Math. Phys.*, 5, 944, 1964.

41. **Stell, G. and Hoye, J. S.**, Dielectric constant and mean polarizability in the critical region, *Phys. Rev. Lett.*, 33, 1268, 1974.

42. **Hocken, R. and Stell, G.**, Index of refraction in the critical region, *Phys. Rev. A*, 8, 887, 1973.

43. **O'Connell, J. P.**, Thermodynamic properties of solutions and the theory of fluctuations, *Fluid Phase Equilibria*, 6, 21, 1981.

44. **Sengers, J. V., Bedeaux, D., Mazur, P., and Greer, S. C.**, Behavior of the dielectric constant of fluids near a critical point, *Physica*, 104A, 573, 1980.

45. **Pestak, M. W. and Chan, M. H. W.**, Dielectric constant anomaly of CO near its liquid-vapor critical point, *Phys. Rev. Lett.*, 46, 943, 1981.

46. **Saad, H. and Gulari, E.**, Diffusion of carbon dioxide in heptane, *J. Phys. Chem.*, 88, 136, 1984.

47. **Wheeler, J. C. and Petschek, R. G.**, Anomalies in chemical equilibria near critical points of dilute solutions, *Phys. Rev. A*, 28, 2442, 1983.

48. **Morrison, G.**, Comment on "anomalies in chemical equilibria near critical points", *Phys. Rev. A*, 30, 644, 1984.

49. **Gouy, G.**, Sur quelqueas phénomènes présentés par les tubes de Natterer, *C. R. Acad. Sci.*, 116, 1289, 1893.

50. **Straub, J.**, Dichtemessungen am Kritischen Punkt mit einer Optischen Methode bei Reinen Stoffen und Gemischen, Ph.D. thesis, Technische Universität, Munich, 1965. (See also Reference 24, p. 13.)

51. **Sengers, J. V.**, Thermal Conductivity Measurements at Elevated Gas Densities, Including the Critical Region, Ph.D. thesis, Municipal University, Amsterdam, 1962.

52. **Michels, A., Sengers, J. V., and van der Gulik, P. S.**, The thermal conductivity of carbon dioxide in the critical region, *Physica*, 28, 1201, 1216, 1238, 1962.

53. **Jacobsen, R. T. et al.**, *Ethylene, International Thermodynamic Tables of the Fluid State*, IUPAC/Blackwell Scientific, Oxford, 1988.

54. **Doiron, T. and Meyer, H.**, Dielectric constant of ^{3}He near the liquid-vapor critical point, *Phys. Rev. B*, 17, 2141, 1978.

55. **Rowlinson, J. S., Ed.**, *J. D. van der Waals: On the Continuity of the Gaseous and Liquid States*, Studies in Statistical Mechanics, North-Holland, Amsterdam, 1989.

56. **Levelt Sengers, J. M. H.**, The state of the critical state of fluids, *Pure Appl. Chem.*, 55, 437, 1983.

57. **Kirkwood, J. G. and Buff, F. P.**, The statistical mechanical theory of solutions. I. *J. Chem. Phys.*, 19, 774, 1951.

58. **Fisher, M. E.**, Renormalization of critical exponents by hidden variables, *Phys. Rev.*, 176, 257, 1968.

59. **Van Gunst, C. A.**, De Oplosbaarheid van Mengsels van Vaste Stoffen in Superkritische Gassen, Ph.D. thesis, Technical University, Delft, The Netherlands, 1950.

60. **Scott, R. L.**, Critical exponents for binary fluid mixtures, in *Chemical Thermodynamics, Specialist Periodical Report*, Vol. 2, McGlashan, M. L., Ed., Chemistry Society, London, 1978, 238.

61. **Leung, S. S. and Griffiths, R. B.**, Thermodynamic properties near the gas-liquid critical line in mixtures of He^3 and He^4, *Phys. Rev. A*, 8, 2670, 1973.

62. **Rainwater, J. C. and Moldover, M. R.**, Thermodynamic models for fluid mixtures near critical conditions, in *Chemical Engineering at Supercritical Conditions*, Paulaitis, M. E. et al., Eds., Ann Arbor Science Publications, Ann Arbor, MI, 1983, chap. 10.

63. **Rainwater, J. C. and Williamson, F. R.**, Vapor-liquid equilibrium of near-critical binary alkane mixtures, *Int. J. Thermophys.*, 7, 65, 1986.

64. **Rainwater, J. C. and Lynch, J. J.**, The modified Leung-Griffiths model for vapor-liquid equilibria: application to polar fluid mixtures, *Fluid Phase Equilibria*, 52, 91, 1989.

65. **D'Arrigo, G., Mistura, L., and Tartaglia, P.**, Leung-Griffiths equation of state for the system CO_2-C_2H_4 near the vapor-liquid critical line, *Phys. Rev. A*, 12, 2587, 1975.

66. **Moldover, M. R. and Gallagher, J. S.**, Critical points of mixtures: an analogy with pure fluids, *AIChE J.*, 24, 267, 1978.

67. **Chang, R. F. and Doiron, T.**, Leung-Griffiths model for the thermodynamic properties of mixtures of CO_2 and C_2H_6 near the gas-liquid critical line, *Int. J. Thermophys.*, 4, 337, 1983.

68. **Chang, R. F., Doiron, T., and Pegg, I. L.**, Decay rate of critical fluctuations in ethane + carbon dioxide mixtures near the critical line including the critical azeotropy, *Int. J. Thermophys.*, 7, 295, 1986.

69. **Chang, R. F., Levelt Sengers, J. M. H., Doiron, T., and Jones, J.**, Gravity-induced density and concentration profiles in binary mixtures near gas-liquid critical lines, *J. Chem. Phys.*, 79, 3058, 1983.

70. **Diepen, G. A. M. and Scheffer, F. E. C.**, The solubility of naphthalene in supercritical ethylene, *J. Am. Chem. Soc.*, 70, 4085, 1948.

71. **Tsekanskaya, Yu.V., Iomtev, M. B., and Mushkina, E. V.,** Solubility of naphthalene in ethylene and carbon dioxide under pressure, *Russ. J. Phys. Chem.,* 38, 1172, 1964.

72. **Cochran, H. D., Lee, L. L., and Pfund, D. M.,** Application of the Kirkwood-Buff theory of mixtures to dilute supercritical mixtures, *Fluid Phase Equilibria,* 34, 219, 1987.

73. **McGuigan, D. B. and Monson, P. A.,** Analysis of infinite dilution partial molar volumes using a distribution function theory, *Fluid Phase Equilibria,* in press.

74. **Debenedetti, P. G.,** Clustering in dilute, binary supercritical mixtures: a fluctuation analysis, *Chem. Eng. Sci.,* 42, 2203, 1987.

75. **Jones, G. L., Kozak, J. J., Lee, E., Fishman, S., and Fisher, M. E.,** Critical point correlations of the Yvon-Born-Green Equation, *Phys. Rev. Lett.,* 46, 795, 1981.

76. **Fisher, M. E. and Fishman, S.,** Critical scattering and integral equations for fluids, *Phys. Rev. Lett.,* 47, 421, 1981.

77. **Stell, G.,** Some critical properties of Ornstein-Zernike systems, *Phys. Rev.,* 184, 135, 1969.

78. **Stell, G.,** Extension of the Ornstein-Zernike theory of the critical region. II., *Phys. Rev. B,* 1, 2265, 1970.

79. **Abbott, M. M. and Nass, K. K.,** Equations of state and classical solution thermodynamics, in *Equations of State, Theory and Applications,* ACS Symp. Ser. 300, Chao, K. C. and Robinson, R. L., Jr., Eds., American Chemical Society, Washington, D.C., 1986, chap. 1.

80. **Redlich, O. and Kister, A. T.,** On the thermodynamics of solutions. VII. Critical properties of mixtures, *J. Chem. Phys.,* 36, 2002, 1962.

81. **Franck, E. U.,** Fluids at high pressures and temperatures, *J. Chem. Thermodyn.,* 19, 225, 1987.

82. **Debenedetti, P. G. and Mohamed, R. S.,** Attractive, weakly attractive and repulsive near-critical systems, *J. Chem. Phys.,* 90, 4528, 1989.

83. **Tsiklis, D. S., Linshits, L. R., and Goryunova, N. P.,** Phase equilibria in the system ammonia-water, *Russ. J. Phys. Chem.,* 39, 1590, 1965.

84. **Wichterle, I. and Kobayashi, R.,** Vapor-liquid equilibrium of methane-propane system at low temperatures and high pressures, *J. Chem. Eng. Data,* 17, 4, 1972.

85. **Bischoff, J. L., Rosenbauer, R. J., and Pitzer, K. S.,** The system NaCl-H$_2$O: relations of vapor-liquid near the critical temperature of water and of vapour-liquid-halite from 300° to 500°C, *Geochim. Cosmochim. Acta,* 50, 1437, 1986.

86. **Bischoff, J. L. and Rosenbauer, R. J.,** Liquid-vapor relations in the critical region of the system NaCl-H$_2$O from 380 to 415°C: a refined determination of the critical point and two-phase boundary of seawater, *Geochim. Cosmochim. Acta,* 52, 2121, 1988.

87. **Biggerstaff, D. R. and Wood, R. H.,** Apparent molar volumes of aqueous argon, ethylene, and xenon from 300 to 716 K, *J. Phys. Chem.,* 92, p. 1988, 1988.

88. **Levelt Sengers, J. M. H. and Gallagher, J. S.,** Generalized corresponding states and high-temperature aqueous solutions, *J. Phys. Chem.,* submitted.

89. **Van Wasen, U., Swaid, I., and Schneider, G. M.,** Physicochemical principles and applications of supercritical fluid chromatography, *Angew, Chem. Int. Ed. Engl.,* 19, 575, 1980.

90. **Eckert, C. A., Ziger, D. H., Johnston, K. P., and Ellison, T. K.,** The use of partial molar volume data to evaluate equations of state for supercritical fluid mixtures, *Fluid Phase Equilibria,* 14, 167, 1983.

91. **Busey, R. H., Holmes, H. F., and Mesmer, R. E.,** The enthalpy of dilution of aqueous sodium chloride to 673 K using a new heat-flow and liquid-flow calorimeter. Excess thermodynamic properties and their pressure coefficients, *J. Chem. Thermodyn.,* 16, 343, 1984.

92. **Wood, R. H.,** private communication, and Ref. 94, 95.

93. **Biggerstaff, D. R., White, D. E., and Wood, R. H.,** Heat capacities of aqueous argon from 306 to 578 K, *J. Phys. Chem.,* 89, 4378, 1985.

94. **White, D. E., Wood, R. H., and Biggerstaff, D. R.,** Heat capacities of 0.0150 mol.kg^{-1} NaCl(aq) from 604 to 718 K at 32 MPa, *J. Chem. Thermodyn.,* 20, 159, 1988.

95. **Tanger, J. C., IV and Helgeson, H. C.,** Calculation of the thermodynamic and transport properties of aqueous species at high pressures and temperatures: revised equation of state for the standard partial molar properties of ions and electrolytes, *Am. J. Sci.,* 288, 19, 1988.

96. **Crovetto, R., Fernández Prini, R., and Japas, M. L.,** Solubilities of inert gases and methane in H$_2$O and D$_2$O in the temperature range of 300 to 600 K, *J. Chem. Phys.,* 76, 1077, 1982.

97. **Crovetto, R. and Fernández Prini, R.,** Solubility of simple apolar gases in light and heavy water: a critical assessment of data, *J. Phys. Chem. Ref. Data,* 18, 1231, 1989.

98. **Japas, M. L. and Levelt Sengers, J. M. H.,** Gas solubility and Henry's law near the solvent's critical point, *AIChE J.,* 35, 705, 1989.

99. **Harvey, A. H. and Levelt Sengers, J. M. H.,** Correlation of aqueous Henry's constants from 0°C to the critical point, *AIChE J.,* 36, 539, 1990.

100. **Connolly, J. F. and Kandalic, G. A.,** Thermodynamic properties of dilute solutions of carbon monoxide in a hydrocarbon, *J. Chem. Thermodyn.,* 16, 1129, 1984.

101. **Harvey, A. H., Crovetto, R., and Levelt Sengers, J. M. H.,** Limiting vs. apparent critical behavior of Henry's constant and K factors, *AIChE J.,* 36, 1901, 1990.

102. **Gitterman, M. and Procaccia, I.,** Quantitative theory of solubility in supercritical fluids, *J. Chem. Phys.,* 78, 2648, 1983.

103. **Morrison, G., Levelt Sengers, J. M. H., Chang, R. F., and Christensen, J. J.,** Thermodynamic anomalies in supercritical fluid mixtures, in *Supercritical Fluid Technology,* Penninger, J. M. L. et al., Eds., Elsevier, Amsterdam, 1985, 25.

104. **Levelt Sengers, J. M. H., Morrison, G., Nielson, G., Chang, R. F., and Everhart, C. M.,** Thermodynamic behavior of supercritical fluid mixtures, *Int. J. Thermophys.,* 7, 231, 1986.

Chapter 2

VAPOR-LIQUID EQUILIBRIUM AND THE MODIFIED LEUNG-GRIFFITHS MODEL

James C. Rainwater

TABLE OF CONTENTS

ABSTRACT

This chapter is a review of vapor-liquid equilibria (VLE) of binary mixtures and their correlation with the Leung-Griffiths model as modified by Moldover and Rainwater. The model is specifically designed for an extended critical region, from the mixture critical pressure down to one half that pressure, and explicitly incorporates scaling-law critical exponents. The various possible types of mixture phase diagrams are reviewed, with emphasis on the class 1 mixture that has a continuous critical locus. Criteria are established for thorough measurement of VLE in the extended critical region, and a comprehensive bibliography is presented of 129 thoroughly measured mixtures composed of 73 pure fluids; the distribution of critical points of those fluids is examined and it is demonstrated that most mixtures of two of these fluids are class 1. The mathematical structure of the model is briefly reviewed and a systematic procedure is established with guidelines on the optimal number of adjustable parameters based on a quantitative measure (α_{2m}) of fluid dissimilarity. Parameters are listed and analyzed for 42 successful nonazeotropic mixture fits to date. The utility of the model is discussed for several related problems including ternary mixtures, liquid-liquid equilibrium, explicit asymptotic expansions, azeotropy, a novel technique for critical density measurement, and interfacial tension.

I. INTRODUCTION

Supercritical fluid (SCF) technology, essentially by definition, involves the exploitation of the unusual variations in solubility with pressure and temperature immediately above the critical point of a volatile solvent. Consequently, optimal development of the technology will ultimately require accurate mathematical correlations of such solubilities. The primary subject of the present review is the thermodynamic model of Leung and Griffiths,[1] as modified by Moldover, Rainwater, and co-workers.[2-6] Among other thermodynamic behaviors, this model accurately describes the mutual solubilities of two fluids in the simplest case, when the only phase transition is VLE and the two-phase region is bounded from above in pressure by a critical locus that extends continuously from the critical point of the solvent to that of the solute.

For many prototypical solutes in the study of supercritical extraction, such as naphthalene in carbon dioxide, the simple description above does not hold; near the critical point of carbon dioxide, the critical line is interrupted by a three-phase solid-liquid-vapor (SLV) or liquid-liquid-vapor (LLV) locus;[7] however, for many other solutes of interest such as toluene[8] or *n*-decane,[9] the critical locus with carbon dioxide is continuous and the simple description applies. Another aspect of SCF technology is the successful and rapidly expanding field of SFC chromatography, as reviewed elsewhere in this volume. While the highly complex pharmaceutical solute molecules typically examined by SFC do not follow the simple pattern of miscibility with carbon dioxide, a desirable enhancement of solubility can often be obtained by adding to the solvent up to several percent of a "modifier" or "entrainer" such as methanol, acetonitrile, or tetrahydrofuran.[10] Binary mixtures of these fluids with carbon dioxide qualitatively display the simple phase equilibrium pattern, although such modifiers are much less volatile than carbon dioxide. The precise description of near-critical dilute solutions of such modifiers in carbon dioxide is at present unavailable and urgently needed by SFC researchers.

Conventional mixture phase equilibrium calculations begin with an equation of state (EOS), pressure (P) as a function of temperature (T), density (ρ), and mole fraction (x), typically adapted from a pure-fluid EOS with various mixing and combining rules. The phase boundary is obtained by numerical solution for the equality of chemical potentials or fugacities. There is currently a widespread recognition that a standard EOS is deficient in the critical region, and the improvement of VLE predictions near the critical locus is a current frontier topic in the field of phase equilibria, with many modifications of the equations or mixing rules having been recently proposed to deal with critical behavior. As is well known, however, any analytic EOS leads to (incorrect) classical critical exponents, and thus is in principle inadequate very close to a critical point. Therefore, for the representation of phase behavior near a critical locus, it is appropriate to turn to a nonconventional model such as that of Leung and Griffiths which does not employ an EOS and which explicitly incorporates nonclassical exponents. Section IX provides a quantitative comparison, for the mixture carbon dioxide + propane, of the deviations between experimental and calculated thermodynamic functions according to the nonclassical model and two classical EOSs as the critical locus is approached.

The success of the modified Leung-Griffiths model is due first to certain important theoretical insights of Griffiths and co-workers. In particular, they noted that the mathematical description of phase boundaries is simplest in a space of "field" variables,[11] by definition variables such as P, T, and chemical potential (μ_i) that are equal in the coexisting phases. By contrast, the conventional EOS, P (T, ρ, x), is a mixed representation involving two "density" variables, ρ and x, which by definition differ across the phase boundary. They also successfully adapted to mixtures the pure-fluid Schofield model[12] with the explicit inclusion of nonclassical exponents, and cleverly introduced a fugacity ratio variable (ζ) that served as a very useful field-space analog of x. Subsequently, Moldover and Gallagher[2,3] introduced some important modifications in the parametric structure for greater ease in developing practical correlations to mixture VLE data.

Despite these important fundamental advances, the successful application of the Leung-Griffiths model (which predates the much more widely used Peng-Robinson equation[13] by 3 years) to a wide variety of fluid mixtures was not attained until the last few years,[4-6,14-18] by means of the following systematic empirical methodology. First, original, unsmoothed critical-region VLE data have been collected, to the extent possible, for all available mixtures and from all sources, as the narrow focus on a particular mixture would not be fruitful. Second, the importance of a certain quantitative parameter[19] has been recognized as a measure of the difficulty of fitting a particular mixture, in particular the function α_{2m} where

$$\alpha_2 = \Delta x/(\Delta\rho/\rho_c) \tag{1}$$

$$\alpha_{2m} = \max|\alpha_2| \tag{2}$$

the symbol Δ denotes the difference between liquid and vapor values in the limit as the mixture critical point is approached, and ρ_c is the mixture critical density. Here, α_{2m} is a measure of the dissimilarity of the pure fluids or, in the language of lattice-gas theories,[20] of the "asymmetry" of the mixture. Third, the mixture VLE data sets have been ordered with increasing α_{2m}, or increasing difficulty of fitting. While many important mixtures have yet to be measured in the critical region, there is a wealth of accurate critical-region VLE data (see Section V) and enough distinct mixtures studied that a near-continuum of α_{2m} values is available. Fourth, the mathematical structure of the model has been carefully analyzed[6] as to what adjustable parameters are possible (given the thermodynamic and critical-behavior constraints) and can produce specific modifications of the predicted VLE surface such as changes in the width or orientation of the dew-bubble curves. Fifth, the overall analysis has begun with the mixtures easiest to correlate, those of lowest α_{2m}. As a point of reference, the algorithm of Moldover and Gallagher[2,3] has been used, wherein, apart from parameters needed to fit the pure fluid coexistence surfaces and the critical locus, no further adjustable parameters are introduced and the mixture VLE surface is determined.

Finally, the progression of mixtures with continuously increasing α_{2m} has been studied sequentially, and new degrees of freedom are added to the model by means of further adjustable parameters, but only as necessary. It has been found that the behavior of non-azeotropic mixtures changes very systematically with α_{2m}, and specific guidelines can be developed for the maximum number of permissible adjustable parameters as a function of that parameter (see Section VII), with certain exceptions made for polarity effects. With this procedure, since at any stage none of the parameters is redundant, the model is capable of fitting good data without overfitting, i.e., without inappropriately fitting noise or error. It does not follow that all types of experimental error are detectable, since data with smoothly varying erroneous deviations in thermodynamic variables will be correlated to apparently high precision; however, abrupt or irregular deviations between the data and the predictions of the model as optimized by the guidelines established by this procedure are highly unlikely, and in such cases the data can legitimately be regarded as suspect, as has been shown explicitly in a case history of carbon dioxide with the butane isomers.[18]

Furthermore, this systematic addition of new degrees of freedom ensures that the parameters are largely independent of one another and have little mutual interference, in a manner similar to the fitting of an arbitrary function to a sum of orthogonal polynomials. The parameters are not orthogonal in a strict mathematical sense, but are so within the more descriptive notion of orthogonality. The "original" Leung-Griffiths formalism, without the modifications of Moldover, Rainwater, and co-workers, suffered from a high degree of coupling and interference among the parameters, and therefore has made actual correlations[21,22] very difficult; additionally, no effort to optimize its parameters by formal nonlinear fitting has met with success. By contrast, recently a formal nonlinear (Simplex) fitting program has been developed for the modified model, and, because of the independence of the parameters, convergence is robust and straightforward. Before this development, successful correlations for many of the simpler mixtures had been obtained with the modified model by purely visual techniques.

The empirical approach just described in principle is clearly applicable to other methods than the modified Leung-Griffiths model and to other problems than VLE; in fact, it is the approach planned in a later study of the problem of supercritical one-phase behavior. Since the model is consistent with both thermodynamics and modern understanding of critical phenomena, however, it provides VLE correlations in the critical region that are, to the knowledge of the author, superior to those of any other method. The model continues to be under development, and at present consistently provides excellent P-T-x-ρ correlations for all simple mixtures such that $\alpha_{2m} < 0.25$ and good P-T-x correlations for $\alpha_{2m} < 0.4$.

Perhaps not surprisingly, the most difficult data to correlate accurately have been those mentioned at the outset, relatively dilute solutions of a nonvolatile compound in a volatile solvent, which also are of greatest interest for SCF technology. The same peculiar physical and mathematical behavior of solubility that is technologically exploitable is also the most difficult to model accurately. This was first seen by Moldover and Gallagher[3] in an attempted correlation of propane + *n*-octane, where the greatest discrepancies by far for their version of the model with a single adjustable parameter were at the dilute octane limit. While this mixture can now be accurately correlated with a six-parameter model, it is found that the dilute octane end is extremely sensitive to small variations in the parameters, and similar limitations are now found in mixtures even more difficult to correlate such as carbon dioxide + methanol.[23] There are also certain fundamental peculiarities such as the "double retrograde vaporization" found by Kobayashi et al.[24,25] that will require special mathematical analysis and will be difficult to model. Nevertheless, rapid progress has been made, and it is realistically hoped that with features judiciously added to the formalism, such problems can be overcome. It is clear that the use of nonclassical exponents is extremely important; a concave region of constant-composition loci in temperature-density space, most pronounced for a dilute nonvolatile solute in a volatile solvent, has been shown to be a direct consequence of the nonclassical exponents.[26]

This review is admittedly written as a frank advocacy of the modified Leung-Griffiths model, and is not meant to be a neutral, unbiased, third-party evaluation of various critical-region mixture theories presently available. The modified Leung-Griffiths model in its present stage of development has its obvious shortcomings, and the author intends to be forthright about acknowledging them. Nevertheless, it is hoped that most of the shortcomings (except for the fundamental restriction to an extended critical region) can be overcome within the philosophy of the model, and that its great advantages in describing mixture critical behavior outweigh such deficiencies.

Since the distinction between a standard EOS and a parametric field-space model is most transparent for the pure-fluid case, Section II begins the discussion with a review of pure-fluid vapor-liquid coexistence. Section III reviews the six possible types of binary mixture phase diagrams according to the classification scheme of Van Konynenberg and Scott,[27] where type 1 mixtures (those with a continuous critical locus and without liquid-liquid equilibrium), in the absence of azeotropy, are of particular interest and are discussed in Section IV. Inasmuch as the research has of necessity been semi-empirical, Section V provides a comprehensive review of the experimental literature for thoroughly measured VLE surfaces of class 1 binary mixtures (those with a continuous critical locus) in the critical region.

A review of the Leung-Griffiths model is given in Section VI, with the basic thermo-dynamic transformations, the parametric equations that introduce the phase transition, and the incorporation of adjustable parameters. A summary of the many successful correlations is presented in Section VII, and the techniques for obtaining these correlations are described in Section VIII. A detailed comparison for the carbon dioxide + propane mixture between the correlation accuracy of the Rainwater-Moldover model, the DDMIX[28,29] computer package, and the Soave-Redlich-Kwong EOS[30,31] is presented in Section IX. Since the model has been used for supercritical solubilities in carbon dioxide, Section X describes the results of such a study in which the pentane isomers are the solutes.

Section XI analyzes the interesting special case of carbon dioxide + methane, which has a solid-liquid-vapor locus at relatively high temperature and a distorted VLE surface. The discussion of azeotropic mixtures in Section XII completes the study of VLE of binary mixtures. Sections XIII and XIV briefly examine some related problems: ternary mixtures, experimental determination of mixture critical densities, interfacial tension, phase equilibrium algorithms for a classical EOS, liquid-liquid equilibrium, and modeling of enthalpies and entropies. The subject is summarized in Section XV.

II. PURE FLUID PHASE TRANSITIONS

Many of the advantages and limitations of a parametric model designed for the critical region of mixtures are most transparent in the simpler case of a pure fluid.[12] For the purpose of clarifying the philosophy of nonclassical models in a simple context, this chapter initially reviews the pure fluid vapor-liquid equilibrium transition. The pressure-temperature diagram

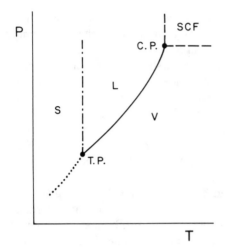

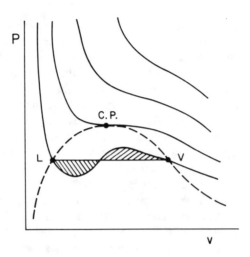

FIGURE 1. Phase diagram in P-T space for a pure fluid (schematic). The vapor pressure curve (solid curve) extends from the triple point (T.P.) to the critical point (C.P.) and separates the liquid (L) and vapor (V) phases. Melting and sublimation loci (broken and dotted curves) separate the solid phase (S) from the liquid and vapor, respectively. The SCF region is the region above the critical pressure and temperature, as indicated by dashed lines.

FIGURE 2. Vapor-liquid transition for a pure fluid in P-v space (schematic). The dashed curve is the phase boundary and its maximum is the critical point. The solid curves are isotherms; the critical isotherm has a point of inflection at the critical point. The Maxwell construction is shown, with equal shaded areas and liquid (L) in coexistence with vapor (V).

is shown in Figure 1, with a vapor pressure curve, critical point, triple point, melting curve, and sublimation curve, a pattern followed qualitatively by all fluids except the helium isotopes. Since neither a standard EOS nor the parametric Schofield model describes the transition to a solid phase, only the vapor pressure curve and critical point are considered. For the purposes outlined here, the supercritical region is the area higher in both pressure and temperature than the critical point, although this region has been defined differently and with some confusion elsewhere; the portion of this region relatively close to the critical point is of greatest present interest.

A pressure-volume plot and the means of constructing the phase boundary from an EOS are displayed in Figure 2. At temperatures above the critical temperature T_c, isotherms as described by the EOS show a monotonic decrease of P with molar volume v, but the critical isotherm shows a point of inflection at the critical point. For isotherms below T_c, according to the EOS there is a region of thermodynamic instability where $(dP/dv)_T$ is positive. As is well known, the correct thermodynamic description is obtained with a Maxwell construction as displayed, where the two shaded regions between the predicted isotherm and the horizontal line are equal in area, thereby yielding liquid and vapor states in coexistence. It is shown in textbooks that the Maxwell construction is equivalent to the determination of states of equal Gibbs free energy or chemical potential.[32] Thus, the fluid is not described by the EOS literally, but by the EOS plus the Maxwell construction; a formal thermodynamic instability has been converted into a phase transition.

The highly skewed phase boundary in P-v space is considerably more symmetric in T-ρ space, as shown in Figure 3, where the phase boundary may be accurately represented within a "simple scaling" approximation by

$$\rho = \rho_c[1 \pm C_1(-t)^\beta + C_2 t] \qquad (3)$$

where

$$t = (T - T_c)/T_c \qquad (4)$$

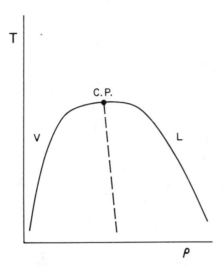

FIGURE 3. Vapor-liquid transition for a pure fluid in T-ρ space (schematic). The solid curve is the phase boundary and its maximum is the critical point; the dashed line is the rectilinear diameter, empirically found to be nearly straight.

and β is a critical exponent; for $\beta = {}^1\!/_3$ this is the equation attributed to Guggenheim,[33] where the rectilinear diameter,

$$\rho_d = {}^1\!/_2(\rho_\ell + \rho_v) \tag{5}$$

is a straight line on a P-T plot. The exponent of approximately one third actually goes back to Verschaffelt[34] at the turn of the 20th century. The fundamental limitation of analytic EOSs is that in the limit of approach to the critical point, the form of Equation 3 is obtained but with β assuming the classical, and incorrect, value of ${}^1\!/_2$. Of less quantitative importance for phase equilibria, but of equal fundamental importance, is that the vapor pressure curve of Figure 1 is in principle represented by the power series

$$P/RT = P_c/RT_c[1 + C_3(-t)^{2-\alpha} + C_4t + C_5t^2 + C_6t^3 + \ldots] \tag{6}$$

where α is the critical exponent characterizing the divergence of the constant-volume specific heat on the one-phase critical isochore,

$$C_v = C_{vo}(-t)^{-\alpha} \tag{7}$$

As calculated from an analytic EOS $\alpha = 0$ so that C_v does not diverge, in contradiction to experiment. Our modern understanding of critical phenomena due to Wilson[35] (by means of renormalization group theory) indicates that the three-dimensional Ising model is in the same universality class as the three-dimensional liquid-vapor transition, and therefore possesses the same critical exponents. Both theoretical studies on the Ising model and careful experiments on fluids extremely close to the critical point ($|t| < 0.001$) have shown[36-38] that $\beta = 0.325 \pm 0.001$ and $\alpha = 0.110 \pm 0.001$, in contrast to the classical values.

The difference in philosophies between conventional phase equilibrium calculations with EOS and the parametric Schofield model[12,39] for accurate critical region behavior must be emphasized at this point. As stated earlier, field variables are by definition continuous

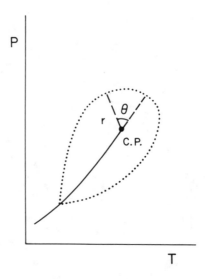

FIGURE 4.   Parametric variables of Schofield in P-T space (schematic). The upper part of the vapor pressure curve in Figure 1 is shown with a locus of constant r (dotted line), which measures a distance from the critical point. θ is an angle-like variable that goes to +1 for the saturated liquid and −1 for the saturated vapor.

functions of each other, even across a phase boundary, but that boundary is characterized by a discontinuity in the partial derivatives of field variables with respect to each other. Such partial derivatives are density variables,[11] for example $(d\mu/dP)_T = \rho^{-1} = V$, the molar volume. The alternative to the standard method is to construct mathematically, in field variable space, the appropriate multidimensional function which is continuous but has discontinuous derivatives. Furthermore, the discontinuity must go to zero at a critical point, and must increase with thermodynamic distance from the critical point, i.e., $|t|$, according to the proper critical exponent. The Schofield model is one such construction; it is neither the only nor necessarily the best such construction, but it has clearly proven to be a very useful technique for correlation of pure fluid properties in the vicinity of the critical point.

The version of the model described here is not the original version or the one most frequently used for pure fluids, but is an acceptable alternate formulation that is more easily generalizable to mixtures. As functions of field variables are clearly themselves field variables, the dependent field variable or thermodynamic potential[11] is

$$\omega = P/RT \tag{8}$$

where R is the gas constant, and the independent field variables are t, defined earlier, and

$$h = \mu/RT - \mu^\sigma/RT \tag{9}$$

where μ^σ is the value of the chemical potential on the vapor pressure curve (or its smooth extension) for a given t. The dependent variable ω is modeled as the sum of a regular or analytic part and a singular part that incorporates the discontinuous derivatives or phase boundary.

The singular part is expressed parametrically in terms of two new variables, r and θ, as shown schematically in Figure 4. They are somewhat like a polar coordinate system centered at the critical point, although loci of constant r are not circular and θ is restricted to the interval $(-1, 1)$ rather than $(-\pi, \pi)$. The explicit model, a variant of the so-called "restricted linear Schofield model"[39], is

$$\omega_{sing} = (P_c C_3/RT_c a_T)r^{2-\alpha}(a_0 + a_2\theta^2 + a_4\theta^4) \tag{10}$$

$$t = r(1 - b_2\theta^2)/(b_2 - 1) \tag{11}$$

$$h = (P_c C_3/RT_c \rho_c C_1 a_T)r^{2-\alpha-\beta}\theta(1-\theta^2) \tag{12}$$

where C_1 and C_3 are defined in Equations 3 and 6 and b, a_0, a_2, a_4, and $a_T = a_0 + a_2 + a_4$ are constants and, within the "restricted" model, functions of α and β; see Reference 6, Equations 51 to 57 for explicit forms. The regular, or analytic, part of the potential is given by

$$\omega_{an} = (P_c/RT_c)(1 + C_4t + C_5t^2 + C_6t^3) + \rho_c h(1 + C_2t) \tag{13}$$

The coexistence curves, Equations 3 and 6, now follow by setting $h = 0$ and $\theta = +1$ (for coexisting liquid) or $\theta = -1$ (for coexisting vapor), and the relation

$$\rho = (\partial\omega/\partial h)_t = C_1\rho_c r^\beta\theta + \rho_c(1 + C_2t) \tag{14}$$

There are now several observations to be made about the advantages and shortcomings of the two respective models, which are most transparent in the pure-fluid case but which are also in effect for the generalization to mixtures:

1. The Schofield model automatically incorporates correct nonclassical exponents or, if desired, effective exponents (which differ slightly from the true asymptotic-limit values but which approximately represent the thermodynamics over a wider region), whereas regardless of the form of the EOS, very close to the critical point the (incorrect) classical exponents $\beta = 1/2$ and $\alpha = 0$ are obtained with any analytic EOS. Since a nonanalytic function can be approximated by a sum of analytic functions to any desired approximation, however, a many-parameter EOS can mimic the nonclassical exponents. In fact, the 32-term equation recently developed by Schmidt and Wagner[40] does mimic the value $\beta = 1/3$ to within about $|t| = 0.001$. Although successful correlations of near-critical pure-fluid thermodynamic properties were performed much earlier with the parametric model, the Schmidt-Wagner equation is at present as accurate as the parametric model for most pure-fluid practical applications; however, its successful generalization to mixtures has not yet been achieved and is by no means straightforward.
2. The phase boundary is much more simply described in the parametric model by the algebraic relations $h = 0$ and $\theta = \pm 1$. By contrast, with the EOS a relatively difficult numerical calculation must be performed to determine the phase boundary.
3. A critical point can be calculated from the EOS, but in general the critical point and the phase boundary cannot both be made simultaneously to agree with experiment. In Equation 3, the classical value $\beta = 1/2$ yields a parabolic shape for the coexistence curve in Figure 3, whereas the true curve, with $\beta = 1/3$, is much flatter on top. Consequently, optimizing the parameters of the EOS to the coexistence boundary at low pressure will cause the predicted critical point, in general, to overshoot and to be too high in pressure and temperature, whereas optimization of the parameters to obtain the experimental critical point will be at the expense of correct phase boundary prediction at low pressures. On the other hand, in the Schofield model the critical point is input rather than output. The model makes no statement of where the critical point is located; it only describes thermodynamic behavior relative to the critical point.

4.   For engineering purposes, the EOS may have a greater intuitive appeal, in that it algebraically relates measurable thermodynamic variables, whereas the field-space parametric model involves quantities such as the chemical potential that are not normally directly measurable; however, the derivation of expressions for measurable variables from the parametric model from numerical methods is relatively straightforward,[41] and with modern computational resources this should not be regarded as a significant drawback.

5.   In our view, perhaps the greatest shortcoming of the scaling-law parametric approach is the following. With the EOS and within the normal domain of the independent variables (for example, the disallowing of negative density or temperature), a thermodynamic instability is converted into a phase transition by means of the Maxwell construction or its appropriate generalization for mixtures; however, in the Schofield picture, an instability is simply an unwanted artifact; for example, if $C_1 \geqslant 1.28$ and $C_2 = 0$, for $t < -0.5$ a negative saturated vapor density is predicted, which is an obviously unphysical result and can be interpreted as an instability.

6.   The two approaches are complementary in that the EOS works best away from the critical point whereas the Schofield model works best near the critical point. Accordingly, global representations have been developed by using both approaches in their respective domains and merging them with a switching function.[42,43] Since there are some technical problems with such switching functions,[44] and since they are perceived as inelegant by some, alternative approaches to the global equation have been proposed. Two of these are the model of Fox,[45] in which the thermodynamic measures are transformed near critical, and the formal crossover theory introduced by Nicoll[46] and recently developed by Sengers and co-workers.[47-50] Although successful crossover EOSs for steam, carbon dioxide, and ethylene by means of the second method have recently been developed,[49,50] neither method to date has been successfully generalized to yield accurate correlations for a variety of mixtures.

The Leung-Griffiths model may be regarded as the (not very straightforward) generalization of the above-described Schofield model from pure fluids to mixtures. Accordingly, one must expect it to have the same advantages and drawbacks for mixtures that the Schofield model has for pure fluids. At present, this generalization is well ahead of competing techniques in its demonstrated capacity to produce accurate mixture VLE correlations in the critical region.

III. TYPES OF BINARY MIXTURE PHASE DIAGRAMS

The currently accepted scheme for classification of binary mixture phase diagrams is that of van Konynenberg and Scott.[27] According to their scheme, there are six distinct types of phase diagrams, as shown in Figure 5. Useful detailed discussions of the thermodynamic behavior of the various types of mixtures, with specific examples, are provided by Rowlinson and Swinton[32] and by Streett.[51]

The simplest case, discussed in the Introduction, is type 1 (Figure 5A), for which there is a continuous critical locus connecting the two pure-fluid critical points and no liquid-liquid immiscibility. For all of the other types, there are certain thermodynamic regions in which the liquid phase separates into two immiscible liquid phases, or liquid-liquid equilibrium (LLE).

For type 2 (Figure 5B), in addition to the VLE critical locus or locus of plait points, there is a separate LLE critical locus or locus of consolute points that normally occurs at temperatures lower than the critical temperature of the more volatile component (although there are exceptions such as ethane + ammonia[52]). On a plane of constant pressure, the

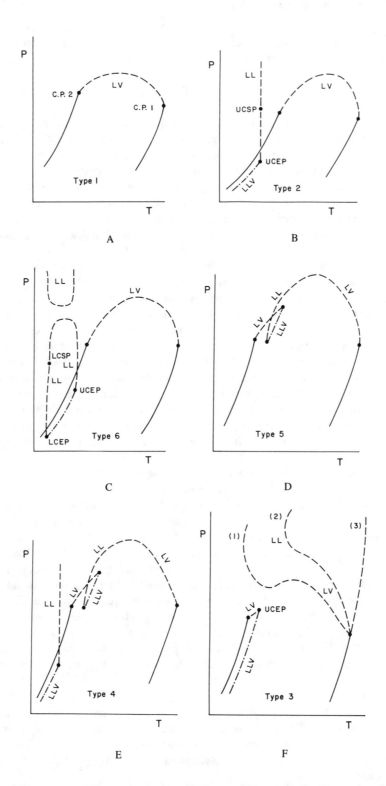

FIGURE 5. Different types of binary mixture phase diagrams according to the Van Konynenberg-Scott classification method (schematic), where solid curves denote pure vapor pressure curves, dashed curves denote critical loci and broken curves denote three-phase loci, L denotes liquid and V denotes vapor: (A) type 1; (B) type 2; (C) type 6; these first three cases are "class 1"; (D) type 5; (E) type 4; (F) type 3, for which three cases of the upper branch of the critical locus are shown in order of increasing fluid dissimilarity: (1) minimum pressure and temperature, (2) minimum temperature only, and (3) no minimum in pressure or temperature.

liquid-liquid two-phase region is described in T-x space by a diagram similar to Figure 3, with ρ replaced by x, so that to leading order

$$(x - x_c) \propto (T - T_c)^\beta \tag{15}$$

where, as before, $\beta \approx {}^1/_3$ in practice but $\beta = {}^1/_2$ if calculated from an analytic EOS.

The slope of the consolute point or upper critical saturation point (UCSP) locus can be either positive or negative, but initially the most noteworthy feature of the consolute point locus is that it is nearly independent of pressure, i.e., almost completely vertical in Figure 5B. The T-x boundary curve at a constant pressure, except at the lowest pressures, is also almost pressure independent. Thus, Francis,[53] in a widely used compendium of LLE data, does not even discuss pressure dependence, while in the more recent compendium of Arlt and Sorensen,[54] the issue of pressure dependence is mentioned but discounted in importance. The LLE surface for all higher pressures can therefore be estimated from data taken only at atmospheric pressure, which is by far the most common case. As shown in Figure 5B, the UCSP locus ends from below in a critical endpoint, and the LLE surface joins the VLE surface along a three-phase locus on which two liquids and one vapor are in equilibrium.

The remaining types, for which nonclassical models have yet to be attempted, are best discussed out of numerical order. Type 6 is similar to type 2 except that on an isobaric T-x plane, the LLE boundary curve does not resemble Figure 3, but rather is a closed loop with a lower critical saturation point (LCSP) as well as the previous UCSP. The region of liquid immiscibility occurs only over a finite temperature range; above or below this range the liquids are miscible, as shown in Figure 5C. Van Konynenberg and Scott[27] have noted that type 6 is the only one of the mixture types not describable qualitatively by the simple van der Waals equation, and depends strongly on polarity of the fluid molecules. The many examples of such behavior in the Arlt and Sorenson[54] compendium all contain either water or glycerol as a component.

Types 4 and 5 are relatively infrequently occurring transition cases. As shown in Figure 5D, for type 5 the locus of plait points is discontinuous and consists of two branches. Each point goes from a pure-fluid critical point to a critical endpoint, and the two critical endpoints are then joined by a three-phase LLV locus. While the lower branch of the plait point locus describes VLE, the upper branch, in the vicinity of the lower critical endpoint first describes LLE, but as the pressure increases, smoothly makes a transition to VLE. The distinction between liquid and vapor here may seem somewhat arbitrary, but might better be understood in terms of the α_2 variable of Equation 1. The essential feature of a vapor-liquid transition near a critical point is that as the two-phase region is entered, the two coexisting phases show an abrupt difference in density and a relatively small difference in composition. In a liquid-liquid transition the reverse is true; there is an abrupt difference in composition and relatively small difference in density for the two phases. Therefore, if a strict dividing point is desired for the upper branch of the critical locus of a type 4 or 5 mixture, we would propose the point at which α_2 changes from a value greater than 1 (LLE) to a value less than 1 (VLE).

The distinction between type 4 and type 5 mixtures is analogous to that between types 2 and 1. As shown in Figure 5E, type 4 mixtures display the same broken critical line structure between the two pure fluid critical points, but in addition there is a second three-phase locus and a separate locus of consolute points, similar to those of type 2 mixtures, at lower temperatures.

Finally, type 3 mixtures occur in the limit of extreme dissimilarity of the components. The structure is shown in Figure 5F. There is a small branch of the critical locus extending from the critical point of the more volatile component to a critical endpoint, to which a three-phase LLV locus is joined. The critical endpoint temperature is usually higher than the critical temperature of the more volatile component, but there are exceptions.[55-57] The LLV locus extends downward in pressure and temperature and is roughly parallel to the vapor pressure curve of the more volatile component. A separate branch of the critical locus starts at the critical point of the less volatile component and ultimately tends toward "indefinitely high pressures" in the P-T plane. Actually, at some extremely high pressure this branch of the critical locus must intersect a solid phase, but such pressures are far above the usual pressure range of most fluid property measurement laboratories.

The upper branch of the critical locus, as shown in Figure 5F, may have sequentially a pressure maximum, a pressure minimum, and a temperature minimum (curve 1). For greater dissimilarity of the components, it will have only a temperature minimum (curve 2) and in the extreme case it will proceed immediately from the critical point of the less volatile component to higher pressures and temperatures (curve 3). The last two cases are often called "gas-gas immiscibility" of the first and second kind, respectively,[32] although such a term is somewhat controversial.[58] As with type 4 and 5 mixtures, there is no unambiguous division between gas and liquid or between VLE and LLE along the upper branch of the critical locus. Again, since the parameter α_2 of Equation 1 is well-defined and finite along the critical locus, if a strict division between VLE and LLE is needed the author recommends the point at which $\alpha_2 = 1$.

Van Konynenberg and Scott introduce the concept of "class" as well as type, where "class 1" mixtures (types 1, 2, and 6) possess a continuous locus of plait points from one pure-fluid critical point to the other, whereas "class 2" mixtures (types 3, 4, and 5) do not. Strictly speaking, van Konynenberg and Scott assign type 6 to a third class, but type 6 meets the requirement of a continuous plait point locus. Also, those authors do not use the term "type 6", but this term has been used by subsequent authors[32] and is now standard. The terminology is confusing in that some authors, for example, Streett,[51] use "class" in the manner that van Konynenberg and Scott use "type". For this discussion, the separate concepts of "type" and "class" are useful and thus both terms are retained.

A. PROGRESSION OF TYPES

A greater appreciation and understanding of the different types of phase equilibrium can be obtained by considering a succession or family of mixtures in which one component is fixed and successive members of a homologous series are taken as the second component.[59] Here, consider the case that the first component, the solvent, is the most volatile, and the solutes are members of a homologous series, for example, carbon dioxide with the n-alkanes. Brunner[56] has provided a comprehensive bibliography of experimental studies of such families. It is assumed for the present example that type 6 mixtures cannot occur.

For the lighter solutes, the mixtures are type 1; there is no LLE and the VLE is bounded from below by some kind of three-phase SLV locus. Either it is a simple triple locus, with both the liquids and the solids completely miscible, or it is a eutectic[60] structure in which the liquids are miscible but the solids in general are not. The type 1 mixture is shown in Figure 6A. In Figures 6A-6G, the SLV loci show the common eutectic structure with a solid-solid-liquid-vapor quadruple point (Q') at a temperature and pressure slightly below those of the triple point of the more volatile fluid.

At some point in the series, an LLE locus emerges, either as a type 2 structure (Figure 6B) or as a type 5 structure (Figure 6C). The limiting case of a type 5 LLV locus of zero length is known as a tricritical point, and is of considerable theoretical interest.[61] For a binary mixture, there is zero measure of probability for obtaining an LLV locus of length zero and thus a tricritical point cannot occur, but it does occur for ternary mixtures and for

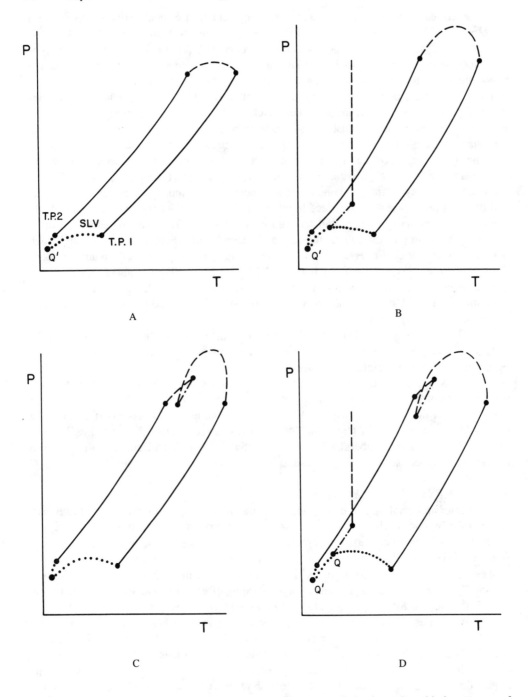

FIGURE 6.   Sequence of phase diagrams for a family of mixtures according to increasing critical temperature of the solute (schematic), where solid curves denote pure vapor pressure curves, dashed curves denote critical loci, dotted curves denote solid-liquid-vapor (SLV) loci, broken curves denote liquid-liquid-vapor (LLV) loci, Q denotes a SLLV quadruple point and Q' denotes a SSLV quadruple point: (A) type 1, (B) type 2, (C) type 5, where either (B) or (C) can emerge first; (D) type 4 with short LLV loci, (E) type 4 with long LLV loci, (F) initial emergence of type 3, (G) final structure in which SLV loci cut off the critical loci.

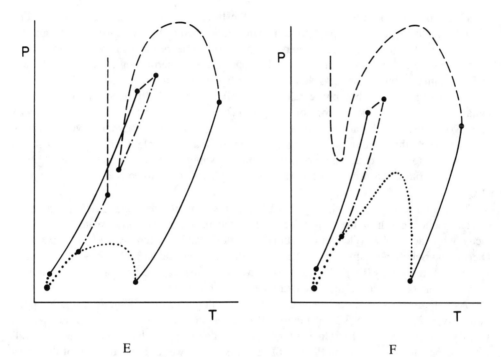

E

F

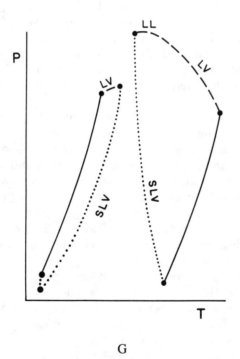

G

FIGURE 6, continued

so-called "pseudobinaries" in which two of the three components are very similar. For example, methane forms a type 1 mixture with 2,2-dimethyl butane but a type 5 mixture with each of the other hexane isomers.[62] Thus, experiments have been conducted with methane as the solvent and a mixture of the appropriate proportion of 2,2-dimethyl butane and 2,3-dimethyl butane as the solute, so that a tricritical point is approached.[63,64]

According to the Gibbs phase rule, the number of degrees of freedom for a thermo-dynamic locus equals the number of components minus the number of coexisting phases plus two. Thus, for a binary mixture there are the one-phase thermodynamic volume, two-phase surfaces, three-phase lines, and the possibility of a quadruple point. Figure 6B shows the first example of a quadruple point (Q) with one vapor, two liquid, and one solid phase all in equilibrium. According to the 180° rule, any two of the three-phase loci must meet at Q at an angle[65,66] less than 180°.

The initial emergence of the type 5 LLV locus is generally near the critical point of the solvent, rather than near the equimolar mixture where α_{2m} is typically greatest. The LLV locus, while never of zero length for a binary mixture, can be extremely short. For example, Diepen and Scheffer[67] found for ethane + 1,4-dichlorobenzene an LLV locus that extended in temperature for only 0.6 K. In order to determine definitively the type of a binary mixture, therefore, the critical locus must be examined in extreme detail.

As one proceeds to higher members of the homologous series of solutes, at some point both a type 2 and a type 5 LLV locus will emerge; in other words, the structure will be type 4. Furthermore, both LLV loci will get longer and will approach two interrupted intervals of the same curve, as shown in Figures 6D and 6E. The lower critical endpoint of the upper (originally type 5) LLV locus will extend to temperatures below the critical temperature of the solvent. Further into the sequence, the two LLV loci will merge, as will the consolute point locus of the original type 2 structure and the upper branch of what originally was the plait point locus, so that the structure is transformed to type 3, as shown in Figure 6F. If conditions were such that the two LLV branches and the two critical locus branches all joined at a single point, that would be a "critical double point". Again, the measure of probability for this to occur is zero for a binary mixture, but it can be made to occur with a "pseudobinary".

For solutes of even larger molecules, the phase diagram proceeds through the different possible cases of type 3 mixtures, as shown earlier in Figure 5F; however, there is a further complication in that as the solute molecules become larger and the solute critical temperature increases, the temperature of the triple point of the solute also increases and eventually becomes substantially larger than the critical temperature of the solvent. Then the SLV locus, which bounds the fluid phase diagram from below, intrudes on the critical loci so that the lower critical endpoint of the LLV locus proceeds to higher pressures and temperatures, until the LLV locus becomes obliterated, as shown in Figure 6G. This is qualitatively equivalent to the carbon dioxide-naphthalene diagram,[7] the prototype of a supercritical extraction process with a solid solvent.

Another way to view this progression is with the diagrams used by Luks and co-workers (see the recent article by Miller and Luks[68] and references therein). These researchers have conducted an extensive project to measure the LLV loci of various families of mixtures; Figures 7 and 8 show schematically the diagrams they employ. The vertical axes, with discrete steps, represent the homologous series of solutes in, for example, carbon number, for fixed solvent; here they are labeled A through H. The horizontal axes denote temperature.

In both cases, for solute A there is no LLV locus, so that mixture is type 1. In Figure 7, an LLV locus first emerges at lower temperature (type 2, Figure 6B) for solute B, while in Figure 8, it first emerges above the critical temperature of the solvent (type 5, Figure

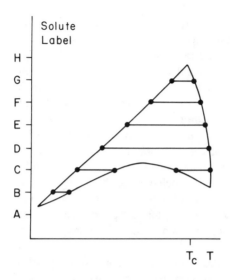

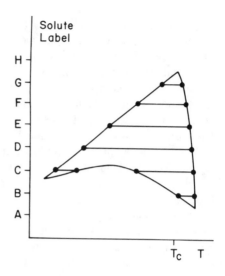

FIGURE 7. Three-phase LLV loci for a family of mixtures according to the diagrams of Luks and co-workers (schematic), where T_c denotes the critical temperature of the solvent and A through H denote the solvents in order of increasing critical temperature or, typically, increasing carbon number: type 2 structure, Figure 5B, first emerges.

FIGURE 8. Diagram similar to Figure 7 except that type 5 structure, Figure 5C, first emerges.

6C). In both cases the mixture with solute C is type 4, as in Figures 6D and 6E. The lower branches of the LLV loci are bounded from below by a solid-liquid-liquid-vapor quadruple point, where the solid is normally rich in the solute but in rare cases (e.g., carbon dioxide + octane[68]) may be rich in the solvent. In both figures the two branches of the LLV loci merge for solutes D through G, and these solutes form type 3 mixtures. The lower critical endpoint continues to increase with temperature until, at solute H, the LLV locus is obliterated as in Figure 6F. These studies do not establish the behavior of the upper branch of the critical locus, but do establish quite unambiguously the mixture type.

The above discussion assumes that the mixture type is determined by the solid, as well as the fluid, phase diagram; however, there is an alternate viewpoint. From the statistical mechanics of pure fluids as well as mixtures, consider both a fluid EOS (in which, by definition, the single-particle position distribution function is uniform) and a solid EOS, in which the average positions of the molecules are constrained to a lattice. In the most general case, consider several possible lattices such as face-centered cubic, body-centered cubic, etc. For a real substance, there is some premelting behavior in the solid such as the buildup of defects and vacancies and some prefreezing behavior in the liquid such as local clusters of lattice-like structures; as a first approximation, however, regard the liquid solution and the various solid solutions to the statistical mechanics of the substance as independent, and at a given temperature and pressure the phase of the substance will then correspond to the solution having the lowest free energy.

Within the alternate viewpoint, the mixture type behavior depends only on the fluid solution, not on the behavior with consideration of both fluid and solid. The second viewpoint must be taken when analyzing mixture behavior with an equation of state, which does not describe the solid transition. The pertinent question then is, if freezing could be suppressed, whether at temperatures below actual freezing would two miscible liquids at some point exhibit liquid-liquid immiscibility? For example, if the left boundary or locus of Q-points was removed from Figure 7 and the LLV loci were allowed to extend to arbitrarily low temperature, the mixture with solute A would be type 2 rather than type 1.

Certain lattice-gas models[69] of mixtures possess a ground state, at T = 0 K, of complete separation of the components. This suggests the hypothesis that most (or, as an extreme viewpoint, all) miscible liquid mixtures (except type 6 mixtures) would undergo liquid-liquid immiscibility, or LLE, at sufficiently low temperature. Within this viewpoint, all type 1 mixtures are "really" type 2 and all type 5 mixtures are "really" type 4; even a mixture such as ethane + propane would exhibit a locus of consolute points in some cryogenic regime. Yet, the diagram of Miller and Luks[68] for ethane as solvent and the *n*-alkane series as solutes shows no tendency toward the "bending-back" behavior of the lower curve in Figures 7 and 8. Thus, LLE for ethane + propane, even it if exists in principle for a fluid solution, would occur at temperatures so far below freezing that its possible occurrence becomes a strictly academic matter. In this sense, it is reasonable to define phase behavior type with inclusion of the solid phase.

Since the Leung-Griffiths model only describes fluid thermodynamics, it is independent of the location of a freezing locus. It is not appropriate to restrict the model completely to type 1 mixtures, since the presence of LLE at very low temperatures should not affect significantly the VLE behavior in the "extended critical region". Indeed, the mixture helium 3 + helium 4, the subject of the original Leung-Griffiths paper,[1] exhibits liquid-liquid immiscibility at very low temperature.[70]

However, the presence of LLE at higher temperatures, as shown later, can lead to distortions in the field-space structure of the model and thus to substantial mathematical complications. Also, the interesting possibility exists that a type 1 mixture, as defined including the solid phase, may have a freezing locus at relatively high temperature and "virtual" LLE just below the freezing locus, which is not directly measurable but which distorts the critical-region VLE structure. Section XI provides evidence that carbon dioxide + methane is, in fact, such a mixture. At present in the literature, there seems to be no consensus as to whether to define mixture types including or excluding the solid phase. In view of the considerations above, such ambiguity may to a certain degree be appropriate.

IV. TYPE 1 MIXTURES: OVERVIEW

Since the modified Leung-Griffiths model in its present form is suitable only for type 1 mixtures, this review shall henceforth (except for Sections XI and XIV) consider only these simple mixtures in which the liquids are always completely miscible. This section reviews the qualitative features of nonazeotropic or "normal" type 1 mixture phase diagrams; see Section XII for the azeotropic case.

The usual domain of the model ranges approximately from the mixture critical pressure down to half that pressure; this region of P-T space shall be referred to as the "extended critical region". VLE data in this region are typically measured either along isopleths (loci of constant composition) or isotherms; isobaric data in this region are extremely rare.[71-74] The four common projections of the coexistence surface are as shown in the following four figures: Figure 9A, isopleths on a P-T plot; Figure 9B, isopleths on a T-ρ plot; Figure 9C, isotherms on a P-x plot, and Figure 9D (which is rarely seen in the experimental literature), isotherms on a ρ-x plot.

In Figure 9A, the two pure vapor pressure curves form the left and right boundaries of the coexistence surface, and both terminate at their highest pressures in pure fluid critical points. A continuous critical locus extends from one pure critical point to the other; this locus typically has a maximum in pressure although there are mixtures for which the critical pressure is monotonically increasing[16] or decreasing.[17] Also shown are four dew-bubble

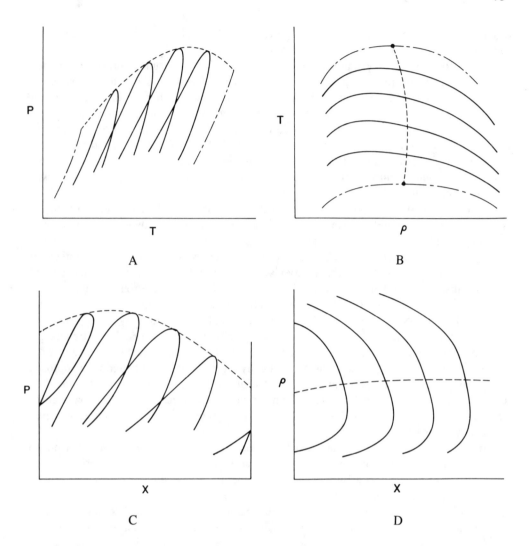

FIGURE 9. Binary mixture vapor-liquid equilibria diagrams showing loci along which experimental measurements are usually taken (schematic); dashed lines denote critical loci and broken lines denote pure-fluid coexistence curves: (A) isopleths (loci of constant composition) in P-T space; (B) isopleths in T-ρ space; (C) isotherms in P-x space; and (D) isotherms in ρ-x space.

curves; the bubble curve is on the left and the dew curve on the right. The dew-bubble curves bound the two-phase region for a given composition and, unlike pure vapor pressure curves, have finite width. These curves are tangent to the critical locus at the mixture critical point, which is (except in special cases) neither the point of highest temperature nor of highest pressure on the curves. The loci of highest temperature and of highest pressure points are called either the cricondentherm and cricondenbar loci, respectively, or the maxcondentherm and maxcondenbar loci, respectively; the author prefers the latter terminology.

A single dew curve and a single bubble curve pass through each (P, T) point of the two-phase region bounded from the sides by the pure vapor pressure curves, from above by the critical locus, and from below by a freezing (SLV) locus as shown in Figure 6A. At the given (P, T) point, liquid of composition characterized by the bubble curve (and rich in the less volatile component) is in coexistence with vapor of composition characterized by the dew curve (and rich in the more volatile component). While a pure fluid boils or condenses at a single point on an isotherm or isobar, a mixture boils or condenses over a finite range of temperature on an isobar or a finite range of pressure on an isotherm. The mathematical form of these dew-bubble curves is discussed in Reference 26.

Figure 9B shows a typical temperature-density plot. The coexistence surface is bounded from above and below by the T-ρ coexistence surfaces of the respective two pure components. The mixture isopleths somewhat resemble the pure-fluid curves, but are distorted. As can be seen from Figure 9A, the maxcondentherm points are typically on the vapor side, so the points of highest temperature on the isopleths of Figure 9B are at a density below the critical point. A result Rainwater has recently shown (see Reference 26) is that around the critical point there is in general a small interval of concave upward curvature, and that the critical point is the point of maximum concave curvature instead of an inflection point as was previously conjectured.[75]

A typical pressure-composition plot is shown in Figure 9C, where convention is that x = 1 is the more volatile component. Again, the critical locus is shown with a maximum in P vs. x. Unlike the earlier example, for isothermal dew-bubble curves the critical point is the point of maximum pressure. This may be understood by noting that there are horizontal tie lines across each isotherm joining coexisting liquid and vapor states, and the critical point occurs where the length of the tie lines vanishes. The curves are rounded at their tops for $T_{c2} < T < T_{c1}$, but for $T < T_{c2}$, as shown at the right of the diagram, the isotherms hit the x = 1 boundary as a wedge shape with a sharp angle, as required by Henry's law.

Finally, Figure 9D shows a density-composition plot, with isotherms. The extreme right point of each isotherm is the maxcondentherm point, which again occurs on the vapor side. The isotherms cross the critical locus not vertically but at an angle, although they approach a vertical crossing in the pure fluid limits. One of the interesting results of the modified Leung-Griffiths theory is that for many mixtures the slopes of the dew-bubble curves at the critical locus in Figures 9B and 9D may be predicted from the critical properties and their composition derivatives.[26]

In Figure 9A the critical locus is the envelope of the dew-bubble curves, but not in Figure 9C. Because the isotherms are horizontal at the critical point while the critical locus (in P and x) in general has a finite slope, the isotherms extend slightly outside the critical locus as shown. It is interesting that on a P-T plot the envelope is the critical locus and the locus of highest pressure points is the maxcondenbar locus, whereas on a P-x plot the reverse is true; the envelope is the maxcondenbar locus and the locus of highest pressure points is the critical locus.

While these observations are for the most part elementary features of simple phase diagrams, it is surprising and rather discouraging to see how frequently researchers in the field apparently confuse or misunderstand these features. For example, Mandlekar et al.[76] in the presentation of their *n*-hexane + diethylamine VLE data, assert that the maxcondenbar and maxcondentherm points of their isopleths are identical to the critical points. This implies that their dew-bubble curves would have a sharp cusp at the critical point, and is in principle incorrect. Dew-bubble curves have a wide variety of widths and curvatures at the critical point (measured by α_2 of Equation 1), but in principle are curved rather than sharply pointed at the intersection of the critical locus. In other articles,[77,78] P-x diagrams such as Figure 9C are presented with the critical locus as the envelope of the isothermal dew-bubble curves. For some mixtures, the region in which the isotherms extend outside the critical locus is not readily visible on the scale of a typical figure, but from a Leung-Griffiths analysis, such is not the case for the articles cited. More seriously in error are several papers from the Soviet State Institute for the Nitrogen Industry,[79-84] in which isothermal dew-bubble curves are shown incorrectly to terminate with a sharp cusp at the critical locus.

Perhaps the most peculiar examples of erroneously illustrated dew-bubble curves are those of Tyvina et al.[85] for ammonia + acetonitrile. Isotherms are shown as having a maximum in pressure on the vapor side, when in fact the maximum must be at the critical point. The authors attribute this behavior to retrograde condensation as defined by Kuenen;[86] however, retrograde condensation (of the first kind) is a very general phenomenon in mixtures that is easily explained in Figure 9C.

Consider a supercritical mixture of fixed composition slightly above its critical point, i.e., at a temperature corresponding to one of the isotherms as shown but at a composition and pressure slightly larger than those of the critical point for that temperature. At the same time, the composition is less than the maximum composition (maxcondentherm point) for that isotherm. The pressure is isothermally lowered until the two-phase region is reached. The phase boundary will be intersected on the vapor side, to the right of the critical point but to the left of the maxcondentherm point for the given temperature, and the coexisting phase that emerges will be described by a tie line to the left, i.e, a liquid state. This transition is "retrograde" in that lowering of pressure results in condensation (the formation of a "liquid" state more dense than the original fluid, or a meniscus forming at the bottom of a cell), in contrast to the usual situation in which isothermal lowering of pressure results in "boiling" or a meniscus forming at the top of the cell. Retrograde condensation certainly does not imply a pressure maximum on the vapor side of an isothermal dew-bubble curve.

V. SURVEY OF THE EXPERIMENTAL LITERATURE

Since the modified Leung-Griffiths model is a semi-empirical technique, the history of its development cannot be discussed fully without examination of the experimental literature. In this section, the body of experimental VLE measurements on class 1 binary mixtures over the extended critical region is reviewed. The author does not intend to review or to comment critically on the various different experimental techniques; these are discussed or referenced in most of the cited articles.

It is first necessary to specify some criteria for what constitutes a "primary" source, or a thorough measurement of a binary coexistence surface from which a reliable correlation can be developed. While the choices are somewhat arbitrary, the basic criteria are that the experiment should include measurements along at least four isotherms (if conducted along constant-temperature loci) or four isopleths (if conducted along constant-composition loci). The requirement of four thermodynamic loci is appropriate in that most critical loci, when fitted to polynomials, require at least three adjustable coefficients; thus, at least four mixture critical points are needed to avoid overfitting.

With this basic criterion and some further stipulations noted below, in a search of the literature through 1989, 129 mixtures have been located, collectively composed of 73 pure fluids, that have been thoroughly measured and therefore are excellent candidate mixtures for examination with the modified Leung-Griffiths model. The pure fluids and their critical temperatures and pressures are listed in Table 1, and the mixtures with their VLE references are listed in Table 2. These sources were obtained with the help of bibliographies,[240-248] textbooks,[32,249] and review articles;[250,251] from inquiries to leaders of experimental groups, from the databases of the National Institute of Standards and Technology (NIST) Thermophysics Division; and through considerable individual, unstructured searching. In the construction of these tables, the only consideration for each reported experiment was whether the data have adhered to the criteria (essentially quantity rather than quality of data) as specified in this section, so that there are papers, for example, with three isotherms not on the list that may be superior to some of those on it.

The source list has not been restricted solely to the archival literature. There are a number of excellent experiments reported in unpublished Masters and Ph.D. theses, and the timely publication of material from the theses evidently often has been a matter of personal circumstance rather than a reflection of the quality of the work. Also, in many cases the original, or raw, data do not appear in the archival publication but do appear in a thesis or

TABLE 1
"Primary" Pure Fluids

	Fluid	Abbrev.	T_c/K	P_c/MPa	No. of mixtures
1.	Helium 3	He_3	3.311	0.115	1
2.	Helium 4	He_4	5.188	0.227	1
3.	Hydrogen	H_2	33.25	1.297	1
4.	Neon[a]	Ne	44.40	2.654	1
5.	Nitrogen[a]	N_2	126.24	3.398	1
6.	Carbon monoxide[a]	CO	132.85	3.494	1
7.	Argon[a]	Ar	150.66	4.860	1
8.	Oxygen[a]	O_2	154.58	5.043	1
9.	Methane[a]	CH_4	190.55	4.595	11
10.	Krypton	Kr	209.46	5.49	5
11.	Carbon tetrafluoride	R14	227.6	3.74	1
12.	Ethylene[a]	C_2H_4	282.35	5.040	10
13.	Xenon	Xe	289.7	5.87	3
14.	Chlorotrifluoromethane	R13	302.0	3.92	1
15.	Carbon dioxide[a]	CO_2	304.17	7.386	26
16.	Ethane[a]	C_2H_6	305.34	4.871	17
17.	Acetylene	C_2H_2	308.70	6.247	3
18.	Nitrous oxide	N_2O	309.15	7.285	3
19.	Sulfur hexafluoride	SF_6	318.82	3.765	1
20.	Hydrogen chloride	HCl	324.55	8.263	3
21.	Bromotrifluoromethane	R13B1	340.08	3.956	1
22.	Propylene	C_3H_6	365.05	4.600	5
23.	Chlorodifluoromethane	R22	369.27	4.967	4
24.	Propane[a]	C_3H_8	369.85	4.247	21
25.	Hydrogen sulfide	H_2S	373.40	8.963	5
26.	Trifluoropropylene	$C_3H_3F_3$	376.2	3.80	1
27.	Dichlorodifluoromethane	R12	385.01	4.129	1
28.	1,1-Difluoroethane	R152a	386.44	4.520	1
29.	Perfluorocyclobutane	$c\text{-}C_4F_8$	388.43	2.785	1
30.	Dimethyl ether	Me_2O	400.10	5.370	5
31.	Ammonia	NH_3	405.45	11.278	5
32.	Isobutane[a]	$i\text{-}C_4H_{10}$	407.84	3.629	5
33.	Dichlorotetrafluoroethane	R114	418.86	3.220	3
34.	Perfluoro-n-pentane	$n\text{-}C_5F_{12}$	420.59	2.045	1
35.	n-Butane[a]	$n\text{-}C_4H_{10}$	425.38	3.809	15
36.	Sulfur dioxide	SO_2	430.75	7.884	2
37.	Vinyl chloride	C_2H_3Cl	432.00	5.670	1
38.	Neopentane	$neo\text{-}C_5H_{12}$	433.75	3.196	2
39.	Perfluorocyclohexane	$c\text{-}C_6F_{12}$	457.29	2.237	1
40.	Isopentane[a]	$i\text{-}C_5H_{12}$	460.51	3.371	2
41.	Diethyl ether	Et_2O	466.56	3.651	1
42.	n-Pentane[a]	$n\text{-}C_5H_{12}$	469.65	3.370	8
43.	Perfluoro-n-heptane	$n\text{-}C_7F_{16}$	474.85	1.636	4
44.	Epoxypropane (propylene oxide)	—	482.25	4.924	1
45.	Perfluoromethylcyclohexane	PFMCH	485.90	2.019	5
46.	2,2-Dimethyl butane	22DMB	489.25	3.102	2
47.	2-Methyl pentane[a]	2MP	498.05	3.035	2
48.	2,3-Dimethyl butane	23DMB	500.23	3.147	2
49.	3-Methyl pentane	3MP	504.62	3.128	2
50.	n-Hexane[a]	$n\text{-}C_6H_{14}$	507.95	3.032	4
51.	Acetone	Me_2CO	508.15	4.758	2
52.	Isopropanol	i-PrOH	508.31	4.764	1
53.	Cyclopentane	$c\text{-}C_5H_{10}$	511.76	4.502	1
54.	Methanol	MeOH	512.58	8.097	4
55.	Ethanol	EtOH	516.25	6.384	2
56.	Perfluorobenzene	C_6F_6	516.71	3.275	2

TABLE 1 (continued)
"Primary" Pure Fluids

	Fluid	Abbrev.	T_c/K	P_c/MPa	No. of mixtures
57.	n-Propanol	n-PrOH	536.70	5.142	1
58.	n-Heptane[a]	n-C_7H_{16}	540.00	2.736	6
59.	Iso-octane (2,2,4-trimethylpentane)	i-C_8H_{18}	543.83	2.564	1
60.	Acetonitrile	CH_3CN	545.50	4.833	1
61.	Cyclohexane	c-C_6H_{12}	553.54	4.075	2
62.	Benzene	C_6H_6	562.24	4.888	9
63.	n-Butanol	n-BuOH	562.89	4.416	3
64.	n-Octane	n-C_8H_{18}	569.20	2.603	3
65.	Methylcyclohexane	MCH	572.2	3.47	1
66.	Toluene	—	591.79	4.109	2
67.	Acetic acid	AcOH	592.71	5.786	1
68.	m-Xylene	—	617.05	3.541	1
69.	n-Decane	n-$C_{10}H_{22}$	619.3	2.096	4
70.	Mesitylene	—	637.3	3.13	1
71.	Water	H_2O	647.29	22.09	3
72.	trans-Decalin	—	687.1	2.615	1
73.	Tetralin	—	719.2	3.515	1

[a] Included in the package DDMIX as of 2/91.

TABLE 2
Primary Mixtures

Mixture	Ref.

Normal Mixtures with Density Measurements
(Small to Moderate α_{2m})

	Mixture	Ref.
1.	He_3 + He_4	87/88
2.	N_2 + CH_4	89, 90*[†], 91*[†], 92*[†]
3.	CH_4 + C_2H_6	93, (94*[†] + 95*[†] + 96*[†])
4.	C_2H_4 + i-C_4H_{10}	84*
5.	C_2H_4 + n-C_4H_{10}	97, 79*
6.	C_2H_4 + $C_3H_3F_3$	98*
7.	CO_2 + N_2O	99, 100
8.	CO_2 + H_2S	101/102, 103/104
9.	CO_2 + C_3H_6	105, 106[†], 107[†]
10.	CO_2 + C_3H_8	108*, 109[†], 110*
11.	CO_2 + i-C_4H_{10}	111*, 112[†]*
12.	CO_2 + n-C_4H_{10}	113*, 112[†]*, 109[†], 114[†]* (115* + 116* + 117[†]*)
13.	C_2H_6 + C_3H_6	118
14.	C_2H_6 + i-C_4H_{10}	119*
15.	C_2H_6 + n-C_4H_{10}	120/121, 122[†]*, (123[†]* + 124[†]*)
16.	C_2H_6 + n-C_5H_{12}	125*
17.	R13B1 + R114	126
18.	R22 + R114	127
19.	R22 + R12	128/129
20.	C_3H_8 + n-C_4H_{10}	130, 131
21.	C_3H_8 + neo-C_5H_{12}	132/133
22.	C_3H_8 + i-C_5H_{10}	134
23.	C_3H_8 + n-C_5H_{12}	130, 135*
24.	C_3H_8 + 22DMB	136/137
25.	C_3H_8 + 2MP	136/137

TABLE 2 (continued)
Primary Mixtures

	Mixture	Ref.

Normal Mixtures with Density Measurements
(Small to Moderate α_{2m})

26.	C_3H_8 + 23DMB	136/137
27.	C_3H_8 + 3MP	136/137
28.	C_3H_8 + n-C_6H_{14}	136/137, 138/139
29.	C_3H_8 + C_6H_6	140*/141*
30.	C_3H_8 + n-C_8H_{18}	142
31.	R152a + R114	143
32.	NH_3 + CH_3CN	85*
33.	n-C_4H_{10} + n-C_5H_{12}	144
34.	n-C_4H_{10} + n-C_6H_{14}	144
35.	n-C_4H_{10} + n-C_7H_{16}	145/146
36.	n-C_4H_{10} + n-C_8H_{18}	142
37.	n-C_4H_{10} + AcOH	147*
38.	SO_2 + H_2O	148
39.	c-C_6F_{12} + C_6F_6	149
40.	Et_2O + n-BuOH	150
41.	MeOH + n-BuOH	150

Normal Mixtures with Density Measurements (Large α_{2m})

42.	C_2H_4 + n-C_7H_{16}	151/152
43.	C_2H_4 + C_6H_6	153
44.	CO_2 + epoxypropane	(154* + 82*)
45.	CO_2 + MeOH	23*
46.	CO_2 + n-C_7H_{16}	155*
47.	CO_2 + c-C_6H_{12}	[156* + 157*]
48.	CO_2 + n-$C_{10}H_{22}$	9*, 158[†]*
49.	C_2H_6 + n-C_7H_{16}	75/159, 160[†]*
50.	C_2H_6 + c-C_6H_{12}	161
51.	C_2H_6 + C_6H_6	162
52.	C_2H_6 + n-$C_{10}H_{22}$	163*
53.	C_3H_8 + n-$C_{10}H_{22}$	164*
54.	n-C_4H_{10} + n-$C_{10}H_{22}$	165*

Azeotropic Mixtures with Density Measurements

55.	C_2H_4 + CO_2	105, 106[†], 166[†]*
56.	CO_2 + C_2H_6	167, 168[†]*
57.	C_2H_6 + C_2H_2	169
58.	C_2H_6 + N_2O	170
59.	C_2H_6 + HCl	171
60.	C_2H_6 + H_2S	172
61.	SF_6 + C_3H_8	173
62.	H_2S + C_3H_8	174
63.	C_3H_8 + c-C_4F_8	175
64.	NH_3 + n-C_4H_{10}	176
65.	NH_3 + i-C_8H_{18}	177
66.	n-C_5F_{12} + n-C_5H_{12}	178
67.	n-C_4H_{10} + n-C_7F_{16}	179/180
68.	n-C_5H_{12} + n-C_7F_{16}	179/180
69.	n-C_5H_{12} + Me_2CO	181
70.	PFMCH + 22DMB	182
71.	PFMCH + 2MP	182
72.	PFMCH + 23DMB	182

TABLE 2 (continued)
Primary Mixtures

Mixture   Ref.

Azeotropic Mixtures with Density Measurements

73. PFMCH + 3MP   182
74. PFMCH + n-C_6H_{14} 182
75. n-C_6H_{14} + n-C_7F_{16} 179/180
76. MeOH + C_6H_6 183
77. EtOH + C_6H_6 183
78. n-PrOH + C_6H_6 183
79. n-C_7H_{16} + n-C_7F_{16} 179/180
80. C_6H_6 + n-BuOH   183

Normal Mixtures without Density Measurements

81. CO + CH_4 184*
82. Ar + Kr   185*, 186*
83. O_2 + N_2O 187*
84. CH_4 + Kr   188*
85. CH_4 + C_2H_4 189*/190*, 191*
86. CH_4 + CO_2 192*
87. CH_4 + R22   193*
88. CH_4 + C_3H_8 194*, 195*
89. CH_4 + i-C_4H_{10} 196*
90. CH_4 + n-C_4H_{10} 197*
91. CH_4 + n-C_5H_{12} [198* + 199††*], [200* + 201*]
92. Kr + Xe   202*
93. Kr + C_2H_6 203*
94. Kr + HCl   204/205
95. R13 + R14   206*
96. C_2H_4 + C_3H_6 207*
97. C_2H_4 + C_2H_3Cl 208*
98. Xe + C_2H_6 209*
99. CO_2 + R22   193*
100. CO_2 + Me_2O 210*, 107
101. CO_2 + neo-C_5H_{12} 211*/212*, 213*
102. CO_2 + i-C_5H_{12} 214*, (215††* + 216*)
103. CO_2 + n-C_5H_{12} 217*, 109, (218††* + 216*)
104. CO_2 + i-PrOH   219*
105. CO_2 + c-C_5H_{10} 219A*/212*
106. CO_2 + n-C_8H_{18} 220
107. CO_2 + MCH   221*
108. CO_2 + toluene   8*
109. CO_2 + m-xylene   222*
110. CO_2 + mesitylene   223*
111. C_2H_6 + C_3H_8 224*, 225*
112. C_2H_2 + C_3H_6 226
113. C_2H_2 + C_3H_8 226
114. HCl + C_3H_8 204/227
115. H_2S + i-C_4H_{10} 228*
116. H_2S + n-C_4H_{10} 228*, 229*
117. C_3H_6 + Me_2O 107
118. C_3H_8 + n-C_7H_{16} 138
119. Me_2O + n-C_4H_{10} 230*
120. Me_2O + MeOH   231*
121. NH_3 + H_2O 78*
122. NH_3 + $trans$-decalin   232*
123. Me_2CO + C_6H_6 233*

TABLE 2 (continued)
Primary Mixtures

Mixture	Ref.

Normal Mixtures without Density Measurements

124. C_6F_6 + C_6H_6	234
125. Toluene + tetralin	235*

Azeotropic Mixtures Without Density Measurements

126. H_2 + Ne	236*
127. C_2H_4 + Xe	237*
128. Me_2O + SO_2	238**
129. EtOH + H_2O	239*

* Isothermal data (otherwise data are on isopleths).
** Negative azeotrope.
† Source without coexisting density measurements.
†† Source with density measurements.

institutional report. Theses are omitted, however, when the authors cast doubt on the accuracy of their own work and evidently have made a deliberate decision not to publish (e.g., Reference 252); however, only sources organized into some sort of formal manuscript (except work currently in progress) were considered and we do not, for example, give credence to a handwritten list of numbers left on a laboratory shelf for many years.

In addition to the requirement of at least four dew-bubble curves, there are other criteria that preclude certain sources from being listed in Table 2. The requirements noted in the following discussion include citations to examples of articles that are omitted due to the failure to meet that particular requirement.

VLE data must be presented in tabular form; sources that present the data only in graphical form are excluded,[253] as well as sources in which the tabulation of data is highly incomplete;[254] however, archival publications with only graphs are included if the data are tabulated in theses,[190] unpublished institutional reports,[88] or auxiliary publication depositories,[148] or are available on request from the authors.[129] Data could be obtained from digitization of the graphs, but the trouble and errors inherent in digitization relegate these sources to a lower priority. For some of the sources without tabulated data, auxiliary publication depository material may satisfy the above requirements, but thus far that material, including sources on the mixtures ethane + ammonia,[255] carbon dioxide + sulfur hexafluoride,[256] and ethanol + cyclohexane,[257] has been elusive.

As the data are required to cover the entire extended critical region, sources are excluded if most of the data are too close to the critical region[76] ($p > 0.75 \, P_c$) or too far removed[258] ($p < 0.75 \, P_c$), in which case the critical locus is not well characterized. Similarly, for data along isopleths a source is excluded if all dew-bubble curves are on one side of the equimolar mixture; therefore, Reference 169 is excluded as a primary source for carbon dioxide + ethane. For isothermal data, typically the geometric mean of the two pure-fluid critical temperatures evenly divides the extended critical region, and a source if excluded if all isotherms are on one side of this geometric mean.[259] A source is excluded if the data are too sparse;[260] the one general requirement is that each dew-bubble curve should contain at least five points in the extended critical region, with a coexisting liquid and vapor state counting as two points. Sources are excluded if the components are by the authors' admission highly impure,[261] if the data in the extended critical region are extrapolated rather than measured,[262] and if the critical locus is measured but the dew-bubble curves are inferred from an equation of state.[263]

Mixtures such as carbon dioxide + oxygen[264] and carbon dioxide + argon,[265] which might be type 1 except that their critical loci are cut off by an SLV locus, are excluded. A source does not qualify if only the liquid[266] or only the vapor[267] has been measured. Tabulated binary mixture data that are actually interpolated from ternary measurements[268] are not included, although true binary measurements in the course of a study primarily of a ternary mixture are acceptable.[124]

Some sources with four or more dew-bubble curves are excluded if the removal of some inadequate curves would result in three or fewer. Reasons for inadequacy include data taken from another source (therefore excluding Reference 194 as a primary reference for methane + ethane), conjectured or extrapolated curves,[269] curves with only the liquid or vapor side measured,[270] curves that do not extend to high enough pressure to define the critical point,[271] and curves for which the temperatures or compositions have not been determined or identified.[272]

The extended critical region covers the temperature interval $0.9\, T_{c2} < T < T_{c1}$, where fluid 2 is the more volatile, but for isothermal data at least three isotherms are required in the range $T_{c2} < T < T_{c1}$ in order to establish the critical locus.[273] A source is excluded if data on both isotherms or isopleths are presented, but neither set by itself satisfies the criteria described here.[274,275] It is sometimes difficult to identify the mixture class by the VLE source alone; for example, the articles by Reamer et al. on methane + cyclohexane[276] and methane + n-heptane[277] are of the same format as their article on methane + n-pentane.[199] From their articles there appear to be no fundamental differences among the three mixtures, but the last mixture is class 1, while the former two are not[62] and thus are excluded.

Finally, there are some calorimetric experiments primarily designed for the measurement of absolute[278-280] or excess[281-283] enthalpy, but which yield dew and bubble points from, for example, the slope discontinuity of isobars on a plot of enthalpy vs. temperature. In principle, such experiments could determine the coexistence surface of binary mixtures. In practice, compared to other experimental techniques, such a method is indirect and inaccurate. With the same degree of care and skill, accurate VLE data are much harder to obtain from enthalpy experiments than from more conventional methods. Thus, such sources are excluded from the primary list. Without this exclusion, the mixtures n-octane + benzene,[278] benzene + cyclohexane,[279] pentane + trans-decalin,[279a] and pentane + benzene[280] could be added to Table 2.

These exclusions demonstrate that VLE experimental sources appear in all kinds of different formats and presentations and that their value as input for correlations cannot be determined from a cursory examination of tabulted pressure and temperature ranges. The reasons for exclusion are intended to be guidelines rather than strict rules; some sources do not technically satisfy the criteria as specified but fall short by small and insignificant amounts, and thus are included.[106,165] As always with a bibliographic project of this nature, it cannot be guaranteed that this list is comprehensive and the author would greatly appreciate calling any inadvertent omissions to his attention.

In the development of a bibliographic system for binary mixture VLE, it is necessary to decide upon a list of pure fluids that are of interest. For example, Hiza et al.[245] limit their tables to 30 fluids of cryogenic interest, while Williamson and Olien[248] consider 24 fluids of interest to the natural gas industry. Knapp et al.[247] considered 46 fluids (all but ten of which are listed in Table 1 of this chapter) in their 1900 to 1973 survey; for their 1973 to 1980 survey they added 72 more fluids but did not recheck the earlier literature for the additions. By contrast, the bibliography of Wichterle et al.[240-244] considers essentially all fluids, but only literature references are given for a specific mixture, whereas the more focused bibliographies also provide pressure and temperature ranges and other information.

For the purposes of this chapter, a compromise is made between the limited component list of the specialized bibliographies and, for example, the DIPPR project with over 1000 pure fluids.[284] The 73 fluids of Table 1 as the range of the bibliography and database is the focus here. In progress is a project to collect all critical region VLE and critical locus information for binary mixtures composed of these fluids. The author also hopes to construct a chart of the mixture types of all possible binaries; other sources have given examples[32,51] of mixtures of types 1 through 6, but to the author's knowledge there is no reference that identifies the mixture type as a function of the two pure components. Since experimentally the mixture type is not often determined defintively, such a chart will of necessity include various caveats.

Admittedly the list of Table 1 excludes some important fluids such as chloroform, and includes some "obscure" fluids, such as perfluorocyclobutane, but by and large, the table includes most of the fluids of greatest interest to the scientific and engineering communities. It includes all noble gases (except radon), all normal and branched alkanes through hexane, and all normal alkanes through *n*-octane. Representative samples are provided for less important homologous series, for example, five alcohols, three normal perfluoroalkanes, two alkenes and ethers, and one alkyne, ketone, carboxylic acid, and nitrile. Also, it seems more systematic to choose the list of fluids on the basis of the collective judgment of previous experimental investigators of critical-region VLE, rather than to make arbitrary and subjective choices.

A. CRITERIA FOR MISCIBILITY

A useful question to raise at this point is whether the type of a binary mixture can be predicted from the properties of its two pure components. The components are most fundamentally characterized by a critical pressure, temperature, and density (P_c, T_c, ρ_c). In effect, only two of these are independent, since the critical compressibility factor is nearly universal for nonassociating fluids,

$$P_c/(R\rho_c T_c) \approx 0.27 \qquad (16)$$

Microscopically, these critical parameters reflect the energy and length scales of the intermolecular potential as optimized, for example, to the Lennard-Jones potential. The critical temperature is proportional to the energy of the potential well, and the cube root of the critical molar volume (inverse critical density) is proportional to the molecular diameter. For nonspherical molecules, these statements apply approximately to the spherically averaged potential. The principle of corresponding states is then the approximate result that in terms of dimensionless reduced units, all fluids exhibit universal thermodynamic behavior.

In principle, the properties of the two pure fluids alone cannot predict the mixture behavior, because microscopically the mixture thermodynamics depends on the forces between unlike molecules (fluid 1 with fluid 2) as well as between like molecules (1 - 1 and 2 - 2). In the absence of other information, however, the unlike interaction can be approximated by combining rules such as those of van der Waals, in which the energy scale and the length scale of the cross interaction are, respectively, the geometric mean of the pure fluid energy scales and the arithmetic mean of the pure fluid length scales. Therefore, from a simple equation of state such as the van der Waals equation, one can map out mixture type boundaries in parameter space; in this case the parameters a and b are proportional to $V_c T_c$ and V_c, respectively. It is important to remember, however, that an EOS alone cannot determine the intrusion of solid phases onto the phase diagram.

Van Konynenberg and Scott,[27] who devised the mixture type system currently in use, also provided the first systematic study of type boundaries in the space of EOS parameters. They considered the van der Waals equation with a_{12} considered a free parameter (rather than determined from combining rules) but with b_{12} fixed to be the arithmetic mean of b_{11} and b_{22}. (This differed from the usual combining rule using the arithmetic mean of the cube roots of the pure fluid b's, but for the most part they considered mixtures of equal size molecules for which the distinction does not matter.) Their parameter space then had three degrees of freedom, and the type boundaries were mapped in planes of fixed size ratio. Among other results, they showed that types 1 through 5 could occur for the appropriate van der Waals parameters, but type 6 could not. Examination of type boundaries in parameter space continues to be an interesting problem; recent work includes further study of the van der Waals equation by Meijer,[285] the Redlich-Kwong equation[30] by Deiters and Pegg,[286] and the Ree equation[287] by Boshkov and Mazur.[288] Romig and Hanley[289] have examined phase type boundaries with a generalized corresponding states method that includes prediction of solid transitions, so the phase types are defined with the inclusion of solid boundaries (although, for real fluids, the solid transition does not occur as regularly as it does for the somewhat idealized corresponding states calculations). The boundaries are displayed in terms of size and well depth ratios, and show the same general pattern as a group of selected real mixtures of nonpolar or weakly polar fluids in the space of critical temperature ratio and critical volume ratio. Again, no type 6 behavior was found from the calculations.

The concept of "immiscibility" is closely connected with the mixture type, in which type 1 mixtures represent a miscibility ideal, type 3 mixtures represent the extreme case of immisicibility, and other types are intermediate. Immiscibility is reflected not only in the structure of the mixture critical and three-phase loci, but also in the behavior of excess functions (free energies, enthalpies, entropies, and volumes), of which there is a wealth of data at both low and high pressures. Immiscibility is determined not only by the relative location of the critical points of the components, but also by a hydrogen-fluorine interaction, molecular polarity, and the presence of ionization. These effects are discussed as they arise in the overview of pure fluid critical points and miscibility.

B. DISTRIBUTION OF CRITICAL POINTS: CRYOGENIC FLUIDS

Table 1 lists the number of primary mixtures in which each pure fluid is a component. In a rough sense, this number can be considered a measure of the importance of that fluid within the VLE database; however, for several entries with unusually high or low critical temperatures or pressures such as hydrogen, nitrogen, carbon monoxide, and water, many additional mixtures have been investigated, but these are class 2 and are thus excluded from this survey.

These "primary" fluids are ordered in Table 1 by critical temperature, in contrast to other ordering methods by normal boiling temperature[247] and by molecular mass.[245] Figure 10 displays the pure-fluid critical points, except for a cluttered region of n-alkane points, in pressure-temperature space. It is instructive to examine this figure in order to understand the various mixture types that occur with fluids of greatest interest; a useful rule of thumb is that a mixture is almost certainly class 2 (with a discontinuous critical locus), and probably type 3, if the critical temperature ratio is greater than 2.5.

The list of pure fluids begins with the helium isotopes, the subject of the original Leung-Griffiths paper.[1] Other than the isotopic mixture, helium forms a type 3 mixture with all primary fluids, including hydrogen.[290] Next listed are hydrogen and neon, which form a class 1 (type 2) mixture with each other[236] but type 3 mixtures with all other primary fluids. This temperature range also includes HD and D_2, and there has been considerable interest[291-295] (partly for purposes of fusion research) of the SLV locus and subcritical VLE of isotopic hydrogen mixtures, but to the author's knowledge no studies of the critical region. Probably the binary mixtures formed from H_2, HD, D_2, and neon would all have simple, continuous critical loci. With the added feature that the homonuclear hydrogens can be in an *ortho, para,* normal, or equilibrium state, there are many interesting possibilities.

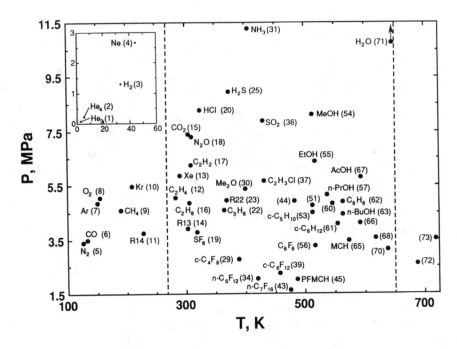

FIGURE 10.   Distribution of primary fluid critical points in P-T space. Fluids are labeled by their numbers (and in most cases, abbreviations) in Table 1. For all but 14 of the 73 primary fluids, the critical points are in the range 280 K < T_c < 650 K and 1.5 MPa < P_c < 11.5 MPa, and almost all mixtures of two fluids with critical points in this range are class 1. Because of lack of space the critical points of the normal and isomeric alkanes are not shown; these lie in a narrow band between that of propane (369.85 K, 4.247 MPa) and that of *n*-decane (619.3 K, 2.096 MPa). Other critical points within this band are also omitted. Note that H_2O (22.09 MPa) is off the scale.

Streett[296,297] has measured LLE and slightly subcritical VLE of H_2 and D_2 with neon, and with the thorough critical region study of the former mixture by Heck and Barrick,[236] hydrogen-neon is an excellent candidate mixture for a complete theoretical model of type 2 phase equilibria.

Next, as opposed to the previous four "extreme cryogenic fluids", are seven important "cryogenic fluids": nitrogen, carbon monoxide, argon, oxygen, methane, krypton, and carbon tetrafluoride (R14). These probably all form class 1 mixtures with each other, although the evidence is not conclusive[298,299] for nitrogen + R14. Critical loci for all binary mixtures of these fluids, except krypton and R14, were measured by Jones and Rowlinson;[300] however, as noted by Rainwater and Jacobsen,[16] the critical region VLE for mixtures of the constituents of air is surprisingly deficient. Nitrogen (and probably carbon monoxide) form type 3 mixtures with all "noncryogenic" fluids described below, including ethylene,[301] ethane,[302] and carbon dioxide,[303] whereas, as noted earlier, methane forms continuous critical loci with alkanes up to 2,2-dimethyl butane[62] and R14 with alkanes up to propane.[304]

As a general rule, mixtures of hydrocarbons and fluorocarbons are considerably more immiscible than would be expected on the basis of critical parameter ratios or polarity. For example, the critical points of methane (T_c = 190.55 K, P_c = 4.595 MPa, ρ_c = 10.14 kmol/m³) and R14 (T_c = 227.6 K, P_c = 3.74 MPa, ρ_c = 7.16 kmol/m³) are quite close, and both molecules are nonpolar and essentially spherical, yet the mixture exhibits LLE at 95 K, a much higher temperature than would be expected from the pure-fluid critical properties.

The anomalous hydrocarbon-fluorocarbon interaction has been a subject of interest and confusion ever since the original review in 1948 by Scott[305] pointing out such effects. His original explanation, later discounted, was an anomalously low Hildebrand solubility parameter for the fluorocarbons. Later, Scott[306] and Jordan[180] reviewed developments during the 1950s. Several additional explanations were proposed, including the greater ease of interpenetration of hydrocarbon molecules,[307] and the relative flexibility of hydrocarbon to fluorocarbon molecules,[308] but these were also discounted. It was realized that a satisfactory microscopic theory would have to describe, in addition to hydrocarbon-fluorocarbon mixtures, mixtures of hydrocarbons and fluorocarbons with each other and with entirely distinct fluids such as argon or krypton. In particular, a satisfactory theory must be consistent with the fact that no such anomalies have been seen in fluorocarbon-fluorocarbon mixtures.

More contemporary approaches have avoided the earlier geometrical interpretations and have tried to fit the hydrocarbon-fluorocarbon mixtures within the framework of conventional mixture theory, with the anomalous features embodied in a cross interaction determined from ionic potentials and multipolar and dispersion forces. Since the review by Swinton[309] in 1977 (summarized in the book of Rowlinson and Swinton[32]), there appears not to have been further comprehensive reviews or breakthroughs in theoretical explanations, although experimental confirmations of large excess functions have continued (see References 310 and 311 and articles cited therein).

Within conventional corresponding-states approaches, anomalous cross interactions are described by representing the cross-interaction energy parameter as the geometric mean of the pure energy scales times $(1 - \gamma)$, where γ is a binary interaction parameter. For CH_4 + CF_4, the simplest case, Swinton notes that $\gamma = 0.12$, an unusually large value which can nevertheless be obtained from the Hudson-McCoubrey combining rules.[312] Ionization parameters of the pure fluids are required by the rules as input. At present such a procedure is not predictive for mixtures of larger hydrocarbon and fluorocarbon molecules, since the ionization potentials of the latter are not known. Also, as shown in Swinton's Table 3, no mixture theory can simultaneously predict with accuracy the excess Gibbs energy, enthalpy, and volume of CH_4 + CF_4.

Swinton points out that similar anomalous behavior occurs, but with diminished intensity, for mixtures of perfluorobenzene with n-alkanes and other hydrocarbons, and with mixtures of hydrocarbons and partially fluorinated hydrocarbons. The perfluorobenzene mixtures display additional anomalous and poorly understood behavior.[32,309]

The hydrocarbon-fluorocarbon anomaly may spark renewed interest because of the current concern about the upper atmospheric ozone layer and the search for alternative refrigerants.[313] Since the ozone depletion is due to a reaction involving chlorine, the alternative refrigerants under consideration primarily are fluorinated methanes or ethanes. The need to understand the thermophysical properties of mixtures of these compounds, and perhaps of the pure compounds themselves, will require a greater understanding of the anomalous interaction.

Additional studies of the type-change boundaries through mixtures of argon and krypton with noncryogenic fluids would be of interest; the work of Orobinskii et al.[314] shows an extensive LLV locus and probably a type 3 structure for argon + propylene. It would also be of interest to locate the type transition in the mixture family of R14 with the normal perfluoroalkanes. Of course, at critical temperatures and pressures oxygen would form explosive mixtures with most organic compounds.[315]

C. DISTRIBUTION OF CRITICAL POINTS: NONCRYOGENIC FLUIDS

The substantial gap between the critical temperature of R14 (227.6 K) and that of ethylene (282.35 K) forms a useful division between "cryogenic" and "noncryogenic" fluids, although there are some rarely studied fluids such as nitrogen trifluoride (233.9 K) or boron trifluoride (260.8 K) within the gap. It is most noteworthy that the vast majority of primary fluids (59 of 73) possess critical points in the range 280 K < T_c < 650 K and 1.5 MPa <

P_c < 11.5 MPa, or essentially the range bounded by the critical temperatures of ethylene and mesitylene and the critical pressures of perfluoro-*n*-heptane and ammonia. Furthermore, within this range almost all mixtures are class 1. The exceptions are anomalous hydrocarbon-fluorocarbon mixtures such as R13 with *n*-heptane[304] or higher alkanes, or mixtures that span the extremes of the region such as sulfur hexafluoride + ammonia[56] or ethane + ethanol.[316,317]

It can therefore be concluded that at least 75% of mixtures of primary fluids are class 1. While some researchers into three-phase loci or critical phenomena at very high pressures assert that class 1 mixtures are "uninteresting", from a practical viewpoint it is evident that class 2 mixtures are overall of a rather secondary importance, and consequently that a model that reliably correlates the phase boundary surfaces of only class 1 mixtures is still of great value. The most important class 2 mixtures are (1) petroleum mixtures with nitrogen (or, to a lesser degree, carbon monoxide or hydrogen) as a component, (2) mixtures of carbon dioxide and heavy organic molecules for supercritical extraction, and (3) aqueous mixtures.

Figure 10 shows a subdivision of this region by type of pure fluid. The alkanes (not shown) lie within a narrow linear strip between the critical points of ethane (T_c = 305.34 K) and of *n*-decane (T_c = 619.3 K). Fluorine compounds, except for perfluorobenzene, lie beneath this strip at lower pressures. It is evident that fluorination of a hydrocarbon lowers the critical pressure without having much effect on the critical temperature, whereas chlorination substantially increases both the critical temperature and pressure. Consequently, the common refrigerants based on methane or ethane lie above the ethane critical point in the range 300 K < T_c < 450 K and 3.0 MPa < P_c < 5.0 MPa.

Within a higher pressure range, 5.0 MPa < P_c < 11.5 MPa and 280 K < T_c < 450 K, are a number of miscellaneous but important fluids: carbon dioxide, hydrogen sulfide, ammonia, sulfur dioxide, nitrous oxide, acetylene, xenon, and dimethyl ether. Above 450 K, as shown, is the domain of the alcohols and single-ring aromatics and cyclic alkanes. The alcohols start with methanol (T_c = 512.58 K, P_c = 8.097 MPa) and proceed abruptly to lower pressures and slowly to higher temperatures, while the aromatics and cyclic alkanes have critical points in the range 500 K < T_c < 650 K and 3.0 MPa < P_c < 5.0 MPa; these are at higher pressure than the *n*-alkane critical points in this temperature range.

Traditionally, 650 K has been the upper limit of critical temperatures of primary fluids, because for T > 600 K, organic molecules decompose rapidly and traditional VLE experimental techniques are not useful. It is, in fact, this upper barrier that has kept the number of primary fluids and mixtures relatively small or at least manageable; however, recently new fast-flow apparatus development has enabled VLE expriments to be conducted for T > 600 K (the flow of mixture constituents is faster than the decomposition rate and an analysis of the mixture after the experiment shows minimal amounts of decomposition products). Such techniques were pioneered by Chao and co-workers,[318,319] and subsequently Inomata et al.[158,320] extended the method to high pressures as well as temperatures. They measured VLE at temperatures as high as 709.9 K for benzene + quinoline.[158] Niesen et al.[321] have developed a flow apparatus with a temperature range extending to 625 K.

Very recently, the first studies qualifying as primary sources with double-ring compounds have been published. Thies et al.[235] measured toluene + tetralin up to 672.3 K, and Maheswari and Lentz[232] published a study of ammonia + *trans*-decalin, although this work was by the authors' admission subject to problems of decomposition and had been presented in graphical form much earlier.[322] With the interest in coal tar derivatives as possibly useful fuels,[323] such studies are likely to continue in the near future.

The last primary fluid is water, a special case because of its extremely high critical pressure (22.1 MPa). Like the cryogenic fluids, but for the opposite reason, water is miscible with very few of the other primary fluids. Those that do form a continuous critical line with water are generally at the regions of high pressure and/or temperature in Figure 10, and include methanol,[324] ethanol,[239] isopropanol,[239] acetone,[325] ammonia,[78] and sulfur dioxide.[148] Also, acetonitrile is usually considered miscible with water (Reference 32, p. 175), although to the author's knowledge VLE for the water + acetonitrile mixture has only been measured at subcritical pressures (p <0.7 MPa).[326] Fluids that do not form a continuous critical locus with water include benzene,[55] dimethyl ether,[326] hydrogen sulfide,[328] and carbon dioxide.[329] Water, as well as all of the fluids mentioned above (except for benzene and carbon dioxide), have strong permanent molecular dipole moments.

The immiscibility of water with benzene relative to that with acetonitrile or acetone demonstrates the effect of polarity on the structure of mixture phase diagrams. Rowlinson and Swinton[32] have summarized the results of such studies for a variety of polar-polar and polar-nonpolar mixtures. The overall picture is quite complicated, but as a general rule, polar-polar mixtures are less miscible than nonpolar-nonpolar mixtures, while polar-nonpolar mixtures are still less miscible than polar-polar mixtures. The quantitative predictions of the effects of polarity will probably be a very challenging problem. For example, recently Jangkamolkulchai et al.[320] measured the LLV loci of the family of mixtures of nitrous oxide with n-alkanes and compared them against those of carbon dioxide with n-alkanes. Nitrous oxide (T_c = 309.15 K, P_c = 7.285 MPa) and carbon dioxide (T_c = 304.17 K, P_c = 7.386 MPa) have almost identical critical points, yet the type structure for nitrous oxide as a solvent is very different from that of carbon dioxide. Nitrous oxide has a nonzero but relatively small permanent dipole moment ($5.57 \cdot 10^{-31}$ C·m or 0.167 D), whereas carbon dioxide has no permanent dipole moment but a strong quadrupole moment. This suggests that diagrams such as Figures 7 and 8 are highly sensitive to the polarity of the solvent. Further systematic studies of this kind would be very informative.

At this point it is appropriate to note some mysterious features of ionic mixtures. While none of the primary mixtures are ionic, there has been considerable recent interest in the mixture water + salt (sodium chloride).[331-338] The critical temperature of salt is estimated to be 2200 K, and thus VLE data are only available at the water-rich end; however, with the absence of an LLV locus such as those of Figures 5D to 5F, and a known SLV locus that clearly does not intersect the locus of plait points as in Figure 6G, it is commonly assumed that (water + salt) has a continuous critical locus, in complete contradiction to the guideline mentioned earlier concerning a critical temperature ratio greater than 2.5. This result would crudely suggest that the presence of ions causes a large enhancement in miscibility.

Another ionic mixture that has been studied in part is water + hydrogen chloride, but here exactly the opposite effect seems to occur. As seen from Figure 10, the critical point of hydrogen chloride is not much further removed from that of water than those of ammonia or sulfur dioxide, both of which display continuous critical loci with water. From the water critical point, however, the critical locus of HCl + H_2O extends to higher pressures and temperatures,[325,339] thereby apparently displaying not only a type 3 structure but "gas-gas equilibrium of the first kind" or extreme immiscibility. Jockers and Schneider[340] have collectively shown in their Figures 9 and 10 the upper critical locus branches for many type 3 aqueous mixtures, and of their examples gas-gas equilibrium of the first kind occurs only for argon (but not for nitrogen or methane). In view of recently taken data,[340a] this conclusion for argon has now been changed.

Phase boundary diagrams to moderately high pressure have been presented by Haase et al.[341] at 298.15 K (with a similar structure for aqueous mixtures of hydrogen bromide and hydrogen iodide) and by Kao[342] at 323.15 K. The structure is consistent with Figure 5F, curve 3, with a three-phase LLV point at T = 298.15 K, P = 4.75MPa. A negative azeotrope, nearly parallel to but at lower pressure than the water vapor pressure curve, probably joins the critical locus tangentially above the critical point of water. The measurement of the critical locus up to 70 MPa by Bach et al.[339] rules out the possibility that the upper branch of the critical locus would turn around toward lower pressures and join a three-phase locus, as in Figure 6E.

In summary, the miscibility of two fluids is determined by four distinct factors:

1. The relative position of, or "distance" between, the critical points in pressure and temperature, where, of course, fluids with less distance between their critical points are more miscible
2. The hydrogen-fluorine interaction; hydrocarbons and fluorocarbons are considerably more immiscible than would otherwise be expected
3. Polarity effects; nonpolar-nonpolar mixtures are the most miscible, then polar-polar, then polar-nonpolar
4. The presence of ions; the apparently paradoxical situation occurs that water is much more miscible with sodium chloride but much less miscible with hydrogen chloride (and probably other hydrogen halides) than would otherwise be expected (there is, to the author's knowledge, no explanation for these ionic effects)

D. BINARY MIXTURE VLE DATA SOURCES

The primary mixture sources that satisfy the criteria described earlier are listed in Table 2. The significance of the asterisks and daggers is explained in the footnote. When two citations are separated by a slash, both sources describe the same experiment, but the first reference is an archival publication not containing a tabulation of the original data or a publication in press and not yet available, while the second reference, typically a thesis or unpublished report, contains such a tabulation.

In several instances a group of experimental papers from the same institution and group leader collectively qualifies as a primary source; these "composite" sources are noted by regular parentheses. For two additional mixtures (methane + *n*-pentane and carbon dioxide + cyclohexane), no output from a single laboratory qualifies as a primary source, but experiments from separate laboratories collectively do qualify; these composite sources from independent groups are noted by square brackets.

For correlations with the modified Leung-Griffiths model it is very helpful to have coexisting density data available as well as P-T-x-y data, and fortunately for 81 of the 129 mixtures there is a primary source with density measurements. For several additional mixtures coexisting density data are available over part of the temperature range; these mixtures include methane + carbon dioxide,[343] methane + propane,[344] methane + isobutane,[345] methane + *n*-butane,[346] methane + *n*-pentane,[199] carbon dioxide + isopentane,[215] and carbon dioxide + *n*-pentane.[218] Globally, VLE data are most commonly taken isothermally and without coexisting densities,[32] but in the high-pressure extended critical region data along isotherms and isopleths are equally common, although the abundance of constant-composition data is mostly due to the efforts of a single experimental group, that of W. B. Kay.

Kay and co-workers are in fact responsible for the measurement of 41 of the 129 primary mixtures in Table 2. In general, Kay's group published smoothed data in their archival publications but retained the original data in industrial reports[121,146,152,159] or theses. Because of space limitations, all of the theses are not listed in the bibliography of the table, but a complete listing is provided in Reference 132. For many additional mixtures Kay and co-workers measured either VLE in a narrow range extremely close to the critical loci[76,347-350] or the critical loci themselves.[179,351-354]

The 73 pure fluids could form a total of 2628 binary mixtures, and, as noted earlier, at least $^3/_4$ of these are class 1. Not all critical-region VLE surfaces of these mixtures call for a thorough measurement; for example, oxygen would form explosive mixtures with the hydrocarbons,[315] and safety problems with nitrous oxide at elevated temperatures[156,355] have also been noted. Other mixtures are chemically reacting at critical conditions, the most noteworthy being carbon dioxide + ammonia,[356] and for many mixtures not listed in Table 2 there are more limited studies of the critical-region VLE or at least some measurements of the critical locus.[251] Nevertheless, the near-critical coexistence surfaces of only a small fraction of important binary mixtures have been thoroughly measured to date, and there is ample cause and opportunity for additional experiments.

As seen from the dates of the sources in Table 2, experimental work in this area has a long history, starting with the pioneering measurements of Kuenen[169,170] and Quint[171] of the Amsterdam laboratory. Kuenen's descriptive writings display a great depth of understanding of phase equilibrium behavior, in contrast to the high incidence of contemporary misunderstanding noted in Section IV, and he was also responsible for such important discoveries as discontinuous critical loci[357] and negative critical azeotropy.[358] At the same time Caubet also conducted VLE experiments up to critical pressures, including a remarkable study of eight isopleths of carbon dioxide + sulfur dioxide[253] that graphically looks excellent to the modern eye but most unfortunately was not tabulated. Caubet also conducted interesting, if perhaps questionable, studies of carbon dioxide + nitrous oxide[100] (in substantial disagreement with Cook's later data[99]), carbon dioxide + methyl chloride,[359] and sulfur dioxide + methyl chloride.[360] (The data of References 253 and 360 were subsequently tabulated in Reference 360a.) Another noteworthy pioneering effort was Sander's measurement[361] of three isotherms of carbon dioxide + diethyl ether together with an investigation of the solubilities of many fluids in carbon dioxide.

From a study of the dates of the articles cited in Table 2, one sees that since 1940, a fairly steady average of two to three new mixtures have been examined each year. Despite technological advances in chromatography, fast-flow apparatus, etc., there has been no noticeable trend toward an increase in the yearly rate of new mixture measurement, although the rate of duplicate measurement of mixtures has increased. While the technology to perform high-pressure VLE experiments has been available for nearly a century, it remains a difficult and tedious undertaking and one resistant to automation. Consequently, the world database of VLE data in the extended critical region can be expected to increase only incrementally in the coming years; it is unrealistic to expect wholesale reexaminations of or additions to that database, and as a correlator one must then largely work with what is available at present.

Attempts to correlate the data of Table 2 with the modified Leung-Griffiths model have revealed a wide range of quality and reliability of the data, although for most mixtures there is at least one reliable source. The model has proven to be an excellent method of data evaluation[18] as well as correlation; poor-quality sources have been identified as much for their inconsistency with reliable sources as for their incompatibility with the model. It is hard to judge the relative merits of the many different experimental techniques from the experimentalists' descriptions, but the model can "circumstantially" detect incompatibility with other mixture data and with modern understanding of critical phenomena.

As a general rule, the collective VLE database seems tainted less by outright and demonstable error in the original measurements than by the presentation of smoothed, cross-plotted, or otherwise "processed" data in place of the original measurements. Unfortunately, the majority of archival literature sources cited in Table 2 do not present the original data. Considerable effort has been made to track down the original measurements, but in many cases success is elusive, and as a result the availability of raw data was not made a requirement for inclusion in the primary source list. Some manipulations of the data are more innocuous than others; such manipulations can be divided into three types. First is smoothing of data, where, for example, dew and bubble points originally measured over irregular pressure intervals are reported at regular pressure intervals by graphical interpolation. Second, and more troublesome, is cross-plotting of data, for example, graphically transforming constant-composition dew-bubble curves into isothermal loci. The worst case is extensive "processing" of the data, or manipulating the original measurements to an extreme degree perhaps with the inclusion of some sort of theoretical or empirical model.

An example of such "processing" is the advocacy by one group that measurements be taken more or less randomly in thermodynamic space, not along loci of constant composition, temperature, or pressure, and of generating further data by repeated interpolation.[362] In a few cases the original, randomly distributed data from that group have been reported,[81,83] but usually the generated data have been published, and clearly there is much systematic error in such a procedure. Also, the extrapolation of such data to the critical locus[84,85] is influenced by an empirical equation of Benson et al.,[363] which states that rectilinear diameters of isotherms on a plot of log P vs. x (Figure 9C with a logarithmic vertical axis) are linear. This is an empirical assumption that works least well near the critical point, and may have generated the incorrect cusps in dew-bubble isotherms, as noted in Section IV. The author has attempted to remove the "Benson distortion" and then to correlate the data on ethylene + isobutane[84] and ammonia + acetonitrile,[85] but without success.

From personal experience in correlating VLE with the modified Leung-Griffiths model, the author has found that it is best to use original or raw data as input; therefore, it is recommended that experimentalists initially tabulate and publish their data as originally measured, rather than as smoothed, cross-plotted, or otherwise processed data.

VI. OVERVIEW OF LEUNG-GRIFFITHS MODEL

In this section the Leung-Griffiths model[1] as modified by Moldover, Rainwater, and co-workers[2-6] is reviewed. The procedure is basically a generalization of the Schofield model[12,39] for pure fluids with nonclassical critical exponents (as described in Section II) to mixtures by a field variable representation. The concepts of field variables and models with nonclassical exponents are independent, and one could be used without the other; in fact, Fox and Storvick[364,365] consider a field-variable mixture model based on the classical Berthelot and BWR EOSs. The author's experience, however, indicates that the advantages of using field variables occur largely in the critical region and thus it is appropriate to incorporate into the mixture model the correct critical exponents.

One related but unsuccessful approach is to use a Schofield or an alternate "scaling-law" model as a reference equation for one pure fluid and then to describe the mixture with classical corresponding states based on that reference equation.[366] Among other difficulties, the critical locus obtained from such an approach in P-T space begins as colinear with the pure-fluid vapor pressure curve, in contradiction to experiment.[32,367]

In a separate report,[6] the author has developed in detail the mathematics of the modified Leung-Griffiths model, with interpretive comments and detailed comparison with the "unmodified" model and with more conventional phase equilibrium techniques. This report is recommended for those unfamiliar with the model or who have questions concerning its relationship to other methods.

For the present, less detailed discussion, the previous outline of the development of the model is retained. First, field variables that represent thermodynamic "distances" are introduced and the basic laws of thermodynamics are expressed in terms of these field variables; this part is without approximation but by itself not useful. Second, a Schofield construction is imposed so that a vapor-liquid phase equilibrium surface is described at the right location in thermodynamic space, with the correct nonclassical critical exponents and with some flexibility or degrees of freedom. Third, the remaining degrees of freedom are analyzed and a minimal number of adjustable parameters are introduced so that the model can be made to agree with experiment. A simple fixed version of the model due to Moldover and Gallagher,[2,3] in which there are in effect no adjustable parameters, is used as a point of reference.

Section VI.C examines how, empirically, the VLE mixture data as collected in the previous section and ordered in α_{2m} are used to introduce the minimum necessary number of adjustable parameters to the model for a given mixture.

A. BASIC THERMODYNAMIC TRANSFORMATIONS

Binary mixture thermodynamics is described by three independent thermodynamic variables and one dependent variable. In the field-space formalism of Griffiths and Wheeler,[11] the dependent field variable is called the "potential"; here, the potential is $\omega = P/RT$ as introduced earlier in Equation 8. The potential is a continuous function of the independent field variables, but on a phase boundary, partial derivatives of the potential with respect to the independent field variables are discontinuous. The fundamental field variables are pressure, temperature, and the chemical potential of each component; functions of field variables are, of course, themselves field variables. Initially chosen are the independent variables

$$\nu_i = \mu_i/RT, \; i = 1, 2 \tag{17}$$

$$B = 1/RT \tag{18}$$

The fundamental thermodynamic relationship, derivable from the first and second laws of thermodynamics,[6] is

$$d\omega = \rho_1 d\nu_1 + \rho_2 d\nu_2 - u dB \tag{19}$$

where u is the internal energy per unit volume. This representation is particularly useful since the differentials are of field variables, whereas the conjugate functions are related to measurable density variables by the following equations

$$\rho = \rho_1 + \rho_2 \tag{20}$$

$$x = \rho_2/(\rho_1 + \rho_2) \tag{21}$$

$$h_m = \rho^{-1}(P + u) \tag{22}$$

where h_m is the molar enthalpy. The present convention is that fluid 2 is the fluid with the lower critical temperature, and $x = 1$ is pure fluid 2. As noted in the experimental VLE literature survey, composition is almost always measured, coexisting densities are measured in many if not all experiments, and there is a much more limited number of experiments that measure coexisting enthalpy data. Normally, such quantities as entropy or free energies are not subject to direct experimental measurement.

The next step is to transform to three independent "distance" variables in field space. Two of these, t and h, are simple generalizations of those variables from the earlier pure-fluid discussion Equations 4 and 9. In addition, a variable is needed to characterize the distance from pure fluid 1. Conventionally, x serves such a purpose, but x is a density variable. Much of the essence of the Leung-Griffiths model is in their very clever introduction of ζ, a field-variable analog of x,

$$\zeta = \frac{K_1(B)e^{\nu_1}}{K_2(B)e^{\nu_1} + K_1(B)e^{\nu_1}} = \frac{1}{K(B)e^{\nu_2 - \nu_1} + 1} \tag{23}$$

where the parameter $K = K_2/K_1$ is explained below. Since the chemical potentials diverge in the respective dilute limits, e.g.,

$$\nu_2 \rightarrow \ell n x \tag{24}$$

it is seen that $\zeta = 1$ for $x = 0$ and $\zeta = 0$ for $x = 1$. Originally, K was assumed to be constant, but the author has shown that it can also be a temperature-dependent function with very little formal change in the equations of the model.[6]

New variables are defined both in terms of the old and the previously defined new variables. In the original version of the model, the other "distance" variables are

$$\tau = \frac{1}{RT_c(\zeta)} - B = B_c(\zeta) - B \tag{25}$$

$$h = \ell n \left[K_1(B)e^{\nu_1} + K_2(B)e^{\nu_2} \right] - \ell n \left[K_1(B)e^{\nu_1^\sigma} + K_2 (B)e^{\nu_2^\sigma} \right]$$

$$= \ell n \left[K_1(B)e^{\nu_1} + K_2(B)e^{\nu_2} \right] - H(\zeta,\tau) \tag{26}$$

Here, in the interval ($0 < \zeta < 1$), which bounds the physically meaningful ζ, there is a critical temperature for each ζ for class 1 mixtures, and this temperature enters into Equation 25. Furthermore, for each ζ and $t < 0$ in the one-phase region, there is a point with the same values of ζ and t on the coexistence surface, and functions evaluated at that coexistence point are denoted by the superscript σ in Equation 26. For $t > 0$, the superscript σ corresponds to a point on the smooth extension of the coexistence surface into the supercritical region, the precise definition of which is model dependent.

These equations illustrate that the Leung-Griffiths model is not merely a description of the VLE surface, but a complete description of thermodynamic space, including the one-phase region, near the critical locus. Within the original version of the model, one-phase data have been correlated in the original paper[1] as well as by D'Arrigo et al.[21] for carbon dioxide + ethylene and by Chang and Doiron[22] for carbon dioxide + ethane. For certain technical reasons discussed below the author has provided no analysis of the mixture supercritical one-phase region with the modified model, though one is being prepared. The expectation is that with some further modifications as yet undetermined, the model that will be developed can provide a superior description of one-phase mixture thermodynamics near the critical locus than that obtainable from a classical EOS.

For a discussion of the differences and relative merits of the original and modified models see Reference 6, Appendix C, where the author strongly advocates the use of the modified model. The first distinct feature of the modified model is the replacement of the dimensioned τ by a dimensionless t as in Equation 4,

$$t = \frac{T - T_c(\zeta)}{T_c(\zeta)} = \frac{\tau}{B_c(\zeta) - \tau} \tag{27}$$

so that the modified model describes ω as a function of ζ, t, and h. The independent variables are thermodynamic distances; ζ is a distance from fluid 1, t is a distance from the critical locus, and h is a distance from the coexistence surface. In three-dimensional field space, the critical locus is $0 < \zeta < 1$, $t = 0$, $h = 0$, and the coexistence surface is $0 < \zeta < 1$, $t < 0$, $h = 0$.

Although it is introduced a bit prematurely, Figure 11 is instructive in showing loci of constant ζ in P-T space for a type 1 mixture. Within the modified model, such loci form a curvilinear grid parallel to the vapor pressure curves, as illustrated. The constant-ζ loci are displayed in steps of 0.2, and the dots on the curves denote intervals in t of 0.02. Because of the steepness of the vapor pressure curves and consequently of the constant-ζ loci, $t = -0.1$ corresponds roughly to half the critical pressure. The "extended critical region" is defined as $-0.1 < t < 0$, which is approximately equivalent to $0.5\ P_c < P < P_c$. Typically

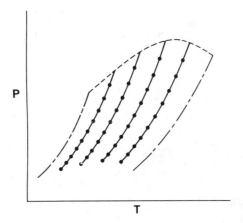

FIGURE 11. Loci of constant ζ (solid curves, with dots) for a binary mixture in P-T space at intervals of $\Delta\zeta = 0.2$. The dots show increments of t where $\Delta t = 0.02$. The dashed line is the critical locus and the broken lines are the vapor pressure curves.

within this region the modified Leung-Griffiths model works well and at lower pressures it breaks down, whereas a classical EOS typically works well at lower pressures but tends to break down within the extended critical region.

Note that for each (P,T) point in the two-phase region there is one value of ζ but two values of x (one for the liquid, one for the vapor). Another characteristic of the modified model for most mixtures is that, on the critical locus,

$$x = 1 - \zeta \tag{28}$$

so that, on Figure 11, the intercepts of the constant-ζ loci with the critical locus are at intervals in x of 0.2.

Originally, Equation 28 was thought to be an approximation, and it indeed is if K is restricted to be a constant; however, if $T_c(x)$ is a monotonic function of x (true for most normal mixtures but not true for most azeotropic ones), a K(T) exists such that Equation 28 can be made exact. To see this, it is noted that a rearrangement of Equation 23 yields

$$\ell n\ K = (\mu_1 - \mu_2)/RT + 2\ tanh^{-1}(1 - 2\zeta) \tag{29}$$

which is consistent with Equation 28 if

$$\ell n\ K = (\mu_1 - \mu_2)/RT + 2\ tanh^{-1}(2x - 1) \tag{30}$$

The values of the chemical potentials along the critical locus must be regarded as unknown, but in principle for each point on the critical locus (i.e., each temperature) K can be chosen so that Equation 30 is satisfied, and therefore Equation 28 is satisfied, so long as the critical temperature is a monotonic function of x. This point of understanding has been most useful in the recent work of the author, who has found that use of Equation 28 works for a wide variety of normal mixtures, but, upon close inspection, is inadequate for the majority of azeotropic ones (see Section XII).

The next step is to express the conjugate density functions of Equation 19 in terms of partial derivatives with respect to the new variables.

$$\left[\frac{\partial \omega}{\partial v_i}\right]_{v_j,B} = \rho_i = \left[\frac{\partial \omega}{\partial \zeta}\right]_{\tau,h}\left[\frac{\partial \zeta}{\partial v_i}\right]_{v_j,B} + \left[\frac{\partial \omega}{\partial \tau}\right]_{\zeta,h}\left[\frac{\partial \tau}{\partial v_i}\right]_{v_j,B}$$

$$+ \left[\frac{\partial \omega}{\partial h}\right]_{\zeta,\tau}\left[\frac{\partial h}{\partial v_i}\right]_{v_j,B} \tag{31}$$

where i = 1 and j = 2 or i = 2 and j = 1, and

$$-u = \left[\frac{\partial \omega}{\partial B}\right]_{v_1,v_2} = \left[\frac{\partial \omega}{\partial \zeta}\right]_{\tau,h}\left[\frac{\partial \zeta}{\partial B}\right]_{v_1,v_2} + \left[\frac{\partial \omega}{\partial \tau}\right]_{\zeta,h}\left[\frac{\partial \tau}{\partial B}\right]_{v_1,v_2}$$

$$+ \left[\frac{\partial \omega}{\partial h}\right]_{\zeta,\tau}\left[\frac{\partial h}{\partial B}\right]_{v_1,v_2} \tag{32}$$

An interesting observation to be made here is that, since ρ_1 and ρ_2 depend on partial derivatives with B (and thus T) held constant, the temperature dependence of K has no bearing on the formal expressions for ρ and x, although u will contain terms involving dK/dT.

At this point the mathematics becomes rather involved, but one important point of simplification occurs in that

$$\left[\frac{\partial \zeta}{\partial v_2}\right]_{v_1,B} = -\zeta(1-\zeta) = -\left[\frac{\partial \zeta}{\partial v_1}\right]_{v_2,B} \tag{33}$$

In other words, the partial derivatives with respect to ζ occurring in Equations 31 and 32 are algebraic functions of ζ itself. Such a structure would not occur for other functions of T, μ_1, μ_2, and K; the required partial derivatives would come out in terms of different combinations of T, μ_1, μ_2, and K not connected algebraically to the original function. Because of this property, it is not required that the value of K be known explicitly when correlating the P-T-ρ-x coexistence surface. Another remarkable simplification is that

$$\rho = \rho_1 + \rho_2 = \left[\frac{\partial \omega}{\partial h}\right]_{\zeta,\tau} \tag{34}$$

which is largely a consequence of the fact that ζ depends only on the difference of μ_1 and μ_2.

The expressions for composition and energy density are more involved. Details are given in Reference 6; the final results according to the original model are

$$x = \rho_2/\rho = 1 - \zeta - \zeta(1-\zeta)Q/\rho \tag{35}$$

$$Q = \left[\frac{\partial \omega}{\partial \zeta}\right]_{\tau,h} + \frac{dB_c(\zeta)}{d\zeta}\left[\frac{\partial \omega}{\partial \tau}\right]_{\zeta,h} - \rho\left\{\left[\frac{\partial H}{\partial \zeta}\right]_{\tau} + \frac{dB_c(\zeta)}{d\zeta}\left[\frac{\partial H}{\partial \tau}\right]_{\zeta}\right\} \tag{36}$$

$$u = \left[\frac{\partial \omega}{\partial \tau}\right]_{\zeta,h} - \rho\left[\frac{\partial H}{\partial \tau}\right]_{\zeta} - \rho_1\frac{d}{dB}\ell n\, K_1 - \rho_2\frac{d}{dB}\,\ell n\, K_2 \tag{37}$$

which were derived by Leung and Griffiths except for the terms of Equation 37 involving the temperature dependence of K_1 and K_2.

It is evident that the temperature dependence of K can be obtained only through a correlation of coexisting enthalpies. As noted in Section V, some experiments have been published[278-283] that provide four dew-bubble curves through enthalpy data. In addition, the groups led by Smith,[368,369] Lenoir,[370,371] Katz,[372, 273] Christensen,[374-381] and Wormald[382] have presented less complete P-T-x-h_m data in the extended mixture critical region, and Sage and co-workers have measured the enthalpy[383] as well as the P-T-x-ρ surface[165] of n-butane + n-decane. The experiments of the Christensen group include a substantial amount of one-phase data, and they have found difficulty in simultaneously correlating P-T-x-ρ data and enthalpies with a single classical equation in the supercritical region, a problem that may call for an extension of a model such as the author's with correct critical exponents. Correlations of coexisting enthalpies are in the planning stage, but no specific results have been obtained at present.

In the modified model, τ is eliminated in favor of t, and some addition and subtraction is performed to yield the expressions

$$x = 1 - \zeta - \zeta(1 - \zeta)\left[\frac{\overline{Q}(\zeta,t)}{\rho} - \frac{\overline{Q}(\zeta,t=0)}{\rho_c(\zeta)} - \overline{H}(\zeta,t)\right] \tag{38}$$

$$\overline{Q} = \left[\frac{\partial\omega}{\partial\zeta}\right]_{t,h} + \frac{1}{B_c(\zeta)}\frac{dB_c(\zeta)}{d\zeta}(1 + t)\left[\frac{\partial\omega}{\partial t}\right]_{\zeta,h} \tag{39}$$

$$\overline{H}(\zeta,t) = \left[\frac{\partial H}{\partial\zeta}\right]_{\tau} + \frac{dB_c(\zeta)}{d\zeta}\left[\frac{\partial H}{\partial\tau}\right]_{\zeta} - \frac{\overline{Q}(\zeta,t=0)}{\rho_c(\zeta)} \tag{40}$$

so that $\overline{Q}$ is obtainable from the critical locus and the model for $\omega(\zeta, t, h)$, while $\overline{H}$ in practice is modeled separately. Use of the liquid or vapor density in Equation 38 yields, respectively, the liquid or vapor composition.

While K(T) does not appear in the formal expression for the P-T-x-ρ surface and only its temperature derivative appears in the expression for enthalpy, for other thermodynamic quantities such as entropy or Helmholtz free energy the explicit knowledge of K is required.[1] Thus, the modified Leung-Griffiths model is "incomplete" compared with an integrable classical EOS; however, Section XIV notes that the entropy difference between dew and bubble points on the same isopleth at the same temperature can be calculated, and from this information one may be able to obtain a good estimate for $\Delta s(\zeta, t)$, where Δs is the difference between vapor and liquid molar entropies.

B. INCORPORATION OF SCHOFIELD MODEL

Having set up the thermodynamics of the binary mixture in terms of the new distance variables, a model is now constructed for ω, the potential, such that the vapor-liquid phase transition is explicitly built in. As discussed in Section II, this involves constructing a multidimensional function of field variables that is continuous but has discontinuous derivatives, and where the discontinuity as a function of distance from a critical point increases with the correct nonclassical exponent. It was noted that the Schofield model was a convenient means of constructing such a multidimensional function, not the only or necessarily the best way, but with certain distinctly advantageous features.

The idea here is to make a Schofield construction upon each surface of constant ζ, together with a background part of the potential that describes the thermodynamics other than the phase transition. Specifically, ω consists of an analytic and a singular part,

$$\omega(\zeta,t,h) = \omega_{an}(\zeta,t,h) + \omega_{sing}(\zeta,t,h) \tag{41}$$

For the singular part, Equations 10 to 12 are employed except that now P_c, T_c ρ_c, C_1, and C_3 are functions of ζ. Therefore, a phase transition is imposed in surfaces of constant ζ starting at the mixture critical point. Differences in coexisting densities increase as distance from the critical locus (i.e., $|t|$) to an exponent equal to the correct nonclassical value of β instead of $\beta = \frac{1}{2}$. The analytic part is equivalent to Equation 13 with ζ-dependent coefficients; in particular,

$$\omega_{an}(\zeta,t,h) = \frac{P_c(\zeta)}{RT_c(\zeta)}\left[1 + C_4(\zeta)t + C_5(\zeta)t^2 + C_6(\zeta)t^3\right]$$
$$+ \rho_c(\zeta)[1 + C_2(\zeta)t]h \tag{42}$$

The value of ω on the coexistence surface is given by setting $h = 0$ and $\theta = \pm 1$, hence $r = -t$. The coexisting densities are obtained from Equation 34, together with the Jacobian representation

$$\left[\frac{\partial\omega_{sing}}{\partial h}\right]_{t,\zeta} = \frac{\partial(\omega_{sing},t)/\partial(r,\theta)}{\partial(h,t)/\partial(r,\theta)} \tag{43}$$

which reduces to

$$\rho(\zeta,t)_{h=0} = \rho_c(\zeta)[1 \pm C_1(\zeta)(-t)^\beta + C_2(\zeta)t] \tag{44}$$

i.e. the analog of Equation 3, but with ζ-dependent coefficients.

The key insight of Leung and Griffiths here is that on surfaces of constant ζ, mixtures behave very similarly to pure fluids. In the usual approaches, mixtures are studied along surfaces of constant composition, on which they behave very differently from pure fluids, e.g., they possess wide dew-bubble curves instead of infintely narrow pure vapor pressure curves. Composition is a density variable, however, and the conceptual breakthrough of Leung and Griffiths was to recognize that in field variable space the behavior of mixtures and pure fluids is very similar. The construction in the original model closely follows the above form, except that τ is used in place of t, an extra function occurs in the Schofield parameterization, and the term proportional to (th) in Equation 42 is absent. This term yields a finite slope in the rectilinear diameter of the temperature-density coexistence curve, and is not needed for (helium 3 + helium 4), but is needed for other mixtures.

The six functions $C_i(\zeta)$ have yet to be specified; there is some convenient freedom in the model within these functions; however, they are subject to well-defined boundary conditions. The vapor pressure curves and coexisting density curves of each pure fluid are fitted to Equations 6 and 3, respectively, thereby yielding pure fluid coefficients $C_i^{(j)}$, $i = 1, \ldots ,6$ and $j = 1, 2$. The ζ-dependent coefficients then must satisfy the boundary conditions

$$C_i(\zeta = 1) = C_i^{(1)} \tag{45}$$

$$C_i(\zeta = 0) = C_i^{(2)} \tag{46}$$

An additional degree of freedom within the model is the specification of the auxiliary function $\overline{H}(\zeta, t)$, as discussed in the following section.

Section II discussed the advantages and disadvantages of the Schofield model relative to classical EOSs, and noted that such patterns would persist for the mixture problem. Let us now return to items A through F in the discussion below Equation 14 and discuss how they generalize to mixtures.

1. As in the pure fluid case, any analytic EOS will yield a phase transition surface with (incorrect) classical critical exponents. Therefore, the Leung-Griffiths model, unlike a classical EOS, has the capacity to describe correctly the thermodynamic behavior in the near-critical region. The Schmidt-Wagner equation[40] has mimicked the correct exponents with substantial success for pure fluids, but to date no comparably successful mimicking of critical exponents for a mixture phase boundary surface has been implemented.

2. The phase equilibrium surface is described in the Leung-Griffiths model with the simple algebraic condition h = 0. Although some numerical inversion is necessary for practical implementation of the model, phase equilibrium appears to be calculable much more quickly with the Leung-Griffiths model than with equations of state where typically time-consuming flash calculations must be performed. In fact, one possible advantage of the model may be simply the savings in computational time and effort. This consideration is becoming less important as faster and more powerful computers are developed, but as some industrial calculations require thousands of phase equilibrium calculations, the savings in computation time with the Leung-Griffiths model for some applications would be worth investigating.

3. Just as the critical point is input rather than output for the pure-fluid scaling-law equation, the critical locus is input for the Leung-Griffiths model, and typically is obtained from experiment. Also, a conventional EOS predicts a critical locus, but because of a trade-off between the optimization of low-pressure VLE and the incorrect parabolic shape of the coexistence loci at the critical point, certain optimizations of the EOS can lead to incorrect critical loci. One procedure strongly recommended against is to calculate a critical locus from an EOS and then to construct a Leung-Griffiths correlation from the calculated critical locus. This was attempted by Al-Sahhaf et al.[384] with generally poor results, and typically brings out the worst aspects of both approaches. A possible alternative for future consideration would be to use a classical EOS reliable up to, for instance, $t = -0.05$ (or $P = {}^3/_4 P_c$) and then, within the band $-0.1 < t < -0.05$, fit the Leung-Griffiths model to the classical representation and extrapolate it to critical conditions. Thus, the critical line would be predicted with inclusion of scaling-law exponents, although the hazards inherent with any extrapolation process would be present.

4. As with pure fluids, the Leung-Griffiths model is an indirect approach that algebraically describes nonmeasurable quantities and has the further mysterious feature of an unspecified function K(T) in the definition of ζ. Again, there is no problem in application so long as the errors inherent in the numerical analysis are controllable and can be made much smaller than the corresponding experimental errors, which is the case with the availability of modern computational resources.

5. Thermodynamic stability remains a serious problem within the Leung-Griffiths model. In a field variable representation, the phase boundaries are explicitly imposed, rather than converted from a formal instability of the EOS as in classical approaches. Furthermore, there is no guarantee that the model is stable everywhere, and if it is not, the instability cannot be converted into a phase transition; it is simply an artifact of, and a taint on, the model. In fact, such an instability has occurred in one of the published Leung-Griffiths correlations. In the propane + n-octane fit of Moldover and Gallagher[3] (see their Figure 10), the dew curve of the 4% octane mixture could not be described; the model according to their solid curve deviated toward high temperatures and the version described by their dashed curve deviated in an opposite manner. Subsequent study by the author has shown that along this dew curve, near T = 410 K and P = 3 MPa, their model predicts a vapor composition greater than 1, a completely unphysical result and a violation of thermodynamic stability criteria.

With subsequent improvements in the model this unphysical artifact has been removed and the dew curve has been correlated accurately, but the example shows that it is easy (inadvertently) to create artificial instabilities with a Schofield-like model. For field-space models of VLE, it becomes the responsibility of the modeler to ensure that the correlation is thermodynamically stable at least over some well-specified domain of the model.

6.   The modified Leung-Griffiths model and the classical EOS complement each other, in that the former is most accurate near the critical locus and the latter is most accurate away from the critical locus. For VLE, it is hoped that over some band, say $-0.1 <$ $t < -0.05$, both methods are accurate enough that they can be joined by a switching function, although this has not yet been attempted. Use of a switching function for the entire thermodynamic space, including the one-phase region, is much more problematic. Research is in progress or in the planning stages by other groups to join the critical and noncritical regions "seamlessly" by crossover theory[46-50] or similar methods.

While type 1 mixtures behave similarly to pure fluids with only the vapor-liquid phase transition, types 2 through 6 have various liquid-liquid transitions that can be described qualitatively by conventional approaches but do not appear in the Leung-Griffiths model as thus far developed. The field variable approach does have some pedagogical utility for the higher mixture types. In P-T-x space, as shown by Streett,[51] mixtures of higher type can have extremely complicated phase boundary surfaces. However, in P-T-ζ space, the figures are considerably simplified because all phase boundaries are continuous single-valued sheets, intersecting each other along three-phase boundaries in accordance with the 180° rule.[65,66] In order to understand mixtures of higher type the effort to comprehend the P-T-x-ρ diagrams is still worthwhile.

To describe higher type mixtures within the spirit of the Leung-Griffiths model, the additional phase boundaries would have to be constructed or imposed. A preliminary attempt to do so for type 2 mixtures is described in Section XIV; such a line of research, although very challenging, may be the most promising means of obtaining a correct nonclassical description of an arbitrary class 1 binary mixture.

The model as thus far described incorporates only "simple scaling"; here brief comments are presented on "revised" and "extended" scaling.[385] Within revised scaling, t is replaced by $t + qh$ in Equation 11 and its generalization to mixtures, where q is a mixing parameter. For pure fluids, this has the effect of producing a "hook" in the rectilinear diameter of the temperature-density coexistence curve, as shown in Figure 12 and in contrast to the straight-line rectilinear within simple scaling. Also, in the supercritical region, the h = 0 locus, after another small hook, proceeds approximately along the critical isochore, in contrast to simple scaling with a sloping rectilinear diameter where the supercritical h = 0 locus continues to slope as in Figure 3.

For pure fluids, the rectilinear diameter hook has been confirmed in the recent experiments of Balzarini and co-workers.[386-388] For hydrogen, the diameter hooks in the opposite direction to that of Figure 12, with interesting thermodynamic ramifications that have yet to be investigated; however, the shape of the VLE surface changes very little from simple scaling. Although revised scaling is in principle inherent in mixtures, we believe it is appropriate not to incorporate it into the model for VLE studies alone. New functions such as a ζ-dependent q would lead to redundancies in the model, i.e., different sets of parameter values that would yield essentially the same P-T-x-ρ coexistence surface. The author's success in correlating a wide variety of mixtures is largely due to a parameter structure that avoids such redundancies, whereas the failures of the original Leung-Griffiths model are largely because of the many redundancies within its parameter structure.

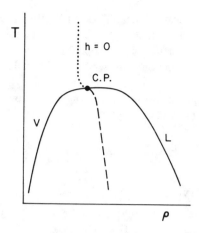

FIGURE 12. Vapor-liquid transition for a pure fluid in T-ρ space according to a thermodynamic representation with revised scaling (schematic, exaggerated). Unlike Figure 3, the rectilinear diameter has a "hook" at the critical point, and the one-phase h = 0 locus also has a hook before approximately following the critical isochore. The difference between the estimated critical density with and without revised scaling is typically $^1/_{10}$ of 1%.

However, for future one-phase studies it will be necessary to incorporate revised scaling because the sloping h = 0 locus for the one-phase region is clearly contrary to experiment. Thus far, the original model with its absence of a rectilinear diameter slope is the only version that has been used in the one-phase region, though the absence of a slope has been at the expense of accurate VLE predictions for carbon dioxide + ethylene[21] and carbon dioxide + ethane.[22] Unpublished works of Chang and of Moldover show some promising potential for correlation of both VLE and the supercritical region with a Leung-Griffiths model incorporating revised scaling.

Extended scaling, a completely independent model refinement, has the effect of adding a term to Equation 3,

$$\rho = \rho_c[1 \pm C_1(-t)^\beta \pm C_7(-t)^{\beta+\Delta} + C_2t] \tag{47}$$

as first proposed by Wegner[389] in which $\Delta \approx 0.5$. Balfour et al.[385] have shown how to incorporate both revised and extended scaling into the Schofield model for a pure fluid; again, the generalization to mixtures in the spirit of Leung and Griffiths would be to make the coefficient ζ-dependent. Extended scaling requires a second, relatively complicated term in ω_{sing} of Equation 10. Unlike revised scaling, extended scaling may be useful in improving the author's VLE correlations. Lynch has developed a code for a generalized model with extended scaling for the purpose of improving the predictions of coexisting densities for carbon dioxide + n-butane[5,18] and similar mixtures with some promising initial results, but at present no firm conclusions can be drawn.

C. ADJUSTABLE PARAMETER STRUCTURE

The history of the Leung-Griffiths model to date has been to use it in a correlative rather than a predictive mode.[20] The predictive ideal is to determine a VLE surface from very minimal information which cannot be achieved with the model at present. Rather, the emphasis has been placed on developing high-accuracy correlations of mixture VLE surfaces for which there is a good quantity of data. Thus, in Section V criteria were developed for such thorough studies, listed in Table 2, but no experiment can cover the complete mixture surface and at most 13 isotherms[212] or 11 isopleths[177] have been measured for a single mixture. A high-accuracy correlation thus serves to fill in large unmeasured regions of thermodynamic space.

To continue the development of the model, an expression for $\overline{Q}$ more specific to the model as developed in the last section is useful,

$$\overline{Q}(\zeta,t) = \frac{dx}{d\zeta} \frac{PT_c(x)}{RTP_c(x)} \frac{d}{dx} [P_c(x)/T_c(x)]$$

$$+ \frac{P_c(x)}{RT_c(x)} \left[C_3'(-t)^{1.9} + C_4't + C_5't^2 + C_6't^3 \right]$$

$$+ \frac{dx}{d\zeta} \frac{P_c(x)}{R} \frac{d[1/T_c(x)]}{dx} (1 + t) \left[-1.9 \, C_3(\zeta)(-t)^{0.9} + C_4(\zeta) \right.$$

$$\left. + 2C_5(\zeta)t + 3C_6(\zeta)t^2 \right] \tag{48}$$

where $C_i' = dC_i/d\zeta$.

Second, explicit representations of the critical locus are required. Such representations for these simple polynomial fits of P_c/T_c, $1/T_c$, and ρ_c are

$$\frac{1}{RT_c(x)} = \frac{1-x}{RT_{c1}} + \frac{x}{RT_{c2}} + x(1-x) \left[T_1 + (1-2x)T_2 \right.$$

$$\left. + (1-2x)^2 T_3 + (1-2x)^3 T_4 \right] \tag{49}$$

$$\frac{P_c(x)}{RT_c(x)} = \frac{(1-x)P_{c1}}{RT_{c1}} + \frac{xP_{c2}}{RT_{c2}}$$

$$+ x(1-x)[P_1 + (1-2x)P_2 + (1-2x)^2 P_3 + (1-2x)^3 P_4] \tag{50}$$

$$\rho_c(x) = (1-x)\rho_{c1} + x\rho_{c2} + x(1-x)[\rho_1 + (1-2x)\rho_2 + (1-2x)^2 \rho_3] \tag{51}$$

For some cases,[4] it was found convenient to replace x by x_T, where

$$x_T = \frac{1/T_{c1} - 1/T_c(x)}{1/T_{c1} - 1/T_{c2}} \tag{52}$$

in Equations 50 and 51, where in our terminology the coefficients P_i and ρ_i are replaced by $\overline{P}_i$ and $\overline{\rho}_i$. Also, in some cases[5] fourth-order terms have been needed in Equations 49 and 50.

There is already a substantial proliferation of parameters, six $C_i^{(j)}$ parameters for each of the two pure fluids and at least nine for the critical locus. While the importance of minimizing the number of parameters is recognized, those introduced so far are regarded more as "boundary conditions", as the properties of pure fluids, and in many cases the critical locus, are typically more accurately determined than the mixture VLE surface.

The key question at this point is how many additional parameters that are truly characteristic of the mixture are needed. Moldover and Gallagher[3] developed a recipe in which no more parameters are needed; in other words, the VLE surface is fully determined by the pure-fluid coexistence properties and the critical locus. This recipe has turned out to be overly simple for most mixtures, but it remains an excellent point of reference. Their recipe is simply that all C_i functions are linearly interpolated, i.e.,

$$C_i(\zeta) = C_i^{(2)} + \zeta C_i' \tag{53}$$

$$\overline{H}(\zeta,t) = 0 \tag{54}$$

which, through Equation 38, yields $x = 1 - \zeta$ on the critical locus.

If all pure fluids were to obey the principle of corresponding states, then each C_i would be a universal number for all fluids. This is not exactly the case, but from fluid to fluid the variation is normally not great. The linear interpolation assumption, together with the approximate equality of $C_i^{(1)}$ and $C_i^{(2)}$, leads to the structure of nearly parallel constant-ζ loci as shown in Figure 11, where these loci are given by the equation

$$\frac{P}{RT} = \frac{P_c(\zeta)}{RT_c(\zeta)} [1 + C_3(\zeta)(-t)^{2-\alpha} + C_4(\zeta)t + C_5(\zeta)t^2 + C_6(\zeta)t^3] \tag{55}$$

However, the Moldover-Gallagher recipe as such provides an optimal correlation for only one mixture, helium 3 + helium 4. This was the subject of the original Leung-Griffiths paper,[1] and is nearly "ideal" in the sense that the dew-bubble curves are very narrow.

As described in the Introduction, the author's empirical procedure is to order the mixtures according to difficulty of fitting or dissimilarity of the two components, for which there is the useful quantitative parameter α_{2m} of Equation 2, but from Equations 1 and 38,

$$\alpha_2 = \rho_c^{-1}\zeta(1 - \zeta)\overline{Q}(\zeta,t = 0)$$

$$= (R\rho_c)^{-1}\zeta(1 - \zeta)\left\{ \frac{d}{d\zeta}\left(\frac{P_c}{T_c}\right) + \left[\frac{d}{d\zeta}\left(\frac{1}{T_c}\right)\right]\left[T_c\left(\frac{\partial P}{\partial T}\right)_{\zeta,t=0} - P_c\right]\right\} \tag{56}$$

which, if $x = 1 - \zeta$ on the critical locus, reduces to

$$\alpha_2 = -(R\rho_c)^{-1}x(1 - x)\left[\frac{d}{dx}\left(\frac{P_c}{T_c}\right) + C_4 P_c \frac{d}{dx}\left(\frac{1}{T_c}\right)\right] \tag{57}$$

As well as providing excellent correlations, the modified Leung-Griffiths model yields an understanding of critical region behavior not found elsewhere in the VLE literature. Equation 57 specifically relates α_2, an independently measurable quantity, to the critical locus and its composition derivatives. One can determine α_2, for example, from the slope of isotherms on a ρ - x plot such as Figure 9D, since

$$\alpha_2 = \rho_c(dx/dP)_T \tag{58}$$

or, as the author has shown both from a chain rule argument[6] and a formal asymptotic analysis,[26] from the slope of isopleths on a T-ρ plot such as Figure 9B,

$$\alpha_2 = \frac{\rho_c dx_c/dT_c}{\dfrac{d\rho_c}{dT_c} - \left(\dfrac{\partial\rho}{\partial T}\right)^c_{x,cxs}} \tag{59}$$

where the subscript cxs means "on the coexistence surface" (see Reference 11).

For the mixtures with coexisting density data listed in Table 2, and with the above assumptions, α_2 and α_{2m} can be calculated. On comparing the simple Moldover-Gallagher recipe with the experimental dew-bubble curves, the model curves show certain systematic deviations such as having a skew toward high temperature or being too wide. The objective at this point is to understand how the degrees of freedom within the model can alter the curves in a controlled manner, and then to utilize these degrees of freedom by means of adjustable parameters to obtain accurate correlations. In such an analysis, it is best to think in terms of averages and differences of calculated density variables.

Average composition: It is evident from Equation 38 that with x kept linear in ζ on the critical locus, at two-phase points variation of $\overline{H}$ will not affect the composition difference between liquid and vapor, but will affect the average value of the composition. Also, $\overline{H}$ is assumed to be analytic in ζ and t, and for Equation 28 to hold, $\overline{H}$ must vanish along the critical locus, i.e., at t = 0. Therefore, to control the average composition a term in $\overline{H}$ linear in t should first be introduced.

Moldover and Gallagher[3] have provided an argument in their Appendix B that such a term also should be proportional to the composition gradient of the critical temperature. In addition, it was found for mixtures with $\alpha_{2m} > 0.17$ that a linear ζ-dependence was necessary. The result is to model the t-dependent part of $\overline{H}$ as

$$\overline{H}(\zeta,t) = -\frac{1}{T_c(\zeta)}\frac{dT_c(\zeta)}{d\zeta}C_H(1 + C_Z\zeta)t \qquad (60)$$

which introduces the adjustable parameters C_H and C_Z. Although the notation was not used, Moldover and Gallagher[3] first introduced C_H in their alternate correlation of propane + *n*-octane, while C_Z was first introduced by Rainwater and Moldover.[4]

Composition and density differences: It is easily seen that the coefficient $C_1(\zeta)$ determines the rate at which composition and density differences between liquid and vapor will increase with $|t|$. At an early stage the author noticed empirically that model dew-bubble curves in both P-T and T-ρ spaces (Figures 9A and 9B) with the simple recipe were almost always wider than the experimental curves. In a preliminary report on this project,[390] it was first suggested that an essential problem with the philosophy of the simple recipe was that a mixture has both density and composition differences, whereas a pure fluid has only composition differences. If the two fluids exactly obeyed corresponding states, i.e., $C_i^{(1)} = C_i^{(2)}$ for all i, the simple recipe would yield the same density difference for the pure fluids and the equimolar mixture, but the equimolar mixture has a pronounced composition difference as well.

The author feels that it is not the density difference but some combination of density and composition difference that should behave "universally" across the mixture surface. At first a quadratic combination[390] was considered, and this was used in the analysis of Al-Sahhaf et al.,[384] but was later changed to a linear combination.[4] In particular, we considered an "amount of phase change" $A_0(t)$ defined as

$$A_0(t) = \frac{\Delta\rho}{\rho_c(\zeta)} + \left[1 + C_X|\alpha_2(\zeta)|\right] \qquad (61)$$

where C_X is an adjustable parameter. This line of reasoning is in keeping with the search for an "order parameter" or density variable that displays the greatest symmetry. As with C_H, it was later found that an additional linear ζ-dependence can be useful. The form for $C_1(\zeta)$ that keeps $A_0(t)$ "universal" to leading order for mixtures of fluids that obey corresponding states, but which is appropriately generalized to other mixtures, is

$$C_1(\zeta) = \frac{C_1^{(2)} + \zeta[C_1^{(1)} - C_1^{(2)}]}{1 + C_X(1 + C_Y\zeta)|\alpha_2(\zeta)|} \tag{62}$$

Thus, two adjustable parameters, C_X and C_Y, are added to the formalism. Rainwater and Moldover introduced C_X in Reference 4 and C_Y in Reference 5.

Average density: The average of the liquid and vapor densities at a given two-phase point is governed by $C_2(\zeta)$, the slope of the rectilinear diameter on a constant-ζ surface. For mixtures of dissimilar n-alkanes such as butane + n-octane, Rainwater and Moldover,[4] as well as Rainwater and Williamson,[15] found with linearly interpolated C_2 that equimolar liquid densities were substantially overpredicted. The same effect for butane + n-octane was found by Storvick and Fox[365] in their field-space model using classical EOSs where corresponding states were assumed to be strictly obeyed along their surfaces of constant ζ (or η in their notation). Rainwater found that such overprediction could be largely corrected by adding the simplest ζ-dependence to C_2 that is consistent with the pure-fluid boundary conditions, i.e.,

$$C_2(\zeta) = C_2^{(2)} + \zeta C_2' + C_R x_c(\zeta)[1 - x_c(\zeta)] \tag{63}$$

This adds the adjustable parameter C_R, first introduced by Moldover and Rainwater,[5] Thus far, $x_c(\zeta) = 1 - \zeta$, but Equation 63 has been found to be the most useful form when the linear relationship between x and ζ on the critical locus is altered, as described next.

Ratio of composition difference to density difference: The alteration of C_1 as in Equation 44 increases or decreases both the composition and density differences while keeping their ratio the same. The dimensionless ratio is α_2, Equations 1 and 57, which within the model can only be altered by changing $C_4(\zeta)$ or by changing the relation between x and ζ on the critical locus. It was shown earlier that for monotonic $T_c(x)$, a K(T) exists for the definition of ζ such that the critical ζ is linear in x. There is no comparable proof that $C_4(\zeta)$ is given by a linear interpolation between the two pure fluids, but it is a remarkable empirical result that such an assumption yields excellent VLE correlations for a fairly wide range of normal mixtures.

When the components become too dissimilar (for example, for n-alkane mixtures at a carbon number ratio greater than 2), however, the composition differences and density differences must be adjusted independently. According to Equation 57 there is the choice of changing C_4 or changing the relationship between x and ζ. The author has elected to retain the linear interpolation for C_i, i = 3, 6, so that the parallel-curve structure of Figure 11 is preserved.

An additional insight into the mathematics of the model is a guide into the alternative choices of the relationship between x and ζ. For an arbitrary analytic relationship satisfying the boundary conditions,

$$1 - \zeta = \epsilon x + O(x^2) \tag{64}$$

where, previously, $\epsilon = 1$. Now, in the dilute limit, α_2 is independent of ϵ; specifically,

$$\alpha_2 = (R\rho_{c1})^{-1} \epsilon x(-1/\epsilon) \left\{ \frac{d}{dx}\left(\frac{P_c}{T_c}\right)_{x=0} \right.$$

$$+ \left[\frac{d}{dx}\left(\frac{1}{T_c}\right)\right]_{x=0} \left[T_c\left(\frac{dP}{dT}\right)^\sigma_{1c} - P_c\right] \right\} + O(x^2)$$

$$= - (R\rho_{c1}T_{c1})^{-1} x \left\{\left(\frac{dP_c}{dx}\right)_{x=0} - \left(\frac{dP}{dT}\right)^\sigma_{1c}\left(\frac{dT_c}{dx}\right)_{x=0}\right\} + O(x^2) \tag{65}$$

It is useful to introduce the quantity A, where

$$A = \left(\frac{dP_c}{dx}\right)_{x=0} - \left(\frac{\partial P}{\partial T}\right)_{1c}^{\sigma} \left(\frac{dT_c}{dx}\right)_{x=0}$$

$$= \left(\frac{dT_c}{dx}\right)_{x=0} \left\{\left(\frac{dP_c}{dT_c}\right)_{x=0} - \left(\frac{\partial P}{\partial T}\right)_{1c}^{\sigma}\right\} \tag{66}$$

Thus, for dilute mixtures and to linear order in x, α_2 is independent of ϵ or of the slope of x vs. ζ, and is proportional to the function A which depends only on the derivatives of the pure vapor-pressure curve and the initial critical locus. The argument applies to both ends of the mixture surface.

The function A was first considered by Rozen[391] and by Krichevskii,[392] who showed that it governed certain properties of dilute near-critical mixtures. This combination has appeared prominently in the more recent work of Levelt Sengers and co-workers.[393-397] Much of their analysis involves thermodynamic paths off the coexistence surface which is beyond the scope of the present work, but two results apply particularly to the phase equilibrium surface. First, Levelt Sengers[393] showed that A is equal to the limiting slope of the "bird's beak" isotherm, i.e., the isotherm at the critical temperature of the more volatile fluid, on a P-x diagram such as Figure 9C. Second, Japas and Levelt Sengers[394] have derived a relationship between A and a Henry's law function that, while proven only in the asymptotic limit, works well over a wide range of temperature and pressure. In particular, where y is the vapor composition, for the limiting value of K = y/x at the pure solvent (as shown by the wedge shapes on the right sides of the slightly subcritical isotherm in Figure 9C) Japas and Levelt Sengers have derived the result

$$\ln K \approx 2A \, (\rho_\ell - \rho_c)/RT\rho_c^2 \tag{67}$$

where ρ_ℓ and ρ_c are the liquid coexistence and critical densities of the solvent. In a recent experimental VLE paper on mixtures of the refrigerant R13 with the two butane isomers, Weber's[398] data have shown excellent agreement with this formula.

Therefore, any alteration in the x vs. ζ relationship will not have an effect on the dilute limits and should have its greatest effect at the equimolar mixture. To obtain the optimal adjustability for x = 0.5,

$$x_c(\zeta) = 1 - \zeta + H_1\zeta(1 - \zeta)(1 - 2\zeta) \tag{68}$$

so that

$$\overline{H}(\zeta, t = 0) = H_1(1 - 2\zeta) \tag{69}$$

or, on combining Equations 60 and 69

$$\overline{H}(\zeta, t) = H_1(1 - 2\zeta) - \frac{1}{T_c(\zeta)} \frac{dT_c(\zeta)}{d\zeta} C_H(1 + C_z\zeta)t \tag{70}$$

thereby introducing a sixth adjustable parameter H_1, first introduced by Moldover and Rainwater.[5] This completes the description of the model at present; further parameters will be added by the incorporation of extended scaling.

It should be made clear that the author does not advocate the proliferation of adjustable parameters. Rather, it is desirable to keep their number as small as possible. The idea here is to analyze mathematically the model with its degrees of freedom to understand how various parameters can lead to specific changes in the model dew-bubble curves, but then to use them only if necessary. The following section orders the well-measured mixtures according to α_{2m}, and a guideline structure is developed so that the number of needed adjustable parameters depends on, and increases with α_{2m}.

VII. SUMMARY OF MIXTURE VLE CORRELATIONS

The ultimate goal of this project has been to develop accurate correlations of all 129 mixtures in Table 2, except those for which it can be convincingly shown that the data are of irreparably poor quality. Although this goal has yet to be attained, the project has reached a certain level of maturity. In particular, of the mixtures in Table 2 listed under the heading "Normal Mixtures with Density Measurements (Small to Moderate α_{2m})" dew-bubble curves have been successfully correlated (all except those with clearly poor data, very narrow curves one containing a highly associating fluid, or those possessing LLE that intrudes into the extended (VLE) critical region).

Table 3 lists all of the nonazeotropic mixtures successfully correlated with the modified Leung-Griffiths model to date. In contrast to Table 2 which was ordered by the critical temperatures of the components, Table 3 is ordered according to α_{2m}. Due to space limitations all parameters characterizing the pure-fluid coexistence curves or the critical loci cannot be listed though these are available on request. The mixture-dependent adjustable parameters for each correlation are listed.

The column labeled "Optimized to Ref." lists the source or sources containing the data to which the model has been optimized. As with Table 2, a slash denotes two sources reporting on the same experiment, where typically the first source is an archival literature article and the second is an unpublished report or thesis containing the original data. Except for ethane + propylene, all correlations agree with these input sources with no substantial systematic deviations in P-T-x space, and they also agree in density space at least up to α_{2m} = 0.25.

The column labeled "Compared with Ref." lists sources with extended critical region data on the same mixture against which the correlation has been compared. Some "secondary" sources[399-419] with three or fewer dew-bubble curves that did not appear in Table 2 are added here; the author has tried to make this list comprehensive with the inclusion of all sources presenting any critical-region VLE data for these mixtures. For the secondary sources, agreement with the correlations ranges from very good to poor, but in most cases the sources that show poor agreement have questionable data that clearly disagree with a primary experimental source as well as with the model. In such cases, the primary source should be considered the more credible in that it reports smoother or more extensive data or is from a laboratory with more overall experience in VLE measurements.

Of particular importance is the pattern of the adjustable parameters. A value of zero for a particular parameter is equivalent to not introducing it into the formalism in the first place, and the use of zeroes for all parameters is equivalent to the simple Moldover-Gallagher recipe.[3] As is evident from Table 3, mixtures of greater α_{2m}, as expected, require more parameters. In fact, from these results the following guidelines can be constructed:

1. If $\alpha_{2m} < 0.1$, only C_H is required.
2. If $0.1 < \alpha_{2m} < 0.155$, C_H and C_X are required.
3. If $0.155 < \alpha_{2m} < 0.25$, five parameters, C_H, C_X, C_Z, C_Y, and C_R are required.
4. If $\alpha_{2m} > 0.25$, at least six parameters [the five listed in (3) and H_1] are required.

Even for the most nearly "ideal" mixtures such as propane + n-butane, inclusion of C_H improves the fit over that obtained from the simple Moldover-Gallagher recipe.

TABLE 3
Mixture Correlations with the Modified Leung-Griffiths Model

Mixture	α_{2m}	Optimized to Ref.	Compared with Ref.	C_H	C_X	C_Z	C_R	C_Y	H_1
$n\text{-}C_4H_{10} + n\text{-}C_5H_{12}$	0.055	144		−6.0	0	0	0	0	0
$He_3 + He_4$	0.074	87, 88		0	.	0	0	0	0
$C_3H_8 + n\text{-}C_4H_{10}$	0.078	130	131, 399	−7.0	0	0	0	0	0
$Et_2O + n\text{-}BuOH$	0.086	150		−9.0	0	0	0	0	0
$CO_2 + H_2S$	0.089	101, 102	103, 104	−12.0	−0.4	−0.7	0	0	0
$C_2H_6 + C_3H_6$	0.096	118, 400	401	−12.0	0	0	0	0	0
$C_3H_8 + neo\text{-}C_5H_{12}$	0.104	132, 133		−7.0	0.25	0	0	0	0
$MeOH + n\text{-}BuOH$	0.104	150		−10.0	1.2	0.0	2.5	−0.5	0
$n\text{-}C_4H_{10} + n\text{-}C_6H_{14}$	0.111	144	402	−9.0	0.15	0	0	0	0
$R22 + R114$	0.114	127	403	−16.0	1.3	0.0	2.3	−0.3	0
$C_3H_8 + i\text{-}C_5H_{12}$	0.137	134		−6.0	0.1	0	0	0	0
$R13B1 + R114$	0.137	126	404	−6.2	0.3	0	0	0	0
$C_3H_8 + n\text{-}C_5H_{12}$	0.141	130	135, 273	−8.0	0.3	0	0	0	0
$CO_2 + C_3H_6$	0.153	105	106, 107, 401	−12.0	0.2	−1.0	0	0	0
$n\text{-}C_4H_{10} + n\text{-}C_7H_{16}$	0.156	145, 146		−12.0	0.55	−0.4	2.0	−0.2	0
$C_2H_6 + i\text{-}C_4H_{10}$	0.169	119		−8.0	0.8	−0.3	2.5	−0.3	0
$N_2 + CH_4$	0.177	89	90—92, 405	−6.0	0.3	0	0	0	0
$C_3H_8 + 22DMB$	0.179	136, 137		−7.0	0.9	0	3.2	−0.4	0
$CO_2 + C_3H_8$	0.180	108	109, 110	−14.0	0.7	−0.7	2.9	−0.1	0
$C_3H_8 + 23DMB$	0.187	136, 137		−8.0	0.8	−0.5	2.5	−0.5	0
$C_3H_8 + 2MP$	0.199	136, 137		−6.0	0.75	0.2	1.5	−0.6	0
$n\text{-}C_4H_{10} + n\text{-}C_8H_{18}$	0.200	142		−10.0	0.9	−0.6	3.3	−0.7	0
$C_2H_6 + n\text{-}C_4H_{10}$	0.203	120, 121	122—124	−7.0	0.9	−0.4	3.3	−0.35	0
$C_3H_8 + 3MP$	0.211	136, 137		−9.0	0.7	−0.7	1.5	−0.6	0
$C_3H_8 + n\text{-}C_6H_{14}$	0.218	136, 137	138, 139	−8.0	0.75	−0.2	2.9	0	0
$C_3H_8 + C_6H_6$	0.228	140, 141		−10.8	0.27	−0.9	1.7	1.0	0
$CH_4 + C_2H_6$	0.243	93	94—96, 259	−7.0	0.95	−1.2	3.3	−0.2	0
$C_2H_4 + n\text{-}C_4H_{10}*$	0.255	97	79	−11.6	0.67	−1.38	2.7	0	0
$CO_2 + i\text{-}C_4H_{10}$	0.272	111, 112	117	−10.0	0.9	−0.5	4.0	−0.2	−0.2
$C_2H_6 + n\text{-}C_5H_{12}$	0.298	125	406	−4.0	1.2	−1.2	3.9	−0.5	−0.2

CO_2 + n-C_4H_{10}	0.304	112, 113	109, 114—117, 407—409	−14.0	0.9	−1.0	4.5	−0.2	−0.2
CO_2 + neo-C_5H_{12}*	0.311	212	213	−9.76	0.838	−0.758	4.084	−0.022	−0.364
C_3H_8 + n-C_8H_{18}	0.314	142	410	−15.0	0.75	−1.4	5.0	−0.1	−0.1
CO_2 + i-C_5H_{12}*	0.359	214	215, 216, 411	−11.0	0.671	−1.016	3.357	−0.078	−0.644
CO_2 + n-C_5H_{12}*	0.370	217	109, 216, 218	−10.41	0.692	−0.959	2.415	−0.38	−0.587
C_2H_6 + C_6H_6**	0.376	162	412	−7.42	0.63	−0.801	3.526	0.058	−0.477
C_2H_6 + c-C_6H_{12}**	0.408	161		−4.69	0.925	−1.206	5.455	0.115	−0.283
CO_2 + c-C_5H_{10}*	0.502	212, 219A	413	−8.22	1.032	−2.026	7.062	0.128	−0.564

Fits to Mixtures with Sparse Data

Ar + O_2	0.010	73	414	−12.0	0	0	0	0	0
N_2 + Ar	0.054	73		−8.0	0.7	0	0	0	0
N_2 + O_2	0.063	415—417	73, 418	−6.0	0.7	0	0	0	0
i-C_4H_{10} + i-C_5H_{12}	0.065	14, 419		−12.0	0	0	0	0	0

* Simplex fit to data on isotherms.
** Simplex fit to data on isopleths.

Al-Sahhaf et al.[384] show the simple-recipe fit for this mixture in their Figure 1; the model curves are in fair agreement with the experimental, but they are systematically shifted toward high temperatures. Setting $C_H = -7$ moves the model curves in line with the experimental curves.

The mixtures in Table 3 that do not follow the guidelines are those subject to polarity effects; most of the mixtures listed are nonpolar-nonpolar. Rainwater and Lynch[17] have discussed in detail the polar mixtures diethyl ether + n-butanol, methanol + n-butanol, R22 + R114, and R13B1 + R114. In general, these mixtures require more parameters, and higher values of the parameters, than mixtures of comparable α_{2m}. Such results are consistent with our earlier discussions of miscibility in which, for the same relative placing of critical points, polar-polar mixtures are more immiscible than nonpolar-nonpolar mixtures. Evidently a decrease in miscibility is reflected by larger values of the modified Leung-Griffiths parameters, although the procedure so far is empirical and it is not understood microscopically how this works.

Weakly polar fluids such as propylene (dipole moment $1.22 \cdot 10^{-30}$ C·m or 0.366 D) show no anomalies when mixed with nonpolar fluids. In Table 3, the only mixture of a nonpolar and a strongly polar fluid is carbon dioxide (no permanent dipole moment, but a substantial quadrupole moment) with hydrogen sulfide (dipole moment $3.24 \cdot 10^{-30}$ C·m or 0.97D). As expected, this mixture shows the greatest deviation from the guidelines in that four parameters are needed while the guidelines call for only one. Also, uniquely among the nonazeoptropic mixtures, carbon dioxide + hydrogen sulfide has a negative value of C_X. This is somewhat contradictory to the philosophy of Equation 61, in that the "amount of phase change", A_0, is expected intuitively to increase, rather than decrease, with Δx. Again no microscopic explanation is possible at present, although the somewhat similar azeotropic mixture ethane + hydrogen sulfide also calls for a negative C_X (see Section XII).

For a particular range of α_{2m}, quality fits of the VLE surface in P-T-x-y space can be obtained with fewer parameters than quality fits of coexisting densities. The correlations of n-butane + n-octane by Rainwater and Moldover,[4] and of ethane + n-butane by Rainwater and Williamson,[15] were accurate in P-T-x-y space, but there were discrepancies, particularly in the liquid densities. For those correlations, only the three parameters C_H, C_X, and C_Z were utilized. The revised correlations of Table 3, with C_Y and C_R also included, yield accurate density predictions as well. At present, problems with coexisting densities remain for the listed correlations that use H_1 and for mixtures such that $\alpha_{2m} > 0.25$; these hopefully can be alleviated with additional parameters that have yet to be chosen.

It is obviously not possible to show graphically in this article the comparison of correlation and experiment for all of the fits in Table 3; the reader is referred to previously published graphs;[4,5,15-18,108] however, as representative samples of such graphs, Figures 13A and 13B show the revised correlation of ethane + n-butane and Figures 14A and 14B show the revised correlation of n-butane + n-octane. These should be compared against the earlier fits of the same mixtures in Figure 4 of Reference 15 and Figure 6 of Reference 4. The revised fits with use of the parameters C_Y and C_R show considerable improvement in T-ρ space, particularly for the liquid densities, while retaining the same high-quality fit in P-T space. It should also be noted that the figures in References 4 and 15 display the smoothed data of the journal publications,[120,142] while the figures in this chapter show the original or raw data. The three dew points lowest in pressure on the $x(n\text{-}C_4H_{10}) = 0.4631$ cannot be correlated and are probably in error; the author posits that some chemical decomposition at this relatively high temperature may have occurred.

As an example of a correlation to isothermal data, Figures 15A and 15B show the mixture carbon dioxide + propane as measured and correlated by Niesen and Rainwater.[108] For many years literature VLE data for this mixture in the critical region were seriously in error, despite the mixture's importance; in Table 1 the two components happen to be those

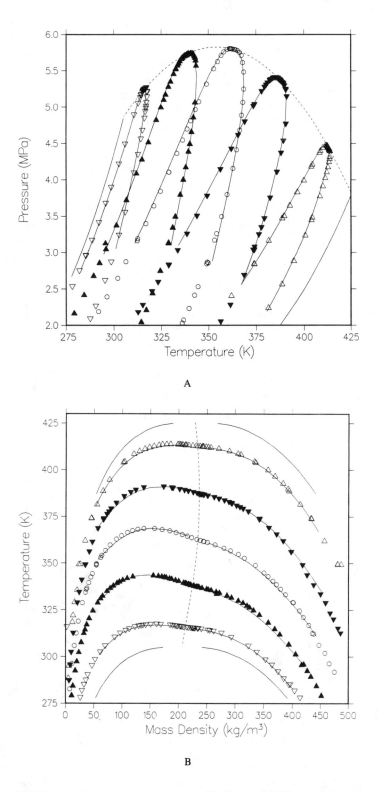

A

B

FIGURE 13. Coexistence surface and modified Leung-Griffiths correlation of
ethane + *n*-butane, with data from Reference 121: (A) P-T space, (B) T-ρ space.
The ethane mole fractions, right to left in (A) and top to bottom in (B), are 0.1749,
0.4510, 0.6577, 0.8218, and 0.9471. The use of the parameters C_R and C_Y greatly
improves the correlation of liquid densities (cf. Figure 4 of Reference 15).

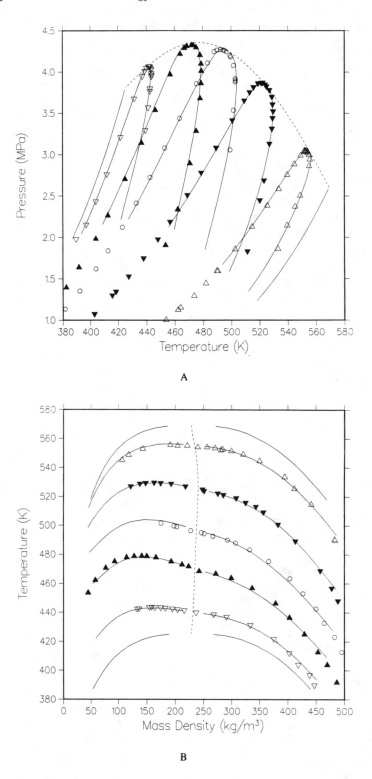

FIGURE 14.   Coexistence surface and modified Leung-Griffiths correlation of
n-butane + *n*-octane, with data from Reference 142: (A) P-T space, (B) T-ρ
space. The *n*-butane mole fractions, right to left in (A) and top to bottom in (B),
are 0.1823, 0.4631, 0.6707, 0.8183, and 0.9461. The use of the parameters C_R
and C_Y greatly improves the correlation of liquid densities (cf. Figure 6 of Reference 4).

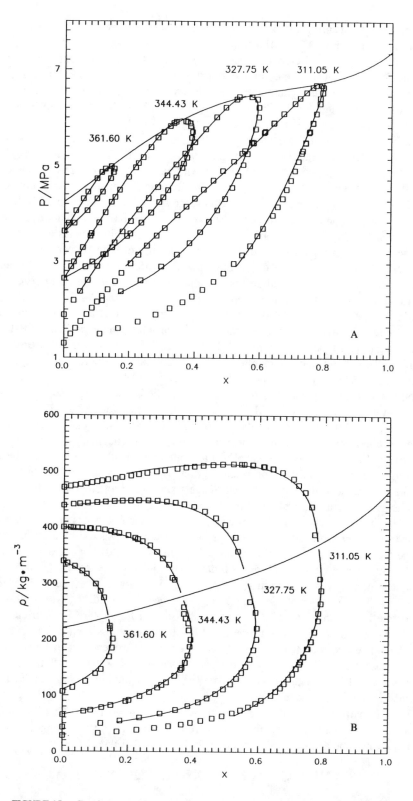

FIGURE 15.   Coexistence surface and modified Leung-Griffiths correlation of carbon dioxide + *n*-propane; data and correlation from Reference 108: (A) Isotherms in P-x space, with temperatures as indicated; (B) Isotherms in ρ-x space.

most frequently appearing in the primary mixture survey. The critical locus is lower in pressure by as much as 0.785 MPa from that of Poettmann and Katz[109] and of Reamer et al.[110] Such a significant discrepancy was pointed out in the critical locus measurements of Roof and Baron,[420] but the remeasurement of this mixture clearly called for by such a large discrepancy was not implemented until the recent work by Niesen and Rainwater. The high quality of the modified Leung-Griffiths fit is evident from the figures; in Section IX, a detailed quantitative analysis is presented.

The list of mixtures in Table 3 through carbon dioxide + *n*-butane contains most mixtures in Table 2 under the heading ''Normal Mixtures with Density Measurements (Small to Moderate α_{2m})''. The exceptions are mixtures with extremely narrow dew-bubble curves or $\alpha_{2m} < 0.05$: carbon dioxide + nitrous oxide,[99] perfluorocyclohexane + perfluorobenzene,[149] and R22 + R12.[128] To correlate these mixtures, the special transformation of Reference 26, as discussed in Section XII, is required. Also exempted are mixtures with highly questionable data, such as ethylene + isobutane[84] and ammonia + acetonitrile.[85] As discussed earlier, the experimental figures for these mixtures are inconsistent with proper critical behavior. Correlations have been developed that minimize the differences between the data and the model predictions; however, unlike the correlations listed in Table 3, these correlations cannot be recommended as accurate representations of the actual VLE surfaces.

Additional exceptions are sulfur dioxide + water[148] with LLE in the proximity of the VLE region, and ethylene + trifluoropropylene[98] for which the experimental P-T diagram strongly suggests a type 2 structure with LLE, even though there are no direct LLE measurements. This is yet another manifestation of the hydrogen-fluorine anomaly. The final exception is *n*-butane + acetic acid;[147] since acetic acid has a strong tendency to form dimers,[421] its coexisting densities do not obey Equation 3 and a special model including association would be necessary. Alcohols also exhibit dimerization but to a lesser degree.

In the course of this research, the author and associates have encountered two opportunities to contribute to projects on mixtures with less thorough measurements. The first was a study of isobutane + isopentane[14] for a geothermal plant design in which the working fluid was a mixture with 10% isopentane. As input to the correlation, VLE measurements very close to the critical locus by Levelt Sengers were used together with dew and bubble points generated by a classical equation of state contributed by Gallagher.[419] The modified Leung-Griffiths calculation yielded excellent agreement with both sets of input, whereas the classical model by itself led to a critical pressure locus that was too high by 0.1 MPa.

The second project was a study of air by R. T. Jacobsen and colleagues at the University of Idaho. Surprisingly, there are little VLE data on nitrogen + oxygen[73,415,418] in the extended critical region, and even less on nitrogen + argon[73,414] and argon + oxygen.[73] Measurements on the dew-bubble curve of air itself have twice been reported,[416,417] as well as critical temperature loci of the three binary mixtures.[300] Several Leung-Griffiths correlations have been developed from this limited input, and ancillary equations for the dew-bubble curve of air have been constructed, first by treating air as a nitrogen-oxygen binary mixture,[16] second by using an explicit ternary correlation[422] (see Section XIV), and third by optimizing the binary correlation to the actual air data[416,417] with a nonlinear Simplex method,[423] as explained in the following section.

While initially the author has chosen to test the modified Leung-Griffiths model only against the most thorough data, the results from the special projects suggest that reliable correlations may be achievable from relatively limited VLE data.

VIII. CORRELATION METHODOLOGY

Most of the correlations in Table 3 have been obtained by graphical methods, in which parameters have been adjusted until agreement between experiment and model predictions has been attained according to visual criteria. The algorithms for constructing isopleths or isotherms as predicted by the model are relatively straightforward. For isopleths, which are somewhat more naturally analyzed within this approach, first a geometric series in $|t|$ is constructed on the interval $-0.1 < t < 0$. Then, for each value of t, Equation 38 is numerically inverted to find values of ζ, one for the liquid and one for the vapor state, so that the composition is predicted to be the experimental value. In some cases, the asymptotic analysis of Reference 26 can be exploited to obtain a good initial guess and to improve convergence. With an array of t and ζ, temperature is calculated from Equation 27, pressure from Equation 55, and density from Equation 44.

For the calculation of isotherms, $\zeta_c(T)$, the value of ζ on the critical locus at the desired temperature, must first be found. This is done by inverting Equation 49 with the Newton-Raphson method to obtain the critical value of x, and then using Equation 28 or 68. Dew and bubble points along an isotherm in the extended critical region are then obtained by choosing an array of ζ values from ζ_c to ζ_{max}, where

$$\zeta_{max} = max[1, \zeta_c(10T/9)] \tag{71}$$

From such an array, t is obtained algebraically for each point from Equation 27, P from Equation 55, and the liquid and vapor densities and compositions from Equations 44 and 35, respectively. Thus, calculation of isotherms is somewhat simpler in that less numerical inversion is required.

Unlike the simple geometric series in $|t|$ for isopleths, however, the array of ζ must be carefully chosen so that all calculated curves are smooth. In P-x diagrams such as Figure 9C, the isotherms have a wide range of curvatures near their maxcondentherm points, and a greater density of calculated points is needed in the region of greatest curvature. The author's technique is to derive and to utilize an approximate expression for ζ_2, defined as the value of ζ other than ζ_c such that y is equal to the critical composition, in other words, the point vertically directly below the critical point on an isotherm of Figure 9C. An analysis similar to that of Reference 26, with the assumption of a critical locus that "locally" is linear in pressure, temperature, and composition, yields

$$\zeta_2 - \zeta_c = (C_1\alpha_2)^{\frac{1}{1-\beta}} \left(\frac{dT_c}{d\zeta}\right)^{\frac{\beta}{1-\beta}} (T_c)^{-\frac{\beta}{1-\beta}} \tag{72}$$

where C_1, α_2, and T_c are evaluated at $\zeta = \zeta_c$.

This approximate expression typically agrees with the exact interval $\zeta_2 - \zeta_c$ to within 10%, which is adequate for the present purposes since only a scale is needed to construct smooth graphs. The array of ζ points is chosen to be most dense in this interval and particularly near ζ_c, and less dense within the interval $\zeta_2 < \zeta < \zeta_{max}$. Care is taken to detect the occasional instance where $\zeta_{max} < \zeta_2$, and for $0.9\,T_{c1} < T < T_{c1}$, a uniform distribution of ζ intervals suffices and is used instead. These techniques led to the smoothly plotted isotherms of figures in previous references.[5,16,18,108]

The adjustable parameters have been incorporated so that their variations lead to predictable changes in the model dew-bubble curves, as discussed in some detail by Rainwater and Moldover.[4] For example, on a P-T plot such as Figure 9A, lowering C_H moves the model curves to the left without change of width, while raising C_X narrows the curves. Also, while one usually starts by linearly fitting the critical locus as stated by the experimentalist to Equations 49 to 51, variation is allowed for in the critical locus parameters, particularly those of the density locus, as measurements extremely close to the critical point are typically the least precise. With some practice and acquired skill, a correlator can obtain quality visual fits with five parameters (all but H_1) if the critical locus is well characterized; however, inclusion of the sixth parameter and the need for additional critical locus parameters test the patience of even an experienced correlator.

Recently, Lynch[424] has developed a version of the code that performs formal nonlinear optimization by the Simplex technique.[425] For the mixtures in Table 3 with largest α_{2m}, correlations were obtained by the Simplex program as noted. In any such analysis, there is a need to define an "objective function" or "distance" between the experimental and model coexistence curves to be minimized. Lynch's initial objective function corresponds to the distance minimized visually. On a P-T plot, there is a scale for pressure, P_{scale}, and a scale for temperature, T_{scale}, corresponding to the scale of a figure such as Figure 9A, including the critical locus, vapor pressure curves, and dew-bubble curves in the extended critical region. Lynch calculates the model (isopleth) dew-bubble curve, finds the two points closest to a given experimental point, and calculates the perpendicular distance (which naturally depends on the figure scale) from the experimental point to the line joining the two closest model points. This is, in effect, the perpendicular distance from the measured point to the model dew-bubble curve. The objective function is then

$$\sum\left[\left(\frac{\Delta P}{P_{scale}}\right)^2 + \left(\frac{\Delta T}{T_{scale}}\right)^2\right]$$

where ΔP and ΔT are the intervals along the perpendicular drop, and the sum is over all data points. Similar objective functions are defined for P-x, T-ρ, and ρ-x plots. The initial guess for the model parameters is obtained by an attempted visual fit. For optimization of the model to P-T-x-ρ data on isopleths, the product of the P-T and T-ρ sums is minimized, and similarly for isothermal data with the product of the P-x and ρ-x sums.

There are many different possible choices for an objective function; a more conventional approach might be to treat T and x as independent variables and P, y, and the densities as dependent variables, and to minimize the sums of the squares or absolute values of the deviations. Some more involved methods incorporate the stated experimental uncertainties and examine deviations in various different projections of thermodynamic space.[426] Further study of the most appropriate objective function is warranted, although careful judgment should be employed as such analyses can easily be overdone. It is hoped that the use of any reasonable objective function will give largely the same result, and that the model is not especially sensitive to the exact choice.

As is evident from Table 2, there are a large number of mixtures for which P-T-x-y data are available but coexisting density data are not. Since the mixture critical density locus is in principle required as input to the model, the author did not try to fit any of these mixtures by visual methods. With the Simplex program, as noted in Table 3, the critical density parameters as well as the mixture parameters can be allowed to float and, in some cases, obtain quality P-T-x-y fits. The coexisting densities obtained from the resulting model, however, are speculative and may be quite wrong.

The correlations listed in Table 3 for $\alpha_{2m} > 0.35$ (as well as that of carbon dioxide + neopentane) were obtained by the Simplex program, while those for $\alpha_{2m} < 0.35$ were obtained by visual methods. As an example of a Simplex correlation, Figure 16 shows the fit to ethane + benzene.[162] At this stage six mixture parameters and four parameters in the critical locus are needed, and visual correlations become too tedious, although an approximate visual correlation is useful as a first guess for the Simplex algorithm. In P-T space, the high standards of correlations of simpler mixtures are retained, while in T-ρ space there are some minor deviations from the experimental shapes of the isopleths. Nevertheless, this represents a substantial improvement over earlier correlations of mixtures with large α_{2m} such as propane + *n*-octane,[3] particularly for mixtures rich in the more volatile component and thus relevant to the study of supercritical solubility.

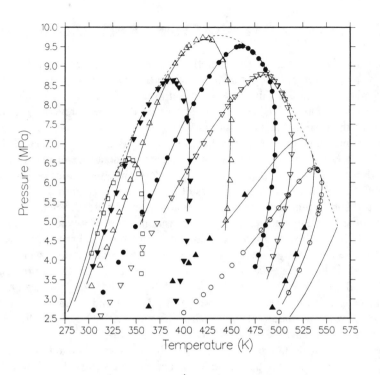

A

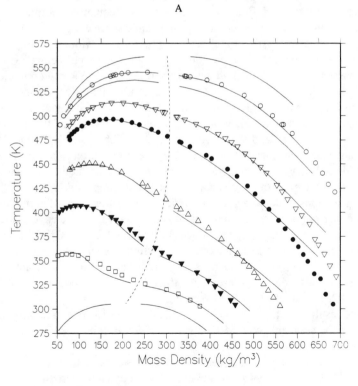

B

FIGURE 16. Coexistence surface and modified Leung-Griffiths correlation generated by the Simplex algorithm for ethane + benzene, with data from Reference 162: (A) P-T space, (B) T-ρ space.

IX. QUANTITATIVE ANALYSIS OF FITTING ACCURACY

A quantitative measure of the overall agreement between experiment and model predictions is useful not only for optimization of the fit, but also to compare the quality of the
correlation against those obtained by other methods. One measure of goodness of fit is the
distance between model prediction and experiment described previously as the objective
function for the current Simplex procedure; however, this is not standard and depends on
an arbitrary pressure, temperature, or density scale. The typical procedure in the recent
correlation literature has been to take T and x as independent variables and P, y, and (when
density data are available) ρ_l and ρ_v as dependent variables. Such an analysis requires
isothermal VLE data with coexisting pairs of x and y, and precludes data on isopleths since
the dew and bubble points do not form coexisting pairs. The tendency to overlook constant-
composition data is unfortunate, in that the measurements of that type from Kay's laboratory
are typically more dense and more self-consistent in the critical region.

Nevertheless, recently Niesen and Rainwater[108] have undertaken such an analysis for
the carbon dioxide-propane data as measured by Niesen. As illustrated previously (Figures
15A and 15B), the modified Leung-Griffiths model can be seen graphically to provide an
excellent fit to this mixture. The comparison by Niesen and Rainwater of correlations of
this mixture with the modified Leung-Griffiths model, the Soave-Redlich-Kwong equation,[30,31] and the package DDMIX[28,29] are reviewed below.

The advantages of the state-of-the-art DDMIX package should be noted here. The
package at present includes the 18 fluids indicated by footnote a in Table 1. Pure-fluid
thermodynamic surfaces are fitted by a 32-term BWR equation; the many-parameter correlation ensures an accurate fit over all thermodynamic space except, to some degree, within
the critical region, since it still is a classical representation without scaling-law exponents.
The extended corresponding states technique with "exact shape factors" is employed[29] so
that the BWR representation of fluid 1 is mapped into the BWR representation of fluid 2.

Unlike the present modified Leung-Griffiths model, DDMIX provides calculations of
the one-phase supercritical region and of one- and two-phase states at low pressure outside
the extended critical region. DDMIX also calculates, for one- or two-phase states, such
thermophysical quantities and the entropy, enthalpy, Gibbs and Helmholtz free energies,
specific heats, and even transport properties, whereas the current modified Leung-Griffiths
codes provide only pressure, temperature, and coexisting densities and compositions. Furthermore, DDMIX can accommodate an arbitrary multicomponent mixture, whereas Leung-
Griffiths calculations to date have been restricted to binaries and a few ternaries.

The use of many-parameter equations of state requires an extensive amount of data for
each pure fluid in the package, and since the modified Leung-Griffiths model for VLE
requires only the coexistence surface of the pure fluid, the author has been able to examine
mixtures of certain compounds such as alcohols, ethers, and refrigerants in advance of
DDMIX using that model. At present, for thermodynamic points too close to the critical
locus, the DDMIX algorithm for phase equilibrium fails to converge, and the program
provides the result after a given number of iterations together with an error message. The
algorithm for the simpler, cubic SRK equation similarly fails to converge if too close to a
critical point.

In practice, DDMIX calculates the P-T-x-y coexistence surface with the Peng-Robinson
equation,[13] also cubic, and densities and other thermodynamic variables are then calculated
by the extended corresponding states method with the BWR equation as reference. A much
more time-consuming phase equilibrium calculation from the mixture BWR equation would
in principle yield slightly different results, but in practice the differences appear not to have
been significant. Another important point of comparison is that in the modified Leung-
Griffiths model a fit of the experimental locus is included as input, whereas in DDMIX the

critical locus (as defined by the EOS) can be predicted from that equation and the mixing rules.

In order to justify development of the modified Leung-Griffiths model as an engineering project, it must be demonstrated to be superior to state-of-the-art classical methods, at least within some regime close to the critical locus. Figures 17A to 17D are deviation plots of correlations of the carbon dioxide-propane data due to Niesen and Rainwater[108] by means of DDMIX, the Soave-Redlich-Kwong equation, and the modified Leung-Griffiths model. These correlations have not been completely optimized; the binary interaction parameter used in DDMIX has not been adjusted to the data (although it agrees well with that used in the Soave-Redlich-Kwong equation, which was optimized for each isotherm), and the modified Leung-Griffiths model has been visually optimized rather than calculated by the Simplex program. Nevertheless, the overall pattern should be representative of the quality of fit expected by these various methods. The ordinate on these figures is $P_c - P$ along each isotherm, a measure of distance from the critical point; the variable t could equally well have been used. As noted earlier, the independent variables are T and x.

Figures 17A and 17B are deviation plots for pressure and vapor composition, respectively. In both plots, the deviations for the modified Leung-Griffiths model are randomly distributed about zero while those of DDMIX and the Soave-Redlich-Kwong equation are more predominantly biased in the negative direction. Again, note that for DDMIX these independent variables are in fact predicted from a Peng-Robinson equation. The negative deviations are expected from a model in which $\beta = 1/2$ rather than $\beta = 1/3$, in that the "parabolic" dew-bubble curves with $\beta = 1/2$ should lie "inside" those with a flatter critical region, as found from nonclassical exponents.

The deviation plot for liquid densities, Figure 17C shows the well-known result that a cubic EOS such as Soave-Redlich-Kwong predicts coexisting liquid densities very poorly, and much lower than experimental results. In other work this problem has been partially solved by a volume shifting technique[427,428] that improves predictions quantitatively but retains the incorrect classical exponents. DDMIX and the modified Leung-Griffiths model both perform well, with the former yielding somewhat better densities for $P_c - P > 2$ MPa, and the latter better close to the critical locus, as expected. Again, within DDMIX the P-T-x-y surface is determined from a Peng-Robinson equation, but the coexisting densities are calculated with a many-parameter EOS.

Figure 17D is the deviation plot for vapor densities. Both the modified Leung-Griffiths model and DDMIX show randomly distributed deviations about zero, the percent deviations being greater for DDMIX. The Soave-Redlich-Kwong equation greatly underpredicts near the critical point and is consistently biased negatively.

This example shows that overall and in a correlative rather than a predictive mode, within the extended critical region the coexistence surface as predicted by the modified Leung-Griffiths model is superior to that found from classical methods currently in use. The predictions of DDMIX are not bad, however, and may be adequate for most practical applications. As the current DDMIX is a "hybrid" method, another possible method, not yet attempted, would be to use the P-T-x-y surface as determined from the modified Leung-Griffiths model (instead of the Peng-Robinson equation) and the many-parameter EOS for other thermodynamic variables. This would probably improve predictions over the current DDMIX near the critical locus while providing a means of calculating quantities such as enthalpy or entropy not available from the modified Leung-Griffiths model in its present form. Such a hybrid method, however, would not describe the divergences in one-phase specific heats that are characterized by nonclassical exponents.

A comprehensive study of the accuracy of modified Leung-Griffiths and classical correlations is in order, where the more abundant data along isopleths as well as those along isotherms should be analyzed. Every new or modified EOS and mixing rule, of which so

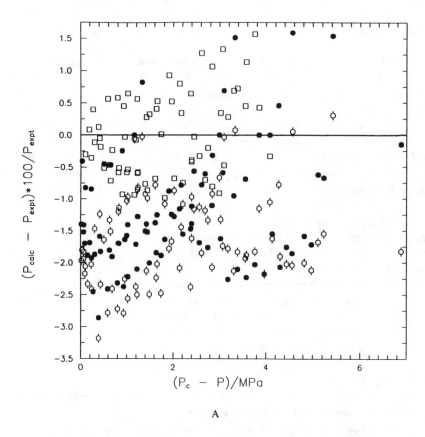

A

FIGURE 17. Deviation plots for the correlation of carbon dioxide + propane: □, Modified Leung-Griffiths model; ●, Soave-Redlich-Kwong equation; ○, DDMIX, with P_c-P (a distance from the critical locus) as the independent variable: (A) percent pressure deviations, (B) absolute vapor composition deviations, (C) percent liquid density deviations, (D) percent vapor density deviations.

many have been recently proposed, cannot be tested, but those that claim the most accurate high-pressure fits can be compared against. Such a study, however, should be deferred until the problems associated with convergence of classical phase-equilibrium algorithms in the critical region have been resolved (see Section XIV).

X. NONLINEAR FITS OF CARBON DIOXIDE-PENTANE MIXTURES

To justify the modified Leung-Griffiths model as even a good preliminary technique for SCF technology, it is necessary to demonstrate its utility for at least some mixtures with carbon dioxide as solvent and with a two-phase region in the supercritical domain of carbon dioxide, i.e., in the region immediately above the critical pressure and temperature. The method of visual fitting is useful only up to carbon dioxide + *n*-butane,[17] which has only a very small two-phase region in the domain of supercritical carbon dioxide. To find the next *n*-alkane solute with a thoroughly measured VLE surface including coexisting densities, one must go to carbon dioxide + *n*-decane,[9] well outside the range of the current six-parameter model.

Fortunately, the recently developed Simplex program can develop good fits beyond carbon dioxide + *n*-butane. The Cornell group of W. B. Streett and J. A. Zollweg have also measured at least eight isotherms each on mixtures of carbon dioxide with neopentane,[212]

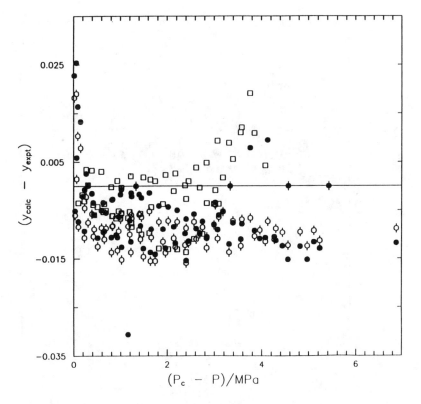

FIGURE 17B.

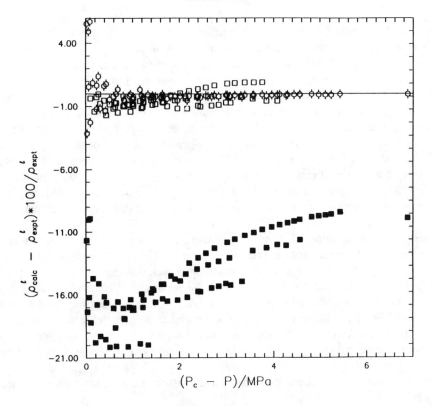

FIGURE 17C.

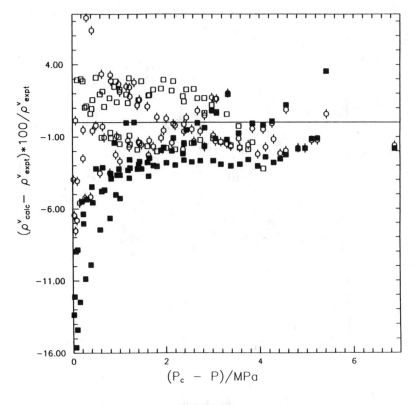

FIGURE 17D.

isopentane,[214] *n*-pentane,[217] and cyclopentane.[212] All of these mixtures have a substantial two-phase region in the supercritical domain of carbon dioxide.

Figures 18, 19, and 20A show the data and modified Leung-Griffiths correlations for carbon dioxide with neopentane, *n*-pentane, and cyclopentane, respectively. The neopentane correlation ($\alpha_{2m} = 0.311$) is quite satisfactory, as is the isopentane correlation ($\alpha_{2m} = 0.359$) which is not shown. Accurate correlation of supercritical solubility is equivalent to accurate representation of the dew curves in the upper right part of the diagram, since the statement "the dew curve of the carbon dioxide-pentane mixture for temperature $T > T_c$ (CO_2) passes through the point [$P > P_c (CO_2)$, x]" is eqivalent to the statement "the solubility of pentane in supercritical carbon dioxide at P, T is $(1 - x)$ mole fraction pentane". This is the saturated solution for bulk supercritical carbon dioxide at that pressure and temperature; any further pentane would lead to a relatively pentane-rich liquid phase that would fail to dissolve.

The correlation to carbon dioxide + *n*-pentane deviates significantly from experiment only at the highest mixture critical pressure and on bubble curves near subcritical carbon dioxide; here $\alpha_{2m} = 0.370$. These correlations compare well with the neopentane, isopentane, and *n*-pentane mixture data of Robinson and co-workers[213,215] except that their highest isotherms are in all cases shifted from those correlated by the author for reasons not presently understood.[18] For the cyclopentane mixture, the correlation is best at high temperature and shows noticeable deviations from experiment in the supercritical carbon dioxide domain; it appears that some "fine-tuning" is needed with the insertion of some appropriate but unknown additional parameter. This represents the upper limit of ($\alpha_{2m} = 0.502$) beyond which the current six-parameter modified Leung-Griffiths model cannot provide adequate fits.

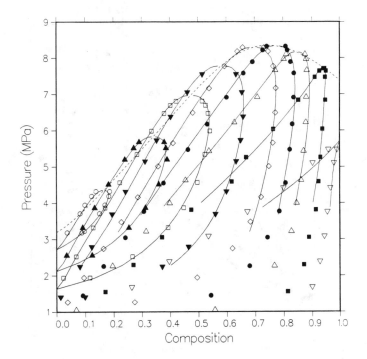

FIGURE 18. Coexistence surface and modified Leung-Griffiths correlation for carbon dioxide + neopentane in P-x space; data from Reference 211. Temperatures of isotherms range from 292.99 K at right to 423.55 K at left.

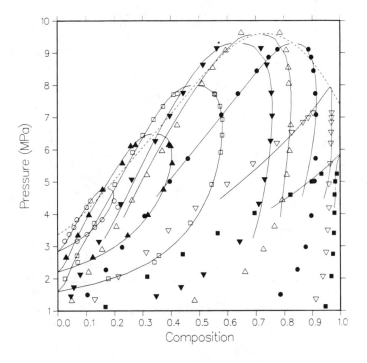

FIGURE 19. Coexistence surface and modified Leung-Griffiths correlation for carbon dioxide + n-pentane in P-x space; data from Reference 216. Temperatures of isotherms range from 294.09 K at right to 458.54 K at left.

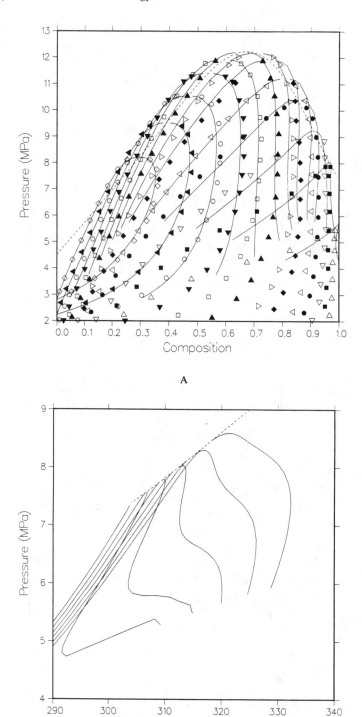

A

B

FIGURE 20. (A) Coexistence surface and modified Leung-Griffiths correlation for carbon dioxide + cyclopentane in P-x space; data from Reference
212. Temperatures of isotherms range from 293.14 K at right to 478.17 K at
left. (B) The same correlation in P-T space near the carbon dioxide critical
point: dew-bubble isopleths, from left to right 1, 2, 3, 4, and 5% cyclopentane.
The more dilute mixtures show a pattern of ''double retrograde vaporization''.

Figure 20B shows the same fit of carbon dioxide + cyclopentane in Figure 20A transformed to P-T space, at the carbon-dioxide-rich side of the diagram. For 5% cyclopentane, the dew-bubble isopleth shows the normal pattern of Figure 9A, with a maxcondenbar point and a maxcondentherm point or point of maximum temperature; however, for more dilute solutions the isopleths develop a point of minimum temperature below the points of maximum temperature. This topology is consistent with the "double retrograde vaporization" found by Chen et al. for methane + n-butane[24] and methane + n-pentane.[25] By contrast, the fits for carbon dioxide with neopentane and isopentane show no pattern of double retrograde vaporization, and that for carbon dioxide + n-pentane shows such a pattern only over an extremely small temperature range.

The significant result here is not that the model can produce such minimum temperature points, since it was shown earlier that the model, if used with insufficient care, can lead to thermodynamic instabilities and unphysical results such as negative vapor densities. What may well be significant is that when optimized by the Simplex program to abundant data, a double retrograde condensation pattern emerges at about the point where it also appears in nature. The reasoning is as follows. Different normal alkane mixtures with the same carbon number ratio, e.g., ethane + n-butane of Figure 13 and n-butane + n-octane of Figure 14, have nearly the same VLE diagram in reduced units and the same α_{2m}, as seen from Table 3. Furthermore, again as seen from Table 3, mixtures of ethane + C_nH_{2n+2} have nearly the same α_{2m} value as mixtures of carbon dioxide with $C_{n-1}H_{2n}$. Therefore, if double retrograde vaporization emerges between methane + propane and methane + n-butane, it should also emerge between ethane + n-hexane and ethane + n-octane, and also between carbon dioxide + n-pentane and carbon dioxide + n-heptane.

This behavior has yet to be fully analyzed. It is hoped that with mathematical analysis similar to that of Section XII, such patterns may be predictable from parameters such as A of Equation 66. Such understanding may be a key to extending the model and to modeling the large enhancement of supercritical solubility in mixtures of more dissimilar components, which is closely related to double retrograde condensation phenomena.

XI. CARBON DIOXIDE-METHANE: A SPECIAL CASE

The only detailed independent critique of the modified Leung-Griffiths model to date has been that of Al-Sahhaf et al.,[384] who tested an earlier, less developed version of the model. They examined the mixtures propane + n-butane, carbon dioxide + n-butane, nitrogen + carbon dioxide, and methane + carbon dioxide. Correlations for the first two mixtures were largely successful, as noted in Table 3. Nitrogen + carbon dioxide is quite clearly a type 3 mixture,[303] and thus in principle cannot be correlated with the model in its present form.

The fourth mixture examined by Al-Sahhaf et al. was carbon dioxide + methane, and since this on first examination appears to be a well-behaved type 1 mixture, the failure of the model to correlate this mixture, as found by them and verified by Rainwater, has been a source of concern. Figure 21 shows a P-x diagram for this mixture with isothermal data and a preliminary correlation using only $C_H = -7$. The only "primary source" for this mixture that spans the entire phase diagram is by Donnelly and Katz,[192] but these data were not used because their critical compositions as a function of critical temperature differed substantially from those of other sources, and there is evidence that in the work of the Michigan group at that time, mixture compositions were not measured accurately. Elsewhere[18] Rainwater et al. have shown that the carbon dioxide + n-butane data of Poettmann and Katz[109] is consistent with many more recent sources only if the experimentally stated compositions are shifted by up to 3%, and Lynch[424] has shown that the data of Williams[97] for ethylene + n-butane can be fitted accurately only if some of the stated compositions are similarly shifted.

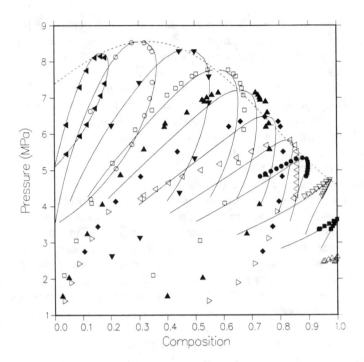

FIGURE 21.   Coexistence surface of carbon dioxide + methane in P-x space
with an unsuccessful attempted correlation by the modified Leung-Griffiths
model. Isotherms, from left to right, are from Reference 343, 288.15 K, 273.15
K, 253.15 K; from Reference 429, 240 K; from Reference 95, 230 K; and
from References 430 and 431, 219.26 K, 210.15 K, 203.15 K, 193.15 K,
183.15 K, and 173.15 K. The correlation yields good agreement at the highest
and lowest temperatures, but yields considerably smaller dew-bubble curve
widths at intermediate temperatures. The experimental curves for T < 230 K
terminate from below on an SLV locus.

Instead, VLE data are taken from a number of different but mutually consistent
sources;[95,343,429-431] other data have been reported by Somait and Kidnay.[432] The preliminary
correlation shows a very unusual pattern not present in any of the other mixtures in Table
3. Carbon dioxide + methane is apparently a normal type 1 mixture, with $\alpha_{2m} = 0.214$
according to the preliminary fit, and indeed from Figure 21 the correlations at the highest
temperatures are accurate and superior to those previously reported from classical EOSs.
(Reference 247, p. 400, and Reference 303); however, at lower temperatures, particularly
in the range 200 to 230 K, the model dew-bubble curves are narrower (by as much as a
factor of 2) than the experimental curves. Therefore, α_2 is seriously underpredicted. The
expected pattern for mixtures of this type is that with only C_H, the model curves are initially
too wide and need to be made narrower by using a positive C_X. This pattern was first noticed
by Al-Sahhaf et al.,[384] and the present correlation has independently confirmed it.

Our conjectured explanation is as follows. Figure 22 shows the P-T diagram for this
mixture. Like Figures 6F and 6G, a solid phase invades the VLE surface; it almost cuts off
the critical locus, as is the pattern for carbon dioxide + oxygen[264] and carbon dioxide +
argon.[265] In Figure 21 the VLE data for isotherms with T ≤230 K terminate as the pressure
is lowered, and these termination points are the points of intersection with SLV locus. This
locus has been measured in detail by Donnelly and Katz[192] and by Davis et al.[433] Such a
pattern is largely due to the well-known, anomalously large triple-point pressure (0.5135
MPa) of carbon dioxide. Of the primary fluids of Table 1, only one other fluid has a triple
point pressure (just slightly) above atmospheric pressure, acetylene (0.126 MPa).

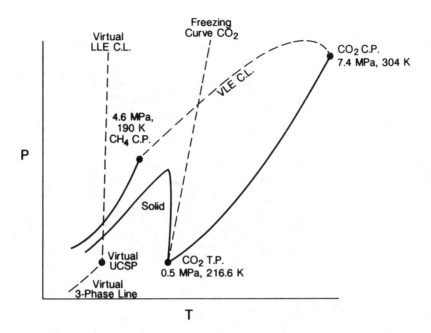

FIGURE 22. Partially conjectured P-T phase diagram for carbon dioxide + methane (schematic). According to References 192 and 433, the vapor pressure curves, plait point locus, and SLV locus are as shown. It is conjectured that if freezing could be suppressed, the mixture would show a type 2 structure with a "virtual" LLV locus, UCEP, and consolute point locus, the latter at about 180 K.

Such an SLV locus leads to a solid-liquid phase transition surface that depends on composition. Our conjecture is that below this surface there is for the fluid description the structure of LLE, with a consolute point locus at about 180 K. This mixture is believed to display liquid-liquid immiscibility as the temperature is lowered, except that it freezes first. To the author's knowledge, no actual LLE has been observed for this mixture and it may not occur in nature. It might be detectable experimentally by studying ternary mixtures, including a judiciously chosen third component, where observable LLE would disappear at some small composition of the third fluid but extrapolation of the phase diagram to zero concentration of the third component in the solid region would yield the "virtual" consolute point.*

As evidence for this conjecture, note the following:

1. According to the DDMIX program,[28,29] after suppression of a warning message about the probable entering of a solid phase, for this mixture at 180 K and 5 MPa there is a state of two-phase equilibrium where the first phase has a density of 28.02 kmol/m^3 and methane composition of 0.166, and the second phase has a density of 23.30 kmol/m^3 and a methane composition of 0.653. From the densities and compositions, clearly this is a state of liquid-liquid rather than liquid-vapor equilibrium.
2. A Peng-Robinson correlation of this mixture by Knapp et al. (Reference 247, p. 400) clearly shows at 199.82 K a tendency of the bubble curves toward the development of an inflection point, and therefore a consolute point locus, as shown in Figure 23.
3. Miller and Luks[68] have presented in their Figure 3 a diagram similar to Figure 7 showing the pattern of LLV loci for the mixture family of carbon dioxide with a normal alkane. Actual LLE is observed only for carbon numbers of 7 and above, but extrap-

* The author thanks Graham Morrison for first pointing out this possibility.

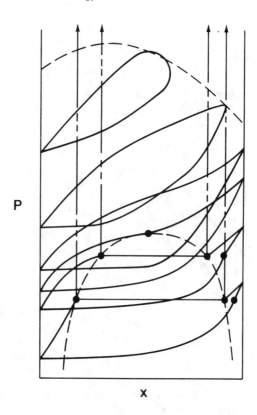

P

X

FIGURE 23. Phase diagram in P-x space for the isotherms of a type 2 mixture. The upper dashed curve is
the plait point locus; the lower dashed curve is the three-
phase locus that terminates from above at the UCEP (●).
The isotherms at high temperature (and pressure) follow
the pattern of Figure 9C, but the isotherm at the temperature of the UCEP has a point of inflection at the
UCEP on the liquid side. At lower temperatures, there
are two liquid phases and one vapor phase in coexistence
(●) on the three-phase locus. The vertical lines rising
"indefinitely" from the liquid points describe liquid-
liquid coexistence in the idealized limit that the LLE
surface is independent of pressure and composition; otherwise these lines would not be exactly vertical. The
dew curve at the three-phase locus has a change of slope
in accordance with the 180° rule (see Reference 65).

olation of their critical endpoint locus to carbon numbers of 1 and 2 yields a consolute
point for the methane mixture at about 180 K and for the ethane mixture at about 185
K. In fact, the possibility of "virtual" LLE for carbon dioxide + ethane has been
pointed out explicitly by Rowlinson and Swinton (Reference 32, p. 209).

The following section presents some evidence from a study of azeotropic mixtures that
the presence of LLE, including "virtual" LLE, can significantly distort the VLE surface in
P-T-ζ space; see the discussion following Equation 83. Furthermore, this distortion is manifested in a nonlinear dependence of x vs. ζ on the critical locus and a strong enhancement
of dx/dζ and α_2 near the value of ζ at which the upper critical endpoint occurs.

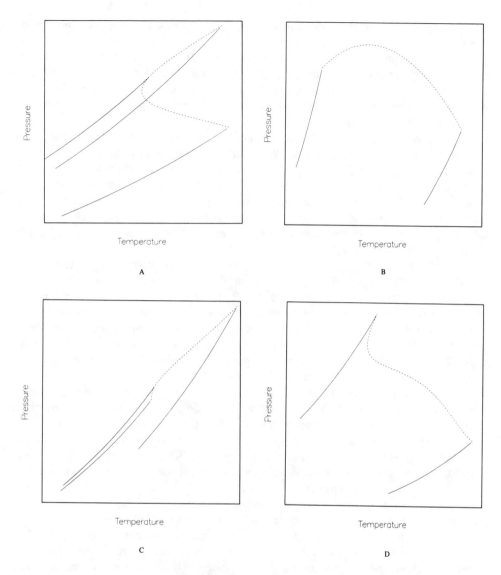

FIGURE 24.   Relation of azeotropy to the presence of a minimum mixture critical temperature, where the dashed curve is the critical locus and the solid curves are vapor pressure or azeotropic loci: (A) azeotropic mixture with minimum in T_c; (B) nonazeotropic mixture with no minimum in T_c; (C) the unusual example of an azeotropic mixture without a minimum in T_c; (D) the unusual example of a nonazeotropic mixture with a minimum in T_c.

XII. AZEOTROPIC MIXTURES

Like the nonazeotropic or normal type 1 diagrams of Figures 9A to 9D, the phase diagrams for azeotropic mixtures have been subject to a fair amount of misunderstanding. The diagram for a prototypical azeotropic mixture is shown in Figure 24A. The two components have nearly the same critical temperature but substantially different critical pressures and the critical locus has a crescent shape with a temperature minimum as shown. The key point to recognize in understanding such diagrams in the critical region is that the azeotrope, or locus of two-phase points at which the liquid and vapor compositions are identical, must be tangent to the critical locus at its point of highest pressure.[3,32] As a first approximation, in P-T space the azeotrope is parallel to the pure-fluid vapor pressure curves as are the loci of constant ζ within the modified Leung-Griffiths model. While the azeotrope is not a locus of constant ζ, it may initially be viewed as such on a P-T diagram. More precisely, the azeotrope is the left envelope of the manifold of constant-ζ loci (for h = 0) on a P-T plot. For the typical normal mixture, as shown in Figure 24B, there is no point on the critical locus at which the slope equals the interpolated slope of the vapor pressure curves, and therefore there is no critical azeotrope.

A common misconception is that a minimum in mixture critical temperature implies an azeotropic structure, and the converse. While this is usually true, there are exceptions as shown in Figures 24C and 24D. The first of these figures shows a mixture with an azeotrope but without a temperature minimum; there is a point on the critical locus with slope comparable to that of the vapor pressure curves, but no point of vertical slope. This structure occurs for hydrogen sulfide + ethane,[172] and was once thought to be the pattern for carbon dioxide + ethylene,[21] although a more recent experiment[434] has clearly established a minimum in critical temperature.

Figure 24D shows the opposite case — a mixture with a minimum critical temperature but without an azeotrope. Here, the minimum T_c occurs near the critical point of the component with the lower T_c, and while dP_c/dT_c is infinite at the minimum, the critical locus is always either sloping in the opposite direction or is steeper than the vapor pressure curve. This pattern applies to carbon dioxide + propane[108,435] and hydrogen sulfide + isobutane.[228] The mixture n-butane + perfluoro-n-heptane[179,180] is also probably of this form, although because of its pronounced temperature minimum, for correlational purposes it has been included with the azeotropic mixtures in Table 2.

The azeotropes described so far are commonly called "positive" or "minimum-boiling" azeotropes, the latter term referring to an isobar on a T-x diagram along which the azeotrope is the two-phase point of minimum temperature. "Negative" or "maximum-boiling" azeotropy is also possible. Here, in contrast to Figure 24A, the critical locus displays a maximum in critical temperature, and the negative azeotrope on a P-T plot lies to the right of both pure-fluid vapor pressure curves. Also, corresponding to but opposite from the case of Figure 5B, a negative azeotrope can form if the critical locus almost forms a temperature maximum near the critical point of the component with the larger critical pressure.

If the critical temperature of the mixture is higher than that of either pure fluid, it follows that the cross (1-2) interaction is more strongly attractive than either the (1-1) or (2-2) interaction. This then increases the probability that the unlike molecules will undergo a chemical reaction, and traditionally the study of negatively azeotropic mixtures has been hindered by such reactions. Kuenen[358] attempted to measure the phase boundary of hydrogen chloride + dimethyl ether, and Khazanova et al.[436,437] attempted the same for triethylamine + acetic acid, but in both cases chemical reactions prohibited a thorough measurement. Recently Noles et al.[238] have reported a thorough critical-region study of dimethyl ether + sulfur dioxide, an associating but stable mixture with a negative azeotrope first noted by Zawisza and Glowka[438] on one critical isotherm. (A study of the similar mixture dimethyl ether + R22 by Noles and Zollweg has also recently been completed.)

Other structures for azeotropic mixtures in P-T space are possible. In some cases, the critical locus displays a small "hook" toward a minimum temperature before turning and proceeding smoothly to the critical point of the component with the higher critical temperature. This pattern, which occurs for sulfur hexafluoride + propane,[173] methanol + benzene,[183] and n-propanol + benzene,[183] is topologically equivalent to Figure 24A, but unlike Figure 24A, the azeotropic mixture is a very dilute solution. Another possibility, occurring for ethanol + benzene,[183] is that an azeotrope exists at lower pressures; however, as pressure is increased, it terminates on a pure-fluid vapor pressure curve and fails to reach the critical locus.

A very interesting and rarely occurring example is a binary mixture with both a positive and a negative azeotrope. Although this pioneering work apparently was never independently confirmed and has been largely overlooked, for sulfur dioxide + methyl chloride Caubet[360] found an S-shaped critical locus with each type of azeotrope extending to the critical point. Within the extended critical region benzene + perfluorobenzene[234] appears to be a normal mixture with small α_{2m}, and is listed as such in Table 2; however, at lower pressures there is a Bancroft point at which the pure vapor pressure curves intersect. Below and slightly above the Bancroft point are both a positive and a negative azeotrope. The double azeotropic pattern was first noticed at lower pressure by Gaw and Swinton;[439] Caubet's work notwithstanding, this mixture has been cited as the only known example of double azeotropy. (In a recent study of sulfur dioxide + methyl chloride, Noles and Zollweg did not find doubly azeotropic behavior.)

Despite the wide variety of possible phase diagram structures, the most common and prototypical azeotropic mixture structure is Figure 24A, with a crescent-shaped critical locus and a pronounced temperature minimum. Between the critical azeotropic composition and the point of minimum critical temperature, there occurs what Kuenen[86] defined as "retrograde condensation of the second kind". Here, in contrast to retrograde condensation of the first kind as discussed at the end of Section IV, the critical point of an isopleth is at a lower pressure than the maxcondentherm point, which in this case is a liquid rather than a vapor point. As another example of a peculiar phase diagram, for some rare examples the critical locus in P-T space is always steeper than the constituent vapor pressure curves and, while there is no azeotrope, all dew-bubble curves display retrograde condensation of the second kind. Kay and Hissong found such a pattern for benzene + *n*-heptane,[347] cyclopentane + *n*-hexane,[348] and cyclohexane + *n*-heptane.[348]

At one time the author believed, as stated in a preliminary report,[390] that azeotropic mixtures were relatively easy to correlate with the modified Leung-Griffiths model. The reasoning was that the parameter α_2, which characterizes the fitting difficulty, is zero at the azeotrope as well as at the pure-fluid limits, and ordinarily is never very large in between. There are exceptions; for ammonia + *n*-butane,[176] $\alpha_{2m} = 0.175$. It was also noted that of the four mixtures successfully fitted according to the standards of that time, three were azeotropic (carbon dioxide + ethylene,[21] carbon dioxide + ethane,[3] and sulfur hexafluoride + propane[3]), while the other[1] (helium 3 + helium 4) was a normal mixture of low α_{2m}.

Upon closer examination, an argument can be made that azeotropic mixtures are difficult to fit according to more exacting standards. Since in most cases $T_c(x)$ is not monotonic, the argument given by Equations 29 and 30 that ζ is linear in x on the critical locus does not necessarily hold. For azeotropy to occur, the two components must have nearly equal T_c but substantially different P_c, which precludes mixtures of two members of the same homologous series. Consequently, the components are likely to differ in important characteristics such as polarity, and are likely to exhibit LLE at relatively high temperatures. Also, of the 30 azeotropic mixtures listed in Table 2, 11 are hydrocarbon-fluorocarbon mixtures subject to the anomalous behavior described in Section V.B.

For the graphical development and checking of azeotropic correlations, it is better to plot isopleths in P-ρ space rather than T-ρ space as in Figures 13B and 14B. For most azeotropic mixtures $P_c(x)$ is monotonic while T_c is not, and the T-ρ diagram will be cluttered with many overlapping isopleths (see Figure 3 of Reference 176). Conversely, for the typical nonazeotropic mixture $T_c(x)$ is monotonic but $P_c(x)$ is not, and the P-ρ diagram will be similarly cluttered.

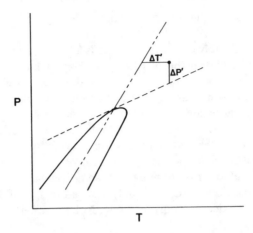

FIGURE 25.   Upper part of dew-bubble isopleth in P-T space (schematic) and geometric definition of the linear transformation to $\Delta T'$ and $\Delta P'$. The dashed line is tangent to the critical locus and the broken line is tangent to the constant-ζ locus at the mixture critical point.

The author's rather stringent conditions for successfully fitting the VLE surface of an azeotropic mixture are as follows. The dew-bubble curves in P-T space from the model should agree with experiment in a relative as well as an absolute sense. Since most dew-bubble curves are very narrow, a model curve with half or twice the width of the experimental curve will yield small absolute pressure or temperature differences with the measured dew and bubble points, but will not be really correct. The model should agree with the P-ρ diagram by the same standards as the T-ρ diagrams of normal mixtures in Table 3. Finally, if the azeotropic locus itself is experimentally measured, the model should accurately predict it.

A. ASYMPTOTIC EXPANSIONS OF DEW-BUBBLE CURVES

In order to analyze the experimental and model-predicted widths of the dew-bubble curves, a transformation Rainwater[26] developed earlier is very useful. Consider a dew-bubble curve in the critical region as shown in Figure 25, and assume that "locally" it is a good approximation that the critical pressure and critical temperature are linear functions of the critical composition (and therefore are linear functions of each other). The transformation in P-T space is

$$\Delta P' = \Delta P - \Delta T \frac{dP_c}{dT_c} \tag{73}$$

$$\Delta T' = \Delta T - \Delta P \left(\frac{\partial P}{\partial T}\right)_{\zeta, t=0}^{-1} \tag{74}$$

where $\Delta P = P_c - P$ and $\Delta T = T_c - T$; note also from Equation 55 that

$$\left(\frac{\partial P}{\partial T}\right)_{\zeta, t=0} \equiv \left(\frac{\partial P}{\partial T}\right)_\zeta = \frac{P_c}{T_c} [1 + C_4] \tag{75}$$

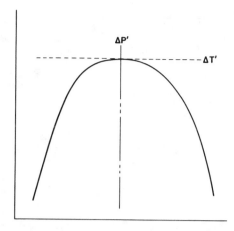

FIGURE 26.   Dew-bubble curve of Figure 25 in transformed space (schematic). The curve now resembles the pure-fluid T-ρ diagram of Figure 3.

The dew-bubble curve in $\Delta P' - \Delta T'$ space is shown in Figure 26. It resembles the pure-fluid T-ρ diagram of Figure 3.

For very narrow dew-bubble curves (small α_2), which occur for mixtures of very similar components, azeotropic mixtures near the azeotrope, and dilute mixtures, graphically it is often difficult to distinguish the bubble side from the dew side in P-T space, and no alteration of pressure or temperature scales can make the distinction clearer. In the transformed space, however, the dew and bubble points can be separated and distinguished by "stretching" the $\Delta T'$ axis.

The author has shown elsewhere[26] that within the modified Leung-Griffiths model, the nonanalytic expansion of a dew-bubble curve about the critical point has the form

$$\Delta T' = \pm a_1 |\Delta P'|^\beta + a_2 |\Delta P'|^{2\beta} \pm a_3 |\Delta P'|^{3\beta}$$
$$+ a_4 |\Delta P'|^{1-\alpha} + a_5 (\Delta P') + \ldots \qquad (76)$$

where plus refers to vapor and minus to liquid, and a_1 is proportional to α_2.

Application of the transformation to experimental dew and bubble points is itself a useful data-analysis technique. Figure 27 shows in transformed space the dew-bubble curve for n-butane + n-hexane as reported by Kay et al.[144] together with the correlated curve, and Figure 28 similarly shows the dew-bubble curve for ethane + propylene as reported by Lu et al.[118] The former data form a smooth curve in a transformed space, whereas the latter data show an inconsistent distortion to the right near the maxcondentherm point, a feature not evident in the P-T plot presented by the authors. A similar distortion is found for the R12 + R22 data of the Keio University group.[128,129]

To compare experimental and model-derived widths of dew-bubble curves for azeotropic mixtures, this transformation is very useful. Caution is necessary in that, exactly at the critical azeotrope, the $\Delta P'$ and $\Delta T'$ axes are parallel and thus the transformation becomes singular. Also, in the region of retrograde condensation of the second kind between the critical azeotrope and point of minimum critical temperature, the pattern of Figure 27 is inverted and the transformed curve has a minimum rather than a maximum in $\Delta P'$. The azeotrope is calculated by a numerical search in (ζ, t) space for the one-dimensional locus such that

$$\overline{Q}(\zeta, t) = 0 \qquad (77)$$

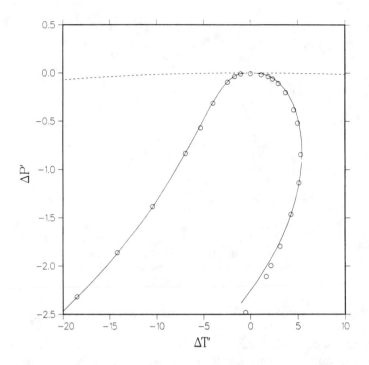

FIGURE 27. Data of Kay et al.[144] in transformed space for the *n*-butane + *n*-hexane mixture (0.4928 mol fraction *n*-butane). The data are smooth and self-consistent and asymptotically follow the shape of Figure 26. The solid line is the modified Leung-Griffiths correlation.

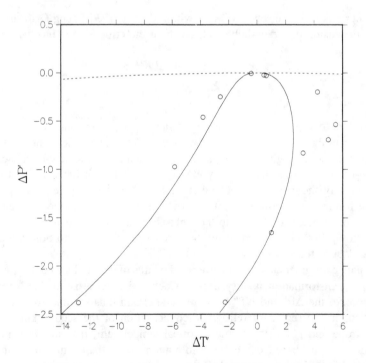

FIGURE 28. Data of Lu et al.[118] in transformed space for the ethane + propylene mixture (0.5285 mol fraction ethane). Unlike Figure 27, the data do not form a smooth curve and are particularly distorted, and hence suspect in the maxcondentherm region.

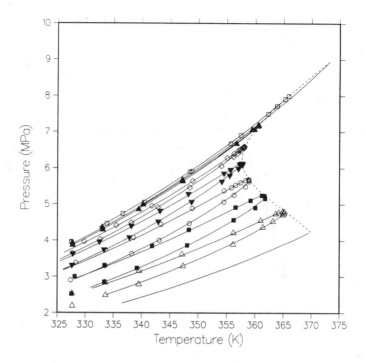

FIGURE 29. Coexistence surface and modified Leung-Griffiths correlation of hydrogen sulfide + propane, with data from Reference 174 in P-T space. The hydrogen sulfide mole fractions, top to bottom, are 0.8984, 0.7817, 0.6755, 0.5641, 0.4342, 0.2986, and 0.1633.

and therefore, from Equation 38, the coexisting vapor and liquid compositions are equal, which is the definition of the azeotrope.

B. CURRENT STATUS OF AZEOTROPIC CORRELATIONS

For the azeotropic mixtures with thorough VLE data including coexisting densities (entries 55 to 80 of Table 2), modified Leung-Griffiths correlations with the linear assumption of Equation 28 and the guidelines of Section VII have been attempted, so that in most cases only the adjustable parameters C_H and C_X are used. The critical density locus for several mixtures appears to be very irregular. As an example, for benzene + methanol the critical density of dilute solutions of benzene appears to increase very rapidly with benzene content, apparently a molecular packing phenomenon, so that the simple polynomial of Equation 51 must be replaced by a form including exponential functions such as $x \exp(-x/a)$, where a is a new parameter.

When the critical density locus is not too irregular, in most cases a reasonably good correlation in P-ρ space and dew-bubble curves in P-T space can be obtained that are correct in an absolute (but usually not in a relative) sense. The only mixtures that have been successfully correlated to the previously described standards on the initial attempts have been propane + hydrogen sulfide and, to a lesser degree, ethane + hydrogen sulfide.

Figures 29 and 30 show the successful correlation of C_3H_8 + H_2S, and Figure 31 shows in transformed space the $x(C_2H_6) = 0.4999$ isopleth for C_2H_6 + H_2S. In Figure 31, the "theoretical" (or model-derived) curve has a slightly smaller width (by about 10%) than the experimental curve.

The leading-order "amplitude" a_1 of the transformed curve (Equation 76) is proportional to α_2, so that Figure 31 implies that the theoretically predicted α_2 is 10% lower. From Equation 56, and given an experimentally determined critical locus $P_c(x)$, $T_c(x)$, and $\rho_c(x)$,

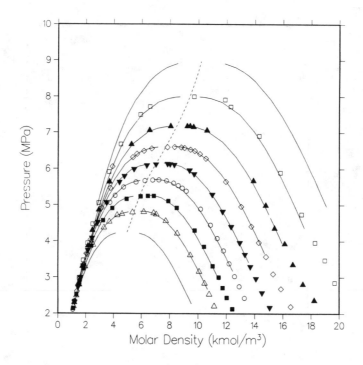

FIGURE 30.   Coexistence surface for the correlation of Figure 29 in P-ρ space.

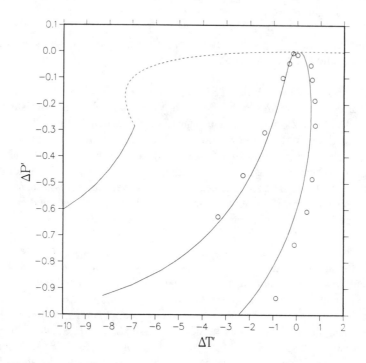

FIGURE 31.   Dew-bubble curves in transformed space for the hydrogen sulfide + ethane mixture of 0.4999 mol fraction ethane. The dashed line is the critical locus and the dew-bubble curve is from the modified Leung-Griffiths model. Data from Reference 172.

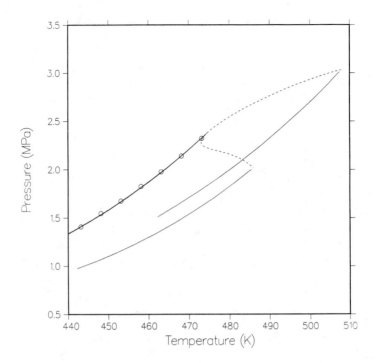

FIGURE 32. Prediction of the azeotrope in P-T space of PFMCH + *n*-hexane: bold solid curve, azeotrope; dashed curve, critical locus; narrow solid curves, vapor pressure loci. Azeotropic data from Reference 182.

clearly there are only two ways to adjust α_2 within the model: by changing the slopes of the constant-ζ curves at t = 0 [or, equivalently, $C_4(\zeta)$], or by changing the functional relationship between x and ζ on the critical locus.

However, there is some experimental information not available in previous correlations for azeotropic mixtures, namely the azeotrope itself, which is the left envelope of constant-ζ loci in P-T space. For those mixtures with measured azeotropic loci, the assumption of linearly interpolated $C_i(\zeta)$, i = 3, . . . ,6 (or "parallel" constant-ζ loci) yields excellent predictions of azeotropes. Figure 32 shows the example perfluoromethylcyclohexane (PFMCH) + *n*-hexane; equally good predictions have been obtained for entries 63, 69, 70 to 73, 78, and 80 of Table 2. To correct the dew-bubble curve widths, therefore, the only option is to alter the dependence of ζ on x along the critical locus.

A general procedure has been developed to do this. The mixture is first correlated under the assumption that for t = 0, x = 1 − ζ. Experimental and "theoretical" curves are examined in transformed space and a width ratio, $w_r(x_1) = w_{th}(x_1)/w_{ex}(x_1)$ is determined, where $x_1 = 1 − x$ and the subscripts denote "theoretical" and "experimental". A new $\zeta(x_1)$ that would reproduce the experimental widths satisfies the differential equation

$$\frac{d\zeta}{\zeta(1 - \zeta)} = \frac{dx_1}{x_1(1 - x_1)} w_r(x_1) \qquad (78)$$

It is presumed that $w_r(x_1)$ is analytic and may be expanded as

$$w_r(x_1) = w_0 + w_1 x_1 + w_2(1 - x_1) + x_1(1 - x_1)[w_3 + w_4 x_1 + \ldots] \qquad (79)$$

but, from Equations 64 to 66 and the accompanying discussion, $w_r(x_1 = 0) = w_r(x_1 = 1)$ = 1, so $w_0 = 1$ and $w_1 = w_2 = 0$. Hence, Equation 79 may be rewritten as

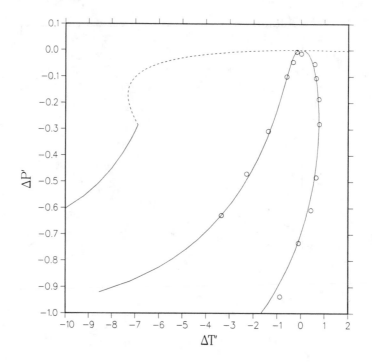

FIGURE 33.   The isopleth shown in Figure 31 with $w_3 = -0.6$ in Equation 81.

$$w_r(x_1) = 1 + x_1(1 - x_1)w_E(x_1) \tag{80}$$

where the "excess width ratio" is

$$w_E(x_1) = w_3 + w_4 x_1 + \ \ldots \tag{81}$$

The solution to the differential equation is then

$$\zeta(x_1) = \frac{Cx_1 \exp\left[\int_0^{x_1} w_E (x_i) \, dx_i\right]}{1 - x_1 + Cx_1 \exp\left[\int_0^{x_1} w_E (x_i) \, dx_i\right]} \tag{82}$$

where C is an arbitrary constant of integration. For the simple choice $w_E(x_1) = w_3 = -0.6$, the revised correlation yields Figure 33 for the $C_2H_6 + H_2S$ isopleths, where the experimental and theoretical widths now coincide.

For the other azeotropic mixtures, there are regions where the theoretical and experimental widths substantially disagree, particularly for the hydrocarbon-fluorocarbon mixtures near the azeotrope. This result calls for an explanation as to the better behavior of the hydrogen sulfide mixtures, and there is evidence that the answer is related to the absence or presence of LLE. Brewer et al.[440] have measured the coexistence surface of propane + hydrogen sulfide down to a three-phase SLV locus without finding LLE, so this mixture is type 1 in the sense of Figure 6A. Robinson et al.[441] measured VLE of ethane + hydrogen sulfide down to 199.9 K; again, no LLE was found, although the bubble curve at 199.9 K may be tending toward a point of inflection and hence a consolute point, as in Figure 23.

For many of the other azeotropic mixtures, however, a consolute point has been directly measured or can be inferred from similar mixtures. Francis[53] reports a consolute point for PFMCH + *n*-hexane at 282.15 K; it can be inferred that mixtures of other hexane isomers with PFMCH have consolute points at similar temperatures. Francis also reports consolute points of perfluoro-*n*-heptane with *n*-heptane at 323.15 K and with *n*-hexane at 302.15 K; by extrapolation the consolute points of perfluoro-*n*-heptane with *n*-pentane and *n*-butane can be inferred. Additionally, as discussed in Section XI, carbon dioxide + ethane probably has a "virtual" consolute point at 185 K. Surprisingly, the apparently greater miscibility of the hydrogen sulfide mixtures contradicts the rule of thumb noted in Section V.C that polar-nonpolar mixtures are less miscible than nonpolar-nonpolar mixtures; there is, to the author's knowledge, no explanation for this behavior.

A tentative explanation can be provided for the deviation from the linear hypothesis of Equation 28 for most of the azeotropic mixtures. A mixture with a consolute point in the neighborhood of $t = -0.4$ is an intermediate case between the normal mixtures of Table 3, where the LLE is either nonexistent or so low in temperature that it has no effect on the (VLE) extended critical region, and the mixtures noted in Section XI, for which a consolute point occurs at $t \geq -0.1$ and an explicit singular part of the potential for LLE as well as VLE is necessary (see Section XIV). For these azeotropic mixtures, the LLE part of the potential has no singularities in the extended critical region $-0.1 < t < 0$, and thus can be represented analytically; however, to some degree it distorts the P-T-ζ coexistence surface from the ideally smooth sheet proposed by Moldover and Gallagher. Specifically, at the liquid-liquid phase boundary x (the liquid composition) changes abruptly and discontinuously at $\zeta = \zeta_c$ where ($\zeta_c t_c$) is the location of the critical end point, so that

$$(\partial x/\partial \zeta)_{\zeta = \zeta_c, \ t<t_c, \ h=o+} = \infty \tag{83}$$

For $t > t_c$, the partial derivative is finite but may still be very large up to $t = 0$, so that at the plait point locus there is a kink in ζ as a function of x. Also, the azeotropic ζ is frequently equal or nearly equal to ζ_c. Such a kink is consistent with the observations of $w_r(x)$ to date.

Such behavior may also be present in the carbon dioxide + methane system as described in the previous section, where "virtual" LLE may lead to a strong enhancement of α_2 and the critical value of dx/dζ along the extension of the (virtual) three-phase locus, and therefore lead to the substantial underprediction of dew-bubble curve widths as noted for that mixture.

Further analysis of azeotropic mixtures is in progress. It is hoped that with judicious choices (not necessarily polynomial) for the functional forms of $\rho_c(x)$ and $\zeta(x)$ at $t = 0$, correlations can be developed that satisfy the rather stringent criteria specified earlier, and perhaps some new fundamental understanding can be achieved for the anomalous hydro-carbon-fluorocarbon behavior.

XIII. OTHER APPLICATIONS OF THE MODIFIED LEUNG-GRIFFITHS MODEL

In addition to the analysis of binary mixture VLE over the extended critical region presented thus far, the modified Leung-Griffiths model has proven useful for the solution of other related problems. Such problems include ternary VLE, a novel technique for meas-uring critical densities,[442,443] and interfacial tension between the liquid and vapor of a mixture.[5] Such work is reviewed briefly in this section; except for the ternary mixture results, which have not yet been published, details may be found in the cited papers.

Rainwater and Van Poolen have successfully generalized the modified Leung-Griffiths model to ternary mixtures. The mathematical derivations were derived independently by them, although it was later found that a separate but simplified independent derivation had

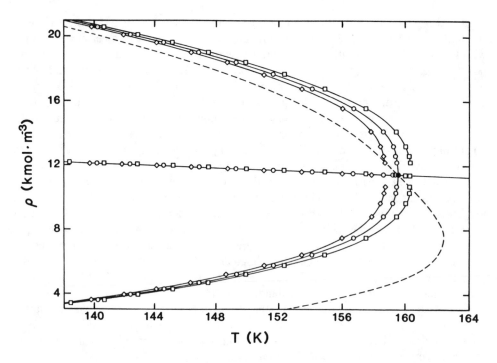

FIGURE 34.   Coexisting densities along two-phase isochores from the correlation of nitrogen + methane of Reference 4, for 0.5088 mol% nitrogen: circles, critical isochore; diamonds, $\rho_T/\rho_c = 1.069$; squares, $\rho_T/\rho_c = 0.935$. The rectilinear diameters for all three isochores are seen nearly to fall onto the same straight line. The dashed line is the dew-bubble curve for this composition.

been presented by Balbuena et al.[444] Significantly, a procedure was developed whereby, once correlations for each of the three binaries have been constructed, no additional adjustable parameters are needed for the ternary. The results for nitrogen + oxygen + argon have been used as input for a recently published ancillary equation of the air coexistence curve.[422] Also, successful correlations have been developed from the data of Thodos and co-workers[445,446] for ethane + *n*-butane + *n*-pentane.

A novel experimental technique for determining mixture critical densities has arisen from a study by Van Poolen and Rainwater of two-phase thermodynamic paths in which an overall, or "system-intensive" variable (in the terminology of Van Poolen[447]) is held constant. Van Poolen[447] has developed a generalization of the Gibbs phase rule to understand such two-phase thermodynamic paths. Of particular interest are two-phase isochores, in which the overall density and overall composition are held constant. Van Poolen[448] has developed an alternate coordinate system in field space for modeling VLE in which loci of constant ζ are replaced by critical two-phase isochores. From a correlation of nitrogen + methane, Van Poolen and Rainwater[442] showed that if the overall density is within 7% of the mixture critical density for the specified composition, the rectilinear diameter (Equation 5) of that isochore is nearly a straight line and nearly independent of overall density, as shown in Figure 34.

This observation led to the suggested experimental procedure of determining the critical density by measuring such an isochore and extrapolating the rectilinear diameter to a (known) mixture critical temperature, either by assuming a straight-line form or using an equation based on the liquid volume fraction.[449] It can be seen from Table 2 that for many mixtures the critical temperature locus is known but the critical density locus is not. Recently, the theoretically proposed method has been confirmed in the laboratory by Van Poolen et al.,[443] who measured two-phase isochores of ethane + *n*-butane and determined critical densities

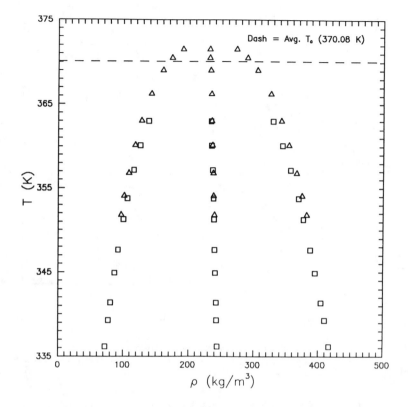

FIGURE 35.   Laboratory verification of the pattern of Figure 34 for the mixture ethane + n-butane: squares, 0.5966 mol% ethane, $\rho_T/\rho_c = 1.0684$; triangles, 0.5943 mol% ethane, $\rho_T/\rho_c = 0.9235$. The rectilinear diameters coincide to within experimental error; for details see Reference 443.

consistent with Rainwater's modified Leung-Griffiths correlation of that mixture. Figure 35 shows the results for the rectilinear diameters of two isochores of different overall densities but nearly the same overall composition; to within experimental error the two rectilinear diameters are straight and coincident.

The model has also led to a technique for predicting interfacial tension. The basis is the principle of "two-scale factor universality" as introduced by Stauffer et al.[450] According to this principle, the singular free energy in a volume of ζ^3 per unit kT is a universal number, where ξ is the correlation length. Also, the surface energy in an area of ξ^2 per unit kT is a second universal number. Elimination of the correlation length from the resulting equations and calculation of the singular free energy from Equation 10 leads to the result along a pure-fluid vapor pressure curve that

$$\sigma(t) = Y^-(k_B T_c)^{1/3}[P_c C_3 \alpha (1 - \alpha)(2 - \alpha)]^{2/3}(-t)^\nu \tag{84}$$

where Y^- is a third universal number.

Moldover[451] has determined empirically from an extensive study of the interfacial tension literature that $Y^- = 3.74 \pm 0.008$. This number differs from theoretically derived values,[452-455] but Moldover recommends it as being the correct value. In a generalization to mixtures, Moldover and Rainwater[5] proposed that Equation 84 holds along loci of constant ζ, where T_c, P_c, and C_3 are ζ-dependent functions. Here, the mixture critical pressures and temperatures, but not the densities, are needed as input. This equation was compared against the interfacial tension data for carbon dioxide + n-butane of Hsu et al.[409] There is an error

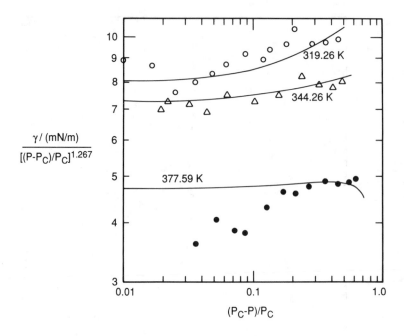

FIGURE 36.   Reduced interfacial tension data for carbon dioxide + *n*-butane from Reference 409 and predicted interfacial tension from Equation 84. This figure corrects an error in Figure 6 of Reference 5.

in Figure 6 of Reference 5 in that the model-derived interfacial tension was inadvertently divided by 1 − t or 1 + |t|. The corrected diagram is given here as Figure 36. The original conclusion remains valid that Equation 84, with ζ-dependent parameters, predicts the interfacial tension to within the scatter of the data.

Nadler et al.[456] applied Equation 84 to their recent interfacial tension data of argon + krypton. They also considered a model in which ζ was determined from the chemical potentials as calculated from a Peng-Robinson correlation, and with K = 1 in Equation 23. Some of their data were outside the extended critical region, but overall the authors' method agreed with their data with an average deviation of 5.6%, and their alternate correlation agreed to 3.9%. Finally, Jones[457] has used the VLE correlation of the Rainwater group and predicted surface tensions of carbon dioxide + *n*-pentane to analyze his flow experiments on dilute *n*-pentane solutions in supercritical carbon dioxide. Jones has found that the portion of heat and mass transfer driven by interfacial tension gradients may be significant.

XIV. WORK IN PROGRESS AND FUTURE WORK

This section discusses some additional problems related to VLE of binary mixtures in the extended critical region that have yet to be solved, but for which the author has developed specific plans based on the ideas presented in this chapter. The first problem, as mentioned at the end of Section IX, is a careful quantitative study of the relative accuracies of the present model and classical equations of state as the critical locus is approached.

At present the difficulties occur not with the Leung-Griffiths model, but with classical methods. In conventional phase equilibrium calculations near critical conditions, there are two problems that are sometimes confused but in fact are quite different and distinct. First, very close to the critical locus the predictions of a classical equation must be inaccurate because of the (incorrect) classical exponents. Second, the algorithms currently used to find coexisting liquid and vapor states, by searching for equality of chemical potentials or fugacities, fail to converge near the critical locus. The many Peng-Robinson correlations presented graphically in Reference 247 are open-ended at high pressure, thus showing the limits of convergence. These convergence failures are a defect of the algorithm, not of the EOS from which a definite critical locus and a coexistence surface extending to that locus can in principle be calculated.

Insights from the Leung-Griffiths theory can be exploited to develop an improved method for near-critical phase equilibrium calculations from a classical EOS. In the author's proposed method for a binary mixture (the multicomponent generalization is unclear), first the critical locus from the equation of state is calculated by a method such as that of Heidemann and Khalil.[458,459] Then, with the usual Leung-Griffiths assumptions about ζ and t, the composition derivatives of the calculated critical locus, and the author's asymptotic expansions,[26] an initial estimate of the point of phase equilibrium is obtained. In the conventional procedure, phase equilibrium calculations are a form of numerical root finding, and the success of any root-finding algorithm depends strongly on a good first guess for the solution. With such a first guess and a judiciously designed iteration procedure, the author expects that a program can be constructed that converges rapidly and is guaranteed not to converge onto the trivial solution. Different proposed methods to solve this problem are presently under investigation by other workers. It this method or another method is successful, it can be established unambiguously how good (or how bad) classical VLE predictions are in the critical region.

Liquid-liquid equilibrium is of interest both for its own sake and for its role in distorting the vapor-liquid equilibrium surface. Several studies have been completed that employ nonclassical critical exponents for LLE in an isobaric plane. Levelt Sengers[460] has developed a parametric Schofield-like model similar to the pure-fluid VLE model of Equations 10 to 13, but with a rearrangement of dependent and independent variables more appropriate for LLE. This model has been generalized by Johnson[461] to better describe the asymmetry of the T-x boundary. Also, dePablo and Prausnitz[462] have applied the method of Fox[45] for rescaling of the pure-fluid vapor-liquid transition to the description of isobaric, near-critical LLE.

However, the author's goal is to describe over the full thermodynamic space a type 2 mixture with both LLE and VLE, and with emphasis on the distortions of the VLE surface caused by the presence of LLE. A tentative thermodynamic potential has been constructed of the form $\omega = \omega_{an} + \omega_{VLE} + \omega_{LLE}$, where ω_{an} and ω_{VLE} are given by Equations 42 and 10, respectively, and ω_{LLE} is a new and separate Schofield construction. The upper critical endpoint is located at $\zeta = \zeta_c$ and $t = t_c$ (a negative constant) and $h = 0$, and the locus of consolute points is $\zeta = \zeta_c$, $t = t_c$, and $h > 0$.

Since this model has not yet been tested on actual data, the explicit equations are omitted. For the present discussion, the philosophy of the type 2 model is more important than the mathematical details. As with VLE, classical EOSs and phase equilibrium algorithms for type 2 mixtures yield an LLE surface, but with incorrect classical exponents. It is therefore in order to construct a model that explicitly builds in LLE and VLE surfaces with correct classical exponents. It should be easier to do this in field variable space, where continuous multidimensional functions with discontinuous derivatives are required, than in the customary mixed field and density variable space that calls for discontinuous functions. The challenge is to construct such a thermodynamic potential in P-T-ζ space with the proper derivative discontinuities on the indicated surfaces, proper (although perhaps "effective") critical exponents, and the sufficient amount of flexibility in the coefficients of the nonanalytic and analytic background parts of the thermodynamic potential so that type 2 mixture data can be fitted without overfitting. A major challenge is to ensure that no unwanted phase boundaries are constructed, such as an artificial and spurious "vapor-vapor equilibrium" transition for $h < 0$.

At present, the other major problems are the correlation of the supercritical single-phase region and the analysis of thermodynamic functions other than P, T, ρ, and x, such as enthalpies and entropies. For the former problem, revised scaling as discussed before Equation 47 should be incorporated into the model; also, there is flexibility in that terms quadratic, cubic, etc., in h can be added to Equation 42 and change the one-phase thermodynamics without changing the coexistence surface in P-T-ρ-x space. There is a significant body of coexisting enthalpy data in the literature, as reviewed in Sections V.D and VI.A, and these data can be correlated starting from Equations 22 and 37, with some integrating and differentiating of the present thermodynamic potential and with some modeling of the temperature derivative of K.

An alternate approach is suggested by a generalization of the Clausius-Clapeyron equation due to King.[463] The difference in molar entropy between a bubble point and a dew point on the same isopleth at the same temperature, i.e, a bubble point and a dew point vertically directly below it in Figure 9A, is

$$ s_{DP} - s_{BP} = v_{DP} \left. \frac{dP}{dT} \right)_x^{DP} - v_{BP} \left. \frac{dP}{dT} \right)_x^{BP} + \frac{d}{dT} \int_{P_{DP}}^{P_{BP}} v \, dP \qquad (85) $$

where the subscripts BP and DP denote bubble point and dew point, respectively. The integral is along a two-phase path of constant composition and temperature with v the overall or "system-intensive" molar volume of the two-phase state with the specified overall composition.

Furthermore, since the integral is the difference in Gibbs free energy per mole between the bubble point and dew point, and the pressures and molar volumes of the phase boundary points are known, the differences in molar enthalpies and molar Gibbs and Helmholtz free energies may be calculated. All the necessary information is in principle contained within the current modified Leung-Griffiths model, although the success of this procedure depends strongly on the statistical accuracy of the slopes of the model bubble and dew curves.

With molar entropy differences from a large number of such pairs of points, a function for $\Delta s(\zeta, t)$ may be fitted where Δs is the molar entropy difference between vapor and liquid. In addition to α_2, Onuki[19] has defined the variable α_1 as

$$ \alpha_1 = \lim_{t \to 0} \Delta s / R (\Delta \rho / \rho_c) \qquad (86) $$

This variable, as well as α_2, is needed within Onuki's theory[19] to calculate the critical thermal conductivity enhancement of a binary mixture. Friend and Roder[464,465] have measured the near-critical thermal conductivity of methane + ethane, and with the correlation of the VLE data of this mixture due to Bloomer et al.,[93] it may be possible to calculate α_1 from Equations 85 and 86 and thus to develop a quantitative analysis of the thermal conductivity results of Friend and Roder.

XV. SUMMARY

This chapter has reviewed the phase equilibrium structures of binary mixtures, the vapor-liquid equilibrium literature, and the success to date in correlating binary mixture VLE with the modified Leung-Griffiths model. A systematic theoretical, computational, correlational, and bibliographic investigation has been described in which "primary" fluids have been chosen as components of at least one thoroughly measured class 1 mixture over an extended critical region at high pressure. The criteria for thorough measurement have been specified in some detail according to the author's correlation experiences. It has been noted that the miscibility of two fluids depends on the relative position of their critical points, polarity effects, and an incompletely understood anomalous hydrogen-fluorine interaction. Of the 73 primary fluids, most are largely miscible with each other as their critical points are mostly in the range 280 K $<$ T_c $<$ 650 K and 2 MPa $<$ P_c $<$ 11.5 MPa. This helps to justify the efforts in constructing a model limited to the application of only type 1 mixtures, although preliminary extensions to type 2 with LLE have been considered.

Of the 129 class 1 mixtures identified as having been thoroughly measured, the Leung-Griffiths model, as modified by Moldover, Rainwater, and co-workers, has successfully correlated 38 of them as listed in Table 3, along with four additional mixtures with sparse data and two azeotropic mixtures. The author has described a history of development of the model requiring careful mathematical and physical analysis of its possibilities, while at the same time its development has been highly interactive with the experimental data. Of the uncorrelated mixtures, most of those without azeotropy and density data probably could be correlated in P-T-x-y space with the newly developed Simplex program. The problems with azeotropic mixtures are becoming understood and good progress on fitting their VLE surfaces with very exacting standards is being made. Also, the project has led to understanding of shapes of the dew-bubble curves and their relations with the critical locus not available elsewhere in the VLE literature.

The most important frontier in the development at present is the application of the model to mixtures of more highly dissimilar fluids, or, in the useful quantitative terminology, of greater α_{2m}. Particular attention is being paid to the problem of a dilute heavy solute in a light solvent, the condition most important for SCF technology. This condition is the most sensitive to parameter variations and the most difficult to model, but this chapter has also shown that nonclassical exponents are important and a truly accurate model of supercritical solubilities will probably require a model such as the author's with specific consideration of such nonclassical exponents. The success in the modeling mixture of carbon dioxide and pentane isomers shows promise in such a direction.

While this chapter has largely been an advocacy of the modified Leung-Griffiths model, some products of this investigation should be useful even for those who doubt the utility of such scaling-law techniques. The bibliographical survey, as summarized in Table 2, has extracted ''benchmark-quality'' experimental sources for a wide variety of mixtures that should be used as input for a comprehensive study of any other high-pressure phase equilibrium model. The author has shown that the best data have been more frequently taken along isopleths rather than isotherms, so that for serious studies of the critical region of mixtures, phase equilibrium algorithms should be adapted to study constant composition as well as isothermal loci. The prevalence of smoothed, cross-plotted, or otherwise processed data in the literature and in current databases has been noted, and the databases the author plans to develop with emphasis on original or raw data should prove useful for many investigators. Also, the distance variables within the model such as t and ζ provide useful axes within any model for the construction of deviation plots and an analysis of how a classical model behaves as the critical locus is approached.

Meanwhile, the research has taken the trouble to bring the model more into line with modern engineering practices by going beyond previous visual and graphical methods and by developing Simplex and deviation-plot programs. For more convenient use of the model by the experimental community it will be the responsibility of the author to develop the appropriate computer packages, and this is gradually being accomplished. At this point, the author can unreservedly recommend the model as a correlation technique and consistency check on thorough VLE experiments such as those listed in Table 2, and he has already entered into several interactions and collaborations with leading experimentalists.[14,108,132,136,384,456,457] The utility of the model for other engineering purposes will depend on the continued development of the model itself and accompanying codes.

As discussed in the comparisons with the DDMIX package, there are many important thermodynamic problems for mixtures that the modified Leung-Griffiths model in its present stage of development cannot address; it should be remembered that the Leung-Griffiths model, with its field variables, scaling-law equations, and adaptations of corresponding states principles, is a very different approach from conventional phase equilibrium calculations based on analytic EOSs. The conventional methods have been developed over many decades by a large number of institutions and investigators. By contrast, although the Leung-Griffiths model had previously shown some promise with a few mixtures,[1-3,21] its successful adaptation and application to a wide range of mixtures has been a nearly single-handed effort of the present author over the last few years, with assistance from the individuals listed in the first paragraph of the acknowledgment. Consequently, we have probably only scratched the surface of the potential applications of the Leung-Griffiths model and related techniques. It is hoped that the promise of the approach, as demonstrated herein, will encourage others to investigate important contemporary problems of mixture thermodynamics by means of these alternative viewpoints.

ACKNOWLEDGMENTS

The author first would like to thank Michael Moldover for suggesting the development of the modified Leung-Griffiths model, for substantial theoretical and computational contributions, and for excellent guidance over many years. He thanks Isaura Vazquez, Frank Williamson, Christina Swanger, Hepburn Ingham, Lambert Van Poolen, John Lynch, and Jerry Chatel for assistance with computer coding, graphics, and correlations, and Rachel Moldover for extensive data entry. Significant contributions to the theory by Lambert Van Poolen and John Lynch are also gratefully acknowledged.

The project has advanced significantly through collaborations on the thermodynamics of air with Richard Jacobsen, Steve Beyerlein, and Paul Clarke; on isobutane + isopentane with Jan and Anneke Sengers; and on a novel method to measure critical densities with Lambert Van Poolen and Vicki Niesen. The author thanks Webster Kay, Vicki Niesen, Lloyd Weber, and John Zollweg for providing data for analysis prior to publication. Finally, the project has benefitted from discussions with those individuals already mentioned as well as with Howard Hanley, James Ely, Daniel Friend, Jeffrey Fox, Graham Morrison, Ren-Fang Chang, Benjamin Widom, Robert Griffiths, John Wheeler, Joseph Magee, Mark McLinden, Arthur Kidnay, Taher Al-Sahhaf, Kraemer Luks, and Dendy Sloan. This work was supported by the Department of Energy, Office of Basic Energy Sciences.

NOTES ON THE BIBLIOGRAPHY

For sources in general and experimental vapor-liquid equilibrium references in particular originally written in a language other than English, an English translation has been cited where available. In most cases the translation refers to the original source, and all sources are in English unless otherwise noted. For non-English references, titles have been translated into English; the translations of *Chemical Abstracts* are used where available. Many M.Sc. and Ph.D. theses are cited; the Ph.D. theses are available from University Microfilms International, Ann Arbor, MI, if the date is after the year that the university in question joined that system, typically between 1950 and 1960. Earlier Ph.D. theses and all M.Sc. theses are available only from the granting university.

REFERENCES

1. **Leung, S. S. and Griffiths, R. B.,** Thermodynamic properties near the liquid-vapor critical line in mixtures of He₃ and He₄, *Phys. Rev. A,* 8, 2670, 1973.
2. **Moldover, M. R. and Gallagher, J. S.,** Phase equilibria in the critical region: an application of the rectilinear diameter and "1/3 power" laws to binary mixtures, in *Phase Equilibria and Fluid Properties in the Chemical Industry Estimation and Correlation,* ACS Symp. 60, Sandler, S. I. and Storvick, T. J., Eds., American Chemical Society, Washington, D.C., 1977, 498.
3. **Moldover, M. R. and Gallagher, J. S.,** Critical points of mixtures: an analogy with pure fluids, *AIChE J.,* 24, 267, 1978.
4. **Rainwater, J. C. and Moldover, M. R.,** Thermodynamic models for fluid mixtures near critical conditions, in *Chemical Engineering at Supercritical Fluid Conditions,* Paulaitis, M. E., Penninger, J. M. L., Gray, R. D., and Davidson, P., Eds., Ann Arbor Science, Ann Arbor, MI, 1983, 199.
5. **Moldover, M. R. and Rainwater, J. C.,** Interfacial tension and vapor-liquid equilibria in the critical region of mixtures, *J. Chem. Phys.,* 88, 7772, 1988.
6. **Rainwater, J. C.,** Vapor-Liquid Equilibrium of Binary Mixtures in the Extended Critical Region I. Thermodynamic Model, NIST Tech. Note 1328, National Institute of Standards and Technology, Washington, D.C., 1989.
7. **McHugh, M. and Paulatis, M. E.,** Solid solubilities of naphthalene and biphenyl in supercritical carbon dioxide, *J. Chem. Eng. Data,* 25, 327, 1980.
8. **Ng, H.-J. and Robinson, D. B.,** Equilibrium phase properties of the toluene-carbon dioxide system, *J. Chem. Eng. Data,* 23, 325, 1978.
9. **Reamer, H. H. and Sage, B. H.,** Phase equilibria in hydrocarbon systems. Volumetric and phase behavior of the n-decane-CO₂ system, *J. Chem. Eng. Data,* 8, 508, 1963.
10. **Smith, R. M.,** *Supercritical Fluid Chromotography,* The Royal Society of Chemistry, London, 1988.
11. **Griffiths, R. B. and Wheeler, J. C.,** Critical points in multicomponent systems, *Phys. Rev. A,* 2, 1047, 1970.
12. **Schofield, P.,** Parametric representation of the equation of state near a critical point, *Phys. Rev. Lett.,* 22, 606, 1969.
13. **Peng, D. Y. and Robinson, D. B.,** A new two-constant equation of state, *Ind. Eng. Chem. Fund.,* 15, 59, 1976.
14. **Diller, D. E., Gallagher, J. S., Kamgar-Parsi, B., Morrison, G., Rainwater, J. C., Levelt Sengers, J. M. H., Sengers, J. V., Van Poolen, L. J., and Waxman, M.,** Thermophysical Properties of Working Fluids for Binary Geothermal Cycles, National Bureau of Standards Interagency Rep. 85-3124, NBS, Washington, D.C., 1985.
15. **Rainwater, J. C. and Williamson, F. R.,** Vapor-liquid equilibrium of near-critical binary alkane mixtures, *Int. J. Thermophys.,* 7, 65, 1986.
16. **Rainwater, J. C. and Jacobsen, R. T.,** Vapour-liquid equilibrium of nitrogen-oxygen mixtures and air at high pressure, *Cryogenics,* 28, 22, 1988.
17. **Rainwater, J. C. and Lynch, J. J.,** The modified Leung-Griffiths model for vapor-liquid equilibria: application to polar fluid mixtures, *Fluid Phase Equilibria,* 52, 91, 1989.
18. **Rainwater, J. C., Ingham, H., and Lynch, J. J.,** Vapor-liquid equilibrium of carbon dioxide with i-butane and n-butane: modified Leung-Griffiths correlation and data evaluation, *J. Res. NIST,* 95, 701, 1990.
19. **Onuki, A.,** Statics and dynamics in binary mixtures near the liquid-vapor critical line, *J. Low Temp. Phys.,* 61, 101, 1985.
20. **Brennecke, J. F. and Eckert, C. A.,** Phase equilibria for process design, *AIChEJ.,* 35, 1409, 1989.
21. **D'Arrigo, G., Mistura, L., and Tartaglia, P.,** Leung-Griffiths equation of state for the system CO₂-C₂H₄ near the liquid-vapor critical line, *Phys. Rev. A,* 12, 2587, 1975.
22. **Chang, R. F. and Doiron, T.,** Leung-Griffiths model for the thermodynamic properties of mixtures of CO₂ and C₂H₆ near the gas-liquid critical line, *Int. J. Thermophys.,* 4, 337, 1983.
23. **Brunner, E., Hultenschmidt, W., and Schlichtharle, G.,** Fluid mixtures at high pressures. IV. Isothermal phase equilibria in binary mixtures consisting of (methanol + hydrogen or nitrogen or methane or carbon monoxide or carbon dioxide), *J. Chem. Thermodyn.,* 19, 273, 1987.
24. **Chen, R. J. J., Chappelear, P. S., and Kobayashi, R.,** Dew point loci for methane-n-butane system, *J. Chem. Eng. Data,* 19, 53, 1974.
25. **Chen, R. J. J., Chappelear, P. S., and Kobayashi, R.,** Dew point loci for methane-*n*-pentane binary system, *J. Chem. Eng. Data,* 19, 58, 1974.
26. **Rainwater, J. C.,** Asymptotic expansions for constant-composition dew-bubble curves near the critical locus, *Int. J. Thermophys.,* 10, 357, 1989.
27. **Van Konynenberg, P. H. and Scott, R. L.,** Critical lines and phase equilibria in binary van der Waals mixtures, *Philos. Trans. R. Soc. London,* 298, 495, 1980.

28. **Ely, J. F., Haynes, W. M., and Bain, B. C.,** Isochoric (p, V_m, T) measurements on CO_2 and on (0.982 CO_2 + 0.018 N_2) from 250 to 330 K at pressures to 35 MPa, *J. Chem. Thermodyn.,* 21, 879, 1989.

29. **Ely, J. F.,** A predictive, exact shape factor extended corresponding states model for mixtures, *Adv. Cryogen. Eng.,* 35, 1511, 1990.

30. **Redlich, O. and Kwong, J. N. S.,** On the thermodynamics of solutions. V. An equation of state, Fugacities of gaseous solutions, *Chem. Rev.,* 44, 233, 1949.

31. **Soave, G.,** Equilibrium constants from a modified Redlich-Kwong equation of state, *Chem. Eng. Sci.,* 27, 1197, 1972.

32. **Rowlinson, J. S. and Swinton, F. L.,** *Liquids and Liquid Mixtures,* 3rd ed., Butterworths, Boston, 1982.

33. **Guggenheim, E. A.,** The principle of corresponding states, *J. Chem. Phys.,* 13, 253, 1945.

34. **Verschaffelt, J. E.,** On the critical isothermal line and the densities of saturated vapour and liquid in isopentane and carbon dioxide, *Commun. Phys. Lab. Leiden,* 55, 1, 1900.

35. **Wilson, K. G.,** Renormalization group and critical phenomena. I. Renormalization group and the Kadanoff scaling picture, *Phys. Rev. B,* 4, 3174, 1971.

36. **Greer, S. C.,** Coexistence curves at liquid-liquid critical points: Ising exponents and extended scaling, *Phys. Rev. A,* 14, 1770, 1976.

37. **Jacobs, D. T., Anthony, D. J., Mockler, R. C., and O'Sullivan, W. J.,** Coexistence curve of a binary mixture, *Chem. Phys.,* 20, 219, 1977.

38. **Edwards, T. J.,** Specific-Heat Measurements Near the Critical Point of Carbon Dioxide, Ph.D. thesis, University of Western Australia, Perth, 1984.

39. **Schofield, P., Lister, J. D., and Ho, J. T.,** Correlation between critical coefficients and critical exponents, *Phys. Rev. Lett.,* 23, 1098, 1969.

40. **Schmidt, R. and Wagner, W.,** A new form of the equation of state for pure substances and its application to oxygen, *Fluid Phase Equilibria,* 19, 175, 1985.

41. **Moldover, M. R.,** Implementation of scaling and extended scaling equations of state for the critical point of fluids, *J. Res. Natl. Bur. Stand.,* 83, 329, 1978.

42. **Chapela, G. A. and Rowlinson, J. S.,** Accurate representation of thermodynamic properties near the critical point, *Trans. Faraday Soc. I,* 70, 584, 1974.

43. **Hill, P. G.,** A unified equation of state for H_2O, *J. Phys. Chem. Ref. Data,* 19, 1233, 1990.

44. **Woolley, H. W.,** A switch function applied to the thermodynamic properties of steam near and not near the critical point, *Int. J. Thermophys.,* 4, 51, 1983.

45. **Fox, J. R.,** Method for construction of nonclassical equations of state, *Fluid Phase Equilibria,* 14, 45, 1983.

46. **Nicoll, J. F.,** Critical phenomena of fluids: asymmetric Landau-Ginzburg-Wilson model, *Phys. Rev. A,* 24, 2203, 1981.

47. **Albright, P. C., Sengers, J. V., Nicoll, J. F., and Ley-Koo, M.,** A crossover description for the thermodynamic properties of fluids in the critical region, *Int. J. Thermophys.,* 7, 75, 1986.

48. **Albright, P. C., Chen, Z. Y., and Sengers, J. V.,** Crossover from singular to regular thermodynamic behavior of fluids in the critical region, *Phys. Rev. B,* 36, 877, 1987.

49. **Chen, Z. Y., Abbaci, A., and Sengers, J. V.,** Nonasymptotic critical thermodynamic behavior of steam, in Proc. 11th Int. Conf. on the Properties of Steam, to be published.

50. **Chen, Z. Y., Albright, P. C, and Sengers, J. V.,** Crossover from singular critical to regular classical behavior of fluids, *Phys. Rev. A,* 41, 3161, 1990.

51. **Streett, W. B.,** Phase equilibria in fluid and solid mixtures at high pressure, in *Chemical Engineering at Supercritical Fluid Conditions,* Paulaitis, M. E., Penninger, J. M. L., Gray, R. D., and Davidson, P., Eds., Ann Arbor Science, Ann Arbor, MI, 1983, 3.

52. **Shvarts, A. V. and Efremova, G. D.,** Liquid-liquid-gas equilibrium in the ethane-ammonia system, *Russ. J. Phys. Chem.,* 46, 1654, 1972.

53. **Francis, A. W.,** *Critical Solution Temperatures,* Advances in Chemistry Series No. 31, American Chemical Society, Washington, D.C., 1961.

54. **Arlt, W. and Sorensen, J. M.,** *Liquid-Liquid Equilibrium Data Collection,* Vol. 1, *Binary Systems,* DECHEMA, Frankfurt, West Germany, 1979.

55. **Rebert, C. J. and Kay, W. B.,** The phase behavior and solubility relations of the benzene-water system, *AIChE J.,* 5, 285, 1959.

56. **Brunner, E.,** Fluid mixtures at high pressures. VII. Phase separations and critical phenomena in 12 binary mixtures containing ammonia, *J. Chem. Thermodyn.,* 20, 1397, 1988.

57. **Brunner, E.,** Fluid mixtures at high pressures. IX. Phase separation and critical phenomena in 23 (n-alkane + water) mixtures, *J. Chem. Thermodyn.,* 22, 335, 1990.

58. **Streett, W. B.,** Gas-gas equilibrium: high pressure limits, *Can. J. Chem. Eng.,* 52, 92, 1974.

59. **Luks, K. D.,** The occurrence and measurement of multiphase equilibria behavior, *Fluid Phase Equilibria,* 29, 209, 1986.

60. **Findlay, A. and Campbell, A. N.,** *The Phase Rule and its Applications,* Dover Publications, New York, 1945, 107.

61. **Widom, B.,** Three-phase equilibrium and the tricritical point, in *Chemical Engineering at Supercritical Fluid Conditions,* Paulaitis, M. E., Penninger, J. M. L., Gray, R. D., and Davidson, P., Eds., Ann Arbor Science, Ann Arbor, MI, 1983, 199.

62. **Davenport, A. J. and Rowlinson, J. S.,** The solubility of hydrocarbons in liquid methane, *Trans. Faraday Soc.,* 59, 78, 1963.

63. **Creek, J. L., Knobler, C. M., and Scott, R. L.,** Tricritical points in ternary mixtures of hydrocarbons, *J. Chem. Phys.,* 67, 366, 1977.

64. **Creek, J. L., Knobler, C. M., and Scott, R. L.,** Tricritical phenomena in "quasibinary" mixtures of hydrocarbons. I. Methane systems, *J. Chem. Phys.,* 74, 3489, 1981.

65. **Wheeler, J. C.,** Geometric constraints at triple points, *J. Chem. Phys.,* 61, 4474, 1974.

66. **Wheeler, J. C.,** The 180° rule at a class of higher-order triple points, *Phys. Rev. A,* 12, 267, 1975.

67. **Diepen, G. A. M. and Scheffer, F. E. C.,** On critical phenomena of saturated solutions in binary systems, *J. Am. Chem. Soc.,* 70, 4081, 1948.

68. **Miller, M. M. and Luks, K. D.,** Observations on the multiphase equilibria behavior of CO_2-rich and ethane-rich mixtures, *Fluid Phase Equilibria,* 44, 295, 1989.

69. **Wheeler, J. C.,** Decorated lattice-gas models of critical phenomena in fluids and fluid mixtures, *Ann. Rev. Phys. Chem.,* 28, 411, 1977.

70. **Tedrow, P. M. and Lee, D. M.,** On the phase diagram of ^{3}He-^{4}He mixtures, in *Proceedings of the 9th International Conference on Low Temperature Physics,* Part A, Daunt, J. G., Edwards, D. O., Milford, F. J., and Yaqub, M., Eds., Plenum Press, New York, 1965, 248.

71. **Gilliland, E. R. and Scheeline, H. W.,** High-pressure vapor vapor-liquid equilibrium for the systems propylene-isobutane and propane-hydrogen sulfide, *Ind. Eng. Chem.,* 32, 48, 1940.

72. **Scheeline, H. W. and Gilliland, E. R.,** Vapor-liquid equilibrium in the system propane-isobutylene, *Ind. Eng. Chem.,* 31, 1050, 1939.

73. **Wilson, G. M., Silverberg, P. M., and Zellner, M. G.,** Argon-oxygen-nitrogen three-component system experimental vapor-liquid equilibrium data, *Adv. Cryogen. Eng.,* 10, 192, 1965.

74. **Zakharov, N. D., Semenov, V. G., and Domnina, E. V.,** Critical parameters in nitrogen-Freon-13 and nitrogen-Freon-14 systems, *Russ. J. Phys. Chem.,* 56, 28, 1982.

75. **Kay, W. B.,** Liquid-vapor phase equilibrium relations in the ethane-n-heptane system, *Ind. Eng. Chem.,* 30, 459, 1938.

76. **Mandlekar, A. V., Kay, W. B., Smith, R. L., and Teja, A. S.,** Phase equilibria in the n-hexane + diethylamine system, *Fluid Phase Equilibria,* 23, 79, 1985.

77. **Kidnay, A. J., Miller, R. C., Sloan, E. D., and Hiza, M. J.,** A review and evaluation of the phase equilibria, liquid phase heats of mixing and excess volumes, and gas phase PVT measurements for nitrogen + methane, *J. Phys. Chem. Ref. Data,* 14, 681, 1985.

78. **Rizvi, S. S. H. and Heidemann, R. A.,** Vapor-liquid equilibria in the ammonia-water system, *J. Chem. Eng. Data,* 32, 183, 1987.

79. **Efremova, G. D. and Sorina, G. A.,** Phase and volume relations in the system ethylene-butane (in Russian), *Khim. Prom.,* No. 6, 503, 1960.

80. **Tsiklis, D. S., Linshits, L. R., and Goryunova, N. P.,** Phase equilibria in the system ammonia-water, *Russ. J. Phys. Chem.,* 39, 1590, 1965.

81. **Pryanikova, R. O., Pleninka, R. M., Kiseleva, N. I., and Efremova, G. D.,** Liquid-gas equilibrium in the cyclohexane-ammonia system, *Russ. J. Phys. Chem.,* 45, 864, 1971.

82. **Shakhova, S. F., Rutenberg, O. L., and Targanskaya, M. N.,** Liquid-gas equilibrium in the epoxy-propane-carbon dioxide system, *Russ. J. Phys. Chem.,* 47, 895, 1973.

83. **Tyvina, T. N., Efremova, G. D., and Pyranikova, R. O.,** Liquid-gas equilibrium in the propene-toluene system, *Russ. J. Phys. Chem.,* 47, 1513, 1973.

84. **Naumova, A. A. and Tyvina, T. N.,** Liquid-gas equilibrium in the isobutane-ethylene system, *J. Appl. Chem. U.S.S.R.,* 54, 2440, 1981.

85. **Tyvina, T. N., Naumova, A. A., and Fokina, V. V.,** Liquid-gas equilibrium in the system acetonitrile-ammonia, *J. Appl. Chem. U.S.S.R.,* 52, 2458, 1979.

86. **Kuenen, J. P.,** On retrograde condensation and the critical phenomena of mixtures of two substances, *Commun. Phys. Lab. Univ. Leiden,* No. 4, 1892.

87. **Wallace, B. and Meyer, H.,** Pressure-density-temperature relations of He_3-He_4 mixtures near the liquid-vapor critical point, *Phys. Rev. A,* 5, 953, 1972.

88. **Wallace, B. and Meyer, H.,** Tabulation of the Original Pressure-Volume-Temperature Data for He_3-He_4 Mixtures and for He_3, Duke University Tech. Rep., Duke University, Raleigh, NC, 1971.

89. **Bloomer, O. T. and Parent, J. D.,** Liquid-vapor phase behavior of the methane-nitrogen system, *Chem. Eng. Prog. Symp. Ser.,* 49(6), 11, 1953.

90. **Cines, M. R., Roach, J. T., Hogan, R. J., and Roland, C. H.,** Nitrogen-methane vapor-liquid equilibria, *Chem. Eng. Prog. Symp. Ser.,* 49(6), 1, 1953.

91. **Stryjek, R., Chappelear, P. S., and Kobayashi, R.,** Low-temperature vapor-liquid equilibria of nitrogen-methane system, *J. Chem. Eng. Data,* 19, 334, 1974.

92. **Kidnay, A. J., Miller, R. C., Parrish, W. R., and Hiza, M. J.,** Liquid-vapor phase equilibria in the N_2-CH_4 system from 130 to 180 K, *Cryogenics,* 15, 531, 1975.

93. **Bloomer, O. T., Gami, D. C., and Parent, J. D.,** Physical-Chemical Properties of Methane-Ethane Mixtures, IGT Res. Bull. No. 22, July 1953.

94. **Miller, R. C., Kidnay, A. J., and Hiza, M. J.,** Liquid + vapor equilibria in methane + ethene and in methane + ethane from 150.00 to 190.00 K, *J. Chem. Thermodyn.,* 9, 167, 1977.

95. **Davalos, J., Anderson, W. R., Phelps, R. E., and Kidnay, A. J.,** Liquid-vapor equilibria at 250.00 K for systems containing methane, ethane, and carbon dioxide, *J. Chem. Eng. Data,* 21, 81, 1976.

96. **Gupta, M. K., Gardner, G. C., Hegarty, M. J., and Kidnay, A. J.,** Liquid-vapor equilibria for the N_2 + CH_4 + C_2H_6 system from 260 to 280 K, *J. Chem. Eng. Data,* 25, 313, 1980.

97. **Williams, B.,** Pressure Volume Temperature Relationships, and Phase Equilibria in the System Ethylene-Normal Butane, Ph.D. thesis, University of Michigan, Ann Arbor, 1949.

98. **Zernov, V. S., Kogan, V. B., Lyubetskii, S. G., and Duntov, F. I.,** Liquid-vapor equilibrium in the system ethylene-trifluoropropylene, *J. Appl. Chem. U.S.S.R.,* 44, 693, 1971.

99. **Cook, D.,** The carbon-dioxide-nitrous-oxide system in the critical region, *Proc. R. Soc. Chem. A,* 219, 245, 1953.

100. **Caubet, F.,** The liquifaction of gas mixtures (in German), *Z. Phys. Chem.,* 49, 101, 1904.

101. **Bierlein, J. A. and Kay, W. B.,** Phase equilibrium properties of system carbon dioxide-hydrogen sulfide, *Ind. Eng. Chem.,* 45, 618, 1953.

102. **Bierlein, J. A.,** Liquid-Vapor Equilibrium Relations in the System Carbon Dioxide-Hydrogen Sulfide, Ph.D. thesis, Ohio State University, Columbus, 1951.

103. **Sobocinski, D. P. and Kurata, F.,** Heterogeneous phase equilibria of the hydrogen sulfide-carbon dioxide system, *AIChE J.,* 5, 545, 1959.

104. **Sobocinski, D. P.,** Phase Behavior of the Hydrogen Sulfide-Carbon Dioxide System, Ph.D. thesis, University of Kansas, Lawrence, 1953.

105. **Haselden, G. G., Holland, F. A., King, M. B., and Strickland-Constable, R. F.,** Two-phase equilibrium in binary and ternary systems. X. Phase equilibria and compressibility of the systems carbon dioxide/propylene, carbon dioxide/ethylene and ethylene/propylene, and an account of the thermodynamic functions of the system carbon dioxide/propylene, *Proc. R. Soc. Chem. A,* 240, 1, 1957.

106. **Haselden, G. G., Newitt, D. M., and Shah, S. M.,** Two-phase equilibrium in binary and ternary systems. V. Carbon dioxide-ethylene. VI. Carbon dioxide-propylene, *Proc. R. Soc. Chem. A,* 209, 1, 1951.

107. **Winkler, C. A. and Maass, O.,** The critical temperatures and pressures of the three two-component systems comprised of carbon dioxide, methyl ether and propylene, *Can. J. Res.,* 6, 458, 1932.

108. **Niesen, V. G. and Rainwater, J. C.,** Critical locus, (vapor + liquid) equilibria, and coexisting densities of (carbon dioxide + propane) from 311 to 361 K, *J. Chem. Thermodyn.,* 22, 777, 1990.

109. **Poettmann, F. H. and Katz, D. L.,** Phase behavior of binary carbon dioxide-paraffin systems, *Ind. Eng. Chem.,* 37, 845, 1945.

110. **Reamer, H. H., Sage, B. H., and Lacey, W. N.,** Phase equilibria in hydrocarbon systems. Volumetric and phase behavior of the propane-carbon dioxide system, *Ind. Eng. Chem.,* 43, 2515, 1951.

111. **Besserer, G. J. and Robinson, D. B.,** Equilibrium-phase properties of i-butane-carbon dioxide system, *J. Chem. Eng. Data,* 18, 298, 1973.

112. **Weber, L. A.,** Simple apparatus for vapor-liquid equilibrium measurements with data for the binary systems of carbon dioxide with n-butane and isobutane, *J. Chem. Eng. Data,* 34, 171, 1989.

113. **Olds, R. H., Reamer, H. H., Sage, B. H., and Lacey, W. N.,** Phase equilibria in hydrocarbon systems. The n-butane-carbon dioxide system, *Ind. Eng. Chem.,* 41, 475, 1949.

114. **Pozo de Fernandez, M. E., Zollweg, J. A., and Streett, W. B.,** Vapor-liquid equilibrium in the binary system carbon dioxide + n-butane, *J. Chem. Eng. Data,* 34, 324, 1989.

115. **Kalra, H., Krishnan, T. R., and Robinson, D. B.,** Equilibrium phase properties of carbon dioxide-n-butane and nitrogen-hydrogen sulfide systems at subambient temperatures, *J. Chem. Eng. Data,* 21, 222, 1976.

116. **Besserer, G. J. and Robinson, D. B.,** A high pressure autocollimating refractometer for determining coexisting liquid and vapor phase densities, *Can. J. Chem. Eng.,* 49, 651, 1971.

117. **Leu, A. D. and Robinson, D. B.,** Equilibrium phase properties of the n-butane-carbon dioxide and isobutane-carbon dioxide binary systems, *J. Chem. Eng. Data,* 32, 444, 1987.

118. **Lu, H., Newitt, D. M., and Ruhemann, M.,** Two-phase equilibrium in binary and ternary systems. IV. The system ethane-propylene, *Proc. R. Soc. London A,* 178, 506, 1941.

119. **Besserer, G. J. and Robinson, D. B.,** Equilibrium phase properties of i-butane-ethane system, *J. Chem. Eng. Data,* 18, 301, 1973.

120. **Kay, W. B.,** Liquid-vapor equilibrium relations in binary systems. The ethane-n-butane system, *Ind. Eng. Chem.,* 32, 353, 1940.

121. **Kay, W. B.,** Liquid-Vapor Equilibrium Composition Relations in Binary Systems of Mixtures. II. Ethane-n-Butane System., Conference Report, Standard Oil Company, Whiting, IN, October 25, 1938.

122. **Lhotak, V. and Wichterle, I.,** Vapour-liquid equilibria in the ethane-n-butane system at high pressures, *Fluid Phase Equilibria,* 6, 229, 1981.

123. **Mehra, V. S. and Thodos, G.,** Vapor-liquid equilibrium in the ethane-n-butane system, *J. Chem. Eng. Data,* 10, 307, 1965.

124. **Dingrani, J. G. and Thodos, G.,** Vapor-liquid equilibrium behavior of ethane-n-butane-n-hexane system, *Can. J. Chem. Eng.,* 56, 616, 1978.

125. **Reamer, H. H., Sage, B. H., and Lacey, W. N.,** Phase equilibria in hydrocarbon systems. Volumetric and phase behavior of ethane-n-pentane system, *J. Chem. Eng. Data,* 5, 44, 1960.

126. **Hosotani, S., Maezawa, Y., Uematsu, M., and Watanabe, K.,** Measurements of PVTx properties for the R13B1 + R114 system, *J. Chem. Eng. Data,* 33, 20, 1988.

127. **Hasegawa, M., Uematsu, M., and Watanabe, K.,** Measurements of PVTx properties for the R22 + R114 system, *J. Chem. Eng. Data,* 30, 32, 1985.

128. **Takaishi, Y., Kagawa, N., Uematsu, M., and Watanabe, K.,** Volumetric properties of the binary mixtures dichlorodifluoromethane + chlorodifluoromethane, *Proc. 8th Symp. Therm. Prop.,* Vol. 2, p. 1982, 387.

129. **Higashi, Y.,** Private communication, 1985.

130. **Kay, W. B.,** Vapor-liquid relations of binary systems. The propane-n-alkane systems, n-butane and n-pentane, *J. Chem. Eng. Data,* 15, 46, 1970.

131. **Nysewander, C. N., Sage, B. H., and Lacey, W. N.,** Phase equilibria in hydrocarbon systems. The propane-n-butane system in the critical region, *Ind. Eng. Chem.,* 32, 118, 1940.

132. **Hissong, D. W., Kay, W. B., and Rainwater, J. C.,** Critical properties and vapor-liquid equilibria of the binary system propane-neopentane, *J. Chem. Eng. Data,* in preparation.

133. **Hissong, D. W.,** Phase Relationships of Binary Hydrocarbon Systems: Propane-Tetramethyl Methane, M.Sc. thesis, Ohio State University, Columbus, 1965.

134. **Vaughan, W. E. and Collins, F. C.,** P-V-T-x relations of the system propane-isopentane, *Ind. Eng. Chem.,* 34, 885, 1942.

135. **Sage, B. H. and Lacey, W. N.,** Phase equilibria in hydrocarbon systems. Propane-n-pentane system, *Ind. Eng. Chem.,* 32, 992, 1940.

136. **Chun, S.-W., Kay, W. B., and Rainwater, J. C.,** Critical properties and vapor-liquid equilibria of binary systems of propane with the hexane isomers, *J. Chem. Eng. Data,* in preparation.

137. **Chun, S. W.,** The Phase Behavior of Binary Systems in the Critical Region: Effect of Molecular Structure (The Propane-Isomeric Hexane Systems), Ph.D. thesis, Ohio State University, Columbus, 1964.

138. **Kay, W. B.,** Vapor-liquid equilibrium relationships of binary systems. Propane-n-alkane systems, n-hexane and n-heptane, *J. Chem. Eng. Data,* 16, 137, 1971.

139. **Porthouse, J. D.,** Phase Relations of Binary Hydrocarbon Systems Propane-n-Hexane, M.Sc. thesis, Ohio State University, Columbus, 1962.

140. **Glanville, J. W., Sage, B. H., and Lacey, W. N.,** Volumetric and phase behavior of propane-benzene system, *Ind. Eng. Chem.,* 42, 508, 1950.

141. **Sage, B. H. and Lacey, W. N.,** Some Properties of the Lighter Hydrocarbons, Hydrogen Sulfide and Carbon Dioxide, API Research Project 37, 1955.

142. **Kay, W. B., Genco, J., and Fichtner, D. A.,** Vapor-liquid equilibrium relationships of binary systems propane-n-octane and n-butane-n-octane, *J. Chem. Eng. Data,* 19, 275, 1974.

143. **Yada, N., Uematsu, M., and Watanabe, K.,** Study of the PVTx properties for Binary R152a + R114 system (in Japanese), *Trans. JAR,* 5, 107, 1988.

144. **Kay, W. B., Hoffman, R. L., and Davies, O.,** Vapor-liquid equilibrium relationships of binary systems n-butane-n-pentane and n-butane-n-hexane, *J. Chem. Eng. Data,* 20, 333, 1975.

145. **Kay, W. B.,** Liquid-vapor equilibrium relations in binary systems. n-Butane-n-heptane system, *Ind. Eng. Chem.,* 33, 590, 1941.

146. **Kay, W. B.,** Liquid-Vapor Equilibrium Composition Relations in Binary Systems of Mixtures. III. n-Butane-n-Heptane System, Rep. No. C40-25, Standard Oil Company, Whiting, IN, November 14, 1940.

147. **Krichevskii, I. R., Efremova, G. D., Pryanikova, R. O., and Polyakov, E. V.,** Phase and volume relations in the system acetic acid-butane (in Russian), *Khim. Prom.,* No. 7, 498, 1961.

148. **Spall, B. C.,** Phase equilibria in the system sulphur dioxide-water from 25-300°C, *Can. J. Chem. Eng.,* 41, 79, 1963.

149. **Davies, D. R. and McGlashan, M. L.,** Phase equilibria and critical constants of fluid (dodecafluorocyclohexane + hexafluorobenzene), *J. Chem. Thermodyn.,* 13, 377, 1981.

150. **Kay, W. B. and Donham, W. E.,** Liquid-vapor equilibria in the iso-butanol-n-butanol, methanol-n-butanol and diethyl ether-n-butanol systems, *Chem. Eng. Sci.,* 4, 1, 1955.

151. **Kay, W. B.,** Liquid-vapor equilibrium relations in binary systems. Ethylene-n-heptane system, *Ind. Eng. Chem.,* 40, 1459, 1948.

152. **Kay, W. B.**, Liquid-Vapor Equilibrium Composition Relations in Binary Systems. The Ethylene-n-Heptane System, Rep. No. F45-12, Standard Oil Company, Whiting, IN, August 22, 1945.

153. **Lyubetskii, S. G.**, Liquid-vapor equilibrium in the system ethylene-benzene, *J. Appl. Chem. U.S.S.R.*, 35, 125, 1962.

154. **Shakhova, S. F., Rutenberg, O. L., and Targanskaya, M. N.**, Volume relations in the epoxypropane-carbon dioxide system, *Russ. J. Phys. Chem.*, 47, 894, 1973.

155. **Kalra, H., Kubota, H., Robinson, D. B., and Ng, H.-J.**, Equilibrium phase properties of the carbon dioxide-n-heptane system, *J. Chem. Eng. Data*, 23, 317, 1978.

156. **Krichevskii, I. R. and Sorina, G. A.**, Liquid-gas phase equilibria in the cyclohexane-carbon dioxide and cyclohexane-nitrous oxide systems, *Russ. J. Phys. Chem.*, 34, 679, 1960.

157. **Shibata, S. K. and Sandler, S. I.**, High-pressure vapor-liquid equilibria of mixtures of nitrogen, carbon dioxide, and cyclohexane, *J. Chem. Eng. Data*, 34, 419, 1989.

158. **Inomata, H., Tuchiya, K., Arai, K., and Saito, S.**, Measurement of vapor-liquid equilibria at elevated temperatures and pressures using a flow type apparatus, *J. Chem. Eng. Jpn.*, 19, 386, 1986.

159. **Kay, W. B.**, Liquid-Vapor Phase Equilibrium Relations in the Ethane-n-Heptane System, Conference Report, Standard Oil Company, Whiting, IN., March 11, 1937.

160. **Mehra, V. S. and Thodos, G.**, Vapor-liquid equilibrium in the ethane-n-heptane system, *J. Chem. Eng. Data*, 10, 211, 1965.

161. **Kay, W. B. and Albert, R. E.**, Liquid-vapor equilibrium relations in the ethane-cyclohexane system, *Ind. Eng. Chem.*, 48, 422, 1956.

162. **Kay, W. B. and Nevens, T. D.**, Liquid-vapor equilibrium relations in binary systems. The ethane-benzene system, *Chem. Eng. Prog. Symp. Ser.*, 48(3), 108, 1952.

163. **Reamer, H. H. and Sage, B. H.**, Phase equilibria in hydrocarbon systems. Volumetric and phase behavior of the ethane-n-decane system, *J. Chem. Eng. Data*, 7, 161, 1962.

164. **Reamer, H. H. and Sage, B. H.**, Phase equilibria in hydrocarbon systems. Volumetric and phase behavior of the propane-n-decane system, *J. Chem. Eng. Data*, 11, 17, 1966.

165. **Reamer, H. H. and Sage, B. H.**, Phase equilibria in hydrocarbon systems. Phase behavior in the n-butane-n-decane system, *J. Chem. Eng. Data*, 9, 24, 1964.

166. **Bae, H. K., Nagahama, K., and Hirata, M.**, Isothermal vapor-liquid equilibria for the ethylene-carbon dioxide system at high pressure, *J. Chem. Eng. Data*, 27, 25, 1982.

167. **Khazanova, N. E., Lesnevskaya, L. S., and Zakharova, A. V.**, Liquid-vapor equilibrium in the ethane-CO$_2$ system (in Russian), *Khim. Prom.*, 42, 364, 1966.

168. **Ohgaki, K. and Katayama, T.**, Isothermal vapor-liquid equilibria data for the ethane-carbon dioxide system at high pressures, *Fluid Phase Equilibria*, 1, 27, 1977.

169. **Kuenen, J. P.**, Experiments on the condensation and critical phenomena of some substances and mixtures, *Philos. Mag. Ser.*, 5, 44, 174, 1897.

170. **Kuenen, J. P.**, On the condensation and the critical phenomena of mixtures of ethane and nitrous oxide, *Philos. Mag. Ser.*, 5, 40, 173, 1895.

171. **Quint, Gzn., N.**, Isotherms of mixtures of hydrogen chloride and ethane (in German), *Z. Phys. Chem.*, 39, 14, 1901.

172. **Kay, W. B. and Brice, D. B.**, Liquid-vapor relations in ethane-hydrogen sulfide system, *Ind. Eng. Chem.*, 45, 615, 1953.

173. **Clegg, H. P. and Rowlinson, J. S.**, The physical properties of some fluorine compounds and their solutions. II. The system sulphur hexafluoride + propane, *Trans. Faraday Soc.*, 51, 1333, 1955.

174. **Kay, W. B. and Rambosek, G. M.**, Liquid-vapor equilibrium relations in binary systems. Propane-hydrogen sulfide system, *Ind. Eng. Chem.*, 45, 221, 1953.

175. **Barber, J. R., Kay, W. B., and Teja, A. S.**, A study of the volumetric and phase behavior of binary systems. II. Vapor-liquid equilibria and azeotropic states of propane perfluorocyclobutane mixtures, *AIChE J.*, 28, 138, 1982.

176. **Kay, W. B. and Fisch, H. A.**, Phase relations of binary systems that form azeotropes. I. The ammonia-n-butane system, *AIChE J.*, 4, 293, 1958.

177. **Kay, W. B. and Warzel, F. M.**, Phase relations of binary systems that form azeotropes. II. The ammonia-isooctane system, *AIChE J.*, 4, 296, 1958.

178. **Aftienjew, J. and Zawisza, A.**, High pressure liquid-vapor equilibria, critical state, and p(V,T,x) up to 503.15 K and 4.560 MPa for n-pentane + n-perfluoropentane, *J. Chem. Thermodyn.*, 9, 153, 1977.

179. **Jordan, L. W. and Kay, W. B.**, Phase relations of binary systems that form azeotropes: n-alkanes-perfluoro-n-heptane systems (ethane through n-nonane), *Chem. Eng. Prog. Symp. Ser.*, 59(44), 46, 1963.

180. **Jordan, L. W., Jr.**, Liquid-Vapor Phase Behavior in the Critical Region in Systems which Form Azeotropes: The Binary Systems Perfluoro-n-Heptane with n-Alkanes, Ph.D. thesis, Ohio State University, Columbus, 1959.

181. **Hajjar, R. F., Cherry, R. H., and Kay, W. B.**, Critical properties of the vapor-liquid equilibria of the binary system acetone-n-pentane, *Fluid Phase Equilibria*, 25, 137, 1986.

182. **Genco, J. M., Teja, A. S., and Kay, W. B.,** Study of the critical and azeotropic behavior of binary mixtures. II. PVT-x data and azeotropic states of perfluoromethylcyclohexane-isomeric hexane systems, *J. Chem. Eng. Data,* 25, 355, 1980.

183. **Skaates, J. M. and Kay, W. B,** The phase relations of binary systems that form azeotropes. N-alkyl alcohol-benzene systems: methanol through n-butanol, *Chem. Eng. Sci.,* 19, 431, 1964.

184. **Toyama, A., Chappelear, P. S., Leland, T., and Kobayashi, R.,** Vapor-liquid equilibria at low temperatures: the carbon monoxide-methane system, *Adv. Cryogen. Eng.,* 7, 125, 1961.

185. **Schouten, J. A., Derenberg, A., and Trappeniers, N. J.,** Vapour-liquid and gas-gas equilibria in simple systems. IV. The system argon-krypton, *Physica,* 81A, 151, 1975.

186. **Nadler, K. C., Zollweg, J. A., Streett, W. B., and McLure, I. A.,** Surface tension of argon + krypton from 120 to 200 K, *J. Coll. Int. Sci.,* 122, 530, 1988.

187. **Zeininger, H.,** Liquid-vapor equilibriums of the binary systems nitrous oxide/nitrogen, nitrous oxide/oxygen, and nitrous oxide/methane at low temperatures and high pressures (in German), *Chem. Ing. Tech.,* 44, 607, 1972.

188. **Calado, J. C. G., Dieters, U., and Streett, W. B.,** Liquid-vapour equilibrium in the krypton + methane system, *J. Chem. Soc. Faraday Trans.,* 77, 2503, 1981.

189. **Guter, M., Newitt, D. M., and Ruhemann, M.,** Two-phase equilibrium in binary and ternary systems. II. The system methane-ethylene. III. The system methane-ethane-ethylene, *Proc. R. Soc. London, A,* 176, 140, 1940.

190. **Guter, M.,** Study of some Binary and Ternary Systems at High Pressure, Ph.D. thesis, Imperial College, London, 1939.

191. **Volova, L. M.,** Equilibrium of coexisting liquid and gas phases in the binary system methane-ethylene (in Russian), *Zh. Fiz. Khim.,* 14(2), 268, 1940.

192. **Donnelly, H. G. and Katz, D. L.,** Phase equilibria in the carbon dioxide-methane system, *Ind. Eng. Chem.,* 46, 511, 1954.

193. **Nohka, J., Sarashina, E., Arai, Y., and Saito, S.,** Correlation of vapor-liquid equilibria for systems containing a polar component by the BWR equation, *J. Chem. Eng. Jpn.,* 6, 10, 1973.

194. **Price, A. R. and Kobayashi, R.,** Low temperature vapor-liquid equilibrium in light hydrocarbon mixtures: methane-ethane-propane system, *J. Chem. Eng. Data,* 4, 40, 1959.

195. **Akers, W. W., Burns, J. F., and Fairchild, W. R.,** Low temperature phase equilibria: methane-propane system, *Ind. Eng. Chem.,* 46, 2531, 1954.

196. **Barsuk, S. D., Skripka, V. G., and Benyaminovich, O. A.,** Liquid-vapor equilibriums in a methane-isobutane system at low temperatures (in Russian), *Gazov. Promst.,* 15, 38, 1970.

197. **Roberts, L. R., Wang, R. H., Azernoosh, A.,and McKetta, J. J.,** Methane-n-butane system in the two-phase region, *J. Chem. Eng. Data,* 7, 484, 1962.

198. **Chu, T. C., Chen, R. J. J., Chappelear, P. S., and Kobayashi, R.,** Vapor-liquid equilibrium of methane-n-pentane system at low temperatures and high pressures, *J. Chem. Eng. Data,* 21, 41, 1976.

199. **Sage, B. H., Reamer, H. H., Olds, R H., and Lacey, W. N.,** Phase equilibria in hydrocarbon systems. Volumetric and phase behavior of methane-n-pentane system, *Ind. Eng. Chem.,* 34, 1108, 1942.

200. **Kahre, L. C.,** Low-temperature K data for methane-n-pentane, *J. Chem. Eng. Data,* 20, 363, 1975.

201. **Reiff, W. E., Peters-Girth, P., and Lucas, K.,** A static equilibrium apparatus for (vapor + liquid) equilibrium measurements at high temperatures and pressures. Results for (methane + n-pentane), *J. Chem. Thermodyn.,* 19, 467, 1987.

202. **Calado, J. C. G., Chang, E., and Streett, W. B.,** Vapour-liquid equilibrium in the krypton-xenon system, *Physica,* 117A, 127, 1983.

203. **Calado, J. C. G., Chang, E., Clancy, P., and Streett, W. B.,** Vapor-liquid equilibrium in the krypton + ethane system, *J. Phys. Chem.,* 97, 3914, 1987.

204. **Glocker, G., Fuller, D. L., and Roe, C. P.,** Binary systems in two phases. I. HCl-Kr. II. HCl-C_3H_8, *J. Chem. Phys.,* 1, 714, 1933.

205. **Roe, C. P.,** The Solubilities of Various Gases in Various Liquids at High Pressures. II. The System HCl-Kr, Ph.D. thesis, University of Minnesota, Minneapolis, 1933.

206. **Kubic, W. L. and Stein, F. P.,** An experimental and correlative study of the vapor-liquid equilibria of the tetrafluoromethane-chlorotrifluoromethane system, *Fluid Phase Equilbria,* 5, 289, 1980.

207. **Kubota, H., Inatome, H., Tanaka, Y., and Makita, T.,** Vapor-liquid equilibrium of the ethylene-propylene system under high pressure, *J. Chem. Eng. Jpn.,* 16, 99, 1983.

208. **Zernov, V. S., Kogan, V. B., and Lyubetskii, S. G.,** Thermodynamic consistency of data on liquid-vapor equilibria in binary systems at high pressures, *J. Appl. Chem. U.S.S.R.,* 50, 1488, 1977.

209. **Nunes da Ponte, M., Chokappa, D., Calado, J. C. G., Clancy, P., and Streett, W. B.,** Vapor-liquid equilibrium in the xenon + ethane system, *J. Phys. Chem.,* 89, 2746, 1985.

210. **Tsang, C. Y. and Streett, W. B.,** Vapor-liquid equilibrium in the system carbon dioxide/dimethyl ether, *J. Chem. Eng. Data,* 26, 155, 1981.

211. **Shah, N. N., Pozo de Fernandez, M. E., Zollweg, J. A., and Streett, W. B.,** Vapor-liquid equilibrium in the system carbon dioxide + 2,2-dimethylpropane from 262 to 424 K at pressures to 8.4 MPa, *J. Chem. Eng. Data,* 35, 278, 1990.

212. **Shah, N. N.,** Vapor-Liquid Equilibrium Study for the Binary Systems Carbon Dioxide/Neopentane and Carbon Dioxide/Cyclopentane, Ph.D. thesis, Cornell University, Ithaca, NY, 1985.

213. **Leu, A.-D. and Robinson, D. B.,** Equilibrium-phase properties of the neopentane-carbon dioxide binary system, *J. Chem. Eng. Data,* 33, 313, 1988.

214. **Cheng, H., Zollweg, J. A., and Streett, W. B.,** in preparation.

215. **Besserer, G. J. and Robinson, D. B.,** Equilibrium phase properties of isopentane-carbon dioxide system, *J. Chem. Eng. Data,* 20, 93, 1975.

216. **Leu, A.-D. and Robinson, D. B.,** Equilibrium phase properties of selected carbon dioxide binary systems: n-pentane-carbon dioxide and isopentane-carbon dioxide, *J. Chem. Eng. Data,* 32, 447, 1987.

217. **Cheng, H., Pozo de Fernandez, M. E., Zollweg, J. A., and Streett, W. B.,** Vapor-liquid equilibrium in the system carbon dioxide + n-pentane from 252 to 458 K at pressures to 10 MPa, *J. Chem. Eng. Data,* 34, 319, 1989.

218. **Besserer, G. J. and Robinson, D. B.,** Equilibrium phase properties of n-pentane-carbon dioxide system, *J. Chem. Eng. Data,* 18, 416, 1973.

219. **Radosz, M.,** Vapor-liquid equilibrium for 2-propanol and carbon dioxide, *J. Chem. Eng. Data,* 31, 43, 1986.

219a. **Shah, N. N., Zollweg, J. A., and Streett, W. B.,** Vapor-liquid equilibrium in the system carbon dioxide-cyclopentane, *J. Chem. Eng. Data,* in press, 1991.

220. **Schneider, G., Alwani, Z., Heim, W., Horvath, E., and Franck, E. U.,** Phase equilibriums and critical phenomena in binary mixed systems to 1500 bars. CO₂ with n-octane, n-undacene, n-tridecane and n-hexadecane (in German), *Chem. Ing. Tech.,* 39, 649, 1967.

221. **Ng, H.-J. and Robinson, D. B.,** The equilibrium phase properties of selected naphthenic binary systems. Carbon dioxide-methylcyclohexane, hydrogen sulfide-methylcyclohexane, *Fluid Phase Equilibria,* 2, 283, 1979.

222. **Ng, H.-J., Huang, S. S.-S., and Robinson, D. B.,** Equilibrium phase properties of selected m-xylene binary systems. m-xylene-methane and m-xylene-carbon dioxide, *J. Chem. Eng. Data,* 27, 119, 1982.

223. **Huang, S. S.-S. and Robinson, D. B.,** The equilibrium phase properties of selected mesitylene binary systems: mesitylene-methane and mesitylene-carbon dioxide, *Can. J. Chem. Eng.,* 63, 126, 1985.

224. **Matschke, D. E. and Thodos, G.,** Vapor-liquid equilibria for the ethane-propane system, *J. Chem. Eng. Data,* 7, 232, 1962.

225. **Miksovsky, J. and Wichterle, I.,** Vapour-liquid equilibria in the ethane-propane system at high pressures, *Collect. Czech. Chem. Commun.,* 40, 365, 1975.

226. **McCurdy, J. L. and Katz, D. L.,** Phase equilibria in the systems propylene-acetylene and propane-acetylene, *Oil Gas J.,* 43(44), 102, 1945.

227. **Fuller, D. L.,** The Solubility of Various Gases in Various Liquids at High Pressures. I. The System Hydrogen Chloride-Propane, Ph.D. thesis, University of Minnesota, Minneapolis, 1933.

228. **Leu, A.-D. and Robinson, D. B.,** Equilibrium phase properties of the n-butane-hydrogen sulfide and isobutane-hydrogen sulfide systems, *J. Chem. Eng. Data,* 34, 315, 1989.

229. **Robinson, D. B., Hughes, R. E., and Sandercock, J. A.,** Phase behavior of the n-butane-hydrogen sulfide system, *Can. J. Chem. Eng.,* 42, 143, 1964.

230. **Pozo de Fernandez, M. E.,** Fluid Phase Equilibria Studies for the Binary Systems Carbon Dioxide/n-Butane, Dimethyl Ether/n-Butane and Dimethyl Ether/1-Butene, Ph.D. thesis, Cornell University, Ithaca, NY, 1986.

231. **Chang, E., Calado, J. C. G., and Streett, W. B.,** Vapor-liquid equilibrium in the system dimethyl ether/methanol from 0 to 180°C and at pressures to 6.7 MPa, *J. Chem. Eng. Data,* 27, 293, 1982.

232. **Maheshwari, R. C. and Lentz, H.,** Phase equilibria and critical curve of the ammonia-trans-decahydro-naphthalene system, *Fluid Phase Equilibria,* 41, 195, 1988.

233. **Campbell, A. N. and Chatterjee, R. M.,** Vapor-liquid equilibrium in the system acetone-benzene, *Can. J. Chem.,* 48, 277, 1970.

234. **Ewing, M. B., McGlashan, M. L., and Tzias, P.,** Phase equilibria, and critical temperatures and pressures, of fluid (benzene + hexafluorobenzene), *J. Chem. Thermodyn.,* 13, 527, 1981.

235. **Thies, M. C., Daniel, W. E., and Todd, M. A.,** Vapor-liquid equilibrium for tetralin/toluene mixtures at elevated temperatures and pressures, *J. Chem. Eng. Data,* 33, 134, 1988.

236. **Heck, C. K. and Barrick, P. L.,** Liquid-vapor phase equilibria of the neon-normal hydrogen system, *Adv. Cryogen. Eng.,* 11, 349, 1966.

237. **Nunes da Ponte, M., Chokappa, D., Calado, J. C. G., Zollweg, J., and Streett, W. B.,** Vapor-liquid equilibrium in the xenon + ethene system, *J. Phys. Chem.,* 90, 1147, 1986.

238. **Noles, J. R., Zollweg, J. A., and Streett, W. B.,** Vapor-liquid equilibrium in the associating system dimethyl ether/sulfur dioxide, presented at the A.I.Ch.E. 1989 Annual Meeting, San Francisco, November 9, 1989.

239. **Barr-David, F. and Dodge, B. F.,** Vapor-liquid equilibrium at high pressures. The systems ethanol-water and 2-propanol-water, *J. Chem. Eng. Data,* 4, 107, 1959.

240. **Wichterle, I., Linek, J., and Hala, E.,** *Vapor-Liquid Equilibrium Data Bibliography,* Elsevier, New York, 1973.

241. **Wichterle, I., Linek, J., and Hala, E.,** *Vapor-Liquid Equilibrium Data Bibliography,* Supplement I, Elsevier, New York, 1976.

242. **Wichterle, I., Linek, J., and Hala, E.,** *Vapor-Liquid Equilibrium Data Bibliography,* Supplement II, Elsevier, New York, 1979.

243. **Wichterle, I., Linek, J., and Hala, E.,** *Vapor-Liquid Equilibrium Data Bibliography,* Supplement III, Elsevier, New York, 1982.

244. **Wichterle, I., Linek, J., and Hala, E.,** *Vapor-Liquid Equilibrium Data Bibliography,* Supplement IV, Elsevier, New York, 1985.

245. **Hiza, M. J., Kidnay, A. J., and Miller, R. C.,** *Equilibrium Properties of Fluid Mixtures,* Vol. 2, *A Bibliography of Experimental Data on Selected Fluids,* Plenum Press, New York, 1982.

246. **Hirata, M., Ohe, S., and Nagahama, K.,** *Computer-Aided Data Book of Vapor-Liquid Equilibria,* Elsevier, New York, 1975.

247. **Knapp, H., Doring, R., Oellrich, L., Plocker, U., and Prausnitz, J. M.,** *Vapor-Liquid Equilibria for Mixtures of Low-Boiling Substances,* DECHEMA, Frankfurt West Germany, 1982.

248. **Williamson, F. R. and Olien, N. A.,** Compilation and Evaluation of Available Data on Phase Equilibria of Natural and Synthetic Gas Mixtures, NBS Int. Rep. NBSIR 83-1692, National Bureau of Standards, Washington, D.C. June 1983.

249. **Kreglewski, A.,** *Equilibrium Properties of Fluids and Fluid Mixtures,* Texas A & M University Press, College Station, 1984.

250. **Kidnay, A. J., Hiza, M. J., and Miller, R. C.,** Liquid-Vapor Equilibria Research on Systems of Interest in Cryogenics — A Survey, *Cryogenics,* 13, 575, 1973.

251. **Hicks, C. P. and Young, C. L.,** The gas-liquid critical properties of binary mixtures, *Chem. Rev.,* 75, 119, 1975.

252. **Magruder, L. B. and Mitchell, G. R.,** High Pressure Rectification, M.Sc. thesis, Massachusetts Institute of Technology, Cambridge, 1940.

253. **Caubet, F.,** On the liquefaction of gaseous mixtures of carbon dioxide and sulfur dioxide (in French), *Comp. Rend.,* 130, 828, 1900.

254. **Zawisza, A. C.,** High-pressure phase equilibria and the compressibility of the sulphur dioxide-diethyl ether system, *Bull. Acad. Polon. Sci. Ser. Sci. Chim.,* 15, 291, 1967.

255. **Khazanova, N. E., Sominskaya, E. E., Zakharova, A. V., and Nechitailo, N. L.,** Phase and volume relations and the critical curves for azeotropic systems (ethane-ammonia), *Russ. J. Phys. Chem.,* 46, 319, 1972.

256. **Khazanova, N. E., Sominskaya, E. E., Zakharova, A. V., and Timofeeva, G. V.,** Systems with an azeotrope at high pressures. I. Phase and volume relations and critical curves in the carbon dioxide-sulphur hexafluoride system, *Russ. J. Phys. Chem.,* 48, 950, 1974.

257. **Konyakhin, V. P.,** Phase equilibrium in the ethanol-cyclohexane system, *Russ. J. Phys. Chem.,* 48, 143, 1974.

258. **Rodrigues, A. B. J., McCaffrey, D. S., and Kohn, J. P.,** Heterogeneous phase and volumetric equilibrium in the ethane-n-octane system, *J. Chem. Eng. Data,* 13, 164, 1968.

259. **Wichterle, I. and Kobayashi, R.,** Vapor-liquid equilibrium of methane-ethane system at low temperature and high pressure, *J. Chem. Eng. Data,* 17, 9, 1972.

260. **Zawisza, A. C.,** High-pressure vapour-liquid equilibria and the critical state of the trifluoroacetic acid-perfluorobutane system, *Bull. Acad. Polon. Sci.,* 15, 307, 1967.

261. **Boomer, E. H., Johnson, C. A., and Piercey, A. G. A.,** Equilibria in two-phase, gas-liquid hydrocarbon systems. II. Methane and pentane, *Can. J. Res. B,* 16, 319, 1938.

262. **Ottenweller, J. H., Holloway, C., and Weinrich, W.,** Liquid-vapor equilibrium compositions in hydrogen chloride-n-butane system, *Ind. Eng. Chem.,* 35, 207, 1943.

263. **Kay, W. B. and Kreglewski, A.,** Critical locus curves and liquid-vapor equilibria in the binary systems of benzene with carbon dioxide and sulfur dioxide, *Fluid Phase Equilibria,* 11, 251, 1983.

264. **Fredenslund, A. and Sather, G. A.,** Gas-liquid equilibrium of the oxygen-carbon dioxide system, *J. Chem. Eng. Data,* 15, 17, 1970.

265. **Kaminishi, G., Arai, Y., Saito, S., and Maeda, S.,** Vapor-liquid equilibria for binary and ternary systems containing carbon dioxide, *J. Chem. Eng. Jpn.,* 1, 109, 1968.

266. **de Loos, Th. W., Poot, W., and de Swaan Arons, J.,** Vapor liquid equilibria and critical phenomena in methanol + n-alkane systems, *Fluid Phase Equilibria,* 42, 209, 1988.

267. **Do, V. T. and Straub, J.,** Surface tension, coexistence curve, and vapor pressure of binary gas-liquid mixtures, *Int. J. Thermophys.,* 7, 41, 1986.

268. **McCurdy, J. L. and Katz, D. L.,** Phase equilibria in the system ethane-ethylene-acetylene, *Ind. Eng. Chem.,* 36, 674, 1944.

269. **Gomez-Nieto, M. and Thodos, G.,** Vapor-liquid equilibrium measurements for the propane-ethanol system at elevated pressures, *AIChE J.,* 24, 672, 1978.

270. **Amick, E. H., Johnson, W. B., and Dodge, B. F.,** P-V-T-x relationships for the system: methane-isopentane, *Chem. Eng. Prog. Symp. Ser.,* 48(3), 65, 1952.

271. **Semenova, A. E., Emel'yanova, E. A., Tsimmerman, S. S., and Tsiklis, D. S.,** Phase equilibria in the methanol-carbon dioxide system, *Russ. J. Phys. Chem.,* 53, 1428, 1979.

272. **Churchill, S. W., Collamore, W. G., and Katz, D. L.,** Phase behavior of the acetylene-ethylene system, *Oil Gas J.,* 41(13), 33, 1942.

273. **Vejrosta, J. and Wichterle, I.,** The propane-pentane system at high pressures, *Collect. Czech. Chem. Commun.,* 39, 1246, 1974.

274. **Mislavskaya, V. S. and Khodeeva, S. M.,** Liquefied acetylene as solvent. II. The system acetylene-n-hexane, *Russ. J. Phys. Chem.,* 43, 1062, 1969.

275. **Campbell, A. N. and Musbally, G. M.,** Vapor pressures and vapor-lquid equilibria in the systems: (1) acetone-chloroform, (2) acetone-carbon tetrachloride, (3) benzene-carbon tetrachloride, *Can. J. Chem.,* 48, 3173, 1970.

276. **Reamer, H. H., Sage, B. H., and Lacey, W. N.,** Phase equilibria in hydrocarbon systems. Volumetric and phase behavior of the methane-cyclohexane system, *Ind. Eng. Chem. Data Ser.,* 3, 240, 1958.

277. **Reamer, H. H., Sage, B. H., and Lacey, W. N.,** Phase equilibria in hydrocarbon systems. Volumetric and phase behavior of the methane-n-heptane system, *Ind. Eng. Chem. Data Ser.,* 1, 29, 1956.

278. **Lenoir, J. M., Hayworth, K. E., and Hipkin, H. G.,** Enthalpies of benzene and mixtures of benzene with n-octane, *J. Chem. Eng. Data,* 16, 280, 1971.

279. **Lenoir, J. M., Hayworth, K. E., and Hipkin, H. G.,** Enthalpies of mixtures of benzene and cyclohexane, *J. Chem. Eng. Data,* 16, 285, 1971.

279a. **Lenoir, J. M., Hayworth, K. E., and Hipken, H. G.,** Enthalpies of decalins and of *trans*-decalin and *n*-pentane mixtures, *J. Chem. Eng. Data,* 16, 129, 1971.

280. **Lenoir, J. M. and Hipkin, H. G.,** Measured enthalpies for mixtures of benzene with n-pentane, *J. Chem. Eng. Data,* 17, 319, 1972.

281. **Pando, C., Renuncio, J. A. R., Schofield, R. S., Izatt, R. M., and Christensen, J. J.,** The excess enthalpies of (carbon dioxide + toluene) at 308.15, 385.25 and 413.15 K from 7.60 to 12.67 MPa, *J. Chem. Thermodyn.,* 15, 747, 1983.

282. **Christensen, J. J., Zebolsky, D. M., and Izatt, R. M.,** The excess enthalpies of (carbon dioxide + toluene) at 470.15 and 573.15 K from 7.60 to 12.67 MPa, *J. Chem. Thermodyn.,* 17, 1, 1985.

283. **Cordray, D. R., Christensen, J. J., Izatt, R. M. and Oscarson, J. L.,** The excess enthalpies of (carbon dioxide + toluene) at 390.15, 413.15, 470.15, and 508.15 K from 7.60 to 17.50 MPa, *J. Chem. Thermodyn.,* 20, 877, 1988.

284. **Selover, T. B.,** DIPPR: Past-Present-Fugure, Paper No. 40k, presented at the American Institute of Chemical Engineers, 1989 Annual Meeting, San Francisco, November 9, 1989.

285. **Meijer, P. H. E.,** The van der Waals equation of state around the van Laar point, *J. Chem. Phys.,* 90, 448, 1989.

286. **Dieters, U. K. and Pegg, I. L.,** Systematic investigation of the phase behavior in binary fluid mixtures. I. Calculations based on the Redlich-Kwong equation of state, *J. Chem. Phys.,* 90, 6632, 1989.

287. **Ree, F. H.,** Analytic representation of thermodynamic data for the Lennard-Jones fluid, *J. Chem. Phys.,* 73, 5401, 1980.

288. **Boshkov, L. Z. and Mazur, V. A.,** Phase equilibria and critical lines of binary mixtures of Lennard-Jones molecules, *Russ. J. Phys. Chem.,* 60, 16, 1986.

289. **Romig, K. D., Jr. and Hanley, H. J. M.,** Phase equilibria from the one fluid model, *Cryogenics,* 29, 65, 1989.

290. **Streett, W. B.,** Phase equilibria in molecular hydrogen-helium mixtures at high pressures, *Astrophys. J.,* 186, 1107, 1983.

291. **Kogan, V. S., Lazarev, B. G., and Bulatova, R. F.,** The phase diagram of the hydrogen-deuterium system, *Sov. Phys. JETP,* 7, 165, 1958.

292. **Simon, M.,** On the thermodynamic properties of H_2-D_2 solid mixtures, *Phys. Lett.,* 9, 122, 1964.

293. **Bereznyak, N. G., Bogoyavlenskii, I. V., Karnatsevich, L. V., and Kogan, V. S.,** Melting diagram of the pH2-oD2, pH2-HD and oD-HD systems, *Sov. Phys. JETP,* 30, 1048, 1970.

294. **Bereznyak, N. G., Bogoyavlenskii, I. V., and Karnatsevich, L. V.,** Vapor pressure of liquid hydrogen-deuterium solutions below 20.4 K, *Sov. Phys. JETP,* 36, 304, 1973.

295. **Blagoi, Yu. P., Zimohlyad, B. N., and Zhun, G. G.,** Selective adsorption of the ortho- and para-modifications and separation of the hydrogen isotopes, *Russ. J. Phys. Chem.,* 43, 691, 1969.

296. **Streett, W. B. and Jones, C. H.,** Liquid phase separation and liquid-vapor equilibrium in the system neon-hydrogen, *J. Chem. Phys.,* 42, 3989, 1965.

297. **Streett, W. B.,** Liquid-phase separation and liquid-vapor equilibrium in the system neon-deuterium, *Proc. 2nd. Int. Cryogenic Engineering Conference,* Science and Technology Publications, Guildford, U.K. 1968, 260.

298. **Van Poolen, L. J. and Graham, L.,** Dew point and gas-phase data for a mixture of 30.6 mole percent nitrogen — 69.4 percent R-14 (CF_4), *ASHRAE Trans.,* 78(Part 2), 92, 1972.

299. **Zakharov, N. D., Matyash, Yu. I., Fisher, E. A., and Domnina, E. V.,** Critical parameters of binary mixtures containing nitrogen and freons, *Russ. J. Phys. Chem.,* 51, 1385, 1977.

300. **Jones, I. W. and Rowlinson, J. S.,** Gas-liquid critical temperatures of binary mixtures. II., *Trans. Faraday, Soc.,* 59, 1702, 1963.

301. **Gasem, K. A. M., Hiza, M. J., and Kidnay, A. J.,** Phase behavior in the nitrogen + ethylene system from 120 to 200 K, *Fluid Phase Equilibria,* 6, 181, 1981.

302. **Ellington, R. T., Eakin, B. E., Parent, J. D., Gami, D. C., and Bloomer, O. T.,** in *Thermodynamic and Transport Properties of Gases, Liquids and Solids,* McGraw-Hill, New York, 1958. 180.

303. **Kreglewski, A. and Hall, K. R.,** Phase equilibria calculated for the systems N_2 + CO_2, CH_4 + CO_2 and CH_4 + H_2S, *Fluid Phase Equilibria,* 15, 11, 1983.

304. **Wirths, M. and Schneider, G. M.,** High pressure phase studies on fluid binary mixtures of hydrocarbons with tetrafluoromethane and trifluoromethane between 273 and 630 K and up to 250 MPa, *Fluid Phase Equilibria,* 21, 257, 1985.

305. **Scott, R. L.,** The solubility of fluorocarbons, *J. Am. Chem. Soc.,* 70, 4090, 1948.

306. **Scott, R. L.,** The anomalous behaviour of fluorocarbon solutions, *J. Phys. Chem.,* 62, 136, 1958.

307. **Simons, J. H. and Dunlop, R. D.,** The properties of n-pentforane and its mixtures with n-pentane, *J. Chem. Phys.,* 18, 335, 1950.

308. **Hildebrand, J. H. and Rotariu, G. J.,** Molecular order in n-heptane and perfluoro-n-heptane, *J. Am. Chem Soc.,* 74, 4455, 1952.

309. **Swinton, F. L.,** Mixtures containing a fluorocarbon, in *Chemical Thermodynamics,* Vol. 2, *Specialist Periodical Reports,* The Chemical Society, London, 1978, 147.

310. **Cooney, A. and Morcom, K. W.,** Thermodynamic behavior of mixtures containing fluoroalcohols. II. (2,2,2-Trifluoroethanol + benzene), *J. Chem. Thermodyn.,* 20, 1469, 1988.

311. **Aracil, J., Rubio, R. G., Caceres, M., Diaz Pena, M., and Renuncio, J. A. R.,** Thermodynamics of fluorocarbon-hydrocarbon mixtures, *J. Chem. Soc. Faraday Trans. I,* 84, 539, 1988.

312. **Hudson, G. H. and McCoubrey, J. C.,** Intermolecular forces between unlike molecules. A more complete form of the combining rules, *Trans. Faraday Soc.,* 56, 761, 1960.

313. **McLinden, M. O. and Didion, D. A.,** Thermophysical-property needs for the environmentally acceptable halocarbon refrigerants, *Int. J. Thermophys.,* 10, 563, 1989.

314. **Orobinskii, N. A., Blagoi, Yu. P., and Semyannikova, E. L.,** Liquid-vapor phase equilibrium of the argon-propylene and krypton-propylene systems at high temperatures, *Ukran. Fiz. Zh. (Russ. Ed.,),* 13, 262, 1968.

315. **McKinley, C. and Wang, E. S. J.,** Hydrocarbon-oxygen systems stability, *Adv. Cryogen. Eng.,* 4, 11, 1958.

316. **Brunner, E.,** Fluid mixtures at high pressures. II. Phase separation and critical phenomena of (ethane + an n-alkanol) and of (ethene + methanol) and (propane + methanol), *J. Chem. Thermodyn.,* 17, 871, 1985.

317. **Brunner, E. and Hultenschmidt, W.,** Fluid mixtures at high pressures. VIII. Isothermal phase equilibria in the binary mixtures: (ethanol + hydrogen or methane or ethane), *J. Chem. Thermodyn.,* 22, 73, 1990.

318. **Sebastian, H. M., Nageshwar, G. D., Lin, H. M., and Chao, K. C.,** Gas-liquid equilibria in mixtures of carbon dioxide and tetralin at elevated temperatures, *Fluid Phase Equilibria,* 4, 257, 1980.

319. **Sebastian, H. M., Lin, H. M., and Chao, K. C.,** Gas-liquid equilibrium of carbon dioxide plus m-cresol and carbon dioxide plus quinoline at elevated temperatures, *J. Chem. Eng. Data,* 25, 381, 1980.

320. **Inomata, H., Arai, K., and Saito, S.,** Vapor-liquid equilibria for CO_2/hydrocarbon mixtures at elevated temperatures and pressures, *Fluid Phase Equilibria,* 36, 107, 1987.

321. **Niesen, V. G., Palavra, A., Kidnay, A. J., and Yesavage, V. F.,** An apparatus for vapor-liquid equilibrium at elevated temperatures and pressures and selected results for the water-ethanol and methanol-ethanol systems, *Fluid Phase Equilibria,* 31, 283, 1986.

322. **Lentz, H. and Franck, E. U.,** Phase equilibria and critical curves of binary ammonia-hydrocarbon mixtures, in *Extraction with Supercritical Gases,* Schneider, G. M., Stahl, E., and Wilke, G., Eds., Verlag Chemie, Weinheim, 1980, 83 (translation of *Angew. Chem.,* 90, 775, 1978, in German).

323. **Oshmyansky, Y., Hanley, H. J. M., Ely, J. F., and Kidnay, A. J.,** The viscosities and densities of selected organic compounds and mixtures of interest in coal liquefaction studies, *Int. J. Thermophys.,* 7, 599, 1986.

324. **Griswold, J. and Wong, S. Y.,** Phase equilibrium of the acetone-methanol-water system from 100°C into the critical region, *Chem. Eng. Prog. Symp. Ser.,* 48(3), 18, 1952.

325. **Marshall, W. L. and Jones, E. V.,** Liquid-vapor critical temperatures of several aqueous-organic and organic-organic solution systems, *J. Inorg. Nucl. Chem.,* 36, 2319, 1974.

326. **Chirikova, Z. P., Galata, L. A., Kotova, Z. N., and Kofman, L. S.,** Liquid-vapour equilibrium in the system acetonitrile-water, *Russ. J. Phys. Chem.,* 40, 493, 1966.

327. **Pozo, M. E. and Streett, W. B.,** Fluid phase equilibria for the system dimethyl ether/water from 50 to 220°C and pressures to 50.9 MPa, *J. Chem. Eng. Data,* 29, 324, 1984.

328. **Selleck, F. T., Carmichael, L. T., and Sage, B. H.,** Phase behavior in the hydrogen sulfide-water system, *Ind. Eng. Chem.,* 44, 2219, 1952.

329. **Toedheide, K. and Franck, E. U.,** Two-phase range and the critical curve in the system carbon dioxide-water up to 3500 bar (in German), *Z. Phys. Chem.,* 37, 387, 1963.

330. **Jangkamolkulchai, A., Lam, D. H., and Luks, K. D.,** Muliphase equilibria of nitrous oxide + n-paraffin mixtures, *Fluid Phase Equilibria,* 50, 175, 1989.

331. **Levelt Sengers, J. M. H., Everhart, C. M., Morrison, G., and Pitzer, K. S.,** Thermodynamic anomalies in near-critical aqueous NaCl solutions, *Chem. Eng. Commun.,* 47, 315, 1986.

332. **Morrison, G.,** Modelling aqueous solutions near the critical point of water, *J. Sol. Chem.,* 17, 887, 1988.

333. **Harvey, A. H. and Levelt Sengers, J. M. H.,** On the NaCl-H_2O coexistence curve near the critical temperature of H_2O, *Chem. Phys. Lett.,* 156, 415, 1989.

334. **Bischoff, J. L., Rosenbauer, R. J., and Pitzer, K. S.,** The system NaCl-H_2O: Relations of vapor-liquid near the critical temperature of water and of vapor-liquid-halite from 300 to 500°C, *Geochim. Cosmochim. Acta,* 50, 1437, 1986.

335. **Pitzer, K. S., Bischoff, J. L., and Rosenbauer, R. J.,** Critical behavior of dilute NaCl in H_2O, *Chem. Phys. Lett.,* 134, 60, 1987.

336. **Bischoff, J. L. and Rosenbauer, R. J.,** Liquid-vapor relations in the critical region of the system NaCl-H_2O from 380 to 415°C: a refined determination of the critical point and the two-phase boundary of seawater, *Geochim. Cosmochim. Acta,* 52, 2121, 1988.

337. **Tangier, J. C. and Pitzer, K. S.,** Thermodynamics of NaCl-H_2O: a new equation of state for the near-critical region and comparisons with other equations for adjoining regions, *Geochim. Cosmochim. Acta,* 53, 973, 1989.

338. **Pitzer, K. S.,** Fluids, both ionic and non-ionic, over wide ranges of temperature and composition, *J. Chem. Thermodyn.,* 21, 1, 1989.

339. **Bach, R. W., Friedrichs, H. A. and Rau, H.,** P-V-T relations for HCl-H_2O mixtures up to 500°C and 1500 bars, *High Temp. High Press.,* 9, 305, 1977.

340. **Jockers, R. and Schneider, G. M.,** Fluid mixtures at high pressures. Fluid phase equilibria in the systems fluorobenzene + water, 1,4-difluorobenzene + water, and 1,2,3,4-tetrahydronaphthalene + decahydro-naphthalene(trans) + water up to 360 MPa, *Ber. Bunsenges. Phys. Chem.,* 82, 576, 1978.

340a. **Wu, G., Heilig, M., Lentz, H., and Franck, E. U.,** High pressure phase equilibrium of the water-argon system, *Ber. Bunsenges. Phys. Chem.,* 94, 22, 1990.

341. **Haase, R., Naas, H., and Thumm, H.,** The thermodynamic behavior of concentrated hydrohalic acids (in German), *Z. Phys. Chem.,* 37, 210, 1963.

342. **Kao, J. T. F.,** Vapor-liquid equilibrium of water-hydrogen chloride system, *J. Chem. Eng. Data,* 15, 362, 1970.

343. **Arai, Y., Kaminishi, G., and Saito, S.,** The experimental determination of the P-V-T-x relations for the carbon dioxide-nitrogen and the carbon dioxide-methane systems, *J. Chem. Eng. Jpn.,* 4, 113, 1971.

344. **Reamer, H. H., Sage, B. H., and Lacey, W. N.,** Phase equilibria in hydrocarbon systems. Volumetric and phase behavior of the methane-propane system, *Ind. Eng. Chem.,* 42, 534, 1950.

345. **Olds, R. H., Sage, B. H., and Lacey, W. N.,** Methane-isobutane system, *Ind. Eng. Chem.,* 34, 1008, 1942.

346. **Sage, B. H., Hicks, B. L., and Lacey, W. N.,** Phase equilibria in hydrocarbon systems. The methane-n-butane system in the two-phase region, *Ind. Eng. Chem.,* 32, 1085, 1940.

347. **Kay, W. B. and Hissong, D.,** The critical properties of hydrocarbons. I. Simple mixtures, *Proc. Am. Petrol. Inst. Refining Div.,* 47, 653, 1967.

348. **Hissong, D. W. and Kay, W. B.,** The critical properties of hydrocarbons. II. Correlational studies, *Proc. Am. Petrol. Inst. Refining Div.,* 48, 397, 1968.

349. **Kay, W. B. and Hissong, D.,** The critical properties of hydrocarbon mixtures. III. Prediction of critical constants, *Proc. Am. Petrol. Inst. Refining Div.,* 49, 13, 1969.

350. **Abara, J. A., Jennings, D. W., Kay, W. B., and Teja, A. S.,** Phase behavior in the critical region of six binary mixtures of 2-methylalkanes, *J. Chem. Eng. Data,* 33, 242, 1988.

351. **Kay, W. B.,** P-T-x diagrams in the critical region. Acetone-n-alkane systems, *J. Phys. Chem.,* 68, 827, 1964.

352. **Kreglewski, A. and Kay, W. B.,** Critical constants of conformal mixtures, *J. Phys. Chem.,* 73, 3359, 1969.

353. **Mousa, A. E. H. N., Kay, W. B., and Kreglewski, A.,** The critical constants of certain perfluorocompounds with alkanes, *J. Chem. Thermodyn.,* 4, 301, 1972.

354. **Pak, S. C. and Kay, W. B.,** The critical properties of binary hydrocarbon systems, *I & EC Fundam.,* 11, 255, 1972.

355. **Lien, R. H.,** Research on the Stability of an N_2O-Metal Powder System, presented at the Northwest Conference on Research in Supercritical Fluids, Moscow, ID, April 1988.

356. **Lemkowitz, S. M., Goedegbuur, J., and van den Berg, P. J.,** Bubble point measurements in the ammonia-carbon dioxide system, *J. Appl. Chem. Biotechnol.,* 21, 229, 1971.

357. **Kuenen, J. P.,** On the mutual solubility of liquids. II., *Philos. Mag.,* 6, 637, 1903.

358. **Kuenen, J. P.,** Mixtures of hydrochloric acid and methylether, *Philos. Mag.,* 1, 593, 1901.

359. **Caubet, M. F.,** On the liquefaction of gaseous mixtures (in French), *Comp. Rend.,* 130, 167, 1900.

360. **Caubet, M. F.,** On the liquefaction of gaseous mixtures. Methyl chloride and sulfur dioxide (in French), *Comp. Rend.,* 131, 108, 1900.

360a. **Kracek, F. C.,** P-T-x relations for systems of two or more components and containing two or more phases (L-V, L-L-V and S-L-V systems), in *International Critical Tables of Numerical Data, Physics, Chemistry, and Technology,* Vol. III, Washburn, E. W., Ed., McGraw-Hill, New York, 1928, 351.

361. **Sander, W.,** The solubility of carbon dioxide in water and some other solvents under high pressures (in German), *Z. Phys. Chem.,* 78, 513, 1911.

362. **Efremova, G. D. and Ioffe, A. I.,** Interpolation section method of studying liquid-gas equilibria at high pressures, *Russ. J. Phys. Chem.,* 43, 1319, 1969.

363. **Benson, S. W., Allen, P. E., and Copeland, C. S.,** A rectilinear diameter law of mole fractions for binary liquid-vapor coexistence regions near the critical locus, *J. Chem. Phys.,* 22, 247, 1954.

364. **Fox, J. R. and Storvick, T. S.,** A field-space conformal solution method, *Int. J. Thermophys.,* 11, 49, 1990.

365. **Storvick, T. S. and Fox, J. R.,** A field-space conformal-solution method: binary vapor-liquid phase behavior, *Int. J. Thermophys.,* 11, 61, 1990.

366. **Hastings, J. R., Levelt Sengers, J. M. H., and Balfour, F. W.,** The critical-region equation of state of ethene and the effect of small impurities, *J. Chem. Thermodyn.,* 12, 1009, 1980.

367. **Levelt Sengers, J. M. H., Morrison, G., and Chang, R. F.,** Critical behavior in fluids and fluid mixtures, *Fluid Phase Equilibria,* 14, 19, 1983.

368. **McCracken, P. G., Storvick, T. S., and Smith, J. M.,** Phase behavior from enthalpy measurements: benzene-ethyl alcohol and n-pentane-ethyl alcohol systems, *J. Chem. Eng. Data,* 5, 130, 1960.

369. **Storvick, T. S. and Smith, J. M.,** Thermodynamic properties of polar substances: enthalpy of hydrocarbon-alcohol systems, *J. Chem. Eng. Data,* 5, 133, 1960.

370. **Lenoir, J. M., Kuravila, G. K., and Hipken, H. G.,** Enthalpies of cyclohexane and mixtures of n-pentane and cyclohexane, *J. Chem. Eng. Data,* 16, 271, 1971.

371. **Lenoir, J. M., Robinson, D. R., and Hipken, H. G.,** Enthalpies of mixtures of n-octane and n-pentane, *J. Chem. Eng. Data,* 15, 26, 1970.

372. **Mather, A. E., Yesavage, V. F., Powers, J. E., and Katz, D. L.,** Enthalpy data for binary mixtures of methane with propane and with nitrogen, Proc. Annu. Conv. Nat. Gas. Proc. Assoc. Tech. Paper 45, 12, 1966.

373. **Yesavage, V. F., Katz, D. L., and Powers, J. E.,** Experimental determinations of several thermal properties of a mixture containing 51 mole percent propane in methane, *AIChE J.,* 16, 867, 1970.

374. **Cordray, D. R., Izatt, R. M., and Christensen, J. J.,** The excess enthalpies of (carbon dioxide + cyclohexane) at 390.15, 413.15, 438.15, 498.15 and 508.15 K from 7.50 to 14.39 MPa, *J. Chem. Thermodyn.,* 20, 225, 1988.

375. **Christensen, J. J., Faux, P. W., Cordray, D., and Izatt, R. M.,** The excess enthalpies of (carbon dioxide + pentane) at 348.15, 373.15, 413.15, 470.15, and 573.15 K from 7.58 to 12.45 MPa, *J. Chem. Thermodyn.,* 18, 1053, 1986.

376. **Christensen, J. J., Cordray, D., and Izatt, R. M.,** The excess enthalpies of (carbon dioxide + decane) from 293.15 to 573.15 K at 12.50 MPa, *J. Chem. Thermodyn.,* 18, 53, 1986.

377. **Christensen, J. J., Zebolsky, D. M., and Izatt, R. M.,** The excess enthalpies of (carbon dioxide + pyridine) at 470.15 and 573.15 K from 7.50 to 12.50 MPa, *J. Chem. Thermodyn.,* 18, 53, 1986.

378. **Christensen, J. J., Walker, T. A. C., Schofield, R. S., Faux, P. W., Harding, P. R., and Izatt, R. M.,** The excess enthalpies of (carbon dioxide + hexane) at 308.15, 358.15 and 413.15 K from 7.50 to 12.50 MPa, *J. Chem. Thermodyn.,* 16, 445, 1984.

379. **Cordray, D. R., Izatt, R. M., Christensen, J. J., and Oscarson, J. L.,** The excess enthalpies of (carbon dioxide + ethanol) at 308.15, 323.15, 373.15, 413.15 and 473.15 K from 5.00 to 14.91 MPa, *J. Chem. Thermodyn.,* 20, 655, 1988.

380. **Christensen, J. J., Cordray, D. R., Oscarson, J. J., and Izatt, R. M.,** The excess enthalpies of (carbon dioxide + an alkanol) mixtures from 308.15 to 573.15 K at 7.50 to 12.50 MPa, *J. Chem. Thermodyn.,* 20, 867, 1988.

381. **Christensen, J. J., Zebolsky, D. M., Schofield, R. S., Cordray, D. R., and Izatt, R. M.,** The excess enthalpies of (ethane + chlorodifluoromethane) from 293.15 to 383.15 at 5.15 MPa, *J. Chem. Thermodyn.,* 16, 905, 1984.

382. **Wormald, C. J. and Yerlett, T. K.,** The enthalpy of $[0.5 \ (CH_3)_2 \ CO \ + \ 0.5 \ C_6H_{14}]$ at temperatures up to 573.2 K and pressures up to 11.3 MPa, *J. Chem. Thermodyn.,* 19, 215, 1987.

383. **Houseman, J. and Sage, B. H.,** Enthalpy changes upon vaporization in the n-butane-n-decane system, *J. Chem. Eng. Data,* 10, 250, 1965.
384. **Al-Sahhaf, T. A., Sloan, E. D., and Kidnay, A. J.,** Calculation method for vapor-liquid equilibrium in the critical region, *AIChE J.,* 30, 867, 1984.
385. **Balfour, F. W., Sengers, J. V., Moldover, M. R., and Levelt Sengers, J. M. H.,** A revised and extended scaled equation for steam in the critical region, in *Proc. 7th Symp. Thermophysical Properties,* Cezairliyan, A., Ed., American Society of Mechanical Engineers, New York, 1977, 786.
386. **De Bruyn, J. R. and Balzarini, D. M.,** Coexistence curve of C_2H_4 in the critical region, *Phys. Rev. A,* 36, 5677, 1987.
387. **De Bruyn, J. R. and Balzarini, D. M.,** Critical behavior of hydrogen, *Phys. Rev. B,* 39, 9243, 1989.
388. **Balzarini, D., DeBruyn, J., Narger, U., and Pang, K.,** Image-plane measurements of coexistence curve diameters, *Int. J. Thermophys.,* 9, 739, 1988.
389. **Wegner, F. J.,** Corrections to scaling laws, *Phys. Rev. B,* 5, 4529, 1972.
390. **Rainwater, J. C.,** Vapor-Liquid Equilibrium of Binary Mixtures Near the Critical Locus, in Phase Equilibria: An Informal Symposium, NBS Tech.Note 1061, National Bureau of Standards, Washington, D.C., 83.
391. **Rozen, A. M.,** The unusual properties of solutions in the vicinity of the critical point of the solvent, *Russ. J. Phys. Chem.,* 50, 837, 1976.
392. **Krichevskii, I. R.,** Thermodynamics of critical phenomena in infinitely dilute binary solutions, *Russ. J. Phys. Chem.,* 41, 1332, 1967.
393. **Levelt Sengers, J. M. H.,** Dilute mixtures and solutions near critical points, *Fluid Phase Equilibria,* 30, 31, 1986.
394. **Japas, M. L. and Levelt Sengers, J. M. H.,** Gas solubility and Henry's law near the solvent's critical point, *AIChE J.,* 35, 705, 1989.
395. **Levelt Sengers, J. M. H., Chang, R. F., and Morrison, G.,** Nonclassical description of (dilute) near-critical mixtures, in *Equation of State — Theories and Applications,* Chao, K. C. and Robinson, R. L., Eds., ACS Symp. Ser. 300, American Chemical Society, Washington, D.C., 1986, 110.
396. **Chang, R. F., Morrison, G., and Levelt Sengers, J. M. H.,** The critical dilemma of dilute mixtures, *J. Phys. Chem.,* 88, 3389, 1984.
397. **Chang, R. F. and Levelt Sengers, J. M. H.,** Behavior of dilute mixtures near the solvent's critical point, *J. Phys. Chem.,* 90, 5921, 1986.
398. **Weber, L. A.,** Vapor-liquid equilibrium in binary systems of chlorotrifluoromethane with n-butane and isobutane, *J. Chem. Eng. Data,* 34, 452, 1989.
399. **Beranek, P. and Wichterle, I.,** Vapor-liquid equilibria in the propane-n-butane system at high pressures, *Fluid Phase Equilibria,* 6, 279, 1981.
400. **McKay, R. A., Reamer, H. H., Sage, B. H., and Lacey, W. N.,** Volumetric and phase behavior in the ethane-propene system, *Ind. Eng. Chem.,* 43, 2112, 1951.
401. **Ohgaki, K., Nakai, S., Nitta, S., and Katayama, T.,** Isothermal vapor-liquid equilibria for the binary systems propylene-carbon dioxide, propylene-ethylene and propylene-ethane at high pressure, *Fluid Phase Equilibria,* 8, 113, 1982.
402. **Cummings, L. W. T.,** High Pressure Rectification, Ph.D. thesis, Massachusetts Institute of Technology, Cambridge, 1933.
403. **Higashi, Y, Uematsu, M., and Watanabe, K.,** Measurements of the vapor-liquid coexistence curve and the critical locus for several refrigerant mixtures, *Int. J. Thermodyn.,* 7, 29, 1986.
404. **Higashi, Y., Kabata, Y., Uematsu, M., and Watanabe, K.,** Measurements of the vapor-liquid coexistence curve for the R13B1 + R114 system in the critical region, *J. Chem. Eng. Data,* 33, 23, 1988.
405. **Chang, S.-D. and Lu, B. C.-Y.,** Vapor-liquid equilibria in the nitrogen-methane-ethane system, *Chem. Eng. Prog. Symp. Ser.,* 63(81), 18, 1967.
406. **Herlihy, J. C.,** Vapor-Liquid Equilibrium Constants for the Ethane-n-Butane-n-Pentane System at 150°F, M.Sc. thesis, Northwestern University, Evanston, June 1962.
407. **Niesen, V. G.,** Vapor-liquid equilibria and coexisting densities of (carbon dioxide + n-butane) at 311 to 395 K, *J. Chem. Thermodyn.,* 21, 915, 1989.
408. **Shibata, S. K., Sandler, S. I., and Kim, H.,** High-pressure vapor-liquid equilibria involving mixtures of nitrogen, carbon dioxide and n-butane, *J. Chem. Eng. Data,* 34, 291, 1989.
409. **Hsu, J. J.-C., Nagarajan, N., and Robinson, R. L., Jr.,** Equilibrium phase compositions, phase densities, and interfacial tensions for CO_2 + hydrocarbon systems. I. CO_2 + n-butane, *J. Chem. Eng. Data,* 30, 485, 1985.
410. **Guillevic, J.-L., Richon, D., and Renon, H.,** Vapor-liquid equilibrium measurements up to 558 K and 7 MPa: a new apparatus, *Ind. Eng. Chem. Fundam.,* 22, 495, 1983.
411. **Fontalba, F., Richon, D., and Renon, H.,** Simultaneous determination of vapor-liquid equilibria and saturated densities up to 45 MPa and 433 K, *Rev. Sci. Inst.,* 55, 944, 1984.
412. **Ohgaki, K., Sano, F., and Katayama, T.,** Isothermal vapor-liquid equilibrium data for binary systems containing ethane at high pressures, *J. Chem. Eng. Data,* 21, 55, 1976.

413. **Eckert, C. J. and Sandler, S. I.**, Vapor-liquid equilibria for the carbon dioxide-cyclopentane system at 37.7, 45.0 and 60.0°C, *J. Chem. Eng. Data*, 31, 26, 1988.

414. **Narinskii, G. B.**, Liquid-vapor equilibrium in the argon-nitrogen system. I. Experimental data and their verification, *Russ. J. Phys. Chem.*, 40, 1093, 1966.

415. **Kuenen, J. P., Verschoyle, J., and Van Urk, A. T.**, Isotherms of diatomic substances and their binary mixtures. XX. The critical curve of oxygen-nitrogen mixtures, the critical phenomena and some isotherms of two mixtures with 50% and 75% by volume of oxygen in the neighborhood of the critical point, *Commun. Phys. Lab. Leiden*, No. 161, 1922.

416. **Blanke, W.**, PVT measurements of air in the two-phase region, in *Proc. 7th Symp. Thermophysical Properties*, Cezairliyan, A., Ed., American Society of Mechanical Engineers, New York, 1977, 461.

417. **Michels, A., Wassenaar, T., Levelt, J. M., and DeGraaff, W.**, Compressibility isotherms of air at temperatures between −25°C and −155°C and at densities up to 560 amagats (pressure up to 1000 atmospheres), *Appl. Sci. Res.*, A4, 381, 1954.

418. **Dodge, B. F. and Dunbar, A. K.**, An investigation of the coexisting liquid and vapor phases of solutions of nitrogen and oxygen, *J. Am. Chem. Soc.*, 49, 591, 1927.

419. **Gallagher, J. S.**, An equation of state for isobutane-isopentane mixtures with corrections for impurities, *Int. J. Thermophys.*, 7, 923, 1986.

420. **Roof, J. G. and Baron, J. D.**, Critical loci of binary mixtures of propane with methane, carbon dioxide and nitrogen, *J. Chem. Eng. Data*, 12, 292, 1967.

421. *Kirk-Othmer Encyclopedia of Chemical Technology*, Vol. 1, 3rd ed., John Wiley & Sons, New York, 1978, 124.

422. **Jacobsen, R. T., Clarke, W. P., Beyerlein, S. W., Rousseau, M. F., Van Poolen, L. J., and Rainwater, J. C.**, A thermodynamic property formulation for air. II. Pressure and density estimation functions for the dew and bubble lines, *Int. J. Thermophys.*, 11, 179, 1990.

423. **Penoncello, S. G., Jacobsen, R. T., Clarke, W. P., and Beyerlein, S. W.**, Ancillary Equations for the Representation of the Bubble and Dew Point Pressures and Densities for Air, Center for Applied Thermodynamic Studies Rep. 89-2, University of Idaho, Moscow, 1989.

424. **Lynch, J. J.**, The Modified Leung-Griffiths Model of Vapor-Liquid Equilibrium: Developments for Binary Mixtures of Dissimilar Fluids, Ph.D. thesis, University of Colorado, Boulder, 1990.

425. **Caceci, M. S. and Cacheris, W. P.**, Fitting curves to data. The Simplex algorithm is the answer, *Byte*, 9, 340, 1984.

426. **Niesen, V. G. and Yesavage, V. F.**, Application of a maximum likelihood method using implicit constraints to determine equation of state parameters from binary phase behavior data, *Fluid Phase Equilibria*, 50, 249, 1989.

427. **Rauzy, E. and Peneloux, A.**, Vapor-liquid equilibrium and volumetric properties calculations for solutions in the supercritical carbon dioxide, *Int. J. Thermophys.*, 7, 635, 1986.

428. **Chou, G. F. and Prausnitz, J. M.**, A phenomenological correction to an equation of state for the critical region, *AIChE. J.*, 35, 1487, 1989.

429. **Al-Sahhaf, T. A.**, Measurement and Prediction of Vapor-Liquid Equilibria for the Nitrogen-Methane-Carbon Dioxide System, Ph.D. thesis, Colorado School of Mines, Golden, 1981.

430. **Mraw, S. C., Hwang, S.-C., and Kobayashi, R.**, Vapor-liquid equilibrium of the CH_4-CO_2 system at low temperature, *J. Chem. Eng. Data*, 23, 135, 1978.

431. **Hwang, S.-C., Lin, H., Chappelear, P. S., and Kobayashi, R.**, Dew point study in the vapor-liquid region of the methane-carbon dioxide system, *J. Chem. Eng. Data*, 21, 493, 1976.

432. **Somait, F. A. and Kidnay, A. J.**, Liquid-vapor equilibria at 270.00 K for systems containing nitrogen, methane, and carbon dioxide, *J. Chem. Eng. Data*, 23, 301, 1978.

433. **Davis, J. A., Rodewald, N. and Kurata, F.**, Solid-liquid phase behavior of the methane-carbon dioxide system, *AIChE J.*, 8, 537, 1962.

434. **Khazanova, N. E., Sominskaya, E. E., Zakharova, A. V., Rozovskii, M. B., and Nechitailo, N. L.**, Azeotropic systems at high pressures. X. Phase and volume relations in the ethylene-carbon dioxide system, *Russ. J. Phys. Chem.*, 53, 902, 1979.

435. **Morrison, G. and Kincaid, J. M.**, Critical point measurements in nearly polydisperse fluids, *AIChE J.*, 30, 257, 1984.

436. **Khazanova, N. E., Sominskaya, E. E., and Zakharova, A. V.**, Systems with azeotropism at high pressures. V. Negative azeotropes, *Russ. J. Phys. Chem.*, 51, 1610, 1977.

437. **Khazanova, N. E., Sominskaya, E. E., and Zakharova, A. V.**, Systems with azeotropism at high pressures. XI. Negative azeotropes, *Russ. J. Phys. Chem.*, 53, 905, 1979.

438. **Zawisza, A. C. and Glowka, S.**, Liquid-vapour equilibria and thermodynamic functions of dimethyl ether-sulfur dioxide system up to 300°C and 77.81 atm., *Bull. Acad. Polon. Sci. Ser. Sci. Chim.*, 18, 549, 1972.

439. **Gaw, W. J. and Swinton, F. L.**, Occurrence of a double azeotrope in the binary system hexafluorobenzene + benzene, *Nature*, 212, 283, 1966.

440. **Brewer, J., Rodewald, N., and Kurata, F.**, Phase equilibria of the propane-hydrogen sulfide system from the cricondentherm to the solid-liquid-vapor region, *AIChE J.*, 7, 13, 1961.

441. **Robinson, D. B., Kalra, M., Krishnan, T., and Miranda, R. D.,** The phase behavior of selected hydrocarbon-nonhydrocarbon binary systems: C_2-H_2S and N_2-iC_4 systems, *Proc. Ann. Conv. GPA,* 54, 25, 1975.

442. **Van Poolen, L. J. and Rainwater, J. C.,** Determination of binary mixture vapor-liquid critical densities from coexisting density data, *Int. J. Thermophys.,* 8, 697, 1987.

443. **Van Poolen, L. J., Niesen, V. G., and Rainwater, J. C.,** An experimental method for obtaining critical densities for binary mixtures: application to ethane + n-butane, *Fluid Phase Equilibria,* in press, 1991.

444. **Balbuena, P. B., Campanella, E. A., Gribaudo, L. M., and Martinelli, E.,** Liquid-vapor equilibrium calculations near the critical point, presented at the CODATA meeting, Paris, September 1985.

445. **Herlihy, J. C. and Thodos, G.,** Vapor-liquid equilibrium constants: ethane-n-butane-n-pentane system at 150° F, *J. Chem. Eng. Data,* 7, 348, 1962.

446. **Mehra, V. S. and Thodos, G.** Vapor-liquid equilibrium constants for the ethane-n-butane-n-pentane system at 200, 250 and 300° F, *J. Chem. Eng. Data,* 8, 1, 1963.

447. **Van Poolen, L. J.,** Extended phase rule for non-reactive, multiphase, multicomponent systems, *Fluid Phase Equilibria,* 58, 133, 1990.

448. **Van Poolen, L. J.,** Representation of binary mixture liquid-vapor equilibria by means of critical isochores, *Ind. Eng. Chem. Res.,* 29, 451, 1990.

449. **Van Poolen, L. J.,** Analysis of Liquid Volume and Liquid Mass Fractions at Coexistence for Pure Fluids, NBSIR 80-1631, 1980.

450. **Stauffer, D., Ferer, M., and Wortis, M.,** Universality of second-order phase transitions: the scale factor for the correlation length, *Phys. Rev. Lett.,* 29, 345, 1972.

451. **Moldover, M. R.,** Interfacial tension of fluids near critical points and two-scale-factor universality, *Phys. Rev. A,* 31, 1022, 1985.

452. **Binder, K.,** Monte Carlo calculation of the surface tension for two- and three-dimensional lattice gas models, *Phys. Rev. A,* 25, 1699, 1982.

453. **Tarko, H. B. and Fisher, M. E.,** Theory of critical point scattering and correlations. III. The Ising model below T_c and in a field, *Phys. Rev. B,* 11, 1217, 1975.

454. **Brezin, E. and Feng, S.,** Amplitude of the surface tension near the critical point, *Phys. Rev. B,* 29, 472, 1984.

455. **Breizin, E., Le Guillou, J. C., and Zinn-Justin, J.,** Universal ratio of critical amplitudes near four dimensions, *Phys. Lett.,* 47A, 285, 1974.

456. **Nadler, K. C., Zollweg, J. A., and Streett, W. B.,** Global representation of interfacial tension in the system (Ar + Kr), *Int. J. Thermophys.,* 10, 333, 1989.

457. **Jones, M. C.,** Two-Phase Heat Transfer in the Vicinity of a Lower Consolute Point, ACS Symp. Ser. 406, American Chemical Society, Washington, D.C., 1989, 396.

458. **Heidemann, R. A. and Khalil, A. M.,** The calculation of critical points, *AIChE J.,* 26, 769, 1980.

459. **Eaton, B. E.,** On the Calculation of Critical Points by the Method of Heidemann and Khalil, NBS Technical Note 1313, National Bureau of Standards, Washington, D.C., 1988.

460. **Levelt Sengers, J. M. H.,** The state of the critical state of fluids, *Pure Appl. Chem.,* 55, 437, 1983.

461. **Johnson, K. A.,** An asymmetric potential to describe the excess enthalpies of binary liquid mixtures near their (liquid + liquid) critical states, *J. Chem. Thermodyn.,* 20, 889, 1988.

462. **De Pablo, J. J. and Prausnitz, J. M.,** Liquid-liquid equilibria for binary and ternary systems including the critical region: transformation to non-classical coordinates, *Fluid Phase Equilibria,* 50, 191, 1989.

463. **King, M. B.,** *Phase Equilibrium in Mixtures,* Pergamon Press, New York, 1969, 178.

464. **Friend, D. G. and Roder, H. M.,** Thermal-conductivity enhancement near the liquid-vapor critical line of binary methane-ethane mixtures, *Phys. Rev. A,* 32, 1941, 1985.

465. **Friend, D. G. and Roder, H. M.,** The thermal conductivity surface for mixtures of methane and ethane, *Int. J. Thermophys.,* 8, 13, 1987.

Chapter 3

MOLECULAR ANALYSIS OF PHASE EQUILIBRIA IN SUPERCRITICAL FLUIDS

Michael P. Ekart, Joan F. Brennecke, and Charles A. Eckert

TABLE OF CONTENTS

I. INTRODUCTION TO SUPERCRITICAL FLUIDS

The very special physical properties of supercritical fluid (SCF) solutions have led to intense attempts to exploit them as solvents for separations and reactions. SCFs are characterized by high density, leading to very high loadings of solutes, low viscosity, and high molecular diffusivity. Moreover, they exhibit compressibilities typically one to three orders of magnitude greater than other fluids. Since solubilities appear to be virtually exponential in density, this means that very small pressure changes can result in enormous solubility variations, giving the design engineer the opportunity to pull chemicals into solution or drop them out very selectively. All of these properties make SCFs ideal candidates for mass transfer solvents.

To make the potential for SCF extraction appear even more promising, small additions of second components, called cosolvents or entrainers, often modify observed phase equilibria dramatically. Again, this special property affords the thermodynamicist the opportunity to "tailor" solvents to achieve specific goals, either in separation or reactions.

In order to achieve all these very lofty goals, it is imperative to have a good mathematical model of SCF solution behavior, capable of giving a good representation of not only the PVT properties, but also of their many derivatives, especially minor component fugacities.

Many investigators have used variations of the classic cubic equations of state (EOS), but in general these are inadequate for anything beyond a gross representation, even with violent perturbations of the ubiquitous k_{12}, or its brethren a_{12} or δ_{12}. Basically all of the cubic EOSs were originally developed to characterize petroleum cuts, and they embody a very strong component of corresponding states theory (CST). CST assumes similarity of molecular size and force constant; however, SCF solutions are highly asymmetric in force constant and size, and CST breaks down. Attempts to modify mixing rules, making them variable with density, temperature, or composition, may be useful, but are stopgap solutions; the real need is to develop better EOSs.

Quite substantial additional evidence now exists for the occurrence of rather unusual phenomena in SCF solutions. In addition to extreme entrainer effects of many orders of magnitude,[49] there are the well-known synergistic effects observed for the solubility of physical mixtures of solids,[65,67] the very unusual partial molal volumes,[30] and now a plethora of spectral observations.[11,13,56,59,60] The burgeoning evidence says ever more strongly that SCF solutions may no longer be treated by the classical continuum methods, but an attempt must be made to understand from a molecular point of view exactly what is occurring.

This review discusses qualitatively the nature of SCF solutions by examining phase diagrams, and then considers modeling techniques to describe this phase behavior. Further addressed is the study of the intermolecular interactions that lead to the unique behavior observed in supercritical fluids.

II. PHASE DIAGRAMS

An understanding of the phase equilibria of mixtures is crucial to the design of supercritical fluid processes. Quality high pressure equilibria data are expensive and time-consuming to obtain; thus, it would be desirable to predict phase equilibria from easily obtained pure component properties. As it is generally not yet possible to make such predictions, it is then useful to correlate and extrapolate limited experimental data. Due to the complexity of high pressure mixtures, quantitative modeling of equilibria is challenging.

Although process mixtures are generally quite complex, it is instructive to study the phase diagrams of binary mixtures. In a binary mixture, the phase rule is

$$F = 4 - p \tag{1}$$

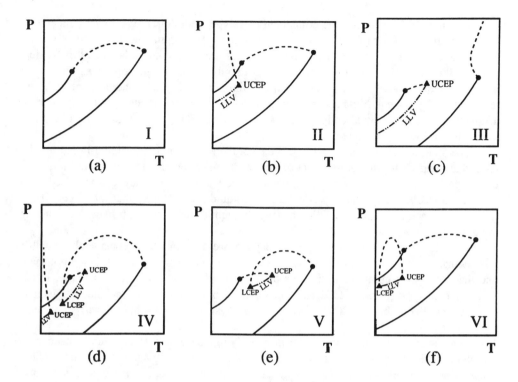

FIGURE 1.   van Konynenburg and Scott classification of phase diagrams based upon PT projections.

where p is the number of phases and F is the degrees of freedom, i.e., the number of intensive variables that must be specified to describe the thermodynamic state of the mixture. A three-dimensional diagram must be used to describe fully the phase behavior of a binary system; the most commonly used are pressure-temperature-mole fraction (PTx) diagrams. In these diagrams, one-phase mixtures are geometrically represented by volumes, two phases by surfaces, three phases by lines, and four phases by points.

Often, binary phase equilibria are depicted by a projection of the PTx surfaces onto the PT plane. These PT projections show pure component phase boundaries (in this chapter shown as solid curves), the loci of the binary mixture critical points (dashed curves), and three phase lines (dashed dotted curves). The critical point of a binary mixture at a given composition is the point at which two phases in equilibrium become identical. Also shown on the PT projections are the pure component critical points (circles), lower critical endpoints (LCEPs, shown as triangles), and upper critical endpoints (UCEPs, also shown as triangles). LCEPs and UCEPs occur at the intersection of three phase lines with critical mixture curves. At critical endpoints, two phases critically merge into a single phase in the presence of another phase.

Although a large diversity in phase behavior is possible, van Konynenburg and Scott[116] have classified PT projections into six fundamental types (Figure 1). Type 1 equilibria are distinguished by the lack of liquid-liquid immiscibility under all conditions and the continuous gas-liquid critical curve between the pure component critical points. This curve can have a variety of shapes, monotonic, or with a maximum or minimum in temperature and/or a maximum in pressure. Type 1 equilibria occur in systems that are not far from ideal. In type 2 equilibria, the gas-liquid critical line is still continuous, but at low temperatures, liquid-liquid phase separation occurs. There is a liquid-liquid-vapor (LLV) equilibrium line that intersects the liquid-liquid critical curve at an UCEP. At this point the two liquid phases become identical in the presence of a vapor phase.

There are two branches of the critical curve in type 3 equilibria. The lower temperature branch, originating from the critical point of the more volatile component, intersects a three-phase line (LLV) at an UCEP. The other branch of the critical curve begins at the critical point of the less volatile component and rises rapidly to very high pressures. This curve may have many shapes, with a maximum or minimum in pressure and/or a minimum in temperature. Type 4 equilibria also have two branches of the critical curve, but both intercept a three-phase line at critical endpoints. Like type 2, there is a region of liquid-liquid immiscibility at low temperatures.

Type 5 equilibria are similar to type 4 except there is no region of liquid-liquid immiscibility at temperatures lower than that of the lower critical endpoint. Type 6 equilibria, the only type that cannot be described by the van der Waals EOS,[116] are characterized by a continuous gas-liquid critical curve and a liquid-liquid critical curve that intersects a LLV line at both an UCEP and a LCEP. This type of equilibrium occurs in hydrogen bonding systems.[103]

Schneider[106] has constructed experimentally phase diagrams for a considerable number of systems, and has shown that the boundaries between the types of equilibria are not sharp and that there are continuous transitions between the different types. For example, in examining binary diagrams of CO_2 and the series of *n*-alkanes, it is seen that CO_2-ethane exhibits type 1 equilibrium, CO_2-octane is a type 2 system, and CO_2-hexadecane shows type 3 behavior.

Although PT projections are useful in classifying phase diagrams, to understand fully the phase behavior of a binary system, three-dimensional PTx diagrams must be studied. To give an overview of the main characteristics of phase diagrams, a type 1 system is explained in detail; then, the more complex type 4 system is discussed. Several reviews give complete discussions of each type of diagram.[80,103,112]

A. TYPE 1 DIAGRAMS

A three-dimensional PTx diagram of a type 1 system is shown in Figure 2. In this diagram are two surfaces enclosing the vapor-liquid envelope; the surfaces intersect at the pure component vapor pressure curves and at the gas-liquid critical curve. The projection onto the PT plane of these curves is shown in Figure 3a. The surface that separates the two-phase envelope from the vapor phase is the assemblage of all possible dewpoints. The other surface separates the single liquid phase from the vapor-liquid region and contains all possible bubble points. At a given temperature and pressure, the compositions of the equilibrium vapor and liquid phases are given by the intersection with the surfaces of a line parallel to the x axis.

Several cross sections of the two phase region are shown in Figure 2 and depicted in two-dimensional diagrams in Figure 3. A Px diagram at temperature T_a (Figure 3b) shows the bubble point and dewpoint curves running smoothly between the pure component vapor pressures. As temperature increases above the critical temperature of the more volatile component, the vapor-liquid envelope separates from the x = 0 axis. This occurs because the more volatile component when pure at this pressure is supercritical; thus, vapor and liquid phases cannot coexist. At temperature T_b (Figure 3c), the mixture critical point occurs at the point of maximum pressure in the vapor-liquid region.

Isobaric Tx diagrams under subcritical (Figure 3d) and supercritical (Figure 3e) conditions are similar to the Px diagrams. When the mixture is subcritical, the dewpoint and bubble point curves run smoothly between the pure component vapor pressures. As pressure increases above the critical pressure of the component with the lowest P_c, the vapor-liquid envelope becomes detached from the axis representing that pure component. If the slope of the PT projection of the gas-liquid critical curve is positive at a given pressure, the vapor-liquid envelope at that pressure has a lower critical solution temperature (LCST) such as the one in Figure 3e. If the slope is negative, then there is an upper critical solution

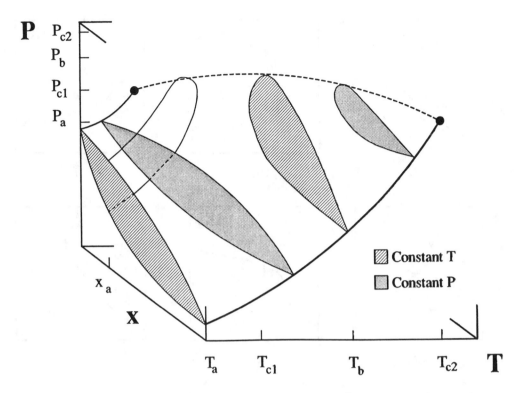

FIGURE 2.   Three dimensional PTx diagram for a type 1 system.

temperature (UCST). In the case of a gas-liquid critical curve with a maximum in pressure, there is both a LCST and an UCST below this pressure and above the critical pressure of both components.

A cross section of the vapor-liquid region at constant composition is shown in Figure 3f. On these representations, the critical point (point C) not only can fall between the points of maximum temperature (point A) and maximum pressure (point B) but on either side of the two points. This means vapor-liquid equilibrium can exist at a pressure and/or temperature higher than those of the critical point, which leads to the unusual behavior known as retrograde condensation. Consider an isotherm at some temperature between those of points B and C. At low pressures, a single vapor phase exists. As pressure is increased, the dewpoint line is intersected, marking the appearance of a liquid phase. Further increasing the pressure initially increases the amount of liquid present, but at some point the liquid begins evaporating, disappearing completely as the dewpoint line is crossed again. This phenomenon demonstrates that peculiarities exist when working in the near critical region.

B. TYPE 4 DIAGRAMS

With the exception of Type 1 equilibria, the phase diagrams all exhibit regions of liquid-liquid immiscibility. Most of the features of these diagrams can be understood by studying a type 4 system. A PT projection (Figure 4) shows the temperatures at which isothermal Px diagrams are examined.

At temperature T_a (Figure 5a), the mixture is a vapor at very low pressures, and there is the usual vapor-liquid envelope intersecting the vapor pressure curve of the less volatile component. As pressure is increased, however, the LLV line is intersected giving rise to three equilibrium phases at the compositions indicated by the points of the diagram. A further increase in pressure can result in vapor-liquid equilibrium if $x_d < x_{mix} < x_e$ or liquid-liquid equilibrium (LLE) if $x_e < x_{mix} < x_f$. The vapor-liquid region intersects the vapor pressure of the more volatile component while the liquid-liquid region ends at a critical point C.

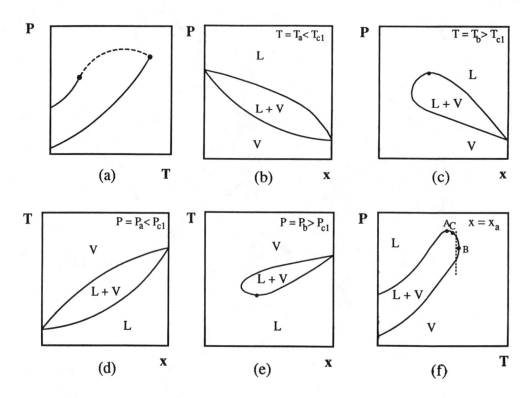

FIGURE 3.   Type 1 system. (a) PT projection. (b) and (c) Isothermal Px diagrams below and above the critical temperature of the lighter component. (d) and (e) Isobaric Tx diagrams below and above the critical pressure. (f) PT diagram at constant composition.

As temperature is increased, the liquid-liquid envelope becomes smaller, until at the upper critical endpoint (T = T_b) the two liquid phases become identical (Figure 5b). As thermodynamically required, the UCEP occurs at a horizontal inflection in the bubble point curve. Between T_b and the temperature of the lower critical endpoint, T_d, Px diagrams have just the single vapor-liquid region, but the LCEP denotes the incipient appearance of another region of liquid-liquid immiscibility. At temperatures between this point and the critical point of the more volatile component, the phase diagrams look qualitatively like Figure 5a.

The vapor-liquid envelope detaches from the x = 0 axis at the pure component critical point, and at temperatures up to the second UCEP there are two mixture critical points (Figure 5c). The critical point on the shaded vapor-liquid envelope is on the branch of the critical curve originating at the critical point of the more volatile component. The other critical point, liquid-liquid, is on the other branch of the critical curve. As temperature approaches T_f, the shaded vapor-liquid envelope shrinks and the liquid-liquid region expands, until at T_f, one of the liquid phases becomes identical to the vapor phase; thus, there is a single vapor-liquid envelope with a horizontal inflection point (Figure 5d). Note that at this temperature, the liquid-liquid critical curve becomes a vapor-liquid critical curve. The phase behavior of type 4 systems is qualitatively identical to type 1 equilibria at temperatures greater than T_f.

C. SUPERCRITICAL FLUID-SOLID EQUILIBRIA

Equilibria between supercritical fluids and solids are of much commercial interest with applications ranging from pharmaceutical processing[68] to environmental control.[28] The solid component exhibits a melting point at a higher temperature than the critical temperature of the SCF. In most cases, the mixtures are highly asymmetric, i.e., the SCF and solid molecules are of highly disparate size and force constant.

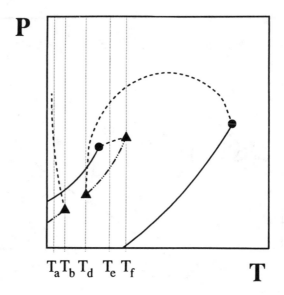

FIGURE 4.   PT projection of a type 4 system.

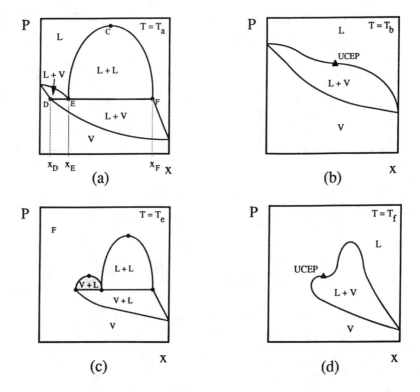

FIGURE 5.   Isothermal Px diagrams of the type 4 system shown in Figure 4.

A typical PT projection of a binary solid-SCF system is shown in Figure 6. The gas-liquid vapor pressure curves of both components are shown along with the solid-liquid and solid-vapor equilibrium curves of the pure solid. A solid-liquid-vapor (SLV) curve originates at the triple point of the solid and terminates at an UCEP where it intersects the upper branch of the gas-liquid critical curve. Note that the SLV curve indicates that the melting point of

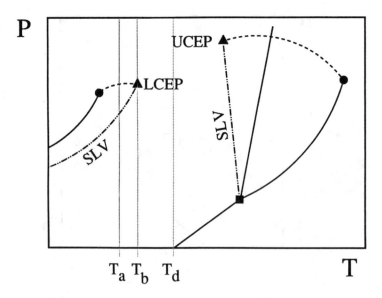

FIGURE 6.   PT projection for a typical solid-SCF system.

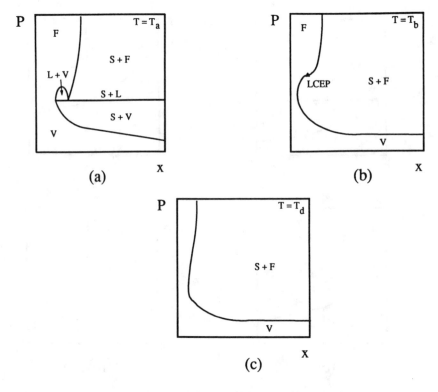

FIGURE 7.   Isothermal Px diagrams of the system shown in Figure 6.

the solid is depressed in the presence of the supercritical fluid. This occurs because the SCF dissolves in the equilibrium liquid phase, lowering the melting point of the heavy component much as salt lowers the melting point of water. The lower branch of the gas-liquid critical curve, which starts at the SCF critical point, meets a SLV curve at a LCEP. Between the LCEP and the UCEP, solid-fluid equilibrium occurs.

At temperature T_a, between the SCF critical point and the LCEP, the phase behavior shown in Figure 7a occurs. At low pressures, there is a region of solid-vapor equilibrium.

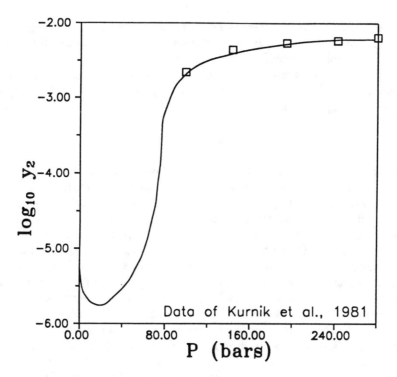

FIGURE 8.   Solubility of 2,3-dimethylnaphthalene(2) in $CO_2(1)$ (T_c = 304.1 K) at 308 K. Solid curve is correlation given by Peng-Robinson EOS.

The solid phase is very nearly pure. As the pressure is increased, the SLV line is encountered. If the composition of the mixture is richer in the solid component than the liquid phase, a further increase in pressure results in solid-liquid equilibrium. This region merges continuously into a region of solid-fluid equilibrium at higher pressures. There is a vapor-liquid envelope at lower compositions of the solid component which is crested by a critical point. At higher pressures, only the region of solid-SCF equilibria remains.

Note that at temperatures between the SCF critical point and the LCEP and at pressures between the SLV and gas-liquid critical curve, solubility measurements can be misleading. Rather than determining the solubility of the solid in the fluid phase as one might expect, the composition of a vapor phase in equilibrium with a liquid phase is being measured. Usually, the LCEP is within a few degrees of the SCF critical point;[80] thus, this region is quite small.

As the temperature is increased, the vapor-liquid envelope shrinks, until at the LCEP it disappears, leaving a horizontal inflection in the solid-SCF region (Figure 7b). Near the LCEP, small increases in pressure result in large changes in solubility of the solid. These dramatic changes near the LCEP can be seen in Figure 8 for the CO_2-2,3-dimethylnaphthalene system. This conveys the usefulness of SCFs as solvents; the solvent properties are widely adjustable over a moderate range of operating conditions.

Increasing the temperature gradually "smooths out" the inflection in the solubility (Figure 7c) until the dramatic changes in solubility with pressure are no longer observed. At the UCEP, another inflection occurs, giving rise to behavior similar to that near the LCEP. The UCEP can be several degrees below the melting point of the solid; for example, in the CO_2-naphthalene system, the UCEP occurs at 60.1°C while the melting point of naphthalene is 80.3°C.[73]

Although knowledge of binary systems is vital if we hope to understand SCF phase equilibria, the multicomponent systems of practical interest pose far greater challenges. SCF extraction from solids involves several solutes. It has been shown that although the solutes may be dilute in the SCF phase, they can exhibit a profound effect on each other (see Section IV). Another example of a multicomponent system is the presence of an entrainer, or cosolvent, in the SCF solvent to enhance the solubility of a particular solute by specific interactions such as hydrogen bonding. The solubility of acridine, a polar compound, in SCF CO_2 was quintupled by adding 2.5% methanol.[115]

III. MODELING

Accurate models of supercritical fluid phase behavior are vital to the design and evaluation of supercritical fluid processes. Unfortunately, the very properties that make supercritical fluids useful prove challenging in modeling efforts — the high compressibility, the asymmetry of the systems, and the mathematically singular nature of the critical point.

An ideal model would use easily measured physical properties to predict phase equilibria at all conditions and would be theoretically based. No current model fits these criteria. Existing correlations of phase equilibria data contain many regressed parameters, are semiempirical at best, and may succeed in fitting the data in portions of the phase diagram with some accuracy. On the other hand, models developed for the purpose of prediction attempt to justify theoretically a link between the model parameters and real physical phenomena. Both methods are useful, but in reality the distinction between the two methods is often lost, as theoretically based models are forced to fit the data better by the introduction of additional adjustable parameters.

There are several categories of models for supercritical phase equilibria. Most often, the SCF phase is treated as a dense gas and an EOS is used to calculate the fugacity coefficient, ϕ_i, of the component i in the fluid phase:

$$f_i^{SCF} = y_i \, \phi_i P \qquad (2)$$

where f_i^{SCF} is the fugacity, y_i is the mole fraction of component i in the SCF phase, and P is the pressure. In this approach, the results are often very sensitive to the composition dependence of the interaction energies and size factors, making mixing rules extremely important. This problem is exacerbated by the typically large asymmetry of the molecules, which causes errors in the corresponding states approach that most mixing rules rely upon; the usual solution is to introduce adjustable parameters that have no theoretical basis. Another drawback to the EOS method is that determination of the fugacity coefficient in a supercritical phase requires an integration through the critical region where most EOSs are less accurate.

Another approach treats the SCF phase as an expanded liquid, thus the fugacity is given by[75]

$$f_i^{SCF} = y_i \gamma_i(y_i,P^o) \, f_i^{oL}(P^o) \exp\int_{P^o}^{P} \frac{\bar{v}_i}{RT} \, dP \qquad (3)$$

where γ_i is the composition-dependent activity coefficient at the reference pressure P^o, f_i^{oL} is the fugacity of the pure reference liquid (hypothetical if component i is solid), and $\bar{v}_i$ is the partial molar volume of component i. If the reference pressure is supercritical, integration through the critical region is not necessary; however, an EOS must be used to calculate $\bar{v}_i$, and the reference activity coefficient must somehow be determined.

There are also many semiempirical correlations to fit phase equilibria data (for example see Reference 20) which have been discussed elsewhere.[13] Here, efforts to model SCF-solid phase equilibria are reviewed, with emphasis on the EOS approach, and attempts to model the more difficult situation of SCF-liquid equilibria are examined.

The fugacity, f_2^s, of a solid in equilibrium with a SCF is given by:

$$f_2^s = P_2^{sat} \exp\left(\frac{v_2^s(P - P_2^{sat})}{RT}\right) \qquad (4)$$

where P_2^{sat} is the vapor pressure and v_2^s is the molar volume of the solid. The reasonable assumptions that have been made in this equation are that the SCF does not dissolve in the solid phase, the solid volume is not a function of pressure, and that the pure saturated solid vapor at temperature T is ideal. As the conditions of equilibrium require that the fugacity of the solute be equal in the solid and the SCF phases, combining Equation 2 and 4, and solving for y_2 gives the solubility of the solid in the SCF phase:

$$y_2 = \frac{P_2^{sat} \exp\left(\frac{v_2^s(P - P_2^{sat})}{RT}\right)}{\phi_2 P} \qquad (5)$$

The exponential term is known as the Poynting correction and is generally less than 2 or 3.[99] The ratio of the actual solubility to the solubility in an ideal gas ($y_{2,ideal} = P_2^{sat}/P$) is known as the enhancement factor. Large enhancement factors in SCFs are thus due to very small values of ϕ_2. In order to predict y_2 accurately, not only are accurate EOSs necessary, but accurate vapor pressure data are required.[53]

The situation is SCF-liquid equilibria is more complicated because substantial amounts of fluid dissolve in the liquid phase. For example, at 35°C and 2000 psia, the solubility of CO_2 in n-hexadecane is almost 85%.[18] Because of this, there is no simple way to determine the fugacities in the liquid. Like the SCF phase, two modeling approaches can be taken — the liquid can be treated with an EOS or as an expanded liquid. Either approach usually requires the introduction of additional adjustable parameters to fit experimental data successfully.

Several different types of EOSs are discussed here, including modification to mixing rules and treatment of the SCF phase as a liquid, and following that is an examination of how computer simulations have furnished insight on the behavior of SCFs. When comparing the models, it is important to remember that the success of a model in correlating phase equilibria is often dependent on the number of adjustable parameters; a model is not necessarily superior because it gives a better representation of a given system than other models.

A. EQUATION OF STATE APPROACH

1. Cubic Equations of State

The most widely used method of analyzing supercritical fluid equilibria data is with cubic EOSs. Their popularity is due not only to the remarkable success in correlating SCF phase behavior, but also the simplicity; cubic equations can be rapidly solved analytically, significantly reducing computation time when trial and error calculations are necessary. Cubic EOSs are easily extended to multicomponent systems, but due to the approximate and somewhat empirical basis of the equations, mixing rules are crucial in determining the quality of the model. Methods of using cubic EOSs to determine thermodynamic properties are described by Reid et al.[102]

There is some theoretical basis to cubic EOSs. They can be derived by considering a first-order perturbation about a hard-sphere reference system and making simplifying assumptions. The earliest cubic EOS, that of van der Waals (vdW), may be obtained by assuming that the integral of the perturbing intermolecular potential for a pair of molecules is a constant. The vdW EOS can predict almost all types of phase behavior qualitatively, but it may not be very good quantitatively. Improvements can be made by assuming that

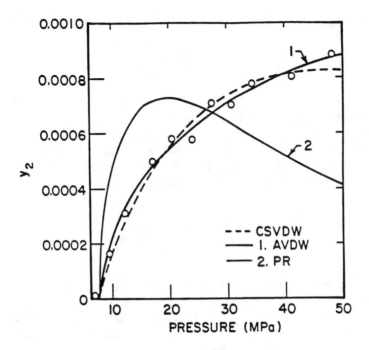

FIGURE 9.   Correlation of the solubility of acridine in SCF CO_2 at 308 K.

the perturbing intermolecular potential is temperature and density dependent, leading to equations such as the Redlich-Kwong (RK) equation (1949), the Soave modification (1972) of the Redlich-Kwong (SRK) equation, and the Peng-Robinson (PR) equation (1976). The parameters in these equations are usually calculated from critical properties, although a better approach may be to optimize the parameters to fit pure component vapor pressure or liquid molar volume data. Morris and Turek[84] determined the optimal temperature dependence of the parameters in the RK form for some common fluids such as CO_2 and ethane.

Many workers have successfully correlated solubility data for solid nonpolar hydrocarbons in SCFs such as ethylene and carbon dioxide with the SRK or PR equations by including a binary interaction parameter, k_{ij}, chosen to fit the data best (see, for example, Kurnik et al.[64]). Unfortunately, k_{ij} is usually found to be temperature or density dependent, necessitating the introduction of more parameters (see Section III.A.3). The empirical nature of k_{ij} makes predictions of solid solubilities by cubic EOSs unreliable. Schmitt and Reid[105] obtained a good fit without the binary interaction parameter by regressing the pure solute parameters from the solubility data, but this approach does not improve the predictive ability of the EOS. It has been found, however, that the PR equation with two adjustable parameters often performed nearly as well as the more complicated perturbed hard-sphere equations for a wide variety of solutes in a diverse field of SCF solvents.[33,43] The solubility of acridine in carbon dioxide at 308 K is shown in Figure 9,[33] along with correlations from the PR (with one adjustable parameter) and the CSvdW and AvdW equations (discussed later). In systems where the solubility is low, the PR typically shows an early false maximum.

Patel and Teja[94] developed a cubic EOS that appears to give improved results for polar fluids. This equation was found to be superior in the correlation of liquid molar volumes.[130] Excess volumes and enthalpies of supercritical mixtures were fitted satisfactorily;[128] such derivative properties are a stringent test for any EOS.

2. Perturbation Equations of State

The cubic equations of state almost invariably use the van der Waals repulsive term $v/(v - b)$; however, for a system of hard spheres, this term is in serious error at moderate to high densities.[42] Carnahan and Starling[15] developed an accurate expression for systems of hard spheres: $(1 + \xi + \xi^2 - \xi^3)/(1 - \xi)^3$ where $\xi = b/v$. Johnston and Eckert[50] used this improved repulsive term in developing the Carnahan-Starling-van der Waals (CSvdW) EOS. Although this equation is not cubic in volume, tabulated pure-fluid densities were used, eliminating the task of solving a higher order polynomial. Additionally, the critical properties of the components were not needed because solute-solute interactions were ignored (valid for very low solubilities). Good results were obtained for the solubility of nonpolar hydrocarbons in ethylene except near the critical region with one nearly temperature independent binary parameter, a_{12}. Because a_{12} correlated well with the enthalpy of vaporization of the solute, the CSvdW is potentially predictive. Bertucco et al.[9] applied this approach to a wider range of systems, including polar solutes, and also obtained good results.

The hard-sphere van der Waals (HSvdW)[127] incorporates an improved repulsive term for *mixtures* of hard spheres[77] in addition to the standard vdW attractive term. In this model, the solute-solute term was not ignored; thus, solubilities greater than 10^{-3} can be considered. Once more, a_{12} was correlated with physical properties leading to the possibility of predictions. The HSvdW was used to correlate solute/SCF/entrainer data[24-26] with good success when the binary parameter was regressed; when the parameter was predicted from physical properties, results were similar to those given by other equations. Again, there was difficulty fitting the critical region.

The failure of these equations in the critical region is due in part to the shortcomings of the mean field attractive term. Clustering (discussed later) of the solvent molecules around the solute in the critical region must be considered. The augmented van der Waals (AvdW)[54] includes second-order perturbation terms based on the results of the square well fluid simulations of Alder et al.[2,3] The AvdW gave improved representation of solubilities in the region where the reduced density was between 1.0 and 1.5.

Most solutes of interest in SCF processes are far from spherical. The hard sphere repulsive terms do not account for interference of vibrational and rotational freedom by nearest neighbors, particularly at high density. This was the impetus for the development of perturbed hard-chain theory,[8,27] which was later extended to supercritical fluids[79] with many variations.[47,48,85,121,123] In this model, the vibrational and rotational degrees of freedom were transformed into effective translational degrees of freedom, allowing the use of the Carnahan-Starling term. The attractive term included the second-order Alder functions. Because a small solvent molecule may interact with only part of a large solute molecule, mixture properties are based on surface or volume fractions along with nonrandom mixing rules. With only one adjustable binary interaction parameter, good results were obtained in fitting the solubility of polycyclic aromatics in nonpolar SCF solvents at high density. The model was not accurate in the subcritical and critical regions.

Another perturbation method[55] uses the Padé approximation, a free energy expansion, which makes the mean field assumption, but includes the effects of higher-order perturbations on the structure of the fluid. With one adjustable parameter, results were obtained for the ethylene-naphthalene system comparable to the RK.

3. Mixing Rules

In order to extend a pure-fluid equation of state to fluid mixtures, mixing rules are required. A typical mixing rule is

$$\Psi_{mix} = \Sigma_i \Sigma_j x_i x_j \Psi_{ij} \qquad (6)$$

where Ψ_{mix} is the mixture parameter used in the EOS, x is some measure of composition (usually mole fractions), and the summations are performed over all of the components in the mixture. When $i = j$, Ψ_{ij} is the parameter Ψ_i from the EOS for the pure fluid i. The unlike pair parameter, $\Psi_{ij, i \neq j}$, is determined from semitheoretical combining rules. Customarily, unlike pair energy parameters are the geometric mean of the pure component parameters, and the unlike size parameters are determined by the arithmetic mean of the pure size parameters. All EOS models are very sensitive to the unlike pair parameters, especially in asymmetric mixtures.[79,107] In most cases, the introduction of one or more adjustable parameters is necessary to compensate for the inadequacies of the combining and mixing rules, particularly for SCF-liquid equilibria. A frequent modification is the addition of an adjustable binary interaction parameter, k_{ij}, to the energy parameter combining rule:

$$a_{ij} = (a_i a_j)^{1/2}(1 - k_{ij}) \qquad (7)$$

Because this is an empirical modification, k_{ij} generally cannot be predicted from physical properties; thus, it must be optimized to fit existing data. Improvement of mixing rules has been the goal of much research. Mixing rule modification represents an advance only if solubility predictions, or perhaps derivative properties, can be improved without the introduction of new adjustable parameters or by introducing parameters that can be predicted from known physical properties.

Many researchers have found that k_{ij} is not constant over all conditions for a given system. To solve this problem, k_{ij} can be made temperature, pressure, density, or composition dependent, effectively increasing the number of adjustable parameters. Note that in SCF-liquid equilibria, a composition or density dependent k_{ij} is different in the two phases. Mohamed and Holder[82] used the PR equation and allowed k_{ij} to vary linearly with density leading to a quartic EOS with two adjustable binary parameters. By fitting one isotherm for CO_2-aromatic liquid systems, they obtained a better prediction of phase behavior at other temperatures than the correlation of the equilibria with one binary parameter fit at each temperature.

Another approach is changing the functional form of the mixing rule.[100] Won[126] suggested an empirical, nonquadratic mixing rule for a modified SRK that improved representation of the naphthalene-CO_2 system, but the mixing rule was incorrect in the limit of very low pressure, where the mixing rule is theoretically required to be quadratic.

The standard mixing rules for cubic EOSs other than the vdW have no theoretical basis. Based upon dimensional arguments, Mansoori and co-workers[5,66] showed that the conventional SRK and PR mixing rules are not equivalent to the vdW mixing rules and derived the appropriate rules. By making the conformal solution assumption (all intermolecular potentials in a solution have the same functional form), a variety of mixing rules can be obtained.[34,76,78] These rules may be temperature, composition, and density dependent adding to the complexity, but comparison with the standard mixing rules yields improved results for asymmetric mixtures. Conformal solution mixing rules for the PR along with an unlike three-body interaction term were applied to systems such as nitrogen/methane/CO_2,[6] giving better representation near the critical point than the original PR. This is not surprising, however, due to the increased number of adjustable parameters.

Other attempts at improving mixing rules have used the local composition concept, i.e., the concentration of molecules in the neighborhood of a central molecule deviates from the bulk. Mollerup[83] showed how a solution model can be incorporated into an EOS. He used the nonrandom, two-liquid approach which recognizes ordering around a solute, unlike the normal one-fluid theory, and applied it to the CSvdW, SRK, and PR equations. Vidal[120] showed that this model correlated the naphthalene/CO_2 system very well. Whiting and Prausnitz[125] used the same type of approach to show that including local composition effects also incorporated a density dependence into the mixing rules. Consequently, the mixing rules can maintain the theoretically required quadratic form at low densities, but smoothly assume more complicated forms at higher densities. Although these models give a better description of the physical situation, the evaluation of a number of system-specific quantities is necessary. Lee and Chao[71,72] incorporated a local composition model obtained from Monte

Carlo simulation in cubic EOSs. Good representation of vapor-liquid equilibria was obtained in polar systems. Johnston and co-workers[52,59] developed a density-dependent local-composition modification of the AvdW equation and applied it to entrainer systems. They correlated the cross-interaction energy with the total configurational internal energy, determined from literature solubility parameters, so no new adjustable parameters were introduced. Qualitative agreement was obtained for the phenol blue/entrainer/CO_2 systems.

4. Lattice-Gas Equations of State

Lattice models have been used successfully to describe liquid phase behavior in highly asymmetric polymer solutions.[36,45] The lattice-gas model is based on a distribution of molecules and holes, or vacant lattice sites, in a three-dimensional lattice. Trappeniers et al.[113] used a mean-field, two-component lattice-gas model to qualitatively explore so-called "gas-gas equilibria" in some noble gas mixtures at high pressure. Kleintjens and Koningsveld[61] extended this treatment to SCF-liquid solutions by including an empirical entropy correction term and allowing molecules to occupy more than one site. Instead of employing a lattice coordination number, interacting surface areas were assigned. This model was extended to binary solid-SCF mixtures[69] with the addition of two mixture parameters. One of the parameters was quite temperature dependent, necessitating the use of additional parameters. The phase diagram of ethylene-naphthalene was reproduced quantitatively. A further extension was made to multicomponent systems.[62]

Vezzetti[118,119] proposed a two-component lattice-gas model for solid solubilities in SCFs incorporating the quasichemical approximation, which attempts to include the effects of nonrandomness on the mixture. This model is attractive because all parameters, including one binary parameter, have physical significance. Unfortunately, the model cannot handle the highly asymmetric systems typical of SCF processes.

Kumar et al.[63] developed a multicomponent model based on the pure lattice-gas model of Panayiotou and Vera.[92] Kumar's model accounts for nonrandomness and includes one temperature-independent binary parameter. Systems such as benzoic acid-CO_2 were correlated well outside the critical region.

Lattice models have also been used to address the fact that behavior in the immediate vicinity of the critical point is nonclassical and cannot be described correctly by classical-type equations such as the vdW equation. At the critical point, the limiting values of the partial molar volume, enthalpy, and heat capacity approach negative or positive infinity, and for some properties, the limiting values are path dependent.[16,17,74,104] In this region, a nonclassical EOS is needed to impose the correct asymptotic behavior at the critical point. Unfortunately, the calculations are terribly complex and in general can be done for only highly idealized binary systems. Usually, consideration of this nonclassical behavior is unnecessary as most SCF processes occur sufficiently far from the critical point (several degrees Kelvin).

Scaling-law models describe thermodynamic properties in terms of the distance from the critical point as governed by the universal "critical exponents".[41] Nonanalytic decorated lattice-gas models have been developed that are accurate in the critical region. A decorated lattice-gas superimposes secondary lattice sites at the midpoints of the regular lattice, an approach more adept at describing asymmetric systems. Wheeler[124] applied a decorated lattice gas model to describe limiting behavior of dilute solutions near the critical point. Gilbert[39,40] described a lattice-gas whose properties can be mapped mathematically from the much-studied three-dimensional Ising model. With this model, partial molar volume and solubility data were fit for highly asymmetric SCF-solid mixtures with two adjustable parameters. Because partial molar volumes are a derivative property, fitting them is a stringent test for any model. Nielson and Levelt Sengers[86] also developed a decorated lattice-gas model for supercritical solutions and were able to reproduce portions of solubility curves with several adjustable parameters. Although they used a more complex nonclassical description of the pure solvent and included the temperature and pressure dependence of the chemical potential of the solute phase, they were unable to represent quantitatively the partial molar volume data with the same parameters used to fit the solubility.

Although not a lattice model, an interesting approach to the nonclassical nature of the critical region was suggested by Fox.[37] In order to make the transition from classical EOSs far from the critical region to the nonclassical scaling laws near the critical point, he proposed the use of a switching function. Though this leads to some internal inconsistencies in the model, the concept may be useful in practice.

5. Association Models

The section on phase diagrams mentioned the importance of entrainers or cosolvents. For instance, the solubility of acridine in carbon dioxide increases several hundred percent when 5% methanol is added to the SCF CO_2. It is most likely that the solubility increase is due to hydrogen bonding between the methanol and the acridine. This type of specific chemical bonding cannot be described by any of the models discussed up to this point; incorporation of a chemical theory term is required. The first efforts in this area were for high-pressure, light hydrocarbon/methanol systems. It was found that dimerization of the methanol was needed to explain the phase behavior.[97]

In a 1:1 or dimerization model the equilibrium constant, K, for complex formation is given by

$$K = \frac{1}{P}\left(\frac{\phi_{AB}}{\phi_A\phi_B}\right)\left(\frac{z_{AB}}{z_A z_B}\right) \tag{8}$$

where AB = the complex
     A = hydrogen bond donor
     B = hydrogen bond acceptor
     z = true mole fractions in solution
     ϕ = fugacity coefficients determined from an EOS

Since the fugacity coefficient is a function of the true mole fractions, the problem is highly coupled. In addition, the equilibrium constant for bond formation is generally not known, so it must be regarded as an adjustable parameter.

Donohue and co-workers[47] sought to eliminate the coupled nature of the association problem by extending their perturbed anisotropic chain theory (PACT) to associated solutions. In this formulation, the only effect of the association is to decrease the number of moles in the ideal gas term. When applied to supercritical entrainer systems, their associated PACT (APACT) model did a reasonably good job of predicting solubilities.[123] They show spectroscopic IR data to support qualitatively the thesis that hydrogen bonds form in the acridine/methanol and benzoic acid/acetone systems, but they do not obtain numerical estimates of the equilibrium constants. As a result, the pure component parameters were fit from pure component data, but two parameters, the equilibrium constant and the nonzero binary interaction parameter, had to be regressed from the solid/SCF/entrainer experimental solubility data.

Two distinct opportunities exist in the area of association models of supercritical fluid solutions. First, experimentally determined equilibrium constants are needed to reduce the number of adjustable parameters in the association model and provide a better test of the ability of the model to predict SCF phase behavior. Second, association models may be an appropriate way to describe the clustering of solvent molecules around a solute in the region near the critical point that is highly compressible, as discussed later. In fact, ideal chemical theory ($\phi = 1$) has been used to describe qualitatively the unusual form of the infinite dilution partial molar volume of solutes in SCFs.[29]

B. EXPANDED LIQUID TREATMENT

The SCF phase can also be treated as an expanded liquid, with the advantage that integration through the critical region is not necessary; however, rather than requiring ϕ, as in the EOS approach, knowledge of two thermodynamic properties is necessary: γ and $\bar{v}$ for the solute. To determine $\bar{v}$, an EOS, thus the pervasive k_{ij}, is required, albeit in a narrower sense. Mackay and Paulaitis[75] assumed that the solid was infinitely dilute in the SCF; thus, the composition dependence of the activity coefficient could be ignored. By adjusting both γ and k_{ij} they were able to correlate successfully the solubility of naphthalene in CO_2 and ethylene. They found that γ was typically on the order of 1000, indicating the considerable nonidealities in the SCF.

Rather than using γ as an adjustable parameter, it can be predicted from a solution model. Unfortunately, in practice, the solution model requires its own adjustable parameters. Giddings et al.[38] proposed the use of the Hildebrand solubility parameter, δ, to characterize the SCF solvent qualitatively. They determined δ from the vdW equation of state and found it to be proportional to the solvent density; thus, the enhancement factor (which removes effects of the solute vapor pressure) should be related to solvent density. This treatment agrees with the experimental observation that in the absence of chemical interactions the log of the enhancement factor is roughly linear with solvent density.[24] Allada[4] showed a generalized method to determine the solubility parameter of an SCF, but found that δ alone could not correlate solubilities of a given solute in diverse solvents.

Indeed, efforts to apply regular solution theory to SCF phase equilibria have had to employ semiempirical modifications. Pang and McLaughlin[93] were able to fit the solubility of several aromatics in CO_2 and ethylene as a function of the solubility parameter, but four adjustable parameters were necessary. Ziger and Eckert[134] developed an expression for the enhancement factor based upon regular solution theory and the vdW equation. By introducing only one temperature-independent parameter per pure component, they were able to correlate the solubilities of a variety of solutes in SCF CO_2, ethylene, and ethane successfully.

C. COMPUTER SIMULATIONS

The advent of high-speed computers has opened new avenues in the study of SCF solutions. They are not yet a good tool for modeling SCF phase equilibria due to the nonanalytic nature of the results and the time-intensive computer use; however, the simulations may provide insight on the molecular level that will form the basis for improved mathematical models. Vogelsang and Hoheisel[122] used molecular dynamics simulations to compare the structure of SCFs and liquids, but the results were inconclusive since the simulations were performed at different densities. Hoheisel et al.[44] obtained "exact" mixture data for a Lennard-Jones system using molecular dynamics, then compared the results with the van der Waals n-fluid approximations. They found that the vdW one-fluid approximation worked best for SCF mixtures, but was invalid for the large size and energy asymmetries typically found in process mixtures. Note that most commonly used EOSs adopt a one-fluid approach.

Monte Carlo simulations have also been used to study SCF phase equilibria. Shing and Chung[108] used the isothermal-isobaric ensemble which accounts for density fluctuations to calculate the solubility of naphthalene in CO_2. Qualitative agreement was obtained using Lennard-Jones potentials plus quadrupole interactions; with optimized potential parameters, improved quantitative agreement would be obtained. In a later study,[88] the grand canonical ensemble (constant temperature, volume, and chemical potential) was employed, which allowed both density and composition fluctuations and gave similar results. In addition, they found that the addition of a small amount of water dramatically increased the solubility of naphthalene.

The phase equilibrium of the CO_2-acetone system was reproduced qualitatively by a canonical ensemble (constant temperature, volume, and number of molecules) Monte Carlo technique.[91] Later, the Gibbs-ensemble Monte Carlo methodology was developed,[89-91] allowing simultaneous simulation of two equilibrium phases (not a solid phase, however). Carbon dioxide-acetone phase behavior was reproduced quantitatively with potential parameters obtained from pure component properties. Surprisingly, the good results were obtained with the simple spherically symmetric Lennard-Jones potential without accounting for the polarity of the solute.

Petsche and Debenedetti[98] have conducted molecular dynamics simulations that have yielded new insight on the behavior of solvent molecules near an infinitely dilute solute (see discussion below).

IV. INTERMOLECULAR INTERACTIONS

Developing an EOS that models PVT and solubility behavior accurately in SCFs is clearly challenging; most of the equations discussed earlier give quantitative results only when several adjustable parameters are introduced. This result points out the pressing need for a better understanding of the interactions in SCFs on a molecular level, upon which one can base EOS models that reflect the true interactions in solution. This need is corroborated by the interesting behavior of some derivative properties of solutes in SCFs, namely, the excess enthalpy of mixing (or partial molar excess enthalpy), and the partial molar volume (discussed below). The impetus provided by these derivative property measurements to investigate intermolecular interactions has been addressed by several investigators with spectroscopic measurements, discussed in subsections B and C. This section further shows how the spectroscopic results are corroborated by theoretical and computational investigations of local structure around solutes in SCFs. Finally, the importance of solute/solute interactions in highly asymmetric SCF solutions is indicated.

A. DERIVATIVE PROPERTIES

Both the excess enthalpy of mixing and the solute partial molar volume exhibit anomalous behavior in the vicinity of the solution critical point. In the context of the decorated lattice-gas model discussed earlier, Wheeler[124] showed a fundamental relationship between these two quantities. The infinite dilution partial molar volume should scale as the infinite dilution partial molar enthalpy arbitrarily close to the critical point. The excess enthalpy of mixing is related to the partial molar enthalpy by

$$h_{mix}^E = x_1(\bar{h}_1 - h_1) + x_2(\bar{h}_2 - h_2) \tag{9}$$

Excess enthalpies of mixing of SCFs, measured by flow calorimetry,[21,35,128] exhibit complex maxima and minima dependent on the proximity of the solution to its critical point, as shown in Figure 10. For this propane/Freon-12 system, the large negative deviations occur at temperatures slightly above the critical temperature of propane (370.0 K) and the large positive deviations occur at temperatures slightly above the critical temperature of Freon 12 (385.2 K).

Christensen et al.[21] have explained this phenomenon as follows: at a temperature just above T_c for the lighter component, the light component is in a denser state in the mixture than it would be as a pure component. Thus, the light component exhibits a lower partial molar enthalpy in the mixture than it would pure. Conversely, near the T_c for the heavy component, the heavy component is in a less dense state than it would be as a pure component, resulting in a higher partial molar enthalpy in the mixture.

Partial molar volumes of solutes in SCFs exhibit even more dramatic behavior and may hold the key to understanding and modeling phase equilibria in SCFs. Although partial molar volume data are extremely scarce, the common thread among those available are large negative values in the region of high compressibility near the critical point of the solution.

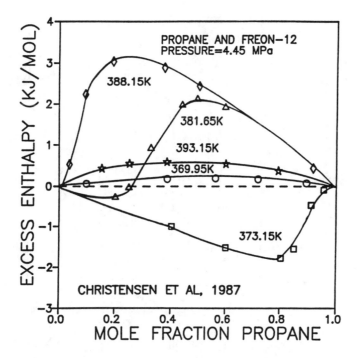

FIGURE 10. Excess enthalpies of mixing of SCF propane and Freon-12.

Several investigators reported values obtained by taking the composition derivative of volume data for mixtures in the critical region.[1,7,19,31,32,114,129] Even at these finite concentrations, the values were large and negative (-100 to -1000 cm^3/mol) near the critical point. These systems involved solutes as diverse as salts and polymers.

Although direct measurement of the partial molar volumes would give better results than the composition derivative of volume data, the experiment is challenging because small changes in temperature and pressure yield large errors in the partial molar volumes; however, Van Wasen and Schneider[117] proposed a chromatographic method to determine infinite dilution partial molar volumes from the isothermal pressure dependence of the solute retention time. Unfortunately, Paulaitis et al.[95] showed that the method made the incorrect assumption that the infinite dilution partial molar volume was not a function of the solute concentration as one approaches the critical point.

The most successful method to measure directly partial molar volumes in SCFs used a high-precision densitometer.[29,30] With this instrument, measurements could be made as close as 2°C from the critical point of the solvent. For several heavy organic solutes in SCF CO_2 and ethylene, the infinite dilution partial molar volume was as large as $-20,000$ cm^3/mol, as shown in Figure 11. Subsequently, Biggerstaff and Wood[10] have used this same method to determine partial molar volumes of argon, ethylene, and xenon in sub- and supercritical water. Since these are repulsive mixtures, they exhibit the interesting phenomenon of large positive values of the partial molar volumes in the vicinity of the critical point of the solvent. While most systems of practical importance involve attractive mixtures, Petsche and Debenedetti[98] clearly demonstrate both attractive and repulsive mixtures in their molecular dynamics simulations, as discussed below.

The large negative values of the infinite dilution partial molar volume of solutes in SCFs physically indicate a large volume decrease when a molecule of solute is added to the pure solvent. This suggests a clustering or agglomeration of the solvent molecules around the solute. Subsequently, researchers have come to view the phenomenon as a local density around the solute that is significantly greater than the bulk density.

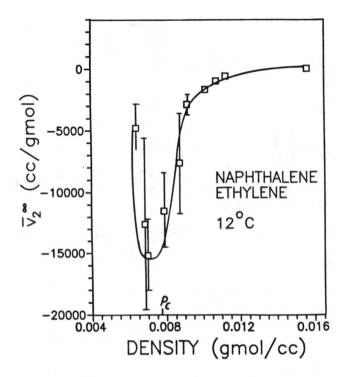

FIGURE 11.   Partial molar volume at infinite dilution of naphthalene in ethylene at 12°C.

The effect of the large negative partial molar volumes may be especially important in entrainer systems, where the increased local density is complicated by the preferential attraction of the entrainer to the solute. This is frequently the case because the entrainer is generally chosen to interact specifically (through hydrogen-bonding, acid-base interactions, or strong dipole-dipole interactions) with the solute. In this situation, one would anticipate increased local densities and local compositions that are significantly different than the bulk. Understanding the local densities and local compositions has been the focus of the spectroscopic investigations discussed below.

The interesting behavior of the partial molar volumes may hold the key to better predictions of the solubility of solutes in SCFs, because these solubilities are governed by the fugacity coefficient of the solute, which is a function of the integral of the solute partial molar volume. The integral over pressure can mask inaccuracies in the expression for the partial molar volume, and in many cases even give reasonable values for the fugacity coefficient (and solubility); however, better understanding of the interactions that produce the partial molar volume behavior and incorporation of that knowledge into EOS models may lead to better overall predictions of SCF phase equilibria.

B. SOLVATOCHROMIC PROBES OF SCF SOLVENT STRENGTH

Spectroscopic probes are an excellent technique to study solute/solvent and solute/entrainer interactions in SCFs; however, the first spectroscopic studies of SCFs were aimed at quantifying the solvent power of the fluids for chromatographic separations.

A popular method for ranking solvent strength is the use of linear solvation energy relationships, such as the Kamlet-Taft equation.[57,58] This equation relates linearly scales of chemical characteristics to some observable quantity [reaction rate constant, position of the maximum absorption peak in the infrared (IR), ultraviolet-visible (UV-vis), nuclear magnetic resonance (NMR), electron spin resonance (ESR), etc.]. The three most commonly included characteristic scales are π^* (measure of polarity and polarizability), α (hydrogen-bond donor strength), and β (hydrogen-bond acceptor strength). A measurable quantity, XYZ, is related to these scales by

$$XYZ = XYZ_0 + a\alpha + b\beta + s\pi^* \qquad (10)$$

where a, b, and s are solute-specific constants, α, β, and π^* are solvent-specific and XYZ_0 is a constant specific to the particular observable quantity.

Hyatt[46] first reported π^* values near -0.5 for CO_2 by measuring the wavelength of maximum absorption, which indicates the degree of solvent-induced stabilization of the electronic states, of a dilute dye in solution. The values increased monotonically with the transition from sub- to supercritical and the values indicated a solvent strength similar to that of a hydrocarbon solvent with low polarizability. In a complementary study, Sigman et al.[109] reported similar values of π^* for CO_2 and values of the hydrogen-bond basicity term near zero; however, they noted that the results showed more variation for different dye indicators than normal liquids. Besides strong specific chemical interactions, the current authors believe this could be due to varying degrees of solvent clustering (discussed in the next section), dependent on the quality of the temperature and pressure control in the experiment.

In more detailed studies of CO_2, Yonker and co-workers[131,133] pointed out the qualitative change in the slope of the π^* parameter as a function of the Onsager reaction field. Later Johnston et al.[51] explained this change in terms of cluster formation, as discussed below. Subsequent data for Xe, SF_6, C_2H_6, and NH_3[110] indicated the same monotonically decreasing solvent strength (lower π^*) with decreasing density; however, most of the change in π^*, as measured by frequency shifts of the UV-vis absorption peak maxima, can be attributed to the changing solvent density. Thermochromic shifts (constant density, varying temperature) are much smaller than solvatochromic shifts (constant temperature, varying pressure and density) and even more dependent on the nature of the solute.[132]

Overall, the investigation of SCFs in the framework of linear solvation energy relationships indicates that solvent strength is proportional to the density. Clearly, this is in agreement with the observation that in the absence of strong chemical interactions the logarithm of the enhancement factor of solid solutes in SCFs are essentially a linear function of the solvent density. The solvatochromic parameters are not linear functions of density, however, and frequently exhibit a change of slope in the vicinity of high compressibility. This observation can be explained by and give added insight into the clustering of the solvent around the solute, which has been the most recent focus of spectroscopic studies of SCF solutions.

C. EXPERIMENTAL PROBES OF LOCAL STRUCTURE

While a general measure of solvent strength (such as the Kamlet-Taft π^* scale) may help discern possible applications of SCFs, the partial molar volume data discussed above suggests the presence of solvent aggregation or clustering that is likely to be important in the quantitative modeling of SCF processes. Kim and Johnston[60] probed the local structure around a solute in a SCF with UV-vis absorption spectroscopy by measuring the wavelength of maximum absorption of a dye. By comparison with the solvent shift theory of McRae,[81] which describes the solvent-induced stabilization of the electronic states in a homogeneous medium, they obtained estimates of the actual local density. These local densities correlated remarkably well with the solvent isothermal compressibility, i.e., the solvent became more tightly packed around the solute when the solvent was very compressible near the critical point. The solvatochromic work of Yonker et al.[131,133] exhibits the same behavior.[51] Instead of describing the change in slope of the data as a discontinuity, one may draw a linear asymptote from the high density data (as suggested by the Onsager reaction field theory; see Figure 12). Then the difference between the data and the prediction represents the degree

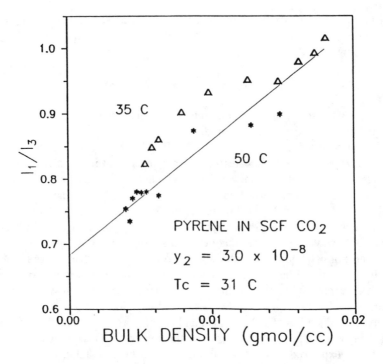

FIGURE 12. I_1/I_3 of pyrene in SCF carbon dioxide at 35 and 50°C.

of clustering and the maximum in deviation corresponds directly with the maximum in the isothermal compressibility.

Kajimoto et al.[56] also looked at frequency shifts, but examined the fluorescence spectra, which is a very sensitive technique for investigating the solvent environment of dilute samples. They looked at both the normal and charge-transfer peaks of (*N*,*N*-dimethyl-amino)benzonitrile in SCF fluoroform, a highly polar solvent where one might expect large spectral shifts. In the slightly SC region, the shifts were greater than those predicted by the theory of McRae[81] and the discrepancy was attributed to an aggregation of the solvent molecules around the solute.

Another measure of the strength of solute/solvent interactions available from the fluorescence spectra of some polycyclic aromatics such as pyrene is the relative intensity of the first to the third peak in the spectrum. The first transition is forbidden by symmetry arguments; however, the stronger the interaction between the solute and the solvent the more that symmetry is disrupted, allowing the transition to take place with higher intensity. For naphthalene and pyrene in SCF CO_2, C_2H_4, and CF_3H, the data indicate much stronger interactions in the compressible region near the critical point, as shown in Figure 12 by the higher value of the ratio I_1/I_3 at the lower temperature (nearer the critical point) when comparing at constant density.[13,14] In fact, the difference between the ratio I_1/I_3 at a temperature near the critical point and I_1/I_3 at a temperature far from the critical point, where one would not expect clustering to occur, gives a measure of the augmented density around the solute. In agreement with the solvatochromic shifts,[51,53,59,60] these data indicate a maximum in this deviation that corresponds directly with the maximum in the isothermal compressibility, as shown in Figure 13. In addition to providing an independent measure of the enhanced local densities, these are the first experimental data to suggest that clustering occurs in nonpolar systems where one would not expect strong specific chemical interactions (e.g., hydrogen bonding or acid-base interactions).

Both Kim and Johnston[59,60] and Yonker and Smith[131] have extended their solvatochromic studies to include systems with entrainers or cosolvents. By measuring the ν_{max} of phenol

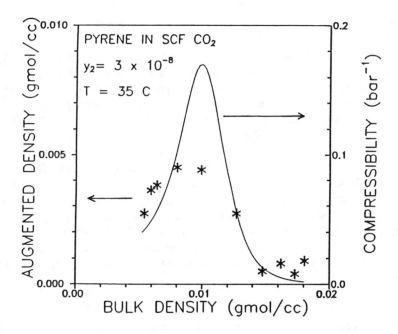

FIGURE 13. The relationship of the augmented density to the isothermal compressibility of pyrene in supercritical carbon dioxide.

blue in several CO_2/entrainer systems and using the dielectric enrichment arguments of Nitsche and Suppan,[87] Kim and Johnston[59,60] estimated local entrainer compositions that were several times the bulk concentration when operating in the compressible region near the critical point. In a slightly varied approach, Yonker and Smith estimated the local compositions by assuming a linear contribution of the entrainer and the solvent to the observed π^* value. Their results for the 2-nitroanisole/2-propanol/CO_2 system were in complete agreement with those of Kim and Johnston. Therefore, in addition to solvent molecules clustering around the solute, in entrainer systems one can adjust the composition of that cluster by careful manipulations of temperature and pressure.

D. THEORY AND MODELING OF LOCAL STRUCTURE

In addition to experimental observations, enhanced local densities have been corroborated by several diverse theoretical and computational efforts. These include Kirkwood-Buff fluctuation theory analysis, molecular dynamics simulations, calculation of pair correlation functions from integral equation theory, and development of an EOS including association for entrainer systems. This subsection briefly summarizes the results of these studies; more detailed discussion of the Kirkwood-Buff formalism and the computer simulations are available elsewhere in this monograph.

Debenedetti[23] used a fluctuation analysis to derive an expression for the solute infinite dilution partial molar volume in terms of an actual cluster size of solvent molecules around a solute. The cluster size is numerically a measure of the statistical correlations between solute and solvent concentration fluctuations, but can take on the mechanistic significance of the number of SCF solvent molecules surrounding the solute. Cluster sizes determined in this way from experimental $\bar{v}_2^\infty$ data indicate as many as 100 solvent molecules around a solute in the highly compressible region.

The clustering is seen more dramatically in molecular dynamics simulations of Lennard-Jones mixtures that represent dilute xenon in SCF neon and dilute neon in SCF xenon.[98] In the attractive mixture (dilute Xe in Ne) clustering clearly occurs in the vicinity of the critical point and a hole or negative cluster forms in the repulsive mixture (dilute Ne in Xe). The latter case corresponds to the large positive partial molar volumes measured by Biggerstaff and Wood[10] of argon, ethylene, and xenon in SCF water.

Lee and co-workers[22,70] have shown that integral equation theories can be very helpful in revealing the local structure around solutes in SCFs. They calculated pair correlation functions of mixtures representing systems as asymmetric as pyrene/CO_2 and naphthalene/CO_2 by solving the Ornstein-Zernike equation. Results indicate strong short-range interactions from the increased first peak height near critical conditions, as well as long-range correlations between the solute and the solvent, consistent with the concept of clustering.

Eckert et al.[29] suggested that chemical theory may also be useful for modeling the clustering of pure solvent around the solute in normal solid/SCF systems. Some additional justification of clustering can be drawn from the relative success or promise of models that include the concept of enhanced local structure. For example, the AvdW equation with density-dependent local composition mixing rules that address the local composition question shows significant promise for modeling entrainer systems.[52,59,60] Similarly, when Donohue and co-workers extended the PACT equation to include a chemical association term for solute/entrainer complexation they were able to obtain the correct form of the solubility curves of an entrainer system.[123]

E. SOLUTE/SOLUTE INTERACTIONS

Besides solvent clustering around the solute, another important consideration is the strength of solute/solute interactions, which appear to be strong even in very dilute solutions. In many of the models of SCF solubilities discussed above,[24,50,54,105,134] investigators assume infinite dilution of all solutes, i.e., no solute/solute interactions, because the mole fractions are generally below 10^{-2} or 10^{-3}; however, there are both experimental measurements and computational evidence to suggest that assumption is not justified.

Solubilities of solutes in ternary systems (two solutes in a SCF) are significantly different than pure component solubilities in the same SCF. In the case of naphthalene and benzoic acid in SCF CO_2 the solubility of both solutes increased several hundred percent in the mixed solute system even though the solubilities were as low as 10^{-3} mol fraction.[65] In the phenanthrene/anthracene/CO_2 system the solubility of anthracene (about 10^{-5} mol fraction) increased, while that of phenanthrene (about 10^{-3} mol fraction) decreased slightly relative to the binary pairs.[67] Clearly, the presence of another compound, even in dilute solutions, influences the solubility behavior.

In addition to this solubility evidence, fluorescence spectroscopy experiments indicate the existence of solute/solute interactions at low concentrations in SCF solutions because of the formation of excimers.[11,13] Excimers are excited state dimers, and their presence in pyrene/SCF CO_2 solutions as low as 10^{-5} mol fraction emphasize that solute/solute interactions may be important in highly asymmetric SCF solutions, even when the solute concentrations are very low.

Finally, Cochran and Lee[22] have calculated solute/solute pair correlation functions of Lennard-Jones mixtures that simulate naphthalene/CO_2 and pyrene/CO_2 systems and found quantitatively stronger solute/solute correlations than the solute/solvent correlations discussed in the context of solvent clustering above. This is in qualitative agreement with the experimental observations of strong solute/solute interactions even in dilute SCF solutions.

V. CONCLUSIONS

At this point it must be recognized that further progress in the characterization and even the exploitation of SCF solution chemistry depends on the achievement of a deeper basic understanding of the molecular processes involved. One must recognize the special properties of such fluids, with noncontinuum behavior, characterized by clustering or spherical anisotropy giving local structures very different from bulk or average structures.

Recognition of the need to determine separately the unusual forces between the molecules and the special ways in which statistical mechanical or other techniques must be used to add these up is vital. The capacity for molecular simulations is ever improving, and these endeavors must be linked with experimental results from spectroscopy and other incisive investigation techniques.

Infinite dilution is a useful composition for the study of pure unlike-pair interactions, but researchers find that infinite dilution may be fantastically dilute for SCF solutions, probably in the parts per million range. Current understanding must be extended to higher compositions and to the effects on the critical point and the resulting properties in the near-critical region.

Finally if one finds success in such efforts, in gaining an appreciation of the detailed molecular processes, the potential for practical applications may exceed even the very optimistic forecasts of its strongest proponents of recent years. The ability will then exist to tailor solvents over an extremely wide range of properties to achieve very difficult separations or carry out very precise reaction chemistry.

NOMENCLATURE

a,b	Equation of state parameters (modeling section)
a,b,s	Solute-specific constants in linear solvation energy relationship (intermolecular interactions section)
F	Degrees of freedom
h	Molar enthalpy
I	Intensity of spectroscopic peaks
K	Equilibrium constant
k	Binary interaction parameter
P	Pressure
p	Number of phases
R	Ideal gas constant
T	Temperature
v	Molar volume
x,y	Mole fraction
XYZ	Observable quantity correlated by linear solvation energy relationship
z	True mole fraction in chemical theory

Greek Letters

α	Kamlet-Taft hydrogen-bond donor strength
β	Kamlet-Taft hydrogen-bond acceptor strength
γ	Activity coefficient
δ	Hildebrand solubility parameter
π^*	Kamlet-Taft polarity/polarizability
ϕ	Fugacity coefficient
Ψ	Equation of state parameter

Superscripts and Subscripts

o	Reference state	i,j	Component identification
∞	Infinite dilution	L	Property in liquid phase
A	Hydrogen bond donor	mix	Mixture property
AB	Hydrogen bonded complex	s	Property in solid phase
B	Hydrogen bond acceptor	sat	Conditions at saturation
c	Critical state	SCF	Property in supercritical phase
E	Excess property		

REFERENCES

1. **Abraham, K. P. and Ehrlich, P.,** Partial molar volume of a polymer in supercritical solution, *Macromolecules,* 8(6), 944, 1975.
2. **Alder, B. J. and Hecht, C. E.,** Studies in molecular dynamics. VII. Hard-sphere distribution functions and an augmented van der Waals theory, *J. Chem. Phys.,* 50(5), 2032, 1969.
3. **Alder, B. J., Young, D. A., and Mark, M. A.,** Studies in molecular dynamics. X. Corrections to the augmented van der Waals theory for the square well fluid, *J. Chem. Phys.,* 56(6), 3013, 1972.
4. **Allada, S. R.,** Solubility parameters of supercritical fluids, *Ind. Eng. Chem. Process Des. Dev.,* 23, 344, 1984.
5. **Benmekki, E.-H., Kwak, T. Y., and Mansoori, G. A.,** Van der Waals mixing rules for cubic equations of state, in *Supercritical Fluids,* Squires, T. G. and Paulaitis, M. E., Eds., American Chemical Society, Washington, D.C., 1987, 101.
6. **Benmekki, E.-H. and Mansoori, G. A.,** The role of mixing rules and three-body forces in the phase behavior of mixtures: simultaneous VLE and VLLE calculations, *Fluid Phase Equilibria,* 41, 43, 1988.
7. **Benson, S. W., Copeland, C. S., and Pearson, D.,** Molal volumes and compressibilities of the system NaCl-H_2O above the critical temperature of water, *J. Chem. Phys.,* 21(12), 2208, 1953.
8. **Beret, S. and Prausnitz, J. M.,** Perturbed hard-chain theory: an equation of state for fluids containing small or large molecules, *AIChE J.,* 21, 1123, 1975.
9. **Bertucco, A., Fermeglia, M., and Kikic, L.,** Modified Carnahan-Starling-van der Waals equation for supercritical fluid extraction, *Chem. Eng. J.,* 32(1), 21, 1986.
10. **Biggerstaff, D. R. and Wood, R. H.,** Apparent molar volumes of aqueous argon, ethylene, and xenon from 300 to 716 K, *J. Phys. Chem.,* 92, p. 1988, 1988.
11. **Brennecke, J. F. and Eckert, C. A.,** Molecular interactions from fluorescence spectroscopy, Proc. Int. Symp. on Supercritical Fluids, Nice, October 1988.
12. **Brennecke, J. F. and Eckert, C. A.,** Phase equilibria for supercritical fluid process design, *AIChE J.,* 35, 1409, 1989.
13. **Brennecke, J. F. and Eckert, C. A.,** Fluorescence spectroscopy studies of intermolecular interactions in supercritical fluids, *Am. Chem. Soc., Symp. Ser.,* 406, 14, 1989.
14. **Brennecke, J. F., Tomasko, D. L., Peshkin, J., and Eckert, C. A.,** Fluorescence spectroscopy studies of dilute supercritical solutions, *Ind. Eng. Chem. Res.,* 29, 1682, 1990.
15. **Carnahan, N. F. and Starling, K. E.,** Intermolecular repulsions and the equation of state for fluids, *AIChE J.,* 18(6), 1184, 1972.
16. **Chang, R. F. and Levelt Sengers, J. M. H.,** Behavior of dilute mixtures near the solvent's critical point, *J. Phys. Chem.,* 90, 5921, 1986.
17. **Chang, R. F., Morrison, G., and Levelt Sengers, J. M. H.,** The critical dilemma of dilute mixtures, *J. Phys. Chem.,* 88(16), 3389, 1984.
18. **Charoensombut-Amon, T., Martin, R. J., and Kobayashi, R.,** Application of a generalized multiproperty apparatus to measure phase equilibrium and vapor phase densities of supercritical carbon dioxide in *n*-hexadecane systems up to 26 MPa, *Fluid Phase Equilibria,* 31, 89, 1986.
19. **Chappelear, D. C. and Elgin, J. C.,** Phase equilibria in the critical region, *J. Chem. Eng. Data,* 6(3), 415, 1961.
20. **Chrastil, J.,** Solubility of solids and liquids in supercritical gases, *J. Phys. Chem.,* 86, 3016, 1982.
21. **Christensen, J. J., Izatt, R. M., and Zebolshy, D. M.,** Heats of mixing in the critical region, *Fluid Phase Equilibria,* 38, 163, 1987.
22. **Cochran, H. D. and Lee, L. L.,** Solvation structure in supercritical fluid mixtures based on molecular distribution functions, *Am. Chem. Soc., Symp. Ser.,* 406, 27, 1989.
23. **Debenedetti, P. G.,** Clustering in dilute, binary supercritical mixtures: a fluctuation analysis, *Chem. Eng. Sci.,* 42, 2203, 1987.
24. **Dobbs, J. M. and Johnston, K. P.,** Selectivities in pure and mixed supercritical fluid solvents, *Ind. Eng. Chem. Res.,* 26, 1476, 1987.
25. **Dobbs, J. M., Wong, J. M., and Johnston, K. P.,** Nonpolar co-solvents for solubility enhancement in supercritical fluid carbon dioxide, *J. Chem. Eng. Data,* 31(3), 303, 1986.
26. **Dobbs, J. M., Wong, J. M., Lahiere, R. J., and Johnston, K. P.,** Modification of supercritical fluid phase behavior using polar cosolvents, *Ind. Eng. Chem. Res.,* 26, 56, 1987.
27. **Donohue, M. D. and Prausnitz, J. M.,** Perturbed hard chain theory for fluid mixtures: thermodynamic properties for mixtures in natural gas and petroleum technology, *AIChE J.,* 24, 849, 1978.
28. **Eckert, C. A., Van Alsten, J. G., and Stoicos, T.,** Supercritical fluid processing, *Environ. Sci. Technol.,* 20(4), 319, 1986.
29. **Eckert, C. A., Ziger, D. H., Johnston, K. P., and Ellison, T. K.,** The use of partial molal volume data to evaluate equations of state for supercritical fluid mixtures, *Fluid Phase Equilibria,* 14, 167, 1983.

30. **Eckert, C. A., Ziger, D. H., Johnston, K. P., and Kim, S.,** Solute partial molal volumes in supercritical fluids, *J. Phys. Chem.,* 90(12), 2738, 1986.
31. **Ehrlich, P.,** Partial molar volume anomaly in supercritical mixtures and the free radical polymerization of ethylene, *J. Macromol. Sci. Chem.,* A5(8), 1259, 1971.
32. **Ehrlich, P. and Fariss, R. H.,** Negative partial molal volumes in the critical region. Mixtures of ethylene and vinyl chloride, *J. Phys. Chem.,* 73(4), 1164, 1969.
33. **Ellison, T. K.,** Supercritical Fluids: Kinetic Solvent Effect and the Correlation of Solid-Fluid Equilibria, Ph.D. thesis, University of Illinois, Urbana-Champaign, 1986.
34. **Ely, J. F.,** Improved mixing rules for one-fluid conformal solution calculations, *Am. Chem. Soc., Symp. Ser.,* 300, 331, 1986.
35. **Faux, P. W., Christensen, J. J., Izatt, R. M., Pando, C., and Renuncio, J. A. R.,** The excess enthalpies of (carbon dioxide + n-hexane + toluene) at 470.15 K and 7.5 and 12.5 MPa, *J. Chem. Thermodyn.,* 20, 1297, 1988.
36. **Flory, P. J.,** Thermodynamics of high polymer solutions, *J. Chem. Phys.,* 10, 51, 1942.
37. **Fox, J. R.,** Method for construction of nonclassical equations of state, *Fluid Phase Equilibria,* 14, 45, 1983.
38. **Giddings, J. C., Myers, M. N., and King, J. W.,** Dense gas chromatography at pressures to 2000 atmospheres, *J. Chromatogr. Sci.,* 7, 276, 1969.
39. **Gilbert, S. W.,** Experimental and Theoretical Studies of Supercritical Fluid Behavior, Ph.D. thesis, University of Illinois, Urbana-Champaign, 1987.
40. **Gilbert, S. W. and Eckert, C. A.,** A decorated lattice-gas model of supercritical fluid solubilities and partial molar volumes, *Fluid Phase Equilibria,* 30, 41, 1986.
41. **Goodstein, D. L.,** *States of Matter,* Dover Publications, New York, 1985.
42. **Henderson, D.,** Practical calculations of the equation of state of fluids and fluid mixtures using perturbation theory and related theories, *Am. Chem. Soc., Adv. Chem. Ser.,* 182, 1, 1979.
43. **Hess, B. S.,** Supercritical Fluids: Measurement and Correlation Studies of Model Coal Compound Solubility and the Modeling of Solid-Liquid-Fluid Equilibria, Ph.D. thesis, University of Illinois, Urbana-Champaign, 1987.
44. **Hoheisel, C., Deiters, U., and Lucas, K.,** The extension of pure fluid thermodynamic properties to supercritical mixtures, *Mol. Phys.,* 49(1), 159, 1983.
45. **Huggins, M. L.,** Some properties of solutions of long-chain compounds, *J. Phys. Chem.,* 46, 151, 1942.
46. **Hyatt, J. A.,** Liquid and supercritical carbon dioxide as organic solvents, *J. Org. Chem.,* 49, 5097, 1984.
47. **Ikonomou, G. D. and Donohue, M. D.,** COMPACT: a simple equation of state for associated molecules, *Fluid Phase Equilibria,* 33, 61, 1987.
48. **Jin, G., Walsh, J. M., and Donohue, M. D.,** A group-contribution correlation for predicting thermodynamic properties with the perturbed-soft-chain theory, *Fluid Phase Equilibria,* 31, 123, 1986.
49. **Johnston, K. P.,** personal communication, 1988.
50. **Johnston, K. P. and Eckert, C. A.,** An analytical Carnahan-Starling-van der Waals model for solubility of hydrocarbon solids in supercritical fluids, *AIChE J.,* 27(5), 773, 1981.
51. **Johnston, K. P., Kim, S., and Combes, J.,** Spectroscopic determination of solvent strength and structure in supercritical fluid mixtures: a review, *Am. Chem. Soc. Symp. Ser.,* 406, 52, 1989.
52. **Johnston, K. P., Kim, S., and Wong, J. M.,** Local composition models for fluid mixtures over a wide density range, *Fluid Phase Equilibria,* 38, 39, 1987.
53. **Johnston, K. P., Peck, D. G., and Kim, S.,** Modeling supercritical mixtures: how predictive is it?, *Ind. Eng. Chem. Res.,* 28, 1115, 1989.
54. **Johnston, K. P., Ziger, D. H., and Eckert, C. A.,** Solubilities of hydrocarbon solids in supercritical fluids. The augmented van der Waals treatment, *Ind. Eng. Chem. Fundam.,* 21, 191, 1982.
55. **Jonah, D. A., Shing, K. S., Venkatasubramanian, V., and Gubbins, K. E.,** Molecular thermodynamics of dilute solutes in supercritical solvents, in *Chemical Engineering at Supercritical Fluid Conditions,* Paulaitis, M. E., et al., Eds., Ann Arbor Science, Ann Arbor, MI, 1983, 221.
56. **Kajimoto, O., Futakami, M., Kobayashi, T., and Yamasaki, K.,** Charge-transfer-state formation in supercritical fluid: (N,N-dimethylamino) benzonitrile in CF_3H, *J. Phys. Chem.,* 92, 1347, 1988.
57. **Kamlet, M. J., Abboud, J. L., Abraham, M. H., and Taft, R. W.,** Linear solvation energy relationships. XXII. A comprehensive collection of the solvatochromic parameter, π^*, α, β, and some methods for simplifying the generalized solvatochromic equations, *J. Org. Chem.,* 48, 2877, 1983.
58. **Kamlet, M. J., Abboud, J. L., and Taft, R. W.,** The solvatochromic comparison method. VI. The π^* scale of solvent polarities, *J. Am. Chem. Soc.,* 99(18), 6027, 1977.
59. **Kim, S. and Johnston, K. P.,** Clustering in supercritical fluid mixtures, *AIChE J.,* 33(10), 1603, 1987.
60. **Kim, S. and Johnston, K. P.,** Molecular interactions in dilute supercritical fluid solutions, *Ind. Eng. Chem. Res.,* 26, 1206, 1987.
61. **Kleintjens, L. A. and Koningsveld, R.,** Lattice-gas treatment of supercritical phase behavior in fluid mixtures, *J. Electrochem. Soc.,* 25, 2352, 1980.

62. **Kleintjens, L. A., Van der Haegen, R., van Opstal, L., and Koningsveld, R.,** Mean-field lattice-gas modelling of supercritical phase behavior, *J. Supercrit. Fluids,* 1, 23, 1988.

63. **Kumar, S. K., Suter, U. W., and Reid, R. C.,** A statistical mechanics based lattice model equation of state, *Ind. Eng. Chem. Res.,* 26, 2532, 1987.

64. **Kurnik, R. T., Holla, S. J., and Reid, R. C.,** Solubility of solids in supercritical carbon dioxide and ethylene, *J. Chem. Eng. Data,* 26, 47, 1981.

65. **Kurnik, R. T. and Reid, R. C.,** Solubility of solid mixtures in supercritical fluids, *Fluid Phase Equilibria,* 8, 93, 1982.

66. **Kwak, T. Y. and Mansoori, G. A.,** Van der Waals mixing rules for cubic equations of state. Applications for supercritical fluid extraction modelling, *Chem. Eng. Sci.,* 41(5), 1303, 1986.

67. **Kwiatkowski, J., Lisicki, Z., and Majewski, W.,** An experimental method for measuring solubilities of solids in supercritical fluids, *Ber. Bunsenges. Phys. Chem.,* 88, 865, 1984.

68. **Larson, K. A. and King, M. L.,** Evaluation of supercritical fluid extraction in the pharmaceutical industry, *Biotech. Progress,* 2(2), 73, 1986.

69. **Leblans-Vinck, A. M., Koningsveld, R., Kleintjens, L. A., Diepen, G. A. M.,** Solubility of solids in supercritical solvents. III. Mean-field lattice-gas description of G-L and S-L equilibria in the system ethylene-naphthalene, *Fluid Phase Equilibria,* 20, 347, 1985.

70. **Lee, L. L. and Cochran, H. D.,** Solvation Structure and Chemical Potentials in Supercritical Fluid Mixtures Based on Molecular Distribution Functions, presented at the AIChE meeting, Washington, D.C., 1988.

71. **Lee, R.-J. and Chao, K.-C.,** Local composition of square-well molecules by Monte Carlo simulation, *Am. Chem. Soc. Symp. Ser.,* 300, 214, 1986.

72. **Lee, R.-J. and Chao, K.-C.,** Local composition embedded equations of state for strongly nonideal fluid mixtures, *Ind. Eng. Chem. Res.,* 28, 1251, 1989.

73. **Lemert, R. M. and Johnston, K. P.,** Solid-liquid-gas equilibria in multicomponent supercritical fluid systems, *Fluid Phase Equilibria,* 45, 265, 1989.

74. **Levelt Sengers, J. M. H., Morrison, G., Nielson, G., Chang, R. F., and Everhart, C. M.,** Thermodynamic behavior of supercritical fluid mixtures, *Int. J. Thermophysics,* 7(2), 231, 1986.

75. **Mackay, M. E. and Paulaitis, M. E.,** Solid solubilities of heavy hydrocarbons in supercritical solvents, *Ind. Eng. Chem. Fundam.,* 18(2), 149, 1979.

76. **Mansoori, G. A.,** Mixing rules for equations of state, *Am. Chem. Soc. Symp. Ser.,* 300, 314, 1986.

77. **Mansoori, G. A., Carnahan, N. F., Starling, K. E., and Leland, T. W., Jr.,** Equilibrium thermodynamic properties of the mixture of hard spheres, *J. Chem. Phys.,* 54(4), 1523, 1971.

78. **Mansoori, G. A. and Ely, J. F.,** Density expansion (DEX) mixing rules: thermodynamic modeling of supercritical extraction, *J. Chem. Phys.,* 82(1), 406, 1985.

79. **Mart, C. J., Papadopoulos, K. D., and Donohue, M. C.,** Application of perturbed-hard-chain theory to solid-supercritical fluid equilibria modeling, *Ind. Eng. Chem. Proc. Des. Dev.,* 25(2), 394, 1986.

80. **McHugh, M. A. and Krukonis, V. J.,** *Supercritical Fluid Extraction: Principles and Practice,* Butterworths, Boston, 1986.

81. **McRae, E. G.,** Theory of solvent effects on molecular electronic spectra frequency shifts, *J. Phys. Chem.,* 61, 562, 1957.

82. **Mohamed, R. S. and Holder, G. D.,** High pressure phase behavior in systems containing CO_2 and heavier compounds with similar vapor pressures, *Fluid Phase Equilibria,* 32, 295, 1987.

83. **Mollerup, J.,** A note on excess Gibbs energy models, equations of state and the local composition concept, *Fluid Phase Equilibria,* 7, 121, 1981.

84. **Morris, R. W. and Turek, E. A.,** Optimal temperature-dependent parameters for the Redlich-Kwong equation of state, *Am. Chem. Soc. Symp. Ser.,* 300, 389, 1986.

85. **Morris, W. O., Vimalchand, P., and Donohue, M. D.,** The perturbed-soft-chain theory: an equation of state based on the Lennard-Jones potential, *Fluid Phase Equilibria,* 32, 103, 1987.

86. **Nielson, G. C. and Levelt Sengers, J. M. H.,** Decorated lattice gas model for supercritical solubility, *J. Phys. Chem.,* 91(15), 4078, 1987.

87. **Nitsche, K. S. and Suppan, P.,** Solvatochromic shifts and polarity of solvent mixtures, *Chimia,* 36, 346, 1982.

88. **Nouacer, M. and Shing, K. S.,** Grand canonical Monte Carlo simulation for solubility calculation in supercritical extraction, *Mol. Simul.,* 2, 55, 1989.

89. **Panagiotopoulos, A. Z.,** Direct determination of phase coexistence properties of fluids by Monte Carlo simulation in a new ensemble, *Mol. Phys.,* 61, 813, 1987.

90. **Panagiotopoulos, A. Z.,** Gibbs-ensemble Monte Carlo simulations of phase equilibria in supercritical fluid mixtures, *Am. Chem. Soc. Symp. Ser.,* 406, 39, 1989.

91. **Panagiotopoulos, A. Z., Suter, U. W., and Reid, R. C.,** Phase diagrams of nonideal fluid mixtures from Monte Carlo simulation, *Ind. Eng. Chem. Fundam.,* 25, 525, 1986.

92. **Panayiotou, C. and Vera, J. H.,** Statistical mechanics of r-mer fluids and their mixtures, *Polymer J.,* 14, 681, 1982.

93. **Pang, T.-H. and McLaughlin, E.**, Supercritical extraction of aromatic hydrocarbon solids and tar sand bitumens, *Ind. Eng. Chem. Proc. Des. Dev.*, 24(4), 1027, 1985.

94. **Patel, N. C. and Teja, A. S.**, A new equation of state for fluids and fluid mixtures, *Chem. Eng. Sci.*, 37, 463, 1982.

95. **Paulaitis, M. E., Johnston, K. P., and Eckert, C. A.**, Measurement of partial molar volumes at infinite dilution in a supercritical-fluid solvent near its critical point, *J. Phys. Chem.*, 85(12), 1770, 1981.

96. **Peng, D.-Y. and Robinson, D. B.**, A new two-constant equation of state, *Ind. Eng. Chem. Fundam.*, 15(1), 59, 1976.

97. **Peschel, W. and Wenzel, H.**, Equation-of-state predictions of phase equilibria at elevated pressures in mixtures containing methanol, *Ber. Bunsenges. Phys. Chem.*, 88, 807, 1984.

98. **Petsche, I. B. and Debenedetti, P. G.**, Solute-solvent interactions in infinitely dilute supercritical mixtures: a molecular dynamics investigation, *J. Chem. Phys.*, 91, 7075, 1989.

99. **Prausnitz, J. M., Lichtenthaler, R. N., and Gomes de Azevedo, E.**, *Molecular Thermodynamics of Fluid-Phase Equilibria*, Prentice-Hall, Englewood Cliffs, NJ, 1986.

100. **Radosz, M., Lin, H.-M., and Chao, K.-C.**, High-pressure vapor-liquid equilibria in asymmetric mixtures using new mixing rules, *Ind. Eng. Chem. Proc. Des. Dev.*, 21, 653, 1982.

101. **Redlich, O. and Kwong, J. N. S.**, On the thermodynamics of solutions. V. An equation of state. Fugacities of gaseous solutions, *Chem. Rev.*, 44, 233, 1949.

102. **Reid, R. C., Prausnitz, J. M., and Poling, B. E.**, *The Properties of Gases and Liquids*, McGraw-Hill, New York, 1987.

103. **Rowlinson, J. S. and Swinton, F. L.**, *Liquids and Liquid Mixtures*, Butterworths, London, 1982.

104. **Rozen, A. M.**, The unusual properties of solutions in the vicinity of the critical point of the solvent, *Russ. J. Phys. Chem.*, 50(6), 837, 1976.

105. **Schmitt, W. J. and Reid, R. C.**, Solubility of monofunctional organic solids in chemically diverse supercritical fluids, *J. Chem. Eng. Data*, 31, 204, 1986.

106. **Schneider, G. M.**, High-pressure phase diagrams and critical properties of fluid mixtures, in *Chemical Thermodynamics, Vol. 2*, Specialist Periodical Reports, The Chemical Society, London, 1978, 105.

107. **Šerbanović, S. P. and Djordjević, B. D.**, Influence of the optimized temperature-dependent interaction parameter on vapor-liquid equilibrium binary predictions of supercritical methane with some alkanes by means of the Soave equation of state, *Ind. Eng. Chem. Res.*, 26(3), 618, 1987.

108. **Shing, K. S. and Chung, S. T.**, Computer simulation methods for the calculation of solubility in supercritical extraction systems, *J. Phys. Chem.*, 91, 1674, 1987.

109. **Sigman, M. E., Lindley, S. M., and Leffler, J. E.**, Supercritical carbon dioxide: behavior of π^* and β solvatochromic indicators in media of different densities, *J. Am. Chem. Soc.*, 107, 1471, 1985.

110. **Smith, R. D., Frye, S. L., Yonker, C. R., and Gale, R. W.**, Solvent properties of supercritical Xe and SF_6, *J. Phys. Chem.*, 91, 3059, 1987.

111. **Soave, G.**, Equilibrium constants from a modified Redlich-Kwong equation of state, *Chem. Eng. Sci.*, 27, 1197, 1972.

112. **Streett, W. B.**, Phase equilibria in fluid and solid mixtures at high pressure, *Chemical Engineering at Supercritical Conditions*, Paulaitis, M. E. et al., Eds., Ann Arbor Science, Ann Arbor, MI, 1983, 3.

113. **Trappeniers, N. J., Schouten, J. A., and Ten Seldam, C. A.**, Gas-gas equilibrium and the two-component lattice-gas model, *Chem. Phys. Lett.*, 5(9), 541, 1970.

114. **Tsekhanskaya, Y. V., Roginskaya, N. G., and Mushkina, E. V.**, Volume changes in naphthalene solutions in compressed carbon dioxide, *Russ. J. Phys. Chem.*, 40(9), 1152, 1966.

115. **Van Alsten, J. G.**, Structural and Functional Effects in Solutions with Pure and Entrainer-Doped Supercritical Solvents, Ph.D. thesis, University of Illinois, Urbana-Champaign, 1986.

116. **Van Konynenburg, P. H. and Scott, R. L.**, Critical lines and phase equilibria in binary van der Waals mixtures, *Philos. Trans. R. Soc. London*, 298, 495, 1980.

117. **Van Wasen, U. and Schneider, G. M.**, Partial molar volumes of naphthalene and fluorene at infinite dilution in carbon dioxide near its critical point, *J. Phys. Chem.*, 84, 229, 1980.

118. **Vezzetti, D. J.**, Solubility of solids in supercritical gases, *J. Chem. Phys.*, 77(3), 1512, 1982.

119. **Vezzetti, D. J.**, Solubility of solids in supercritical gases. II. Extension to molecules of differing sizes, *J. Chem. Phys.*, 80(2), 868, 1984.

120. **Vidal, J.**, Phase equilibria and density calculations for mixtures in the critical range with simple equations of state, *Ber. Bunsenges. Phys. Chem.*, 88, 784, 1984.

121. **Vimalchand, P. and Donohue, M. D.**, Thermodynamics of quadrupolar molecules: the perturbed-anisotropic-chain theory, *Ind. Eng. Chem. Fundam.*, 24, 246, 1985.

122. **Vogelsang, R. and Hoheisel, C.**, Structure and dynamics of a supercritical fluid in comparison with a liquid. A computer simulation study, *Mol. Phys.*, 53(6), 1355, 1984.

123. **Walsh, J. M., Ikonomou, G. D., and Donohue, M. D.**, Supercritical phase behavior: the entrainer effect, *Fluid Phase Equilibria*, 33, 295, 1987.

124. **Wheeler, J. C.,** Behavior of a solute near the critical point of an almost pure solvent, *Ber. Bunsen-Gessellsch.,* 76(3/4), 308, 1972.
125. **Whiting, W. B. and Prausnitz, J. M.,** Equations of state for strongly nonideal fluid mixtures: application of local compositions toward density-dependent mixing rules, *Fluid Phase Equilibria,* 9, 119, 1982.
126. **Won, K. W.,** Phase equilibria of high-boiling organic solutes in compressed supercritical fluids. Equation of state with new mixing rule, in *Chemical Engineering at Supercritical Fluid Conditions,* Paulaitis, M. E. et al., Eds., Ann Arbor Science, Ann Arbor, MI, 1983, 323.
127. **Wong, J. M., Pearlman, R. S., and Johnston, K. P.,** Supercritical fluid mixtures: prediction of the phase behavior, *J. Phys. Chem.,* 89, 2671, 1985.
128. **Wormald, C. J. and Eyears, J. M.,** Excess enthalpies and excess volumes of (0.5 CO_2 and 0.5 C_2H_6) in the supercritical region, *J. Chem. Soc., Faraday Trans. 1,* 84, 1437, 1988.
129. **Wu, P. C. and Ehrlich, P.,** Volumetric properties of supercritical ethane-*n*-heptane mixtures: molar volumes and partial molar volumes, *AIChE J.,* 19(3), 533, 1973.
130. **Yerlett, T. K. and Wormald, C. J.,** The enthalpy of acetone, *J. Chem. Thermophys.,* 18, 371, 1986.
131. **Yonker, C. R. and Smith, R. D.,** Solvatochromic behavior of binary supercritical fluids: the carbon dioxide/2-propanol system, *J. Phys. Chem.,* 92, 2374, 1988.
132. **Yonker, C. R. and Smith, R. D.,** Thermochromic shifts in supercritical fluids, *J. Phys. Chem.,* 93, 1261, 1989.
133. **Yonker, C. R., Frye, S. L., Kalkwarf, D. R., and Smith, R. D.,** Characterization of supercritical fluid solvents using solvatochromic shifts, *J. Phys. Chem.,* 90, 3022, 1986.
134. **Ziger, D. H. and Eckert, C. A.,** Correlation and prediction of solid-supercritical fluid phase equilibria, *Ind. Eng. Chem. Proc. Des. Dev.,* 22, 582, 1983.

Chapter 4

FLUCTUATION THEORY OF SUPERCRITICAL SOLUTIONS

Lloyd L. Lee, Pablo G. Debenedetti, and Henry D. Cochran

TABLE OF CONTENTS

I. INTRODUCTION

One of the most significant developments in the theory of solutions occurred in 1951, when Kirkwood and Buff[1] published their paper entitled "The Statistical Mechanical Theory of Solutions". The Kirkwood-Buff (KB) formalism yields rigorous expressions for key thermodynamic derivatives (such as isothermal compressibility, partial molar volumes, and chemical potential compositional derivatives) in terms of microscopic quantities (correlations between pairwise concentration fluctuations, or, equivalently, integrals of pair correlation functions). Because of its theoretical foundation in the statistics of concentration fluctuations in the grand canonical ensemble, the formalism is also frequently referred to as the fluctuation theory of mixtures.

The KB theory has been fruitfully applied to a variety of situations, including electrolyte solutions,[2,3] solubility of gases in liquids,[4] analysis of mixing rules,[5] etc. The formalism has also been recast in terms of direct correlation function integrals.[6] In recent years, there has been considerable interest in the application of fluctuation theory as a valuable tool for the molecular-based interpretation of supercritical thermodynamics. It is this body of work that is reviewed here.

In 1983, Eckert and co-workers[7] reported their measurements of partial molar volumes of highly dilute organic solutes in supercritical solvents. They obtained very large, negative values, entirely in agreement with classical[8,9] and nonclassical[10] predictions as to the divergence of this quantity for an infinitely dilute solute at the solvent's critical point. The word "clustering" was first used by Eckert and co-workers[7] to interpret these extraordinary volume contractions per added solute molecule. Furthermore, these authors used chemical theory to estimate the number of solvent molecules influenced by the presence of the infinitely dilute solute. The rigorous quantification of negative partial molar volume experiments in molecular terms was achieved through the application of the KB formalism,[11,12] which allowed the calculation of "cluster sizes" from experimental volumetric data.

The KB "cluster size" calculations[11,12] naturally raised questions regarding the nature of solute-solvent interactions in dilute supercritical mixtures. This review discusses not only the KB approach as applied to supercritical systems, but also the experiments in which that work originated, as well as the numerical and computational calculations that have since been carried out to explore the implications of the theory in greater detail.

The chapter begins with a brief summary of the KB approach (Section II). The partial molar volume experiments[7,13] and their quantification via fluctuation theory[11,12] are reviewed in Sections III and IV, respectively. Also discussed is the classification of mixtures into attractive, weakly attractive, and repulsive,[14] which also follows directly from fluctuation arguments. Section V summarizes recent spectroscopic experiments[12,15-17] aimed at investigating molecular interactions in dilute supercritical mixtures. The objective of these investigations is to study the local environment surrounding solute molecules in supercritical solvents. Such knowledge of events on the local length scale provides a natural complement to the global information (i.e., integrated over all length scales) resulting from the theory developed in Section IV. Section VI discusses integral equation[18-20] and computer simulation[21] studies of molecular interactions in dilute supercritical mixtures. The former approach yields information on both long- and short-range effects, the latter, primarily on the local environment surrounding solute molecules in dilute supercritical mixtures. Integral equation calculations of quantities (such as the cluster size) are presented in Section VI. Computer simulation evidence substantiating theoretical predictions as to the difference between attractive and repulsive near-critical mixtures,[14,21] which are defined in Section IV from fluctuation-based criteria is also presented in Section VI. Section VII is a discussion of solubility enhancement, as well as the temperature and pressure dependence of solubility in supercritical mixtures, in the light of the clustering phenomenon. Finally, the chapter includes a review of the theoretical expressions[22-24] for solubility enhancement in supercritical mixtures whose basis is, again, the KB theory.

II. THE KIRKWOOD-BUFF FORMALISM

The statistical mechanical theory of solutions based on concentration fluctuations originated from the seminal paper of Kirkwood and Buff in 1951.[1] Since then, it has become clear that the Kirkwood-Buff approach has established a bridge between the molecular interactions on the microscopic level and the bulk (thermodynamic) properties of solutions. A number of studies ensued,[25] and several excellent reviews and monographs are now available.[26] Here, the theory is described in terms of the statistical mechanics of the grand canonical ensemble, in order to clarify the microscopic connections. Without loss of generality, consider a binary mixture of N_1 molecules of type 1 and N_2 molecules of type 2 (with $N = N_1 + N_2$) in an open system of fixed volume. The partition function is defined as[27]

$$\Xi \equiv \sum_{N1=0}^{\infty} \sum_{N2=0}^{\infty} \frac{z_1^{N_1} z_2^{N_2}}{N_1! \, N_2!} \int d\mathbf{r}^N \exp[-\beta V_N(\mathbf{r}^N)] \tag{1}$$

where z_i is the activity $\exp(\beta\mu_i)/\Lambda_i^3$ of species i; Λ_i is its De Broglie wavelength, V_N is the N-body potential energy, $\beta = 1/kT$; k is Boltzmann's constant; and $\mathbf{r}^N$ is the 3N-vector of the positions of the N particles in the system. The average number of particles $\langle N_i \rangle$ in the system is obtained from the well-known relation in ensemble theory as the derivative[27]

$$\left[\frac{\partial \ln \Xi}{\partial \beta \mu_i} \right]_{T,V,\mu_j} = \langle N_i \rangle, \, j \neq i \tag{2}$$

and the fluctuations of the number of particles are given by the second derivative

$$\frac{\partial^2 \Xi}{\partial \beta \mu_i \partial \beta \mu_j} = \frac{\partial \langle N_i \rangle}{\partial \beta \mu_j} = \frac{\partial \langle N_j \rangle}{\partial \beta \mu_i} = \langle N_i N_j \rangle - \langle N_i \rangle \langle N_j \rangle \tag{3}$$

The last equality gives the statistical correlation between the fluctuations in the number of molecules of species i and j. These correlations of fluctuations can be expressed in terms of the total correlation functions $h_{ij}(r) = g_{ij}(r) - 1$ through the normalization conditions appropriate to the latter quantities in the grand canonical ensemble[1]

$$\frac{1}{V} \int d\mathbf{r}[g_{ij}(r) - 1] = \frac{\langle N_i N_j \rangle - \langle N_i \rangle \langle N_j \rangle}{\langle N_i \rangle \langle N_j \rangle} - \frac{\delta_{ij}}{\langle N_i \rangle} \tag{4}$$

Comparison with Equation 3 shows that

$$\frac{kT}{V} \left[\frac{\partial \langle N_i \rangle}{\partial \mu_j} \right]_{T,V,\mu_k} = \rho_i \delta_{ij} + \rho_i \rho_j \int d\mathbf{r}[g_{ij}(r) - 1] = \rho_i \delta_{ij} + \rho_i \rho_j G_{ij} \tag{5}$$

where ρ_i is the number density of species i, and the Kirkwood-Buff factor is defined as

$$G_{ij} \equiv \int d\mathbf{r}[g_{ij}(r) - 1] \tag{6}$$

The remarkable feature of Equation 5 is the expression of the chemical potential derivative of the number of molecules (or concentration, since V is fixed) in terms of the microscopic pair distribution function (PCF) $g_{ij}(r)$. Thus, a link is established between the measurable bulk thermodynamic quantities and the microscopic spatial distributions of molecules. This connection explains not only the power but also the utility of the KB approach. Using conventional thermodynamic relations, it is also possible to express the isothermal compressibility, and the partial molar volumes (PMV) in terms of the KB factors:[1]

$$K_T \equiv -\frac{1}{V}\left[\frac{\partial V}{\partial P}\right]_T = \frac{D}{kT\Delta} \tag{7}$$

where

$$D \equiv 1 + \rho_1 G_{11} + \rho_2 G_{22} + \rho_1\rho_2(G_{11}G_{22} - G_{12}{}^2)$$

$$\Delta \equiv \rho_1 + \rho_2 + \rho_1\rho_2(G_{11} + G_{22} - 2G_{12}) \tag{8}$$

The PMV becomes

$$\bar{V}_1 = \Delta^{-1}[1 + \rho_2(G_{22} - G_{12})] \tag{9}$$

In standard texts, the derivatives in Equation 5 are identified as elements (B_{ij}) of a B matrix. Its inverse is called the A matrix, with elements

$$A_{ij} \equiv \frac{V}{kT}\left[\frac{\partial \mu_i}{\partial \langle N_j \rangle}\right]_{T,V,<N_k>} , \text{ and } A \cdot B = I \tag{10}$$

where I is the identity matrix. The A matrix plays an important role in the reformulation of the KB theory in terms of short-range quantities, due to O'Connell.[6] If the DCFI (direct correlation function integral) is defined as

$$C_{ij} \equiv \int dr\, C_{ij}(r) \tag{11}$$

where $c_{ij}(r)$ are the direct correlation functions (DCF), the A matrix elements are related to DCFI by[6]

$$\rho_i\rho_j A_{ij} = \rho_i\delta_{ij} - \rho_i\rho_j C_{ij} \tag{12}$$

where use has been made of the Ornstein-Zernike relation. Since the DCFs are short-range compared to g_{ij}, numerically it is more advantageous to use the DCFIs.

The elements of the KB theory of solutions have been outlined thus far. Research in this area is still very active, covering both theoretical and applied aspects. Debenedetti[28] has recently generalized the KB formalism using Legendre transform theory. Fluctuation theory has been applied to a variety of thermodynamic systems. Of particular significance is the work of O'Connell and co-workers, who applied the KB theory to electrolyte solutions,[2,3] and to the solubility of supercritical gases in compressed liquids.[4] This work builds upon the previously mentioned recasting of the KB formalism in terms of direct correlation function integrals due to O'Connell.[6] Since direct correlation functions are, by definition, short ranged, their use offers numerical advantages over that of the full pair correlation functions. For a detailed discussion of this topic, the interested reader is referred to the paper by O'Connell[6] and references therein.

TABLE 1
Experimental Conditions in Eckert and Co-workers'
Partial Molar Volume Measurements[13]

System	P_r[a]	T_r[b]
Naphthalene/carbon dioxide	1.01—5.06	1.01—1.05
Naphthalene/ethylene	1.06—5.00	1.01—1.13
Tetrabromomethane/carbon dioxide	1.05—3.40	1.01
Tetrabromomethane/ethylene	1.05—4.95	1.01—1.06
Camphor/ethylene	1.05—4.93	1.01

[a] $P_r = P/P_{c2}$.
[b] $T_r = T/T_{c2}$.

Hamad et al.[5] have recently used the KB formalism in their treatment of mixing rules. By assuming conformality of the pair correlation functions, they were able to scale the KB factors, and to obtain a size parameter for the conformal fluid, which they used in conjunction with van der Waals mixing for the energy parameter. They applied this approach to mixtures of model fluids. Cochran et al.[23] have proposed a scaled distribution function approximation for the calculation of KB factors by introducing both size and energy scaling. This is the basis of these authors' fluctuation approach to supercritical solubility, which is discussed in detail in Section VII.

McGuigan and Monson[29] have used the KB formalism to derive relationships between the long-range behavior of pair correlation functions in supercritical mixtures. In addition, they investigated the density dependence of G_{12}/G_{22} in the critical region using integral equations and model potentials.

III. PARTIAL MOLAR VOLUME EXPERIMENTS

A significant amount of the theoretical work described in this chapter was inspired by the experimental measurements of PMVs of very dilute, nonvolatile organic solutes in a variety of supercritical solvents, due to Eckert and co-workers.[7,13] Accordingly, this section discusses Eckert's experiments, their interpretation, and their significance.

In 1983, Eckert and co-workers[7] published measurements of PMVs of naphthalene in carbon dioxide, naphthalene in ethylene, tetrabromomethane in carbon dioxide, and camphor in ethylene. Later, these authors reported more extensive data for the above systems, as well as for tetrabromomethane in ethylene.[13] Experimental conditions are summarized in Table 1.

The method used by Eckert and co-workers[13,30] to measure infinite dilution PMVs is described elsewhere in detail. Basically, the technique consisted of measuring the density of an almost pure supercritical solvent at several values of solute mole fraction, and using the relationship

$$\overline{v}_1^\infty = v + \left(\frac{\partial v}{\partial y_1}\right)_{T,P}^\infty \tag{13}$$

where v is the molar volume of the solvent, y_1, the mole fraction of the solute, superscript ∞ denotes infinite dilution, and the last term was found to be linear at sufficiently high dilution.[13] Density measurements were performed using a vibrating tube densitometer.[31]

Figure 1, which is typical of the resulting data, shows the PMV of infinitely dilute naphthalene in supercritical ethylene, at 12°C (T_{c2} = 9.2°C; throughout this chapter, subscript 1 denotes the solute and subscript 2, the solvent). At a pressure of 53.7 bars (P_{c2} = 50.4

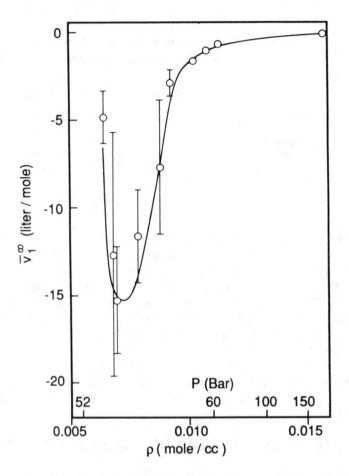

FIGURE 1. PMVs of infinitely dilute naphthalene in supercritical ethylene at 12°C.[7] (From Eckert, C. A., Ziger, D. H., Johnston, K. P., and Ellison, T. K., *Fluid Phase Equilibria*, 14, 167, 1983. With persmission.)

bar), the PMV of the solute, which is negative, is roughly 100 times larger in absolute value than the bulk molar volume of the solvent. In the words of the investigators, " . . . these numbers indicate that the solution process represents a disappearance of solvent of more than 100 moles solvent/mole solute. This suggests some sort of clustering process, perhaps akin to electrostriction about an ion in a protic solvent".[7] These authors then used chemical theory arguments to estimate the number of solvent molecules "complexed" to the dilute solute.[7]

The PMV of an infinitely dilute solute is given by the identity,

$$\bar{v}_1^{\infty} = vK_T \left[N \left(\frac{\partial P}{\partial N_1} \right)_{T,V,N_2} \right]^{\infty} = vK_T \delta \qquad (14)$$

where K_T is the isothermal compressibility of the pure solvent, v, its molar volume, N_i denotes moles of component i ($N = N_1 + N_2$), and δ is necessarily finite. It thus follows straightforwardly that the left side of Equation 14 will diverge at the critical point of the solvent, with a sign given by that of δ. Wheeler[10] quantified this phenomenon, taking into account its nonclassical aspects via a decorated lattice model. A classical discussion of the same topic can be found, for example, in Rozen[8] and in the early pioneering work of Krichevskii and his school.[9]

Thus, although they did not involve the discovery of hitherto unknown phenomena, the experiments of Eckert and co-workers were important, firstly, because they represented the first extensive study of infinite dilution solute PMVs in the near-critical region, and confirmed theoretical predictions as to the very large value of this quantity. Nevertheless, what makes these experiments especially significant in the present context is their interpretation in terms of an effective number of solvent molecules influenced by the presence of the solute. The original attempts at quantifying, in molecular terms, these extraordinary volume contractions upon solute addition[7] elicited the interest of theoretical investigators and eventually gave rise to the more fundamental treatments that are reviewed in this chapter.

IV. KIRKWOOD-BUFF THEORY AND SUPERCRITICAL MIXTURES

Although chemical theory-based calculations[7] yielded some insight into the interpretation of partial molar experiments, the quantitative substantiation of the clustering concept was achieved through the application of fluctuation theory.[1] Since the Kirkwood-Buff formalism has already been discussed in detail in Section II, what is given here are the relevant equations. One starts with the relationship between PMV and fluctuation integrals,

$$\bar{v}_i = [1 + \rho_j(G_{jj} - G_{ij})][\rho_i + \rho_j + \rho_i\rho_j(G_{ii} + G_{jj} - 2G_{ij})]^{-1} \tag{15}$$

Upon taking the limit of infinite dilution, this equation can be rearranged to read

$$\bar{v}_1^\infty\rho = \rho kTK_T - \Gamma \tag{16}$$

where ρ and K_T denote the number density and isothermal compressibility of the pure solvent, $\bar{v}_1^\infty$ is the PMV of the solute at infinite dilution, and Γ, henceforth referred to as the cluster size, is given by

$$\Gamma = \rho \int \left(g_{21}^\infty - 1 \right) d\mathbf{r} \tag{17}$$

where g_{12}^∞ is the solute-solvent pair correlation function at infinite dilution. The physical significance of Γ is therefore the excess number of solvent molecules, with respect to a uniform distribution at the prevailing bulk density, ρ, surrounding the infinitely dilute solute. Equation 16 was derived by Debenedetti,[11] and by Kim and Johnston.[12]

Upon combining Equations 14 and 16,[32]

$$\Gamma = K_T(\rho kT - \delta) \tag{18}$$

Equation 16 shows that the growth in Γ is inseparable from the existence of large, negative solute PMVs at infinite dilution in compressible solvents (since K_T is large and positive). Since, via Equation 18, Γ diverges as K_T, the growth in Γ is necessarily a long-range effect, a fact that also follows from elementary geometric and energetic considerations, since local densities of real molecules cannot possibly diverge.

Long-range correlations have a dramatic effect upon the pressure[11] and temperature[32] dependence of solubility in dilute supercritical mixtures (see Section VII). On the other hand, it is clear that solubility is primarily determined by the nature of the local environment surrounding solute molecules in a given solvent. Thus, an understanding of the solvent power of supercritical fluids necessarily involves the consideration of solute-solvent interactions over a spectrum of length scales. Short-range effects are discussed in detail in Sections V and VI; long-range behavior, in Sections VI and VII.

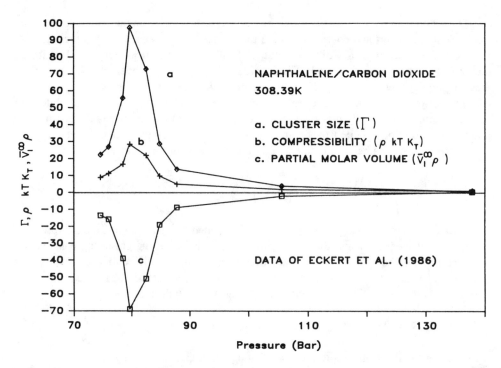

FIGURE 2. Cluster size calculations from infinite dilution PMV measurements. Naphthalene-carbon dioxide system. Experimental data from Reference 13. (From Debenedetti, P. G., Clustering in dilute, binary supercritical mixtures: a fluctuation analysis, *Chem. Eng. Sci.*, 42(9), 2203, 1987. With permission.)

Returning now to Equation 16, it must be noted that it can be used to calculate the excess number of solvent molecules per solute molecule (i.e., Γ) from knowledge of solute PMVs at infinite dilution. Thus, the introduction of fluctuation theory makes the measurement of PMVs equivalent to the determination of solvent enrichment (or depletion) around dilute solutes, and gives quantitative substance to the clustering concept originally proposed by Eckert et al.[7] It should be emphasized here that Γ is, in general, an order 1 quantity. Its calculation via Equation 16 is applicable to any infinitely dilute mixture, but Γ acquires special significance in the near-critical region, where it becomes very large (see Equation 18).

Figure 2 shows an example of a calculation based on Equation 16, in which experimental measurements of solute PMVs at infinite dilution have been used to calculate solvent enrichment around the solute. It can be seen that in the highly compressible region, each naphthalene molecule is surrounded by an excess of roughly 100 solvent molecules over and above a uniform (ideal gas) distribution at bulk conditions. Note that Γ is a small quantity outside of the highly compressible region.

Equations 16 and 18 show that both $\bar{v}_1^\infty$ and Γ diverge as K_T, but with signs given by those of δ and $(\rho kT - \delta)$, respectively. Since these equations are thermodynamic identities, this implies that any infinitely dilute near-critical mixture must necessarily exhibit one of three types of behavior[14]

$$\delta < 0 \Rightarrow \bar{v}_1^\infty \to -\infty, \ \Gamma \to +\infty \text{ (attractive)} \tag{19}$$

$$\rho kT > \delta > 0 \Rightarrow \bar{v}_1^\infty \to +\infty, \ \Gamma \to +\infty \text{ (weakly attractive)} \tag{20}$$

$$\delta > \rho kT \Rightarrow \bar{v}_1^\infty \to +\infty, \ \Gamma \to -\infty \text{ (repulsive)} \tag{21}$$

The weakly attractive regime requires explanation. In an ideal gas mixture, PMVs are positive

$(\bar{v}_i = \rho_i^{-1})$ and Γ vanishes. Therefore, the case $\rho kT > \delta > 0$ implies that the attraction between solute and solvent (Γ large and positive) is insufficient to overcome the purely osmotic effect associated with the introduction of a solute molecule ($\bar{v}_1^\infty$ positive), hence the origin of the term "weakly attractive".

Equations 19 to 21 are significant in the present context in that all of the actual or proposed applications of supercritical fluids (SCFs) to processes as diverse as coffee decaffeination,[33] precipitation polymerization,[34] or powder formation[35] involve attractive mixtures. As discussed in Section VII, the negative divergence of $\bar{v}_1^\infty$ has a pronounced influence upon the rate of change of solubility with respect to pressure in dilute supercritical mixtures. It is therefore of interest to discuss the theoretical prediction of attractive (or repulsive) behavior. It follows from Equations 16 and 18 that this involves the consideration of finite and well-behaved quantities, such as δ and $\rho kT - \delta$.

The criteria according to which a given mixture is predicted to exhibit attractive, repulsive, or weakly attractive behavior differ, of course, among the various mixture models. As an example, consider a van der Waals mixture with van der Waals-1 mixing rules[36] and Lorentz-Berthelot combining rules. According to this model,

$$\delta < 0 \Rightarrow \gamma < \frac{3(3 - \rho_r)^2}{4} \cdot \frac{\alpha^{1/2}}{T_r} + \left(1 - \frac{3}{\rho_r}\right) \tag{22}$$

$$\rho kT > \delta > 0 \Rightarrow \frac{3(3 - \rho_r)^2}{4} \cdot \frac{\alpha^{1/2}}{T_r} + \left(\frac{\rho_r}{3} - 1\right) >$$

$$\gamma > \frac{3(3 - \rho_r)^2}{4} \cdot \frac{\alpha^{1/2}}{T_r} + \left(1 - \frac{3}{\rho_r}\right) \tag{23}$$

$$\delta > \rho kT \Rightarrow \gamma > \frac{3(3 - \rho_r)^2}{4} \cdot \frac{\alpha^{1/2}}{T_r} + \left(\frac{\rho_r}{3} - 1\right) \tag{24}$$

where $\alpha = a_{11}/a_{22}$, $\gamma = b_1/b_2$, $\rho_r = \rho/\rho_{c2}$, and $T_r = T/T_{c2}$ (a and b are the van der Waals attractive and repulsive parameters). Consider now the "symmetric" near-critical mixture obtained by exchanging solute and solvent (in other words, the infinitely dilute solute becomes the near-critical solvent, and vice-versa). Along the solvent's critical isochore, for an attractive system, it must be true that

$$\gamma < \frac{3\alpha^{1/2}}{T_r} - 2 \tag{25}$$

whereas, for an attractive "symmetric" system,

$$\gamma > \frac{\alpha^{1/2}}{\dfrac{3}{T_r} - 2\alpha^{1/2}} \tag{26}$$

Similarly, if both the given and "symmetric" mixtures are repulsive,

$$\gamma > \frac{3\alpha^{1/2}}{T_r} - \frac{2}{3} \tag{27}$$

and, simultaneously,

$$\gamma < \frac{\alpha^{1/2}}{\dfrac{3}{T_r} - \dfrac{2\alpha^{1/2}}{3}} \tag{28}$$

Equations 25 and 26 cannot be simultaneously satisfied.[14] Thus, according to the van der Waals model, attractive behavior cannot be preserved upon interchanging solute and solvent at near-critical conditions. As for Equations 27 and 28, the conditions under which they can both be simultaneously satisfied are discussed elsewhere in detail.[14] Here, it suffices to mention that outside a very limited region in (α, γ) parameter space, repulsive behavior cannot be preserved upon interchanging solute and solvent at near-critical conditions. Extension of these considerations to mixtures that do not obey the Lorentz-Berthelot combining rules (as well as to different mixture models) is discussed in Reference 14.

From the above discussion, it can be concluded that fluctuation theory allows the quantification of cooperative behavior whose macroscopic manifestation are the large and negative PMVs of dilute organic solutes in supercritical solvents.[7,13] The sign of diverging properties such as Γ and $\bar{v}_1^\infty$ are determined by finite and well-behaved quantities, such as δ and $(\rho kT - \delta)$. That δ is finite means that it is a local quantity. In fact, Eckert et al.[13] have given expressions for δ in terms of direct correlation function integrals. A systematic study of these local quantities, their dependence on details of molecular architecture and energetics, and their behavior in the near-critical region, important as it is within the context of solubility in supercritical solvents, has not yet been attempted.

V. SPECTROSCOPIC MEASUREMENTS

Sections III and IV discussed the implications of PMV measurements[7,13] as to the long-range solvent enrichment around solute molecules in attractive mixtures. Recently there have been direct spectroscopic probes of the corresponding local solvent structure around the solutes. These include using the solvatochromic scales as in the ultraviolet (UV)-visible absorption of phenol blue in ethylene[12] and the fluorescence spectroscopy of pyrene in carbon dioxide.[16] Some interesting examples are provided below to illustrate the point. The use of spectroscopic methods to study the local environment surrounding solute molecules in dilute supercritical mixtures has been reviewed recently.[37]

The absorption spectrum of a dye, e.g., phenol blue, is significantly affected by the solvent environment, a phenomenon called "solvatochromism". This is indicated by the transition energy E_T, which is defined as hc/λ_{max}, λ_{max} being the wavelength corresponding to maximum absorption. Solvatochromic scales, based on the absorption wavelengths, are commonly employed to analyze solubility behavior in infrared (IR), nuclear magnetic resonance (NMR), ESR, and UV-visible spectroscopy.[38,39] Phenol blue (N,N-dimethylindoaniline) has an excited state (carbonyl$^-$N$^+$(Me)$_2$) with a higher dipole moment (2.5 D greater than the ground state). The transition energy for this excited state is $E_T = 51.96$ kcal/mol in n-hexane as solvent, and is reduced to 47.03 in methanol as solvent, since methanol is a strongly dipolar and hydrogen bonding fluid that interacts with and stabilizes the excited state. Therefore, the transition energy decreases as the solvent strength is increased. The λ_{max} shifts from 550 nm in hexane to 608 nm in methanol. This spectroscopic method is capable of yielding highly accurate results since the experimental resolution is ± 0.2 nm. When E_T is plotted against bulk solvent density for phenol blue in ethylene,[37] Figure 3 results. The experimental data fall on curve A, while the calculated behavior of E_T based on the assumption of a homogeneous solvent is given by line B.[37] Figure 3 therefore shows an enrichment in the local solvent density (with respect to bulk conditions) around the solute, especially in the region where the solvent is highly compressible. Experiments were also carried out on (N,N-dimethylamino)benzonitrile in fluoroform with similar interpretation of

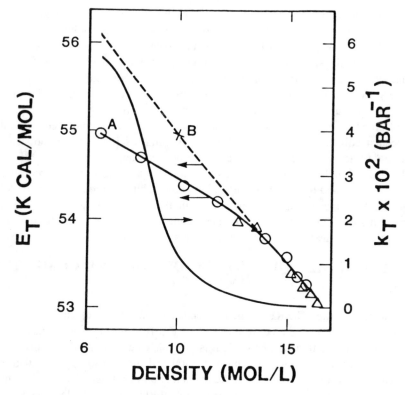

FIGURE 3. Transition energy (E_T) of phenol blue in ethylene and isothermal compressibility of ethylene vs. density.[37] (Circles) 25°C; (triangles) 10°C; (dotted line) calculated transition energy at 25°C based on the assumption of a homogeneous solvent.[37] (Reprinted with permission from ACS Symp. Ser. No. 406, p. 55. ©1989 American Chemical Society.)

solvent clustering.[40] Effects of the addition of cosolvents (e.g., *n*-octane, acetone, methanol) in carbon dioxide were also determined by Kim and Johnston.[15] Yonker et al.[41] studied the shift in absorption wavelength of a chromophore in a number of SCFs as a function of density.

Additional evidence of clustering was put forward in the fluorescence spectroscopy experiments of Brennecke and Eckert.[16,17] They carried out measurements on pyrene in several SCFs: CO_2, C_2H_4, and CHF_3. For CO_2, they further studied the solutes naphthalene, dibenzofuran, and carbazole. For pyrene, the ratio of intensity of the first to the third fluorescence spectra peaks, I_1/I_3, is an indication of the strength of solute-solvent interactions.[16] The transition corresponding to the first peak is not allowed by symmetry arguments.[16] If strong solute-solvent interactions occur, however, symmetry is disrupted and the transition occurs with greater intensity.[16] In the case of naphthalene, the corresponding ratio is I_1/I_4. The peak ratios observed by Brennecke and Eckert were much higher than predictions based on correlations of organic liquids.[41] In addition, upon plotting I_1/I_3 (pyrene) or I_1/I_4 (naphthalene) isotherms against bulk solvent density, Brennecke and Eckert observed higher (but finite) intensity ratios the closer the temperature was to the critical point of the solvent. Thus, the fluorescence spectroscopy experiments[16,17] provide evidence of a substantial solvent enrichment around the solute in attractive near-critical mixtures.

A new and interesting observation from Brennecke and Eckert's experimental work was the recognition of strong solute-solute interactions accompanying supercritical behavior. The solute-solute interactions are revealed by formation of excimers: excited state dimers. Their formation is indicated by a broad structureless band at longer wavelengths than the normal fluorescence. In general, one would not anticipate a high degree of solute-solute aggregation in dilute solutions (mole fraction $<10^{-5}$, i.e., well below saturation); however, Brennecke and Eckert's data indicated significant (approximately 50% of total solute) excimer concentration in dilute (approximately 10^{-5} mol fraction) solutions of pyrene in supercritical ethylene. This is highly suggestive of solute-solute aggregation according to the following argument: one expects excimer formation only when two or more solute molecules are located very close to one another (i.e., within less than about two molecular diameters), and when the probability of electronic excitation is high (e.g., by UV absorption). The lifetime of an excimer (of the order of picoseconds) is very large compared to the duration of a molecular collision (of the order of 10^{-15} s); finite excimer concentrations can exist and can be detected by their fluorescence. The fact that approximately 50% of the solute molecules at 10^{-5} mol fraction should exist as excimers suggests a very high probability of excimer formation, and hence a high degree of solute-solute propinquity. Section VI shows that, indeed, this unusual aggregation occurs in the calculated solute-solute correlation functions of highly dilute supercritical Lennard-Jones mixtures, concurrently with the long-range growth of G_{11} near the critical point of the solvent.

VI. INTEGRAL EQUATION MODELING AND COMPUTER SIMULATION

As mentioned earlier, the KB formalism lends naturally to a molecular interpretation of the macroscopic behavior of mixtures. Since the KB factors, G_{ij}, are defined in terms of the pair correlation functions (PCF) $g_{ij}(r)$, it is possible to use the PCF not only to interpret the bulk properties, but also to show the structural and mechanistic pictures that give rise to such behavior. A notable example is the phenomenon of clustering of solvent molecules around the solute molecule whose macroscopic manifestations are the large negative solute PMVs discussed in Section III, and whose relationship to solubility enhancement is discussed in Section VII.

The PCF can be obtained from several sources: computer simulation on model potentials, integral equations, and scattering experiments. As there are currently no scattering experiments for supercritical mixtures, only the first two approaches are discussed. Each method has its merits and demerits; thus, this chapter takes the proper perspective in treating the results obtained.

There are currently a number of highly accurate integral equations for simple spherical potentials,[27] such as the RHNC (reference hypernetted chain), HMSA (hybrid mean spherical approximation), and some more conventional closures: PY (Percus-Yevick) and HNC (hypernetted chain). In this chapter, the results generated by the HMSA and RHNC methods,[18,19] as well as those from PY are examined. Real molecules in supercritical mixtures (e.g., CO_2 and naphthalene) do not interact with simple spherically symmetrical potentials; however, the essential aspects of the clustering phenomenon illustrated here manifest themselves even with much simpler, spherically symmetric potentials. Consequently, attention is focused on integral equation calculations based on simple model potentials (i.e., the Lennard-Jones potential). In the following we shall discuss several topics:

1.   Changes in the local environment surrounding solute molecules in dilute mixtures
2.   Growth and divergence of the KB factor $G_{12}(r)$ near the critical point as a long-range induced behavior
3.   Solute-solute interactions in highly dilute mixtures
4.   Structural peculiarities in repulsive and attractive supercritical mixtures

In recent studies by Cochran et al.[18,19] the mixture CO_2-naphthalene was modeled by the Lennard-Jones potentials shown in Table 2. The force constants were determined by scaling the true critical constants with the known reduced values of LJ fluids ($T_c^* = kT_c/\epsilon$

TABLE 2
Lennard-Jones Parameters for
CO_2-Naphthalene Mixtures

		ϵ/k (K)	σ(Å)
CO_2-CO_2	(2-2)	225.3	3.794
$C_{10}H_8$-$C_{10}H_8$	(1-1)	554.4	6.199
$C_{10}H_8$-CO_2	(1-2)	353.4	4.997

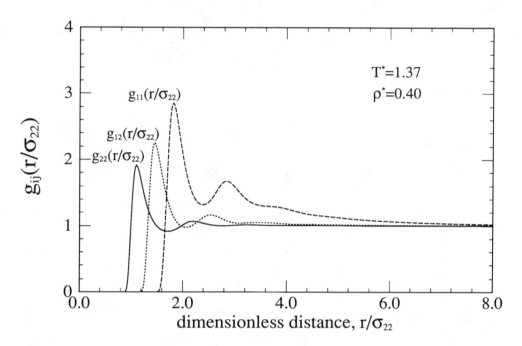

FIGURE 4. Pair correlation functions for solute-solute (1-1), solute-solvent (1-2), and solvent-solvent (2-2) interactions at $T^* = 1.37$, $\rho^* = 0.4$. Potential parameters as per Table 2 (naphthalene-carbon dioxide). HMSA closure. Solute mole fraction = 10^{-9}. (Reprinted with permission from ACS Symp. Ser. No. 406, p. 31. ©1989 American Chemical Society.)

$\cong 1.35$ and $\rho_c^* = \rho_c\sigma^3 \approx 0.35$). The HMSA equation was solved for this simulated mixture (details of the solution can be found in Reference 20). The PCFs are shown in Figure 4 for g_{11}, g_{12}, and g_{22} at $T^* = 1.37$ and $\rho^* = 0.4$ (throughout this review, ρ^* denotes $\rho\sigma^3$, and T^*, kT/ϵ, where σ and ϵ are intermolecular potential size and energy parameters. In dilute mixture calculations, σ and ϵ are solvent quantities). Although the actual numerical data show long-range behavior for the three curves ($g_{ij} \neq 1$ and nonoscillatory for large r), this trend is only apparent, on the figure scale, for g_{11}. Note the greatly elevated first valley (>1) for this curve. Figure 5 singles out the solvent-solute correlations at $T^* = 1.37$ and three different densities. At $\rho^* = 0.6$, the peak height is ~2.15, at 0.5, it decreases to ~2.10. This is normal, since lower density favors rarefaction of neighboring pairs; however, when density is further decreased toward the critical value, the peak height increases to ~2.25. Increased affinity between solvent and solute molecules is in evidence. The simulation data discussed below clearly support this trend. Another interesting feature occurs in the development of long-range behavior in the solvent-solute PCF (see Figure 5b: g_{12} shows persistent correlations between solute and solvent at values above 1 for very long ranges of separations, up to 40 molecular diameters, approaching the limit of the numerical integration. In contrast, this behavior is absent at higher densities (away from the critical value of 0.35).

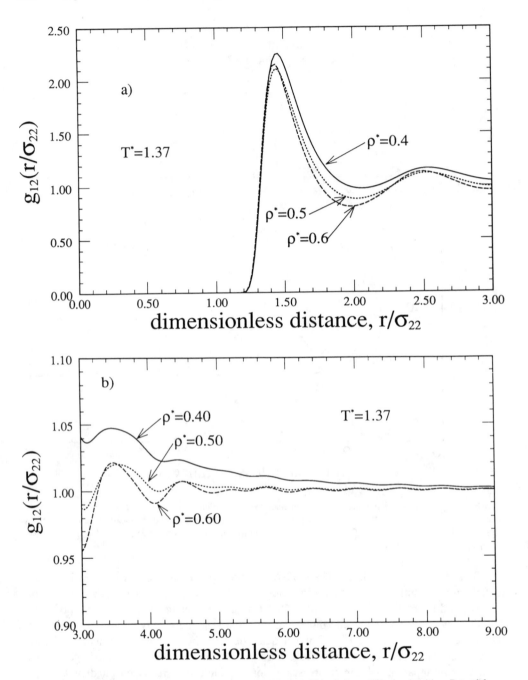

FIGURE 5. Solute-solvent pair correlation functions at T* = 1.37 and three different densities. Potential parameters as per Table 2. HMSA closure. Solute mole fraction = 10^{-9}. Short-range structure (a); long-range structure (b). (Reprinted with permission from ACS Symp. Ser. No. 406, p. 32. ©1989 American Chemical Society.)

At high densities, alternating peaks and valleys fluctuate around unity, indicating typical liquid-like behavior (see the curves for $\rho^* = 0.5$ and 0.6), and absence of long-range correlations. To further ascertain clustering, Cochran and co-workers integrated the PCF g_{12} to get the excess coordination number,

$$\rho_2 G_{12}(L) = \rho_2 \int_0^L dr\, 4\pi r^2 [g_{12}(r) - 1] \qquad (29)$$

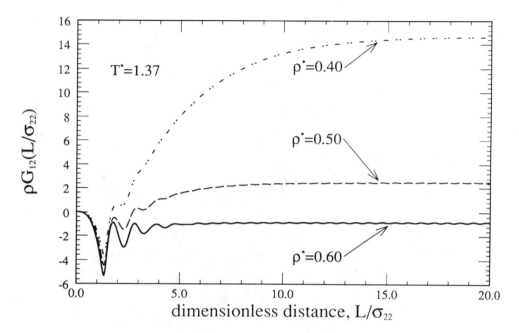

FIGURE 6.   Number of excess solvent molecules within a sphere of radius L around a central solute molecule. Same conditions as in Figure 5. (Reprinted with permission from ACS Symp. Ser. No. 406, p. 33. ©1989 American Chemical Society.)

This integral (whose infinite length limit is the quantity Γ, or cluster size, defined in Equation 17) counts the "excess" (or surplus) number of solvent molecules 2 surrounding the central solute molecule 1 as the coordination distance L, measured from the center of the solute molecule, increases. "Excess" refers to the number of solvent molecules above and beyond the normal level established by the bulk density ρ_2 (i.e., $(4\pi/3)L^3\rho_2$). This quantity is plotted in Figure 6 for three densities. At $\rho^* = 0.6$, the function shows a "deficit" at small L derived from the excluded volume, a behavior common to all correlations at short ranges. At long range, there still remains a deficit of around -1 solvent molecules. The deficit is made up at $\rho^* = 0.5$, to a surplus of $+2$; however, as the critical density is approached, the surplus increases to 15 solvent molecules ($\rho^* = 0.4$). This buildup is gradual, not all of it coming from the first neighbors. In fact, the first neighbors contribute only ~4 excess solvent molecules. Cluster size calculations based on PMV measurements[11] (see also Figure 2) indicate that this excess number of solvent molecules can easily exceed 100 under conditions relevant to supercritical processes. This should be interpreted as resulting from long-range accumulations. The first neighbors do experience enhancement, shown by the increased peak heights. Due to geometrical constraints, there are only so many solvent molecules that can be packed into the space adjacent to a central molecule. Thus, the drastic increase in the cluster size, eventually translated into the large negative PMVs, derives most of is value from long-range accumulations. Figure 7 shows the changes of the correlation functions at fixed density ($\rho^* = 0.28$) as the temperature is lowered. Near the critical temperature, the long-range quality of g_{12} again appears.

As noted in Figure 4, the solute-solute correlation shows a prominent second peak with a very shallow intermediate valley. By a geometric examination of the radial distribution, the structure indicates that the solvent molecules are excluded from a "sandwich" or "wedge" position between two solute molecules, i.e., forbidding structures such as "solute-solvent-solute". The solute molecules are, on average, very close together, so that there is no room for insertion of a solvent molecule. Physically, this is consistent with the picture whereby the solute molecules come together and form excimers (as discussed above in fluorescence spectroscopy), thus excluding the solvent molecules from entering the first valley.

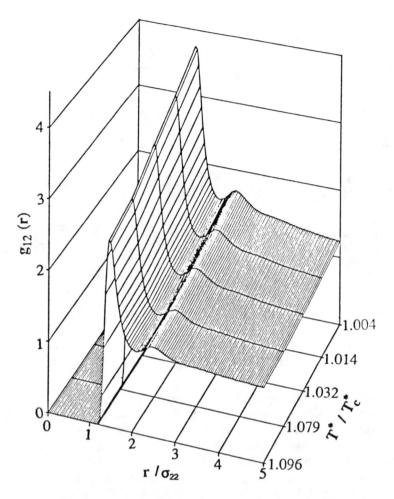

FIGURE 7.   Behavior of the solute-solvent correlation function at fixed density (ρ^* = 0.28) and several near-critical temperatures. Same conditions as in Figure 5.

TABLE 3
Lennard-Jones Parameters for
CO_2-Pyrene Mixtures

	$\epsilon/k(K)$	$\sigma(\text{Å})$
CO_2-CO_2	225.3	3.794
Pyrene-pyrene	662.8	7.14
CO_2-pyrene	386.4	5.467

To study this question in more detail, Wu et al.[42] solved the RHNC equation for LJ molecules simulating the interaction energies between pyrene and CO_2 (Table 3). To find the temperature effects, the PCFs at T^* = 1.37, 1.50, and 2.00 were obtained and are exhibited in Figures 8 to 10.

An important feature of these correlation functions is the increasing depth of the first minimum of $g_{pyr-pyr}$ as the temperature is increased, allowing the gradual intervention of a solvent (CO_2) molecule between two pyrene molecules, and the concurrent disappearance of long-range correlations.

The integral equation calculations discussed up to now pertain to attractive systems. Wu

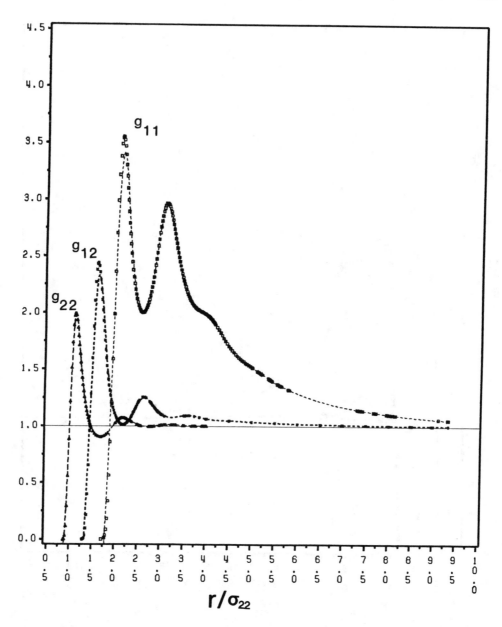

FIGURE 8. Pair correlation functions for solute-solute (squares), solute-solvent (stars), and solvent-solvent (triangles) interactions at $T^* = 1.37$, $\rho^* = 0.4248$. Potential parameters as per Table 3 (pyrene-carbon dioxide). RHNC closure. Solute mole fraction = 3×10^{-7}. (From Wu, R.-S., Lee, L. L., and Cochran, H. D., *Ind. Eng. Chem. Res.*, 29, 977, 1990. With permission.)

et al.[42] have also solved the PY integral equations for a model repulsive system, infinitely dilute Ne in near-critical Xe. In these calculations, Wu and co-workers used the same Lennard-Jones parameters employed by Petsche and Debenedetti[21] in their molecular dynamics study of attractive and repulsive systems (these Molecular Dynamics calculations are discussed below). Figure 11 presents the solute-solvent correlations, $g_{12}(r)$. Three state conditions were studied: $T^* = 1.4$, $\rho^* = 0.8$; $T^* = 1.4$, $\rho^* = 0.35$; and $T^* = 2.0$, $\rho^* = 0.35$. The striking feature is the damped oscillatory behavior in the two low density cases, in contrast to the liquid-like behavior at high density. Temperature seems to have little effect on structure at this density. At $T^* = 1.4$ and $\rho^* = 0.35$ (near the critical point of the solvent), after the first peak and valley, g_{12} barely recovers above unity (at the second

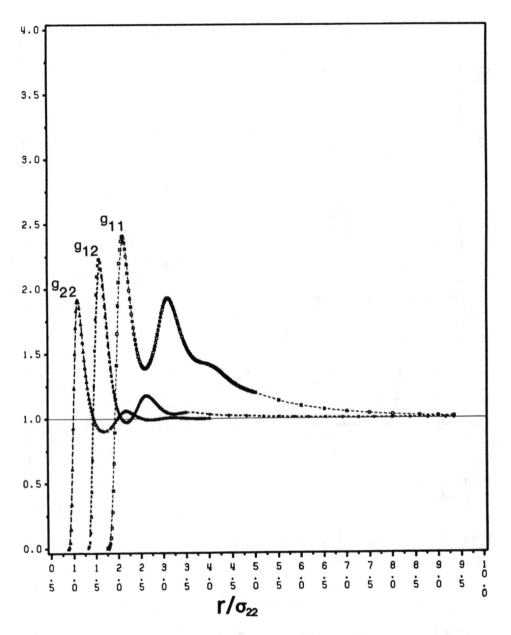

FIGURE 9.   Pair correlation functions for solute-solute (squares), solute-solvent (stars), and solvent-solvent (triangles) interactions at T* = 1.50, ρ* = 0.4248. Potential parameters as per Table 3 (pyrene-carbon dioxide). RHNC closure. Solute mole fraction = 3×10^{-7}. (From Wu, R.-S., Lee, L. L., and Cochran, H. D., *Ind. Eng. Chem. Res.*, 29, 977, 1990. With permission.)

peak, g_{12} is 1.007). Thereafter, the curve is consistently below unity, indicating long-range negative correlations between solute and solvent. This behavior is the exact opposite of that exhibited by the attractive systems discussed earlier.

To further examine this long-range solvent depletion, Wu and co-workers[42] calculated the excess coordination number defined in Equation 29 at T* = 1.4, ρ* = 0.35 for the pure Lennard-Jones solvent (P), the repulsive system (R), and the attractive system (A) obtained by interchanging solute and solvent (i.e., in this case, infinitely dilute Xe in near-critical Ne; see Equations 25 to 28 and the accompanying discussion). Results are shown in Figure 12, which clearly highlights the fact that the repulsive system (R) is characterized

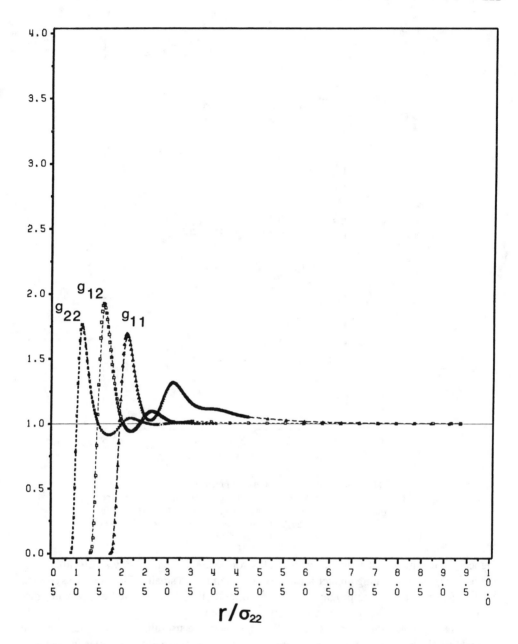

FIGURE 10. Pair correlation functions for solute-solute (squares), solute-solvent (stars), and solvent-solvent (triangles) interactions at $T^* = 2.00$, $\rho^* = 0.4248$. Potential parameters as per Table 3 (pyrene-carbon dioxide). RHNC closure. Solute mole fraction = 3×10^{-7}. (From Wu, R.-S., Lee, L. L., and Cochran, H. D., *Ind. Eng. Chem. Res.*, 29, 977, 1990. With permission.)

by a deficit of solvent molecules around the solute. Note that most of the rarefaction results from long-range negative correlations, despite the simultaneous presence of short-range effects (lowered first peak). This says that on average, the solvent molecules (Xe) reduce not only their encounters with Ne (weak first peak), but also stay away from the solute at longer distances (weaker second peak and consistent undercorrelation thereafter). At lower temperatures, the second peak does not even rise above unity,[42] and the solvent depletion around the solute is even more evident. This interesting phenomenon is discussed below in connection with the computer simulation study of Petsche and Debenedetti[21] on attractive and repulsive systems. These authors have interpreted repulsive behavior in terms of dif-

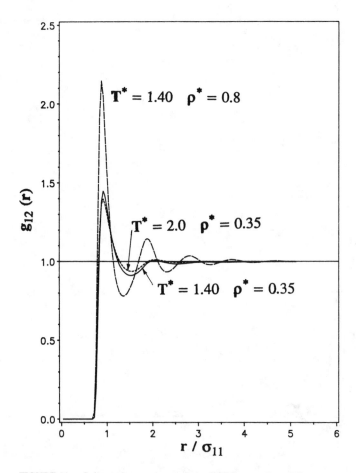

FIGURE 11.   Solute-solvent pair correlation functions at three different state
conditions for dilute Ne in Xe (repulsive mixture). Potential parameters as per
Table 4 (mixture 2). PY closure. Solute mole fraction = 10^{-9}. (From Wu,
R.-S., Lee, L. L., and Cochran, H. D., *Ind. Eng. Chem. Res.*, 29, 977, 1990.
With permission.)

ferences in magnitude of potential well depths for solute-solvent vis-à-vis solvent-solvent
interactions, the latter being greater for repulsive systems, and smaller for attractive systems,
these differences becoming progressively more important (i.e., correlated over greater distances) as the critical point of the solvent is approached.

The results that have been discussed thus far show the usefulness of integral equation
methods in elucidating important structural features associated with molecular interactions
in dilute supercritical mixtures. Particularly evident were the importance of long-range
correlations, as well as of solute-solute interactions at high dilution. The latter results are
consistent with, and shed additional light on, recent experimental observations[16] which
suggest that the very concept of infinite dilution needs to be critically evaluated in the vicinity
of the critical point of the solvent. Of course, the fact that solute-solute interactions become
important in this region even at high dilutions is simply a manifestation of the growth of
the correlation length.

The limitations of integral equation approaches are their approximate nature, the increased computational difficulty associated with realistic molecular geometries, and, in
general, their failure to describe nonclassical behavior in the immediate vicinity of the critical
point. Most situations of technological interest for supercritical applications are not confined
to this region.

Attention now turns to the study of supercritical mixtures via molecular-based computer

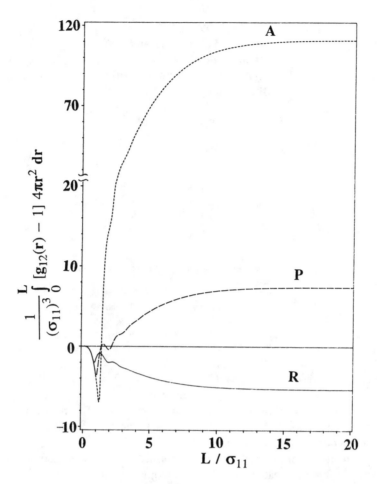

FIGURE 12.   Number of excess solvent molecules within a sphere of radius L around a central solute, for the pure Lennard-Jones solvent (P), an attractive mixture (A; Xe in Ne), and a repulsive mixture (R; Ne in Xe), at T* = 1.4, ρ* = 0.35. Potential parameters as per Table 4 (A = Mixture 1; B = Mixture 2). PY closure. Solute mole fraction = 10^{-9}. (From Wu, R.-S., Lee, L. L., and Cochran, H. D., *Ind. Eng. Chem. Res.*, 29, 977, 1990. With permission.)

simulations. This topic is discussed elsewhere in detail in this volume; the particular work of interest here is a recent study[21] of infinitely dilute mixtures aimed at verifying the predictions of the van der Waals theory as to the nonpreservation of attractive behavior upon solute-solvent interchange, as well as to investigating the structure and dynamics of the local environment surrounding infinitely dilute solute molecules in the vicinity of the critical point of the solvent, both for attractive and repulsive systems.

It was shown earlier in the chapter that fluctuation theory allows the classification of near-critical systems into three classes: attractive, weakly attractive, and repulsive (see Equations 19 to 21). Furthermore, a van der Waals-Lorentz-Berthelot model predicts that attractive ($\bar{v}_1^\infty \to -\infty$) behavior cannot be preserved upon going from one pure component critical point to the other, whereas preservation of repulsive behavior ($\Gamma \to -\infty$), although theoretically possible, can only occur for systems whose α and γ parameters fall within extremely limited ranges.

In order to test the above predictions, Petsche and Debenedetti[21] studied model mixtures of noble gases, which are reasonably well represented both by the van der Waals model and by the Lennard-Jones potential. Table 4 lists the potential parameters and state conditions investigated by Petsche and Debenedetti.

TABLE 4
**Potential Parameters and State
Conditions in Petsche and
Debenedetti's Simulations of Infinitely
Dilute Supercritical Mixtures[21]**

Mixture	m_1/m_2	σ_2/σ_2 [a]	ϵ_1/ϵ_2 [a]
1	6.506	1.435	7.04
2	0.154	0.697	0.142

State point[b]	ρ^* [c]	T^* [d]
A	0.35	2
B	0.35	1.4
C	0.80	1.4

[a] Lorentz-Berthelot combining rules used for 1–2 interactions.
[b] The critical point of the Lennard-Jones fluid occurs (approximately) at $\rho^* = 0.35$ and $T^* = 1.35$.
[c] $\rho^* = \rho\sigma^3$ ($\rho\sigma_2^3$ in mixture simulations).
[d] $T^* = kT/\epsilon$ (kT/ϵ_2 in mixture simulations).

Mixture 1 is representative of Xe (1; solute) in Ne (2; solvent); mixture 2, of Ne (1; solute) in Xe (2; solvent). Potential parameters were obtained from viscosity data.[43] According to the van der Waals model[14,43] (see Equations 22 to 24), mixture 1 is predicted to exhibit attractive behavior (solvent enrichment around the solute), and mixture 2, repulsive behavior (solvent depletion around the solute).

Three simulations per state point were performed, using an isothermal, isochoric (N, V, T) ensemble: 864 identical (Lennard-Jones) atoms, 863 solvent atoms plus one solute, with potential parameters corresponding to mixture 1, and 863 solvent atoms plus one solute, with potential parameters corresponding to mixture 2. Use of periodic boundary conditions precluded solute-solute interactions, and the systems were therefore at infinite dilution (this applies to all quantities reported here, but not to properties such as energy and pressure, whose value is sensitive to the presence of the solute atom).

It is essential, in simulating the behavior of fluids in the relative vicinity of critical points, to use sample sizes large enough to guarantee that the correlation length is smaller than the computational box size. These and other technical aspects are discussed in detail by Petsche and Debenedetti.[21]

Figure 13 shows the pair correlation functions corresponding to the distribution of Ne atoms around infinitely dilute Xe (curve A); curve P is for the pure solvent, in both cases at state point B (critical isochore of the solvent; $T_r = T/T_{c2} = 1.037$). A pronounced solvent enrichment around the solute is evident for the attractive mixture, where the local solvent density in the nearest neighbor shell is almost four times higher than at bulk conditions. A detailed analysis of the correlation functions in the range $2 \leq r/\sigma_2 \leq 4$ (not shown in Figure 13) shows evidence of long-range decay.[21]

Upon interchanging solute and solvent at state point B, the behavior shown in Figure 14 resulted. Each curve is an average of 5000 snapshots and spans an interval of 22.1 ps. The noise is indicative of extremely weak solute-solvent correlations. The environment surrounding the Ne atom fluctuates continuously, and it exhibits the pronounced solvent depletion predicted by the van der Waals theory. This depletion is a consequence of enhanced solvent-solvent interactions with respect to solute-solvent interactions, magnified by long-range correlations in the highly compressible region. The term "repulsive" actually refers

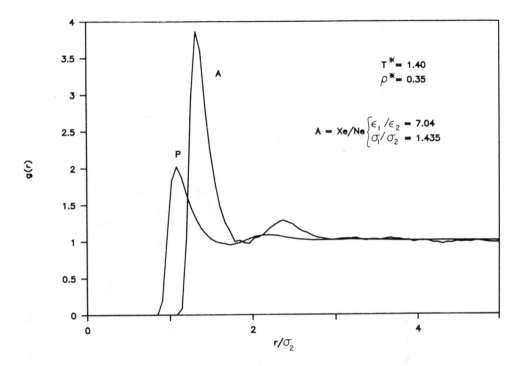

FIGURE 13.   Pair correlation functions for the pure (P) and attractive (A; 1-2 correlation) systems at T* = 1.4, ρ* = 0.35. Potential parameters as per Table 4. Molecular dynamics simulation. (From Petsche, I. B. and Debenedetti, P. G., *J. Chem. Phys.*, 91, 7075, 1989. With permission.)

more to competitive attractions than to true repulsions. In a repulsive near-critical system, the environment surrounding solute molecules is dominated by strong solvent interactions (and, therefore, by pronounced solvent density fluctuations). In an attractive near-critical system, solute-solvent interactions are dominant.

Upon compressing isothermally to point C (ρ* = 0.8), the correlation functions shown in Figure 15 were obtained. The attractive first peak decreased from 3.86 (state point B) to 2.98, whereas the repulsive peak increased. Differences between attractive and repulsive behavior tend to disappear away from the compressible region. In the former case, the solvent enrichment around the solute (with respect to bulk conditions) decreases; in the latter, the noise associated with weakened solute-solvent interactions disappears, as does any evidence of solvent depletion around the solute.

Hill[44] provided a quantitative framework for the study of cluster formation in nonideal systems. In this approach, two molecules are "bound" if their relative kinetic energy is less than their attractive energy, and not "bound", either if this condition is not satisfied or if the pair potential is repulsive.[21,44] An energetically defined cluster, in this context, consists of all pairs directly or indirectly connected at a given instant in time. Thus, if pairs α-β and β-γ are "bound", but α-γ is not, α-β-γ is a triplet. No molecule can belong to more than one such cluster at a given instant in time. Using Hill's cluster definition and an elegant cluster-counting algorithm due to Sevick et al.,[45] Petsche and Debenedetti[21] compared cluster statistics for the pure, attractive, and repulsive systems.

Figure 16 shows the cumulative relative frequency of occurrence of energetically defined clusters at state point B (Table 4). In the mixture cases, only those clusters including the solute were considered; for the pure Lennard-Jones fluid, clusters including a randomly selected molecule were tracked. Reported cluster sizes include the solute; a cluster size of 1 indicates no solvent particles bound to the solute (mixture case) or randomly selected solvent atom (pure case). The repulsive (Ne in Xe) mixture exhibited a cluster size of 1 (no

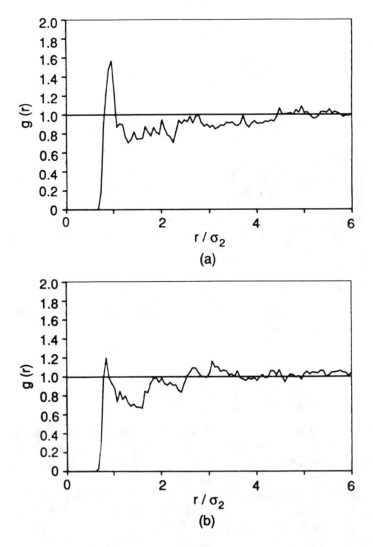

FIGURE 14.   Solute-solvent pair correlation functions for the repulsive mixture at $T^* = 1.4$, $\rho^* = 0.35$. Each curve is an average over 5000 steps, spanning 22.1 ps. Noise is indicative of weakened solute-solvent correlations and enhanced density fluctuations near the critical point of the solvent. Potential parameters as per Table 4. Molecular dynamics simulation. (From Petsche, I. B. and Debenedetti, P. G., *J. Chem. Phys.*, 91, 7075, 1989. With permission.)

Xe atom bound to the Ne probe) during 89.98% of the observations. The curvature of the distribution shows that larger clusters occurred with decreasing frequency: clusters involving more than five atoms (including the Ne probe) accounted for only 5.18% of the observations. The pure near-critical system exhibited a cluster size of 1 during 36.98% of the observations, and dimers, 12.62% of the time. The curvature shows, as in the repulsive case, decreasing frequency of occurrence with increasing cluster size. By contrast, the attractive (Xe in Ne) mixture exhibited a vanishing slope at the origin (unbound probe never observed), and maximum slope for cluster sizes between 20 and 40 (cluster sizes within this range occurred most frequently). Dimers were observed during only 0.09% of the observations, and clusters involving more than five atoms (including the Xe probe), accounted for 99.52% of the observations. Median cluster sizes (cumulative relative frequency = 0.5) were 2 (pure solvent) and 59 (attractive mixture).

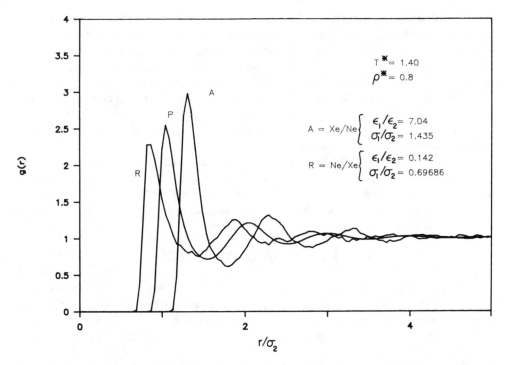

FIGURE 15. Pair correlation functions for the pure system (P), attractive (A; 1-2 correlation), and repulsive (R; 1-2 correlation) mixtures at T* = 1.4, ρ* = 0.8. Potential parameters as per Table 4. Molecular dynamics simulation. (From Petsche, I. B. and Debenedetti, P. G., *J. Chem. Phys.*, 91, 7075, 1989. With permission.)

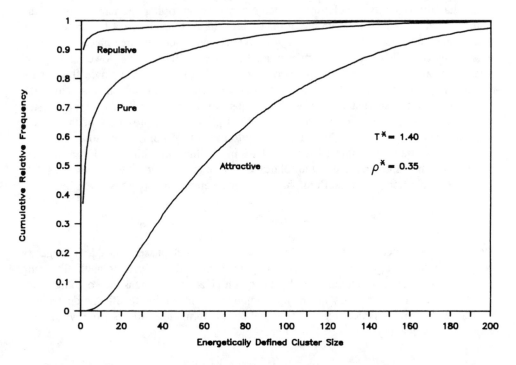

FIGURE 16. Cumulative relative frequency of occurrence of energetically defined clusters surrounding the Xe atom (attractive mixture), the Ne atom (repulsive mixture), and a randomly chosen atom (pure system), at T* = 1.4, ρ* = 0.35. For a given abscissa value (n), the ordinate is the sum over all clusters of size ≤ n of the number of times such clusters were observed, divided by the total number of snapshots (5000). (From Petsche, I. B. and Debenedetti, P. G., *J. Chem. Phys.*, 91, 7075, 1989. With permission.)

The simulation results shown in Figures 13 to 16 confirm the predictions of the van der Waals model. Attractive near-critical behavior (pronounced solvent enrichment around dilute solutes) is not preserved upon interchanging solute and solvent. The predictions of the van der Waals model do not necessarily apply to mixtures with more pronounced asymmetries, or whose components depart significantly from spherical geometry.

Simulations also show that in addition to the long-range effects that underlie the divergence in $\overline{v}_1^\infty$ and Γ, the local environment around dilute solutes in attractive mixtures ($\overline{v}_1^\infty \rightarrow -\infty$) is significantly enriched in solvent with respect to bulk conditions. This is in agreement with experimental evidence, as discussed in Section V.

The limitations associated with the use of computer simulations to study molecular interactions in dilute supercritical mixtures are the practical impossibility of including solute-solute interactions at high dilution, and the need to use very large systems in the vicinity of the critical point, which makes the study of polyatomic systems at these state conditions particularly demanding from a computational viewpoint.

VII. THERMODYNAMICS

This section interprets well-known thermodynamic relationships for the enhancement factor, as well as for the pressure and temperature dependence of solubility in dilute supercritical mixtures in the light of the clustering phenomenon. Subsequently, solubility enhancement is recast in fluctuation-explicit form, thereby arriving at molecular-based expressions for supercritical solubility.

The phenomena thus far discussed influence both the extent to which solutes dissolve in SCFs, as well as the sensitivity of this solvent power with respect to changes in temperature and pressure. Mixtures here are considered composed of a solid solute and a supercritical solvent that differ greatly in volatility in such a way that the triple point temperature of the solute exceeds the critical temperature of the solvent. In many such cases, the critical locus joining pure-component critical points can be intersected twice by three-phase loci. Such intersections are called lower and upper critical endpoints. Between the lower and upper-critical endpoint temperatures, the pure solid solute can coexist with a single fluid phase (the solubility of the SCF in the solid phase is neglected). In low-solubility systems, the critical point of the solvent is normally very close to the lower critical endpoint of the mixture. This means that the behavior of the fluid phase in the proximity of the critical endpoint is strongly influenced by the proximity to the critical point of the solvent.

For a pure, incompressible condensed phase in equilibrium with a fluid at a pressure far exceeding the vapor pressure of the solid, the ratio of actual to ideal-gas-predicted solute equilibrium mole fraction in the fluid phase (or enhancement factor, E) is given by

$$E = [\exp(Pv_1^s/kT)]/\hat{\phi}_1 \qquad (30)$$

where v_1^s is the molecular volume of the solute in the solid phase, and $\hat{\phi}_1$, its fugacity coefficient in the fluid phase. The exponential term is normally referred to as the Poynting factor. For the systems of interest here, E can be as high as 10^6, whereas the Poynting factor is an order 1 quantity. Thus, by far the most important contribution to E is the decrease in $\hat{\phi}_1$. The relationship between this decrease and cluster formation follows from writing the thermodynamic identity

$$kT \ln \hat{\phi}_1 = \int_0^P \left(\overline{v}_i - \frac{kT}{P'} \right) dP' \qquad (31)$$

which, at infinite dilution, becomes

$$\ln \hat{\phi}_1^\infty = \int_0^P \left[\left(K_T - \frac{1}{P'} \right) - \frac{\Gamma}{\rho kT} \right] dP' \tag{32}$$

where Equation 16 has been used to relate $\bar{v}_1^\infty$ and Γ. The first term in the integrand is the difference between the compressibility of the solvent and the ideal-gas value of this quantity. This difference increases monotonically along a supercritical isotherm, until the pressure corresponding to the critical isochore of the solvent, whereupon K_T attains its maximum value for that particular temperature.[32] Thus, the decrease in $\hat{\phi}_1^\infty$ can only be caused by the increase in Γ. Of course, in an actual mixture, y_1 is not zero and Equation 32 does not lead to quantitatively exact values of E. Nevertheless, it has been shown[32,46] that in those dilute supercritical mixtures where y_1 is inherently low (<0.001), the essential features of solubility enhancement can be described in terms of solute-solvent interactions at infinite dilution. In such cases, Equation 32 shows that the increase in E is a macroscopically observable consequence of the corresponding long-ranged increase in Γ that characterizes attractive near-critical behavior.

The relationship between the growth in Γ and the pressure and temperature dependence of the solubility of nonvolatile solids in supercritical solvents is another important topic. The relevant thermodynamic identity in this case is

$$\left(\frac{\partial y_1}{\partial P} \right)_{T,\sigma} = \frac{v_1^s - \bar{v}_1}{\left(\dfrac{\partial \mu_1}{\partial y_1} \right)_{T,P}} \tag{33}$$

where subscript σ denotes differentiation along a phase equilibrium locus. The denominator, being proportional to a stability coefficient,[47] is non-negative. Thus, the sign of the left side (i.e., the very fact that solubility can increase upon increasing the pressure isothermally) is determined solely by that of the right side numerator. Consequently, in attractive supercritical systems in which solubility increases dramatically with pressure in the vicinity of the lower critical endpoint of the system, a volume contraction accompanies the dissolution of a pure solid into a fluid ($v_1^s > \bar{v}_1$).

Once again, for dilute mixtures infinite dilution arguments can be invoked. It follows at once that the large, negative values of $\bar{v}_1^\infty$ that characterize attractive near-critical behavior underlie the pressure dependence of the solvent power of SCFs. Of course, the vanishing of the denominator in Equation 33 at the lower critical endpoint largely determines the magnitude of this effect, but its occurrence (i.e., the fact that $\partial y_1/\partial P$ is positive), which requires that the dissolution of a pure solid into a fluid be accompanied by a volume contraction, is a direct manifestation of the growth in Γ.

The above considerations apply unchanged to the analysis of the temperature dependence of the solvent power of SCFs. In this case, the relevant thermodynamic identity is

$$\left(\frac{\partial y_1}{\partial \ln T} \right)_{P,\sigma} = \frac{\bar{h}_1 - h_1^s}{\left(\dfrac{\partial \mu_1}{\partial y_1} \right)_{T,P}} \tag{34}$$

which relates the isobaric change in equilibrium solubility due to temperature variations to the enthalpy change associated with the dissolution of the pure solute. The phenomenon described by Equation 34 is commonly known as retrograde solubility, the isobaric decrease in y_1 with temperature that characterizes dilute supercritical mixtures slightly above the lower critical endpoint. Chimowitz and co-workers[48,49] have implemented an elegant separation process based on this phenomenon. It follows from Equation 34 that retrograde behavior implies the exothermic dissolution of the pure solid solute into the SCF.

For an infinitely dilute near-critical system,

$$\frac{\bar{h}_1^{\infty}}{\bar{v}_1^{\infty}} \sim T \left(\frac{\partial P}{\partial T}\right)_{\sigma} \tag{35}$$

where σ now denotes differentiation along the vapor-liquid coexistence curve of the pure solvent. This implies that the solute PMV and enthalpy become asymptotically proportional to each other, with a positive proportionality constant. Therefore, in dilute mixtures, to which infinite dilution arguments may be applied with qualitative accuracy, retrograde solubility is also a macroscopically observable consequence of attractive behavior ($\bar{v}_1^{\infty} \rightarrow -\infty$, $\bar{h}_1^{\infty} \rightarrow -\infty$, $\Gamma \rightarrow \infty$). Once again, the vanishing of the denominator in Equation 34 at the lower critical endpoint enhances the magnitude of a phenomenon whose occurrence (sign) is solely dependent upon the exothermicity that accompanies solute dissolution.

It follows from what has been said that long-range, cooperative behavior has a profound influence upon the temperature and pressure dependence of solubility in dilute, attractive supercritical mixtures.

Fluctuation-explicit expressions for E are now discussed. From Equation 30

$$\ln E = \beta v_1^s \cdot \Delta P - \ln \hat{\phi}_1 \tag{36}$$

where $\beta = 1/kT$, and $\Delta P = P - P_{vp}(T)$ has been used instead of P (see Equation 30); P_{vp} is the vapor pressure of the solute at T. For the systems of interest here, $P \gg P_{vp}$. Equation 36 is valid at any concentration. In fact, it is the basis of the compressed gas approach.[50] This chapter is concerned with the KB approach. Debenedetti and Kumar[46] examined the behavior of ϕ_1 near the infinite dilution limit. Upon expanding $\ell n \phi_1$ about its infinite dilution limit, and truncating higher order terms, they obtained

$$\hat{\phi} = \hat{\phi}_1^{\infty} \exp[-Kx_1] \tag{37}$$

where K, which is a function of T and P only, is the second derivative of the fugacity coefficient of the solvent with respect to x_1 at infinite dilution. Later, Cochran and Lee[22] showed the K could be expressed in terms of the KB factors. Their result is

$$K = \rho_2^0[G_{11}^{\infty} + G_{22}^0 - 2G_{12}^{\infty}] \tag{38}$$

where superscript 0 denotes pure solvent.

For infinite dilution, Cochran et al.[23] formulated an equivalent form of Equation 36 in terms of the KB factors,

$$\lim_{\rho 1 \rightarrow 0} \ln E = \beta v_1^s \Delta P + \beta \int_{P_{vp}}^{P} dP \, G_{12}^{\infty} + \ln Z_2^0 \tag{39}$$

where Z_2^0 is the compressibility factor of the solvent. Pfund et al.[24] proposed an approximation for G_{12} in terms of G_{22} via a scaled distribution function assumption (i.e., the correlation functions were assumed to scale by ratios of energy and size parameters) by introducing a scaling parameter $\bar{\alpha}$

$$\bar{\alpha} \equiv \frac{G_{12} + V_{12}}{G_{22} + V_{22}} \tag{40}$$

where V_{ij} is the excluded (hard core) volume for the interactions of ij molecules. In general, this parameter is a function of the same variables as G_{ij}, namely temperature, density, and composition. For simplicity, Pfund and associates took $\bar{\alpha}$ to be a function of temperature only and estimated its magnitude using critical constants. Then the fugacity, f_1^∞, of the solute could be expressed in terms of the fugacity, f_2^0, of the pure solvent as

$$\ln \frac{f_1^\infty}{\rho kT} \approx \bar{\alpha} \ln \frac{f_2^0}{\rho kT} - \beta[\bar{\alpha}V_{22} - V_{12}] \cdot \Delta P \tag{41}$$

Thus, Equation 36 becomes

$$\ln E \approx \beta v_1^s \Delta P - \bar{\alpha} \ln \left(\frac{f_2^0}{\rho kT}\right) + \ln Z_2^0 + \beta[\bar{\alpha}V_{22} - V_{12}]\Delta P \tag{42}$$

The advantage of Equation 42 over Equation 36 is obvious: except for the solid molar volume, one needs only solvent properties in order to evaluate the enhancement factor. Data on common supercritical solvents such as CO_2 are readily available in the literature. The price to pay is the determination of the scaling parameter $\bar{\alpha}$ as a function of temperature (note that $\bar{\alpha}$ is also a function of pressure; however, this dependence is weak). $\bar{\alpha}$ can be estimated from the pure component critical constants. Application of this approach to the CO_2-naphthalene system is shown in Figure 17. In this case $\bar{\alpha}$ was approximated by

$$\bar{\alpha} \approx 0.5\left[1 + \frac{V_{c,naph}}{V_{c,CO_2}}\right] \cdot \sqrt{T_{c,naph}/T_{c,CO_2}} = 4.38 \tag{43}$$

where T_c and v_c refer to the critical temperature and volume of solute (naphthalene) and solvent (CO_2). The excluded volumes were assumed to be given by the critical volumes $V_{22} = v_{c,CO_2} = 90.036$ cm³/mol, and $V_{12} = 0.5(v_{c,CO_2} + v_{c,naph}) = 251.5$ cm³/mol. With these rough estimates, the agreement seems reasonable (see Figure 17). Later developments employed local composition arguments[24] and yielded a modified excluded volume

$$\ln E \approx \beta v_1^s \Delta P - \bar{\alpha} \ln \left(\frac{f_2^0}{\rho kT}\right) + \ln Z_2^0 + \beta[\bar{\alpha}V_{22}e^{\bar{b}/RT} - V_{12}]\Delta P \tag{44}$$

A temperature-dependent characteristic volume $V_{22}\exp(\bar{b}/RT)$ was introduced (with one additional parameter $\bar{b}$). Agreement with the same data was better.

Equation 42 is inexact. It would be, nonetheless, interesting to examine its implications, for example, for the van der Waals (vdW) equation of state (EOS) with the usual mixing rules: $a = \Sigma\Sigma x_i x_j a_{ij}$ and $b = \Sigma x_i b_i$. The KB factors G_{11} and G_{22} for a binary mixture of van der Waals gases of species 1 and 2 can be easily derived. For solute 1 at infinite dilution in solvent 2 ($\lim \rho_1 \to 0$)

$$\lim_{\rho_1 \to 0} G_{12} = -\frac{1}{D}\left[\frac{(b_1 + b_2)}{(1 - b\rho)} + \frac{\rho b_1 b_2}{(1 - b\rho)^2} - 2\beta a_{12}\right]$$

$$\lim_{\rho_1 \to 0} G_{22} = -\frac{1}{D}\left[\frac{2b_2}{(1 - b\rho)} + \frac{\rho b_2^2}{(1 - b\rho)^2} - 2\beta a_{22}\right] \tag{45}$$

where b_i and a_{ij} are the covolumes and energy parameters of the van der Waals equation, $\rho = \rho_1 + \rho_2$, and D is given by

$$\lim_{\rho_1 \to 0} D = 1 + \frac{2\rho_2 b_2}{(1 - b\rho)} + \frac{\rho\rho_2 b_2^2}{(1 - b\rho)^2} - 2\beta\rho_2 a_{22} \tag{46}$$

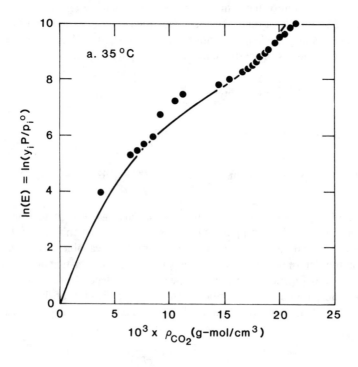

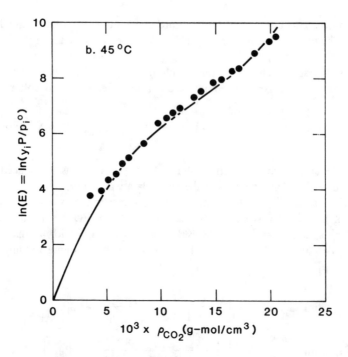

FIGURE 17.   Logarithm of the enhancement factor vs. experimental fluid density for naphthalene in supercritical carbon dioxide. Data from Reference 51. Curve is the calculation from the Kirkwood-Buff model (Equation 42). Temperatures are 35°C (a), 45°C (b), and 55°C (c). (From Cochran, H. D., Lee, L. L., and Pfund, D. M., in *Fluid Phase Equilibria*, 34, 228, 1987. With permission.)

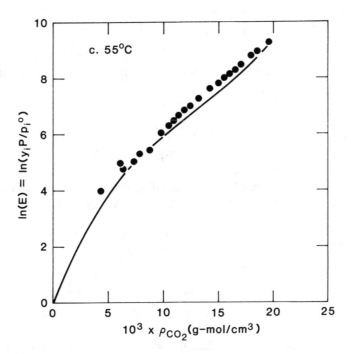

FIGURE 17, continued

Rearrangement gives

$$G_{12} + \frac{1}{D}\left[\frac{(b_1 + b_2)}{(1 - b\rho)} + \frac{\rho b_1 b_2}{(1 - b\rho)^2}\right] = [2\beta a_{12}]/D \tag{47}$$

Similarly,

$$G_{22} + \frac{1}{D}\left[\frac{2b_2}{(1 - b\rho)} + \frac{\rho b_2^2}{(1 - b\rho)^2}\right] = [2\beta a_{22}]/D \tag{48}$$

From Equations 47 and 48 the following relations are obtained

$$\bar{\alpha} \equiv \frac{G_{12} + V_{12}}{G_{22} + V_{22}} = \frac{a_{12}}{a_{12}} = \left(\frac{\epsilon_{12}}{\epsilon_{12}}\right)\left(\frac{\sigma_{12}^3}{\sigma_{22}^3}\right) \approx \left(\frac{T_{c,12}}{T_{c,2}}\right)\left(\frac{v_{c,12}}{v_{c,2}}\right) \tag{49}$$

with

$$V_{ij} \equiv \frac{b_i + b_j}{D(1 - b\rho)} + \frac{\rho b_i b_j}{D(1 - b\rho)^2}, \; i,j = 1,2 \tag{50}$$

The van der Waals parameters a_{ij} were split into $\epsilon_{ij} \cdot \sigma_{ij}^3$. A recommendation was to estimate these parameters by the critical temperatures, T_c, and the critical volumes, v_c, as mentioned earlier. The so-called excluded volumes, V_{ij}, are shown here to be functions of density and temperature (through D); however, in the integration of Equation 39, the V_{ij} were assumed to be constant. Thus, Equation 42 is, as mentioned previously, inexact.

Returning to Equation 36, the fugacity coefficient ϕ_1 for the vdW gas can be evaluated exactly as

$$\lim_{\rho_1 \to 0} \ln \hat{\phi}_1 = -\ln(1 - b\rho) + \frac{\rho b_1}{(1 - b\rho)} - 2\beta\rho a_{12} - \ln Z_2^0 \qquad (51)$$

By contrast, Formula 39 could also be evaluated using the vdW EOS (incorporating $\bar{\alpha}$ from Equation 40)

$$\lim_{\rho_1 \to 0} \ln \hat{\phi}_{1,KB} = -\bar{\alpha} \ln(1 - b\rho) + \bar{\alpha} \frac{\rho b_2}{(1 - b\rho)}$$
$$- 2\bar{\alpha}\beta\rho a_{22} - \ln Z - \beta\Delta P[\bar{\alpha}V_{22} - V_{12}] \qquad (52)$$

The constancy of V_{ij} was assumed in the integration, which was at best an approximation; thus, the results are inexact. The role of $\bar{\alpha}$ was to enforce, some times irreconcilably, the equalities

$$\bar{\alpha}b_2 = b_1, \quad \bar{\alpha}a_{22} = a_{12} \qquad (53)$$

and

$$\bar{\alpha} \ln(1 - b\rho) + \beta\Delta P[\bar{\alpha} V_{22} - V_{12}] = \ln(1 - b\rho) \qquad (54)$$

Not all these equalities could be satisfied by a single $\bar{\alpha}$ value. In addition, real fluids are not always of the van der Waals type. Caution must be exercised in those cases.

VIII. CONCLUSION

The KB theory naturally relates macro- and microscopic quantities. Its application to the study of molecular interactions in SCFs has allowed the rigorous quantification of the cluster concept,[11] led to the classification of near-critical systems into distinct categories,[14] and resulted in a molecular-based approach to supercritical solubility.[23] In addition, the implications of the theoretical predictions have inspired computational studies of molecular interactions in dilute supercritical mixtures,[18-21] which in turn have provided detailed information on the static[18-21] and dynamic[21] nature of the environment surrounding solute molecules in supercritical solvents. Furthermore, a fruitful dialogue has been established between experimentalists and theoreticians as a consequence of the complementary nature of the information generated in solvatochromic[12,15] and fluorescence[16,17] studies of dilute supercritical mixtures, on the one hand, and integral equation[18,20] and computer simulation[21] calculations, on the other.

That the concept of infinite dilution needs to be carefully applied when the correlation length is very large hardly needs to be emphasized; however, the fluorescence and integral equation evidence for excimer formation raises new and fascinating questions about the nature of solute-solute interactions in dilute supercritical systems on a short-length scale, not commonly associated with critical anomalies. The structural stability of the very dense aggregates of solvent molecules surrounding solutes in attractive near-critical mixtures also needs to be carefully investigated. The dynamic implications of the observed differences between the environment surrounding solute molecules in attractive and repulsive systems (i.e., solute mobility) also deserves further study via molecular dynamics simulations. Little is known about how details of molecular architecture and energetics determine the attractive or repulsive nature of a given binary system, except for van der Waals mixtures.[14]

These are just some of the interesting questions that remain unanswered. A detailed understanding of molecular interactions in supercritical mixtures is now commonly accepted to be a necessary prerequisite to the formulation of a rigorous, molecular-based thermodynamic theory of these highly asymmetric mixtures close to the critical point of a solvent. Experimental data, theoretical developments, and computer simulations can all contribute significantly to such an understanding. In addition to providing the quantitative insights discussed in this chapter, the application of the KB theory to supercritical mixtures has provided a common language to experimental, theoretical, and computational researchers in this field. The very diversity of the topics discussed here is, in the opinion of the authors, the best proof of the field's current vitality.

REFERENCES

1. **Kirkwood, J. G. and Buff, F. P.,** The statistical mechanical theory of solutions. I., *J. Chem. Phys.,* 19(6), 774, 1951.
2. **Perry, R. L. and O'Connell, J. P.,** Fluctuation thermodynamic properties of reactive components from species correlation function integrals, *Mol. Phys.,* 52, 137, 1984.
3. **Perry, R. L., Cabezas, H., and O'Connell, J. P.,** Fluctuation thermodynamic properties of strong electrolyte solutions, *Mol. Phys.,* 63, 189, 1988.
4. **Mathias, P. M. and O'Connell, J. P.,** Molecular thermodynamics of liquids containing supercritical compounds, *Chem. Eng. Sci.,* 36, 1123, 1981.
5. **Hamad, E. Z., Mansoori, G. A., and Ely, J. F.,** Conformality in the Kirkwood-Buff solution theory of statistical mechanics, *J. Chem. Phys.,* 86, 1478, 1987.
6. **O'Connell, J. P.,** Thermodynamic properties of solutions based on correlation functions, *Mol. Phys.,* 20, 27, 1971.
7. **Eckert, C. A., Ziger, D. H., Johnston, K. P., and Ellison, T. K.,** The use of partial molal volume data to evaluate equations of state for supercritical fluid mixtures, *Fluid Phase Equilibria,* 14, 167, 1983.
8. **Rozen, A. M.,** The unusual properties of solutions in the vicinity of the critical point of the solvent, *Russ. J. Phys. Chem.,* 50(6), 837, 1976.
9. **Krichevskii, I. R.,** Thermodynamics of critical phenomena in infinitely dilute binary solutions, *Russ. J. Phys. Chem.,* 41(10), 1332, 1967.
10. **Wheeler, J. C.,** Behavior of a solute near the critical point of an almost pure solvent, *Ber. Bunsenges. Phys. Chem.,* 76(3,4), 308, 1972.
11. **Debenedetti, P. G.,** Clustering in dilute, binary supercritical mixtures: a fluctuation analysis, *Chem. Eng. Sci.,* 42(9), 2203, 1987.
12. **Kim, S. and Johnston, K. P.,** Molecular interactions in dilute supercritical solutions, *Ind. Eng. Chem. Res.,* 26, 1206, 1987.
13. **Eckert, C. A., Ziger, D. H., Johnston, K. P., and Kim, S.,** Solute partial molal volumes in supercritical fluids, *J. Phys. Chem.,* 90, 2738, 1986.
14. **Debenedetti, P. G. and Mohamed, R. S.,** Attractive, weakly attractive, and repulsive near-critical systems, *J. Chem. Phys.,* 90(8), 4528, 1989.
15. **Kim, S. and Johnston, K. P.,** Clustering in supercritical fluid mixtures, *AIChE J.,* 33(10), 1603, 1987.
16. **Brennecke, J. and Eckert, C. A.,** in *Proc. Int. Symp. on Supercritical Fluids,* Perrut, M., Ed., Nice, France, 1988, 263.
17. **Brennecke, J. and Eckert, C. A.,** Fluorescence Spectroscopy Studies of Intermolecular Interactions in Supercritical Fluids, in *Supercritical Fluid Science and Technology,* Johnston, K. P. and Penninger, J., Eds., ACS Symp. Ser. No. 406, American Chemical Society, Washington, D.C., 1989, 14.
18. **Cochran, H. D., Pfund, D. M., and Lee, L. L.,** in *Proc. Int. Symp. on Supercritical Fluids,* Perrut, M., Ed., Nice, France, 1988, 245.
19. **Cochran, H. D. and Lee, L. L.,** Structure and properties of supercritical fluid mixtures from Kirkwood-Buff fluctuation theory and integral equation methods, in *Fluctuation Theory of Mixtures,* Matteoli, E. and Mansoori, G. A., Eds., Taylor & Francis, New York, 1990, 69.
20. **Cochran, H. D. and Lee, L. L.,** Solvation structure in supercritical fluid mixtures based on molecular distribution functions, in *Supercritical Fluid Science and Technology,* Johnston, K. P. and Penninger, J., Eds., ACS Symp. Ser. No. 406, American Chemical Society, Washington, D.C., 1989, 27.
21. **Petsche, I. B. and Debenedetti, P. G.,** Solute-solvent interactions in infinitely dilute supercritical mixtures: a molecular dynamics investigation, *J. Chem. Phys.,* 91, 7075, 1989.
22. **Cochran, H. D. and Lee, L. L.,** General behavior of dilute binary solutions, *AIChE J.,* 33, 1391, 1987.
23. **Cochran, H. D., Lee, L. L., and Pfund, D. M.,** Application of the Kirkwood-Buff theory of solutions to dilute supercritical mixtures, *Fluid Phase Equilibria,* 34, 219, 1987.

24. **Pfund, D. M., Lee, L. L., and Cochran, H. D.,** Application of the Kirkwood-Buff theory of solutions to dilute supercritical mixtures. II. The excluded volume and local composition models, *Fluid Phase Equilibria,* 39, 161, 1988.

25. **Buff, F. P. and Schindler, F. M.,** *J. Chem. Phys.,* 29, 1075, 1958.

26. **Ben Naim, A.,** *Water and Aqueous Solutions,* Plenum Press, New York, 1974.

27. **Lee, L. L.,** *Molecular Thermodynamics of Nonideal Fluids,* Butterworths, Boston, 1988.

28. **Debenedetti, P. G.,** The statistical mechanical theory of concentration fluctuations in mixtures, *J. Chem. Phys.,* 87(2), 1256, 1987.

29. **McGuigan, D. B. and Monson, P. A.,** Analysis of infinite dilution partial molar properties using distribution function theories, *Fluid Phase Equilibria,* 57, 227, 1990.

30. **Ziger, D. H. and Eckert, C. A.,** Simple high-pressure magnetic pump, *Rev. Sci. Instrum.,* 53, 1296, 1982.

31. **Kratky, O., Leopold, H., and Stabinger, H.,** in *Methods in Enzymology,* Hirs, C. H. W. and Timasheff, S. W., Eds., Vol. 27, Part D, Academic Press, London, 1973.

32. **Debenedetti, P. G. and Kumar, S. K.,** The molecular basis of temperature effects in supercritical extraction, *AIChE J.,* 34(4), 645, 1988.

33. **Rozelius, W., Vizthum, O., and Hubert, P.,** Method for the Production of Caffeine-Free Extract, U.S. Patent 3,843,824, 1974.

34. **Kumar, S. K., Suter, U. W., and Reid, R. C.,** Fractionation of polymers with supercritical fluids, *Fluid Phase Equilibria,* 29, 373, 1986.

35. **Mohamed, R. S., Debenedetti, P. G., and Prud'homme, R. K.,** Effects of process conditions upon crystals obtained from supercritical mixtures, *AIChE J.,* 35(2), 325, 1989.

36. **Prausnitz, J. M., Lichtenthaler, R., and de Azevedo, E. G.,** *Molecular Thermodynamics of Fluid-Phase Equilibria,* 2nd ed., Prentice-Hall, Englewood Cliffs, NJ, 1986, chap. 3.

37. **Johnston, K. P., Kim, S., and Combes, J.,** Spectroscopic determination of solvent strength and structure in supercritical fluid mixtures: a review, in *Supercritical Fluid Science and Technology,* Johnston, K. P. and Penninger, J., Eds., ACS Symp. Ser. No. 406, American Chemical Society, Washington, D.C., 1989.

38. **Kamlet, M. J., Abraham, M. H., Doherty, R. M., and Taft, R. W.,** Solubility properties in polymers and biological media. IV. Correlation of octanol/water partition coefficients with solvatochromic parameters, *J. Am. Chem. Soc.,* 106, 464, 1984.

39. **Kamlet, M. J., Abraham, M. H., Doherty, R. M., and Taft, R. W.,** The molecular properties governing solubilities of organic nonelectrolytes in water, *Nature,* 313, 384, 1985.

40. **Kajimoto, O., Futakami, M., Kobayashi, T., and Yamasaki, K.,** Charge-transfer state formation in supercritical fluid: (*N,N*-dimethylamino) benzonitrile in CF_3H, *J. Phys. Chem.,* 92, 1347, 1988.

41. **Yonker, D. R., Frye, S. L., Kalkwarf, D. R., and Smith, R. D.,** Characterization of supercritical solvents using solvatochromic shifts, *J. Phys. Chem.,* 90, 3022, 1986.

42. **Wu, R. S., Lee, L. L., and Cochran, H. D.,** The structure of supercritical fluid mixtures: clustering of solvent and solute molecules and their thermodynamic effects, *Ind. Eng. Chem. Res.,* 29, 977, 1990.

43. **Reid, R. C., Prausnitz, J. M., and Poling, B.,** *The Properties of Gases and Liquids,* 4th ed., McGraw-Hill, New York, 1987, 733.

44. **Hill, T. L.,** Molecular clusters in imperfect gases, *J. Chem. Phys.,* 23, 617, 1955.

45. **Sevick, E. M., Monson, P. A., and Ottino, J. M.,** Monte Carlo calculations of cluster statistics in continuum models of composite morphology, *J. Chem. Phys.,* 88, 1198, 1988.

46. **Debenedetti, P. G. and Kumar, S. K.,** Infinite dilution fugacity coefficients and the general behavior of dilute binary systems, *AIChE J.,* 32(8), 1253, 1986.

47. **Modell, M. and Reid, R. C.,** *Thermodynamics and its Applications,* 2nd ed., Prentice-Hall, Englewood Cliffs, NJ, 1983, chap. 9.

48. **Chimowitz, E. H. and Pennisi, K. P.,** Process synthesis concepts for supercritical gas extraction in the crossover region, *AIChE J.,* 32, 1665, 1986.

49. **Chimowitz, E. H., Kelley, F. D., and Munoz, F. M.,** Analysis of retrograde behavior and the crossover effect in supercritical fluids, *Fluid Phase Equilibria,* 44, 23, 1988.

50. **MacKay, M. E. and Paulaitis, M. E.,** Solid solubilities of heavy hydrocarbons in supercritical solvents, *Ind. Eng. Chem. Fundam.,* 18, 149, 1979.

51. **Tsekhanskaya, Yu. V., Iomtev, M. B., and Mushkina, E. V.,** Solubility of naphthalene in ethylene and carbon dioxide under pressures, *Russ. J. Phys. Chem.,* 38, 1173, 1964.

Chapter 5

APPLICATION OF MOLECULAR SIMULATION TO THE STUDY OF SUPERCRITICAL EXTRACTION

K. S. Shing

TABLE OF CONTENTS

ABSTRACT

This chapter describes the computational techniques commonly used in the simulation of supercritical extraction systems. Conventional Monte Carlo and molecular dynamics techniques are normally used to simulate supercritical phases for the purpose of gaining understanding and insight. Macroscopic property or behavior such as divergence in partial molar properties are traced to microscopic properties such as clustering at the molecular level. Specialized techniques designed for phase equilibria studies and the application of these techniques to a few model and real mixtures are described. Qualitative agreement with experimental results can be obtained when fairly simple effective potential models are used.

I. INTRODUCTION

Molecular simulation is a computational method used to study the behavior of (primarily) fluids. In a typical simulation, a model for the intermolecular forces is first proposed and the macroscopic properties are found numerically. Therefore, simulations are ideally suited to studies designed to establish the relationship between microscopic (or molecular) characteristics of systems and their macroscopic properties or behavior. Two major techniques can be used: the Monte Carlo (MC) method or the molecular dynamics (MD) method. In the former, various arrangements (or configurations) of the molecules consistent with the prescribed thermodynamic state are generated according to statistical mechanical distribution laws through the use of random number generators. Thermodynamic properties are obtained by averaging over the configurations. In the MD method, the trajectories of the molecules are tracked as functions of time by solving numerically the coupled Newton's equations of motion for each molecule. Thermodynamic, transport, and time-dependent properties can be found. (For detailed discussion of the simulation methods, more specialized techniques and applications, the interested reader is referred to Reference 1.)

In thermodynamic and phase equilibria studies, molecular simulation can serve several useful purposes:

1.   To develop accurate effective intermolecular potential models for real molecules of interest. This is accomplished by comparing experimental and simulation results.
2.   To provide unambiguous tests for the approximations in various molecular theories by comparing simulation results to molecular theory predictions using the same intermolecular potential models.
3.   To provide physical insight (particularly at the molecular level) in systems that are poorly understood and/or difficult to study experimentally. Examples include fluids in microporous materials, supercritical extraction systems, etc. In these cases, it is often necessary to use idealized models for the intermolecular potentials as well as for the system geometries.

Most of the simulation work in supercritical extraction studies are carried out to accomplish objective (3). This is because adequate intermolecular potential models are not yet available for the highly asymmetric mixtures involved in these systems. These studies can be further divided into two groups:

1.   Conventional simulations for supercritical phases performed to gain physical insight. Examples include studies of the differences between supercritical fluids and normal fluids, of the solvent structure and clustering phenomena in the vicinity of infinitely dilute solutes, and the calculation of partial molar properties. The effect of varying the intermolecular force models is also of particular interest.

2. Specialized simulations designed primarily to calculate the equilibrium solubilities. Phase equilibria involve free energies and chemical potentials which are nonmechanical properties. These properties are not routinely accessible from conventional MC or MD methods, therefore special techniques are needed.

In the following sections, the simulation methods used in supercritical extraction studies are briefly discussed and major results and conclusions obtained from these studies are summarized.

II. CONVENTIONAL MOLECULAR SIMULATIONS

A. DIFFERENCES BETWEEN A LIQUID AND A SUPERCRITICAL FLUID

Vogelsang and Hoheisel[3] used MD simulation to study the differences between a normal Lennard-Jones (LJ) liquid and a supercritical LJ fluid. Various static and dynamic properties were calculated, including the internal energy, pressure, diffusion coefficient, static pair and triplet correlation functions, mean-square displacement, and velocity autocorrelation functions. The two states chosen were $T^* \equiv kT/\epsilon = 0.7611$, $\rho^* \equiv \rho\sigma^3 = 0.8184$ for the liquid and $T^* = 2.5023$ and $\rho^* = 1.0736$ for the supercritical fluid. (Here, T is the temperature; ρ, the density; k, the Boltzmann constant, ϵ and σ, the LJ energy and size parameters.) According to the authors this choice was made such that g_{max}, the first peak in the radial distribution functions (RDF) for these two states have the same height of 2.98. (The product of the RDF g(r) and the bulk density ρ is a measure of the local density in a spherical shell of radius r centered on a selected molecule.)

These authors reported that the RDF for the liquid and the supercritical fluid (SCF) were similar except that in the SCF case the peaks were shifted to smaller distances. This was consistent with the fact that the SCF state simulated had a higher density. Of greater interest were the differences in dynamic behavior. The simulated self-diffusion coefficients of the liquid and the SCF were 2.3 and 3.1×10^5 cm^2/s, respectively. The difference was relatively small (38%). At first sight, this may be surprising because greatly enhanced diffusion is one of the advantages of operating in the supercritical rather than in the liquid phase. In this case, however, the chosen density for the SCF was so much higher than that of the liquid (ρ^* of 1.0736 vs. 0.8184) that the diffusion process was dominated by the density effect. Therefore, this particular comparison is of limited relevance to supercritical extraction processes. The large density difference may be reduced by selecting a lower T^* for the SCF. $T^* = 2.5$ is too supercritical ($T_c^* \simeq 1.36$ for the LJ fluid). A near supercritical state would be more interesting and more relevant since most SCF processes would probably operate close to the critical point to minimize the compression cost and operating pressure. Furthermore, a better comparison between diffusion in the liquid and SCF phases would be achieved by selecting the states such that the product of $\rho^* g_{max}$ are equal for the liquid and SCF (g_{max} is the height of the first peak in the RDF), rather than basing the selection on the equality of the g_{max} itself as was done here. The former choice is also more reasonable on physical grounds because $\rho^* g_{max}$ is a measure of the local density in the first neighbor shell, whereas g_{max} itself only measures the enhancement of the local density over the average bulk density.

B. PARTIAL MOLAR PROPERTIES AND EQUILIBRIUM STRUCTURE OF SUPERCRITICAL INFINITELY DILUTE LENNARD-JONES MIXTURES

Shing and Chung[4] used MC methods to calculate the various partial molar properties of infinitely dilute LJ solutes in supercritical LJ solvent. The quantities evaluated include the solute partial molar volume $\overline{V}_1^\infty$, partial molar internal energy $\overline{U}_1^\infty$, and the partial molar Gibbs' free energy (or the chemical potential) μ_1^∞. Several types of solute were used in the simulations, including large or small size as well as strongly or weakly attractive solutes. At a

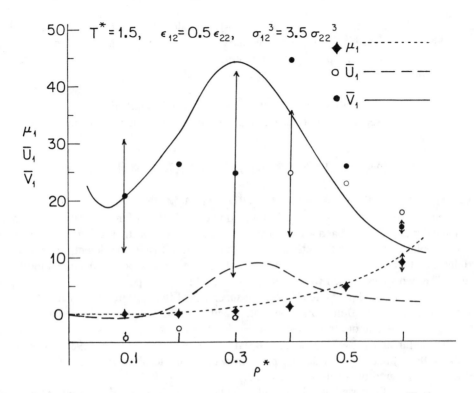

FIGURE 1.   Infinite-dilution partial molar Gibbs' free energy (μ_i), internal energy ($\overline{U}_i$), and volume ($\overline{V}_i$) for LJ mixtures for a supercritical isotherm.[4] The solute is weakly attractive and large. ϵ_{12}, σ_{12}, ϵ_{22}, σ_{22} are energy and size parameters for the solute-solvent and solvent-solvent interactions. $T^* \equiv kT/\epsilon_{22}$, $\rho^* \equiv \rho\sigma_{22}^3$.

moderately supercritical temperature ($T^* = 1.5$), very large and positive partial molar volumes (PMVs) and internal energies were reported for weakly attractive large solutes when the solvent density was near the critical density (Figure 1). Corresponding partial molar properties at either lower or higher densities were much smaller. The positive $\overline{V}_1^\infty$ and $\overline{U}_1^\infty$ were attributed to the fact that the weakly attractive large solute acted as an insulator in the solvent and screened the (negative) attractive interactions between solvent molecules, thus resulting in reduced cohesive energy. The large magnitudes of $\overline{V}_1^\infty$ and $\overline{U}_1^\infty$ at the solvent critical density were associated with the divergence of the isothermal compressibility at the critical point. Similar observations were reported for strongly attractive solutes except that the $\overline{V}_1^\infty$, $\overline{U}_1^\infty$ were now all negative due to increased attractive interaction (Figure 2).

Solute chemical potentials were found to be weaker functions of density. No large deviations were observed near the critical point. This is because unlike volume and internal energy which are discontinuous in vapor-liquid transition, chemical potentials are continuous in coexisting phases. These authors also reported that at a more supercritical temperature of $T^* = 2.0$, the observed trends were similar to those for $T^* = 1.5$ but that the magnitudes of $\overline{V}_1^\infty$ and $\overline{U}_1^\infty$ were greatly reduced. This was attributed to the smaller isothermal compressibilities at the elevated temperature.

The observed behavior of the macroscopic partial molar properties all originated from the perturbation of the solvent fluid structure resulting from the addition of the lone solute molecule. This can be clearly demonstrated in the solute-solvent radial distribution function, g_{12}. For example, g_{12} for the weakly and strongly attractive solutes are shown in Figures 3 and 4. The much higher peaks for the attractive solute (Figure 4) indicate enhanced solvent density in the neighborhood of the solute. The negative $\overline{V}_1^\infty$ and $\overline{U}_1^\infty$ are macroscopic manifestations of this local density enhancement. Similarly, the much lower peaks for the weakly

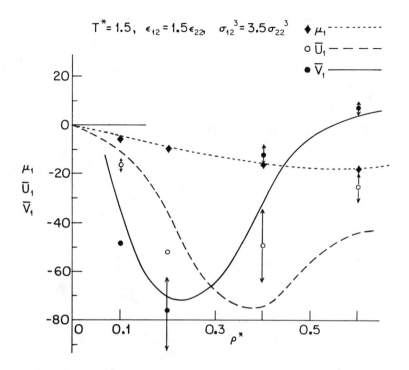

FIGURE 2.   As for Figure 1, but for a large and strongly attractive solute.[4]

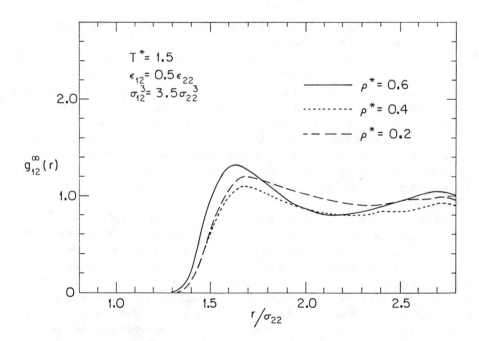

FIGURE 3.   Solute-solvent radial distribution function g_{12} for weakly attractive large solutes in infinitely dilute supercritical LJ mixtures.[4]

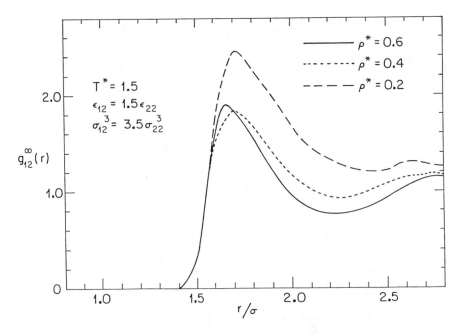

FIGURE 4.   As for Figure 3, but for strongly attractive solutes.[4]

attractive solute indicate depletion of solvent molecules around the solute leading to positive $\overline{V}_1^\infty$ and $\overline{U}_1^\infty$.

C.   INFINITELY DILUTE LENNARD-JONES MIXTURES — CLUSTERING PHENOMENON AND DYNAMIC BEHAVIOR

In a recent study, Petsche and Debenedetti[5] used MD simulations to study infinitely dilute supercritical LJ mixtures, placing particular emphasis on the dynamic solvent structure in the vicinity of the single solute atom. The effect of varying the solute-solvent potential was also reported. It was found that solvent density was greatly enhanced in the neighborhood of an attractive solute (Figure 5) and depleted in the neighborhood of a repulsive solute (Figure 6) (attractive/repulsive means that the solute-solvent interactions are attractive/repulsive compared to the solvent-solvent interactions). The fluid structure around the attractive solute was found to be less chaotic than that around the repulsive solute. This observation was attributed to the stabilizing effect of the stronger attractive solute-solvent interactions in the former case.

Petsche and Debenedetti also analyzed the solvent density enhancement around the attractive solute in terms of aggregation or cluster formation. Two types of clusters were defined and studied. The first was based on energetic considerations, namely, a pair of particles with relative kinetic energy less than the intermolecular attractive potential energy was considered bound. The second type of clusters was defined in geometric terms, i.e., the particles within a specified distance from a central particle (solute in this case) were considered members of the same cluster. Most of the results reported were for the energetically defined clusters, although in interpreting experimental results such as solvatochromic effects and diffusion processes, the geometrical definition was preferred. This is because in an energetically defined cluster, member particles, regardless of whether they are in the periphery or near the center of the cluster constantly fluctuate between being ''bound'' or ''unbound'' due to collisions with neighboring particles. If the clusters are large, dense, and energetic, such fluctuations may make the interpretation of cluster persistence or stability somewhat ambiguous, especially if the persistence time is relatively short.

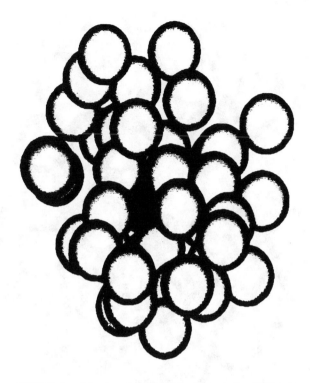

FIGURE 5. Solvent molecules (white spheres) surrounding an infinitely dilute attractive solute (black sphere) showing clustering and high local density.[5]

The major conclusions regarding the clustering phenomenon reported in this work, however, are valid for either definition. For the repulsive solute, the median cluster size was 2 whereas that for the attractive solute was 59. Although the numerical values of the cluster size reported were specific to the system simulated, they nevertheless illustrate the greatly enhanced/reduced solvent density surrounding the attractive/repulsive solute particle. By counting the number of particles belonging to a particular cluster at time t that also belonged to the same cluster at time zero, the persistence of cluster identity may be inferred. No correlation was observed between the initial size of the attractive clusters and the persistence of the identity. This was attributed to the fact that clusters were constantly growing and shrinking when solvent particles near the solute molecule fluctuated between being "bounded" and "unbounded" so that the instantaneous (or initial) size of a cluster had no particular significance. The persistence time of an attractive cluster was, however, longer than that of a pure cluster because the "average" size of an attractive cluster was larger than that in the pure case. One may conclude, then, that attractive clusters are dynamically stable, but not static entities, i.e., solvent particles continuously diffuse in and out of the "sphere of influence" of the strongly attractive solute, without affecting significantly the enhanced solvent density in the cluster or its structure.

III. PHASE EQUILIBRIA SIMULATIONS IN SUPERCRITICAL EXTRACTION

These simulations are performed primarily to calculate the equilibrium solubilities, although in many cases fluid structure is also studied simultaneously. The determination of phase equilibria involves the calculation of free energies and chemical potentials, which are nonmechanical properties and are hence not routinely accessible from conventional MC or

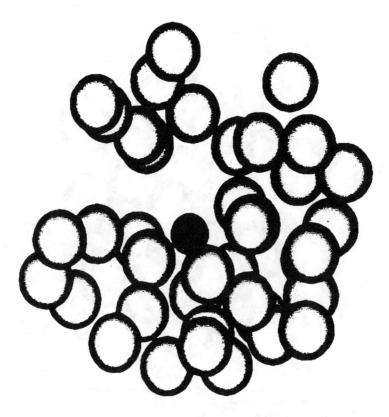

FIGURE 6. Solvent molecules (white spheres) surrounding an infinitely dilute repulsive solute (black sphere) showing lower local density.[5] The macroscopic or bulk density is the same as in Figure 5.

MD techniques. Because of the importance of phase equilibria in chemical processes, active research in recent years resulted in the development of several specialized techniques for phase equilibria calculations. These include the "direct" method of Gibbs' ensemble MC technique,[6] the "indirect" methods of grand canonical MC,[7] the test particle method,[8] and the thermodynamic integration method.[9] In the "direct" method, phases in equilibrium are directly simulated, but the equilibrium chemical potentials are usually not calculated. In the "indirect" methods, the chemical potentials for various states are calculated and the equilibrium phases determined by equating the chemical potentials. In the following sections, these methods and their application to supercritical extraction are briefly described.

A. THE GIBBS' ENSEMBLE MONTE CARLO

This simulation method was first suggested by Panagiotopoulos[10] and has several variations. In the most useful form of this technique, each of the equilibrium phases is simulated in a central cell surrounded by replicas of itself. No interface is simulated. Molecules are displaced in the usual MC manner with trial random displacements accepted with the probability

$$P_{disp} = \min\ [1, \exp(-\Delta E/kT)] \tag{1}$$

where ΔE is the configurational energy change due to the trial displacement. The volume of each phase is varied independently to ensure that the pressure of each phase simulated is equal to the prescribed pressure. Each trial volume change is accepted with probability

$$P_{vol} = \min \left\{ 1, \exp \left[-\left[\Delta E^I - N^I kT \ln \left(\frac{V^I + \Delta V^I}{V^I} \right) + P \Delta V^I \right] /kT \right] \right\} \qquad (2)$$

where ΔE^I is the configurational energy change in phase I due to volume change ΔV^I and V^I is the volume of phase I before change. Composition of the coexisting phases can be varied by interchanging molecules of different species between the phases or by moving molecules from one phase to the other according to the statistical probability distributions derived based on the equality of the chemical potentials. For example, when a particle of species i is transferred from phase II to phase I, this trial transfer is accepted with probability

$$P_{transfer} = \min \left\{ 1, \exp \left[-\left(\Delta E^I + \Delta E^{II} + kT \ln \frac{V^{II}(N_i^I + 1)}{V^I N_i^{II}} \right) /kT \right] \right\} \qquad (3)$$

where N_i^I and N_i^{II} are, respectively, the number of particles of species i in phases I and II before the trial move. Total number of molecules in all phases is kept constant.

Panagiotopoulos[6] used this technique to study the ternary CO_2/acetone/water system at room temperature by simulating the equilibrium between the CO_2-rich supercritical phase and the H_2O-rich liquid phase. The Lennard-Jones potential model was used for all interactions in the ternary system. In a ternary system, three like pair potentials and three unlike pair potentials are needed. The pure component parameters (for the like pair potentials) were obtained from either critical properties (for CO_2) or room temperature vapor-liquid equilibrium (for acetone) or both (for H_2O). The unlike pair interaction parameters were based on the Lorenz-Berthelot rules with the energy parameter given by $\epsilon_{ij} \equiv \delta_{ij}(\epsilon_{ii}\epsilon_{jj})^{1/2}$. δ_{ij} was found by fitting to experimental data for the binary ij. For the CO_2/acetone binary, good agreement was obtained with $\delta_{ij} \equiv 1$. Several simulations with artificial values of δ_{ij} of 0.9, 0.8, and 0.7 were performed to study the effect of varying this unlike binary interaction parameter on the binary phase equilibria. It was found that for δ_{ij} of 0.8 and 0.7 (weak unlike interactions) the immiscibility persists to high pressures. The solubility of the less volatile acetone in the supercritical CO_2 increased with δ_{ij} as expected, since increased ϵ_{ij} results in greater attractive interactions. The observed solubility increase is large (factor of 5) for the small increase in δ_{ij} (from 0.7 to 0.8). Due to the inadequacy of the LJ model for the strongly polar and hydrogen-bonded H_2O and acetone, the determination of δ_{ij} in the H_2O/CO_2 and the H_2O/acetone binaries presented some problems. The most severe problem was the large simulated density at the experimental temperature. For the CO_2/H_2O binary, δ_{ij} was therefore determined by simulating at an elevated temperature. For the acetone/H_2O binary, δ_{ij} was left as an arbitrary parameter.

Ternary simulations at a single state point were reported in Reference 6. Two simulations were reported for acetone/water with δ_{ij} of 1.0 and 0.9. The results are shown in Figure 7. It can be seen that the coexistence compositions (shown by tie lines) are very sensitive to δ_{ij}. Furthermore, it is not possible to obtain quantitative agreement by manipulation of δ_{ij} alone. This underscores the difficulty of performing simulations for real systems where adequate potential models are not available. It is also anticipated that temperature-dependent LJ parameters would be necessary to represent data at different temperatures. This, however, is a deficiency of our understanding of intermolecular forces rather than a deficiency of the simulation technique.

B. TEST PARTICLE METHODS AND THERMODYNAMIC INTEGRATION

The techniques described in this and the following sections are the so-called indirect techniques. In these techniques the chemical potentials of particular fluid mixtures are evaluated at various temperatures, pressures, and compositions. The conditions of equilibria, namely, the equality of temperature, pressure, and chemical potentials, are then imposed to

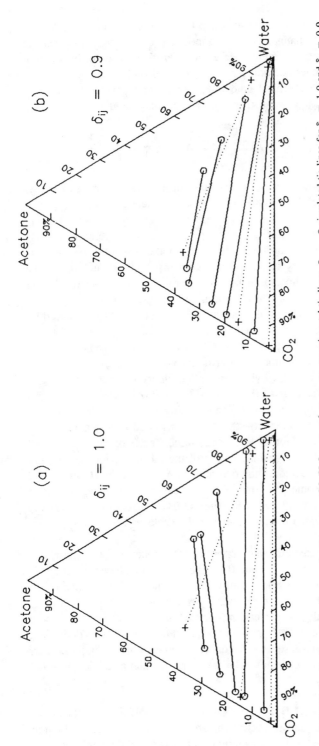

FIGURE 7. Ternary phase diagrams for the acetone/CO_2/H_2O system.[6] $\times$ ...... $\times$ experimental tie lines, $\circ$—$\circ$ simulated tie lines for $\delta_{ij} = 1.0$ and $\delta_{ij} = 0.9$.

determine the equilibrium phases. Compared to the direct method in Section III.A, this method is usually an order of magnitude more time consuming because a series of simulations is needed to locate the equilibrium phases. The advantages are that numerical values for the chemical potentials as well as the variation of chemical potential with other state variables, particularly with composition, are also available from the simulations. Such information can be used in the development of excess Gibbs' free energy-based solution theories. This method can in principle be applied to any mixture regardless of the extent of asymmetry, whereas the direct method described in Section III.A cannot be used for highly asymmetrical mixtures due to the failure of molecular exchanges. In the case where the chemical potentials in one of the equilibrium phases can be evaluated simply without resorting to simulations (such as the case of solid phases), then only the fluid phase needs to be simulated. The direct method, on the other hand, cannot be used when solid phases are present.

Shing and Chung[8] used the test particle and thermodynamic integration method to calculate the solubilities of a solid solute (naphthalene) in supercritical CO_2. At equilibrium, solid naphthalene coexists with a SCF containing naphthalene dissolved in CO_2. The solid naphthalene chemical potential can be evaluated using the vapor pressure and Poynting correction. To reduce the number of simulations required, the naphthalene chemical potential in the fluid phase was assumed to obey Henry's law since the solubilities were not too high for most pressures. With this assumption, only the infinite-dilution chemical potentials were evaluated. At low to moderate pressures where the densities were not too high, the isothermal-isobaric (NPT) ensemble test particle method was used. In this method, the residual chemical potential of the solute μ_1^r is evaluated using

$$\mu_1^r = - kT \ln \left[\frac{\langle V \exp (- \beta \psi_1) \rangle}{\langle V \rangle} \right] \tag{4}$$

where V is the system volume, ψ is the potential energy experienced by an invisible solute particle placed randomly in V, $\beta = 1/kT$, k = Boltzmann constant, T = temperature, and $\langle \ \rangle$ is an NPT ensemble average. NPT ensemble was chosen instead of the canonical (NVT) ensemble because the latter tends to suppress density fluctuations that are important in near critical systems.

For high pressures (and densities), the test particle method is unreliable due to difficulties in adequately sampling significant values of ψ_1. A thermodynamic integration method based on Kirkwood's charging method[8,9] was adopted. In this method, one of the solvent molecules is chosen and its interactions with the rest of the particles in the system varied by using a charging parameter ξ such that when $\xi = 0$, this particle acts as a solvent molecule, while when $\xi = 1.0$, it acts as solute molecule. The difference between the solute and solvent residual chemical potential is given by

$$\mu_1^r - \mu_2^r = \int_0^1 \langle \Delta\psi \rangle_\xi d\xi \tag{5}$$

where $\Delta\psi = \psi_1 - \psi_2$ is the potential energy difference experienced by the specially charged particle between $\xi = 1$ and $\xi = 0$. $\langle \ \rangle_\epsilon$ is a NPT ensemble average when the special particle is charged to the extent ξ.

The integral in Equation 5 can be obtained by performing several simulations at various values of ξ and numerically integrating the ensemble averages. The solvent residual chemical potential μ_2^r is found by using the test particle method with $\xi = 0$. μ_1^r is obtained by adding μ_2^r to the integral in Equation 5.

Simulations for the CO_2/naphthalene system was reported along two supercritical isotherms at 320 and 332 K. All pair interactions were of the quadrupolar LJ type. Experimental

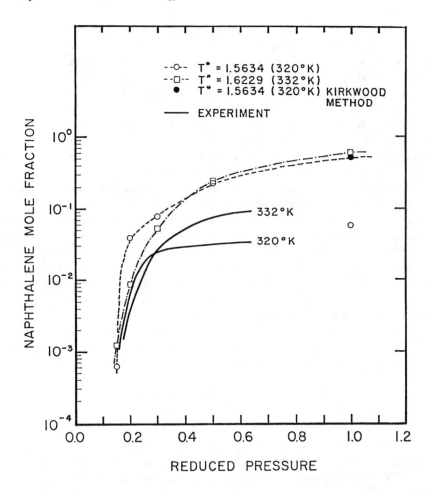

FIGURE 8.   Solubility of naphthalene in supercritical CO_2 as function of reduced pressure.[8]
$P^* \equiv P\sigma^3/\epsilon$ where ϵ,σ are energy and size parameters for the LJ part of the CO_2 pair potential.
$T^* \equiv kT/\epsilon$. Points are MC simulation results.

quadrupoles were used, with approximate LJ parameters obtained from previous perturbation theory studies. The simulated results are shown in Figure 8. The test particle method was found to fail at high pressures (reduced pressure = 1.0, corresponding to reduced density of about 0.7). Simulation results reproduced the qualitative behavior, namely the rapid initial rise in the solubility and a subsequent leveling off at high pressures, as well as the well-known crossover behavior where the solubility increased with temperature above a certain pressure while decreasing with temperature below this pressure. The quantitative representation was not adequate, the temperature dependence of the solubility was particularly weak. The authors attributed this to the fact that the potential models and parameters were not optimized. Chung[11] later optimized the CO_2 LJ parameters by fitting simulation results to the experimental compressibility factor at 320 K. The solute-solvent LJ parameters were fitted to solubility data. The experimental quadrupole moments were retained. With the optimized parameters, the solubility representation was improved (Figure 9), although the higher temperature high pressure solubilities were still too low. This may be attributed to the assumption of Henry's law which is not expected to be valid at high temperature and high pressure where solubility may exceed 10%. To further optimize the potential models without assuming Henry's law, many more simulations will have to be performed at various concentrations with the introduction of additional parameters (probably in the form of com-

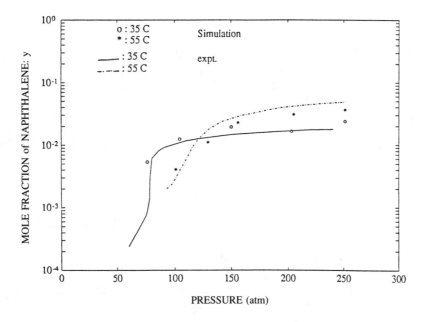

FIGURE 9. As for Figure 8, but using optimized pair potentials.[11]

bining rules). While this will undoubtedly lead to better agreement between simulation and experiments, one would not expect to gain much more additional insight in the process; therefore, the additional computing expense is probably not justified.

C. THE GRAND CANONICAL ENSEMBLE MONTE CARLO (GCEMC) METHOD

In the GCEMC method, the volume, temperature, and chemical potential of all components are specified. The molecules are randomly moved in the usual way. In addition, the concentrations and densities constantly fluctuate when molecules of every component present are continuously added to and removed from the system according to the grand canonical probability distribution. For example, when a molecule of species i is added, this trial addition is accepted with probability

$$P_{add} = \min \left\{ 1, \frac{1}{N_i} \exp \left[\frac{\mu_i^r}{kT} + \ln \langle N_i \rangle - \left(\frac{\Delta E}{kT} \right) \right] \right\} \tag{6}$$

where N_i is the number of molecules of species i before trial addition, μ_i^r is the residual chemical potential of i, and $\langle N_i \rangle$ is the average number of molecules of species i in the system. This method shares many similarities with the Gibbs' ensemble method described in Section II.A, however, only one phase is simulated at a time. Therefore, the simulated phase may not be in equilibrium with any other phase except in the case where the specified chemical potentials correspond to those for a phase such as a solid for which the chemical potentials are known. Thus, the GCEMC is also an "indirect" method and locating fluid phases in equilibrium requires several simulations.

Nouacer and Shing[7] used this technique to study the solubility of naphthalene in supercritical CO_2. The simulations were performed along a slightly supercritical isotherm. Very large and correlated composition and density fluctuations were observed in the vicinity of the critical density (Figure 10). The particularly large fluctuations of the solute concentration relative to the mean value meant that in order to obtain reliable statistics these simulations were quite long. In addition, it implied that simulations in ensembles where

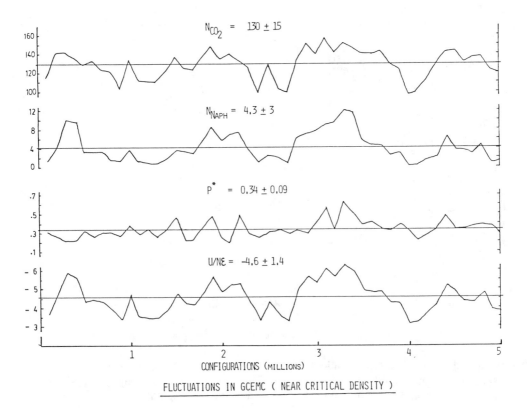

FIGURE 10.   Large and correlated fluctuations in pressure, energy, density, and composition in GCEMC simulations near the vapor-liquid critical point.[7]

either the concentration fluctuations are suppressed (as in NPT ensemble) or both concentration and density fluctuations are suppressed (as in the canonical or microcanonical ensembles) may lead to erroneous results.

These authors also reported the RDFs g_{ij}. Since $\rho_i g_{ij}$ gives the local density of component i in the neighborhood of a molecule of component j, g_{ij} provides information for the local fluid structure. (Here, ρ_i is the bulk density of component i.) Figure 11 shows that there is preferential aggregation of CO_2 about naphthalene since the first peak in g_{12}, the naphthalene-CO_2 RDF, is higher than that in g_{22}, the CO_2-CO_2 RDF. Furthermore, the larger first peak in g_{12} at $\rho^* = 0.15$ compared to that at $\rho^* = 0.42$ implies that the enhancement of local density due to the presence of the attractive naphthalene solute is more pronounced when the fluid is more compressible. Such a trend was also observed in Shing and Chung's[4] simulation study of infinitely dilute near-critical LJ mixtures (Figures 3 and 4). This decreased tendency to aggregate at higher densities is a manifestation of the competition between energetic and entropic effects. At lower densities, energetic effects dominate, aggregation enhances the strong solute-solvent interactions, and it is therefore favored. At higher densities, the entropic effects also become important. Aggregation leads only to a limited increase in attractive interactions (since the molecules are already close together) as a result, the incentive to aggregate is decreased.

Failure of the GCEMC technique was reported for fluid densities in excess of 1.5 times the CO_2 critical density due to unsuccessful attempts to add the large naphthalene molecules. This failure resulted in artificially low calculated solubilities. Calculated pressure and solubilities were reported to be particularly sensitive to system size when the total number of molecules was less than about 150 for this system. This was attributed to the fact that the

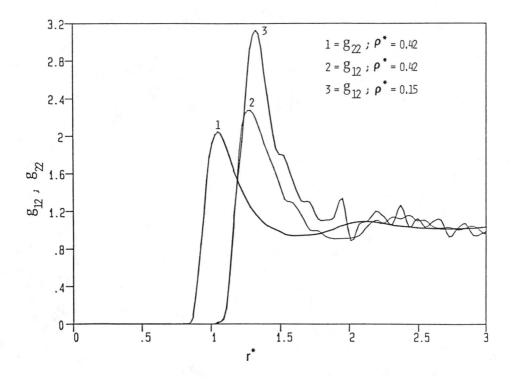

FIGURE 11. Typical RDFs in a CO_2/naphthalene system for a slightly supercritical temperature.[7] g_{22} is the CO_2-CO_2 RDF. g_{12} is the naphthalene/CO_2 RDF. $\rho^* \equiv \rho\sigma^3$, ρ is bulk density, σ is the size parameter for the LJ part of the CO_2 pair potential, $r^* \equiv r/\sigma$, r is the distance between pairs of molecules.

solubility is exponentially dependent on the density and the fact that in small systems, density fluctuations are artificially suppressed.

Nouacer and Shing also studied the effect of the introduction of a few molecules of water (an "entrainer"). It was found that the addition of water dramatically increased the solubility if the fluid density prior to water addition was below the solvent critical density and the fluid was highly compressible. The highly polar water molecules caused aggregation of CO_2 as well as the naphthalene molecules around it, increasing the fluid density in its vicinity, which in turn further enhanced solubility because of the density effect. The aggregation of naphthalene molecules around the added water can be clearly seen in the radial distribution functions g_{ij} (Figure 12). The first peak in g_{13}, the naphthalene-water RDF, is very large compared to those for either g_{12} or g_{22} (Figure 11), indicating that there is a very high degree of preferential aggregation of naphthalene molecules about the H_2O molecules.

An interesting related work is that of van Megan and Snook,[12] who used the GCEMC method to study the adsorption of slightly supercritical ethane on a graphite surface. The graphite surface can be considered to act like an infinitely dilute, large and strongly attractive solute. Therefore, many of the observations in that study are similar to and consistent with those reported for supercritical extraction systems. For example, van Megan and Snook observed that as the pressure (or bulk density) increased, the adsorption (as measured by Γ, the adsorption excess) passed through a maximum at about the ethane critical pressure (Figure 13). This is consistent with the observation described in Section III.C that the tendency of solvent to aggregate about an attractive solute is maximum at about the critical density. The structure of the adsorbed layer reported by these authors also exhibited the large first peaks similar to those shown in Figure 11 for solute-solvent radial distribution functions in supercritical extraction systems.

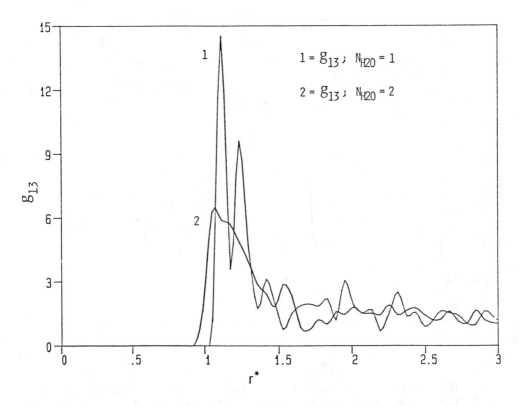

FIGURE 12.   Distribution of solute naphthalene molecules about the entrainer water molecules in the naphthalene/ CO_2/H_2O system.[7] g_{13} is the naphthalene-H_2O RDF. The very large first peaks (compared to those in Figure 11) implies that there is a high degree of preferential aggregation of naphthalene about H_2O molecules.

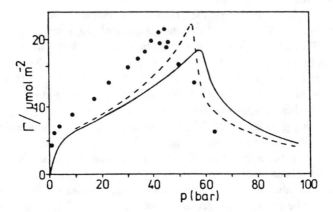

FIGURE 13.   Experimental and computed adsorption excesses Γ for ethane adsorbed on graphite at various pressures.[12] Points are simulated Γ at $T/T_c \simeq$ 1.03, —— experimental Γ at $T/T_c = 1.04$, ----- experimental Γ at $T/T_c = 1.02$.

IV. CONCLUSION

Most of the simulations relevant to supercritical extraction systems reported in the literature were performed to either demonstrate new simulation techniques as described in Sections II.A and III.B or to gain physical insight into the molecular origin of observed phenomena peculiar to these near-critical, highly nonideal mixtures. One of the most informative ways to study these systems is to monitor the structure (particularly the local structure in the vicinity of the dilute solute molecules) while systematically varying state variables such as temperature, pressure, and composition, as well as varying molecular parameters which govern the type and strength of intermolecular forces.

These simulation studies confirmed the large excess enthalpies and excess volumes observed experimentally. Simulations near the critical point also exhibit large and correlated fluctuations in density, solubility, and internal energies. Large enhancements in local density in the vicinity of strongly attractive solutes and entrainer molecules, and the reduction in local density in the vicinity of repulsive solutes were clearly demonstrated. Monitoring of the dynamic behavior of solvent aggregates around attractive solutes indicated that although there was always density enhancement or clustering around the solute, the aggregates were not static objects. Rather, they were shown to be dynamic entities with solvent molecules continuously entering or leaving the clusters. While these generic qualitative behavior were adequately demonstrated using simple potentials, quantitative representation required the use of more realistic and complex potential functions with parameters fitted to experimental solubilities.

Simulation of solubility or phase equilibria requires specialized techniques. The direct method of Gibbs ensemble Monte Carlo simulates the coexisting phases directly by incorporating the phase equilibrium constraints into the probability distribution. This method is efficient and straightforward and is particularly useful for multicomponent simulations. There are a few limitations in the application of this method. Systems containing solid phases or highly asymmetric mixtures cannot be studied using this method. Also, numerical values of the chemical potentials are not obtained routinely. The indirect methods simulate each coexisting phase independently. The phases in equilibrium are located iteratively by applying the phase equilibrium constraints to the ensemble-averaged thermodynamic properties. This is more time consuming than the direct method, but is partially compensated for by the fact that the chemical potentials are always calculated as part of the simulation, so that as a by-product, one obtains information regarding the composition dependence of the chemical potentials. Solid phases and highly asymmetric mixtures can be accommodated. For highly asymmetric mixtures, thermodynamic integration involving a series of simulations may be necessary.

Supercritical extraction systems normally involve highly asymmetric complex molecules near critical points. Therefore, simulations of these systems are normally lengthy and require relatively large system size. Quantitatively accurate results consistent with experimental data also require sophisticated and complex intermolecular potential models. In principle, if the potential models are known *a priori,* simulations can yield predictive results. In reality, since accurate potentials for these complex molecules are often not available, such predictive calculations are rarely possible. This is the most severe limitation of the use of molecular simulation in engineering applications. Until this situation improves significantly, it is likely that molecular simulations will continue to be used primarily in studies of model systems designed to gain physical insight and to discern qualitative trends.

REFERENCES

1. **Allen, M. P. and Tildesley, D. J.,** *Computer Simulation of Liquids,* Clarendon Press, Oxford, 1987.
2. **Gubbins, K. E.,** in *Advances in Chemical Engineering,* Academic Press.
3. **Vogelsang, R. and Hoheisel, C.,** *Mol. Phys.,* 53, 1355, 1984.
4. **Shing, K. S. and Chung, S. T.,** *AIChE J.,* 35, 1973, 1988.
5. **Petsche, I. B. and Debenedetti, P. G.,** *J. Chem. Phys.,* 91, 7075, 1989.

6. **Panagiotopoulos, A. Z.,** *Supercritical Fluid Science and Technology,* ACS Symp. Ser., Johnston, K. P. and Penninger, J., Eds., American Chemical Society, Washington, D.C., 1989.

7. **Nouacer, M. and Shing, K. S.,** *Mol. Sim.,* 2, 55, 1989.

8. **Shing, K. S. and Chung, S. T.,** *J. Phys. Chem.,* 91, 1674, 1987.

9. **Shing, K. S., Gubbins, K. E., and Lucas, K.,** *Mol. Phys.,* 1235, 1989.

10. **Panagiotopoulos, A. Z.,** *Mol. Phys.,* 61, 813, 1987.

11. **Chung, S. T.,** Ph.D. thesis, University of Southern California, Los Angeles, 1988.

12. **van Megan, W. and Snook, I. K.,** *Mol. Phys.,* 45, 629, 1982.

Chapter 6

TRANSPORT PROPERTIES OF SUPERCRITICAL FLUIDS AND FLUID MIXTURES

V. Vesovic and W. A. Wakeham

TABLE OF CONTENTS

I. INTRODUCTION

The increased interest in the use of supercritical fluids (SCFs) technology brings with it renewed requirements for the thermophysical properties of such fluids and fluid mixtures for the determination of the feasibility and subsequent design of chemical process plant. The equilibrium properties of fluids generally determine the feasibility of a given process, whereas the transport properties of the same fluids dictate the sizing of process plant at the design stage. This chapter is concerned with the state of knowledge regarding the transport properties of fluids under supercritical conditions. Although a large amount of experimental data of these properties has been compiled, the diversity of the fluids and of the conditions of interest means that all of the information likely to be required in practice can never be generated by experimental means alone.

Thus, all the other mechanisms for the provision of the necessary information must be involved to supplement experiment if present and future requirements are to be met. These additional mechanisms may include theoretical calculations from first principles as well as sound predictive methods based either on theory or on a limited set of reliable experimental data or on a combination of the two.

The aim of this chapter is to outline the basic principles of the measurement and theory of the transport properties of SCFs outside the critical region. The behavior of transport properties near the critical point is examined elsewhere in this book. In addition, a critical review of some methods for the prediction of these properties is presented, together with a summary of reliable correlations. No attempt has been made to provide a comprehensive coverage of all possible prediction methods or to review and present all the available correlations. Rather, the chapter is intended as a guide to the background to the topic and its current state with a concentration upon those techniques of measurement, prediction, and correlation that have a firm theoretical foundation. As a consequence, if the transport property data for specific SCFs or fluid mixtures are required, the reader is advised to consult a specialized database.[1]

The three major transport properties of interest in SCF technology are the coefficient of the shear viscosity, η, the diffusion coefficient, D, and the thermal conductivity coefficient, λ. These transport properties of the fluid characterize the nonequilibrium process of momentum, mass, or energy transport, respectively, which is manifest when the fluid is perturbed from equilibrium by the application of a velocity, composition, or temperature gradient.[2,3] From the phenomenological point of view the three transport property coefficients of interest here constitute the proportionality factors when the fluxes of momentum, mass, and heat, respectively, are related to the velocity, composition, and temperature gradient in a linear manner. The chapter is not concerned with nonlinear effects such as are revealed by non-Newtonian fluids or indeed in the secondary transport coefficients, such as the thermal diffusion factor, which are of small practical significance in supercritical technology. From the microscopic point of view, the transport coefficients arise from departures of the molecular velocity distribution from that characteristic of equilibrium. The perturbed system relaxes toward equilibrium upon removal of the perturbation by means of the motion of, and collisions among, its constituent molecules.

The measurement of transport properties poses a fundamental problem of principle.[4] In order to obtain a value of the transport property at a specific temperature and density the state of the fluid should be maintained as near equilibrium as possible. It follows that under such a minimal disturbance the effect to be measured will be inherently small and usually near the sensitivity limit of an instrument. Conversely, if a large, easily measurable effect is aimed for, the fluid has to be removed rather far from the equilibrium state and the measured transport property then refers to some average temperature and density which is not always easily identified. The most accurate measurements of the transport properties are therefore performed in instruments in which the departure from equilibrium is small and where thermodynamic stability is maintained in the fluid. At the same time the highest resolution in the detection of the major effects, as well as a rigorous mathematical model of the process involved, must be secured.

The rigorous theoretical calculation of transport properties[3] also poses severe problems, albeit of a different kind. The transport properties are a result of the motion of molecules and as such can only be understood in terms of molecular theory. The kinetic theory of gases[2] seeks to explain the observed transport property coefficients in terms of molecular encounters and the forces between molecules. The theory is most secure for the description of phenomena in a dilute, monatomic gas, where only binary interaction among molecules occur. Any increase in the complexity of the fluid studied either by virtue of more complex molecules or because of an increase in density, leads to a less secure and less rigorous kinetic theory description. Furthermore, the starting point for any calculation of transport properties from first principles must be the intermolecular potential, which is available only for a limited number of systems. For these reasons, practical applications of the rigorous kinetic theory are usually limited to the dilute gas limit. In other regions of the thermodynamic surface a more heuristic approach is usually taken, and qualitative results of the kinetic theory are supplemented by empirical observation.

II. VISCOSITY

The dynamic viscosity of an isotropic Newtonian fluid is defined by the linear phenomenological relationship

$$\tau = - \eta \frac{\partial u}{\partial z} \tag{1}$$

between the applied velocity gradient $\partial u/\partial z$, and the resulting shear stress τ. The dynamic viscosity defined by this relationship is a function only of the thermodynamic state of the system so that for a particular fluid $\eta = \eta(T, P)$ or $\eta = \eta(T, \rho)$, in which T is the temperature of the system, P its pressure, and ρ its density. In practice it is often easier to work with pressure as the independent variable; however, from the theoretical viewpoint, density is the more significant variable so that it is used consistently here. The relationship between P, T, and ρ is, of course, contained in the equation of state (EOS) for the fluid, discussed elsewhere in this book.

In general, the viscosity of an SCF increases with increasing temperature and density, whereas for a liquid the effect of a temperature increase at constant pressure leads to a lower viscosity. Table 1 presents viscosity data for some fluids encountered in supercritical applications for a variety of conditions.[5-9]

A. THE MEASUREMENT OF THE VISCOSITY COEFFICIENT

Measurements of the viscosity of fluids have been performed over a long period with a steadily improving level of accuracy.[4] It is, however, remarkable that given the apparent simplicity of the problem, it was not until 1952 that the viscosity of liquid water at 20°C was determined[10] with a precision of ±0.1%, and not until 1970 that the viscosity of the monatomic gases at a pressure of 0.1 MPa was determined over a range of temperature with comparable precision.[11] It is clear from these facts that the apparent simplicity belies the real problem of measurement, as indicated earlier. Here, attention is focused only upon those experimental techniques of the highest accuracy. While the accuracy attained is not directly necessary for technological applications it will be seen that the use of such accurate data in conjunction with theory leads to superior predictive capabilities.

TABLE 1

**Viscosity Values of Some Common Fluids Used in Supercritical Fluid
Technology**

	$\eta/\mu Pas$					
	T = T$_c$ + 20 K		**T = T$_c$ + 100 K**			
	P = 0.1 MPa	**P = 10 MPa**	**P = 0.1 MPa**	**P = 10 MPa**	**T$_c$**	**Ref.**
CO_2	16.2	27.5	19.9	22.3	304.1	5
N_2	9.71	30.8	14.3	17.8	126.0	6
H_2O	24.2	24.2	27.5	27.8	647.1	7
C_3H_8	10.6	46.6	12.7	20.8	369.8	8
NH_3	15.0	16.4	18.0	18.7	405.4	9

Note: T$_c$ is the critical temperature of the fluid.

1. Capillary Viscometer

This is the earliest type of viscometer used for measurements of viscosity coefficients.[4]
The method is based on the flow of an incompressible fluid through a tube of a uniform,
circular cross section. If the conditions are such that the flow is laminar in the tube, then
the simplest possible analysis leads to a relationship between the pressure drop, ΔP, along
a length L of the tube given by the Hagen-Poiseuille equation[4]

$$\frac{\Delta P}{\dot{m}} = \frac{8L}{\pi \rho R^4} \eta \tag{2}$$

where L is the length of the tube, over which ΔP is measured, $\dot{m}$ is the mass flow rate, ρ
is the fluid density, and R is the radius of the tube.

In most practical instruments the pressure difference is measured not within the tube
but between the reservoirs at the ends of a capillary tube, so as not to disturb the velocity
profile in the tube by the introduction of pressure measuring devices. This practical modification requires a more rigorous analysis of the flow in the system which has been summarized by Kestin et al.[12] The resulting working equation of the modern capillary viscometer
is

$$\frac{\Delta P}{\dot{m}} = \frac{8(L + n_0 R)}{\pi R^4 \rho} \eta + \frac{m_0}{\pi^2 R^4 \rho} \dot{m} \tag{3}$$

where n_0 and m_0 are characteristics of the viscometer.

Theoretical estimates of n_0 and m_0 have been made for a variety of viscometer configurations, and when these are combined with a number of other small corrections (detailed
in Reference 12) and precise measurements of the capillary dimensions, accurate absolute
measurements of the viscosity of the fluid are possible. More commonly, however, n_0 and
m_0 are determined by calibration of the viscometer with a fluid of known viscosity so that
relative measurements can be performed.

In fact, the viscosity can also be derived from measurements of $\Delta P/\dot{m}$ against $\dot{m}$ which
takes the form shown in Figure 1. The viscosity can be determined from the intercept on
the $\Delta P/\dot{m}$ axis without a knowledge of m_0, although the latter can be estimated from the
slope of the same plot.

Figure 2 illustrates the capillary viscometer used by Nagashima and collaborators[13,14]
for measuring the viscosity of fluids at high temperature and high pressure. It is therefore
a suitable design for investigating the viscosity of SCFs in the range of technological interest.
The test fluid was circulated at a steady rate through the capillary tube (1) by means of a

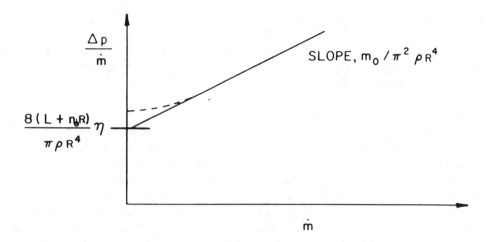

FIGURE 1. The experimental determination of viscosity from a plot of $\Delta P/\dot{m}$ against $\dot{m}$.

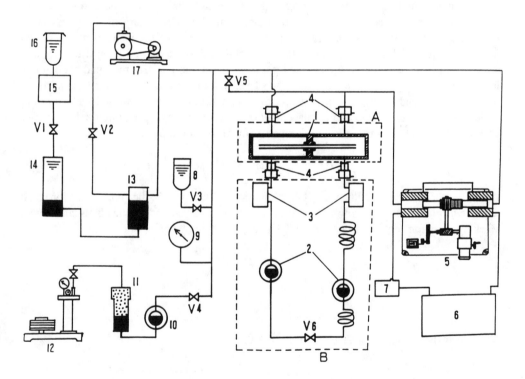

FIGURE 2. A capillary viscometer used for measurements at high temperature and high pressure.[14]

dual piston pump (17), driven by a precisely controlled motor so that the total volume of the fluid system is constant. The total pressure of the system is also maintained constant but with a separate pump. The pump (17) was calibrated very precisely so that its discharge at a particular driving frequency was known. The circulating pump operated at room temperature and, in the application of Nagashima et al.,[13,14] pumped the liquid phase of the fluid. For measurements of the viscosity in the supercritical region the fluid was preheated above its critical temperature before access to the capillary, maintained at an elevated temperature by means of a steam preheater, and then cooled again after flow through the capillary. The pressure difference between the chambers at either end of the capillary was measured with a high precision manometer (2), operated in a null-mode. In applications at high temperatures (up to 900°C) the capillary employed by Nagashima et al.,[13,14] was made of a platinum alloy with a length of 300 mm and an inner diameter of 0.3 mm. It was surrounded by the high pressure fluid (up to 100 MPa) inside and out within an autoclave to avoid distortion. The viscometer was operated in a relative mode and the accuracy obtained was estimated to be ±0.7% over most of the range of conditions.

2. Oscillating Disc Viscometer

The oscillating disc viscometer belongs to a large family of viscometers based on the torsional oscillations of bodies of revolution.[4] These instruments are capable of very high accuracy since a rigorous theory is available for them and because the essential measurements can be reduced to those of length and time.[4] The principle of the measurement in the most popular instrument is that of observing the decay of the amplitude of the damped, torsional oscillations of a disc suspended by a thin strand and immersed in the fluid. The changes in the oscillatory frequency, ω, and the decrement of the oscillation, Δ, compared with those in the vacuum (ω_0 and Δ_0, respectively) depend upon the density and viscosity of the surrounding fluid.

The theory and design of oscillating-body viscometers has recently been reviewed;[4] it is therefore not necessary here to provide more than a brief summary of the type that is most suitable for application in SCFs. The essential elements of the instrument are shown in Figure 3 in which a disc of radius R and thickness d is suspended from a thin torsion strand from its center and an upper fixed point so that it can perform torsional oscillations in its own plane. The disc lies between two fixed plates separated from the disc surfaces by distances b_1 and b_2 (usually $b_1 = b_2 = b$). The entire instrument is filled with the fluid under test. Following an angular displacement of the disc, its angular amplitude decays after an initial transient, according to the damped, simple harmonic equation

$$\theta(t) = \theta_0 \sin \omega t \exp(-\Delta\omega t) \tag{4}$$

A solution of the linearized Navier-Stokes equations of the flow in the fluid, valid for small angular displacements and low frequencies, has been obtained and this leads to complete a working equation for the viscometer.[15,16] Under conditions when

$$2b + d << R \tag{5}$$

and

$$2b + d < \delta \tag{6}$$

where δ is the boundary layer thickness

$$\delta^2 = \eta/\rho \, \omega_0 \tag{7}$$

it is possible to make absolute measurements of viscosity with the instrument. The working equation usually employed for viscosity measurements is then the imaginary part of the complex equation derived by Newell,[16] which reads

$$C_N = \left[\frac{2I}{\pi\rho bR^4} \left(\frac{\Delta}{\omega^*} - \Delta_0 \right) + e_1 \frac{\Delta}{\omega^*} \right] \beta^2 + e_2 \frac{3\Delta^2 - 1.0}{\omega^*} \beta^4 + e_3 \frac{\Delta(\Delta^2 - 1.0)}{\omega^*} \beta^6 \tag{8}$$

where for the usual case of $b_1 = b_2 = b$,

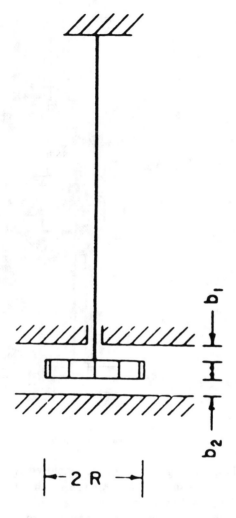

FIGURE 3. Oscillating disc viscometer.

$$\beta = b/\delta$$
$$\omega^* = \omega_0/\omega \tag{9}$$

I is the moment of inertia of the suspension system, and e_1, e_2, and e_3 are $^2/_3$, $^1/_{45}$, $^8/_{945}$, respectively. C_N is an instrument constant determined entirely from the dimensions of the viscometer or from a calibration with fluids of known viscosity.

The working equation evidently requires a value of the fluid density, ρ, before the viscosity can be derived from measurements of Δ, ω, Δ_0, and ω_0, although the result is not especially sensitive to the value of density in this case. In fact, new work has shown that both density and viscosity can be obtained in the same instrument under favorable circumstances.[17]

For SCFs it is not always possible to satisfy the conditions (Equations 5 and 6) required to perform absolute measurements. Nevertheless, relative measurements can still be performed using an analysis due originally to Kestin and Wang[18] and refined by Kestin and Shankland.[19] Under this scheme the working equation takes a form that requires a calibration over a range of boundary layer thicknesses on the disc. Such a calibration can be performed with the aid of a fluid(s) whose viscosity is known over a range of densities. When operated in an absolute or relative manner, the accuracy of the results attained with this type of instrument are of the order of $\pm 0.2\%$ or better.

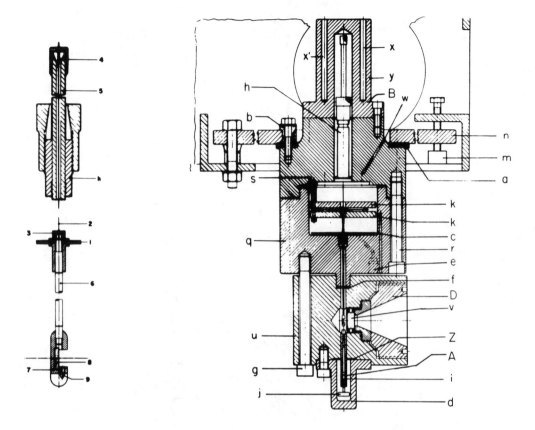

FIGURE 4.   An oscillating disc viscometer used for measurements at high pressures and ambient temperatures.[4]

Figure 4 illustrates the oscillating-disc viscometer employed by Kestin and collaborators[20] for measurements in gases at pressures up to 15 MPa near ambient temperature. Very similar viscometers have been employed by Iwasaki and Takahashi[21] for measurements at higher temperatures in the same pressure range and by Vogel and Strehlow[22] for measurements at low pressures and high temperatures.

The oscillating-disc assembly comprises a platinum/tungsten alloy suspension strand (2) carrying a quartz disc (1) with a radius of 70 mm and a thickness of 0.84 mm between two fixed plates. Below the disc a thin stainless steel rod (6) carries a small mirror (3) used for detection of the torsional motion of the disc. The entire assembly is placed in a high-pressure vessel fitted with an optical window (v) aligned with the mirror. An oscillation was initiated by engaging a rotor (i) in the bottom of the mirror housing. The motion of the disc was then observed by means of the mirror which reflected the image of an illuminated centimeter scale to an observation telescope. The decaying amplitude of the motion and its period could then be determined with a sufficiently high precision to permit viscosity measurements with a precision of ±0.1%. In more modern versions of the instrument all measurements have been reduced to those of time. For example, in the arrangement described by Kestin and Khalifa,[23] a laser beam, reflected from the mirror, is detected by a number of photodetectors placed at various angular displacements from the rest position. The timing of the passage of the light beam between these detectors allows both the decrement and period of the oscillation to be determined.

3. Correlations of Experimental Viscosity

For a small number of fluids, including some of those of importance in SCF technology, a substantial body of experimental viscosity data is available over a wide range of temper-

TABLE 2
Correlations of the Viscosity of Some Fluids

	Temp. range/K	Pressure range/MPa	Claimed accuracy	Ref.
CO_2	200—1500	0—100	±1—±5%	5
H_2O	273—1173	0—300	NA	7
N_2	80—1100	0—100	NA	6
CH_4	95—500	0—75	±3%	25
C_2H_6	200—500	0—75	±5%	26
C_3H_8	140—500	0—50	±5%	8
C_2H_4	110—500	0—50	±5%	27
NH_3	196—680	0—50	±1—±4%	9

atures and pressures.[5-9,24-27] In order to permit interpolation within these data it has become common practice to seek to represent all the information by means of a single equation expressing the viscosity as a function of temperature and density (or pressure). This is an especially useful formulation for the presentation of data for computer-aided design applications. In the best of these representations special care is used in selecting the data for the development of the correlation. The criteria for selection place emphasis on data produced in instruments of high precision and for which a complete working equation and a detailed knowledge of all corrections are available. Only under exceptional circumstances, when such instruments have not yet been applied to a particular range of conditions, have data obtained in inferior instruments been included.

For some fluids of potential interest in supercritical technology, Table 2 summarizes the available viscosity correlations. Naturally, when they are available for the fluid of interest these correlations are to be preferred to any method of prediction.

B. THE THEORY OF THE VISCOSITY COEFFICIENT

The coefficient of the shear viscosity of an SCF at temperature, T, and density, ρ, can be written as the sum of three contributions

$$\eta (\rho, T) = \eta^0(T) + \Delta\eta(\rho, T) + \Delta\eta^c(\rho, T) \qquad (10)$$

where $\eta^0(T)$ is the viscosity of the fluid in the dilute gas limit, $\Delta\eta(\rho, T)$ is the excess viscosity, and $\Delta\eta^c(\rho, T)$ is the critical enhancement arising from long-range velocity correlations in the fluid near its critical point. In this chapter, only the first two terms are examined. The behavior of viscosity near the critical point is dealt with elsewhere in this book. For completeness, it is worth pointing out that for most engineering purposes the term $\Delta\eta^c(\rho, T)$ can be safely neglected. There exist well-developed, formal, theoretical descriptions of the first term. Although the theory in most cases does not allow *ab initio* evaluation of the viscosity from molecular properties, a modest number of high-quality measurements, when combined with the theory, is often sufficient to enable accurate predictions. For the second term there is as yet no adequate theory and empiricism has usually been the only recourse for the estimation of the excess property. Because each contribution to Equation 10 arises from different physical mechanisms and the theoretical description of each is at a different stage of the development, each are examined in turn.

1. The Dilute Gas Limit
a. Pure Fluid

The kinetic theory treatment of the behavior of dilute gas is based on two hypotheses:[2,3]

1. Molecules spend most of the time moving freely, interacting only through binary encounters, which are governed by the intermolecular force field between the two molecules
2. The time between collisions is so large that the velocities of the two molecules which are about to collide are uncorrelated

It is this second assumption that introduces time irreversibility into the theory and relates the reversible microscopic world to the irreversible macroscopic one. Based on these two hypotheses, the Boltzmann equation,[2,3] which describes statistically the behavior of the molecular velocity distribution of the system, can be established and solved for monatomic species. For polyatomic gases there are two additional features to be added to the description. First, the molecules may possess internal energy in rotational and vibrational modes and second, the intermolecular pair potentials are dependent on the orientation of the molecules. Taken together, these two features permit the occurrence of inelastic collisions in which internal energy can be exchanged with translational energy. Despite these complications, and the fact that a quantum-mechanical treatment is then strictly necessary, it has also been possible to solve a Boltzmann-like equation for such systems.[28,29] Thus, for all gases at low density a formal theory exists for the viscosity and its result may be written as

$$\eta^0(T) = \frac{5}{16} (\pi m k T)^{1/2} \frac{f_\eta(T)}{\Omega_\eta(T)} \tag{11}$$

where m is the mass of a molecule, k is the Boltzmann constant, and Ω_η is the viscosity collision integral. In addition, $f_\eta(T)$ is a factor near unity arising from the fact that the theoretical development takes the form of an expansion so that $f_\eta(T)$ accounts for expansion terms beyond the first. The above expression can be rewritten in a more convenient form for numerical calculation as

$$\eta^0(T) = 0.083868 \sqrt{MT} \frac{f_\eta(T)}{\Omega_\eta(T)} \tag{12}$$

where M is the molecular weight in units of grams per mole, Ω_η is in units of square nanometers, and η^0 is in units of micropascal seconds. The viscosity collision integral, Ω_η, contains all the information concerning the statistically averaged dynamics of binary collisions that occur in the gas. Since the dynamics of each collision depends on the binary intermolecular potential $V(\mathbf{r})$, the collision integral, Ω_η, is a functional of the entire potential. For polyatomic molecules the relationship between the intermolecular potential and the collision integral is complex,[30] owing to the nonspherical nature of the intermolecular forces; however, for a monatomic gas the atoms interact through a spherically symmetric potential, $V(r)$, illustrated in Figure 5. Such a potential can be characterized by the well depth, ϵ, and the distance, σ, where the attractive potential is equal to the repulsive potential. The case of a monatomic gas, while not of direct practical importance, is instructive because it has provided the basis of a predictive methodology that is widely applicable. For such a monatomic gas, the collision integral Ω_η is related to the potential function $V(r)$ through the equations[2,3]

$$\Omega_\eta = \frac{1}{6} \int_0^\infty Q_\eta(E) \exp(-E/kT) \left(\frac{E}{kT}\right)^3 d\left(\frac{E}{kT}\right) \tag{13}$$

$$Q_\eta(E) = 3\pi \int_0^\infty [1 - \cos^2 \psi(b, E)] b \, db \tag{14}$$

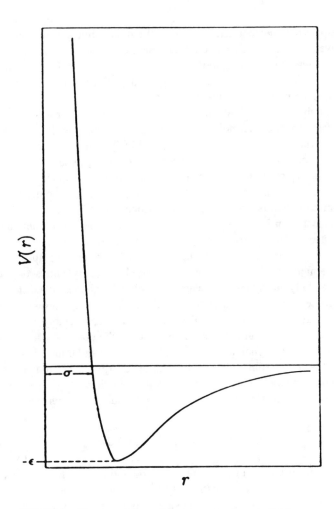

FIGURE 5. The general form of the intermolecular potential for monatomic systems.

$$\psi(b, E) = \pi - 2b \int_{r_0}^{\infty} \frac{dr/r^2}{[1 - b^2/r^2 - V(r)/E]^{1/2}} \qquad (15)$$

where E is the relative kinetic energy in a binary collision, b is the impact parameter of the collision, $\psi(b, E)$ is the deflection angle after the collision has taken place, and r_0 is the distance of the closest approach. For some very simple potentials Equations 13 to 15 can be integrated analytically;[3] however, for more realistic potentials, Equations 13 to 15 must be integrated numerically, and this is achieved both rapidly and routinely using standard computer codes.[3] Furthermore, for monatomic systems, many of the relevant pair potentials are known quite accurately,[31] so that an evaluation of the viscosity from first principles is possible.

As the complexity of the molecules increases, so does the complexity of equivalent relations between the pair potential and the viscosity. The additional features of polyatomic molecules mentioned earlier mean that the dynamics of the binary encounter are very complicated so that except for the very simplest molecules, it is impossible to treat the problem analytically or indeed numerically, without approximation.[32] Furthermore, the statistical averages over binary encounters, although conceptually simpler than the dynamics, turn out to be inordinately time consuming, since one has not only to average over translational energy and impact parameters characteristic of the monatomic system, but to also carry out summations over all possible initial and final rotational and vibrational states of the molecules. As an example, whereas the evaluation of the viscosity from the deflection function for argon (Equations 13 and 14) involves only two integrations, the equivalent classical calculation for nitrogen involves 12 integrations.[33]

Thus, although the expressions for the viscosity collision integral, Ω_η, for the polyatomic molecules in terms of the intermolecular potential are available,[30] their evaluation for any but the most trivial molecular system is prohibitively time consuming with the present generation of computers,[34] and will not become routine for some considerable time.

It follows that in order to be able to evaluate viscosity collision integrals for such a system, it is necessary to perform approximate calculations.[35,36] The most accurate classical mechanics approximation is due to Mason and Monchick,[37,38] and it renders the calculation of the viscosity collision integrals only slightly more demanding of computer time than for monatomic systems. The approximation was originally based on two reasonable physical assumptions that turn out to be equivalent to a subsequent, but approximate, quantum mechanical treatment.[32,34,36] First, it is supposed that the dynamics of the collision are determined essentially during the time when the molecules are near the distance of the closest approach. Because this time is small compared with the rotational or vibrational period of the molecules it is likely that the molecules will not appreciably change their relative configuration. Thus, it is assumed that all encounters take place at a fixed orientation. This assumption simplifies the dynamics by uncoupling internal and translational modes. Second, Monchick and Mason[37] assumed that on average the energy transferred between internal and translational modes is much smaller than the relative kinetic energy. This assumption greatly simplified the statistics of the encounter. The final result can be written in the form

$$\Omega_\eta(T) = \int \Omega_\eta(T; \underline{w}) \, d\underline{w} \tag{16}$$

where $\underline{w}$ is a vector of all angles describing the orientation and $\Omega_\eta(T; \underline{w})$ is the monatomic-like collision integral defined by Equations 13 to 15. Thus, in order to evaluate the viscosity collision integral for polyatomic molecules, it is only necessary to perform a monatomic calculation of the collision integral for the pair potential corresponding to each orientation and subsequently to average over all possible orientations. The Mason-Monchick method has proved successful for molecules with a large number of rotational states available at a given temperature. Its accuracy increases with increasing temperature. For those systems where a check against exact calculations has been possible,[34-36] it yields results within 1 to 2%.

b. Fluid Mixtures

The kinetic theory of gases can be extended to deal with the viscosity of a N-th component mixture.[2,3] For brevity, only the results for a binary mixture in the first order approximation are presented, although the results for higher order are available.[3] The viscosity of a binary mixture, η^0_{mix} is given by

$$\eta^0_{mix} = \left(\frac{x_1^2}{H_1} + \frac{x_2^2}{H_2} - \frac{2x_1 \, x_2 \, H_{12}}{H_1 \, H_2} \right) \Big/ \left(1 - \frac{H_{12}^2}{H_1 \, H_2} \right) \tag{17}$$

where H_1 is given by

$$H_1 = \frac{x_1^2}{\eta_1^0} + \frac{2x_1 \, x_2}{\eta_{12}^0} \frac{m_1 \, m_2}{(m_1 + m_2)^2} \left[\frac{5}{3 \, A_{12}^*} + \frac{m_2}{m_1} \right] \tag{18}$$

while H_2 is obtained by exchange of subscripts. The expression for H_{12} is given by

$$H_{12} = H_{21} = - \frac{2x_1 \, x_2}{\eta_{12}^0} \frac{m_1 \, m_2}{(m_1 + m_2)^2} \left[\frac{5}{3 \, A_{12}^*} - 1 \right] \qquad (19)$$

x_1 and x_2 are mole fraction of species 1 and 2. Here, η_{12}^0 is the interaction viscosity coefficient defined in terms of collision integral $\Omega_{\eta_{12}}$ for the unlike interaction by

$$\eta_{12}^0(T) = 0.083868 \, \sqrt{T} \left[\frac{2 \, M_1 \, M_2}{M_1 + M_2} \right]^{1/2} \frac{1}{\Omega_{\eta_{12}(T)}} \qquad (20)$$

and A_{12}^* is the ratio of the viscosity collision integral to the diffusion collision integral given by

$$A_{12}^* = \Omega_{\eta_{12}}/\Omega_{D_{12}} \qquad (21)$$

Again, in principle, the calculation of the viscosity of a mixture is possible given a knowledge of all the various intermolecular potentials; however, for the reasons given above, such a calculation is practicable only for monatomic species[3] and the simplest polyatomic-monatomic systems.[34-36] Naturally, the same approximate procedure discussed above for pure species is available for mixtures.

c. Estimation Techniques

In cases where little experimental data exist and for which rigorous theoretical calculations are impractical, the estimation of the viscosity coefficient at zero-density is the only recourse. There is a plethora of estimation techniques,[39] some of which are firmly based on theory and others, albeit more numerous, which have little theoretical foundation but are based on empirical observation alone. In this section two estimation techniques are outlined that allow most of the needs for the viscosity of SCFs to be covered.

They are both based to some extent upon the principle of corresponding states. If the intermolecular pair potentials of any set of monatomic species are conformal in the sense that their reduced forms $V^*(r^*) = V \frac{U}{\epsilon}(r/\sigma)$ are identical, then it follows from Equations 13 to 15 that the reduced viscosity collision integral, Ω_η^*

$$\Omega_\eta^* = \Omega_\eta/\pi\sigma^2 \qquad (22)$$

is a universal function of the reduced temperature $T^* = kT/\epsilon$. For estimation purposes, the principle is of great value because it means that the viscosity of one gas in one temperature range may be employed to evaluate the viscosity of another gas in quite a different temperature range.

The conformality of the pair potentials among the noble gases, while not complete, is sufficient, that over an intermediate range of temperatures, the viscosity collision integral is universal with a deviation of less than 1% and may be represented[40,41] by the equation

$$\ln \Omega_\eta^* = 0.46641 - 0.56991 \ln T^* + 0.19591 \ln T^{*2} - 0.03879 \ln T^{*3} + \\ 0.00259 \ln T^{*4} \qquad 1.2 \le T^* \le 10.0 \qquad (23)$$

over a wide range of reduced temperature corresponding to 170 to 1400 K for argon, for example. Departures from conformality in the pair potentials at short and long ranges reveal

TABLE 3
The Scaling Parameters ϵ and
σ for Fluids of Interest in
Supercritical Fluid
Technology

	σ/nm	ϵ/K
N_2	0.3652	98.4
CO_2	0.3769	245.3
CH_4	0.3721	161.4
CF_4	0.4579	156.5
SF_6	0.5252	207.7
C_2H_4	0.4071	244.3
C_2H_6	0.4371	241.9
C_3H_8	0.4992	268.5
$n\text{-}C_4H_{10}$	0.5526	285.6
$i\text{-}C_4H_{10}$	0.5629	260.9
$C Cl_3 F$	0.5757	267.4
$CH Cl F_2$	0.4647	283.3

themselves as deviations from simple two-parameter universality[41] outside of this temperature range. More complicated results that extend to higher and lower temperatures are available.[41]

Although derived from a study of monatomic systems there is a substantial body of evidence that at least for the purposes of estimation, the reduced collision integral for a number of fluids of interest in SCF technology, necessarily polyatomic, is well represented by the same function.[42] The scaling parameters ϵ and σ for some fluids of interest in SCF technology are given in Table 3. More extensive tables are given in Reference 42, where an expression for the higher-order correction factor, $f_\eta(T^*)$, is also given. The law of corresponding states for viscosity has been extensively tested[40-42] on a number of diverse fluids at low density and has been found to predict the viscosity with an accuracy of better than $\pm 2\%$.

Lack of entry for a particular interaction in Table 3 indicates the lack of reliable experimental data rather than the inapplicability of the corresponding states principle. If accurate experimental data are obtained in a limited temperature range for a material other than those listed in Table 3, Equation 23 constitutes not only a preferred means of representing the data over the limited temperature range in which it has been measured, but also a means of generating data over a much wider temperature range. It also follows that since the reduced collision integral is characteristic of a binary interaction, its universality should extend to the quantity $\Omega^*_{\eta_{12}} = \Omega_{\eta_{12}}/\pi\sigma_{12}^2$. This has been verified in a number of studies.[3,40,41] This allows the viscosity of gas mixtures to be calculated from Equations 17 to 20 and Equation 23 given an estimate of A^*_{12} from Equation 21. Tests of this procedure on a wide range of fluid mixtures have been carried out and they confirm that the accuracy is usually one of $\pm 2\%$. Extensive tables of the unlike scaling parameters ϵ_{12} and σ_{12} are given in Reference 3. The only fluids for which the principle of corresponding states is expected to fail drastically are those with large dipole moments because the pair potential is then far from conformal with that of nonpolar fluids.

For fluids for which no scaling parameters are available and for which there are insufficient experimental data to estimate them, including polar fluids, a more heuristic method is recommended. It is also based on the corresponding states principle, but is much less securely founded on theory. In place of scaling parameters characteristic of the intermolecular pair potential ϵ and σ, suitable parameters are constructed from the critical constants of the gas, T_c, P_c, and Z_c and, for polar compounds the dipole moment, μ. Such a method proposed by Lucas[39,43] is based on empirical representations of the zero-density viscosity of a wide range of fluids by the expression.

$$\eta^0(T) = \frac{F_P}{\xi} \left[0.807\, T_r^{0.618} - 0.357\, \exp\left(- 0.449\, T_r\right) \\ + 0.340\, \exp\left(- 4.058\, T_r\right) + 0.018 \right] \tag{24}$$

where T_r is the reduced temperature

$$T_r = T/T_c \tag{25}$$

and

$$\xi = 1.76 \left(\frac{T_c}{M^3 P_c^4}\right)^{1/6} \tag{26}$$

F_p is the correction factor to account for the polarity,

$$
\begin{aligned}
F_P &= 1.0 & 0 \le \mu_r < 0.022 \\
F_P &= 1.0 + 30.55(0.292 - Z_c)^{1.72} & 0.022 \le \mu_r < 0.075 \\
F_P &= 1.0 + 30.55(0.292 - Z_c)^{1.72} \left| 0.96 + 0.1(T_r - 0.7) \right| & 0.75 \le \mu_r
\end{aligned} \tag{27}
$$

μ_r is the reduced dipole moment

$$\mu_r = 52.46\, \frac{\mu^2\, P_c}{T_c^2} \tag{28}$$

The viscosity coefficient $\eta^0(T)$ is in units of micropascal seconds, P_c is in bars, μ is in Debyes, T_c is in degrees Kelvin, and M is in grams per mol. The method is claimed[39] to be accurate to $\pm 3\%$ for nonpolar fluids, but the accuracy may diminish for polar fluids.

2. The Viscosity of a Dense Fluid
a. Pure Fluid

There is at present no rigorous theory that allows an exact evaluation of the viscosity, or for that matter any transport property of a dense fluid, in terms of a realistic intermolecular potential. The reasons are manifold, but center upon the fact that it is no longer possible in the dense gas to involve the assumption of molecular chaos so that molecular velocity correlations can no longer be neglected. Furthermore, the volume of the molecules themselves become significant, and both it and the occurrence of collisional transfer[44,45] must be added to the problem of the dilute gas. Thus, the solution of a rigorous, generalized Boltzmann equation for the state is not available even for the simplest molecular model.[44,45]

The treatment that continues to enjoy the greatest success is based on an approximate theory for hard-sphere molecules[2,44] proposed by Enskog. The basis of Enskog theory is a Boltzmann-like equation whose collision term is modified to account for the increase in density. The collision frequency is in fact increased by a density- and temperature-dependent function, χ, which is the radial distribution function at contact for hard spheres. Account is also taken of the finite volume of the hard spheres, but not of any velocity correlations.[2,44] Within this theory the viscosity coefficient of the dense fluid is given by

$$\eta = \frac{1}{\chi}\left(1 + \frac{n\alpha\chi}{2}\right)^2 \eta^0 + \frac{3}{5}\kappa \tag{29}$$

where κ is the bulk viscosity

$$\kappa = 1.5915 \, \alpha^2 n^2 \chi \, \eta^0 \tag{30}$$

and

$$\alpha = 0.8(2\pi\sigma^3/3) \tag{31}$$

where σ is the diameter of the hard spheres and n is the molecular number density.

The value of Enskog's theory lies not in its ability to make *a priori* predictions of the viscosity of pure gases, because its use of the hard-sphere model and the effects it neglects preclude that. Rather it is important in that it suggests a density expansion for the viscosity of the form

$$\Delta\eta = \eta - \eta^0 = \sum_{i=1}^{N} a_i \, \rho_i \tag{32}$$

which is confirmed in practice when $\Delta\eta^c(\rho, T)$ is negligible and because it is useful in a powerful formulation for fluid mixtures as discussed later. In the latter role and in some applications for pure gases some of the limitations of the hard-sphere model are overcome by the use of a temperature-dependent diameter. In this form the theory is known as the modified Enskog theory,[44] but it has little predictive power without at least some experimental information on the property of interest.

b. Fluid Mixtures

An extension of Enskog's theory can be applied to yield the viscosity of mixtures of hard-sphere molecules at high density.[46,47] Here, only the results for a binary mixture are given in the interest of brevity, although expressions for N-component mixtures are available.[47] The binary viscosity is given by

$$\eta_{mix} = \left(\frac{y_1^2}{H_1} + \frac{y_2^2}{H_2} - \frac{2y_1 \, y_2 \, H_{12}}{H_1 \, H_2}\right) \bigg/ \left(1 - \frac{H_{12}^2}{H_{11} \, H_{22}}\right) + \frac{3}{5} \, \kappa_{mix} \tag{33}$$

where the bulk viscosity contribution is

$$\kappa_{mix} = \frac{25}{5\pi} \, n^2(x_1^2 \, \chi_{11} \, \alpha_1^2 \, \eta_1^0 + x_2^2 \, \chi_{22} \, \alpha_2^2 \, \eta_2^0 + 2x_1 \, x_2 \, \chi_{12} \, \alpha_{12} \, \eta_{12}^0) \tag{34}$$

while y_1, H_1 are given by

$$y_1 = x_1 \left(1 + \frac{1}{2} \, x_1 \, \alpha_1 \, \chi_{11} \, n + \frac{m_2}{m_1 + m_2} \, x_2 \, \alpha_{12} \, \chi_{12} \, n\right) \tag{35}$$

$$H_1 = \frac{x_1\chi_{11}}{\eta_1^0} + \frac{2x_1 \, x_2 \, \chi_{12} \, m_1 \, m_2}{\eta_{12}^0 \, (m_1 + m_2)^2} \left(\frac{5}{3 \, A_{12}^*} + \frac{m_2}{m_1}\right) \tag{36}$$

The corresponding results for y_2 and H_2 are obtained by an exchange of subscripts. The term H_{12} is given by

$$H_{12} = \frac{-2x_1 \, x_2 \, \chi_{12}}{\eta_{12}^0} \, \frac{m_1 \, m_2}{(m_1 + m_2)^2} \left(\frac{5}{3A_{12}^*} - 1\right) \tag{37}$$

In addition to the symbols introduced earlier, χ_{11}, χ_{12}, and χ_{22} are the radial distribution functions (RDF) at contact for the three possible encounters of hard spheres in the presence of all the remaining molecules. Although this formulation is not useful for the evaluation of the viscosity of a mixture directly owing to its reliance on the hard-sphere model, it does provide the basis of a valuable interpolation tool for real fluid mixtures.

c. Estimation Techniques

Most of the theoretically based estimation methods for the fluid viscosity at elevated densities rely on the Enskog theory. In one common technique (the modified Enskog theory[44]) estimates of χ are obtained from the thermal pressure of the gas

$$P_t \equiv T(\partial P/\partial T)_n = nkT(1 + \frac{5}{4} \alpha \chi n) \tag{38}$$

because this is supposed to account for the attractive forces between the molecules.[2,44] The volume factor, α, can then be obtained from the relationship

$$\alpha = 4(B + TdB/dT)/5N_A \tag{39}$$

where B is the second virial coefficient of the real gas and N_A is the Avogadro's constant. There have been a number of extensive comparisons[48] of the predictions of the theory with experimental data that showed that the viscosity predictions were seldom as accurate as the experimental data, even at moderate densities.

An alternative method, also using the Enskog theory, is due to Dymond and collaborators.[49,50] It has been extensively used for liquids, mostly for the monatomic systems, but an extension to fluids of interest in supercritical technology is possible.

The method casts the results of the Enskog theory in the form[49]

$$\eta^* = \frac{\eta}{\eta^0} \left(\frac{V}{V_0}\right)^{2/3} = 2.093 \times 10^8 \, \eta \, V^{2/3}/(MT)^{1/2} \tag{40}$$

where V_0 is the core volume of the molecules and all the quantities are measured in SI units.

The important result following from this formulation is that the dimensionless group η^* should be a function of (V/V_0) only for a hard-sphere system, and this applies even when molecular velocity correlations are introduced to the theory via molecular dynamics (MD) results.[49] Dymond and collaborators, using the ideas of van der Waals, postulated that the same would be true of real fluids at high density and high temperature where the attractive forces between the molecules are weak. The only concession necessary to the existence of real intermolecular forces seems to be the introduction of a weak temperature dependence into V_0 to account for the finite steepness of the true repulsive branch of the potential.

Figure 6 illustrates the behavior of the reduced viscosity, η^*, for hydrocarbons in the dense gas phase as a function of V/V_0. The data are compared with the theoretical results for hard spheres.[51] It can be seen that the data for the hydrocarbons are well represented by the hard-sphere results at high density but depart from the theory at the lowest densities; however, even then the universality of the function $F(V/V_0)$ among them is maintained, although the functionality is different from that of the hard-sphere system. Using this universality and data for each hydrocarbon at a few isotherms, the evaluation of the viscosity along other isotherms, at which $V_0(T)$ can be determined from one datum, is straightforward.[49]

A further means of estimating the viscosity coefficient at elevated densities in the fluid phase is based on the observation that the excess viscosity, of the SCF $\Delta\eta$ defined by

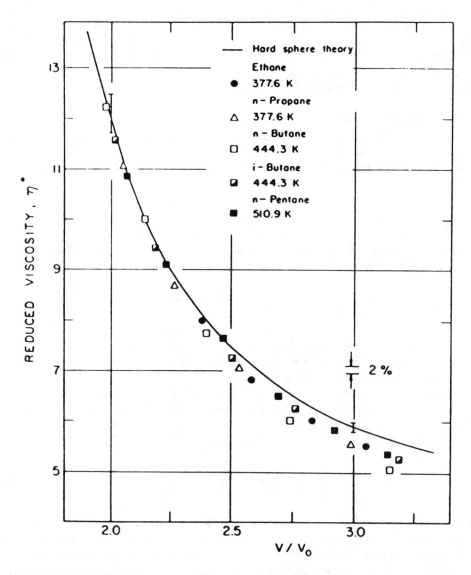

FIGURE 6. The behavior of η^* as a function of V/V_0 for hydrocarbons in the dense gas phase.[51]

Equation 10, is often found to be independent of temperature.[5,39,52,53] Figure 7 illustrates this observation for carbon dioxide.[5] Consequently, for many practical purposes, a knowledge of $\Delta\eta$ at one temperature (preferably experimental) can be used to estimate the total viscosity at another temperature outside the critical region using Equation 10, as long as $\eta^0(T)$ is known at the second temperature. It is noteworthy that this observation is formally at variance with the Enskog theory (Equations 29 and 30), which indicates that it is the ratio $\eta(\rho, T)/\eta_0(T)$ that is temperature independent.

Jossi, Stiel, and Thodos[52,53] have utilized this observation to produce a correlation between the excess viscosity and the reduced density, applicable to a large number of fluids. Their method belongs to a group of much more generally applicable estimation techniques based on the corresponding states principle.[39] The only disadvantage of these methods is that they are purely empirical and as such are prone to errors when applied to fluids and conditions for which they were not tested. An extensive compilation of different predictive methods is given in Reference 39.

The correlation of Jossi et al.[52] for nonpolar gases is given by the following expression

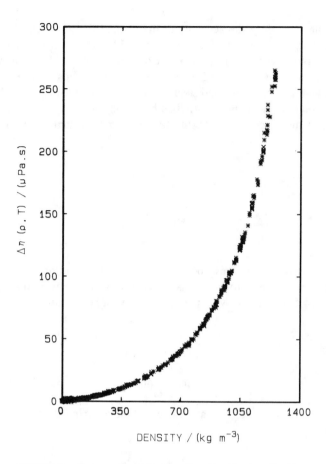

FIGURE 7. The experimental excess viscosity of carbon dioxide as a function of density in the temperature range $T_c < T < 800$ K (for details on experimental data, see Reference 5, Appendix II).

$$\left(\frac{\xi\Delta\eta}{0.176} + 1\right)^{1/4} = 1.0230 + 0.23364\ \rho_r + 0.58533\ \rho_r^2$$
$$- 0.40758\ \rho_r^3 + 0.093324\ \rho_r^4 \quad (41)$$

where ξ is given by Equation 26 and ρ_r is

$$\rho_r = \rho/\rho_c \quad (42)$$

This correlation is valid in the range $0.1 \leq \rho_r < 3$. The same authors have also extended the estimation procedure to polar gases.[52] The correlation then reads

$$\xi\Delta\eta = 0.0106832\ (9.045\ \rho_r + 0.63)^{1.739} \qquad 0.1 \leq \rho_r \leq 0.9$$
$$\log\left(4 - \log\left(\frac{\xi\Delta\eta}{0.176}\right)\right) = 0.6439 - 0.1005\ \rho_r - s \qquad 0.9 < \rho_r < 2.6 \quad (43)$$

where s is

$$s = 0 \qquad\qquad 0.9 < \rho_r \leq 2.2$$
$$s = 4.75 \times 10^{-4}(\rho_r^3 - 10.65)^2 \qquad 2.2 < \rho_r < 2.6 \quad (44)$$

It is very difficult to estimate the accuracy of the method, because reliable data for testing have not always been available. Errors in the range 5 to 10% are to be expected.

The principle of corresponding states has also been employed in estimating the viscosity of dense gases. A method, developed by Hanley and Ely,[54,55] can predict the viscosity of nonpolar fluids and their mixtures and requires only critical constants and Pitzer's acentric factor for each component as the input. It is based on the extended corresponding states principle in which the viscosity of the fluid of interest is related to that of the reference fluid. Thus,

$$\eta_x(\rho, T) = \eta_{ref}(\rho_{ref}, T_{ref}) \, F \, X_\eta \tag{45}$$

where

$$F = \left(\frac{M_x}{M_{ref}}\right)^{1/2} g_x^{1/2} \, h_x^{-2/3} \tag{46}$$

where x refers to the fluid of interest, be it pure or a mixture, and ref refers to the reference fluid, which has been chosen to be methane because its viscosity is readily available[15,54] over a wide range of conditions. M is the molecular weight, T_{ref} and ρ_{ref} are defined by the ratios

$$T_{ref} = T/g_x \tag{47}$$

$$\rho_{ref} = \rho h_x \tag{48}$$

In general for a mixture, g_x, h_x, and M_x are obtained using the following mixture and combining rules

$$g_x = \frac{\displaystyle\sum_i \sum_j x_i \, x_j \, g_{ij} \, h_{ij}}{h_x} \tag{49}$$

$$g_{ij} = (g_i \, g_j)^{1/2}$$

$$h_x = \sum_i x_i x_j h_{ij} \tag{50}$$

$$h_{ij} = \frac{1}{8}\left[h_i^{1/3} + h_j^{1/3}\right]^3$$

$$M_x = \left[\sum_i \sum_j x_i \, x_j \, g_{ij}^{1/2} \, h_{ij}^{4/3} \, M_{ij}^{1/2}\right]^2 / g_x h_x^{8/3} \tag{51}$$

$$M_{ij} = 2M_iM_j/(M_i + M_j)$$

where x_i and M_i are the mole fraction and the molecular weight, respectively, of species i, while g_i and h_i are related to the shape factors of thermodynamics by[54,55]

$$g_i = \frac{T_c}{190.4}\,\theta \tag{52}$$

$$h_i = \frac{V_c}{99.2}\,\phi \tag{53}$$

where T_c is the critical temperature in degrees Kelvin, while V_c is the critical volume in cubic centimeters per mole of the fluid at interest. Finally, the shape factors for the fluid relative to methane are given by

$$\Theta = 1 + (\omega_a - 0.011) \left[0.09057 - 0.86276 \ln T^+ \right.$$

$$\left. + (0.31664 - \frac{0.46568}{T^+}) (V^+ - 0.5) \right] \tag{54}$$

$$\phi = \left\{ 1 + (\omega_a - 0.011) \left[0.39490 (V^+ - 1.02355) - 0.93281 (V^+ - 0.75464) \right. \right. \tag{55}$$

$$\left. \left. \ln T^+ \right] \right\} \frac{0.288}{Z_c}$$

where

$$T^+ = \begin{cases} = 0.5 & T/T_c \leq 0.5 \\ = T/T_c & 0.5 < T/T_c < 2.0 \\ = 2.0 & T/T_c \geq 2.0 \end{cases} \tag{56}$$

$$V^+ = \begin{cases} = 0.5 & V/V_c \leq 0.5 \\ = V/V_c & 0.5 < V/V_c < 2.0 \\ = 2.0 & V/V_c \geq 2.0 \end{cases} \tag{57}$$

and ω_a is the acentric factor.[38]

The quantity X_η is the correction factor for noncorrespondence, but for only one mixture[54] was it necessary to assume a function different from unity. The method has been tested[54] for a number of compounds and their mixtures including n-paraffins, i-paraffins, alkenes, cycloalkenes, alkylbenzenes, carbon dioxide, and others. The viscosity has been predicted to within an absolute percent deviation of about 8%.

If data for the pure fluids comprising a mixture are available, a method based on Enskog's mixture theory for predicting the viscosity of a mixture is preferred to the one just described. The method[47,56-58] uses Equations 33 to 37 to interpolate between the viscosity of the pure components. The additional information required to treat a binary mixture are the RDFs, χ_{ij}, and the characteristic size parameter, α_{ij}, for each pair interaction. The Enskog theory is applicable to hard-sphere systems. Consequently, in its application to real gases the exact radial distribution of the hard-sphere system must be replaced by a pseudo-radial distribution function for the real gas. Accordingly, one derives the pseudo-radial distribution functions for the pure gases from the viscosity of the gas by solving Equations 29 and 30 for $\bar{\chi}_i$ at each temperature and the mixture density of interest so that

$$\bar{\chi}_i = \frac{B_\eta}{2} \frac{(\eta_i - n \alpha_i \eta_i^0)}{n^2 \alpha_i^2 \eta_i^0} \pm B_\eta \left[\left(\frac{\eta_i - n \alpha_i \eta_i^0}{2n^2 \alpha_i^2 \eta_i^0} \right)^2 - \frac{1}{B_\eta n^2 \alpha_i^2} \right]^{1/2} \tag{58}$$

where

$$B_\eta = 0.82994 \tag{59}$$

There are two solutions, $\bar{\chi}_i^+$ and $\bar{\chi}_i^-$, corresponding to the $\pm$ in Equation 58, only one of which is physically meaningful at a particular density. Thus, one constructs the total function $\overline{\chi}_i(n; T)$ so that below a density of n^*, $\overline{\chi}_i^-$ is employed and above n^* $\overline{\chi}_i^+$ is used. The crossover density n^* is obtained from the solution of the equation[56-58]

$$\left[\frac{d\eta_i(n, T)}{dn}\right]_T = \frac{\eta_i(n, T)}{n} \tag{60}$$

It follows that[56]

$$\frac{\eta_i(n^*, T)}{\eta_1^0 \alpha_i n^*} = 3.1954 \tag{61}$$

and a value of α_i can also be determined. Having determined α_i for all the pure components at a given temperature in this manner as well as the corresponding $\overline{\chi}_i$, the interactions pseudo-RDFs, $\overline{\chi}_{ij}$, are generated by means of the combination rules[57-59]

$$\overline{\chi}_{ij}(n, T) = 1 + \frac{2}{5} \sum_{k=1}^{N} x_k(\overline{\chi}_k - 1)$$

$$+ \frac{\frac{6}{5}(\overline{\chi}_i - 1)^{1/3}(\overline{\chi}_j - 1)^{1/3} \sum_{k=1}^{N} x_k(\overline{\chi}_k - 1)^{2/3}}{(\overline{\chi}_i - 1)^{1/3} + (\overline{\chi}_j - 1)^{1/3}} \tag{62}$$

and the requisite α_{ij} from the mixing rule

$$\alpha_{ij} = \frac{1}{8}\left(\alpha_i^{1/3} + \alpha_j^{1/3}\right)^3 \tag{63}$$

The viscosity of the gas mixture is then evaluated from Equations 33 to 37 using $\overline{\chi}_{ij}$ in place of χ_{ij} at the mixture molecular density required. The procedure must be repeated at each temperature of interest because all of the $\overline{\chi}_{ij}$ are temperature dependent. As for low-density mixtures, the quantities A_{ij}^* and η_{ij}^0 are best taken from the representation of the principle of corresponding states.

This scheme has been tested[57,58] against the somewhat limited amount of experimental data available for the viscosity of gas mixtures at high density. Generally, the accuracy of the calculated viscosity has proved commensurate with the experimental uncertainty.

III. DIFFUSION

Phenomenologically, the diffusion coefficient is the proportionality constant between the molecular flux of a species and its composition gradient; however, in order to define the diffusion flux, it is necessary to specify the frame of reference with respect to which the flux is measured. Various reference frames may be employed,[4,60] but here a volume-fixed frame of reference is adopted that allows the diffusion coefficients to be most easily related to those measured in an experiment.[4] For a binary, isothermal, isobaric mixture in the absence of an external force field, the flux equations are given by Fick's law as

$$\underline{J}_1 = -D_{12}\frac{\partial c_1}{\partial z}$$

$$\underline{J}_2 = -D_{21}\frac{\partial c_2}{\partial z} \tag{64}$$

where c_1 and c_2 are the molar concentrations of two components, $\underline{J}_1$ and $\underline{J}_2$ are molar fluxes related by a conservation equation

$$v_1 J_1 + v_2 J_2 = 0 \tag{65}$$

where v_1 and v_2 are the PMVs. For the circumstances prescribed, it can be easily shown that $D_{12} = D_{21}$. The diffusion coefficient, D_{12}, is a positive constant with units of square meters per second, characterized by the thermodynamic state of the fluid — in other words, a function of temperature, composition, and density or pressure. In general, the diffusion process is usually accompanied by small temperature and pressure gradients,[2,4,60,61] but the magnitude of the effects is usually very small and often negligible.[4,62]

Unlike other transport properties, the generalization of Equation 64 to multicomponent mixtures is neither straightforward nor convenient. The molar flux in the multicomponent mixture depends on the composition gradients of all species. The resulting multicomponent diffusion coefficients cannot in general be simply related to the binary diffusion coefficients. The only simplifications of the general expressions[60,61,63] that are available are for low pressures, when the mixture still behaves as an ideal gas. The equations for the multicomponent diffusion coefficients, whereby the composition gradient of the species is related to the difference in fluxes of gas pairs, are then given by[62]

$$\frac{\partial c_i}{\partial z} = \sum_{j=1}^{N} \frac{c_i c_j}{c} \frac{1}{D'_{ij}} \left(\frac{J_j}{c_j} - \frac{J_i}{c_i} \right) \tag{66}$$

where c is the total molar concentration. The newly defined diffusion coefficient $_{ij}'$ is weakly composition dependent and is not equivalent to the binary diffusion coefficient, D_{ij};[62] however, the numerical difference seems to be at most a few percent, and with experimental accuracy available at present, it is reasonable to take $D'_{ij} \sim D_{ij}$.

If a trace of one species is diffusing through the multicomponent mixture Equation 66 simplifies further to

$$\underline{J}_1 = - \left(\sum_{j=2}^{N} \frac{x_{1j}}{D_{1j}} \right)^{-1} \frac{\partial c_1}{\partial z} \tag{67}$$

where x_j is the mole fraction of species j, while subscript 1 denotes the trace species.

In the rest of this section the diffusion coefficients of the greatest importance for SCF technology are considered. In other words, the emphasis is on the binary diffusion coefficient and not, for example, on self- or multicomponent diffusion. Furthermore, special emphasis is placed on mixtures where one of the components is in trace conditions. Table 4 lists representative values[39,64-66] of the binary diffusion coefficient for some supercritical mixtures of technological interest and compares them with the values of the self-diffusion coefficient at low pressures in the gas and liquid phases.

In general, the binary diffusion coefficient in the SCFs decreases with increasing pressure at constant temperature and increases with temperature at constant pressure. At low pressures, the diffusion coefficient is nearly independent of composition, whereas at higher densities the composition dependence becomes more marked,[67] and in the liquid phase is a dominant feature of the behavior of the diffusion coefficient.[39]

A. THE MEASUREMENT OF THE DIFFUSION COEFFICIENT

There exists a variety of methods of measuring the diffusion coefficients;[4] however, the experimental accuracy achieved in many of them is quite modest[4,62] by comparison with what can be achieved for other transport properties, with a few notable exceptions.[68,69]

TABLE 4

The Diffusion Coefficient for Some Fluid Mixtures at T = 313 K

Second Substance is in Trace Concentration

		$D_{12} \times 10^8/m^2/s$	Ref.
Typical D_{12} in dilute gaseous systems, P = 0.1 MPa		1000—4000	39
CO$_2$-benzene, P = 10 MPa		1.67	64
CO$_2$-propylbenzene, P = 10 MPa		1.39	64
CO$_2$-naphthalene, P = 10 MPa		1.52	65
	T = 283		
SF$_6$-1,4 dimethylbenzene, P = 4 MPa	0.58	1.14	66
SF$_6$-1,4 dimethylbenzene, P = 12 MPa	0.48	0.73	66
C Cl F$_3$-2-propanone, P = 8 MPa		1.45	66
Approximate D_{12} in liquid systems		~0.1—0.4	39

Furthermore, methods that give reliable results in the low pressure regime are not especially suitable for high pressures because the inverse dependence of the diffusion coefficient on pressure means that the experiments take an inordinately long time. As an example, the measurement of the diffusion coefficient in a gas mixture at atmospheric pressure might take 5 h with a low pressure technique, whereas at a pressure of 10 MPa it might take 20 d with the same technique. Hence, the choice of a reliable method is not as straightforward as it is for viscosity. In fact, the most appropriate method for measurements of diffusion coefficinets in the SCF state[64,65,70] seems to be one based on Taylor dispersion, which has not proved as reliable at lower pressures.

1. Taylor Dispersion, the Gas Chromatographic Method

The method and its applications are reviewed in detail elsewhere,[4,71] so that here it is necessary only to describe it briefly and indicate those features that make it especially suitable for the supercritical state.[64,70] A trace amount of one material (the solute) is injected as a pulse into a carrier gas (the solvent) flowing through a long circular section tube in laminar flow. The combined action of the parabolic velocity profile across the tube and molecular diffusion disperse the pulse as it travels through the tube. At some distance, L, downstream from the point injection, the spatial distribution of the solute averaged over the tube cross section is a Gaussian whose variance depends upon the diffusion coefficient and the flow characteristics. A standard chromatographic detector is used to observe the cross section-averaged concentrations at the outlet of the tube. The variance of the temporal distribution of the eluted solution, τ, is in the simplest model of the method, given by[62,64,71]

$$\tau^2 = \frac{2 D_{12}L}{\bar{u}^3} + \frac{R^2L}{24 D_{12}\bar{u}} \tag{68}$$

where R is the radius of the tube, and $\bar{u}$ is the average velocity which is given, at the same level of approximation, by

$$\bar{u} = L/\bar{t} \tag{69}$$

where $\bar{t}$ is the time at which the maximum in the eluted distribution occurs. This simple mathematical analysis is valid only subject to a number of constraints on the flow conditions that can seldom all be satisfied. Consequently, for careful measurements a more rigorous

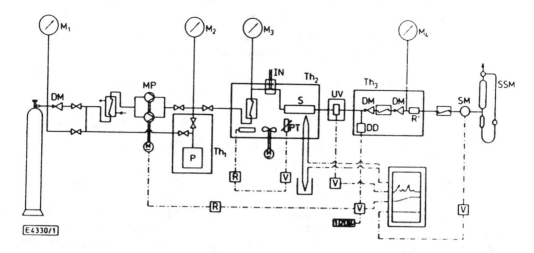

FIGURE 8.   Block diagram of the fluid chromatograph for measurements of binary diffusion coefficients.[64]

analysis is necessary.[71] Nevertheless, the fundamental measurements to be made remain those of the time at which the maximum eluted concentration occurs and the variance of the distribution. Both are readily determined in a variety of ways from the recorded output of the detector.

The Taylor dispersion technique is especially well suited for high pressure applications since the circular section tube is easily pressurized. Additionally, and most importantly, a single measurement can be performed in a matter of minutes and is only slightly affected by an increase of pressure. If the proper care is taken to eliminate the most important sources of systematic errors, then the measurements by the use of a carefully designed instrument and the full working equation, the method is capable of an accuracy[4] of better than ±1%, although in many applications an accuracy of ±5% is more routinely achieved.[64] Figure 8 shows the instrument employed by Swaid and Schneider[64] for extensive studies[64-66,72] of diffusion coefficients of interest in SCF technology. The apparatus is suitable for operation at temperatures up to 100°C and pressures up to 20 MPa. It consists of the tubular diffusion column, around in the form of a helix, which together with an injection system, is mounted in an air thermostat. Detection is carried out using a high-pressure ultraviolet (UV) detector, whose optical cell is held at the column temperature. The effects arising at the two ends of the column were treated experimentally, rather than by theory, by using two columns of different length.

2. Other Methods

The only other methods that can be utilized for the measurement of the diffusion coefficient in SCF are photon correlation spectroscopy[73,74] and the nuclear magnetic resonance (NMR) technique.[75,76] They are not widely used, principally because the equipment required is very expensive and because their application to measurements of diffusion coefficients of industrially important SCF is still in its infancy. Photon correlation spectroscopy, which uses the light scattered by concentration fluctuation in a sample to measure the diffusion coefficients, has so far only been applied to solutions of hydrocarbons in carbon dioxide,[73,74] while the use of the NMR technique is confined to measurements of self-diffusion coefficients.[75,76]

3. Correlations of Experimental Diffusion Coefficient Data

The scarcity of measurements of diffusion coefficients at elevated densities in the supercritical region means that there are no correlations of wide ranging applicability. At low

TABLE 5
The Correlation Parameters A_1, A_2, and A_3 of
Equation 70 for Some Fluid Mixtures

System	$10^5 A_1$/atm cm²/s K^{A_2}	A_2	A_3/K	T/K range
CH_4-SF_6	1.10	1.657	69.2	298—10^4
CH_4-N_2	1.00	1.750	0.0	298—10^4
CO_2-C_3H_8	0.177	1.896	0.0	298—550
CO_2-SF_6	0.140	1.886	0.0	328—472

densities there are many more measurements obtained both directly and indirectly through, for example, measurements of the viscosity of binary gas mixtures.

All of the low density measurements available up to 1972 have been comprehensively reviewed by Marrero and Mason.[62] They included in their analysis information from molecular beam scattering to extend the temperature range and developed representations of the dilute-gas diffusion coefficient as a function of temperature for some mixtures. Their representation takes the form

$$\ln P\, D_{12} = \ln A_1 + A_2 \ln T - A_3/T \qquad (70)$$

where D_{12} is in units of square centimeters per second, P is in units of atmospheres, and T is in degrees Kelvin. The parameters A_1, A_2, and A_3 are tabulated by Marrero and Mason[62] and D_{12} evaluated from Equation 70 refers to an equimolar mixture of the two components. A separate representation of the composition dependence allows correction of the data to any desired composition.[62] Since 1972, very few measurements of the diffusion coefficients have been made so that the correlation[62] is still the most comprehensive available. Table 5 gives the parameters for a few mixtures and the reader is referred to the work of Marrero and Mason[62] for a more comprehensive list.

B. THE THEORY OF THE DIFFUSION COEFFICIENT

Although the diffusion coefficient has not been formally divided into three distinct contributions, as have the other transport properties, it is still useful for the purposes of discussion to deal separately with the diffusion coefficient at low and high pressures. Hence, the first section deals with the diffusion coefficient in the dilute gas limit while the subsequent sections address the theory of a diffusion coefficient of a dense fluid.

1. The Dilute Gas Limit

The theory describing the behavior of the diffusion coefficient in the limit of the dilute gas is well developed. As noted earlier, in the section on viscosity, the solution of the Boltzmann equation leads to the expressions relating each transport property to a collision integral which in turn is related to the intermolecular potential.[2,3] Thus, the diffusion coefficient in a binary mixture, D_{12}^0, is related to a collision integral, $\Omega_{D_{12}}$, by

$$D_{12}^0 = \frac{3}{16n} \left[\frac{2\,(m_1 + m_2)}{m_1\,m_2} \right]^{1/2} (k\,T\,\pi)^{1/2}\, \frac{f_{D12}}{\Omega_{D_{12}}} \qquad (71)$$

where n is the molecular number density, and $f_{D_{12}}$ is the higher order correction term.

The above expression can be rewritten in a more convenient form for numerical calculation

$$D_{12}^0 = 0.083679 \frac{1}{P} T^{3/2} \left[\frac{M_1 + M_2}{2\ M_1\ M_2} \right]^{1/2} \frac{f_{D_{12}}}{\Omega_{D_{12}}} \tag{72}$$

where D_{12}^0 is in units of 10^{-8} m²/s, the pressure, P, is in units of megapascals, T is in degrees Kelvin, M is in grams per mole, and $\Omega_{D_{12}}$ is in units of square nanometers.

In the first-order approximation, $f_{D_{12}} = 1$, the diffusion coefficient is independent of composition. The higher-order correction factor, $f_{D_{12}}$, introduces a composition dependence and departs more from unity than those for the pure gas transport properties. Expressions for $f_{D_{12}}$ are available[3,62] and calculations demonstrate that the composition dependence is strongest for systems of dissimilar mass and size; however, even in these cases it does not usually amount to more than ±5% of the value in the first-order approximation.

In principle, it is possible to evaluate the binary diffusion coefficient from a knowledge of the intermolecular potential between the unlike molecules. The theoretical framework is the same as that set out in Section II.B.1 for viscosity, and formal results exist for monatomic and polyatomic fluids.[2-4] In practice, the lack of a detailed knowledge of the intermolecular forces and the computational complexity renders this approach possible only for simple systems. As for viscosity, the complete direct calculation of the diffusion coefficient in this way can only be performed routinely for monatomic species and has occasionally been carried out for a binary mixture of atoms and diatoms.[34-36] The Mason-Monchick approximation,[37,38] discussed in Section II.B.1.a, has also been applied to diffusion coefficients, but there is some evidence[37,38] that the diffusion coefficient is more strongly affected by the details of the collision dynamics. Consequently, even the approximate theoretical approach to the evaluation of diffusion coefficients is not very useful.

An alternate route to the evaluation of the diffusion coefficient is the use of kinetic theory expressions to relate the diffusion coefficients to other properties in a manner that is nearly independent of a knowledge of the intermolecular potential. Such a procedure leads to the following relationship in the first-order approximation between the binary diffusion coefficient and the interaction viscosity of binary mixture

$$D_{12}^0 = \frac{3}{5} A_{12}^* \frac{kT}{P} \frac{(m_1 + m_2)}{m_1\ m_2} \eta_{12}^0 \tag{73}$$

Here, A_{12}^*, defined by Equation 21, is a weakly temperature-dependent ratio of collision integrals that is very insensitive to the intermolecular potential used for its evaluation and for which a suitable representation exists.[3,40-42] The reliability of the diffusion coefficient evaluated indirectly from Equation 73 and mixture viscosity data is almost the same as the best direct experimental evaluation of D_{12}^0. This is because, notwithstanding the uncertainty in A_{12}^*, the measurements of viscosity are almost an order of magnitude more accurate than measurements of the diffusion coefficient.

a. Estimation Techniques

The most accurate way of estimating the binary diffusion coefficient at low density is based upon the principle of corresponding states.[40-42] By analogy with the discussion on the viscosity, the assumption of conformality of the pair potentials among interacting molecules renders the reduced form of the diffusion collision integral, $\Omega_{D_{12}}^* = \Omega_{D_{12}}/\pi\sigma_{12}^2$, universal within the gas mixtures considered. A correlation of this function developed originally for the monatomic gases and their mixtures, but applicable to polyatomic mixtures as well, is given by[40-42]

$$\ln \Omega^*_{D12} = 0.357588 - 0.472513 \ln T^* + 0.0700902 (\ln T^*)^2$$
$$+ 0.0165741 (\ln T^*)^3 - 0.005929022 (\ln T^*)^4$$
$$1.2 \leq T^* \leq 10.0 \qquad (74)$$

where $T^* = kT/\epsilon_{12}$. When estimating the binary diffusion coefficients, the scaling parameters ϵ_{12} and σ_{12} refer to intermolecular potential between unlike species of the mixture. For most of the molecules listed in Table 3, the unlike scaling parameters are available.[3,40-42] If not, the simple, but frequently adequate, empirical mixing rules

$$\epsilon_{12} = \sqrt{e_1 \, e_2} \qquad (75)$$

$$\sigma_{12} = \frac{1}{2} (\sigma_1 + \sigma_2) \qquad (76)$$

should suffice for many purposes. The scaling parameters for pure components ϵ and σ are the same as those used for viscosity. The corresponding states expression for $f_{D_{12}}$, also exists and is readily available.[41] As an alternative, if an interaction viscosity is available for the system of interests, Equation 73 can be used to obtain D^*_{12} where A^*_{12} should be obtained using corresponding states principle and Equation 21.

In the cases where no scaling parameters are available and no experimental data exist, one of the empirical methods described elsewhere[39] can be used. Unfortunately, for all the methods recommended there are some systems for which large errors have been observed,[39] so that values calculated in this way should be used with caution.

2. The Diffusion Coefficient of a Dense Fluid

The behavior of the binary diffusion coefficients at high density is even more difficult to evaluate than that of other transport properties.[2-4,61] This is partly because the problem concerns a fluid mixture and partly because the velocity correlations discussed in the section on viscosity have a more profound influence on diffusion. The Enskog theory,[2,44] which gave a useful qualitative picture for viscosity, yields the relationship between the diffusion coefficient in a dense binary gas mixture in terms of its dilute gas value

$$D_{12} = D^0_{12}/\chi_{12} \qquad (77)$$

However, the evaluation of χ_{12} for a hard-sphere fluid leads to values of D_{12} that are significantly different from those observed experimentally.[4] This difference is, to a large extent, a result of the velocity correlations among molecules.

In an attempt to avoid the difficulty, at least for the self-diffusion coefficient of a gas, Dymond[49] has made use of the Enskog theory in a different way. He writes

$$D^* = \frac{nD^E}{(nD)^0} \left(\frac{V}{V_0}\right)^{2/3} \frac{D^{MD}}{D^E} \qquad (78)$$

where $(nD)^0$ is the zero-density limit of the product of density and self-diffusion coefficient; V and V_0 have been defined before. In addition, D^{MD} is the self-diffusion coefficient of a hard-sphere system obtained by molecular dynamics and D^E the Enskog result for hard spheres.[2,61] The effects of correlation of molecular velocities are incorporated in D^* through D^{MD}. The dimensionless ratio, D^*, emerges as a function of V/V_0 only.

It has been shown[49] that such a function describes the self-diffusion coefficients of the monatomic fluids at high density quite well, as long as V_0 is properly chosen and allowed to be a weak function of temperature. This approach has been extended[49] to deal with mixtures where one of the components is present in trace quantities. To make the procedure more widely applicable to polyatomic fluids it has been necessary to invoke the rough hard sphere model[77,78] instead of the smooth hard-sphere model of the Enskog theory. This is because the rough hard-sphere model allows coupling between internal and translational energy which occurs in polyatomic molecules. Then, D_{12} is expressed in terms of the smooth hard-sphere theory diffusion coefficients D_{12}^{SHS} as

$$D_{12} = D_{12}^{SHS} \, l_{12} = \frac{D_{12}^{SHS}}{D_{12}^{E}} \, D_{12}^{E} \, l_{12} \tag{79}$$

where l_{12} is a correction factor that deals with effects of translational-rotational coupling.[77,78] There is no theoretical guidance on the value of l_{12}, the so-called coupling constant, beyond the fact that it is unity for smooth hard-spheres and less than unity for real systems;[49] however, when this parameter has been determined from some experimental data and applied more widely in the liquid phase it has proved moderately successful.[49]

Recently, a similar theory[79] has been applied to the tracer diffusivity in a binary mixture where the solvent is an SCF. The term, $(D_{12}^{SHS}/D_{12}^{E})$, on the right of Equation 79, has been approximated using the Sung-Stell formulation[80] of molecular velocity correlations while a value of 0.714 for l_{12} has been adopted.[81] The resulting expressions had two unspecified parameters, the effective hard-sphere diameter of the solute and that of the solvent. Using self-diffusion data, the hard-sphere diameter of the solvent has been established[79] and the solute diameter has been treated as an adjustable parameter. The model has been found to reproduce the experimental diffusion coefficients within $\pm 4\%$; however, the tests have been carried out only over a very limited temperature and pressure range for just one solvent and four solutes. It is therefore premature to speculate on the application of the procedure more generally and more work is evidently necessary.

a. Estimation Techniques

The lack of an easily usable formal theory on even a theoretically based correlation makes the estimation of binary diffusion coefficients of dense fluid rather difficult. Fortunately, there are some qualitative trends that can be utilized to advantage. For example, it has been experimentally observed that at low to moderate pressures (even up to 0.5 P_c) the diffusion coefficient varies inversely with the density.[39,82-84] It should be noted that this is, in fact, a prediction of the dilute gas kinetic theory (Equation 71); however, this result does not imply the applicability of the theory at such high densities, but rather, that there are a number of compensating but unaccounted effects. At higher pressures deviations from the constancy of the product (nD_{12}) behavior are observed as Figure 9 illustrates for fluids where the solutes are in trace quantities. The two sets of data shown for pressures greater than the critical pressure exhibit quite different behavior[64,79] and the scarcity of experimental information means that it is not possible to distinguish between the two types of behavior even qualitatively.

Takahashi[85] has suggested a very simple scheme for predicting the binary diffusion coefficient of a dense gas. The correlation is given by

$$\frac{D_{12}P}{D_{12}^{0}P^{0}} = f(T_r, P_r) \tag{80}$$

where

$$T_r = T/(x_1 \, T_{c1} + x_2 \, T_{c2}) \tag{81}$$

$$P_r = P/(x_1 \, P_{c1} + x_2 \, P_{c2}) \tag{82}$$

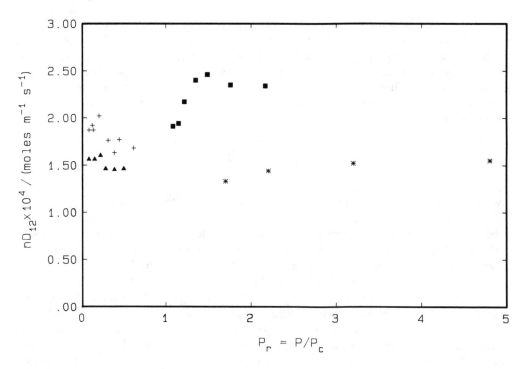

FIGURE 9.   The behavior of (nD_{12}) product as a function of reduced pressure.   + CO_2-naphthalene $T_r = 1.03$;[84] ▲ CO_2-naphthalene $T_r = 0.99$;[84] ■ CO_2-benzene $T_r = 1.03$;[64] * 2,3-dimethylbutane-benzene $T_r = 1.05$.[79]

where T_{c1}, P_{c1}, T_{c2}, and P_{c2} are critical temperature and critical pressure of components 1 and 2, respectively, P^0 is the low pressure at which the value of D_{12}^0 has been taken, and $f(T_r, P_r)$ is illustrated in Figure 10. The scheme is moderately successful as judged by the limited number of tests thus far conducted.[39,85]

IV. THERMAL CONDUCTIVITY

In SCF technology, the need for thermal conductivity data is markedly less than that for viscosity and diffusion. Hence, only a brief summary of modern techniques of measurement and theories for evaluating the thermal conductivity coefficients of the SCFs is presented here.

The thermal conductivity coefficient is defined as the proportional constant between the heat flux and the temperature gradient that exists in the fluid. The relationship is summarized by Fourier's law as

$$q = -\lambda \frac{\partial T}{\partial z} \tag{83}$$

where q is the heat flux, $\partial T/\partial z$ is the temperature gradient, and λ is the coefficient of the thermal conductivity. The thermal conductivity is a function of the temperature and density or pressure of the fluid, and for most SCFs it increases with increasing temperature and increasing density. Table 6 lists some values of the thermal conductivity for a few common fluids used in SCF technology.

A. THE MEASUREMENT OF THE THERMAL CONDUCTIVITY COEFFICIENT

There are two techniques that have been proven to be of lasting value in the measurement

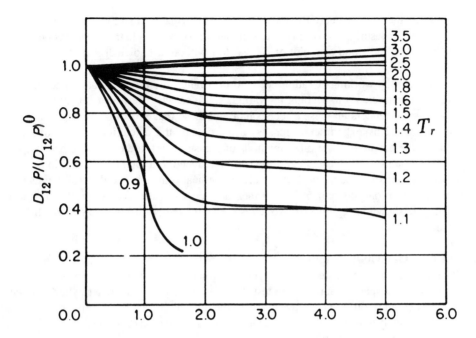

FIGURE 10. Takahashi correlation for evaluating function $f(T_r, P_r)$.[85]

TABLE 6
Thermal Conductivity Values of Some Fluids Used in Supercritical Fluid Technology

	λ/mW/m/K					
	T = T$_c$ + 20 K		T = T$_c$ + 100 K			
	P = 0.1 MPa	P = 10 MPa	P = 0.1 MPa	P = 10 MPa	T$_c$/K	Ref.
CO_2	18.8	51.1	25.5	31.9	304.1	5
N_2	14.1	53.0	20.7	28.9	126.0	6
H_2O	54.0	67.8	63.7	73.0	647.1	7
C_3H_8	29.0	69.9	39.6	54.4	369.8	8

of the thermal conductivity of SCFs over a wide range of conditions. Both have been described in detail elsewhere[4] so that a brief summary is all that is necessary here.

1. The Transient Hot-Wire Technique

The modern versions of this technique[4,86-88] yield the most accurate values of thermal conductivity of fluids. In essence, the instrument consists of a thin (~7 μm) diameter platinum wire suspended vertically in the test fluid. The initiation of a current in the wire at an instant t = 0 initiates dissipation and the wire acts almost as a line source of heat, q(W/m), which is conducted away radially in the surrounding fluid. The platinum wire is also used as a thermometer to determine the temperature change of the fluid in contact with it by virtue of its resistance change. The simplest theory[4] of the device shows that the temperature rise of the wire of radius R, $\Delta T(R, t)$ is related to the time elapsed since initiation of the flux by the equation

$$\Delta T(R,t) = \frac{q}{4\pi\lambda} \ln \frac{4\kappa t}{R^2 C_\lambda} \tag{84}$$

where λ is the thermal conductivity of the fluid, κ its thermal diffusivity, and C_λ a numerical constant. Thus, measurements of $\Delta T(R, t)$ as a function of time yield the thermal conductivity of the fluid from the slope of a plot of ΔT vs. ln t without a knowledge of the wire radius R.

In practice, the working equation needs to be slightly more complicated than Equation 84 because it is necessary to account for various small effects;[89] however, the proper design of an instrument renders all these effects small and those that cannot be treated in this way can be eliminated by experimental means. In particular, the deleterious effects of natural convection, the plague of most other methods of measurement, are eliminated because with modern electronics, the measurement time can be reduced to a total of 1 s, and in this time the fluid is not accelerated sufficiently to effect the heat transfer. The thermal conductivity cells employed in one instrument[86-88] for measurements in gases for 177 to 450 K and up to 35 MPa are shown in Figure 11. Two wires are employed of different lengths to eliminate end effects.

2. The Coaxial Cylinder Method

This technique[4] is of a steady-state type, in which two coaxial cylinders are separated by a small annular space containing the test fluid. A heater within the inner cylinder induces a steady temperature difference between it and the outer cylinder. Measurement of the heat input and the steady temperature difference then allows evaluation of the thermal conductivity of the fluid from the equation

$$q = \frac{2\pi \lambda (T_1 - T_2)}{\ln (R_2/R_1)} \tag{85}$$

where R_1 and R_2 are the radii of the inner and outer cylinders, while T_1 and T_2 are their respective temperatures.

The satisfactory operation of this type of equipment requires considerable attention to detail. For example, the two cylinders must be concentric to a high degree of tolerance because the annular space is a fraction of a millimeter. Furthermore, their surfaces must be silver coated to reduce radiation and the thermocouples employed to determine the temperature difference should be as close to the solid/fluid interfaces as possible. A refined instrument of this type is that described by Le Neindre et al.[90] and Tufeu[91] and Figure 12 contains a schematic diagram of the thermal conductivity cell. This device has been used for measurements of gases[92] at pressures up to 100 MPa and temperatures up to 600°C.

3. Correlations of Experimental Thermal Conductivity

Correlations based entirely on experimental data for a number of fluids of interest to SCF technology exist. Table 7 summarizes them as well as the range of their validity and claimed accuracy. As for viscosity, these correlations are preferred to any method of prediction.

B. THE THEORY OF THE THERMAL CONDUCTIVITY COEFFICIENT

The coefficient of the thermal conductivity can be written as a sum of the contributions of three separate parts

$$\lambda(\rho, T) = \lambda^0(T) + \Delta\lambda(\rho, T) + \Delta\lambda^c(\rho, T) \tag{86}$$

where $\lambda^0(T)$ is the thermal conductivity in the dilute gas limit, $\Delta\lambda^c$ is the critical enhancement, and $\Delta\lambda$ is the excess thermal conductivity. For the reasons given in Section II (Viscosity), it is preferable to deal separately with the first two contributions, while the third contribution is discussed elsewhere in this book.

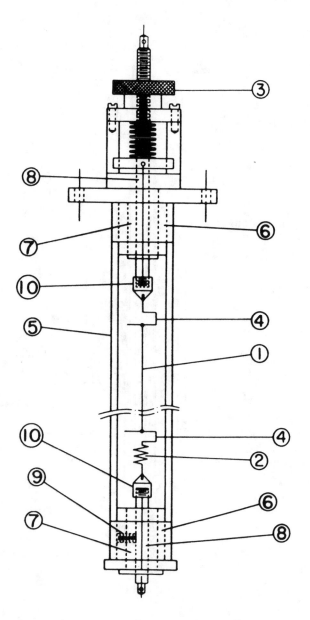

FIGURE 11. A transient hot wire cell.[86]

1. The Dilute Gas Limit

Using the kinetic theory framework,[2,3] the coefficient of the thermal conductivity can be related to the collision integral by a single relationship

$$\lambda^0 = \frac{75}{64} \left[\frac{k^3 \, T\pi}{m} \right]^{1/2} \frac{f_\lambda}{\Omega_\lambda} \qquad (87)$$

where f_λ is the correction factor near unity accounting for higher order approximations than the first and Ω_λ is the collision integral. In the preferred form for numerical calculation, the above expression reads

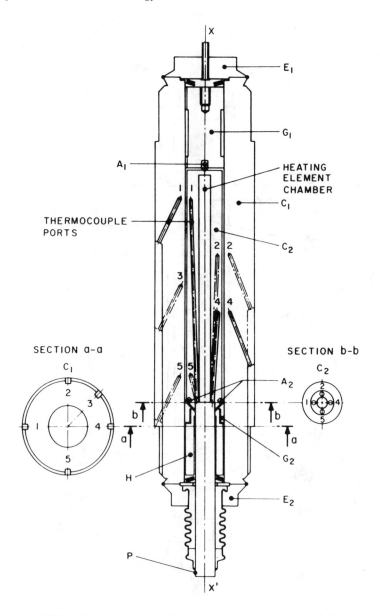

FIGURE 12.   A concentric-cylinder thermal conductivity apparatus.[91]

TABLE 7
Correlations Available for the Thermal
Conductivity of Some Fluids

	Temp. range/K	Pressure range/MPa	Claimed accuracy	Ref.
CO_2	200—1000	0—100	±1—±5%	5
H_2O	273—1073	0—100	NA	7
N_2	70—1100	0—100	NA	6
CH_4	95—500	0—75	±5%	25
C_2H_6	200—500	0—75	±8%	26
C_3H_8	140—500	0—50	±8%	8
C_2H_4	110—500	0—50	±5%	27

$$\lambda^0 = 2.6149 \left(\frac{T}{M}\right)^{1/2} \frac{f_\lambda}{\Omega_\lambda} \tag{88}$$

where λ is in units of milliwatts per meter per degrees Kelvin, T is in degrees Kelvin and Ω_λ is in units of square nanometers. Through a series of expressions it is possible to relate the collision integral Ω_λ to the intermolecular potential for both monatomic and polyatomic species.[3]

For monatomic species

$$\Omega_\lambda = \Omega_\eta \tag{89}$$

and in the first-order approximation one can derive an exact simple relationship between viscosity and the thermal conductivity known as the Eucken factor

$$Eu = \frac{4}{15} \frac{m}{k} \frac{\lambda^0}{\eta^0} = 1 \tag{90}$$

A higher order correction arising from the f_λ/f_η ratio amounts to at most 2% and is readily available.[3] The Eucker relationship is of considerable importance, since it essentially eliminates the need for knowledge of the intermolecular potential in the evaluation of λ^0 from η^0. Furthermore, the result may be used to check the consistency of experimental data on the two properties.

For polyatomic fluids, using the semiclassical theory, the Eucken factor is in the first-order approximation given by[3,27,38]

$$Eu = 1 - \frac{2}{3} Y + \frac{4}{15} \frac{\rho D_{int}^0}{\eta^0} \left[\frac{C_{vint}}{k} + Y\right] \tag{91}$$

where

$$Y = \frac{\dfrac{C_{vint}}{k} \dfrac{2}{\pi\zeta} \left[\dfrac{5}{2} - \dfrac{\rho D_{int}^0}{\eta^0}\right]}{\left[1 + \dfrac{2}{\pi\zeta}\left[\dfrac{5}{3}\dfrac{C_{vint}}{k} + \dfrac{\rho D_{int}^0}{\eta^0}\right]\right]} \tag{92}$$

where C_{vint} is the internal contribution per molecule to the isochoric heat capacity of the ideal gas, ζ is the collision number for internal energy relaxation, and D_{int}^0 is the diffusion coefficient for the internal energy. Of the three new quantities appearing in the expression for polyatomic gases, two are experimentally accessible. The heat capacity at low density is known quite accurately and is readily available from calculation or experiment. In addition, the collision number, ζ, which expresses the number of collisions that must occur in the gas to relax the internal modes of motion to equilibrium, may be determined in sound absorption experiments. In the case of both the rotational and the vibrational modes of a molecule being excited and relaxed at different rates, simple combination rules provide the total collision number ζ_{int}.[38,93] This leaves only a quantity, D_{int}^0, which is not accessible to measurements, since it represents the diffusion of the internal energy through gas. It has to be obtained[3,38] by direct calculation from the intermolecular potential. Hence, unlike the case of monatomics, the thermal conductivity of a polyatomic gas cannot be solely obtained from measurements of other thermophysical properties.

The formal semiclassical kinetic theory expressions for the thermal conductivity of a polyatomic mixture[94] are too lengthy to be given here in full. Instead, an abridged version which involves several physically reasonable approximations is presented.[3,95] In this form, the mixture thermal conductivity is given by

$$\lambda_{mix}^0 = \lambda_{mix}^0(mon) + \lambda_{mix}^0(int) + \lambda_{mix}^0(inel) \tag{93}$$

where $\lambda_{mix}^0(mon)$ is the result for a monatomic system, which for a binary mixture reads

$$\lambda_{mix}^0(mon) = \left(\frac{x_1^2}{L_{11}} + \frac{x_2^2}{L_{22}} - \frac{2x_1x_2L_{12}}{L_{11}L_{22}}\right) \bigg/ \left(1 - \frac{L_{12}^2}{L_{11}L_{22}}\right) \tag{94}$$

where

$$L_{12} = -\frac{x_1x_2m_1m_2}{2\,A_{12}^*\,\lambda_{12}^0\,(m_1 + m_2)^2}\left[\frac{55}{4} - 3\,B_{12}^* - 4A_{12}^*\right] \tag{95}$$

$$L_{11} = \frac{x_1^2}{\lambda_1^0} + \frac{x_1x_2}{2\,A_{12}^*\,\lambda_{12}^0\,(m_1 + m_2)^2}\left[\frac{15}{2}\,m_1^2 + \frac{25}{4}\,m_2^2 - 3m_2^2\,B_{12}^* + 4m_1m_2A_{12}^*\right] \tag{96}$$

and L_{22} can be obtained by exchange of subscripts. The quantity B_{12}^* is a further ratio of collision integrals,[3] which, like A_{12}^*, is weakly dependent on the intermolecular potential. The interaction thermal conductivity λ_{12}^0 is given by

$$\lambda_{12}^0 = \frac{75}{64}\left(\frac{k^3(m_1 + m_2)\,T\pi}{2\,m_1m_2}\right)^{1/2}\frac{1}{\Omega_{\lambda_{12}}} \tag{97}$$

For a monatomic system it is evident that a knowledge of the pure component viscosities and the interaction collision integral $\Omega_{\lambda_{12}}$ as well as A_{12}^* and B_{12}^* suffice to evaluate the thermal conductivity. The quantity $\lambda_{mix}^0(int)$ arising directly for the internal energy transport for the binary mixture is given by[3,95]

$$\lambda_{mix}^0(int) = (\lambda_1^0 - \lambda_{1tr}^0)\bigg/\left[1 + \frac{x_2\,D_{11\,int}^0}{x_1D_{12\,int}^0}\right] + (\lambda_2^0 - \lambda_{2tr}^0)\bigg/\left[1 + \frac{x_1\,D_{22\,int}^0}{x_2\,D_{21\,int}^0}\right] \tag{98}$$

where

$$\lambda_{itr}^0 = \frac{15}{4}\frac{k}{m}\,\eta_i^0 \tag{99}$$

is an approximation to the translational part of the thermal conductivity of the gas. The expression for $\lambda_{mix}^0(inel)$ is too cumbersome to reproduce here and is given in Reference 3. It accounts for the various relaxation processes occurring in the gas mixture.

For monatomic systems, the thermal conductivity can be evaluated from known intermolecular pair potentials with the same ease as described earlier for the viscosity. On the other hand, the evaluation for any system containing a polyatomic molecule is beyond the fastest computers available today owing to its length. The various approximate methods discussed earlier in connection with the viscosity fail more seriously for the thermal conductivity owing to its sensitivity to internal energy exchange.

a. Estimation Techniques

The sensitivity of the thermal conductivity of polyatomic fluids to the anisotropy of the potential and the exchange in internal energy[34,35] mentioned above precludes the use of the corresponding states principle, so successfully applied to viscosity and diffusion.

The best way of approximating the thermal conductivity of a pure polyatomic gas is by means of the theoretical equations (96 and 97) and a series of estimates of some of the quantities occurring in them. The total collision number, ζ_{int}, can be obtained from the simple mixing rule[38]

$$\frac{C_{vint}}{\zeta_{int}} = \frac{C_{vrot}}{\zeta_{rot}} + \frac{C_{vvib}}{\zeta_{vib}} \tag{100}$$

using the information on the independent modes of internal energy, rotation (rot), and vibration (vib). The vibrational collision number, ζ_{vib}, may usually be estimated[93] without the need for high accuracy because it is very large and consequently contributes very little to the total collision number ζ_{int}. Rotational numbers, on the other hand, are much smaller and must be estimated more accurately. Values of ζ_{rot} are not available over a wide range of temperatures, hence it is often necessary to approximate their temperature dependence given a knowledge of ζ_{rot} at one temperature. For diatomics, the temperature function below has been proposed which is reliable and can also be used as an approximation for more complex molecules[96,97]

$$\frac{\zeta_{rot}^{\infty}}{\zeta_{rot}} = 1 + \frac{\pi^{3/2}/2}{T^{*1/2}} + \frac{2 + \pi^2/4}{T^*} + \frac{\pi^{3/2}}{T^{*3/2}} \tag{101}$$

where ζ_{rot}^{∞} is the high temperature limit and $T^* = kT/\epsilon$. Values of ϵ can be obtained from Table 3. The estimation of D_{int}^0 is more difficult. The group $\rho \, D_{int}^0/\eta$, which actually occurs in Equation 91, can be rewritten as

$$\frac{\rho \, D_{int}^0}{\eta^0} = \frac{6}{5} A^* \frac{D_{int}^0}{D^0} \tag{102}$$

Only a few calculations[34,35] and estimates[98-100] of D_{int}^0/D^0 are available and they show that the ratio varies from 0.7 to 1.1. In a recent paper,[101] it has been proposed that for simple molecules the vibrational energy diffuses at the same rate as the molecules. Hence, $\rho \, D_{int}^0/\eta$ should be approximately equal to $\rho D_{rot}^0/\eta^0$. Using the best experimental data available, D_{rot}^0 values have been extracted from the thermal conductivity expressions for a number of molecules and an empirical correlation proposed of the form[101]

$$\frac{1}{(\zeta_{rot}^{\infty})^{1/4}} \left(\frac{\zeta_{rot}^{\infty}}{\zeta_{rot}}\right) \frac{\rho \, D_{rot}^0}{\eta^0} = 1.222 + \frac{4.552}{T^*} \qquad T \leq T_{cross}$$

$$\frac{\rho D_{rot}^0}{\eta_0} = \frac{\eta D^0}{\eta^0} \left[1 + \frac{0.27}{\xi_{rot}^{\infty}} - \frac{0.44}{\xi_{rot}^{\infty 2}} - \frac{0.90}{\xi_{rot}^{\infty 3}}\right] \qquad T > T_{cross} \tag{103}$$

where T_{cross} is a molecule-dependent temperature given in Reference 101. Equation 103 is valid only for nonpolar gases, even though some suggestions have been made on how to include polar gas effects.[101] The correlations have been tested only on the gases, whose data have been used to establish the empirical relationship (Equation 103); however, the uncertainty of the resulting thermal conductivity is estimated to be $\pm 1.5\%$ in the range 300 to 500 K, rising to 3% at lower and higher temperatures.

For more complex fluids, empirical correlations are yet again the only alternative. For polar molecules the Roy-Thodos method[39,102] is recommended. It is based on reducing the thermal conductivity to a dimensionless number using the critical parameters of the system and then using group contributions for various classes of compounds. The method is fully described in Reference 39, together with tables of group contributions. It generally yields the smallest errors of all the available methods, but it is not applicable to inorganic compounds. For nonpolar compounds the method recommended by Ely and Hanley[103] and fully described in the section on the dense fluid (IV.B.2), can also be used.

The calculation of the thermal conductivity of the polyatomic gas mixture using kinetic theory expressions Equations 93 to 99 is less reliable than for pure gases, because the number of quantities that have to be estimated is much larger. For most of the gases of interest the estimation of all the necessary parameters is not possible. Hence, it is better to use another method that has its origins in the full theoretical results but requires less information. The most satisfactory is the Wassilijewa equation[104] and the Mason and Saxena[105] modification of it. For binary mixtures

$$\lambda^0_{mix} = \frac{x_1 \lambda^0_1}{x_1 G_{11} + x_2 G_{12}} + \frac{x_2 \lambda^0_2}{x_1 G_{21} + x_2 G_{22}} \tag{104}$$

where

$$G_{ij} = \frac{\left[1 + \left(\frac{\eta^0_i}{\eta^0_j} \right)^{1/2} \left(\frac{M_j}{M_i} \right)^{1/4} \right]^2}{[8(1 + M_i/M_j)]^{1/2}} \tag{105}$$

Hence, in order to estimate the mixture thermal conductivity, one needs the values of both the viscosities and thermal conductivities of pure components. This equation is capable of representing the thermal conductivity of mixtures that display either a maximum or a minimum as the composition is varied. Errors for nonpolar mixtures are generally less than 3%, whereas for polar mixtures errors of 8% are to be expected.[39]

2. The Thermal Conductivity of a Dense Fluid

The Enskog theory[2,44] of a dense fluid yields the following expression for the thermal conductivity of a polyatomic fluid

$$\lambda = \lambda(mon) + \lambda(int) \tag{106}$$

$$\lambda(mon) = \lambda^0_{tr} \left[1 + \frac{1}{2} \gamma \eta \chi \right]^2 \frac{1}{\chi} + \frac{3}{2} \frac{k}{m} \kappa \tag{107}$$

$$\lambda(int) = \frac{m \rho D^0_{int} C_{vint}}{\chi} \tag{108}$$

where λ^0_{tr} is the value of monatomic low density thermal conductivity given by Equation 99, χ is the radial distribution function, γ is related to rigid sphere co-volume, $\gamma = 1.2(2\pi\sigma^2/3)$, and κ is the bulk viscosity given by

$$\kappa = 0.707355 \, n^2 \, \gamma^2 \, \chi \, \eta^0 \tag{109}$$

The extension of the theory to binary mixture results in the following relationships for the polyatomic fluid[106]

$$\lambda_{mix} = \lambda_{mix}(mon) + \lambda_{mix}(int) \tag{110}$$

where for binary mixture $\lambda_{min}(mon)$ is given by

$$\lambda_{mix}(mon) = \left(\frac{y_1^2}{L_{11}} + \frac{y_2^2}{L_{22}} - \frac{2y_1 y_2 L_{12}}{L_{11} L_{22}}\right) \Big/ \left(1 - \frac{L_{22}^2}{L_{11} L_{22}}\right) + \kappa'_{mix} \tag{111}$$

$$y_1 = x_1 \left[1 + \frac{2m_1 m_2}{(m_1 + m_2)^2} n(x_1 \gamma_1 \chi_{11} + x_2 \gamma_{12} \chi_{12})\right] \tag{112}$$

$$L_{11} = \frac{x_1^2 \chi_{11}}{\lambda_{tr}^0} + \frac{x_1 x_2 \chi_{12}}{2\lambda_{12tr}^0 A_{12}^* (m_1 + m_2)^2} \times$$

$$\left[\frac{15}{2} m_1^2 + \frac{25}{4} m_2^2 - 3m_2^2 B_{12}^* + 4m_1 m_2 A_{12}^*\right] \tag{113}$$

y_2 and L_{22} can be obtained by exchange of subscripts. The quantities L_{12} and κ'_{mix} are given by

$$L_{12} = -\frac{m_1 m_2 x_1 x_2 \chi_{12}}{2\lambda_{12tr}^0 A_{12}^* (m_1 + m_2)^2} \left[\frac{55}{4} - 3B_{12}^* - 4A_{12}^*\right] \tag{114}$$

and

$$\kappa'_{mix} = -\frac{16}{5\pi} \frac{10}{9} n^2 \left(\frac{x_1^2}{4} \lambda_{1tr}^0 \gamma_1^2 \chi_{11} + \frac{x_2^2}{4} \lambda_{2tr}^0 \gamma_2^2 \chi_{22} + \frac{2 m_1 m_2}{(m_1 + m_2)^2} x_1 x_2 \gamma_{12}^2 \chi_{12}\right) \tag{115}$$

The internal contribution to the thermal conductivity is written as

$$\lambda_{mix}(int) = \sum_{i=1}^{2} \left[\frac{\lambda_i^0 - \lambda_{itr}^0}{\chi_{ii}}\right] \Big/ \left[1 + \sum_{\substack{j=1 \\ j \neq 1}}^{2} \frac{x_j \lambda_{itr}^0 \chi_{ij}}{x_i \lambda_{ijtr}^0 \chi_{ii}}\right] \tag{116}$$

which is a simplified form of the Hirschfelder-Euken expression (Equation 98), modified so as to apply to elevated density.

a. Estimation Techniques

As it stands, the Eucken expression for pure dense fluids can, in principle, be used to predict the thermal conductivity, so long as the proper choice of the rigid sphere diameter and the RDF is made.[2,4] The section on viscosity showed that such a choice is possible, but that the resulting viscosity is not accurately predicted. The same is true of the thermal conductivity.

The use of correlations based on the work of Dymond[49] is still at a rudimentary stage for thermal conductivity. Although some empirical success has been achieved[107] in correlating the thermal conductivity of liquids, the scheme is not at the stage where it can be used over all of the thermodynamic surface.

A number of workers[113-118] have reported that the excess thermal conductivity of the SCF is independent of temperature. Like the excess viscosity, this independence is not at the moment substantiated by any theory. In spite of the lack of theoretical justification, this finding can be used as the basis for an empirical prediction scheme. If the thermal conductivity is known along one isotherm and $\lambda^0(T)$ can be calculated, it is possible to predict the thermal conductivity in the entire region of thermodynamic space above the critical temperature. If, on the other hand, the thermal conductivity at high pressures is not available along any isotherm, one has to resort to more general schemes that are based on the above observation. One such method developed by Stiel and Thodos[114] and applicable to nonpolar fluids yields a formula for the excess thermal conductivity in the form

$$\Gamma \, Z_c^5 \Delta\lambda \; (\rho) \; \begin{array}{ll} = 0.0122 \; [\exp \; (0.535 \; \rho_r) - 1] & \rho_r < 0.5 \\ = 0.014 \; [\exp \; (0.67 \; \rho_r) - 1.069] & 0.5 \le \rho_r < 2.0 \\ = 0.0026 \; [\exp \; (1.155 \; \rho_r) + 2.016) & 2.0 \le \rho_r < 2.8 \end{array} \qquad (117)$$

where λ is in units of milliwatts per meter per degree Kelvin, Z_c is the critical compressibility, ρ_r is the reduced density, $\rho_r = \rho/\rho_c$, and Γ is given by

$$\Gamma = 0.21 \left[\frac{T_c \, M^3}{P_c^{\,4}} \right]^{1/6} \qquad (118)$$

where Γ is in units of meters per degree Kelvin per milliwatt, M is in grams per mole, T_c is in degrees Kelvin, and P_c is in bars. Errors as large as 10 to 20% can be expected for some fluids.

In order to estimate the thermal conductivity of dense gas mixtures, the preferred scheme is similar to that used for the viscosity. It is based on Enskog's dense mixture theory summarized by Equations 110 to 116. In a fashion analogous to that for viscosity, the pseudo-RDFs are estimated from the thermal conductivity of pure components using[59,115]

$$\overline{\chi}_i(n, T) = \frac{B_\lambda}{2} \left[\frac{\lambda_i - \lambda_{itr} \, \gamma_i n}{n^2 \, \gamma_i^2 \, \lambda_{itr}^0} \right]$$

$$\pm \frac{B_\lambda}{2} \left[\left(\frac{\lambda_i - \lambda_{itr}^0 \gamma_i n}{2 \, n^2 \, \gamma_i^2 \, \lambda_{itr}^0} \right)^2 - \frac{\lambda_i^0}{B_\lambda \, n^2 \, \gamma_i^2 \, \lambda_{itr}^0} \right]^{1/2} \qquad (119)$$

where

$$B_\lambda = 1.8764 \qquad (120)$$

Again a switch from a solution $\overline{\chi}_i^{\,-}$ to $\overline{\chi}_i^{\,+}$ is performed at density n* such that[59,115]

$$\left[\frac{d\lambda_i(n, T)}{dn} \right]_T = \frac{\lambda_i(n, T)}{n} \qquad (121)$$

Then, using the equation

$$\frac{\lambda_i \, (n^*, T)}{n^* \, \gamma_i \, \lambda_{itr}^0} = 1 + 1.4600 \, \sqrt{\frac{\lambda_i}{\lambda_{itr}}} \qquad (122)$$

a value of γ_i can be determined and a unique pseudo-RDF $\overline{\chi}_i$ can be constructed over the density range. The mixing rule given in Equation 62 can be used to construct $\overline{\chi}_{ij}$ while γ_{ij} is given by

$$\gamma_{ij} = \frac{1}{8} \, (\gamma_i^{1/3} + \gamma_j^{1/3})^3 \qquad (123)$$

The thermal conductivity can then be evaluated using Equations 110 to 116. The scheme has been tested[115] on a number of mixtures, and is able to predict the thermal conductivity within ±5%.

If thermal conductivity data for the pure components are unavailable, the correlation scheme proposed by Ely and Hanley[103] can be used. The scheme is applicable to pure gases as well as mixtures and can be used to supplement the estimates given by the Stiel and Thodos[114] method. The prediction scheme is based on the principle of corresponding states and is analogous to that for viscosity. The thermal conductivity of the mixture is given by

$$\lambda_{mix}(\rho, T) = \lambda'_{ref} (\rho_{ref}, T_{ref}) FX_\lambda + \lambda''_{mix} (T) \tag{124}$$

where F is given by Equations 46 to 57 and X_λ is the correction for noncorrespondence[98]

$$X_\lambda^{2/3} = \left[1 - \frac{T}{g_x} \left(\frac{\partial g_x}{\partial T} \right)_{V_{ref}} \right] \frac{0.288}{Z_x^c} \tag{125}$$

The quantity λ''_{mix} is the contribution of the internal degrees of freedom

$$\lambda''_{mix} (T) = \sum_i \sum_j x_i x_j \lambda''_{ij} \tag{126}$$

$$\frac{2}{\lambda''_{ij}} = \frac{1}{\lambda''_i} + \frac{1}{\lambda''_j} \tag{127}$$

$$\frac{\lambda''_i M_i}{\eta_i^0} = 1.32 N_A \left[C_{vi}^0 - \frac{3}{2} k \right] \tag{128}$$

where i and j refer to the components of mixture, T_{ref} and ρ_{ref} are given by Equations 47 and 48, and $\lambda'(\rho, T)$ is the thermal conductivity of the reference fluid to which everything has been scaled. The reference fluid in this scheme is taken to be methane and its thermal conductivity correlation is readily available.[25]

The calculation of thermal conductivity therefore requires critical parameters, molecular weights, Pitzer's acentric factors,[39] ideal heat capacities, and viscosity of the pure components. It has been tested[103] mainly on organic, nonpolar mixtures and average absolute deviations of less than 7% were found from experimental data.

The thermal conductivity exhibits a strong divergence in the critical region. The magnitude of the critical enhancement $\Delta\lambda^c$ is much larger than that for any other transport property and extends over a considerable range of density and temperature. Because this contribution is discussed elsewhere in this volume, it is omitted here.

REFERENCES

1. **Edmonds, B.,** *PPDS — Physical Property Data Service,* CODATA 58, Pergamon Press, Oxford, 1985; **Laesecke, A., Stephan, K., and Krauss, R.,** The MIDAS data bank system for the transport properties of fluids, *Int. J. Thermophys.,* 7, 973, 1986.
2. **Chapman, S. and Cowling, T. G.,** *The Mathematical Theory of Non-Uniform Gases,* Cambridge University Press, Cambridge, U.K., 1964.
3. **Maitland, G. C., Rigby, M., Smith, E. B., and Wakeham, W. A.,** *Intermolecular Forces: Their Origin and Determination,* Clarendon Press, Oxford, 1981.
4. **Kestin, J. and Wakeham, W. A.,** *Transport Properties of Fluids: Thermal Conductivity, Viscosity and Diffusion Coefficients,* Vol. 1, CINDAS Data Ser. Material Properties, Ho, C. Y., Ed., Hemisphere Publishing, New York, 1988.
5. **Vesovic, V., Wakeham, W. A., Olchowy, G. A., Sengers, J. V., Watson, J. T. R., and Millat, J.,** The transport properties of carbon dioxide, *J. Phys. Chem. Ref. Data,* 19, 763, 1990.
6. **Stephan, K., Krauss, R., and Laesecke, A.,** Viscosity and thermal conductivity of nitrogen for a wide range of fluid states, *J. Phys. Chem. Ref. Data,* 16, 993, 1987.

7. **Kestin, J., Sengers, J. V., Kamgar-Parsi, B., and Levelt Sengers, J. M. H.,** Thermophysical properties of fluid H_2O, *J. Phys. Chem. Ref. Data,* 13, 175, 1984.

8. **Holland, P. M., Hanley, H. J. M., Gubbins, K. E., and Haïle, J. M.,** A correlation of the viscosity and thermal conductivity data of gaseous and liquid propane, *J. Phys. Chem. Ref. Data,* 8, 559, 1979.

9. **Watson, J. T. R.,** The dynamic viscosity of ammonia, *J. Phys. Chem. Ref. Data,* in preparation.

10. **Swindels, J. F., Coe, J. R., and Godfrey, T. B.,** Absolute viscosity of water at 20°C, *J. Res. Natl. Bur. Stand.,* 48, 1, 1952.

11. **Kestin, J., Ro, S. T., and Wakeham, W. A.,** Viscosity of the noble gases in the temperature range 25°—700°C, *J. Chem. Phys.,* 56, 4119, 1972.

12. **Kestin, J., Sokolov, M., and Wakeham, W. A.,** Theory of capillary viscometers, *Appl. Sci. Res.,* 27, 241, 1973.

13. **Nagashima, A. and Tanishita, I.,** Viscosity measurements of water and steam at high temperatures and high pressures, *Bull. JSME,* 12, 1467, 1969.

14. **Kinoshita, H., Abe, S., Nagashima, A., and Tanishita, I.,** Viscosity of D_2O at high temperatures and high pressures, in *Proc. 8th Int. Conf. Properties of Water and Steam,* Vol. 1, Giens, France, 1974.

15. **Kestin, J. and Newell, G. R.,** Theory of oscillating-type viscometer: the oscillating cup, *Z. Angew. Math. Phys.,* 8, 433, 1957.

16. **Newell, G. F.,** Theory of oscillating-type viscometer. V. Disc oscillating between fixed plates, *Z. Angew Math. Phys.,* 10, 160, 1959.

17. **Retsina, R., Richardson, S. M., and Wakeham, W. A.,** The theory of a vibrating rod densimeter and viscometer, *Appl. Sci. Res.,* 43, 127, 1986; 43, 325, 1987.

18. **Kestin, J. and Wang, H. E.,** Corrections for the oscillating-disc viscometer, *J. Appl. Mech.,* 79, 197, 1957.

19. **Kestin, J. and Shankland, I. R.,** A re-evaluation of the relative measurement of viscosity with oscillating-disc viscometers, *Z. Angew. Math. Phys.,* 32, 533, 1981.

20. **Kestin, J. and Leidenfrost, W.,** The effect of pressure on the viscosity of five gases, in *Thermodynamic and Transport Properties of Gases, Liquids and Solids,* Touloukin, Y. S., Ed., American Society of Mechanical Engineers/McGraw-Hill, New York, 1959, 321.

21. **Iwasaki, H. and Takahashi, M.,** Viscosity of carbon dioxide and ethane, *J. Chem. Phys.,* 74, 1930, 1981.

22. **Vogel, E. and Strehlow, T.,** Temperature dependence and initial density dependence of the viscosity of *n*-hexane vapour, *Z. Phys. Chem. Leipzig,* 269, 897, 1987.

23. **Kestin, J. and Khalifa, H. E.,** Measurement of logarithmic decrements through the measurement of time, *Appl. Sci. Res.,* 32, 483, 1976.

24. **Golubev, I. F.,** *Viscosity of Gases and Gas Mixtures,* Israel Program for Scientific Translations, Jerusalem, 1970.

25. **Hanley, H. J. M., Haynes, W. M., and McCarty, R. D.,** The viscosity and thermal conductivity coefficients for dense gaseous and liquid methane, *J. Phys. Chem. Ref. Data,* 6, 597, 1977.

26. **Hanley, H. J. M., Gubbins, K. E., and Murad, S.,** A correlation of the existing viscosity and thermal conductivity data of gaseous and liquid ethane, *J. Phys. Chem. Ref. Data,* 6, 1167, 1977.

27. **Holland, P. M., Eaton, B. E., and Hanley, H. J. M.,** A correlation of the viscosity and thermal conductivity data of gaseous liquid ethylene, *J. Phys. Chem. Ref. Data,* 12, 917, 1983.

28. **Snider, R. F.,** Quantum mechanical modified Boltzmann equation for degenerate internal states, *J. Chem. Phys.,* 32, 1051, 1960.

29. **Waldmann, L.,** *Quantum Theoretical Transport Equations for Polyatomic Gases in Statistical Mechanics of Equilibrium and Non-Equilibrium,* Meixner, J., Ed., North Holland, Amsterdam, 1965.

30. **McCourt, F. R. W., Beenakker, J. J. M., Köhler, W. E., and Kuščer, I.,** *Non-Equilibrium Phenomena in Polyatomic Gases,* Oxford University Press, 1990.

31. **Aziz, R. A.,** *Interatomic Potentials for Rare Gases: Pure and Mixed Interactions in Inert Gases,* Springer Ser. Chem. Phys., Vol. 34, Klein, M. L., Ed., Springer-Verlag, Berlin, 1984.

32. **Bernstein, R. B., Ed.,** *Atom-Molecule Collision Theory: A Guide for the Experimentalist,* Plenum Press, New York, 1979.

33. **Nyeland, C. and Billing, G. D.,** Transport property coefficients of diatomic gases: internal-state analysis for rotational and vibrational degrees of freedom, *J. Phys. Chem.,* 92, 1752, 1988.

34. **Maitland, G. C., Mustafa, M., Wakeham, W. A., and McCourt, F. R.,** An essentially exact evaluation of transport cross-sections for models of the helium-nitrogen interaction, *Mol. Phys.,* 61, 359, 1987.

35. **Dickinson, A. S. and Lee, M. S.,** Classical trajectory calculations for anisotropy-dependent cross-sections for He-N_2 mixtures, *J. Phys. B: Atom. Mol. Phys.,* 19, 3091, 1986.

36. **Wong, C. C. K., McCourt, F. R. W., and Dickinson, A. S.,** A comparison between classical trajectory and IOS calculations of transport and relaxation cross-sections for N_2-He mixtures, *Mol. Phys.,* 66, 1235, 1989.

37. **Monchick, L. and Mason, E. A.,** Transport properties of polar gases, *J. Chem. Phys.,* 35, 1676, 1961.

38. **Mason, E. A. and Monchick, L.**, Heat conductivity of polyatomic and polar gases, *J. Chem. Phys.*, 36, 1622, 1962.

39. **Reid, R. C., Prausnitz, J. M., and Poling, B. E.**, *The Properties of Gases and Liquids*, 4th ed., McGraw-Hill, New York, 1987.

40. **Kestin, J., Ro, S. T., and Wakeham, W. A.**, An extended law of corresponding states for the equilibrium and transport properties of the noble gases, *Physica*, 58, 165, 1972.

41. **Kestin, J., Knierim, K., Mason, E. A., Najafi, B., Ro, S. T., and Waldman, M.**, Equilibrium and transport properties of the noble gases and their mixtures at low density, *J. Phys. Chem. Ref. Data*, 13, 229, 1984.

42. **Boushehri, A., Bzowski, J., Kestin, J., and Mason, E. A.**, Equilibrium and transport properties of eleven polyatomic gases at low density, *J. Phys. Chem. Ref. Data*, 16, 445, 1987.

43. **Lucas, K.**, *VDI-Warmeatlas, Abschmitt DA Berechnungsmethoden fur Staffeigen-Schaften*, Verin Deustcher Ingenieure, Dusseldorf, 1984.

44. **Ferziger, J. H. and Kaper, H. G.**, *Mathematical Theory of Transport Processes in Gases*, North Holland, Amsterdam, 1972.

45. **Dorfman, J. R. and Van Beijeren, H.**, The kinetic theory of gases, in *Statistical Mechanics, Part B: Time-Dependent Processes*, Berne, B. J., Ed., Plenum Press, New York, 1977.

46. **Thorne, H. E.**, quoted in Ref. 2, p. 292.

47. **Di Pippo, R., Dorfman, J. R., Kestin, J., Khalifa, H. E., and Mason, E. A.**, Composition dependence of the viscosity of dense gas mixtures, *Physica*, 86A, 205, 1977.

48. **Hanley, H. J. M., McCarty, R. D., and Cohen, E. D. G.**, Analysis of the transport coefficients for simple dense fluids: application of the modified Enskog theory, *Physica*, A60, 322, 1972.

49. **Dymond, J. H.**, Hard-sphere theories of transport properties, *Chem. Soc. Rev.*, 14, 317, 1985.

50. **Dymond, J. H.**, Corrections to the Enskog theory for viscosity and thermal conductivity, *Physica*, 144B, 267, 1987.

51. **Dymond, J. H.**, Modified hard sphere theory for transport properties of fluids over the whole density range, *Chem. Phys.*, 17, 101, 1976.

52. **Jossi, J. A., Stiel, L. I., and Thodos, G.**, The viscosity of pure substances in the dense gaseous and liquid phases, *AIChE J.*, 8, 59, 1962.

53. **Stiel, L. I. and Thodos, G.**, The viscosity of polar substances in the dense gaseous and liquid regions, *AIChE J.*, 10, 275, 1964.

54. **Ely, J. F. and Hanley, H. J. M.**, Prediction of transport properties. I. Viscosity of fluids and mixtures, *Ind. Eng. Chem. Fundam.*, 20, 323, 1981.

55. **Hanley, H. J. M.**, Prediction of transport properties application of basic theory, *Rev. Port. Quim.*, 25, 27, 1983.

56. **Sandler, S. I. and Fiszdon, J. K.**, On the viscosity and thermal conductivity of dense gases, *Physica*, 95A, 602, 1979.

57. **Vesovic, V. and Wakeham, W. A.**, The prediction of the viscosity of dense gas mixtures, *Int. J. Thermophys.*, 10, 125, 1989.

58. **Vesovic, V. and Wakeham, W. A.**, Prediction of the viscosity of fluid mixtures over wide ranges of temperature and pressure, *Chem. Eng. Sci.*, 44, 2181, 1989.

59. **Kestin, J. and Wakeham, W. A.**, Calculation of the influence of density on thermal conductivity of gaseous mixtures, *Ber. Bunsenges. Phys. Chem.*, 84, 762, 1980.

60. **Bird, R. B., Stewart, W. E., and Lightfoot, E. N.**, *Transport Phenomena*, John Wiley & Sons, New York, 1960, 498.

61. **Hirschfelder, J. O., Curtiss, C. F., and Bird, R. B.**, *Molecular Theory of Gases and Liquids*, John Wiley & Sons, New York, 1964.

62. **Marrero, T. R. and Mason, E. A.**, Gaseous diffusion coefficient, *J. Phys. Chem. Ref. Data*, 1, 1, 1972.

63. **Cussler, E. L.**, *Diffusion: Mass Transfer in Fluid Systems*, Cambridge, Cambridge, U.K., 1984.

64. **Swaid, I. and Schneider, G. M.**, Determination of binary diffusion coefficients of benzene and some alkylbenzenes in supercritical CO_2, *Ber. Bunsenges. Phys. Chem.*, 83, 969, 1979.

65. **Feist, R. and Schneider, G. M.**, Determination of binary diffusion coefficients of benzene, phenol, naphthalene and caffeine in supercritical CO_2 with supercritical fluid chromatography, *Sep. Sci. Technol.*, 17, 261, 1980.

66. **Kopner, A., Hamm, A., Ellert, J., Feist, R., and Schneider, G. M.**, Determination of binary diffusion coefficients in supercritical chlorotrifluoromethane and SF_6 with supercritical fluid chromatography, *Chem. Eng. Sci.*, 42, 2213, 1987.

67. **Takahashi, S. and Hongo, M.**, Diffusion coefficients of gases at high pressures in the CO_2-C_2H_6 system, *J. Chem. Eng. Jpn.*, 15, 57, 1982.

68. **Robjohns, H. L. and Dunlop, P. J.**, Diffusion and thermal diffusion in some binary mixtures of the major components of air, *Ber. Bunsenges. Phys. Chem.*, 88, 1239, 1984.

69. **Dunlop, P. J. and Bignell, C. M.**, Diffusion and thermal diffusion in binary mixtures of methane with noble gases, *Physica A*, 145, 584, 1987.

70. **Balenovic, Z., Myers, M., and Giddings, J. C.,** Binary diffusion in dense gases to 1360 atm by the chromatographic peak-broadening method, *J. Chem. Phys.,* 52, 915, 1970.
71. **Alizadeh, A., Nieto de Castro, C. A., and Wakeham, W. A.,** The theory of the Taylor dispersion technique for liquid diffusivity measurements, *Int. J. Thermophys.,* 1, 243, 1980.
72. **Wilsch, A., Feist, R., and Schneider, G. M.,** Capacity ratios and diffusion coefficients of low volatile organic compounds in supercritical carbon dioxide from supercritical fluid chromatography (SFC), *Fluid Phase Equilibria,* 10, 299, 1983.
73. **Saad, H. and Gulari, E.,** Diffusion of liquid hydrocarbons in supercritical CO_2 by photon correlation spectroscopy, *Ber. Bunsenges. Phys. Chem.,* 88, 834, 1984.
74. **Saad, H. and Gulari, E.,** Diffusion of carbon dioxide in heptane, *J. Phys. Chem.,* 88, 136, 1984.
75. **Jonas, J. and Lamb, D. M.,** Transport and intermolecular interactions in compressed supercritical fluids in *Supercritical Fluids,* ACS Symp. Ser. No. 329, Squires, T. G. and Paulaitis, M. E., Eds., American Chemical Society, Washington, D.C., 1987.
76. **Lamb, W. J., Hoffmann, G. A., and Jonas, J.,** Self-diffusion in compressed supercritical water, *J. Chem. Phys.,* 74, 6875, 1981.
77. **Chandler, D.,** Translational and rotational diffusion in liquids. I. Translational single particle correlation functions, *J. Chem. Phys.,* 60, 3500, 1974.
78. **Chandler, D.,** Translational and rotational diffusion in liquids. II. Orientational single particle correlation function, *J. Chem. Phys.,* 62, 1358, 1975.
79. **Sun, C. K. J. and Chen, S. H.,** Diffusion of benzene, toluene, naphthalene and phenanthrene in supercritical 2,3-dimethylbutane, *AIChE J.,* 31, 1905, 1985.
80. **Sung, W. and Stell, G.,** Theory of transport in dilute solutions, suspensions and pure fluids. I. Translational diffusion, *J. Chem. Phys.,* 80, 3350, 1984.
81. **Baleiko, M. O. and Davis, H. T.,** Diffusion and thermal diffusion in binary dense gas mixtures of loaded spheres and rough spheres, *J. Phys. Chem.,* 78, 1564, 1974.
82. **Iomtev, M. B. and Tsekhanskaya, Yu. V.,** Diffusion of naphthalene in compressed ethylene and carbon dioxide, *Russ. J. Phys. Chem.,* 38, 485, 1964.
83. **Tsekhanskaya, Yu. V.,** Diffusion of naphthalene in carbon dioxide near the liquid-gas critical point, *Russ. J. Phys. Chem.,* 45, 744, 1971.
84. **Vinkler, E. G. and Morozov, V. S.,** Measurement of diffusion coefficients of vapours of solids in compressed gases, *Russ. J. Phys. Chem.,* 49, 1404, 1975; 49, 1405, 1975.
85. **Takahashi, S.,** Preparation of generalized chart for the diffusion coefficients of gases at high pressure, *J. Chem. Eng. Jpn.,* 7, 417, 1974.
86. **Kestin, J., Paul, R., Clifford, A. A., and Wakeham, W. A.,** Absolute determination of the thermal conductivity of the noble gases at room temperature and up to 35 MPa, *Physica A,* 100, 349, 1980.
87. **Haran, E. N., Maitland, G. C., Mustafa, M., and Wakeham, W. A.,** The thermal conductivity of Ar, N_2, CO in the temperature range 300—430 K at pressures up to 10 MPa, *Ber. Bunsenges. Phys. Chem.,* 87, 657, 1983.
88. **Millat, J., Ross, M. J., and and Wakeham, W. A.,** Thermal conductivity of nitrogen in the temperature range 177 K to 270 K, *Physica A,* 159, 28, 1989.
89. **Healy, J. J., de Groot, J. J., and Kestin, J.,** The theory of transient hot wire method for measuring thermal conductivity, *Physica,* 82C, 392, 1976.
90. **Le Neindre, B., Johannim, P., and Vodar, B.,** Conductibilite thermique de l'eau en phase liquide, jusqu'au point critique, *Hebd. Seances Acad. Sci.,* 258, 3277, 1964.
91. **Tufeu, R.,** Etude Experimentale en Function de la Temperature et de la Pression de la Conductivite Thermique de l'Ensemble des Gaz Rares et de Melanges Heliumargon, Ph.D. thesis, University of Paris IV, 1971.
92. **Tufeu, R., Le Neindre, B., and Bury, P.,** Determination du coefficient de conductibilite thermique du methane de 25° a 450°C et jusqu'a 1000 bars, *Physica,* 44, 81, 1969.
93. **Lambert, J. D.,** *Vibrational and Rotational Relaxation in Gases,* Clarendon Press, Oxford, 1977.
94. **Monchick, L., Yun, K. S., Mason, E. A.,** Formal kinetic theory of transport phenomena in polyatomic gas mixtures, *J. Chem. Phys.,* 39, 654, 1963.
95. **Monchick, L., Pereira, A. H. G., and Mason, E. A.,** Heat conductivity of polyatomic and polar gases and gas mixture, *Chem. Phys.,* 42, 3241, 1965.
96. **Parker, J. G.,** Rotational and vibrational relaxation in diatomic gases, *Phys. Fluids,* 2, 449, 1959.
97. **Brau, C. A. and Jonkman, M.,** Classical theory of rotational relaxation in diatomic gases, *J. Chem. Phys.,* 52, 477, 1970.
98. **Millat, J., Vesovic, V., and Wakeham, W. A.,** Theoretically based data assessment for the correlation of the thermal conductivity of dilute gas, *Int. J. Thermophys.,* 10, 805, 1989.
99. **Assael, M. J., Millat, J., Vesovic, V., and Wakeham, W. A.,** The thermal conductivity of methane and tetrafluoromethane in the limit of zero density, *J. Phys. Chem. Ref. Data,* 19, 1137, 1990.
100. **Millat, J. and Wakeham, W. A.,** The thermal conductivity of nitrogen and carbon monoxide in the limit of zero density, *J. Phys. Chem. Ref. Data,* 18, 565, 1989.

101. **Uribe, F. J., Mason, E. A., and Kestin, J.,** A correlation scheme for the thermal conductivity of polyatomic gases at low density, *Physica A,* 156, 467, 1989.
102. **Roy, D. and Thodos, G.,** Thermal conductivity of gases, *Ind. Eng. Chem. Fundam.,* 7, 529, 1968; 9, 71, 1970.
103. **Ely, J. F. and Hanley, H. J. M.,** Predictions of transport properties. II. Thermal conductivity of pure fluids and mixtures, *Ind. Eng. Chem. Fundam.,* 22, 90, 1983.
104. **Wassiljewa, A.,** Conductivity of heat in gas mixtures, *Phys. Z.,* 5, 737, 1904.
105. **Mason, E. A. and Saxena, S. C.,** Approximate formula for the thermal conductivity of gas mixture, *Phys. Fluids,* 1, 361, 1958.
106. **Mason, E. A., Khalifa, H. E., Kestin, J., and Di Pippo, R.,** Composition dependence of the thermal conductivity of dense gas mixtures, *Physica,* 91A, 377, 1978.
107. **Li, S. F. Y., Maitland, G. C., and Wakeham, W. A.,** The thermal conductivity of liquid hydrocarbons, *High. Temp.-High. Press.,* 17, 241, 1985.
108. **Vesovic, V. and Wakeham, W. A.,** Thermal conductivity of carbon dioxide, *High Temp.-High Press.,* 21, 225, 1989.
109. **Carmichael, L. T., Berry, V., and Sage, B. H.,** Thermal conductivity of fluids: ethane, *J. Chem. Eng. Data,* 8, 281, 1963.
110. **Carmichael, L. T., Reamer, H. H., and Sage, B. H.,** Thermal conductivity of fluids: methane, *J. Chem. Eng. Data,* 11, 52, 1966.
111. **Richter, G. N. and Sage, B. H.,** Thermal conductivity of fluids. Ammonia, *J. Chem. Eng. Data,* 9, 75, 1964.
112. **Le Neindre, B., Garrabos, Y., and Tufeu, R.,** Thermal conductivity in supercritical fluids, *Ber. Bunsenges. Phys. Chem.,* 88, 916, 1984.
113. **Owens, E. J. and Thodos, G.,** Thermal conductivity-reduced state correlation for inert gases, *AIChE J.,* 3, 454, 1957.
114. **Stiel, L. I. and Thodos, G.,** The thermal conductivity of non-polar substances in the dense gaseous and liquid regions, *AIChE J.,* 10, 26, 1964.
115. **Vesovic, V. and Wakeham, W. A.,** The prediction of the thermal conductivity of dense gas mixtures, *Int. J. Thermophys.,* in press.

Part II:
Experimental Work and Applications

Chapter 7

THERMOPHYSICAL PROPERTY DATA FOR SUPERCRITICAL FLUID EXTRACTION DESIGN

Thomas J. Bruno

TABLE OF CONTENTS

ABSTRACT

The design of any industrial-scale separation process requires some degree of knowledge about the thermophysical and chemical properties of the materials to be separated. This knowledge can take the form of experimentally measured data, particular for that system, or predictions obtained from a suitable mathematical model such as an equation of state (EOS) or empirical data correlation. It is generally considered better, from the process designer's point of view, to use predicted properties, since calculations are far more economical to perform than are experimental studies. Unfortunately, there are relatively few reliable predictive methods currently available that have sufficient accuracy to be used in the design of supercritical fluid extraction (SFE) processes. This is because such processes usually involve high pressures and large, often polar, solutes. In addition, SFE processes often require operation near solvent critical points, where even good models of well-studied systems become marginal. This lack of data is slowly improving as more experimental and applied theoretical studies are completed. In this chapter, the more important thermophysical properties needed for supercritical fluid extraction design are discussed from the experimental point of view. Both the solvent and solute are discussed in terms of equilibrium and transport properties.

I. INTRODUCTION

One will usually find a common philosophical thread running through the evolutionary progression that characterizes the design of an industrial chemical process. This philosophy requires the existence of a clear, organized structure, from the inception of an idea to the production of a commodity.[1,2] The first stage of chemical process development often occurs in the research laboratory, outside of the engineering community. This is the basic chemical research phase, which can include work on reaction chemistry, experimental thermodynamics and kinetics, and basic theory. Indeed, the history of technology is full of examples of how basic research, often taking a tortuous path, has led to the introduction of a successful new product into the marketplace.

The second step, after the recognition of some of the applicable results of basic research, is usually the market research phase. This is a critical stage in the development of any new process, because there is usually a built-in bias toward the marginal improvement of an old process rather than the development of a new one.[3] This understandable caution or inertia is rooted in the fact that a given firm will have technological, market, and organizational experience in some areas but not in others. In addition, one must consider the business reality that while one company may take risks and cover research and development costs, many other (competing) companies may share widely in the resulting benefits. The patent process and industrial secrecy can sometimes be only marginally effective in protecting the competitive position of a company. The market research stage of process development must therefore carefully consider not only economic factors (for example, supply and demand) but also the technological state-of-the-art.

The next stage in the progression is process engineering and design. This stage of process development incudes the specification of individual unit operations, the determination of the optimal conditions of each unit operation, and the effective control of each operation. To efficiently perform this task, the engineer must have available a reliable and extensive database of thermophysical properties of the commodity compounds to be produced and of the chemicals that are to be used in the process. These properties include both equilibrium properties (such as PVT relations, vapor-liquid equilibria, and solubility) and transport properties (such as thermal conductivities, diffusivities and viscosities). These properties may be measured experimentally for each given system, or predicted from a mathematical model. The availability of measured data can never begin to satisfy the need for knowledge on the millions of chemical compounds of interest to science and engineering.[4] The relatively high cost of experimental measurements and the sheer lack of enough technically trained personnel make this situation an inescapable aspect of technology. The most common resource that the engineer will therefore use to obtain thermophysical properties is predictive (or estimation) methods.

Predictive methods are mathematical models that describe how matter and energy behave on a large scale. These methods are the result of work on basic physical theory, the correlation of existing experimental data with theory, the empirical (no first-principle theoretical basis) description of the behavior of matter, or a combination of these three. These models will often provide estimates that are only approximate and sometimes only accurate in the order of magnitude range. The predictive accuracy will be dependent upon the quality and quantity of experimental data used in the development of the model, and upon the physical and mathematical assumptions inherent in the model. In some engineering applications, marginally accurate models may be adequate for the design task at hand. An example to consider is the use of the ideal gas law to relate mass of air and volumetric flow rate in a large air conditioning system.

The design of industrially viable supercritical fluid extraction (SFE) processes usually represents the opposite extreme of design problems. Many of the existing simple predictive models are not sufficiently accurate to use in the design of these processes.[5] Even the more sophisticated models are subject to serious errors when used for calculations near critical points, an area of great interest in terms of supercritical extraction. An additional complication is that many of the solute molecules of interest in SFE are large and polar, while the solvent molecules (such as carbon dioxide) tend to be small and of low polarity. This makes the thermodynamics less amenable to the usual modeling methods.

Approximate values for the needed thermophysical properties are often inadequate for both the design and control of SFE processes.[6] The phase equilibrium of a system near the critical point is very sensitive to changes in temperature and pressure, and the distribution of components in the phases can be very difficult to predict. This is especially true for multicomponent mixtures, where the relative distribution of the desired product or products (often called the solute or solutes) between a solid phase and a supercritical fluid (SCF) solution phase must be utilized and controlled. This will often constrain SFE processes to operate within narrow regions of economic feasibility, and require very sensitive and robust control schemes. It is therefore desirable at the present time that viable SFE process design be based on an integrated approach of both experimental measurement and correlation with a suitable theoretically based model having good predicitive accuracy.

II. THERMOPHYSICAL PROPERTIES FOR SUPERCRITICAL FLUID EXTRACTION

There are a number of thermophysical properties that are needed in the design of SFE processes. They can be broadly divided into two classifications: pure-component properties and mixture properties. Pure-component properties are needed in the individual characterization of both the solute and the SCF solvent. Mixture properties are needed to describe how the solute and solvent behave when perturbed by one another.

Perhaps the most important pure-component property of the solvent is its PVT surface.[4] The measured PVT surface and correlation of the data using an accurate equation of state (EOS) are needed to provide a global knowledge of the solvent density, one of the most important solvation parameters in the design of efficient SFE processes. If the data are sufficiently accurate and extensive, correlation with either the Schmidt-Wagner[7] or modified Benedict-Webb-Rubin equation[8] can provide density values approaching an accuracy of 1%, even in the vicinity of the critical region. The PVT surface can also be the source of many other important derived thermodynamic parameters through Maxwell's equations. While some limited PVT data do exist for solvent-cosolvent mixtures, the application of mathematical mixing rules to combine the properties of pure solvent components is currently the predominant route to predicting mixed-solvent system densities. It is therefore important to at least have adequate PVT data for each solvent component to allow an accurate correlation to be performed.

The vapor pressure of the pure solute is an important property for a variety of reasons. It is directly related to the solubility of the solute in the supercritical solvent at given density. An increase in temperature will increase the vapor pressure of the solute, and will often lead to a higher solute solubility at a given solvent density. The solute vapor pressure is often needed to estimate the critical parameters of a solute if there are no measured values available. This is especially important for solutes that may be expected to decompose before a critical point is reached. The vapor pressure is also needed in determining the acentric factor of a solute, which is a parameter in many useful EOSs.

The last pure-component property may be more correctly considered as a chemical property rather than as a thermophysical property. The stability of both the solute and the solvent is an important consideration in process design, since chemical degradation can often occur at moderate and high temperatures and pressures. Degradation can result from several causes. The operating temperatures and pressures, process residence times, and materials of construction all play a role in affecting decomposition of solutes and solvents, and therefore have a strong influence on product quality.

The most important mixture property discussed here is the solubility of the solute in the solvent. The solubility is needed for the determination of EOS interaction coefficients to allow prediction of the phase distribution of the solute. This information is used to determine the appropriate conditions of operation for process components such as extractors and separators. This chapter reviews techniques that are appropriate for the measurement of pure solute solubilities and the solubilities of components in a mixture. The latter is an important aspect of solute solubility, since very often the solubility of one component is influenced by the presence of other components in the matrix.

The binary interaction diffusion coefficient is also of value in design. This parameter is useful in sizing extractors, especially in the specification of height. The diffusivity enters through the hydrodynamic Schmidt number, Sc, defined as the kinematic viscosity multiplied by the binary interaction diffusion coefficient.

III. PURE-COMPONENT PROPERTIES

A. DETERMINATION OF THE PVT SURFACE

Perhaps the single most important property of a compressible fluid is its PVT behavior. An accurate and precise description of this three-dimensional surface can, in principle, be used to calculate most equilibrium properties of the fluid. In addition, the PVT surface of a fluid is needed in the interpretation and analysis of data for many other experimental measurements (for example, thermal conductivity and viscosity).[4] It is also useful in engineering calculations on fluid handling systems. Extensive PVT data at elevated temperatures and pressures for technically important fluids are often scarce or unavailable. While it may be possible to arrive at estimates of PVT surfaces (and the derived thermodynamic properties obtained from the PVT surface) using, for example, the principle of corresponding states, it is far more accurate to perform the measurements. This is especially true in the near-critical region.

In the case of mixtures, the historical approach has been the mathematical combination of the pure fluid properties through the pure fluid EOS parameters. This is done using a set of equations called mixing rules, provided that the PVT surfaces of the fluid components have been measured in the temperature and pressure range of interest, and then correlated

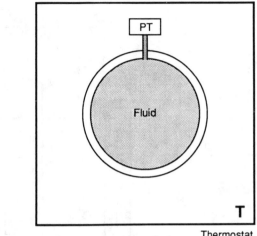

(a)

Thermostat

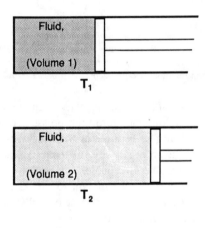

$T_2 > T_1$

FIGURE 1. A schematic diagram illustrating conceptually the measurement of the pressure-volume temperature behavior of a fluid.

with an accurate EOS. The EOS coefficients for each fluid are then mathematically "mixed", often using the mole fractions of the component fluids as the weighing factors.

The measurement of the PVT surface of a fluid is a procedure that seems very simple in concept, but in practice it can be deceptively complex. Figure 1 shows in schematic form two possible approaches. In Figure 1a, a vessel of known volume containing the fluid is held at a desired temperature, while the pressure is measured. This system is called an isochoric or constant volume experiment. In Figure 1b, a piston-cylinder system containing the fluid is changed in temperature, and the change in volume is noted. While the piston-cylinder concept is useful to visualize the experiment, there are many complexities that make it unworkable in practice. The isochoric measurement also has a number of experimental pitfalls, despite the inherent simplicity of the concept. These difficulties, such as temperature gradients and system volume changes as a function of temperature and pressure, must be addressed by design or by correction factors if high accuracy PVT data are to be obtained.

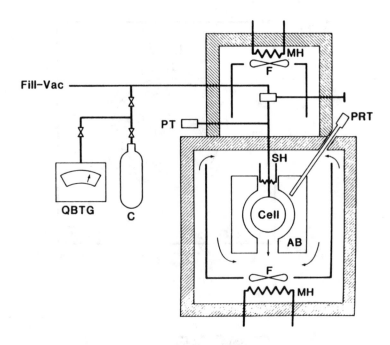

FIGURE 2.   High temperature, high pressure isochoric P-V-T apparatus used at NIST. (Abbreviations: AB = aluminum block, F = fan, MH = main heater, SH = shimming heater, PRT = platinum resistance thermometer, C = sample cylinder, QBTG = quartz Bourdon tube gauge, PT = pressure transducer.)

In order to illustrate some of the techniques involved in PVT measurements, a schematic diagram of the automated high temperature high pressure isochoric apparatus of the Thermophysics Division of the National Institute of Standards and Technology (NIST) is shown in Figure 2. The apparatus incorporates several features that make it well suited for measurements on SCFs.[9] The apparatus is capable of operation over a wide range of conditions, covering temperatures from just above ambient to 600°C, and fluid pressures to over 35 MPa. The ultimate operating pressure is at least partially determined by the temperature, due to the temperature dependence of construction material strength parameters. Density measurements can be made in two separate ways, and are independent of the measured pressure-temperature data that are obtained.

While mixed liquid baths (containing water or a thermally stable silicone fluid) are usually used to provide precise temperature control in PVT measurements, the higher temperatures involved with SCFs makes this approach impossible. Instead, temperature control is usually provided using modified commercially available forced-air ovens. Air is usually used as the convective heat transfer fluid in the ovens, but when toxic or flammable fluids are studied, the oven must be continuously purged with a flow of nitrogen that is exhausted into a surrounding fume hood. The flow rate of the nitrogen purge line is determined by the explosive or flammability limits of the fluid being studied.

The fluid vessel is a spherical cell that has a nominal internal volume of 120 ml, and is constructed from 316L stainless steel (AISI designation). The spherical geometry was adopted to allow reliable corrections to be applied for the effect of increasing fluid pressure on the internal volume of the cell.[10,11] While this effect may, at first glance, appear to be small, the change in cell dimensions can have a very strong effect on the perceived or apparent density of the fluid being characterized. Although such corrections are possible on other cell shapes (such as cylindrically shaped vessels), corrections applied to spherical cells are generally the most accurate and well characterized. The cell is encased in a large aluminum block which serves to "integrate" out large temperature differences that may be present in the oven. The cell is thermally decoupled from the block by an air space that allows the free flow of the heat transfer gas around the cell. This arrangement reduces temperature differences around the cell to a few tenths of a degree. Further reduction of temperature differences is accomplished using a very low power shimming heater located near the top of the cell.

The temperature of the cell is measured using a NIST-calibrated platinum resistance thermometer (PRT) and a set of standard, thermostatted resistors. Differential thermocouples (referenced to the main PRT) placed in various locations in and around the aluminum block provide an indication of thermal gradients. The internal pressure of the cell is measured by determining the vibrational frequency of a commercial quartz-crystal transducer which is accurate to $\pm 0.05\%$ full scale (0 to 40 MPa). This transducer is calibrated using a dead-weight pressure balance traceable to the NIST primary standard. The volumes of the cell, valves, transfer lines, and pressure transducer were determined by low pressure nitrogen gas expansions into calibrated volume cells, a procedure known as gasometry.

The oven containing the isochoric cell is capable of producing a temperature as high as 600°C. Since it is necessary to locate the pressure transducer and manifold valve in a less hostile environment, a second, smaller forced-air oven maintained at a lower temperature is located above the main oven. The temperature of this upper oven is set at a value for which the fluid density has already been determined (the density measurement is discussed later). This allows more accurate corrections to be made along the measured isochores due to the lower temperature in the upper oven.

The density of the fluid can be determined in two ways. The first is a gravimetric method, in which the fluid is condensed into a sampling cylinder held in a cold trap (in liquid nitrogen, 77 K) while the cell and transfer lines are heated to a known, elevated temperature. The quartz Bourdon tube pressure gauge (shown in Figure 2) provides a measure of the residual pressure of uncondensed fluid remaining in the system. Knowledge of the temperature, pressure, and volume of the system thus allows corrections to be applied for the small amount of fluid not condensed. Corrections to the internal volume of the cell at different pressures allow a density to be assigned to each P-T pair. An alternate route to the density is to perform a separate series of experiments called Burnett expansions. This measurement involves the expansion (at constant temperature) of the fluid from an initial cell volume, V_1, into a second, evacuated cell of volume, V_2, through an isolation valve. After the pressures are measured, V_1 is again isolated from V_2, V_2 is evacuated, and the process is repeated. A Burnett apparatus containing three expansion volumes, part of the NIST facility, is shown schematically in Figure 3. For any given expansion, we can write:

$$\frac{P_i}{P_{i+1}} = \frac{z_i}{z_{i+1}} N_i \qquad (1)$$

In this equation, p is the pressure, Z is the compressibility factor, and N_i is a volume ratio for the ith expansion. The compressibility contains the density information (ρ) by definition ($Z = P/\rho RT$). The details of the data analysis techniques are discussed elsewhere,[12,13] and are beyond the scope of this chapter. The main advantage of the Burnett method of determining density values lies in the use of volume ratios (available from the expansion data), rather than an absolute calibration of system volumes.

Fortunately, a number of fluids important to SFE have been the subject of accurate PVT surface measurements. This list will have to be expanded to include other, less common fluids, as well as cosolvents. Among those fluids for which data are available is pure carbon dioxide,[14] and correlations with accurate EOSs have been performed.[15] An example is the modified Benedict-Webb-Rubin equation, presented in Table 1, along with the coefficients for carbon dioxide.[16]

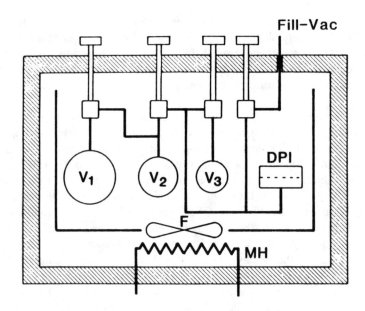

FIGURE 3.   The NIST high-temperature high-pressure Burnett expansion apparatus, showing three vessel volumes. (Abbreviations: F = fan, MH = main heater, DPI = differential pressure indicator.)

B. SOLVENT AND SOLUTE CHEMICAL STABILITY

Although not strictly to be viewed as a thermophysical property, knowledge of the chemical stability of both the solvent system and the solutes to be extracted is of major importance in the proper design of SFE processes. Unfortunately, this aspect of the design process has not received the attention it deserves. It clearly makes little sense to design a separation process that can degrade the product, or in which the solvent itself undergoes decomposition and therefore cannot be recycled. This possibility must be addressed for both solvent and solute early in the design phase. This is especially true for processes operated well above fluid critical points, and those involving fluids other than pure carbon dioxide. There are four main causes of solvent/solute instability: the chemical nature of the solvent or solute(s), the conditions (pressure and temperature) prevailing in the process, the construction materials of the process components, and the residence time of the solvent/solute(s) in the process components.

The chemical decomposition of a number of solvents (and cosolvents) has been noted during the course of other thermophysical property measurements such as the PVT behavior.[17-19] It is therefore desirable to screen fluids for possible decomposition regions before they are specified in extractions. One possible approach is to use a reaction screening apparatus such as that shown in Figure 4.[20] The heart of the apparatus is a small, thick-walled pressure vessel made from 316L stainless steel, which serves as a reaction chamber. The vessel is located in an airtight oven which is maintained in an inert atmosphere (usually a boil-off stream from a liquid nitrogen Dewar flask) under a slightly negative pressure. The apparatus is capable of operation at temperatures up to 250°C and pressures to 130 MPa. The test fluid is held in the vessel for residence times which approximate the conditions of the process. Sampling of the fluid in the vessel can be done in several ways. An on-line high pressure chromatographic sampling valve can be used below test pressures of 30 MPa, and an external sampling cylinder and cold trap can be used at pressures above this value.

The catalytic effects of common component construction materials can also be assessed using an apparatus of this type. Many of the construction materials, such as the stainless steels and glasses, can be obtained as finely divided powders. A test fluid can be held inside

TABLE 1

Functional Form of the Modified BWR Equation of State

$$P = \sum_{n=1}^{9} a_n(T)\rho^n + e^{-\gamma\rho^2} \sum_{n=10}^{15} a_n(T)\rho^{2n-17}$$

$a_1 = RT$	$a_9 = b_{19}/T^2$
$a_2 = b_1T + b_2T^{1/2} + b_3 + b_4/T + b_5/T^2$	$a_{10} = b_{20}/T^2 + b_{21}/T^3$
$a_3 = b_6T + b_7 + b_8/T + b_9/T^2$	$a_{11} = b_{22}/T^2 + b_{23}/T^4$
$a_4 = b_{10}T + b_{11} + b_{12}/T$	$a_{12} = b_{24}/T^2 + b_{25}/T^3$
$a_5 = b_{13}$	$a_{13} = b_{26}/T^2 + b_{27}/T^4$
$a_6 = b_{14}/T + b_{15}/T^2$	$a_{14} = b_{28}/T^2 + b_{29}/T^3$
$a_7 = b_{16}/T$	$a_{15} = b_{30}/T^2 + b_{31}/T^3$
$a_8 = b_{17}/T + b_{18}/T_2$	$+ b_{32}/T^4$

Coefficients for Carbon Dioxide

Parameter	CO_2
P_c	$0.7384325000E+02$
ρ_c	$0.1060000000E+02$
T_c (K)	$0.3042100000E+03$
b(1)	$-0.9818510658E-02$
b(2)	$0.9950622673E+00$
b(3)	$-0.2283801603E+02$
b(4)	$0.2818276345E+04$
b(5)	$-0.3470012627E+04$
b(6)	$0.3947067091E-03$
b(7)	$-0.3255500001E+00$
b(8)	$0.4843200831E+01$
b(9)	$-0.3521815430E+06$
b(10)	$-0.3240536033E-04$
b(11)	$0.4685966847E-01$
b(12)	$-0.7545470121E+01$
b(13)	$-0.3818943540E-04$
b(14)	$-0.4421929339E-01$
b(15)	$0.5169251681E+02$
b(16)	$0.2124509852E-02$
b(17)	$-0.2610094748E-04$
b(18)	$-0.8885333890E-01$
b(19)	$0.1552261794E-02$
b(20)	$0.4150910049E+06$
b(21)	$-0.1101739675E+08$
b(22)	$0.2919905833E+04$
b(23)	$0.1432546065E+08$
b(24)	$0.1085742075E+02$
b(25)	$-0.2477996570E+03$
b(26)	$0.1992935908E-01$
b(27)	$0.1027499081E+03$
b(28)	$0.3776188652E-04$
b(29)	$-0.3322765123E-02$
b(30)	$0.1791967071E-07$
b(31)	$0.9450766278E-05$
b(32)	$-0.1234009431E-02$
γ	$0.8899964400E-02$

Note: Range of application: 215 to 1100 K; pressures to 300 MPa.

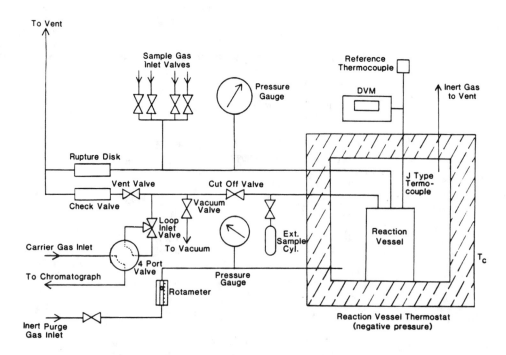

FIGURE 4.   NIST high pressure, high temperature reaction screening apparatus.

the pressure vessel in contact with an appropriate quantity of the powder to simulate worst-case conditions. Surprisingly, the stainless steels have an appreciable catalytic effect and will often be a major cause of chemical decomposition. Materials such as nickel, Teflon®, glasses, and quartz can also be tested by this means.

Several examples can be provided to illustrate the importance of studying the chemical stability of both solvents and solutes in SFE processes. Methanol, which has frequently been used as a modifier or cosolvent in SFE and in SCF chromatography, can be very unstable in the presence of stainless steels.[21] In the course of measuring the PVT properties (using the apparatus described in the previous section), appreciable quantities of hydrogen, carbon monoxide, formaldehyde, as well as acetal and hemiacetal species were detected in the measurement cells. The onset of severe decomposition appeared to be approximately 200°C at any pressure, although other workers have noted the decomposition of methanol/water mixtures in contact with stainless steel at temperatures as low as 100°C.[22] At this temperature, the main reaction products were formaldehyde, hydrogen, and carbon monoxide. Toluene, another fluid which has been used and suggested in higher temperature SFE applications, can also show serious decomposition problems when in contact with stainless steels. Appreciable decomposition is first noticed near the critical temperature of 320.8°C, and becomes quite severe at higher temperatures. A list of decomposition products detected during the measurement of the PVT properties of toluene is provided in Table 2. These measurements were performed using the same high temperature high pressure isochoric PVT apparatus described earlier.

Most of the serious decomposition of SFE solvents occurs at relatively high temperatures. The obvious exception, of course, is methanol. Since methanol is used mainly as a cosolvent or modifier, however, the concentration is usually only a few percent. Ideally, SFE processes are operated at as low a temperature as possible to minimize cost. Many of the solvent decomposition difficulties can therefore be minimized by design. This still leaves the possibility of solute decomposition to be addressed. Many natural products that would be amenable to extraction with an SCF solvent are highly unstable. An example is β-carotene, an extraction process which is described in Chapter 11 of this volume.

TABLE 2
Decomposition Products
Detected in Toluene

Compound	Mol%
Dimethylbiphenyl	15.0
Stilbene	5.7
Diphenylmethane	4.8
Methylbiphenyl	4.8
Benzene	1.4
Benzyltoluene	1.3
Dibenzyl	0.5
Biphenyl	0.1
Fluorene	0.1
Anthracene	0.1
Phenanthrene	0.1
Unidentified components	0.1
Toluene	Balance

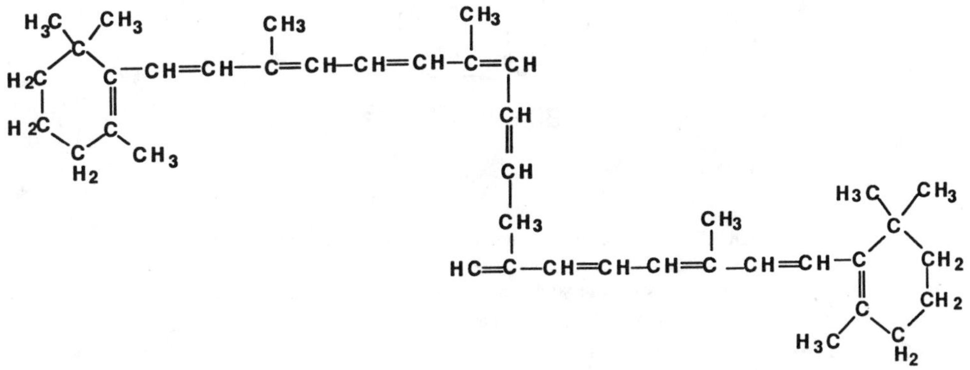

FIGURE 5. The chemical structure of β-carotene.

β-Carotene is a member of the terpene family, and is widely found in plant, marine, and animal life forms. The structure of β-carotene, shown in Figure 5, is that of a tetraterpene (as are all of the approximately 400 known carotenoids), containing eight isoprenoid units. Because of the multiple conjugated double bond system of the molecule, it is very sensitive to light and air oxidation and thermal degradation. The high purity research grade material must be maintained at $-40°C$ under vacuum or an inert atmosphere to prevent decomposition or undesired isomerization (such as between the *cis-* and *trans-* forms). On an industrial scale, it can be stabilized when stored or prepared in oil solutions (especially cottonseed oil) or emulsions, and in a spray-dried form. It is used primarily as a nutritional supplement, especially in the feeds of poultry and cattle. This is due to its ability to enhance the color of beef, poultry, and eggs and because of vitamin A activity that it introduces in these foods. In this respect, β-carotene is also used as a pharmaceutical, since it is the most potent vitamin-A precursor in the human body. It can be converted into as many as two retinol units by an oxidative enzyme system present in the intestinal mucosa of humans and animals. Although 40 of the approximately 400 known carotenoids have some vitamin A activity, the most potent vitamin-A precursor is β-carotene.

In the SCF extraction process discussed in Chapter 11, the extraction of β-carotene from a fermentation broth containing Blakesea Trispora is proposed. The broth consists of a water suspension containing the bacteria and a low-cost soapstock surfactant system. One of the

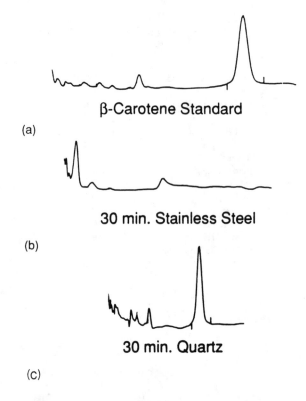

β-Carotene Standard

(a)

30 min. Stainless Steel

(b)

30 min. Quartz

(c)

FIGURE 6.   HPLC chromatograms showing (a) a control sample of β-carotene, (b) β-carotene-water slurry after exposure to powdered stainless steel at 80°C, and (c) β-carotene-water slurry after exposure to powdered quartz, at 80°C.

economically feasible designs that was calculated uses pure carbon dioxide in an extractor maintained at 80°C, at a pressure of 101 MPa, for a residence time of up to 2 h. While such relatively severe conditions are not uncommon in SFE, for example in the extraction of soybean oil,[23] one must address the potential of decomposition or adulteration of the product during the extraction step before proceeding with such a design.

The reaction screening apparatus shown in Figure 4 can be used for this test, either using the pressure vessel as is or using liners inserted into the vessel well in order to minimize the fluid contact with stainless steel. A β-carotene-water slurry that had an approximate composition of 1/99% b.w. was placed in a quartz liner in the well, and the vessel was assembled and placed in the oven. The β-carotene was research grade, and the water was HPLC (high pressure liquid chromatography) grade, and both were used without further purification. The temperature was increased to 80°C, and liquid carbon dioxide was added until approximately 101 MPa pressure was attained. This was done using a refrigerated hand-operated screw pump filled with liquid carbon dioxide. The carbon dioxide flashed to the SCF state in a heat exchanger preceding the pressure vessel (not shown in Figure 4). In addition to tests with pure carbon dioxide, tests were run with powdered quartz and powdered 316 stainless steel in the vessel with the fluid. After the required residence time was simulated, the vessel was cooled and the contents analyzed using an HPLC method.[24] A 25-cm column of octadecylsilane was used for the separation, and an isocratic solvent system of methanol/water (90/10 vol%) was used as the sample solvent and mobile detection was done using a UV-visible (UV-vis) detector (the most common detection method) set at 450 nm, the absorption maximum for β-carotene.

Several representative chromatograms are reproduced in Figure 6. The first chromatogram is a control sample of the pure β-carotene, showing peaks for both the *cis-* and *trans-* isomers. The second chromatogram was measured for the slurry with added 316 stainless steel powder, maintained at 80°C and 101 MPa pressure under carbon dioxide. There is clearly extensive β-carotene decomposition under these circumstances. The extent of decomposition is far less pronounced in the presence of quartz, as shown by the third chromatogram. Other tests were done with Teflon® wool and ground Teflon® added to the vessel, and produced results simlar to those obtained with powdered quartz. These tests with added powdered construction material provide an idea of the most serious decomposition that may be expected. They also serve as a guide in selecting process components. The tests on β-carotene indicate that quartz or Teflon®-lined components would be preferable to stainless steel, although a small amount of decomposition must still be expected under the proposed conditions of extraction.

Tests using a reaction screening apparatus such as the one used at NIST should be included in any process design involving SFE at relatively severe conditions. The definition of the word "severe" cannot be arbitrary, but rather must be determined by the sensitivity of the solvent and solutes to the conditions that prevail. There is very little predictive capability available to allow the elimination of the reaction screening step from the overall design process. This is especially true for many of the large and highly polar natural product solutes of interest to the pharmaceutical industry.

C. SOLUTE VAPOR PRESSURE

The vapor pressure of a pure substance is an important equilibrium quantity that is useful in the description of one-component phase behavior. Vapor pressure can be defined as the pressure above a condensed phase or system when that system is at equilibrium. The equilibrium condition implies a dynamic balance between the number of molecules escaping the condensed phase and those recaptured by it. One normally thinks of vapor pressure only with vapor-liquid systems, the process of sublimation usually being associated with vapor-solid systems. This review adopts the convention of the literature and uses the term "vapor pressure" in the case of solids as well as liquids.

The vapor pressure of the solute is of great interest in applications of SFE for a variety of reasons. The solubility of a solute in an SCF at a given density will generally increase as the temperature increases, due primarily to the increase in the vapor pressure of the solute. A knowledge of the temperature dependence of the vapor pressure is therefore of value in predicting the marginal increase in solute solubility that may be expected with an increase in temperature. This is most easily provided by the correlation of experimental values with a simple expression such as the Antoine equation,[4] provided that the measurements are extensive enough over the temperature range of interest to ensure reliability. While it is generally not acceptable to extrapolate beyond the range of the measured values upon which the correlation is based, the lack of experimental work or the difficulties involved in making accurate experimental measurements often make this necessary for engineering calculations.

The vapor pressure is also needed in EOS modeling. Most reasonably accurate EOSs require knowledge of the critical conditions for all the components being described.[4] Recent measurements of critical properties of many of the solutes of interest in SFE are scarce, and measurements for large, polar solutes are usually not possible. The material will often decompose long before the critical point is attained. The only alternative is to predict the critical conditions using a structural contribution method such as that proposed by Lydersen:[25]

$$T_c = T_b[0.567 + \Sigma \Delta_T - (\Sigma \Delta_T)^2]^{-1} \tag{2a}$$

$$P_c = M(0.34 + \Sigma \Delta_P)^{-2} \tag{2b}$$

$$V_c = 40 + \Sigma \Delta_v \tag{2c}$$

In the above equations, P_c and V_c are the critical point parameters, T_b is the normal boiling point, and M is the relative molecular mass of the substance of interest. The delta quantities are tabulated incremental structural quantities determined from the functional groups of the molecule. The only inputs needed to use this estimation method are the molecular mass and normal boiling point of the solute. Since the boiling point of many large molecules is as difficult to measure as the critical temperature, it must be estimated by extrapolation from the vapor pressure behavior. While this procedure seems inherently inaccurate, since it involves two successive estimation methods, each of questionable reliability, it is often the only way to estimate the critical temperature of high molecular mass or highly polar solutes.

Many EOSs used in engineering phase equilibria prediction require knowledge of the acentric factor for each component. An example is the Peng-Robinson EOS.[26] The acentric factor, proposed by Pitzer[27,28] in 1955, can be considered the degree of nonsphericity of a molecule. For monoatomic gases, it is essentially zero. Nonsphericity increases with molecular mass, and often increases with increasing polarity of the molecule. It is therefore a measure of molecular complexity in terms of both geometry and polarity. The acentric factor, is, by definition,

$$\omega = -\log P_v \, (T_r = 0.7) - 1 \tag{3}$$

In this equation, ω is the acentric factor, P_v is the vapor pressure, and T_r is the reduced temperature, defined as the system temperature divided by the critical temperature. In order to predict the acentric factor for solutes, both the vapor pressure behavior and a predicted critical temperature are needed. As stated before, this estimation method may be only marginally accurate, but often no alternative exists for work involving large and polar species.

1. Effusion Manometry

The measurement of vapor pressures of pure substances is an endeavor that extends well back into the early part of the 19th century, as practiced by such notables as Sir William Ramsey. This review is concentrated more on recent methods applied to the measurement of the vapor pressure of solids such as those often encountered as solutes in SFE. The first technique to be considered is effusion manometry. This method, described by Balson in 1947,[29] has been used to measure vapor pressures as low as 5×10^{-6}. The experiment is of interest today only in its historical context, but is described to illustrate both the simplicity and elegance of the method.

The apparatus used for effusion manometry is shown in Figure 7. A glass vessel consisting of a central tube and two dumbbell-shaped cross-member spheres contains the sample. The spherical cross-members each have a small hole in the center at opposite sides from one another with respect to the central tube. This arrangement is much the same as Hero's steam engine. This vessel, called the effusion vessel, is suspended in the thermostatted flask by a small diameter quartz fiber, which also suspends a mirror that is positioned outside of the thermostatted zone. The heater shown in the figure is used to distill sample from the central tube of the effusion vessel into the spherical cross-members. The thermostatic bath is then set to the desired temperature, and the flask is evacuated to approximately 1.3×10^{-8} atm. The sample will then effuse from the small holes in the spheres, and the resulting reactive force will impose a twisting moment on the supporting fiber. The extent of the twist, which can be up to 180°, is measurable by observing the position of the mirror. The twist is related to the vapor pressure of the sample using a calibration equation. The sensitivity of the apparatus is determined by the hole size and the geometry of the effusion vessel.

The method suffers from several experimental difficulties. The thermostatted flask must be evacuated to reduce the number of collisions of sample molecules (exiting the effusion holes) with residual air molecules. This makes heat transfer to the effusion vessel difficult, and the temperature of the sample in the effusion vessel is therefore uncertain. The effect

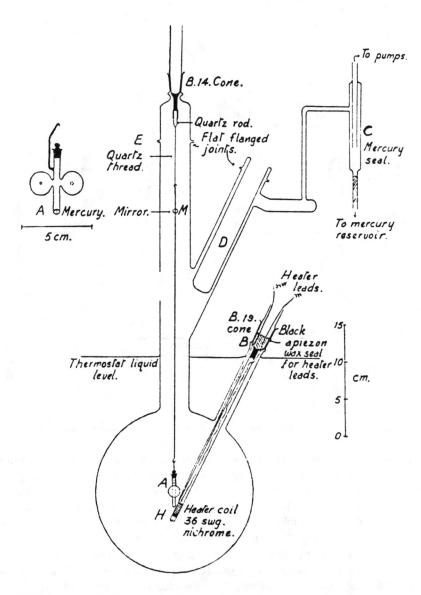

FIGURE 7. An effusion manometer for the measurement of vapor pressures. (From Balson,
E. W., *Trans. Faraday Soc.*, 43, 54, 1947. Reprinted with permission of the Royal Society
of Chemistry.)

of sample cooling upon expansion into the flask is also uncertain, as well as the effect of
collisions of sample molecules with residual air. Despite these problems, the vapor pressures
of many industrially important materials (such as DDT and mercury) were measured with
reasonable accuracy using effusion manometry.

2. Gas Saturation Method

Most of the recent vapor pressure measurements performed on low volatility solids have
used some variation of the gas saturation method. The technique is even specified as a
standard EPA test method in the U.S. code,[30] and has been applied extensively in pesticide
and herbicide work.[31] Although the method dates from 1845, little practical use was derived
from it until the availability of modern analytical techniques, primarily the chromatographic
techniques involving highly sensitive detectors. A schematic diagram of a typical apparatus
(that in use in the Thermophysics Division of NIST) is shown in Figure 8.

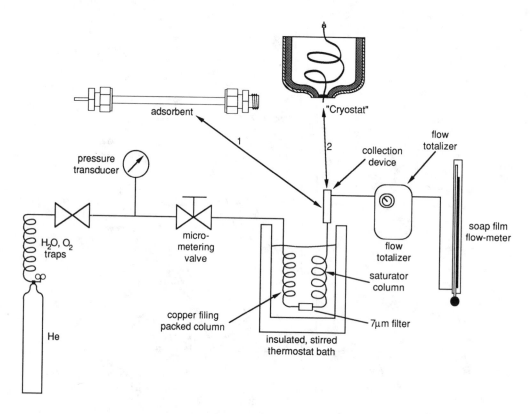

FIGURE 8.   Gas saturation apparatus used for vapor pressure measurements at NIST.

The principle of the measurement is very simple. The sample is contained in a tube that is placed in a thermostatic bath. The sample can be coated on an inert support such as small glass beads, or simply placed neat (without a diluent) into a tube. At any given temperature, sample vapor will form in the head space above the solid. An inert carrier or sweep gas is allowed to flow through the tube and become saturated with the sample vapor. The flow rate of the gas is low enough to ensure that equilibration takes place. This saturated flow then passes out of the sample tube and into some kind of collection vessel. Knowledge of the mass of sample transported as vapor (m) at a given temperature, and of the volume of inert gas needed for transport (V) allows the calculation of the vapor density, ρ_v:

$$\rho_v = m/V \tag{4}$$

The vapor pressure follows upon application of the ideal gas law, which is valid for helium in the temperature range of the measurements (ambient to approximately 100°C):

$$P_v = \rho_v(RT/M) \tag{5}$$

where M is the relative molecular mass of the solute and R is the gas constant.

In the apparatus of Figure 8, a manifold consisting of a diaphragm pressure regulator, mass flow controller, and needle valve provides a uniform flow of helium to the thermostat area. The temperature is regulated using a water bath under proportional control. The pressure is monitored using a strain gauge pressure transducer and readout. Before arriving at the sample saturation tube, the carrier gas is passed through a column of copper powder followed by a fine filter, to ensure temperature equilibration. After leaving the saturation tube, the saturated gas is passed directly into the collection device. This is either a cryogenic trap or

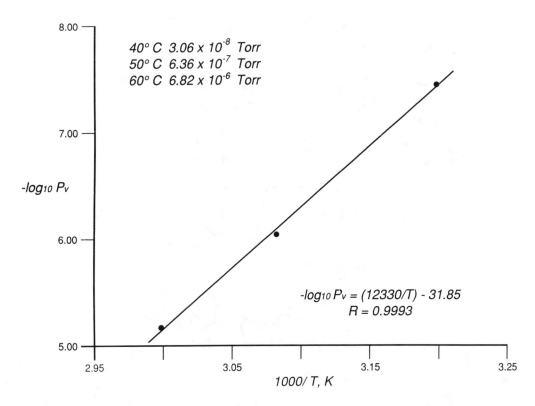

FIGURE 9. The measured vapor pressure of β-carotene as a function of temperature.

an adsorbent cartridge. The cryogenic trap is a length of stainless steel tubing (0.16 cm o.d., 0.05 cm i.d.) wound in an insulated container that is packed with dry ice. The adsorbent cartridge is a 15-cm long stainless steel tube (0.65 cm o.d., 0.48 cm i.d.) packed with 60/80 mesh Tenax-GC* or some other suitable adsorbent. The volume of the carrier is determined using a tin-bellows dry testmeter which follows the collection device. In practice, several saturator tubes are swept simultaneously, with the volumes being totalized using separate dry testmeters.

The sample deposited in the cryogenic or sorbent trap is flushed out using a suitable solvent, and may be analyzed directly or concentrated to a smaller volume of solution. Thermal desorption from the adsorbent should not be used, since the recovery will not be high enough, and sample degradation can occur. For samples that can be analyzed using gas chromatography (GC), flame ionization (FI), or mass selective detection provide high enough sensitivity for vapor pressures in the 10^{-11} atm range.[32] If the sample is fluorescent, spectrofluorimetry provides a similar level of sensitivity, often without a preconcentration step. UV-vis spectrophotometry can be used for samples that have chromophoric groups, although the sensitivity is much lower than the GC methods and somewhat lower than the spectrofluorimetric method. Preconcentration of the sample must usually precede analysis by UV-vis techniques.

Figure 9 shows measurements of the vapor pressure of β-carotene, performed using the apparatus shown in Figure 7, using the adsorbent trap and analysis by UV-vis spectrophotometry after preconcentration of the solute extracted from the trap with hexane. The main

---

* Certain commercial equipment, instruments, or materials are identified in this chapter in order to adequately specify the experimental procedure. Such identification does not imply endorsement by the Natinal Institute of Standards and Technology, nor does it imply that the materials or equipment that are identified are necessarily the best available for the purpose.

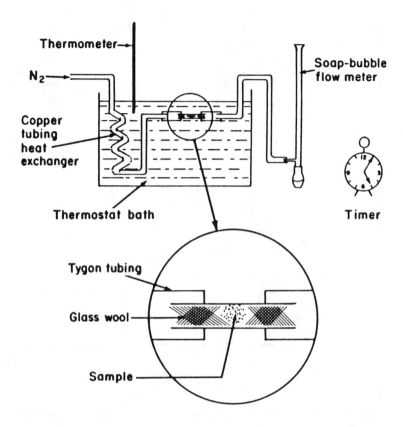

FIGURE 10.   A simple gravimetric gas saturation apparatus suitable for student use.
(Reprinted with permission from Razavi, H., *J. Chem. Educ.*, 63, 639, 1986.
© 1986 American Chemical Society.)

drawback of the method is that very often, long collection times are needed to provide enough sample for analysis. This is especially true if the sample is not easily analyzed by GC. Days and sometimes even weeks are required for very low vapor pressure solids, and even then preconcentration is usually required before analysis.

3. Modifications to the Gas Saturation Method

There are several variations of the gas saturation method that are worthy of mention. A method based on gravimetric determination of the transported vapor is useful for the determination of vapor pressures (and molecular masses, if the vapor pressure is already available or measurable by another means) of moderately volatile solids. Such an apparatus is shown in Figure 10, from a description of an undergraduate laboratory experiment.[33]

A modification to the gas saturation method proposed by Sinke[34] to allow faster measurements on organic materials (especially hydrocarbons) involves the oxidation of the saturated vapor and analysis of the resulting carbon dioxide using infrared (IR) detection. To implement this method, high-purity oxygen is used as the transport gas, which is thermally equilibrated and passed into the saturator column. The saturated oxygen stream is then passed into a reaction chamber containing a platinum gauze maintained at approximately 1450°C. The reacted stream is then passed into the IR analyzer, which is capable of quantitating parts per million levels of CO_2. An important advantage of the method is that the effect of transport gas flow rate is easily assessed, and conditions of nonsaturation can be avoided. It is not clear, however, what effect heteroatoms may have on the reaction process, and how they will ultimately affect the results. Another difficulty concerns measurements on

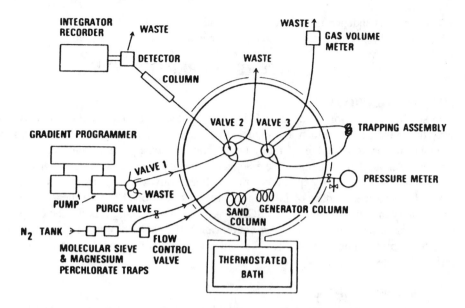

FIGURE 11. A modification of the gas saturation method in which the sample collection device is the sample loop of an HPLC system. (Reprinted with permission from Sonnefeld, W. J., Zoller, W. H., and May, W. E., *Anal. Chem.*, 55, 275, 1983. © 1983 American Chemical Society.)

partially oxidized or reactive compounds. The method is not usable if significant oxidation begins during the saturation process, before the sample enters the reaction chamber. This would be the case if the technique were applied to β-carotene, since the formation of apo- (or oxidative decomposition) products would begin immediately upon contact with the oxygen stream.

A simple modificaiton to Sinke's approach, proposed by Bender et al.,[35] involves the use of purified research grade argon as the saturation gas. The saturated stream is then mixed with a purified oxygen stream in a manifold positioned before the reaction chamber. Using this modified apparatus, the vapor pressures of anthracene, hydroquinone, and resorcinol were determined.[34] These materials would have begun to oxidize upon exposure to pure oxygen as in Sinke's original apparatus.

In an effort to eliminate the problems associated with sample recovery from the trapping system, Sonnefeld et al.[36] proposed a method in which the collection device is an integral part of the sampling system of a liquid chromatograph. A schematic diagram of this apparatus is provided in Figure 11. The saturator column, preceded by a thermal equilibration column containing sand, is connected to the trapping system using a six-port HPLC sampling valve. The collection device then forms the sample loop of the valve. Two collection systems were used: a cryotrap and a column packed with superficially porous octadecylsilane (ODS). Detection is provided using a UV-vis spectrophotometer or a spectrofluorimeter equipped with flow cells, and the data are collected in the form of area counts on an electronic integrator. Quantitation is, as usual, accomplished using a series of standard solutions of the sample.

After the appropriate volume of carrier gas has been passed through the saturator and into the collector, the HPLC eluent is allowed to flush the collector and dissolve the trapped sample. The sample then becomes focused onto the head of a 25-cm column of chemically bonded ODS. This method of sampling is advantageous since it results in no decrease in sensitivity due to the conventional sample handling methods required, for example, with the apparatus of Figure 8. There is also a significant time savings because of this method. The authors report that the cryotrap is the preferred method of sample collection since a slight flow rate dependence is noticed when the adsorbent is used. This flow rate effect is probably due to the pressure drop caused by the packing in the sorbent trap.

IV. MIXTURE PROPERTIES

A. SOLUTE SOLUBILITY

The solubility of a solute in an SCF is probably the most important thermophysical property that must be determined and modeled in order to design effective SFE processes. In particular, the pressure and temperature (and therefore, density) dependence of solubility must be understood. This will allow the engineer to specify the operating conditions of unit operations such as extractors, separators, transfer lines, valves, and process controllers.

There have been a number of different approaches developed for the measurement of solute solubility in both pure and mixed SCF phases. These generally fall into the categories of dynamic (or flow) methods, static (or equilibrium) methods, chromatographic methods, and spectroscopic methods. In addition, several investigators have applied various combinations of these techniques.

1. Dynamic or Flow Methods

The simplest and most straightforward methods of solubility measurement are performed using a flow apparatus. A basic version of such an apparatus is depicted schematically in Figure 12. The solvent fluid is supplied to a compressor from a pressure cylinder. Following the compressor is a surge tank, which minimizes pulsations in pressure caused by the compressor. At the desired pressure, the fluid then passes into the thermostatted extractor cell. The extractor contains the solute, which can be packed neatly into the cell, coated on a support, or placed between multiple layers of glass wool. Ideally, if the solute is a heavy solid, the glass wool should be precoated with the solute to minimize the effects of adsorption. The fluid dissolves solute in the extractor, and is expanded through a heated metering valve, where the solute precipitates from solution. The solute is collected in a device such as a "U" tube or cold finger, and is determined either gravimetrically or using some appropriate analytical technique. The volume of decompressed solvent leaving the collection device is measured using a totalizer such as a wet or dry testmeter or integrating anemometer, or by measuring the flow rate (with, for example, a soap-film flowmeter) and noting the time that collection was allowed to proceed.

The advantages of this method stem mainly from simplicity. Off-the-shelf equipment can be used to assemble an apparatus, and reproducible measurements can be done rapidly. The method suffers from a number of disadvantages, many of which can be addressed by modifications in design. When solute is allowed to expand through the metering valve, the possibility exists for valve clogging and solute holdup in the valve. This difficulty was addressed by Dobbs and Johnston,[38] who developed the microsampling apparatus shown in Figure 13. This apparatus incorporates a chromatographic sampling valve that cuts out two extractor volumes of solution, which are then expanded between valves A and B. The volume of the loop and the density of the solution allow calculation of the sample weight. After the sampling valve has been switched to the sampling position, valve A is opened to allow the effluent to bubble through an appropriate liquid solvent. Valve B is then opened to allow the syringe injection of between 10 and 15 ml of the liquid solvent, to flush solute from the valve and transfer lines. The concentration of the solute is then determined by a suitable analytical technique such as GC with flame ionization detection (FID).

This modification offers a number of advantages in addition to eliminating the sample holdup problem. Mixed solutes can be studied easily, since after the sample is collected and dissolved in a liquid solvent, chromatographic separation can precede quantitation. The flow rate or total volumetric flow measurement is not required since the solubility is determined from the density of the fluid and the volume of the sampling system. A disadvantage is that

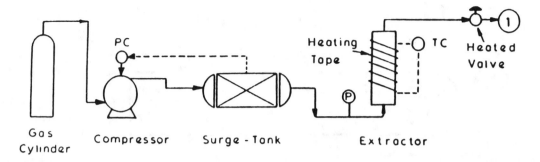

FIGURE 12. A simple dynamic solubility apparatus. (Reprinted with permission from Krukonis, V. J. and Kurnik, R. T., *J. Chem. Eng. Data,* 30, 247, 1985. © 1985 American Chemical Society.)

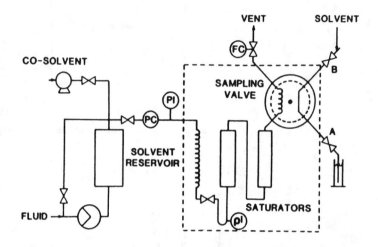

FIGURE 13. A modified dynamic solubility apparatus for use with small sample sizes. (Reproduced by permission of the American Institute of Chemical Engineers © 1986, AIChE.)

the sample should be amenable to analysis by GC to allow the use of the most sensitive detectors. The method could not be used easily in the study of β-carotene, for example, and other large natural products for this reason.

Another modification of the dynamic flow method incorporates an adsorbent trap cartridge into which the SCF solution is expanded. The expanded fluid is then passed to a dry testmeter to provide volume totalization. This is shown in Figure 14, which is a schematic diagram of the NIST solubility apparatus. This approach allows larger volumes of solution to be processed through the extractor, a feature necessary for the study of low solubility solutes and solutes that cannot be sufficiently volatilized for analysis by GC. After an appropriate volume of solution has flowed, the sorbent column is removed and extracted with a suitable liquid solvent. Analysis by IR or UV-vis spectrophotometry or spectrofluorimetry can then be done.

In all dynamic or flow methods, it is important to ensure that the solute and SCF solvent can reach equilibrium during the time that the solvent floods the extractor column. This is usually accomplished by making the measurement at a number of solvent flow rates. If the calculated solubility is independent of flow rate, equilibrium is usually assumed. This is somewhat of a disadvantage of the method, since this test must be performed for each system at each temperature studied.

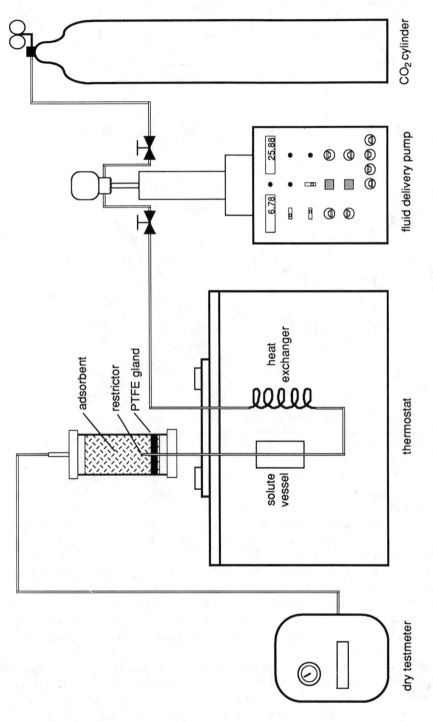

FIGURE 14.   NIST dynamic solubility apparatus using an adsorbent trap to collect solutes of low volatility.

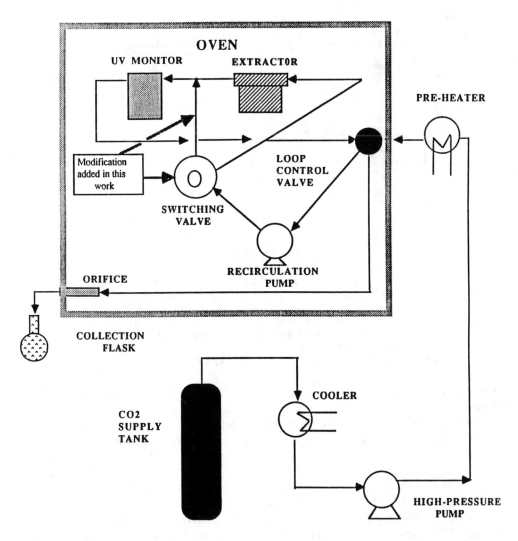

FIGURE 15. An equilibrium or static solubility apparatus constructed from a chromatographic sample preparation accessory.

2. Static (Equilibrium) Methods

Static or equilibrium solubility measurement methods are often used to eliminate the need to sample the SCF solution. A representative instrument is shown schematically in Figure 15, consisting of a modified SCF sampling accessory.[5] The components of the instrument are contained in a small forced air oven that provides temperature control to within $\pm 0.5°C$. The sample is contained in the extractor cell, which consists of a small stainless steel pressure vessel having an inlet and an outlet port, both of which accommodate filter frits. A high-pressure UV-vis cell is placed in the flow circuit to monitor the solution process by variable wavelength spectrophotometry. The solvent fluid is pumped into the extractor cell, and the switching valve isolates the pressurized system to allow the recirculation pump to achieve solute-solvent equilibrium. The approach to equilibrium is monitored directly by the change in absorbance or transmittance of the solution. Fluid recirculation is continued until the absorbance reading becomes constant, at which time it can be assumed that equilibrium has been attained. The concentration of solute is determined from the absorbance using the Beer-Lambert law[15] and a calibration table generated using standard solutions prepared in a suitable liquid solvent. Hexane is usually chosen because it has

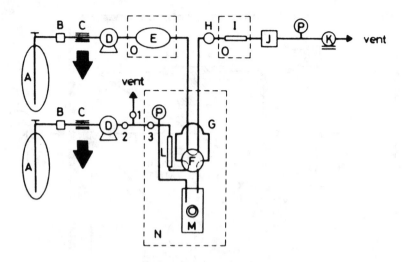

FIGURE 16.   A static solubility apparatus consisting of a recirculation loop interfaced to a SCF chromatograph. (A) Gas cylinder, (B) filter, (C) heat exchanger, (D) high pressure pump, (E) buffer, (F) 6-port valve, (G) sampler, (H) injector, (I) analytical column, (J) UV spectrophotometric detector, (K) back pressure regulator, (L) hand-operated recirculation pump, (M) equilibrium cell, (N) water bath, (O) air bath, (P) pressure sensor. (From Sako, S., Ohgaki, K., and Katayama, T., *J. Supercritical Fluids*, 1, 1, 1988. Reprinted with permission of PRA Press.)

approximately the same polarity as carbon dioxide, and no unexpected solvatochromic effects should disturb the measurement.

This approach has a number of important advantages in addition to those already mentioned. Verification of solute-solvent equilibration is much easier than with the flow methods. Temperature-dependent data are easily obtained without reloading the extractor. The variable wavelength spectrophotometric method can provide for the determination of more than one component if their absorptions are adequately resolved on the wavelength scale. In using this method with UV-vis detection, one must verify that the apparatus is operating under conditions in which the Beer-Lambert law can be considered valid. The wavelength must be carefully chosen to provide accurate absorbance readings. This means that the absorbance readings should cluster around 0.432, and the transmittance around 0.368, where the theoretical absolute error is approximately 1.36%.[15] In addition, areas of very high and low absorbance must be avoided. Another common difficulty is leaking around the windows of the detector cell, which can be persistent until the window gasket material (usually Teflon®) has swelled under the influence of the SCF solvent. An interesting modification of the static method is the combination of the equilibrium cell with an SCF chromatograph.[39] A schematic diagram of such an apparatus is shown in Figure 16. Although the authors describe an SCF chromatograph having UV-vis detection, more sensitive detectors such as the FID can be used as well.

3. Chromatographic Solubility Methods

The instrumentation of SCF chromatography can be adapted to perform solubility measurements, as shown schematically in Figure 17.[40] A relatively short length (approximately 2 m) of fused silica capillary tubing containing the sample replaces the usual capillary column. Sample is coated into this tube using the same techniques that are used to coat a stationary phase on an analytical capillary column. The capillary sample tube is inserted between the injector and detector interface. In the instrument shown in Figure 17, a mass spectrometer serves as the detector.

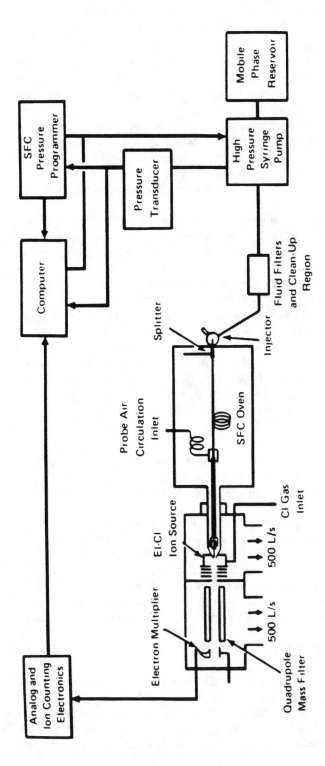

FIGURE 17. A chromatographic solubility apparatus using a mass spectrometer for detection of the dissolved solutes. (Reprinted from Ref. 40, p. 1065, by courtesy of Marcel Dekker.)

Isothermal solubility measurements are made by raising the solvent pressure stepwise from a selected starting point. At each step, the ion current in the mass spectrometer is monitored to determine the quantity of solute in solution. Sufficient time is provided within each step to allow an equilibrium ion current value to be attained. It is possible to monitor the signal from a single ion or the total ion current, and to use both chemical ionization and electron impact methods in detection. Calibration of the signal can be done using response factors obtained from standard solutions which are injected through a valve and splitter arrangement. Alternatively, one can monitor the change in ratio between the signal of the solute and solvent ions.

This method allows relatively rapid measurements on a wide range of solid samples. It is of limited value, however, for heavy liquid samples, which are also of interest in SFE. A serious disadvantage is that this is clearly the most costly method of measuring solute solubility. The instrument described here is necessarily a multipurpose unit used for various kinds of SCF work, including chemical analysis, and is not dedicated exclusively to solubility measurement.

Another interesting application of chromatographic techniques involves the combination of SFE with thin layer chromatography (TLC).[41] A small extractor cell containing the solute of interest is charged with the SCF solvent at the desired temperature and pressure. A stream of the solution controlled by a metering valve is then allowed to escape through a restrictor capillary to impinge directly on a moveable TLC plate coated with a suitable adsorbent. A known quantity of the solvent fluid is allowed to pass through the capillary, and the solute precipitates out upon decompression and is adsorbed onto the plate. The solvent pressure in the extraction vessel is then increased, and the TLC plate is shifted accordingly to collect the new sample at the higher fluid density. The chromatogram is developed after all desired densities have been eluted onto the plate. While this method is not necessarily suited for high accuracy solubility measurements for EOS correlations, it is useful for determining threshold solubility pressures. It can also be used to assess how strongly increasing the density of the supercritical solvent affects the solute solubility.

4. Spectroscopic Methods

Several of the solubility measurement methods discussed earlier have utilized spectroscopic detection as a tool applied in some static or dynamic apparatus. In this section, two methods are discussed in which spectroscopy is the main feature, and the SCF solution is present in an accessory.

The high pressure thermostatted cell shown in Figure 18 has been used with Fourier-transform infrared spectrophotometry (FTIR) to measure the concentration of a solute using the Beer-Lambert law.[42] The solid sample is inserted through the thermowell into a cavity, and the cell is purged with carbon dioxide to exclude atmospheric gases and moisture. The cell is then pressurized with the solvent, and enough time is allowed for the solution to reach equilibrium. Equilibration times are on the order of 1 h, but occasionally much longer times are required. Quantitation is based on the absorption of a strong band such as the C-H stretching frequency.

Nuclear magnetic resonance (NMR) spectroscopy can be used for solubility work using a sample cell such as that shown schematically in Figure 19. The cylindrical sample probe was made from a high temperature polyamide, and consists of a piston-cylinder arrangement. Solid sample is loaded into the well, and the piston is installed and held with an indexing pin to provide a seal. The solvent is introduced from small holes in the cylinder wall. After an equilibration period, the absorption signal of an appropriate nucleus (usually proton) is measured. The signal from the solution is easily distinguished from that of solid in the cell by noting the large difference in spin-spin relaxation times.

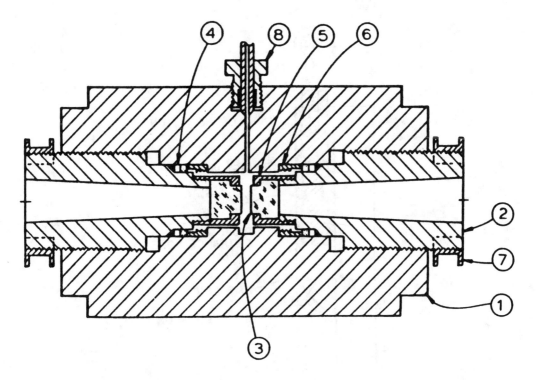

FIGURE 18.   A solubility cell for FTIR measurements. (1) Cell body, (2) plug, (3) sapphire window, (4) Bridgeman rings, (5) window cup, (6) extraction ring, (7) electrical coil, (8) high pressure tube with connection to the cell. (Reprinted with permission from Zerda, T. W., Wiegand, B., and Jonas, J., *J. Chem. Eng. Data,* 31, 274, 1986. © 1986 American Chemical Society.)

B. SOLUTE DIFFUSIVITY

Measurement of the solute-solvent binary interaction diffusion coefficient, D_{12}, for SCF solutions is most often done using the Taylor-Aris (or peak broadening) method.[44,45] This dynamic method uses chromatographic instrumentation that has been adapted for physico-chemical measurements. There are also several static methods that are based on the diffusion of the solute into capillary tubes, and also methods based on light scattering experiments. Examples of the application of these other methods in SCF work are relatively sparse, thus only the peak broadening technique is discussed in this chapter.

Since many authors have treated the assumptions and theoretical consequences of the Taylor-Aris method in detail,[46] only a brief description is provided here. The necessary experimental data are contained in the chromagraphic peak, which can be considered a concentration profile of a solute with respect to the elapsed time (or carrier solvent volume). A sharply defined spike of solute (approximately like a δ-function) is introduced into a laminar stream of carrier fluid flowing in an uncoated, deactivated tube. A stainless steel tube may be deactivated by acidic passivation, while a glass or quartz tube may be deactivated by silation. The solute will be subject to both convective action along the axis of the tube and molecular diffusion in the radial direction. If the tube can be considered straight, the mathematical treatment of Taylor and Aris gives a simple expression for the plate height, H:

$$H = \frac{2D_{12}}{u} + \frac{r^2u}{24D_{12}} \qquad (6)$$

In Equation 6, u is the carrier solvent linear velocity, r is the internal radius of the tube,

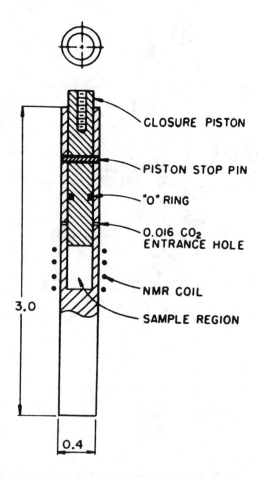

FIGURE 19.   A high pressure NMR tube for solubility
measurements. (Reprinted with permission from Lamb,
D. M., Barbara, T. M., and Jonas, J., *J. Phys. Chem.*,
90, 4210, 1986. © 1986 American Chemical Society.)

and D_{12} is the diffusion coefficient of the solute (1) in the solvent (2). The assumption of
a straight tube may appear difficult to satisfy in practice, since most chromatographic columns
must be coiled to fit into the oven, and a very long, straight tube would be difficult to
properly thermostat. One should therefore consider the assumption satisfied if the ratio of
the coil diameter to the internal diameter of the tube is at least 10:1, and even larger ratios
are preferable. This is not difficult to achieve with small diameter capillary columns.

The plate height H is an important and experimentally accessible quantity in chromatography, since it describes the efficiency of a chromatographic system. It can be defined
as the width of an emergent peak (as designated by its variance, σ^2, in length units) relative
to the distance (L) traveled inside of the tube:

$$H = \sigma^2/L \tag{7}$$

The peak width is directly measurable from a chromatogram, and is often provided directly
by modern chromatographic integrators and commercially available peak processing software
for personal computers. In a straight tube, the concentration profile of the solute in the
carrier will become Gaussian-like when $H < 0.02$ m, a necessary condition for the validity
of Equation 6. At carrier fluid velocities encountered in normal practice, Equation 6 reduces
to:

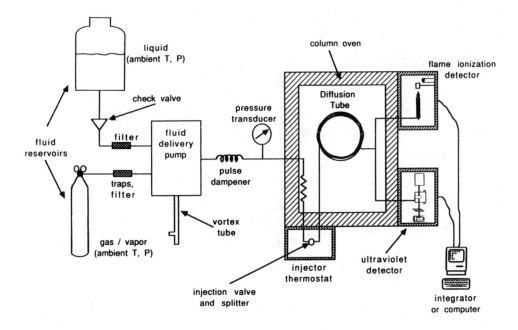

FIGURE 20.   A physicochemical SCF chromatograph used at NIST for the measurement of properties such as the binary interaction diffusion coefficient.

$$H = \frac{r^2u}{24D_{12}}$$

which allows calculation of the diffusion coefficient. This simple result is presented for clarity; a number of corrections to the various parameters must be applied to obtain good precision. There are many sources of error in this measurement, although most of these can be addressed to some extent by proper experimental design.[46]

A representative apparatus is shown schematically in Figure 20.[47,48] The solvent delivery system consists of a pump that is cooled using a Ranque-Hilsch vortex tube.[49,50] The solvent is distilled from a high pressure cylinder and is liquefied in a heat exchanger that precedes the pump. The sample introduction system, based upon SCF extraction, is shown in Figure 21. This arrangement allows sample to be introduced without an additional solvent that would interfere with the measurement. The diffusion tube, which can either be a stainless steel or fused quartz capillary, is housed in a heavy aluminum "racetrack" that integrates out large variations in temperature. The principle here is the same as the aluminum block surrounding the isochoric cells in the high temperature, high pressure PVT apparatus discussed earlier. The chromatographic oven is the main thermostat of the appartus, although very low power shimming heaters are placed in various locations to provide fine tuning of the temperature. Temperature measurement is provided by a PRT inserted in the center of the aluminum racetrack. Differential thermocouples, referenced to the PRT, indicate temperature gradients among the major components. Detection is provided either by an FID or a variable wavelength UV-vis spectrophotometer. Figure 22 shows the results of measurements performed on the β-carotene + carbon dioxide system using this apparatus.

In an interesting extension of the Aris-Taylor method, Matthews and Akgerman[51] developed a technique for making simultaneous density measurements on the carrier fluid. Since the solute moves through the diffusion tube at the mean speed of carrier flow, the center of gravity or normalized first moment of the peak can be related to the steady-state mass flow rate of the carrier. In simplified terms, the mass flow rate is the product of the

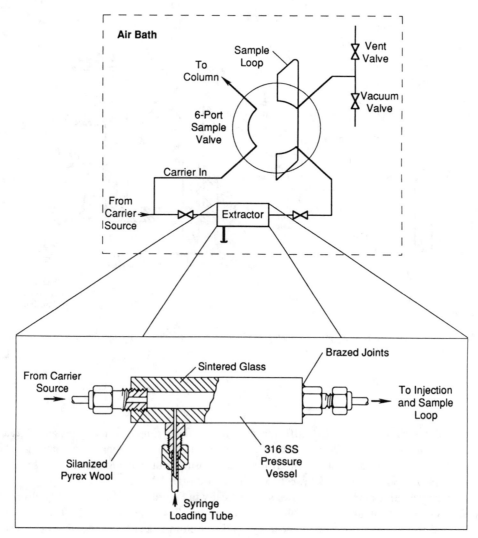

Extractor

FIGURE 21.   The extraction system used for solvent free injection for diffusion coefficient measurements.

fluid velocity, density, and cross-sectional area of the diffusion tube. The mass flow rate of a reference fluid of known density is determined using a tracer solute (at infinite dilution). The unknown density of another fluid is then determined by computing the volume ratio of fluid of known density with that of unknown density which, at steady state, is inside the diffusion tube.

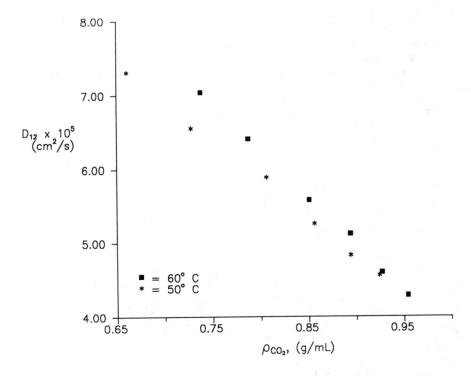

FIGURE 22. Measured diffusion coefficients of β-carotene in carbon dioxide, as a function of density.

REFERENCES

1. **Himmelblau, D. M.**, in *Basic Questions of Design Theory*, Spillers, W. R., Ed., North Holland, Amsterdam, 1974.
2. **Motard, R. L.**, in *Basic Questions of Design Theory*, Spillers, W. R., Ed., North Holland, Amsterdam, 1974.
3. **Mansfield, E.**, in *Economics of Research and Development*, Ohio State University Press, Columbus, 1965.
4. **Reid, R. C., Prausnitz, J. M., and Poling, B. D.**, *The Properties of Gases and Liquids*, 4th ed., McGraw-Hill, New York, 1987.
5. **Cygnarowicz, M. L., Maxwell, R. J., and Seider, W. D.**, *Fluid Phase Equilibria*, in press.
6. **Cygnarowicz, M. L. and Seider, W. D.**, Design and control of supercritical extraction processes—a review, in *Supercritical Fluid Technology: Reviews in Modern Theory and Applications*, Bruno, T. J. and Ely, J. F., Eds., CRC Press, Boca Raton, FL, 1991, chap. 11.
7. **Schmidt, R. and Wagner, W.**, *Fluid Phase Equilibria*, 19, 175, 1985.
8. **Jacobsen, R. T. and Stewart, R. J.**, *J. Phys. Chem. Ref. Data*, 2, 757, 1973.
9. **Straty, G. C. and Palavra, A. M. F.**, *J. Res. Natl. Bur. Stand.*, 89, 375, 1985.
10. **Roark, R. I.**, *Formulas for Stress and Strain*, 5th ed., McGraw-Hill, New York, 1975.
11. **Oberg, E., Jones, F. D., and Horton, H. L.**, *Machinery's Handbook*, 20th ed., Industrial Press, New York, 1975.
12. **Hall, K. R. and Canfield, F. B.**, *Physica*, 33, 481, 1967; 47, 99, 1970.
13. **Wielopolski, P. and Warowny, W.**, *Physica*, 66, 91A, 1978.
14. **Angus, S., Armstrong, B., and deReuck, K. M.**, *International Thermodynamic Tables of the Fluid State-3-Carbon Dioxide*, Pergamon Press, Oxford, 1976.
15. **Bruno, T. J. and Svoronos, P. D. N.**, *Handbook of Basic Tables for Chemical Analysis*, CRC Press, Boca Raton, FL, 1989.
16. **Ely, J. F.**, *Proc. 63rd Gas Processors Assoc. Ann. Conv.*, New Orleans, 9, 1984.
17. **Straty, G. C., Ball, M. J., and Bruno, T. J.**, *J. Chem. Eng. Data*, 32, 133, 1987.

18. **Straty, G. C., Palavra, A. M. F., and Bruno, T. J.,** *Int. J. Thermophys.,* 7, 1077, 1986.
19. **Straty, G. C., Ball, M. J., and Bruno, T. J.,** *J. Chem. Eng. Data,* 33, 115, 1988.
20. **Bruno, T. J. and Hume, G. L.,** *J. Res. Natl. Bur. Stand.,* 90, 255, 1985.
21. **Bruno, T. J. and Straty, G. C.,** *J. Res. Natl. Bur. Stand.,* 91, 135, 1986.
22. **Mayrath, J. E.,** unpublished data, private communication, National Institute of Standards and Technology, Boulder, CO, 1985.
23. **Friedrich, J. P.,** U.S. Patent 4,466,923, 1984.
24. **Bushway, R. J.,** *J. Liq. Chromatogr.,* 8, 1527, 1985.
25. **Lydersen, A. L.,** Univ. Wisc. Coll. Eng. Exp. Stn. Rep. No. 3, Madison, WI, April 1955.
26. **Peng, D.-Y. and Robinson, D. B.,** *AIChE J.,* 23, 137, 1977.
27. **Pitzer, K. S.,** *J. Am. Chem. Soc.,* 77, 3427, 1955.
28. **Pitzer, K. S., Lippmann, D. Z., Curl, R. F., Huggins, C. M., and Peterson, D. E.,** *J. Am. Chem. Soc.,* 77, 3433, 1955.
29. **Balson, E. W.,** *Trans. Faraday Soc.,* 43, 54, 1947.
30. *Fed. Reg.,* 45(227), Friday, November 21, 1980, 40 CFR, Part 772.
31. **Spencer, W. F. and Claith, M. M.,** *Environ. Sci. Technol.,* 3, 670, 1969.
32. **Wong, J. M. and Johnston, K. P.,** *Biotech. Prog.,* 2, 29, 1986.
33. **Razavi, H.,** *J. Chem. Educ.,* 63, 639, 1986.
34. **Sinke, G. C.,** *J. Chem. Thermodynam.,* 6, 322, 1974.
35. **Bender, R., Bieling, V., and Maruer, G.,** *J. Chem. Thermophys.,* 15, 585, 1983.
36. **Sonnefeld, W. J., Zoller, W. H., and May, W. E.,** *Anal. Chem.,* 55, 275, 1983.
37. **Krukonis, V. J. and Kurnik, R. T.,** *J. Chem. Eng. Data,* 30, 247, 1985.
38. **Dobbs, J. M. and Johnston, K. P.,** *Ind. Eng. Chem. Res.,* 26, 1476, 1987.
39. **Sako, S., Ohgaki, K., and Katayama, T.,** *J. Supercritical Fluids,* 1, 1, 1988.
40. **Smith, R. D., Udseth, H. R., and Wright, R. W.,** *Sep. Sci. Technol.,* 22, 1065, 1987.
41. **Stahl, E., Schilz, W., Schütz, E., and Willing, E.,** *Angew. Chem. Int. Ed. Engl.,* 17, 731, 1978.
42. **Zerda, T. W., Wiegand, B., and Jonas, J.,** *J. Chem. Eng. Data,* 31, 274, 1986.
43. **Lamb, D. M., Barbara, T. M., and Jonas, J.,** *J. Phys. Chem.,* 90, 4210, 1986.
44. **Taylor, G. I.,** *Proc. R. Soc. London,* A219, 186, 1953.
45. **Aris, R.,** *Proc. R. Soc. London,* A253, 67, 1956.
46. **Alizadeh, A., Nieto de Castro, C. A., and Wakeham, W. A.,** *Int. J. Thermophys.,* 1, 243, 1980.
47. **Bruno, T. J.,** *J. Res. Natl. Inst. Stand. Technol.,* 94, 105, 1989.
48. **Bruno, T. J.,** *J. Res. Natl. Inst. Stds. Technol.,* 93, 655, 1988.
49. **Bruno, T. J.,** *Anal. Chem.,* 58, 1596, 1986.
50. **Bruno, T. J.,** *J. Chem. Educ.,* 64, 987, 1987.
51. **Matthews, M. A. and Akgerman, A.,** *Int. J. Thermophys.,* 8, 363, 1987.

Chapter 8

THERMOPHYSICAL PROPERTIES OF CO_2 AND CO_2-RICH MIXTURES

Joe W. Magee

TABLE OF CONTENTS

I. INTRODUCTION

Thermophysical properties data, including thermodynamic and transport properties, are fundamental to the clear understanding of supercritical fluid (SCF) processes and technology. Benchmark experimental measurements provide the best check of theoretically based predictive models for the thermophysical properties of SCF, including CO_2 and CO_2-rich mixtures. For this reason, a review of published data is given herein. Only highly accurate thermophysical property data, capable of testing predictive models, have been considered in this context.

II. PURE CARBON DIOXIDE

A comprehensive review of CO_2 thermodynamic properties, covering published results through 1973, was prepared by Angus et al.[1] This chapter covers the period 1973 to mid-1990 with an emphasis on densities, heat capacities, viscosities, and thermal conductivities. Highlighted work includes that work performed at the National Institute of Standards and Technology (NIST), by this author, or by colleagues including James F. Ely, J. G. Sherman, John B. Howley, Dwain E. Diller, Vicki G. Niesen, James C. Rainwater, Howard J. M. Hanley, and W. M. Haynes.

A. DENSITIES

Ely et al.[2] have measured densities (ρ) from 250 to 330 K at pressures to 35 MPa with an isochoric PVT instrument. These densities range from dilute vapor to compressed liquid and are accurate to within 0.1%. Holste et al.[3] have measured densities from 217 to 448 K with Burnett-isochoric apparatus. An analysis of propagated error is complicated by the three separate apparatus used for density measurements, yet one may conclude that the uncertainty is less than 0.1%. Haynes[4] has measured saturated liquid densities with a magnetic suspension densimeter. The data range from 220 to 300 K and are accurate within 0.1%. Magee and Ely[5] have measured densities on selected isochores from 250 to 330 K with an automated version of the same apparatus of Ely et al.,[2] while incurring no loss of accuracy.

B. HEAT CAPACITIES

Magee and Ely[6] have produced measurements of the molar heat capacity at constant volume (C_v) covering broad ranges of density and temperature. The measurements were conducted on both the saturated liquid and on single-phase states using an adiabatic constant volume calorimeter. The estimated uncertainty of the heat capacities does not exceed 2.0%. Extensive two-phase results $C_v^{(2)}$ were obtained, ranging from 220 to 303 K. These results were used to obtain values of the saturated liquid heat capacity and to verify the internal consistency of the measurements by applying the test suggested by the relation of Yang and Yang.[7] This relation given by

$$C_V^{(2)}/T = -d^2\,G/dT^2 + V\,d^2P_{sat}/dT^2 \qquad (1)$$

where G is Gibbs free energy, V is molar volume, and P_{sat} is vapor pressure, requires that $C_v^{(2)}$ vs. V be linear. Magee and Ely found that their results were linear. All accurate C_v data should pass this test for internal consistency. In addition to the two-phase results, single-phase C_v data were measured on 12 isochores ranging form 0.2 to 2.5 times the critical density. The temperatures extended from 233 to 330 K at pressures to 35 MPa. These C_v data overlap and extend published data to both higher and lower densities. Figure 1 shows the behavior of these C_v measurements in the SCF region when they are cross-plotted vs. density on curves of constant temperature. This drawing illustrates the rapid rate of increase in this quantity as the critical point (304.107 K, 467.69 kg m^{-3}, 7.3721 MPa) is approached.

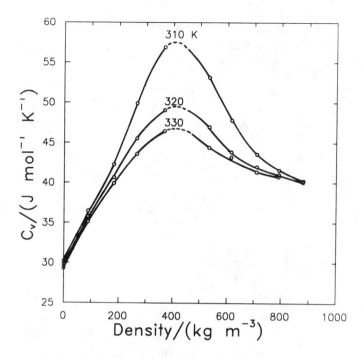

FIGURE 1.   Interpolated C_v values for supercritical CO_2. (From Magee, J. W. and Ely, J. F., *Int. J. Thermophys.*, 7, 1163, 1986.)

Ernst and co-workers[8,9] have recently published measurements of the specific heat capacity at constant pressure (c_p) at temperatures ranging from 303 to 393 K at pressures up to 90 MPa. The maximum uncertainty of the c_p values is stated by the authors to be 1.0%, a claim supported by their close agreement (RMS c_p deviation = 0.5%) with the equations of Angus et al.[1] at 303 K.

C. TRANSPORT PROPERTIES

Transport properties are essential for the design of efficient mass and heat transfer processes. In particular, SCF processes utilize the anomalous behavior of these properties in the vicinity of the critical point. It has been demonstrated experimentally by Michels et al.[10] that the thermal conductivity (λ) of CO_2 diverges as the critical point is approached. In this seminal work, λ was measured with a specially constructed parallel plate apparatus operated between 298 and 348 K. The majority of the λ values reported are accurate to within 1.0%, with the exceptions occurring on the 304.35 K isotherm near the critical density and also at all temperatures when pressure was elevated. Later measurements of λ by le Neindre[11] extended the range of temperatures to 951 K, at pressures up to 120 MPa. A concentric cylinder apparatus was used for the measurements. At temperatures up to 723 K the uncertainty was estimated by le Neindre to be 2.5%. Additional λ measurements were reported by le Neindre et al.,[12] extending the temperature range to 961 K and showing satisfactory agreement with Michels et al.[10] in the region of overlap. In the work, le Neindre and co-workers used the high temperature concentric cylinder results to deduce the background contribution to the $\lambda(\rho,T)$ function. When the background part was subtracted from the parallel plate results of Michels et al., le Neindre et al. obtained the anomalous contribution to the thermal conductivity. When plotted on isotherms vs. density, the anomalous part is seen to rise to a maximum at the critical density as observed for other pure substances. This effect is shown in Figure 2, from a recent review of the transport properties of CO_2 by Vesovic et al.[13] In their critical assessment, Vesovic and co-workers concluded that all

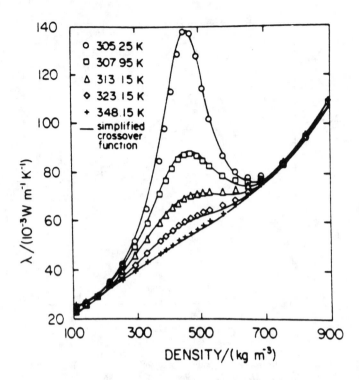

FIGURE 2.   Thermal conductivity of supercritical CO_2 as a function of density at various temperatures. (a) Data are from Michels et al.[10] (b) Solid curves were calculated from a model due to Vesovic et al.[13]

published λ data at high temperatures and low density are inconsistent by about $\pm 5\%$ with theoretical expectation. Further, they concluded that liquid phase λ data are both scarce and of poor quality.

Shear viscosity coefficients (η) of CO_2 have been measured by several workers[14-17] prior to 1985. More recently Diller and co-workers[18,19] have used a torsional piezoelectric crystal method with an uncertainty of better than 2%. Two viscometers were used to cover temperatures from 220 to 500 K and pressure as high as 50 MPa; however, once again, Vesovic et al.[13] have concluded that in the liquid phase, the inconsistencies (up to about 15%) between the published sources of η measurements cannot be resolved. It is further recommended that additional measurements of the best attainable accuracy are needed to resolve the observed discrepancies.

Like thermal conductivity, viscosity of supercritical CO_2 has an identifiable enhancement; however, it is much smaller than the observed enhancement of thermal conductivity. Figure 3, also from Vesovic et al.,[13] shows the viscosity enhancement as a function of density at various temperatures, deduced from the data of Iwasaki and Takahashi.[17] The viscosity of CO_2 is seen to rise to a maximum as the critical point is approached, a behavior observed for all pure fluids.

III. CO$_2$-RICH MIXTURES

Measurements of accurate thermophysical properties for mixtures are always scarcer than for the pure substances. It comes as no surprise that this is true for CO_2-rich mixtures as well. For such work this author has sorted through numerous references to compile a list of references containing data that are suitable for testing predictive models.

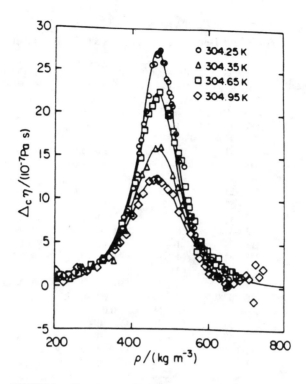

FIGURE 3. The viscosity enhancement $\Delta_c\eta$ of supercritical CO_2. (a) Deduced from the data of Iwasaki and Takahashi.[17] (b) Solid curves were calculated from a model due to Vesovic et al.[13]

A. DENSITIES

Table 1 presents references which give density (ρ) results for CO_2-rich mixtures. An exhaustive literature search yielded many more references than those listed in Table 1. The references tabulated in this review are data sources of sufficient breadth and accuracy to be useful for testing predictive models or theories. Though Table 1 is not comprehensive, it may be considered a starting point for a broad survey. The mixture components considered in Table 1 are alkane or alkene compounds up to C_{33}, water, and selected low molecular weight gases. Entries include the ranges of temperature, pressure, and composition, respectively, followed by an estimate of the maximum uncertainty of the density measurements. The uncertainties were obtained from the authors when possible. Otherwise, they have been estimated from knowledge of generally similar apparatus and procedures.

Many of the density data sets in Table 1 have been used by Ely et al.[47,48] to test a model for thermophysical properties of mixtures containing CO_2, hydrocarbons up to C_7, and other low molecular weight gases. This model designated DDMIX,[48] is based on extended corresponding states theory. In most cases, DDMIX uses propane as a reference fluid; however, it can also use a Schmidt-Wagner type equation of state (EOS) for pure CO_2 as Sherman et al.[24] have demonstrated. The Schmidt-Wagner EOS leads to improved predictions because of its flatter representation of near-critical pressure-density isotherms. Table 2 summarizes some of the comparisons of the experimental density data with predictions of DDMIX. This table lists the number of data values (N) and the root mean square (RMS) deviation of experimental density from that predicted by DDMIX. With the possible exceptions of CO_2 + n-butane and CO_2 + i-butane, the author concludes that DDMIX is a good predictive tool. More mixture density data of the highest attainable accuracy is needed to resolve whether one should suspect the data or the model, but it should also be emphasized that there is never too much good experimental data, since advances in theory depend on advances in the laboratory.

TABLE 1
Selected Published Density Results for CO$_2$-Rich Mixtures

Second component	T Range (K)	P Range (MPa)	X$_{CO_2}$ Range	ρ Uncert. (%)	Ref.
+ Methane	225—400	2—36	0.98	0.1	5
	250—400	2—35	0.90	0.1	20
	206—320	0.08—48	0.48	0.1	21
	311—511	0—70	0.15—1.00	0.2	22
	303—333	0.6—13	0—1.00	0.3	23
+ Ethane	245—400	3—35	0.99	0.1	24
	249—283	1—5	0—1.00	0.2	25
	311—511	1—70	0.17—0.82	0.2—0.4	26
+ Ethylene	312—373	0.3—26	0—1.00	0.2	27
	313—398	0.6—50	0—0.80	0.2	28
+ Propane	311—361	1—4	0—0.67	0.1—0.2	29
	278—511	1—70	0.20—0.79	0.2—0.4	30
+ n-Butane	311—395	0.3—8	0—0.62	0.1—0.2	31
	311—511	1—70	0.17—0.83	0.2—0.4	32
	319—378	2—8	0.09—0.87	0.1—1.0	33
	200—383	0.03—14	0.01—0.93	1	34
+ i-Butane	278—378	0.04—9	0—1.00	1	35
+ n-Pentane	278—378	0.03—10	0—1.00	1	36
+ i-Pentane	311—394	0.5—7	0—0.88	1	37
+ n-Heptane	311—477	0.2—13	0.02—0.95	1	38
+ n-Decane	278—511	0.3—19	0.65—1.00	0.2—0.4	39
	344—378	6—16	0.46—0.92	0.1—0.8	40
+ n-Tetradecane	344	7—16	0.69—0.90	0.1—0.7	41
+ n-Nonadecane	313—333	0.9—8	0.09—0.68	0.4—0.8	42
+ n-Heneicosane	318—338	0.9—8	0.10—0.65	0.3—0.7	42
+ n-Docosane	323—373	1—7	0.08—0.59	0.3—0.5	43
+ n-Octacosane	308—325	8—28	0.99—1.00	1—3	44
+ n-Dotriacontane	348—398	1—7	0.10—0.56	0.2—0.4	43
+ Nitrogen	250—330	2—33	0.98	0.1	2
	209—320	0.1—48	0.45	0.1	22
+ Water	323—498	0.03—10	0.50—0.98	0.05	45
+ Hydrogen sulfide	300—500	0—33	0.49	0.1	46

TABLE 2
Summary of Comparisons of DDMIX with Published Density Data for CO$_2$-Rich Mixtures

Second component	N	Deviation RMS (%)	Ref.
+ Methane	91	0.29	5
	107	0.75	20
	560	0.30	22
+ Ethane	97	0.41	24
	49	0.53	25
	770	0.57	26
+ Propane	72	2.49	29
	691	0.57	30
+ n-Butane	96	3.53	31
	750	0.56	32
	35	3.62	34
+ i-Butane	56	4.87	35
+ Nitrogen	79	0.26	2

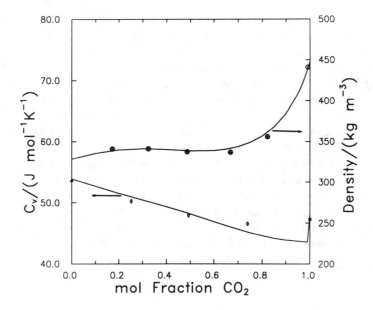

FIGURE 4.   Variation of C_v and ρ with composition for supercritical CO_2 + ethane mixtures at 320 K and 10 MPa. (a) Solid curves were calculated from the DDMIX model.[51] (b) Experimental C_v values from Roder,[50] (▲); Magee,[49] (♦); Magee and Ely,[6] (■) (c); experimental ρ values from Sherman et al.,[24] (●); Reamer et al.[26] (⊕).

B. HEAT CAPACITIES

Magee[49] has measured molar heat capacities at constant volume (C_v) for three compositions of the CO_2 + ethane system. These data complement earlier C_v measurements for pure CO_2[6] and for pure ethane[50] made with the same calorimeter. The data range from 210 to 340 K at pressures to 35 MPa and have an estimated accuracy of 0.5 to 2.0%.

Ernst and Hochberg[9] have measured specific heat capacity at constant pressure (c_p) for an equimolar CO_2 + ethane mixture at temperatures from 303 to 393 K and pressures to 53 MPa. The uncertainty of the c_p results ranges from 0.3 to 2.5%.

Figure 4 shows some predictions (solid lines) and some experimental values (symbols) of densities (ρ) and heat capacities (C_v) for the CO_2 + ethane system. The figure depicts behavior of these properties at a typical state in the supercritical region (320 K, 10 MPa) for compositions ranging from pure ethane to pure CO_2. The experimental values, interpolated by means of spline fits, agree very well with the predictions of Ely,[51] but as expected they show a nonlinear composition dependence C_v and ρ.

C. TRANSPORT PROPERTIES

Accurate and validated experimental measurements of transport coefficients are particularly scarce for mixtures, including those that contain appreciable quantities of CO_2. In fact only a few such data sources were located during this compilation. For viscosity coefficients of CO_2-rich mixtures, two references were located: Diller and co-workers at NIST for the CO_2 + ethane binary system, and that of Hobley et al. for the CO_2 + Ar binary. Hobley et al.[52] measured viscosities with a new capillary flow viscometer from 301 to 521 K at 0.1 MPa. These workers claim an accuracy of 1% for their viscosities. The reported data does not overlap the SCF region of CO_2. Diller et al.[53] measured shear viscosity coefficients from 210 to 320 K at pressures up to 30 MPa. A later companion study[19] of the same compositions ranged from 295 to 500 K at pressures from 2 to 55 MPa. The overall uncertainty of the experimental viscosity values is estimated by the authors not to exceed

3%. The results of Diller and co-workers confirm the accuracy of DDMIX predictions of mixture viscosities. These authors found that the values calculated by DDMIX agree with most of the data to within 10%.

One set of wide-range, validated measurements of thermal conductivity coefficients have been located for CO_2-rich mixtures. Johns et al.[54] have reported measurements with a transient hot-wire apparatus for the CO_2 + N_2 system from 302 to 470 K at pressures to 25 MPa. The authors claim an accuracy of 1% for their thermal conductivities and also found very good ($\pm 3\%$) agreement with the predictions of the corresponding states method of Ely and Hanley.[55,56] In summary, these authors have found very good agreement of their measured transport coefficients with the corresponding states model (DDMIX) most recently revised by Ely. Until a broad range of experimental measurements becomes available, it is recommended that DDMIX be used for predictions of viscosities and thermal conductivities of CO_2-rich mixtures. This model has been cast into an interactive microcomputer program[51] which is available with its supporting documentation from the NIST Office of Standard Reference Data (telephone: 301-975-2208) in Gaithersburg, MD.

REFERENCES

1. **Angus, S., Armstrong, B., and deReuck, K. M.,** *Carbon Dioxide International Thermodynamic Tables of the Fluid State,* Pergamon Press, Oxford, 1976.
2. **Ely, J. F., Haynes, W. M., and Bain, B. C.,** Isochoric (p, V_m, T) measurements on CO_2 and on $(0.982CO_2 + 0.018N_2)$ from 250 to 330 K at pressures to 35 MPa, *J. Chem. Thermodyn.,* 21, 879, 1989.
3. **Holste, J. C., Hall, K. R., Eubank, P. T., Esper, G., Watson, M. Q., Warowny, W., Bailey, D. M., Young, J. G., and Bellomy, M. T.,** Experimental (p, V_m, T) for pure CO_2 between 220 and 450 K, *J. Chem. Thermodyn.,* 19, 1233, 1987.
4. **Haynes, W. M.,** Orthobaric liquid densities and dielectric constants of carbon dioxide, in *Advances in Cryogenic Engineering,* Fast, R. W., Ed., Plenum Press, New York, 1986, 1199.
5. **Magee, J. W. and Ely, J. F.,** Isochoric (p,v,T) measurements on CO_2 and $(0.98CO_2 + 0.02CH_4)$ from 225 to 400 K and pressures to 35 MPa, *Int. J. Thermophys.,* 9, 547, 1988.
6. **Magee, J. W. and Ely, J. F.,** Specific heats (C_v) of saturated and compressed liquid and vapor carbon dioxide, *Int. J. Thermophys.,* 7, 1163, 1986.
7. **Yang, C. N. and Yang, C. P.,** Critical point in liquid-gas transitions, *Phys. Rev. Lett.,* 13, 303, 1964.
8. **Ernst, G., Maurer, G., and Wiederuh, E.,** Flow calorimeter for the accurate determination of the isobaric heat capacity at high pressures; results for carbon dioxide, *J. Chem. Thermodyn.,* 21, 53, 1989.
9. **Ernst, G. and Hochberg, U. E.,** Flow-calorimetric results for the specific heat capacity c_p of CO_2, of C_2H_6, and of $(0.5\ CO_2 + 0.5\ C_2H_6)$ at high pressures, *J. Chem. Thermodyn.,* 21, 407, 1989.
10. **Michels, J., Sengers, J. V., and van der Gulik, P. S.,** The thermal conductivity of carbon dioxide in the critical region. II. Measurements and conclusions, *Physica,* 28, 1216, 1962.
11. **Le Neindre, B.,** Contribution a l'etude experimentale de la conductivite thermique de quelques fluides a haute temperature et a haute pression, *Int. J. Heat Mass Transfer,* 15, 1, 1972.
12. **Le Neindre, B., Tufeu, R., Bury, P., and Sengers, J. V.,** Thermal conductivity of carbon dioxide and steam in the supercritical region, *Ber. Bunsengesellsch.,* 77, 262, 1973.
13. **Vesovic, V., Wakeham, W. A., Olchowy, G. A., Sengers, J. V., Watson, J. T. R., and Millat, J.,** The transport properties of carbon dioxide, *J. Phys. Chem. Ref. Data,* 19, 763, 1990.
14. **Michels, A., Botzen, A., and Schuurman, W.,** The viscosity of carbon dioxide between 0°C and 75°C and at pressures up to 2000 atmospheres, *Physica,* 23, 95, 1957.
15. **Herreman, W., Grevendonk, W., and De Bock, A.,** Shear viscosity measurements of liquid carbon dioxide, *J. Chem. Phys.,* 53, 185, 1970.
16. **Ulybin, S. A and Makarushkin, W. I.,** The viscosity of carbon dioxide at 220-1300 K and pressures up to 300 MPa in *Proc. 7th Symp. on Thermophysical Properties,* American Society of Mechanical Engineers, New York, 1977, 678.
17. **Iwasaki, H. and Takahashi, M.,** Viscosity of carbon dioxide and ethane, *J. Chem. Phys.,* 74, 1930, 1981.
18. **Diller, D. E. and Ball, M. J.,** Shear viscosity coefficients of compressed gaseous and liquid carbon dioxide at temperatures between 220 and 320 K and at pressures to 30 MPa, *Int. J. Thermophys.,* 6, 619, 1985.

19. **Diller, D. E. and Ely, J. F.,** Measurements of the viscosities of compressed gaseous carbon dioxide, ethane, and their mixtures, at temperatues up to 500 K, *High-Temp. High-Pressures,* 21, 613, 1989.
20. **Howley, J. B., Ely, J. F., and Magee, J. W.,** Isochoric PVT measurements on $0.9 CO_2 + 0.1 CH_4$ from 250 to 400 K and pressures to 35 MPa, in press.
21. **Esper, G. J., Bailey, J. C., Holste, J. C., and Hall, K. R.,** Volumetric behavior of near-equimolar mixtures for $CO_2 + CH_4$ and $CO_2 + N_2$, *Fluid Phase Equilibria,* 49, 35, 1989.
22. **Reamer, H. H., Olds, R. H., Sage, B. H., and Lacey, W. N.,** Methane-carbon dioxide system in the gaseous region, *Ind. Eng. Chem.,* 36, 88, 1944.
23. **McElroy, P. J., Battino, R., and Dowd, M. K.,** Compression-factor measurements on methane, carbon dioxide, and (methane + carbon dioxide) using a weighing method, *J. Chem. Thermodyn.,* 21, 1287, 1989.
24. **Sherman, G. J., Magee, J. W., and Ely, J. F.,** PVT relationships in a carbon dioxide-rich mixture with ethane, *Int. J. Thermophys.,* 10, 47, 1989.
25. **Gugnoni, R. J., Eldridge, J. W., Okay, V. C., and Lee, T. J.,** Carbon dioxide-ethane phase equilibrium and densities from experimental measurements and the B-W-R equation, *AIChE J.,* 20, 357, 1974.
26. **Reamer, H. H., Olds, R. H., Sage, B. H., and Lacey, W. N.,** Volumetric behavior of ethane-carbon dioxide system, *Ind. Eng. Chem.,* 37, 688, 1945.
27. **Ku, S. K. and Dodge, B. F.,** Compressibility of the binary systems: helium-nitrogen and carbon dioxide-ethylene, *J. Chem. Eng. Data,* 12, 158, 1967.
28. **Sass, A., Dodge, B. F., and Bretton, R. H.,** Compressibility of gas mixtures carbon dioxide-ethylene system, *J. Chem. Eng. Data,* 12, 168, 1967.
29. **Niesen, V. G. and Rainwater, J. C.,** Critical locus, (vapor + liquid) equilibria, and coexisting densities of (carbon dioxide + propane) at temperatures from 311 K to 361 K, *J. Chem. Thermodyn.,* 22, 777, 1990.
30. **Reamer, H. H., Sage, B. H., and Lacey, W. N.,** Volumetric and phase behavior of the propane carbon dioxide system, *Ind. Eng. Chem.,* 43, 2515, 1951.
31. **Niesen, V. G.,** (Vapor + liquid) equilibria and coexisting densities of (carbon dioxide + n-butane) at 311 to 395 K, *J. Chem. Thermodyn.,* 21, 915, 1989.
32. **Olds, R. H., Reamer, H. H., Sage, B. H., and Lacey, W. N.,** The *n*-butane-carbon dioxide system, *Ind. Eng. Chem.,* 41, 475, 1949.
33. **Hsu, J. J.-C., Nagarajan, N., and Robinson, R. L., Jr.,** Equilibrium phase compositions, phase densities, and interfacial tensions for CO_2 + hydrocarbon systems. I. CO_2 + *n*-butane, *J. Chem. Eng. Data,* 30, 485, 1985.
34. **Kalra, H., Krishnan, T. R., and Robinson, D. B.,** Equilibrium-phase properties of carbon dioxide-*n*-butane and nitrogen-hydrogen sulfide systems at subambient temperatures, *J. Chem. Eng. Data,* 21, 222, 1976.
35. **Besserer, G. J. and Robinson, D. B.,** Equilibrium-phase properties of i-butane-carbon dioxide system, *J. Chem. Eng. Data,* 18, 298, 1973.
36. **Besserer, G. J. and Robinson, D. B.,** Equilibrium-phase properties of *n*-pentane-carbon dioxide system, *J. Chem. Eng. Data,* 18, 416, 1973.
37. **Besserer, G. J. and Robinson, D. B.,** Equilibrium-phase properties of isopentane-carbon dioxide system, *J. Chem. Eng. Data,* 20, 93, 1975.
38. **Kalra, H., Kubota, H., Robinson, D. B., and Ng, H.-J.,** Equilibrium phase properties of the carbon dioxide-*n*-heptane system, *J. Chem. Eng. Data,* 23, 317, 1978.
39. **Reamer, H. H. and Sage, B. H.,** Volumetric and phase behavior of the *n*-decane - CO_2 system, *J. Chem. Eng. Data,* 8, 508, 1963.
40. **Nagarajan, N. and Robinson, R. L., Jr.,** Equilibrium phase compositions, phase densities, and interfacial tensions for CO_2 + hydrocarbon systems. II. CO_2 + *n*-decane, *J. Chem. Eng. Data,* 31, 168, 1986.
41. **Gasem, K. A. M., Dickson, K. B., Dulcamara, P. B., Nagarajan, N., and Robinson, R. L., Jr.,** Equilibrium phase compositions, phase densities, and interfacial tensions for CO_2 + hydrocarbon systems V. CO_2 + *n*-tetradecane, *J. Chem. Eng. Data,* 34, 191, 1989.
42. **Fall, D. J., Fall, J. L., and Luks, K. D.,** Liquid-liquid-vapor immiscibility limits in carbon dioxide + *n*-paraffin mixtures, *J. Chem. Eng. Data,* 30, 82, 1985.
43. **Fall, D. J. and Luks, K. D.,** Phase equilibria behavior of the systems carbon dioxide + *n*-dotriacontane and carbon dioxide + n-docosane, *J. Chem. Eng. Data,* 29, 413, 1984.
44. **McHugh, M. A., Seckner, A. J., and Yogan, T. J.,** High-pressure phase behavior of binary mixtures of octacosane and carbon dioxide, *Ind. Eng. Chem. Fundam.,* 23, 493, 1984.
45. **Patel, M. R. and Eubank, P. T.,** Experimental densities and derived thermodynamic properties for carbon dioxide - water mixtures, *J. Chem. Eng. Data,* 33, 185, 1988.
46. **Bailey, D. M., Liu, C.-H., Holste, J. C., Eubank, P. T., and Hall, K. R.,** Energy functions for H_2S. III. Near-equimolar mixture with CO_2, *Hydrocarbon Processing,* 66, 73, 1987.
47. **Ely, J.F., Magee, J. W., and Haynes, W. M.,** *Thermophysical Properties for Special High CO$_2$ Content Mixtures RR-110,* Gas Processors Association, Tulsa, OK, 1987, 103.

48. **Ely, J. F. and Magee, J. W.,** Experimental measurement and prediction of thermophysical property data of carbon dioxide rich mixtures, *Proc. 68th Gas Processors Assoc. Annu. Conv.*, Gas Processors Association, Tulsa, OK, 1989, 89.

49. **Magee, J. W.,** Molar heat capacity (C_v) measurements on [xCO_2 + (1 − x) $C_2 H_6$] from 220 to 340 K at pressures to 35 MPa, in press.

50. **Roder, H. M.,** Measurements of the specific heats, C_σ and C_v of dense gaseous and liquid ethane, *J. Res. Natl. Bur. Stand.*, 80A, 739, 1976.

51. **Ely, J. F.,** *Computer Program DDMIX*, Office of Standard Reference Data, NIST, Gaithersburg, MD, 20899, 1989.

52. **Hobley, A., Matthews, G. P., and Townsend, A.,** The use of a novel capillary flow viscometer for the study of the argon/carbon dioxide system, *Int. J. Thermophys.*, 10, 1165, 1989.

53. **Diller, D. E., Van Poolen, L. J., and dos Santos, F. V.,** Measurements of the viscosities of compressed fluid and liquid carbon dioxide + ethane mixtures, *J. Chem. Eng. Data*, 33, 460, 1988.

54. **Johns, A. I., Rashid, S., Rowan, L., Watson, J. T. R., and Clifford, A. A.,** The thermal conductivity of pure nitrogen and of mixtures of nitrogen and carbon dioxide at elevated temperatures and pressures, *Int. J. Thermophys.*, 9, 3, 1988.

55. **Ely, J. F. and Hanley, H. J. M.,** Prediction of transport properties. I. Viscosity of fluids and mixtures, *Ind. Eng. Chem. Fundam.*, 20 323, 1981.

56. **Ely, J. F. and Hanley, H. J. M.,** Prediction of transport properties. II. Thermal conductivity of pure fluids and mixtues, *Inc. Eng. Chem. Fundam.*, 22, 90, 1983.

56. **Ely, J. F. and Hanley, H. J. M.,** Prediction of transport properties. II. Thermal conductivity of pure fluids and mixtues, *Ind. Eng. Chem. Fundam.*, 22, 90, 1983.

Chapter 9

THERMAL CONDUCTIVITY AND THERMAL DIFFUSIVITY IN SUPERCRITICAL FLUIDS

C. A. Nieto de Castro

TABLE OF CONTENTS

ABSTRACT

This chapter is a short review of the actual situation of measuring and correlating thermal conductivity and thermal diffusivity data in supercritical fluids (SCFs).

I. INTRODUCTION

The transport properties of fluids have been recognized as very important for the chemical industry, whenever heat, mass, or momentum transfer are involved. Their values, together with those of thermodynamic properties condition the design of heat exchange equipment, compressors, flow meters, distillation columns, extractors, etc.

The thermal conductivity of fluids varies considerably from zone to zone in the phase diagram, from dilute gas to compressed liquid, from vapor to supercritical dense fluid, and therefore it is impossible to understand and predict the behavior of the thermal conductivity, especially of SCFs, without a detailed experimental and theoretical study of well-chosen fluids, and with scientific and/or industrial interest. The knowledge of this property is very important in the development of molecular theories of fluids and fluid mixtures, where no complete theory of molecular transport is available that can reproduce experiment within very tight confidence limits.

The thermal diffusivity of fluids also has some importance in both scientific and industrial aspects, not only because it can be a source of heat capacity data in wide fluid ranges, when used with thermal conductivity, but also because it is the property most important when the critical point is approached and fluctuations become relevant.

A typical behavior of the thermal conductivity of a fluid is shown in Figure 1, where the thermal conductivity of *n*-butane is represented as a function of density,[1] from dilute gas to liquid, including the supercritical enhancement. This figure shows the unique behavior in the transport properties, in which as the density ρ approaches the critical density ρ_c, near the critical temperature (but $T > T_c$), the thermal conductivity shows a strong enhancement, diverging to infinity for $T = T_c^+$. For *n*-butane $\rho_c = 288$ kg m^{-3} and $T_c = 425.16$ K.

Since there is no complete molecular theory that can describe the behavior of the thermal conductivity in the fluid phases, it is useful to divide the thermal conductivity in three contributions, a zero density or dilute gas contribution, $\lambda_0(T)$, a density- and temperature-dependent excess thermal conductivity, $\Delta\lambda_{exc}(T,\rho)$, and a critical enhancement contribution, $\Delta\lambda_{crit}(T,\rho)$, only important in the supercritical zone, i.e.,

$$\lambda(T,\rho) = \lambda_0(T) + \Delta\lambda_{exc}(T,\rho) + \Delta\lambda_{crit}(T,\rho) \tag{1}$$

There are two separate regions where the critical enhancement is important, and therefore, $\Delta\lambda_{crit}(T,\rho)$ significant. The first region, called critical region proper, is the region where a nonanalytical equation of state (EOS), a scaled equation, must be used, and was defined by the Sengers[2] in terms of the reduced coordinates

$$\Delta T^* = (T - T_c)/T_c, \quad \Delta\rho^* = (\rho - \rho_c)/\rho_c \tag{2}$$

as

$$|\Delta T^*| \leq 0.03 \text{ and } |\Delta\rho^*| \leq 0.25 \tag{3}$$

The second region, herein designated by extended critical region, covers the zone in densities and temperatures for which the experimental measurements reveal an anomalous increase in the thermal conductivity and it can extend to quite high temperatures ($T/T_c \sim 2$). In this region an analytical EOS can be used.

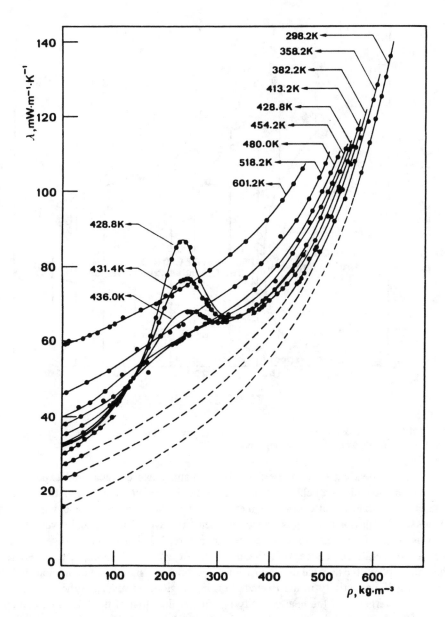

FIGURE 1. The thermal conductivity of *n*-butane. (From Nieto de Castro, C. A., Tufeu, R., Tufeu, B., and le Neindre, B., *Int. J. Thermophys.*, 4, 11, 1983. With permission.)

These regions are illustrated in Figure 2 with the density temperature diagram of oxygen,[3] together with the dilute gas, dense gas, vapor, and compressed liquid zones. It is in the critical region proper that the thermal conductivity can increase ten times or more than the dilute gas values; however, this is a very limited zone in temperature and pressure, where the density is changing very rapidly (the compressibility of the fluid in this zone is very high), and therefore of limited technological applicability. It is therefore in the extended critical region that most of the applications in SCF technology are implemented.

The thermal diffusivity a is defined as

$$a = \frac{\lambda}{\rho C_p} \qquad (4)$$

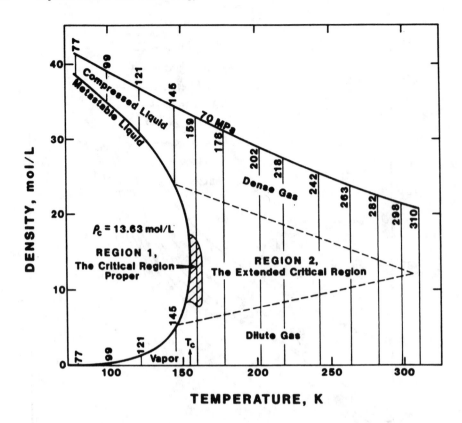

FIGURE 2.   The density vs. temperature diagram for oxygen. (From Roder, H. M., *J. Res. Natl. Bur. Stand.*, 87, 279, 1982, With permission.)

where C_p is the heat capacity of the fluid. This quantity can be measured experimentally and has a complete distinct behavior from that of the thermal conductivity. At low densities (dilute and moderately dense gas) the thermal diffusivity is enormous and varies sharply with density, while for dense gas and liquid densities it decreases and its variation with density is small. This behavior is illustrated in Figure 3 with thermal diffusivity data for argon[4] for temperatures between 171.6 and 321 K. Because the heat capacity also diverges in the critical point, the thermal diffusivity does not show any significant anomalous behavior as the thermal conductivity in the extended critical region; however, in the critical region proper, the divergence of the heat capacity is stronger than that of the thermal conductivity and therefore the thermal diffusivity goes to zero, as illustrated in Figure 4, where the results obtained for sulfurhexafluoride are shown,[5] both for sub- and supercritical data around the critical point.

This chapter presents a short review of the actual situation of thermal conductivity and thermal diffusivity data in SCF, with special emphasis on the work done by the author. A brief description of the best experimental methods of measuring these properties is outlined in the next section, followed by a discussion on the data available, the critical enhancement behavior in the extended critical region, and overall surface fits to correlate thermal conductivity data. Mention is also made of the thermal diffusivity data, and consequently of the heat capacity values obtained from simultaneous measurements of thermal conductivity and diffusivity. On the behavior of the thermal conductivity in the critical region proper, the reader is referred to Chapter 1.[6]

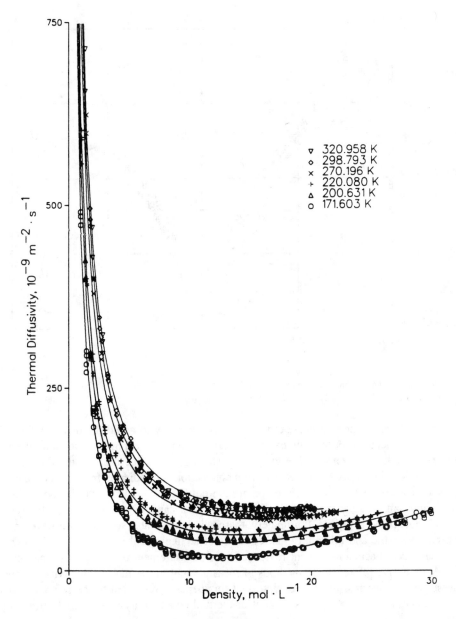

FIGURE 3. The thermal diffusivity of supercritical argon. (From Roder, H. M. and Nieto de Castro, C. A., in *Thermal Conductivity 20,* Hasselman, D. P. H. and Thomas, J. R., Jr., Eds., Plenum Press, New York, 1989, p. 173. With permission.)

II. EXPERIMENTAL METHODS

The experimental measurement of thermal conductivity and thermal diffusivity has received great attention in the last 2 decades, when it became clear that most of the measurements done in the past were incorrect. This situation developed because of the fact that it is not easy to separate completely the particular phenomena under study from other parallel processes generated by the perturbation induced in the system.

This can be easily illustrated with the measurement of thermal conductivity. When a temperature gradient is imposed on a fluid, three different types of heat transfer occur: conduction, convection, and radiation. To isolate the phenomena of pure conduction, and

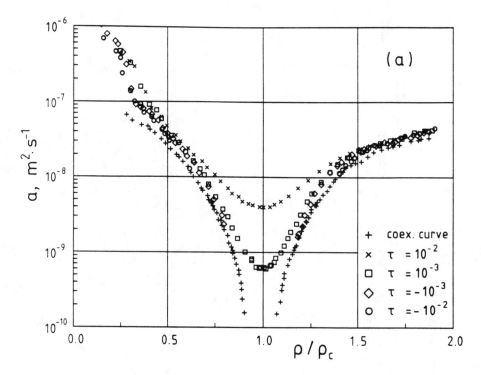

FIGURE 4.   The thermal diffusivity of sulfurhexaflouride. $\alpha = (T - T_c)/T_c$. (From Jany, P. and Straub, J., *Int. J. Thermophys.*, 8, 165, 1987. With permission.)

therefore obtain the thermal conductivity λ, it is necessary to avoid the onset of free convection and to account for the radiative contribution, with a correct mathematical description. Only recently has it been possible to propose standard reference data for the thermal conductivity of liquids at 0.1 MPa, derived from convection and radiation free measurements.[7,8]

The measurements in the SCF region are more difficult to make with high accuracy and they require an accurate value of the density and heat capacity of the fluid in this region, which is not always available. Fortunately, there was considerable success in the development of highly versatile and reliable EOSs for fluids of technological importance in the 1980s.

There have been some recent reviews on the measurement of transport properties that include extensive references to methods of measurement of thermal conductivity,[9-11] and to a lesser extent, thermal diffusivity,[11] so that this paper is restricted to the reference of the best absolute techniques of measurement of thermal conductivity and thermal diffusivity.

A. THE MEASUREMENT OF THERMAL CONDUCTIVITY

The thermal conductivity of a fluid measures its propensity to dissipate energy, when disturbed from equilibrium, by the imposition of a temperature gradient. For isotropic fluids, the thermal conductivity is defined by Fourier's law

$$Q = -\lambda \nabla T \tag{5}$$

where Q is the instantaneous flux of heat, response of the médium to the instantaneous temperature gradient, ∇T.

The thermal conductivity depends on the thermodynamic state of the fluid prior to the perturbation and must be related to a reference thermodynamic state, $\lambda(T_r, \rho_r)$, not necessarily equal to the initial state. As it is impossible to measure local fluxes and local temperature gradients, Equation 5 cannot be used directly to measure the thermal conductivity of fluids.

Assuming that the perturbation of the equation for the conservation of energy* for a viscous isotropic and compressible fluid, with temperature-dependent properties, is small and that a local thermodynamic state is created, such that $K_T\left(\dfrac{\partial P}{\partial T}\right) << \alpha_p$, where K_T is the isothermal compressibility and α_p is the isobaric thermal expansion coefficient, we can write

$$\rho C_p \frac{DT}{Dt} = \lambda \nabla^2 T + \Sigma \tag{6}$$

where D/Dt is the substantial or space derivative, ∇^2 is the symbol for the Laplacian, and Σ represents the additional terms in the energy equation due to the rate of internal energy increase caused by viscous dissipation and the rates of absorption and emission of radiative energy by the fluid.

A general solution of Equation 6 is not possible because of the complicated differential and integrodifferential terms that constitute Σ. Its application to the measurement of thermal conductivity must be restricted.

Small temperature gradients must be used and any fluid movement be avoided or reduced to a negligible level. Under these conditions, the substantial derivative can be replaced by the partial derivative and the rate of internal energy increase due to viscous dissipation is rendered negligible. Finally, it is important to recognize that the transport of energy by radiation is always present, and therefore correct for it in any experimental techniques. Equation 6 is finally reduced to

$$\rho C_p \frac{\partial T}{\partial t} = \lambda \nabla^2 T \tag{7}$$

which is the basis of any experimental method for the measurement of thermal conductivity.

In the last 20 years, a variety of experimental methods have been developed for all the regions of the phase diagram. These methods, according to the use of Equation 7 can be classified in two main categories:

1. Unsteady state on transient techniques, in which the full Equation 7 is used and where the principal measurement is the temporal history of the fluid temperature at a point or line in space, $T(t)$.
2. Steady-state techniques for which the steady state of heat transfer is attained, reducing Equation 7 to

$$\nabla^2 T = 0 \tag{8}$$

which can be integrated for a given geometry (parallel plates, concentric cylinders, etc.).

1. The Transient Hot Wire Technique

Several techniques can be devised, making use of Equation 7; however, the only technique that has received general acceptance for fluids is the transient hot wire technique,[10,11] where the perturbing heat flux is generated by the flow of an electric current through a thin, cylindrical wire, immersed in the fluid. This technique was introduced almost 60 years ago by Stalhane and Pyk.[12] Only in the last 20 years, due to the fast development of electronic and measuring devices, was it possible, however, to design instruments that can take full

* Details of the derivation can be found in References 10 and 11.

advantage of the potential benefits of the technique. The foundations of the modern experimental method were laid by Haarman,[13] Kestin, Wakeham, and co-workers,[14-21] McLaughlin and Pittman,[22] and Mani and Venart.[23] A complete description of the technique, including the mathematical modeling of the instruments and their operation can be found in Reference 24. Instruments based on these principles are capable of measuring the thermal conductivity of fluids to within 0.3% in the gaseous region, 0.5% in the vapor and liquid phases, and 1% for fluids in the extended critical region ($T/T_c > 1.05$). Values of the thermal conductivity of toluene and benzene[7,8] obtained only with this technique have been reported recently as standard reference data.

A brief description of the method is given here, however, because it is the most accurate method for the measurement of thermal conductivity in the zones reported, and it forms the basis of a method for the determination of the thermal diffusivity, described in Section II.B.

The principal reason the transient hot wire technique is so accurate is the existence of a mathematical idealization of the hot-wire cells and a corresponding working equation. In the ideal model, an infinitely long, vertical, line source of zero heat capacity and infinite thermal conductivity is immersed in an infinite isotropic fluid, with properties independent of temperature and in thermodynamic equilibrium with the fluid at time zero. The transfer of energy from the line source, when a stepwise heat flux, q, per unit length is applied is purely conductive and it generates a temperature rise ΔT at the surface of the wire of radius r_0 at time t given by[24]

$$\Delta T_{id}(r_0,t) = \frac{q}{4\pi\lambda} \left[\ell n \frac{4at}{r_0^2 C} + \frac{r_0^2}{4at} + \ldots \right] \tag{9}$$

with $C = 1.781\ldots$ being the exponential of Euler's constant.

If the wire radius is chosen such that the second term on the right side of Equation 9 is less than 0.01% of ΔT_{id}, $r_0^2/4$ at $<< 1$, and Equation 9 is reduced to

$$\Delta T_{id}(r_0,t) = \frac{q}{4\pi\lambda} \ell n \frac{4at}{r_0^2 C} \tag{10}$$

Equation 10 is the fundamental working equation of the transient hot-wire technique. The thermal conductivity of the fluid can be obtained from the slope of the line ΔT_{id} v.s. $\lambda n(t)$, while the thermal diffusivity may be obtained from its intercept (please see previous discussion).

In any practical implementation of this technique, the source of heating is provided by a finite length of metallic wire, which also serves as a resistance thermometer. As a consequence of this experimental arrangement, the measured temperature rise, ΔT_w, departs from the value of Equation 10, even at the surface of the wire, due to its finite properties, its finite length, the finite dimension of the fluid medium, and eventually the occurrence of convection and radiation. The success of the experimental method rests, however, upon the fact that by proper design it is possible to construct an instrument that will match as closely as possible its ideal description, rendering some of the corrections negligible and others very small and accountable by linearizing the mathematical expressions used to describe them.

Most of the possible causes of departure between ΔT_w and ΔT_{id} have been investigated by Healy et al.[15] and posteriorly improved by several authors.[19-21] Their results were expressed in the form of small corrections, δT_i, to be applied to the measured temperature rise ΔT_w, so that

$$\Delta T_{id} = \Delta T_w + \sum_i \delta T_i \tag{11}$$

The thermal conductivity must be assigned to a reference thermodynamic state, defined by T_r and ρ_r given by

$$T_r = T_0 + \sum_i \delta T_i^* \tag{12}$$

$$\rho_r = \rho \, (T_r, P_0) \tag{13}$$

where P_0 and T_0 are the initial equilibrium pressure and temperature and δT_i^* are corrections derived mainly from the fact that the properties of the fluid are temperature and density dependent (see Reference 24 for full details of all the δT_i and δT_i^*).

To obtain the thermal conductivity with this technique, the principal measurements are those of the temperature rise of a thin wire, ΔT_w; time, t; and heat flux per unit length of the wire, q. The solution for the ideal transient hot wire requires a constant q, and the experiment should last less than 1 s for most fluids, to render convection negligible. The accuracy of the measurement of ΔT_w and q limits the overall accuracy of the thermal conductivity obtained. Two methods have been employed thus far. In the first, the hot wire is equipped with potential taps located at some distance from the end. The temperature increase in the wire is deduced from the increase in the potential difference between the taps or from the out of balance of the resistance measuring bridge, as a function of time. This experimental setup has the advantage of having the central portion of the wire free of convection and free of an axial conduction to the wire supports; it was employed by several authors.[22-25] Its main difficulty lies, however, in the mechanical soldering and finishing of the potential taps, which supply an additional path for heat conduction. Furthermore, in these old instruments, the use of a digital voltmeter for data acquisition imposed a time resolution of the measurement approximately equal to its integration time. For all these reasons, an alternate method, first proposed by Haarman,[13] is preferable. This method makes use of an automatic Wheatstone bridge that contains two hot-wire cells of different length with wire resistance ratios R_L/R_S between 1.5 and 3.

This arrangement was found to compensate for the distortion of the temperature field at ends of a single wire and its use is equivalent to the study of a finite section of an infinite wire of resistance $R_w = R_L - R_S$.[26]

The bridge is prepared to determine the times at which R_w attains several predetermined resistance values and by using wire of known temperature-resistance characteristics, such as platinum or tantalum, it is possible to translate the variation $\Delta R(t)$ into $\Delta T(t)$. Details of the Wheatstone bridge used in the past can be found in References 16 and 17 and can be found for a computer-controlled automatic bridge in References 27 and 28. The experimental technique achieves a precision of 0.1 to 0.2% and accuracies between 0.3 and 0.5% in the best instruments. A slightly different approach was used by Roder[29] and Venart et al.,[30] by using the same type of bridges, but measuring the out of balance of the bridge as a function of time. A recent review by Nieto de Castro[31] shows that this technique has been used successfully to measure the thermal conductivity of gases, liquids, vapors, SCFs, molten salts, and molten semiconductors, in wide ranges of temperature and pressure. Most of the experimental problems emanating from the measurement in polar and electrically conducting liquid have been solved, and extensions of this technique to very high temperatures, where radiation heat transfer plays a significant role, are under careful current investigation.

Figure 5 shows the thermal conductivity of a fluid used in SCF technology, ethane, as a function of density,[32] while Figure 6 shows the thermal conductivity of nitrogen[33] as a function of density, for the vapor, the liquid, and the SCF. Figure 7 displays the thermal conductivity of binary mixtures of methane and ethane as a function of density,[34] evidence for the existence of the same type of enhancement in the thermal conductivity of nonquantum binary mixtures.

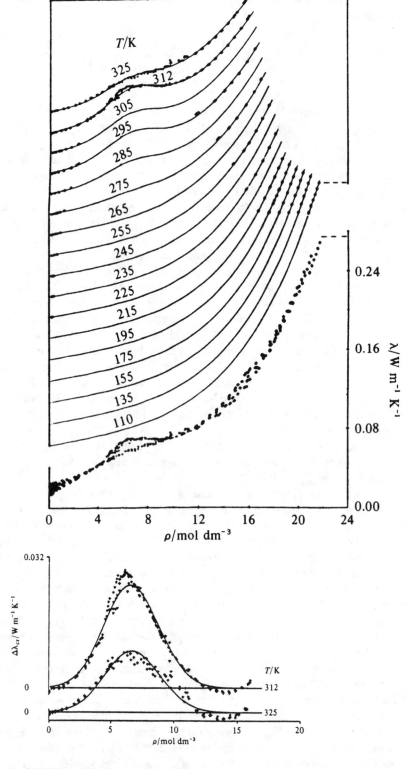

FIGURE 5. The thermal conductivity of ethane. (The isotherms are displaced by 0.02 W/(·K.) The bottom part of the figure shows the critical enhancement for two temperatures, displaced by 0.006 W/(m·K). (From Roder, H. M. and Nieto de Castro, C. A., *High Temp. High Pressure,* 17, 453, 1985. With permission.)

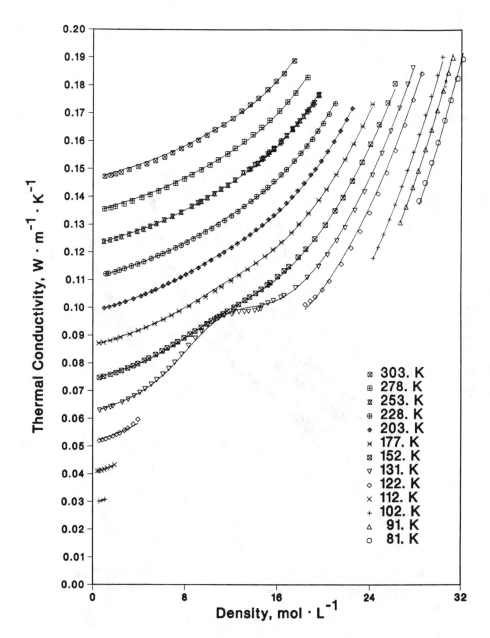

FIGURE 6. The thermal conductivity of nitrogen. The isotherms are separated by 0.010 W/(m·K). (From Perkins, R. A., Roder, H. M., and Nieto de Castro, C. A., *Physica A,* 1991, in press.)

2. The Concentric Cylinders Method

As already stated, the steady-state techniques for the measurement of thermal conductivity use Equation 8 integrated for a given geometry. This is the case for one of the most successful methods in the SCF zone, the concentric cylinder method.

In the ideal model the fluid is constant in an annulus between two infinitely long coaxial cylinders. The inner cylinder, of radius b_1, is maintained at a uniform steady temperature T_1 by the supply of the heat at rate Q, whereas the outer cylinder of diameter b_2 is kept at a uniform temperature T_2. If the cylinders have a length λ, the integration (Equation 8) for a fluid of constant thermal conductivity gives the working equation:

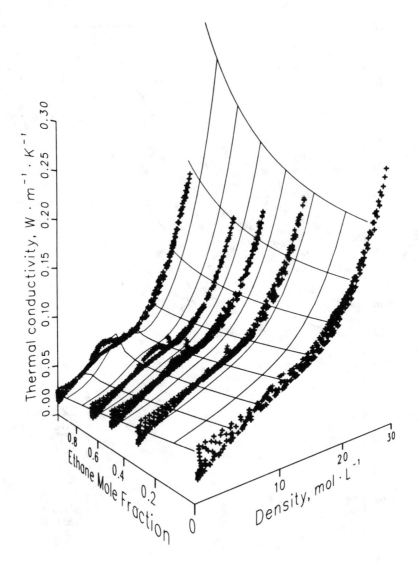

FIGURE 7.   The thermal conductivity of methane-ethane mixtures.[34] (From Friend, D. G. and Roder, H. M., *Int. J. Thermophys.*, 8, 13, 1987. With permission.)

$$Q = \frac{2\pi\ell\lambda \, (T_1 - T_2)}{\ell n \, (b_2/b_1)} \tag{14}$$

Equation 14 can be written in the form

$$\lambda_{app} = \frac{Q}{(T_1 - T_2)} \, K \tag{15}$$

with

$$K = \frac{1}{2\pi\ell} \, \ell n \left(\frac{b_1}{b_2}\right) \tag{16}$$

being the cell constant, that can be determined from geometric or capacitance measurements. The value λ_{app} is an apparent thermal conductivity that has to be corrected for the conduction through the cylinders, electric leads, ceramic supports, usually designated by parallel conduction, λ_p, and for radiation in the fluid, λ_r.

The value of λ_p can be estimated by measuring fluids of known thermal conductivity, while λ_r, in the absence of a complete treatment for conduction and radiation in the cylindrical geometry of participating fluids, must be evaluated assuming the fluid to be transparent.[35] That being the case, one can write

$$\lambda = \lambda_{app} - \lambda_p - \lambda_r \tag{17}$$

$$\frac{\lambda_r}{\lambda} = \frac{Q_r}{Q_c} = \frac{Q_r}{Q - Q_p - Q_r} \simeq \frac{4\sigma\epsilon T_1^3 (T_1 - T_2)}{Q} = \alpha \tag{18}$$

as $Q_p + Q_r \ll Q$. In this equation S is the surface of the emitting inner cylinder, ϵ the emmissivity of the cell material, and σ the Stefan-Boltzmann constant.

The simultaneous use of Equations 17 and 18 gives

$$\lambda = \frac{(\lambda_{app} - \lambda_p)}{(1 + \alpha)} \tag{19}$$

Usually Q is supplied by an electric heater present inside the cylinder and is obtained from the measurement of the voltage supplied and the measured current (Q = VI). The temperature T_1 and the temperature difference $T_1 - T_2$ is usually measured with thermocouples.

This technique is capable of obtaining thermal conductivity values with an accuracy better than 2% when careful measurement of λ_p exists. Due to the fact that the radiative contribution is calculated assuming transparent fluids, larger errors can be found at very high temperatures; however, it does have the advantage over the transient hot-wire technique of using very small gradients, when the fluid is close to the critical point. Temperature differences typically of 0.1 to 2 K, measured to 3 mK, have been used for fluid annulus of 5×10^{-4} m.[1]

Figure 1 shows a plot of the thermal conductivity of n-butane obtained with an instrument developed by le Neindre.[36] Extreme care must be used in the construction of these type of cells to avoid any eccentricity. The readers are referred to References 9 and 37 for details of all the corrections. Several other gases have been measured, including carbon dioxide, ethane, propane, ammonia, and water, in the supercritical region, and the measurements have been recently reviewed by le Neindre et al.[38] (see Section III). A final comment must be made, however, about the possibility of having convection in a concentric cylinder cell as it is very important close to the critical point. In order to evaluate the effect of convection in the fluid layer, one needs to solve the hydrodynamic equations subject to the appropriate boundary conditions for the particular temperature distribution of the experiment. Such an analysis is difficult, especially in the critical region where some of the fluid properties vary considerably with density and temperature and, therefore, with the position in the fluid layer. If the convection is minimized by using small temperature difference and small fluid gaps, a Taylor series expansion in terms of the temperature difference $\Delta T = T_1 - T_2$ can be used and a heat transfer coefficient proportional to the Rayleigh number

$$N_{Ra} = \frac{g\alpha_p\rho^2 C_p \Delta T \Delta b^3}{\lambda\eta} \tag{20}$$

is obtained.[39,40] In this equation, q is the acceleration of gravity and η the shear viscosity. The Rayleigh number diverges strongly when the critical point is approached.[39] The absence of convection in the cells for a given fluid can be verified experimentally by conducting a series of measurements at the same temperature and density, with different values of ΔT and hence different values of N_{Ra}.

3. The Parallel Plates Method

It follows from Equation 8 that when steady-state heating is applied to one of two parallel plates, and the plates are maintained at temperature T_1 and T_2, the heat flow Q across an area A through a fluid layer of thickness d is given by the linearized version of the Fourier's law

$$Q = \lambda A \frac{(T_2 - T_1)}{d} \tag{21}$$

In this solution it is assumed that the thermal conductivity of the fluid is constant, the plate dimensions are infinite, and that the heat flow is unidimensional. The plates must be maintained at a uniform and constant temperature ($T_2 > T_1$). In practice the upper plate is normally surrounded by a guard plate, sufficiently close to the upper plate to eliminate the distortion of the temperature profile at the edges of its lower surface. From Equation 21 one can see that it is very important to know Q, $\Delta T = T_2 - T_1$, d, and A very accurately. Also, the plates must be well parallel and this is only achieved in short areas. In the experiment slightly different quantities than the ideal ones are measured, namely, Q_{exp}, the power developed in the upper plate, and ΔT_{exp} the difference in temperature between the upper plate and the lower plate thermometers. The ratio d/A is the cell constant.

Following Mostert et al.[41] analysis, three main groups of corrections can be defined:

1.   Corrections to the power developed in the upper plate that involve a parasitic heat transfer, Q_p, between the upper plate and the guard plate, the presence of a radiative heat transfer component, Q_r, and a contribution from convection. The value of Q_p can be calculated from Q_{exp}, Q_{rad} is assumed to be given by the Stefan-Boltzmann law, as the optical thickness of the fluid layer is small. The contribution from convection to the heat transfer is more difficult to account for and it is preferable to design the instrument and to perform careful measurements that have no convection. Experience and some mathematical modeling show that if λ is independent of ΔT, convection is absent. If not, the experimental point must be discarded. After these corrections (Q_c = O) one obtains

$$Q = Q_{exp} - Q_p - Q_r \tag{22}$$

2.   The second type of correction is related to the temperature difference, ΔT. These corrections involve the thermometer locations in the upper and lower plate, as they cannot be placed exactly at the surfaces, ΔT_s, and a correction for the temperature discontinuity between the plates and the fluid, which is very important at low densities, must be made. This effect is called the accommodation effect, and the correction, ΔT_a, depends on the mean free path of the molecules of the gas and therefore is inversely proportional to density. The deficient knowledge of the accommodation coefficient that is necessary to calculate ΔT_a does not recommend measurements at very low densities. We therefore obtain

$$\Delta T = \Delta T_{exp} - \Delta T_s - \Delta T_a \tag{23}$$

where ΔT_a must be <0.1% of ΔT_{exp}.

3. Finally, corrections to the cell dimension have to be done, as the cell constant d/A depends directly on them. These corrections involve the edge effects and the variations of d and A with temperature and pressure. In the case of the edge effects the geometrically measured area of the heat transfer surface, A_g, can be corrected by making use of the mathematical analogy between the temperature field caused by conduction and the electric field in a Thomson capacitor,[41] giving $A = cA_g$, where c is a constant that needs to be evaluated for each apparatus.[42] The cell constant d/A has to be corrected for thermal expansion and compression, i.e.,

$$\frac{d}{A}(T,P) = [1 + \alpha(T - T_0)][1 + \beta(P - P_0)]\frac{d}{a}(T_0,P_0) \qquad (24)$$

Using Equations 21 to 24 one may obtain thermal conductivities with an accuracy of 0.5%, decreasing to 2 or 5% as the critical point is approached (at 0.2 K from it).

A final comment must be made regarding the reference state of the fluid. Usually $T_{ref} = (T_1 + T_2)/2$; however, the evaluation of the reference density ρ_{ref} is more complicated, especially when the critical point is approached, where density gradients develop under the influence of the gravity field due to the divergence of the isothermal compressibility (see Reference 41 for details).

A preliminary version of the apparatus described by Mostert et al.[41] was used by Michels et al.[43] to measure the thermal conductivity of CO_2 near the critical point, demonstrating for the first time the phenomenon of critical enhancement. Figure 8 shows the anomalous thermal conductivity of CO_2, $\Delta\lambda_{crit}$, obtained by these authors. Figure 9 shows the measurements of Mostert et al.[44] on ethane, plotted in terms of the thermal diffusivity a, as a function of the density of the fluid, with the heat capacity evaluated from the EOS. An apparatus of this type has been used by Roder and Diller[45] to measure dense gaseous and liquid hydrogen, between 17 to 200 K and pressure up to 15 MPa. Part of the supercritical region was covered.

B. THE MEASUREMENT OF THERMAL DIFFUSIVITY

The thermal diffusivity can be measured by several techniques, utilizing different ways of perturbing the fluid. The discussion is restricted to those applicable to optically transparent fluids, using either light scattering or interferometric techniques, and to the simultaneous measurement of thermal conductivity and thermal diffusivity with the transient hot wire technique.

The thermal diffusivity can be an excellent complement to thermal conductivity data near the critical point, as some of the techniques can be used in the critical region proper. These techniques use very small perturbations to the fluid, where thermal conductivity measurements are unobtainable with adequate accuracy.

1. Dynamic Light Scattering

The dynamic light scattering is one of the best methods to investigate the thermal diffusivity of optically transparent fluids. Previously used only near the critical point, where the correlation length of the fluctuations is big and produces a scattered light of reasonable intensity, it was recently extended to a wide range of states by Jany and Straub[5] in their measurements of sulfurhexafluoride, ethane, and nitrous oxide.

The theory of light scattering is fully described in References 46 and 47, but this subsection is limited to a description of the basic principles. In an equilibrium state, from a thermodynamic point of view, the molecules with their permanent motion create local heterogeneities that scatter incident light, because of the spatial variations of the dielectric constant of the fluid. This scattered light has an angular and a frequency distribution. Because

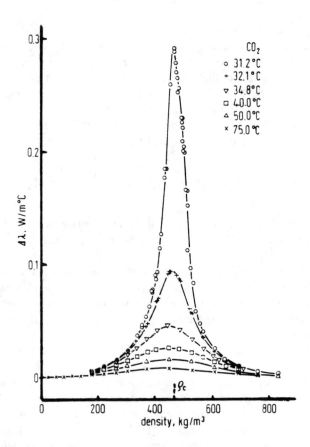

FIGURE 8.   The critical enhancement of carbon dioxide. (From Michels, A., Sengers, J. V., and van der Gulik, P. S., *Physica*, 28, 1216, 1962. With permission.)

the spontaneous fluctuations in the density vary with time, the scattered signal is modulated and therefore has a frequency distribution. In the case of simple fluids, the spectrum consists of three Lorentzian shaped lines, the central one being the Rayleigh line. The half width of the Rayleigh line is given by[5,46-49]

$$\Gamma_R = ak^2 = \frac{\lambda}{\rho C_p} k^2 \tag{25}$$

where k is the scattering vector, the difference between the wave vectors of the incident, and scattered light. Its magnitude is given by

$$|k| = \frac{4\pi n}{\Lambda_0} \sin\left(\frac{\theta}{2}\right) \tag{26}$$

where n is the refractive index of the fluid, Λ_0 the wavelength of the incident light in vacuum, and θ the scattering angle. Details of instruments of this kind can be found in References 5 and 48. The accuracy of the experiments data varies between 5 to 10% as reported by Weber,[48] for his measurements in oxygen and by Ackerson and Straty[49] for methane. Equation 25 has to be modified to be used very near the critical point, as the correlation length ζ becomes very large. Using mode-coupling theory, Weber[48] obtained

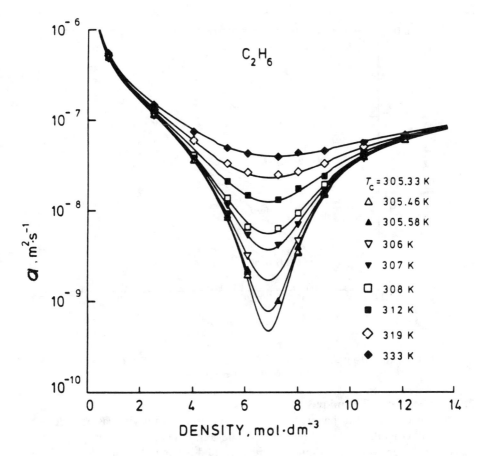

FIGURE 9. The thermal diffusivity of ethane. (From Mostert, R., van der Berg, H. R., and van der Gulik, *Int. J. Thermophys.*, 10, 407, 1989. With permission.)

$$\Gamma_R = \frac{\Delta\lambda_{exc}k^2}{\rho C_p}(1 + k^2 \xi) + \frac{\Delta\lambda_{crit}}{\rho C_p}\frac{K_0(k\xi)}{k^2\xi^2} \tag{27}$$

$$K_0(x) = \frac{3}{4}[1 + x^2 + (x^3 - x^{-1})\arctan(x)] \tag{28}$$

Figure 4 shows the results obtained by Jany and Straub[5] for the thermal diffusivity of SF_6 as a function of reduced density for several values of $\tau = T/T_c - 1$. Figure 10 shows the thermal diffusivity results on the critical isochore and one near-critical isochore of oxygen obtained by Weber[48] as a function of $T - T_c$, obtained with both forward and backscattering.

One of the modifications of the Rayleigh scattering experiment is the forced Rayleigh scattering experiment. Replacing the weak and random fluctuations in a fluid in equilibrium by strong and coherent excitations from a laser-induced grating, the forced scattering of a probing beam becomes much stronger and coherent. Nagasaka et al.[50] summarize the advantages of this technique for thermal diffusivity measurements. It is a contact-free measurement, very fast (typically within 1 ms), avoids convection interference, and it uses very low temperature rises during a measurement (less than 0.1 K), which decreases the effect of radiative heat transfer. Departures from one-dimensional heat conduction and inaccuracies on setting the optical conditions introduce systematic errors that make the accuracy of the results 3%. Unfortunately, this technique has not yet been applied to measurements near the critical point.

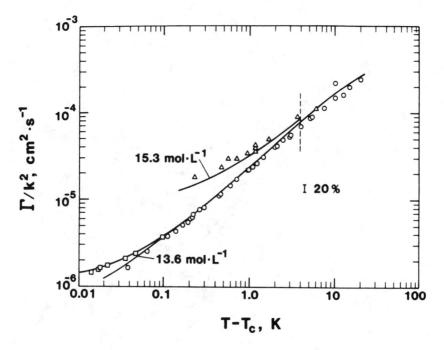

FIGURE 10. The thermal diffusivity of oxygen along two isochores. (From Weber, L. A., *Int. J. Thermophys.*, 3, 117, 1982. With permission.)

2. Interferometric Techniques

A special transient technique to measure the thermal diffusivity has been developed by Becker and Grigull.[51,52] An infinitesimally thin, uniform source of heat q is located at the junction between two semi-infinite materials, a solid heating plate, and the fluid. Initiation of the heat source at time t = O causes the temperature of the fluid, at a distance Z from the heat source, to rise in transient mode, according to

$$\Delta T(Z,t) = \frac{2q}{\lambda} \sqrt{at} \; \text{ierfc} \left(\frac{Z}{2\sqrt{at}} \right) \tag{29}$$

This temperature field causes a change in the refractive index of the fluid, which is proportional to a change in density. Because $\partial \rho / \partial T$ tends to infinity as one approaches the critical point, the temperature change necessary to produce a given change in Δn becomes smaller and smaller as the critical point is approached, and therefore the possibility of convection occurrence decreases. Using optical interferometry and two laser beams, a fringe pattern can be produced and from it, the thermal diffusivity can be obtained, if $|\partial n / \partial T|$ is known for the fluid. Holographic interferometry was used by Becker and Grigull to diminish the errors of additional sources of perturbation to the fluid. The fringe order k, for a light of wavelength Λ, is given by

$$k = \frac{2q}{\lambda} \sqrt{at} \; B \; \text{ierfc} \left(\frac{Z}{2\sqrt{at}} \right) \tag{30}$$

$$B = \frac{q\ell}{\lambda \Lambda} \left| \frac{\partial n}{\partial T} \right| \tag{31}$$

and λ is the distance through the fluid over which the temperature perturbation takes place. Nonlinear regression of Equation 30 permits one to obtain the thermal diffusivity a.

There are several corrections employed to correct the observed fringe pattern that can be found in Reference 51. Although this method has several advantages, especially very close to the critical point, they can only be realized if a high degree of temperature stability and uniformity (10^{-6} K) can be maintained in the cell. This was achieved in CO_2,[51] where the critical temperature is near ambient, but it would be obviously more difficult to achieve at lower or higher temperatures.

3. The Transient Hot-Wire Technique

The transient hot-wire technique has been recognized since its first application as being capable of producing values for the thermal diffusivity simultaneously with the thermal conductivity. Equation 10 can be solved to obtain the thermal diffusivity

$$a = \frac{r_0^2 C}{4t'} \exp\left[\frac{4\pi}{q} \lambda \ (T_r,\rho_r) \ \Delta T_{id}(t')\right] \tag{32}$$

where t' is the specific time at which ΔT_{id} is evaluated. From the regression line of ΔT as a function of ℓnt, it can be demonstrated that when $t' = 1s$, a_0 can be given by

$$a_0 = \frac{r_0^2 C}{4.1s} \exp\left(\frac{T}{S}\right) \tag{33}$$

where I is the intercept of the regression line, $S = q/4\pi\lambda(T_R,\rho_r)$ is its slope, and a_0 refers to the thermal diffusivity at zero time conditions, i.e.,

$$a_0 = a(T_0,P_0) = \frac{\lambda(T_0,P_0)}{\rho_0(C_p)_0} \tag{34}$$

$$\rho_0 = \rho(T_0,P_0) \tag{35}$$

The corrections, δT_i, to the experimental temperature rise, analyzed by several authors[15-21] to obtain thermal conductivity, were reanalyzed in order to obtain thermal diffusivity[53,54] and operational changes in one instrument were made to improve the accuracy of ΔT_w.[4,55,56] It has been shown that it was possible to obtain the thermal diffusivity with an accuracy of 4%. Figure 3 shows the results obtained for argon,[5] while Figure 11 shows the heat capacity of nitrogen obtained by Perkins et al.[33] simultaneously with the data presented in Figure 6. The lines were obtained from the EOS with an accuracy of 5%. The measurements are capable of covering wide ranges of temperatures and pressure and have accuracies comparable to other methods in most of the ranges. As already stated, the transient hot wire cannot be used very close to the critical point, and never in the critical region proper, as Equation 7 is no longer valid there.

III. SUPERCRITICAL THERMAL CONDUCTIVITY SURFACES

This section begins with an overview of the data on thermal conductivity and thermal diffusivity available in the SCF region. Table 1 shows a summary of publications in this field. Sixteen pure fluids and two binary mixtures were studied. There are more data available for thermal conductivity than for thermal diffusivity. Ethane and carbon dioxide, two of the most commonly used fluids in supercritical extractions, have been the subject of very complete studies. Figures 1 to 11 show a reasonable overview of the behavior of thermal conductivity and thermal diffusivity surfaces, as functions of temperature and density.

The analysis of the thermal conductivity surface is very complicated and it has been the

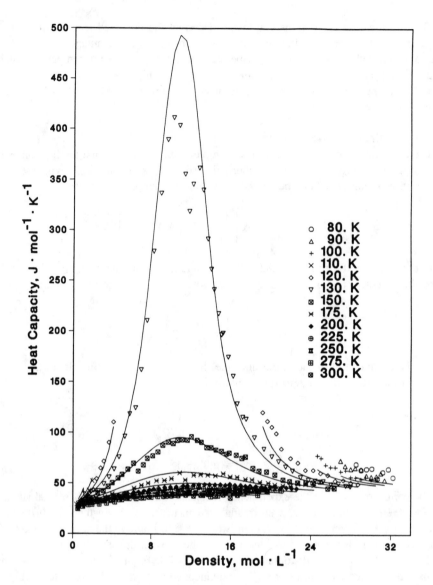

FIGURE 11.   The heat capacity of nitrogen, derived from thermal conductivity and thermal diffusivity simultaneous measurements. (From Perkins, R. A., Roder, H. M., and Nieto de Castro, C. A., *Physica A*, 1991, in press.)

object of several studies. As already explained in the Introduction, it is useful to divide the thermal conductivity into three contributions, usually considered independent. A zero density or dilute gas contribution, λ_0 is a function of temperature only and inaccessible experimentally. This function is obtained, at each temperature, by graphical or numerical extrapolation, or theoretically for monatomic gases[88] or very simple polyatomics such as nitrogen[89] and carbon dioxide.[90] These theoretical calculations or extrapolations to zero density of low density data degrade the accuracy of the best experimental methods. The readers are referred to Chapter 6. Several empirical fits have been tried, the best one having the form

$$\lambda_0 = \sum_{i=1}^{9} a_i T^{(i-4)/3} \tag{36}$$

as used by IUPAC (International Union of Pure and Applied Chemists) correlation for argon.[92]

TABLE 1
Available SCF Data for Thermal Conductivity and Thermal Diffusivity

Fluid	Thermal conductivity (ref.)	Thermal diffusivity (ref.)
He	57	—
Ar	58—62	4, 61
Xe	58, 63	—
$n\text{-}H_2$	45, 64—66	—
N_2	33	33
O_2	3, 48, 67	48
CH_4	60, 68, 69	49
C_2H_6	32, 44, 70—73	5, 44
C_3H_8	74, 75	—
$n\text{-}C_4H_{10}$	1	—
$i\text{-}C_4H_8$	76	—
CO_2	38, 43, 51—52	51, 52, 77—80
H_2O	43	—
N_2O	—	5
NH_3	81	—
SF_6	82—84	5
$He^3 + He^4$	85	—
$CH_4 + C_2H_6$	34, 86, 87	—

The second contribution, designated by excess thermal conductivity, $\Delta\lambda_{exc}$, translates the increase of thermal conductivity over its zero density limit, due to the increase in density, free from critical enhancement contributions. There has been much discussion about the dependence of $\Delta\lambda_{exc}$ on temperature, and the functions used to correlate this part of the thermal conductivity reflect the authors' belief about this subject and the data analysis they perform. Most of the previous attempts have used temperature-dependent excess functions. An example is the form used for methane[68] and argon,[61]

$$\Delta\lambda_{exc} = \alpha(T)\rho + \beta(T)\rho^{n(T)} \tag{37}$$

where the parameters α, β, and η are functions of temperature

$$\alpha = \alpha_1 + \alpha_2 T \tag{38}$$

$$\ell n\beta = \beta_1 + \beta_2 T + \beta_3 T^2 + \beta_4 T^3 \tag{39}$$

$$n = n_1 + n_2 T + n_3 T^2 \tag{40}$$

For nontemperature-dependent excess functions, an example is the form used for butane,[1] nitrogen,[33] and argon[62]

$$\Delta\lambda_{exc} = \sum_{i=1}^{4} B_i \rho^i \tag{41}$$

The combination of the dilute gas thermal conductivity and the excess thermal conductivity is often designated the background thermal conductivity in the context of critical enhancement studies.

The third contribution to the thermal conductivity is $\Delta\lambda_{crit}$, or the critical enhancement contribution. The experimental critical enhancement can be obtained by subtracting the background thermal conductivity from the experimental data; however, this is a difficult task because the extension of the critical enhancement in the temperature and density space has been the object of some controversy. Until 1981, it was believed that no critical enhancement was found for $T/T_c > 1.5$, even at the critical density. Nieto de Castro and Roder[93] have shown, however, that argon thermal conductivity data at 300.65 K, about twice the critical temperature, displayed a critical enhancement three times larger than the estimated accuracy of the data. This controversy affected what should be used as background thermal conductivity, and therefore, the critical enhancement amplitude. These results were recently confirmed with new data on argon[61] and nitrogen.[33]

The empirical representation of the critical enhancement can be demonstrated by a function of the type

$$\Delta T_{crit}(\rho, T) = A\, e^{-x^2} \tag{42}$$

$$A = \frac{C_1}{T + C_2} + C_3 + C_4 T \tag{43}$$

$$\rho_{center} = \rho_c + C_5 (T - T_c)^a \tag{44}$$

$$x = C_6(\rho - \rho_{center}) \text{ for } \rho \leq \rho_{center} \tag{45}$$

$$x = C_7(\rho - \rho_{center}) \text{ for } \rho \geq \rho_{center} \tag{46}$$

Equation 44 translates the fact that the critical enhancement is only centered at ρ_c when $T = T_c$. Equations 45 and 46 show that the critical enhancement is slightly asymmetric about ρ_{center}. Figure 12 shows the critical enhancement function for argon. The lines are given by Equations 42 to 46 with coefficients described in Reference 61. The asymmetry found for argon was also found for methane[68] and oxygen,[3] but the nitrogen data[33] did not show a significant asymmetry. When this is found, the variable x in Equations 45 and 46 is replaced by

$$x = C_5(\rho - \rho_c) \tag{47}$$

and $\rho_{center} = \rho_c$. Also, in Equation 43, T is replaced by $T' = 2T_c - T$ for $T < T_c$.

There have been several attempts to predict theoretically the values of $\Delta\lambda_{exc}$ and $\Delta\lambda_{crit}$, however, this chapter restricts its analysis to the best current analysis of those values.

In considering a moderately dense gas, its thermal conductivity can be expanded in terms of density to the first order, i.e.,

$$\lambda(\rho, T) = \lambda_0(T) + \lambda_1(T)\rho + \ldots \tag{48}$$

The most recent and comprehensive treatment of calculating $\lambda_1(T)$ from first principles using a Lennard-Jones (LJ) 12-6 potential has been given by Friend and Rainwater[94,95] and posteriously extended to polyatomic fluids by Nieto de Castro et al.[96] This model includes two-monomer, three-monomer, and monomer-dimer contributions. Details of the application of this model to argon and nitrogen can be found in References 33 and 62. We define $B_\lambda = \lambda_1/\lambda_0$. The reduced value B_λ^*, given by

$$B_\lambda^* = \frac{3B_\lambda\lambda_0}{2\pi N_A \sigma^3 \lambda_{0_{tr}}} + 0.625\left[\left(\frac{\lambda_0}{\lambda_{0_{tr}}} - 1\right)\right] \tag{49}$$

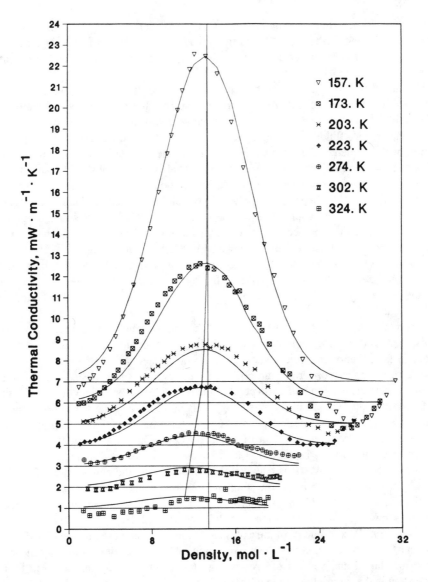

FIGURE 12. The critical enhancement for argon. The lines are generated by the empirical surface fitting described in Equations 37 to 40. Isotherms are separated by 0.001 W/(m·K). The curve to the left of ρ_c is the locus of values of ρ_{center}. (From Roder, H. M., Perkins, R. A., and Nieto de Castro, C. A., *Int. J. Thermophys.*, 10, 1141, 1989. With permission.)

$$\lambda_{0_1 tr} = \frac{15R\eta_0}{4M} \qquad (50)$$

equal to the transitional contribution to the dilute gas thermal conductivity. In these equations, η_0 is the dilute gas viscosity, M the molecular mass of the gas, R the universal gas constant, and σ the LJ 12-6 potential distance parameter. It has been proven that $B_\lambda{}^*$ is a universal function of $T^* = k_B T/\epsilon$, where ϵ is the LJ 12-6 potential energy parameter,[96] given by

$$B_\lambda{}^* = \frac{a_1 + a_2 T^*}{a_3 + T^*} \qquad (51)$$

with $a_1 = 2.9749$, $a_2 = 0.1140$, and $a_3 = -0.04953$. The application of this theory allowed the prediction of the thermal conductivity within 2% over the range 0 to 2 mol dm^{-3} for nitrogen[33] and up to 4.5 mol dm^{-3} for argon. The results for these fluids showed that λ_1 is almost independent of temperature. No theory is available for higher-order coefficients (B_2, B_3, etc., in Equation 41).

The divergence of the thermal conductivity at the critical point arises from the long-wavelength fluctuations of the order parameter and of other relevant hydrodynamic modes of the fluid.[97,98] The theory describing this divergence in the critical region has been well developed for pure fluids,[99] and more recently, Olchowy and Sengers[100] have proposed a solution of the mode-coupling equations that allows the calculation of $\Delta\lambda_{crit}$ throughout the fluid state. A simplified version of this theory has also been published,[101] and it has been shown that the theory describes transport data of carbon dioxide, methane, ethane, helium-3, and water.[99-102] This theory was applied by the author and co-workers to nitrogen[33] and argon[62] and details of its application can be found there. There were some simplifications done by evaluating the mode-coupling integral in closed algebraic form. The matrix inversion algorithm, which must use the results reported by Olchowy and Sengers,[100-102] is no longer necessary. Figure 13 shows the critical enhancement of nitrogen,[33] with the lines translating Olchowy and Sengers[101] theory simplified by Perkins et al.[33] The agreement between the data and the theory is good, and also proves that the background thermal conductivity was well chosen.[33]

At this point, the author can recommend thermal conductivity surface fits, over wide ranges of density and temperature, from dilute gas to liquid, including the supercritical region. In fact, theory can give a very good insight to the overall fit.

In a recent paper concerning argon,[62] the authors have used the extended law of corresponding state developed by Kestin et al.,[88] the first density coefficient theory of Rainwater and Friend,[94,95] and the Olchowy and Sengers theory[100-102] to generate a global surface fit for argon, using a temperature-independent excess function (Equation 41). Figure 14a shows the deviation plot between the data of Nieto de Castro and a surface,[61] where the $\lambda_0(T)$ function was replaced by the Kestin et al.[88] function, and $\Delta\lambda_{exc}$ and $\Delta\lambda_{crit}$ were empirical. Figure 14 b shows the deviations between the same data and the new surface. The improvement is gratifying. At a 2σ level, the standard deviation of the first fit is $\pm 3.4\%$, with large deviations in the vapor and liquid isotherms close to the critical temperatures. The theoretical based fit shows a standard deviation at a 2σ level of $\pm 2.2\%$. It has only five adjustable parameters, four B_i coefficients of the excess function, and the wave number cutoff parameter q_D of the Olchowy and Sengers theory. In addition, the deviations along the near-critical temperature liquid and vapor isotherms are no longer present. It can therefore be recommended that in the future, the fitting of thermal conductivity surfaces shall be done, following the process outlined in Reference 61.

As a final comment, it can be said that we have come a long way since the first surface fits to thermal conductivity were done. New and more experimental data have appeared, and new and better theories were shown to be reliable. The use of these surfaces in transport properties calculations for SCF technology can now be enforced, providing an inestimable tool for optimum technological design and operation.

ACKNOWLEDGMENTS

The author wishes to thank those colleagues who introduced him to the measurement of thermal conductivity in supercritical fluids, namely B. le Neindre (Paris) and H. M. Roder (Boulder).* Special thanks are given to the Thermophysics Division of the National Institute of Standards and Technology (ex-NBS) for financial support as permanent guest scientist.

---

* This chapter is dedicated to Hans Roder who has recently retired.

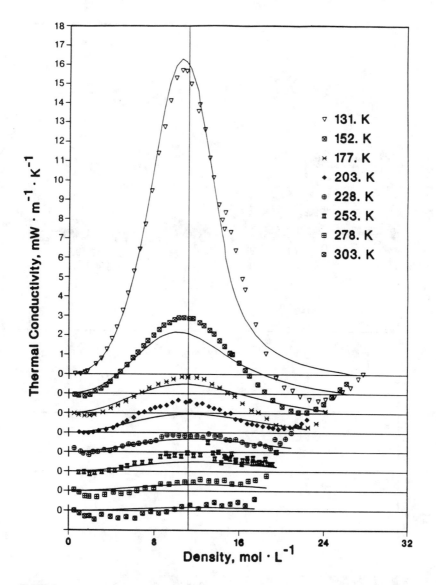

FIGURE 13.   The critical enhancement for nitrogen.[33] The lines were generated by the Olchowy and Sengers theory.[100-102] (From Perkins, R. A., Roder, H. M., and Nieto de Castro, C. A., *Physica A*, 1991, in press.)

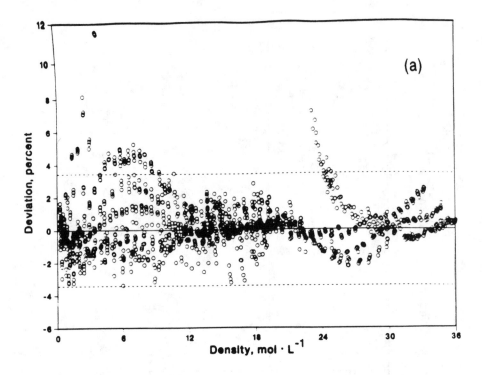

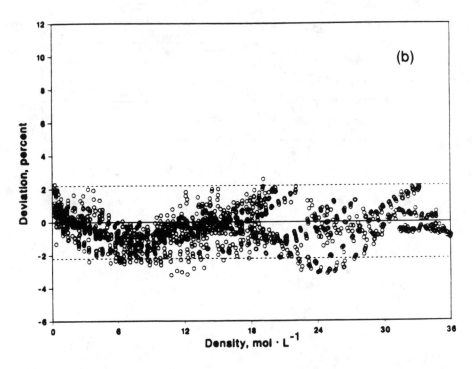

FIGURE 14.   Deviations between experimental data for argon[62] and a correlation where λ_o was taken from Kestin et al.[88] (a) $\Delta\lambda_{exc}$ and $\Delta\lambda_{crit}$ are empirical; (b) $\Delta\lambda_{exc}$ is arranged to be temperature independent following the theory of Rainwater and Friend[94,95] and $\Delta\lambda_{crit}$ is from the Olchowy and Sengers theory.[100-102]

REFERENCES

1. **Nieto de Castro, C. A., Tufeu, R., and le Neindre, B.,** *Int. J. Thermophys.,* 4, 11, 1983.
2. **Sengers, J. V. and Levelt Sengers, J. M. H.,** Concepts and methods for describing critical phenomena in fluids, in *Progress in Liquid Physics,* Croxton, C. A., Ed., John Wiley & Sons, New York 1978, 103.
3. **Roder, H. M.,** *J. Res. Natl. Bur. Stand.,* 87, 279, 1982.
4. **Roder, H. M. and Nieto de Castro, C. A.,** in *Thermal Conductivity 20,* Hasselman, D. P. H. and Thomas, J. R., Jr., Eds., Plenum Press, New York 1989, 173.
5. **Jany, P. and Straub, J.,** *Int. J. Thermophys.,* 8, 165, 1987.
6. **Sengers, J. V.,** this volume, chap. 1.
7. **Nieto de Castro, C. A., Li, S. F. Y., Nagashima, A., Tengrove, R. D., and Wakeham, W. A.,** *J. Phys. Chem. Ref. Data,* 15, 1073, 1986.
8. **Assael, M. J., Ramires, M. L. V., Nieto de Castro, C. A., and Wakeham, W. A.,** *J. Phys. Chem. Ref. Data,* 19, 113, 1989.
9. **Kestin, J. and Wakeham, W. A.,** *Transport Properties of Fluids, Thermal Conductivity, Viscosity and Diffusion Coefficients, Cindas Data Series on Material Properties,* Vol. I/1, Hemisphere Publishing, New York, 1988.
10. **Nieto de Castro, C. A.,** *JSME Int. J. II,* 31, 387, 1988.
11. **Wakeham, W. A., Nagashima, A., Sengers, J. V., Eds.,** *Measurement of the Transport Properties of Fluids,* IUPAC/Blackwell Scientific, Oxford, 1991.
12. **Stalhane, B. and Pyk, S.,** *Tek. Tidskr.,* 61, 389, 1931.
13. **Haarman, J.W.,** *Physica,* 52, 603, 1971.
14. **de Groot, J. J., Kestin, J., and Sookiazian, H.,** *Physica,* 75, 454, 1974.
15. **Healy, J. J., de Groot, J. J., and Kestin, J.,** *Physica,* 82C, 392, 1976.
16. **Anderson, G. P., de Groot, J. J., Kestin, J., and Wakeham, W. A.,** *J. Phys. E., Sci. Instrum.,* 7, 948, 1974.
17. **Nieto de Castro, C. A., Calado, J. C. G., Wakeham, W. A., and Dix, M.,** *J. Phys. E., Sci. Instrum.,* 9, 1073, 1976.
18. **Assael, M. J., Dix, M., Lucas, A., and Wakeham, W. A.,** *J. Chem. Soc. Faraday Trans. I,* 77, 439, 1980.
19. **Clifford, A. A., Kestin, J., and Wakeham, W. A.,** *Physica,* 100A, 370, 1980.
20. **Nieto de Castro, C. A., Li, S. F. Y., Maitland, G. C., Wakeham, W. A.,** *Int. J. Thermophys.,* 4, 311, 1983.
21. **Calado, J. C. G., Fareleira, J. M. N. A., Nieto de Castro, C. A., and Wakeham, W. A.,** *Rev. Port. Quim.,* 26, 173, 1984.
22. **McLaughlin, E. and Pittman, J. F. T.,** *Philos. Trans. Soc. London, Ser. A,* 270, 557, 1971.
23. **Mani, N. and Venart, J. E. S.,** in *Proc. 6th Symp. on Thermophysical Properties,* American Society of Mechanical Engineers, New York 1973, 1.
24. **Assael, M. J., Nieto de Castro, C. A., Roder, H. M., and Wakeham, W. A.,** Transient methods for thermal conductivity, in *Measurement of the Transport Properties of Fluids,* IUPAC/Blackwell Scientific, Oxford, 1991.
25. **Nagasaka, Y. and Nagashima, A.,** *Rev. Sci. Instrum.,* 52, 222, 1981.
26. **Nieto de Castro, C. A.,** Ph.D. thesis, IST, Technical University, Lisbon, 1977.
27. **Charitidou, E., Dix, M., Assael, M. J., Nieto de Castro, C. A., and Wakeham, W. A.,** *Int. J. Thermophys.,* 8, 511, 1987.
28. **Dix, M., Wakeham, W. A., and Zalaf, M.,** in *Thermal Conductivity 20,* Hasselman, D. P. H. and Thomas, J. R., Jr., Eds., Plenum Press, New York, 1989, 195.
29. **Roder, H. M.,** *J. Res. Natl. Bur. Stand.,* 86, 457, 1981.
30. **Buyukbicer, E. F., Venart, J. E. S., and Prasad, R. C.,** *High Temp. High Pressure,* 18, 55, 1986.
31. **Nieto de Castro, C. A.,** in *Thermal Conductivity 21,* Cremers, C. and Fine, A., Eds., Plenum Press, New York, 1990.
32. **Roder, H. M. and Nieto de Castro, C. A.,** *High Temp. High Pressure,* 17, 453, 1985.
33. **Perkins, R. A., Roder, H. M., Nieto de Castro, C. A.,** *Physica A,* 1991, in press.
34. **Roder, H. M. and Friend, D. G.,** *Int. J. Thermophys.,* 6, 607, 1985.
35. **Ozisik, M. N.,** *Radiative Transfer,* Wiley-Interscience, New York, 1973.
36. **Le Neindre, B.,** Ph. D. thesis, Universite de Paris VI, Paris, 1969.
37. **Snell, J. A. A., Trappeniers, N. J., and Botzen, A.,** *Proc. Kon. Ned. Acad. Wet.,* 82B, 303, 1979.
38. **Le Neindre, B., Garrabos, Y., and Tufeu, R.,** *Ber. Bunsenges. Phys. Chem.,* 88, 916, 1984.
39. **Michels, A. and Sengers, J. V.,** *Physica,* 28, 1238, 1962.
40. **Guildner, L. A.,** *J. Res. Natl. Bur. Stand.,* 66A, 333, 1962.
41. **Mostert, R., van den Berg, H. R., and van der Gulik, P. S.,** *Rev. Sci. Instrum.,* 60, 3466, 1989.

42. **Maxwell, J. C.**, *A Treatise on Electricity and Magnetism*, Dover, New York, 1954.
43. **Michels, A., Sengers, J. V., and van der Gulik, P. S.**, *Physica*, 28, 1216, 1962.
44. **Mostert, R., van den Berg, H. R., and van der Gulik, P. S.**, *Int. J. Thermophys.*, 10, 409, 1989.
45. **Roder, H. M. and Diller, D. E.**, *J. Chem. Phys.*, 52, 5928, 1970.
46. **Berne, B. J. and Pecora, R.**, *Dynamic Light Scattering*, John Wiley & Sons, New York, 1976.
47. **Chu, B.**, *Laser Light Scattering*, Academic Press, New York, 1974.
48. **Weber, L. A.**, *Int. J. Thermophys.*, 3, 117, 1982.
49. **Ackerson, B. J. and Straty, G. C.**, *J. Chem. Phys.*, 69, 1207, 1978.
50. **Nagasaka, Y., Hatakeyama, T., Okuda, M., and Nagashima, A.**, *Rev. Sci. Instrum.*, 59, 1156, 1988.
51. **Becker, H. and Grigull, U.**, *Waerme Stoffuebertrag.*, 10, 233, 1977; 11, 9, 1978.
52. **Becker, H. and Grigull, U.**, in *Proc. 7th Symp. Thermophysical Properties*, American Society of Mechanical Engineers, New York, 1977, 814.
53. **Nieto de Castro, C. A., Taxis, B., Roder, H. M., and Wakeham, W. A.**, *Int. J. Thermophys.*, 9, 293, 1988.
54. **Fareleira, J. M. N. A. and Nieto de Castro, C. A.**, *High Temp. High Pressure*, 21, 363, 1989.
55. **Roder, H. M. and Nieto de Castro, C. A.**, *Cryogenics*, 27, 312, 1987.
56. **Perkins, R. A., Roder, H. M., and Nieto de Castro, C. A.**, *J. Res. Nat. Inst. Standards Technol.*, 1991, in press.
57. **Kerrisk, J. F. and Keller, W. E.**, *Phys. Rev.*, 177, 341, 1969.
58. **Trappeniers, N. J.**, in *Proc. 8th Symp. Thermophysical Properties*, Sengers, J. V., Ed., American Society of Mechanical Engineers, New York, 1982, 232.
59. **Bailey, B. J. and Kellner, K.**, *Physica*, 31, 444, 1968.
60. **Kenberry, L. D. I. and Rice, S. A.**, *J. Chem. Phys.*, 39, 156, 1963.
61. **Roder, H. M., Perkins, R. A., and Nieto de Castro, C. A.**, *Int. J. Thermophys.*, 10, 1141, 1989.
62. **Perkins, R. A., Friend, D. G., Roder, H. M., and Nieto de Castro, C. A.**, *Int. J. Thermophys.*, 1991, in press.
63. **Swinney, H. L., Henry, D. L., and Cummins, H. Z.**, *J. Phys. C*, 1, 81, 1972.
64. **Roder, H. M.** in *Thermal Conductivity 17*, Hust, J. G., Ed., Plenum Press, New York, 1983, 257.
65. **Sengers, J. V.**, *Int. J. Heat Mass Transf.*, 8, 1103, 1965.
66. **Roder, H. M.**, *Int. J. Thermophys.*, 5, 322, 1984.
67. **Ziebland, H. and Burton, J. T. A.**, *Br. J. Appl. Phys.*, 6, 416, 1955.
68. **Roder, H. M.**, *Int. J. Thermophys.*, 6, 119, 1985.
69. **Prasad, R. C., Mani, N., Venart, J. E. S.**, *Int. J. Thermophys.*, 5, 265, 1984.
70. **Prasad, R. C. and Venart, J. E. S.**, *Int. J. Thermophys.*, 5, 367, 1984.
71. **Le Neindre, B., Tufeu, R., Bury, P., Johannin, P., and Vodar, B.**, in *Proc. 8th Int. Thermal Cond. Conf.*, Ho, C. Y. and Taylor, R. D., Eds., Plenum Press, New York, 1969, 229.
72. **Tufeu, R., Garrabos, Y., and le Neindre, B.** in *Thermal Conductivity 16*, Larsen, D. C., Ed., Plenum Press, New York 1983, 605.
73. **Desmarest, P. and Tufeu, R.**, *Int. J. Thermophys.*, 8, 293, 1987.
74. **Tufeu, R. and le Neindre, B.**, *Int. J. Thermophys.*, 8, 27, 1987.
75. **Prasad, R. C., Wang, G., and Venart, J. E. S.**, *Int. J. Thermophys.*, 10, 1013, 1989.
76. **Nieuwoudt, J. C., le Neindre, B., Tufeu, R., and Sengers, J. V.**, *J. Chem. Eng. Data*, 32, 1, 1987.
77. **Swinney, H. L. and Henry, D. L.**, *Phys. Rev. A*, 8, 2586, 1973.
78. **Maccabee, B. S. and White, J. A.**, *Phys. Lett.*, 35A, 187, 1971.
79. **Garrabos, Y., Tufeu, R., le Neindre, B., Zalczer, G., and Beysens, D.**, *J. Chem. Phys.*, 4637, 1980.
80. **Reile, E., Jany, P., and Straub, J.**, *Waerme Stoffuebertrag.*, 18, 99, 1984.
81. **Needham, D. P. and Ziebland, H.**, *Int. J. Heat Mass Transf.*, 8, 1307, 1965.
82. **Lis, J. and Kellard, P. O.**, *Br. J. Appl. Phys.*, 16, 1099, 1965.
83. **Kin, T. K., Swinney, H. L., Langley, K. H., and Kachnowski, Th. A.**, *Phys. Rev. Lett.*, 27, 1776, 1971.
84. **Feke, G. T., Hawkins, G. A., Lastovka, J. B., and Benedek, G. B.**, *Phys. Rev. Lett.*, 27, 1780, 1971.
85. **Cohen, L. H., Dingus, M. L., and Meyer, H.**, *Phys. Rev. Lett.*, 50, 1058, 1983.
86. **Friend, D. G. and Roder, H. M.**, *Phys. Rev. A*, 32, 1941, 1985.
87. **Friend, D. G. and Roder, H. M.**, *Int. J. Thermophys.*, 8, 13, 1987.
88. **Kestin, J., Knierim, K., Mason, E. A., Najafi, B., Ro, S. T., and Waldman, M.**, *J. Phys. Chem. Ref. Data*, 13, 229, 1984.
89. **Millat, J. and Wakeham, W. A.**, *J. Phys. Chem. Ref. Data*, 18, 565, 1989.
90. **Vesovic, V., Wakeham, W. A., Olchowy, G. A., Sengers, J. V., Watson, J. T. R., and Millat, J.**, *J. Phys. Chem. Ref. Data*, 19, 763, 1990.
91. **Vesovic, V. and Wakeham, W. A.**, this volume, chap. 6.
92. **Younglove, B. and Hanley, H. J. M.**, *J. Phys. Chem. Ref. Data*, 15, 1323, 1986.
93. **Nieto de Castro, C. A. and Roder, H. M.**, in *Proc. 8th Symp. Thermophysical Properties*, Sengers, J. V., Ed., American Society of Mechanical Engineers, New York 1982, 241.

94. **Friend, D. G. and Rainwater, J. C.,** *Chem. Phys. Lett.,* 107, 590, 1984.
95. **Rainwater, J. C. and Friend, D. G.,** *Phys. Rev. A,* 36, 4062, 1987.
96. **Nieto de Castro, C. A., Friend, D. G., Perkins, R. A., and Rainwater, J. C.,** *Chem. Phys.,* 145, 19, 1990.
97. **Kawasaki, K.,** in *Phase Transitions and Critical Phenomena,* Vol. 5A, Domb, C. and Green, M. S., Eds., Academic Press, New York, 1976, 165.
98. **Hohenbert, P. C. and Halperin, B. I.,** *Rev. Mod. Phys.,* 49, 435, 1977.
99. **Sengers, J. V., Basu, R. S., and Levelt Sengers, J. M. H.,** NASA Contractor Report 3424 National Aeronautics and Space Administration, 1981.
100. **Olchowy, G. A. and Sengers, J. V.,** *Phys. Rev. Lett.,* 61, 15, 1988.
101. **Olchowy, G. A. and Sengers, J. V.,** *Int. J. Thermophys.,* 10, 417, 1989.
102. **Olchowy, G. A.,** Ph.D. thesis, University of Maryland, College Park, 1989.

Chapter 10

MASS TRANSFER IN SUPERCRITICAL EXTRACTION FROM SOLID MATRICES

Michael C. Jones

TABLE OF CONTENTS

ABSTRACT

A review is presented of mass transfer in the extraction of materials from solid matrices using supercritical fluids (SCFs) as solvents. The purpose is to assemble the body of knowledge relevant to the quantitative aspects of mass transfer leading to mass-transfer fluxes and extraction times. Experimental results are available from studies of extraction of natural substances and fossil fuels and from the regeneration of industrial adsorbents. Factors relevant to extraparticle mass transfer are also discussed and, in addition, an outline of mathematical models of intraparticle mass transfer is presented leading to estimates of single particle extraction times.

I. INTRODUCTION

In many proposed extractions using SCFs as solvents, the material to be extracted is fixed in a porous matrix by physical, chemical, or even mechanical bonds. In addition to extractions of natural products such as caffeine from coffee beans or oils from seeds, there is interest in using SCF extraction for many other purposes for which this situation holds. For example, activated carbon used for removal of trace organics from water may be regenerated by extraction with SCFs; porous catalyst particles used in reforming reactions may be reactivated by treatment at supercritical conditions, indeed, it may be advantageous to carry out the reaction under supercritical conditions.[1] Hydrotreating of crude shale oil has been tried successfully, improved separation of n-paraffins from petroleum distillate using zeolites has been demonstrated,[3] and kerogen may be extracted from oil shale under supercritical conditions.[4] Apparently, research in some laboratories on supercritical processing of coal is still active.

In all of these examples, the soluble constituent—the extract—must be released from its bound state, diffuse through the porous structure and, finally, diffuse through the stagnant external fluid layer. What is needed in design and scale-up to industrial scale is a knowledge of the mass transfer rate for the overall process. In general, such rates are controlled by the combination of internal and external diffusional resistances, and possibly by the kinetics that describe the release of the extract from its bound state. They may be enhanced by the favorable equilibrium solubility in the dense SCF.

The objectives of extraction studies fall into two general categories. In the first, one wishes to determine the slate of compounds extractable under various conditions as a function of the extent of extraction. In the second category, the object is to observe the dynamics of the process so as to be ultimately able to predict contact times. This chapter focuses primarily on the second objective and thus considers only work in which mass-transfer parameters were measured or derived from experiments. In a final section, mathematical models which may be used to interpret extraction data and which form the basis of process modeling are discussed.

While conventional language and the acronym "SCF" to describe this field have been adopted, some research is included for which the solvent is subcritical with respect to either its critical temperature or its critical pressure. What is important is that the solvent under discussion is a gas in a compressed state where its density is liquid-like.

II. THE EFFECT OF EXTRACTION CONDITIONS

In the design of an extractor, one should minimize the contact time needed for a given degree of extraction. Given the surface area of particles, this comes down to maximizing the mass transfer flux from particle to bulk fluid. One important question refers to what solvent conditions will best achieve this. A unique paper by Rance and Cussler[5] addresses this question directly, although for a nonporous system. In their experiments, these authors

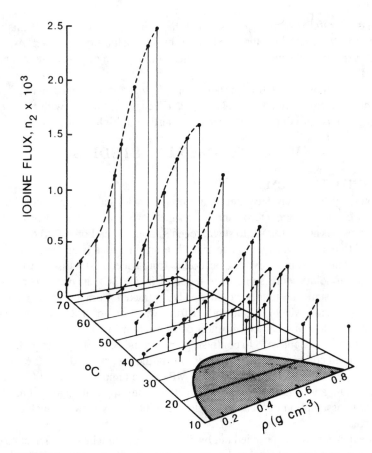

FIGURE 1. Mass-transfer fluxes from a spinning disc of iodine dissolving in supercritical CO_2. (From Rance, R. W. and Cussler, E. L., *ALChE J.*, 20, 353, 1974. Reproduced by permission of the American Institute of Chemical Engineers.)

observed mass transfer rates from a spinning disk of pure iodine dissolving in supercritical CO_2. While these experiments did not per se involve diffusion in porous matrices, the mass transfer conditions were well defined and the results led to an interesting conclusion. The initial solute molar flux n_2 at a simple solid-solvent interface is written by Rance and Cussler as

$$n_2 = k_m \rho_1 \, y_2^s / M_1 \tag{1}$$

where k_m is the mass transfer coefficient, M_1 is the solvent molecular weight, ρ_1 is the solvent density, and y_2^s is the mole fraction solute at saturation. For the spinning disk, k_m is proportional to $D^{2/3}$, where D is the solute diffusivity. The product of diffusivity of a solute in a SCF and density is roughly constant,[6] provided a small region in the vicinity of a critical solution point is avoided. On the other hand, y_2^s is known to increase with density up to a maximum according to the usual enhancement factor formulation[7] for solid solubility. The question is which of these effects dominates as density increases? The experimental results, reproduced in Figure 1, show an increase in mass flux with density at all temperatures investigated. Thus, in this instance, the rate of increase in solubility with density outweighs the reduction in diffusivity. This agrees with a calculation of the enhancement factor according to classical corresponding states theory presented in the same paper.

Similar results may apply to extractions from porous matrices whenever the driving force for mass transfer is phase equilibrium at the solid-solvent interface. If this is the case, one should look for conditions that lead to the highest solubility; however, when the condensed state is that of surface-absorbed molecules, the situation may be different, as discussed in Section III.

Figure 1 also illustrates the important principle that for a given density, one can take advantage of the increase in solute volatility, and hence, solubility, with temperature. The cost of doing this is, of course, that operation must be at higher pressure.

III. EXPERIMENTAL STUDIES

A. ADSORPTIVE SYSTEMS

Adsorptive systems may have much in common with natural products, but appear to stand a better chance of characterization. Activated carbon is used to adsorb organics from aqueous streams. As an alternative to steam regeneration, the carbon may then be regenerated using an SCF to extract the organic material. The original idea is attributed to DeFilippi et al.,[8] who were interested in recovering pesticides from aqueous streams; however, quite low concentrations of solute were sometimes required in the SCF in order to reverse the adsorption. In fact, some pesticides could not be extracted from the carbon at all. As an example, the pesticide Alachlor is adsorbed from water at up to 25 mass% of the carbon adsorbent. At this loading, the equilibrium concentration in CO_2 is only 0.4 mass% even though the solubility is 2 mass%. In contrast, phenol could be readily extracted. Experiments with Alachlor used commercial activated carbons in the size range 0.4 to 2.6 mm. CO_2 was used as a solvent at a pressure typically of 27 MPa and a temperature of 100°C. The authors derived Langmuir adsorption isotherm parameters from application of a one-dimensional bed model with assumption of local adsorption equilibrium. No mass-transfer parameters as such were obtained.

Similar experiments were carried out by Tan and Liou[9] in which ethyl acetate or toluene, previously adsorbed from aqueous solution, were extracted at constant CO_2 pressures (8.83 and 13.1 MPa) and a series of temperatures. Experiments with mixtures of benzene and toluene were also recently reported by these authors.[10] In another paper,[11] they reported on extractions carried out at constant densities for a series of temperatures. Tan and Liou interpreted their data with a one-dimensional bed model in which mass-transfer resistances were neglected and desorption was taken to be irreversible. It is noteworthy that extraction rates at constant temperature increased with density in accordance with the results of Rance and Cussler described above.

Recasens et al.[12] showed how the data of Tan and Liou could be analyzed on the basis of a more complete model. Two parameters were determined from extraction-vs.-time data: the adsorption equilibrium constant, K, and the fluid mass transfer coefficient, k_{eff}. From a van't Hoff plot of ρK vs. 1/T for constant pressures, these authors inferred the existence of a region of retrograde desorption where the heat of desorption would be negative.

The availability of adsorption equilibrium data is essential to the prediction and analysis of extraction rates. Equilibrium adsorption of phenol from water onto activated carbon and regeneration with supercritical CO_2 were investigated by Kander and Paulaitis.[13] Regeneration experiments were conducted at 13.7 MPa, 36°C and 17.1 MPa, 60°C. They found good correlation between adsorption isotherms for aqueous solution and those for SCF CO_2 when the expanded liquid model of Mackay and Paulaitis[14] was used to represent the fugacity of phenol in the supercritical phase. This approach, of course, requires the additional set of aqueous data.

The subject has also been addressed in the related field of supercritical chromatography. A predictive method by Yonker et al.,[15] based on thermodynamic analysis, was shown to predict the retention of naphthalene and biphenyl successfully on cross-linked phenyl polymethylphenyl-siloxane over a range of pressures and constant temperatures. A single choice of the partial molar volume of solute in the stationary phase was sufficient to fit the data, but here, too, an additional set of data was required in the form of solubilities of adsorbate in the supercritical solvent. Insights have also come from the retention volume experiments for simple hyrocarbons reported by King.[16] As far as these experiments are generally applicable, with alumina and porous organic resins as the stationary phase, they seem to suggest that for a given temperature, enhancement of desorption equilibrium levels out at about 10 MPa for CO_2. According to King, the mechanism is that of competitive adsorption of the CO_2, which forms a liquid-like film on the adsorbent surface at pressures below P_c for temperatures slightly above T_c. Consequently, there is no advantage in exceeding this pressure if desorption is to be maximized. Further work relating these results to desorption equilibria for industrial adsorbents would be welcome.

B. NATURAL PRODUCTS

Several studies are concerned with the rates of extraction from natural products. In general, the nature of the bound state of the extract and the structure of the matrix are not well characterized and researchers have relied a good deal on empiricism.

In Gil et al.[17] and King et al.,[18] oils from crushed rapeseeds were extracted by marginally subcritical CO_2 ($P_r = 0.95$, $T_r = 0.98$) at 7.0 MPa in fixed beds of varying height. The total oil content of the seeds was determined to be 39.7 mass%. Particle sizes were in the range 500 to 600 μm. The effluent was monitored for total oil concentration as a function of time and the experiments were repeated at a number of solvent velocities. The effluent concentration in both equilibrium (recirculation) and nonequilibrium (once-through) tests were essentially constant until about 65% of the available oil had been extracted. It then invariably dropped abruptly to a new, constant level — evidence of at least two different states of "bonding" of the oil. The constant rate of extraction exhibited up to the 65% level is reminiscent of drying processes and may indicate that an excess of essentially free oil fills the macropores of the matrix, possibly released by the mechanical process of crushing (see also the discussion under Section IV). The authors' data for both states were consistent with a simple mass-transfer model for the bed in which, for a given flow rate and bed, the equilibrium concentration and overall mass transfer coefficient are constant along the bed. The details of particle internal mass transfer are ignored at this level of modeling. The quantity determined was the product of an overall mass-transfer coefficient, k_{eff}, and surface area per unit volume of bed, a, or, the volumetric mass-transfer coefficient. This quantity increased with velocity, showing that external resistance was at last partially controlling; however, with this type of measurement, no breakdown of mass-transfer resistances can be obtained.

Brunner[19] studied extraction of caffeine from whole coffee beans by humidified SCF nitrous oxide (20.0 MPa; 100°C) and crushed rapeseed by dry CO_2 (20.5 to 75.0 MPa; 41.5 to 81.7°C). The results in both cases were again characterized by an initial constant rate of extraction which was pronounced in the case of rapeseed. The results were interpreted differently for each seed type. For the coffee beans (7 mm diameter equivalent sphere), it was assumed that internal resistance was the controlling factor and that a lumped mass balance on a particle was valid. Bed dynamics were ignored, however. With Sherwood numbers (see Nomenclature for definitions) obtained from correlations, the latter part of the extraction/time curve could be predicted reasonably well. In a second model, similarly ignoring bed dynamics, a single-sphere, unsteady diffusion solution was applied in approximate form to obtain an exponential decay of extractable solid mass fraction. For the rapeseed extractions, analysis was confined to predicting the constant rate portion of the process. Mass transfer was assumed to be controlled by the external fluid mechanical resistance, and appropriate Sherwood numbers were obtained from packed-bed flow correlations. Bed dynamics were again ignored.

Similar experiments were performed by Rao and Mukhopadhyay[20] on crushed cumin seeds using CO_2. Pressures ranged from 8.0 to 14.0 MPa and extractions were performed at 40 and 60°C. In contrast to rapeseed, the oil content of cumin seeds is only 3.4 mass%. Furthermore, equilibrium concentrations of cumin oil in CO_2 were not constant over wide ranges of solid oil content, as seen in the rapeseed. Unsteady-state differential equations for solid and fluid phases in the bed were solved numerically and the solution iterated on the volumetric mass transfer coefficient to match experimental concentration-vs-time data. Intraparticle diffusional resistance was ignored and the effect must be considered as lumped in with the mass-transfer coefficient. When these data are plotted against (particle) Reynolds number, the result is an unexpected decline with Re; however, the variation in Re comes not only from the fluid velocity but also from variations in viscosity as solvent T and P are varied. The authors state that this result may have to do with changes in internal resistance brought about by reductions in viscosity and, presumably, diffusivity. The Sherwood number would be more instructive to plot, but once again the shortcomings of experiments that only yield overall coefficients are evident.

In studying the extraction of the alkaloid, monocrotaline, from the crushed seeds of *Crotalaria spectabilis,* mass-transfer experiments were reported by Schaeffer et al.[21] for the extraction of the soluble lipid material. Extractions were carried out in CO_2 at pressures in the range 18.46 to 27.41 MPa and at temperatures from 308 to 328 K. These seeds have only 4.4 mass% extractable material. Volumetric mass-transfer coefficients were obtained from regression of the effluent concentration-vs.-time data using the same model as Rao and Mukhopadhyay with an equilibrium solubility assumed proportional to the local solid phase concentration. Separate studies were made of solubilities of the lipid material and the desired product, monocrotaline, in both CO_2 and in CO_2 with ethanol as cosolvent.[22]

In studies of extraction of natural products, the level of modeling used to infer physical parameters was primitive. In most cases, the parameter derived from the experiments was a lumped volumetric mass transfer coefficient. Such a quantity is likely to be difficult to correlate. Furthermore, parameters were derived from the application of approximate models and are inescapably model dependent. Even when the model is a good representation of the physical problem, it is still advisable to carry out sensitivity analyses to attach error bands to the inferred parameter values. It is also preferable to supply as many parameters as possible from independent sources of information. This would be particularly important for equilibrium data, where errors have a direct effect on inferred mass-transfer coefficients. What seems to be missing from many of the studies quoted is a rational approach to model selection, as used, for instance, by Recasens et al.[12] for the inference of model parameters.

The main value of these experimental results is that they permit a rough scale-up to the next largest size, for example, pilot plant, provided the designer adheres to conditions approximating the tests. Problems that will be encountered in going much beyond the conditions of the experiments deal with: (1) the complex and not always reproducible chemical and physical structure of the natural material; (2) the difficulty of fully characterizing the mass-transfer resistance, and (3) the difficulty of characterizing equilibria for the extract between substrate and solvent.

C. OTHER SYSTEMS

The decomposition and dissolution of kerogen in oil shale by SCF toluene extraction was investigated by Triday and Smith.[4] One clear advantage of supercritical extraction over conventional retorting was found to be the absence of coke or gas formation. Extractions were carried out in a differential reactor at 3.69 to 5.96 MPa, temperatures 533 to 672 K, and particle diameters of 2.6 and 6.4 mm. For toluene, $T_c = 591.7$ K and $P_c = 4.11$ MPa; thus, runs were both super- and subcritical. The superiority of yield under super- as compared to subcritical conditions was clearly established; it appears to be primarily a temperature effect.

This work has several distinguishing features as an example of SCF extraction:

1. The process takes place at elevated temperature, 652 K. The release of soluble bitumen under this condition is a combined chemical decomposition and activated solvation.
2. The removal of kerogen decomposition products causes an increase in porosity in the course of extraction.
3. In the heating process, the rapid change in density at near-critical conditions results in an intraparticle convective flow.
4. The process is influenced by heat as well as by mass transfer rates.

These features were taken into account in writing equations for the mathematical model of the process. Since experiments were performed on a differential reactor, the equations could be written for a single particle and were solved numerically. Three parameters were found by fitting model predictions to experimental reactor effluent concentrations. These were the pre-exponential factor and activation energy for the kerogen decomposition and a scaling factor for the correlation of diffusivity of bitumen in toluene.

D. THE EXTERNAL MASS-TRANSFER RESISTANCE

A single study reported by Lim et al.[23] focuses on the external, hydrodynamically controlled mass-transfer resistance. Mass-transfer rates were obtained for the initial dissolution rate of two layers of cylindrical naphthalene cylinders (4.5 mm length $\times$ 4.5 mm diameter) embedded in a packed bed of glass beads with CO_2 throughflow. Experiments were carried out at pressures from 1.0 to 20.0 MPa. Lim and colleagues' paper is quite important because it establishes that such mass-transfer coefficients in fixed beds, in addition to their flow-rate dependence, are also affected by buoyancy. This had previously been observed by Debenedetti and Reid[24] in their measurements of diffusivities by a boundary layer diffusion method on a flat plate. The mass-transfer Grashof number (see Nomenclature) must therefore figure in the correlation of external mass-transfer coefficients along with the particle Reynolds number. The authors' correlation for the entire pressure range is

$$Sh/(Sc.Gr)^{1/4} = 1.692 \, [Re/Gr^{1/2}]^{0.356} \qquad (2)$$

with an absolute average relative deviation of 15.3%. The range of the particle Reynolds number was 2 to 70, while the range of the mass-transfer Grashof number was 78 to 3.25 $\times$ 10^7.

When the buoyant forces are unimportant, i.e., $Re/Gr^{1/2} \gg 1$, a large body of experimental evidence favors a correlation such as that given by Wakao and Kaguei:[25]

$$Sh = 2 + 1.11 \, Re^{0.6}Sc^{1/3} \qquad (3)$$

E. EXPERIMENTAL BED CHARACTERISTICS

An important feature of the studies reviewed is the state of aggregation of particles when contacted with the SCF. The majority of studies used a fixed bed of size-fractionated particles continuously fed with pure SCF at the inlet. They differ in particle size and the relative bed length. If the bed residence time is short and axial gradients are negligibly small, it may be considered a differential bed, as in Triday and Smith.[4] There, all particles are in the same state of extraction. If it is long, equilibrium conditions are approached at the bed exit and only a fraction of the bed is active at a given time. When it is difficult to obtain uniform flow distribution through a fixed bed, or when the particle Reynolds number is low (Re < 1) the external particle-to-fluid mass-transfer coefficient may be anamalously low.[26] Other arrangements such as a fluidized bed or slurry reactor may then be advantageous in view of the high density and low viscosity of SCFs. A study of comparative extraction rates in fixed and fluidized beds and agitated baths would be useful.

IV. MASS-TRANSFER MODELS

Mass-transfer models are used to predict transfer rates and the dynamics of extraction as portrayed, for example, by breakthrough curves. They also find use as a means of estimating model parameters by fitting predicted dynamics to experimental results. The problem of mass transfer from the entire packed bed or agitated bath is complex, but may be solved once a description of the internal desorption and diffusion of extract is complete. Since what distinguishes one system from another is this internal process, there is no discussion here of such aggregate behavior of particles. This has, in any case, received much attention and can be found in texts such as Ruthven[27] and Wakao and Kaguei.[25]

In predicting extraction times, the use of models for single particles may be considered the most optimistic estimate for an extraction by whatever process and provides a useful benchmark. Note that although internal mass-transfer resistance may be eliminated by grinding the raw material to powder form, when a fixed-bed configuration is chosen for an industrial scale process, pressure drop considerations dictate the use of relatively large particles.

A. SHRINKING-CORE LEACHING MODEL

When the material to be extracted constitutes a large part of the particle mass it is possible that much of it is held in place as bulk condensed phase in the macropores by mechanical or capillary forces. Further, the porosity of the matrix available to the solvent is expected to be a function of the degree of extraction. A simple way of accounting for this considers a sharp boundary to exist between a core, in which all porosity is filled with extract material, and an outer region, in which only partially saturated solvent exists in the pores. This is a plausible model of SCF extraction where the diffusivity of the solvent in the solid or liquid extract is orders of magnitude less than the diffusivity of extract in solvent, however, there is no evidence of it having been exploited in the current context. The model is similar to the well-known shrinking-core model of heterogeneous fluid-solid reactions. Since it is a moving boundary problem, it is nonlinear and its rigorous solution poses some analytical problems; however, the essence of the model may be preserved in making the pseudo-steady-state approximation for diffusion of extract in the outer porous layer, neglecting the velocity of the interface, while the radius of the core as a function of time is determined by a simple mass balance. For present purposes, heat of solution is also neglected and isothermal extraction is treated. Assume, as in Equation 1, that the extract concentration in the fluid phase at the interface is at its saturation value. Thus, in dimensionless variables, the diffusion problem for the outer layer is

$$\frac{1}{r^2}\frac{\partial}{\partial r}\left[r^2\frac{\partial c}{\partial r}\right] = 0, \; r_c < r < 1 \tag{4}$$

$$t > 0, \; r = r_c(t), \; c = 1 \tag{5}$$

$$r = 1, \; -(\partial c/\partial r) = Bi(c(1) - C) \tag{6}$$

A mass balance at the interface gives

$$N\frac{dr_c}{dt} - \frac{\partial c}{\partial r}\bigg|_{r=r_c} \tag{7}$$

$$t = 0, r_c = 1 \tag{8}$$

In these equations, time is measured in diffusion time units, R^2/D_e, and distance in particle radius units, R. Concentrations are defined as the excess over C_0 normalized relative to $(c^* - C_0)$ where c^* is the equilibrium solubility of extract in solvent and C_0 is the initial concentration of solvent in the bath or bed to which the particle is exposed. The important parameters of the model are the Biot number Bi, defined as the ratio of internal to external mass transfer resistance, and the dimensionless group N, the ratio of mass concentration of extract material in the core to that in the solvent phase at equilibrium. The approximations require $N \gg 1$. Other symbols are defined in Nomenclature.

Straightforward integration of Equation 4, subject to the boundary conditions of Equations 5 and 6, and of Equation 7, subject to Equation 8, gives the following results in dimensionless variables:

- Mass flux into fluid at particle boundary:

$$n_2 = \frac{Bi(1 - C)}{Bi(1/r_c - 1) + 1} \tag{9}$$

- Initial mass flux:

$$n_2(t=0) = Bi \tag{10}$$

- Core radius as a function of time:

$$N\left[\frac{Bi - 1}{3}(r_c^3 - 1) - \frac{Bi}{2}(r_c^2 - 1)\right] = Bi(1 - C)t \tag{11}$$

Setting $r_c = 0$ in Equation 11 gives

- Time for complete extraction:

$$t(r_c=0) = \frac{N}{Bi(1 - C)}\left[\frac{1}{3} + \frac{Bi}{6}\right] \tag{12}$$

A parametric plot of Equations 9 and 11 for $C = 0$ is given in Figure 2. This plot illustrates the evolution of constant rate behavior as the external film begins to control when Bi is reduced. Note the remarkable "on-off" behavior for $Bi < 1$.

It is instructive to compare the predictions of this model with experimental extraction results. The experimental results of King et al.[18] discussed earlier are documented in sufficient detail to permit such a comparison. Using the leaching model, an attempt is made to calculate the time to extract 65% of the extractable material—the initial constant-rate extraction—on the assumption that this portion of the extract is free in the sense that it is held in place by mechanical or capillary forces. This extraction time may be easily obtained from the experimental data given since the extract loading in the effluent, the bed superficial velocity, the residence time, and the oil content of the seeds are known. An experimental extraction time of 8.8 h is found. In these experiments, fluid phase concentrations remain constant during the course of the experiment, making the application of Equation 12 straightforward. The results of the calculation are summarized in Table 1. An extraction time is calculated both for bed inlet and outlet conditions to bracket the observed time. The two values are given in the table. Listed in the table are experimental conditions and calculated quantities

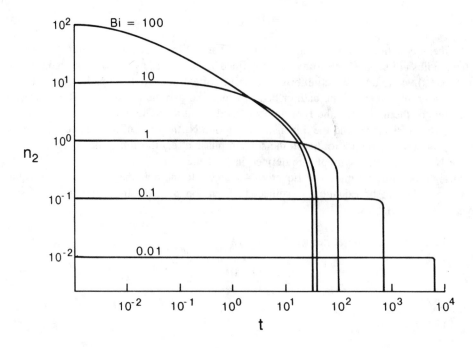

FIGURE 2.   Extract flux as a function of time for the shrinking core leaching model from Equations 9 and 11. N = 200.

TABLE 1
Summary of Calculation of Extraction Time

Quantity	Value	Ref.
Pressure	7.0 MPa	18
Temperature	25°C	18
CO_2 density	749 kg/m³	28
CO_2 viscosity	63.3(10^{-6}) Pa·s	28
Equilibrium loading	1.11 kg/m³	18
Dynamic loading	0.475 kg/m³	18
Superficial velocity	0.8 (10^{-3}) m/s	18
Residence time	91 s	18
Bed height	7.28 cm	Inferred from 18
Bulk density of seeds	640 kg/m³	19
Oil content of seeds	39.7 mass%	18
Particle diameter	0.5 (10^{-3}) m	18
Diffusivity of GTO in CO_2	1.68 (10^{-9}) m²/s	Takahashi method, 29
(porosity/tortuosity)	0.1	Assumed, typical
Re	4.7	
Sc	50	
Sh	12.2	25
k_m	4.11 (10^{-5}) m/s	
Bi	61.2	
N	200	
C at bed inlet	0 (dimensionless)	
C at bed outlet	0.522 (dimensionless)	
t (inlet)	3.6 h	Equation 12
t (outlet)	7.0 h	Equation 12
t (experimental)	8.8 h	Inferred from data

leading to the predicted extraction times. The diffusivity is estimated for the molecule glyceroltrioleate (GTO).

Extraction times calculated from Equation 12 are somewhat less than the experimental value; however, some of the needed quantities such as particle porosity and tortuosity are not known, and the author resorted to "typical" values, which could be considerably in error. Another explanation is possible. If one assumes uniform spherical particles, the surface area per unit volume of the bed can be calculated and the mass-transfer coefficient k_m can be inferred. Based on this calculation, the experimental coefficient is smaller than predicted from the Wakao-Kaguei correlation by a factor of about 50, as also noted by King et al.[18] Thus, it is quite possible that bed conditions were far from ideal and that such extractions may suffer from low contact efficiency caused perhaps by channeling. The Biot number for the run examined here is probably much closer to 1 which, according to Figure 2, would indeed result in the observed constant rate behavior. This attempt at calculating extraction times based on the shrinking-core leaching model illustrates typical difficulties that can be expected in any attempt to give precision to this quantity.

B. ADSORPTION MODEL

When the material to be extracted constitutes only a small fraction of the total mass, or where adsorbents such as activated carbon or zeolites were specifically treated, it is often possible to take the porosity as a constant, with the solvent permeating the entire porous structure from time zero. At low surface coverage, the physical or chemical bonding of extract material gives rise to adsorption-desorption kinetics. Then, the process of extraction may be described as one in which material is simultaneously desorbed and diffuses through the solvent in the macropores to the external stagnant fluid layer and then to the bulk fluid. In the case of chemisorption or accompanying chemical reaction, the enthalpy of desorption or reaction may result in a significantly nonisothermal particle and the mass transfer problem is coupled to the heat transfer problem through the temperature dependence of the adsorption-desorption rate constants. Further complicating the issue is the fact that the solvent density may be strongly temperature dependent. The convective transport must, therefore, be included in the diffusion equation. In these cases, one is usually driven to an entirely numerical solution, as in the case of oil shale extraction modeled by Triday and Smith.[4] Nonlinear equilibrium isotherms also contribute to the difficulty of analytical solution. This topic is reviewed by Ruthven.[27]

Confining attention to the isothermal case with first order adsorption kinetics in order to arrive at analytical solutions, the set of conservation equations can be written for both adsorbed and desorbed states of the adsorbate taking account of finite rates. Following Recasens et al.,[12] and numerous previous authors, the adsorbed phase is assumed to be nondiffusing and the equations are written in nondimensional form using the same scalings as for the leaching model:

$$\frac{\partial c}{\partial t} = \frac{1}{\epsilon r^2} \frac{\partial}{\partial r} \left[r^2 \frac{\partial c}{\partial r} \right] - P(c - c^*), \tag{13}$$

$$0 < r < 1$$

$$\frac{\partial c^*}{\partial t} = Q(c - c^*) \tag{14}$$

$$t = 0, c = c^* = 1 \tag{15}$$

$$t > 0, r = 0, \partial c/\partial r = 0 \tag{16}$$

$$r = 1, \ \partial c / \partial r = -Bi(c - C) \tag{17}$$

Additional parameters occurring in these equations are separate Thiele-like moduli for adsorption and desorption, respectively: $P = (\rho k_a R^2 / \epsilon D_e)$ and $Q = (k_a / K D_e)$. c^* is now the dimensionless concentration in the fluid phase in adsorptive equilibrium with the solid phase. In dimensional units, $c^* = C_a / K$, where c_a is the adsorbed phase concentration on unit mass basis.

Tomida and McCoy[30] and Suzuki and Smith[34] add an additional chemical reaction rate to Equation 14 for greater generality. Since the solution procedure is unaffected, the term is omitted for the sake of clarity. Recasens et al.[12] show how this set of equations leads to several routes to solution, depending on what further approximations one is prepared to make.

If adsorption is physical, the equilibrium approximation is in order due to the very large values of the rate constants k_a and $k_d = k_a / K$. In this case, replace the source term in Equation 13 by the value $(P/Q) \partial c^* / \partial t$ from Equation 14 and let $c^* = c$. Note that the order here is important; the source term cannot be set to zero. For the equilibrium model, then, Equation 13 becomes

$$\frac{\partial c}{\partial t} = \frac{1}{r^2} \frac{\partial}{\partial r} \left[r^2 \frac{\partial c}{\partial r} \right] \tag{18}$$

where the time scale is revised to $R^2(\epsilon + \rho K)/D_e$.

Alternatively, one can make the linear driving force approximation which enables one to lump the internal diffusion resistance with the external mass-transfer resistance. A rational approach to this approximation is detailed in the papers by Do and Rice,[31] Tomida and McCoy,[30] and Recasens et al.[12] It hinges on approximating the pore concentration by a parabolic profile and volume-averaging Equations 13 and 14 over the particle volume. The result is that Equation 13 becomes an ordinary differential equation in t, incorporating the boundary condition Equation 17 in a linear term with an effective mass-transfer coefficient given by $k_{eff} = 5k_m/(5 + Bi)$. This approximation is valid only if pore diffusion is not controlling. Such an approximation is implicit in the model used by both Rao and Mukhopadhyay[20] and Schaeffer et al.,[21] and their mass-transfer coefficients should be considered to be this k_{eff}.

Finally, simplifications result when desorption may be assumed to be irreversible and Equation 14 becomes

$$\frac{\partial c^*}{\partial t} = -Qc^* \tag{19}$$

Recasens et al.[12] argue that this cannot be the case when the fraction desorbed at a given time decreases with temperature, as observed in the data of Tan and Liou.[9] This process must therefore be governed by equilibrium.

The following illustrates the use of the method of moments to calculate an extraction time from any sorption model whose dynamics may be represented by linear partial differential equations. Unlike the leaching model, extraction time in the present case is not uniquely defined due to the exponential nature of the solution. The strategy is to Laplace transform the equations with respect to time and derive an expression for the transform of the mass flux $n_2(s)$ from the surface of a particle into the bulk fluid. Even when the inversion of this transform presents insurmountable difficulties, it may still be possible to calculate the needed moments. These may be obtained from the van der Laan[32] relation

$$M_j = \lim_{s=0}[(-1)^j (\partial^j n_2(s)/\partial s^j)] \tag{20}$$

The zeroth moment M_0 is the total mass adsorbed in the particle. The normalized first moment M_1/M_0 is the mean time, and the second central moment, $M_2/M_0 - (M_1/M_0)^2$, is the variance. As a representative time for extraction take

$$t_e = M_1/M_0 + [M_2/M_0 - (M_1/M_0)^2]^{1/2} \tag{21}$$

—the mean time plus standard deviation of the flux-vs.-time curve. This time certainly accounts for a majority of the extract material. For example, the area under the curve between 0 and t_e for the distributions e^{-t}, te^{-t}, t^4e^{-t} is in the range 0.847 to 0.864 of the total. For the normal distribution $\exp(t - a)^2$, the value is 0.841. Thus, considering uncertainties in properties, one may expect that t_e will provide a very useful measure of needed engineering extraction times. To illustrate, the author provides the results for the equilibrium desorption approximation. The exact solution to Equation 18 subject to Equations 16 and 17 is given by Crank.[33] With C set to zero, moment analysis leads to

·   The Laplace transform of the mass flux:

$$n_2(s) = \frac{1/s}{1/(\sqrt{s}\coth\sqrt{s} - 1) + 1/Bi} \tag{22}$$

·   Zeroth moment:

$$M_0 = 1/3 \tag{23}$$

·   First moment:

$$M_1 = \frac{Bi + 5}{45\ Bi} \tag{24}$$

·   Second moment:

$$M_2 = \frac{2\ (Bi^2 + 14\ Bi + 35)}{945\ Bi^2} \tag{25}$$

·   Extraction time:

$$t_e = \frac{[7(13\ Bi^2 + 70\ Bi + 175)]^{1/2}}{105\ Bi} + \frac{Bi + 5}{15\ Bi} \tag{26}$$

In order to bring out the complete dependence of the extraction time, note that in real time units Equation 26 must be multiplied by the time scale $R^2(\epsilon + \rho K)/D_e$. A plot of Equation 26 is given in Figure 3, which illustrates how the extraction time becomes independent of the Biot number for large values because diffusion is then controlled by the internal resistance. Unfortunately, to test the predictions of extraction time of the model, the author does not have data for either single particle or differential bed. In some cases of incorporating the particle analysis into a bed model, it may be possible to obtain moments (along the lines of Suzuki and Smith[34]) of the flux of extract in the bed effluent. Then extraction times may again be computed.

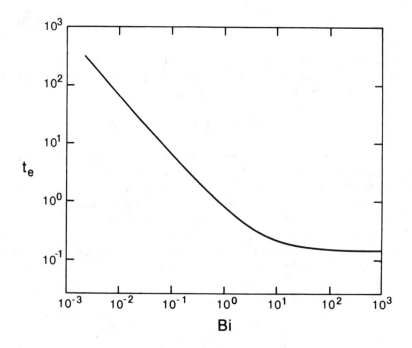

FIGURE 3.   Extraction time as a function of Biot number for the equilibrium desorption model.

V. CONCLUSIONS

This chapter was devoted to the quantitative aspects of SCF extraction with emphasis on the prediction of fluxes, dynamics, and extraction times. Several studies treat natural products of vegetable origin as well as synthetic systems exemplified by attempts to regenerate activated carbon used to adsorb organic molecules from aqueous streams. The barrier to treating natural products quantitatively lies in the difficulty of characterizing the structure of the matrix and the equilibrium partitioning of the multicomponent extract between solvent and substrate. On the other hand, synthetic systems such as industrial adsorbents may be studied under ideal conditions with single adsorbates in reproducible structures; these can be subjected to characterization as regards pore size distribution and the effective diffusivity. While it is not yet possible to establish adsorption equilibria in supercritical solvents from first principles, progress has nevertheless been made using semiempirical methods. Finally, it is noted that a diverse set of applications appears to share a commonality brought out by mathematical models of particle transport. A suggestion for future work is that progress in being able to predict extraction times for complex systems such as the natural products discussed may benefit from precise measurements on a few basic synthetic systems where the matrix structure and phase equilibria are established independently.

NOMENCLATURE

Bi   Biot number $= Rk_m/D_{eff}$
c   Concentration in macropores, kmol/m³ or dimensionless
c*   Equilibrium concentration in macropores, kmol/m³ or dimensionless
C   Bulk fluid concentration, kmol/m³ or dimensionless
C_0 Initial bulk fluid concentration, kmol/m³ or dimensionless
D   Diffusivity of extract in solvent, m²/s
D_e Effective diffusivity in porous matrix, m²/s

d	Particle diameter, m
Gr	Mass-transfer Grashof number $= d^3g\rho\Delta\rho/\mu^2$
g	Gravitational acceleration, m/s^2
k_a	Adsorption rate constant, $m^3/(kg\cdot s)$
k_d	Desorption rate constant, s^{-1}
k_m	External mass-transfer coefficient at particle-fluid boundary, m/s
k_{eff}	Effective mass-transfer coefficient, m/s
K	Adsorption equilibrium constant $= k_a/k_d$, m^3/kg
M_1	Molecular weight of solvent
M_2	Molecular weight of solute (extract)
M_j	jth moment of flux vs. time curve
N	Dimensionless concentration ratio $= \epsilon\rho_c/M_2(c^* - C_0)$
n_2	Flux of solute (extract) at solid-fluid boundary, $kmol/(m^2\cdot s)$
P	Thiele modulus for adsorption
Q	Thiele modulus for desorption
R	Radius of particle, m
Re	Reynolds number $= \rho vd/\mu$
r	Dimensionless radial coordinate
r_c	Dimensionless radius of unleached core
Sc	Schmidt number $= \mu/\rho D$
Sh	Sherwood number $= k_m d/D$
s	Laplace transform variable
t	Dimensionless time,
v	Superficial velocity, m/s
y_2	Mole fraction solute

Greek

ϵ	Porosity of solid matrix
μ	Viscosity of fluid phase, $Pa\cdot s$
ρ	Density of particles, kg/m^3
ρ_c	Density of condensed phase in leaching model, kg/m^3
ρ^*	Density of solvent at solid interface, kg/m^3
$\Delta\rho$	Density difference in Grashof number $= \rho^* - \rho$, kg/m^3

REFERENCES

1. **Saim, S., Ginosar, D. M., and Subramaniam, B.,** Phase and reaction equilibria considerations in the evaluation and operation of supercritical fluid reaction processes, in *Supercritical Fluid Science and Technology,* ACS Symp. Ser. 406, Johnston, K. P. and Penninger, J. M. L., Eds., American Chemical Society, Washington, D.C., 1988, chap. 20.
2. **Low, J. Y.,** Hydrotreating in supercritical media, in *Supercritical Fluids. Chemical Engineering Principles and Applications,* ACS Symp. Ser. 329, Squires, T. G. and Paulaitis, M. E., Eds., American Chemical Society, Washington, D.C., 1987, chap. 22.
3. **Barton, P. and Kasun, T. J.,** Reducing capillary condensation of paraffinic distillate in zeolite with supercritical fluid, in *Supercritical Fluid Technology,* Penninger, J. M. L., Radosz, M., McHugh, M. A., and Krukonis, V. J., Eds., Elsevier, Amsterdam, 1985, 435.
4. **Triday, J. and Smith, J. M.,** Dynamic behavior of supercritical extraction of kerogen from shale, *AIChE J.,* 34, 658, 1989.
5. **Rance, R. W. and Cussler, E. L.,** Fast fluxes with supercritical solvents, *AIChE J.,* 20, 353, 1974.

6. **Vinkler, E. G. and Morozov, V. S.,** Measurement of diffusion coefficients of vapors of solids in compressed gases. II. Diffusion coefficients of naphthalene in nitrogen and in carbon dioxide, *Russ. J. Phys. Chem.,* 49, 1405, 1975.

7. **Prausnitz, J. M., Lichtenthaler, R. N., and Gomes de Azevedo, E.,** *Molecular Thermodynamics of Fluid-Phase Equilibria,* Prentice-Hall, Englewood Cliffs, NJ, 1986, 171.

8. **DeFilippi, R. P., Krukonis, V. J., Robey, R. J., and Modell, M.,** Supercritical Fluid Regeneration of Activated Carbon for Adsorption of Pesticides, Rep. No. EPA-600/2-80-054, U.S. Environmental Protection Agency, Washington, D. C., March 1980.

9. **Tan, C.-S. and Liou, D.-C.,** Desorption of ethyl acetate from activated carbon by supercritical carbon dioxide, *Ind. Eng. Chem. Res.,* 27, 988, 1988.

10. **Tan, C.-S. and Liou, D.-C.,** Supercritical regeneration of activated carbon loaded with benzene and toluene, *Ind. Eng. Chem. Res.,* 28, 1222, 1989.

11. **Tan, C.-S. and Liou, D.-C.,** Modelling of desorption at supercritical conditions, *AIChE J.,* 35, 1029, 1989.

12. **Recasens, F., McCoy, B. J., and Smith, J. M.,** Desorption processes: supercritical fluid regeneration of activated carbon, *AIChE J.,* 35, 951, 1989.

13. **Kander, R. G. and Paulaitis, M. E.,** The adsorption of phenol from dense carbon dioxide onto activated carbon, in *Chemical Engineering at Supercritical Fluid Conditions,* Paulaitis, M. E., Penninger, J. M. L., Gray, R. D., and Davidson, P., Eds., Ann Arbor Science Publishers, Ann Arbor, MI, 1983, 461.

14. **Mackay, M. E. and Paulaitis, M. E.,** Solid solubilities of heavy hydrocarbons in supercritical solvents, *Ind. Eng. Chem. Fundam.,* 18, 149, 1979.

15. **Yonker, C. R., Wright, R. W., Frye, S. L., and Smith, R. D.,** Mechanism of solute retention in supercritical fluid chromatography, in *Supercritical Fluids. Chemical Engineering Principles and Applications,* ACS Symp. Ser. 329, Squires, T. G. and Paulaitis, M. E. Eds., American Chemical Society, Washington, D.C., 1987, 172.

16. **King, J. W.,** Supercritical fluid adsorption at the gas-solid interface, in *Supercritical Fluids. Chemical Engineering Principles and Applications,* ACS Symp. Ser. 329, Squires, T. G. and Paulaitis, M. E., Eds., American Chemical Society, Washington, D.C., 1987, chap 13.

17. **Gil, M. G. B., King, M., and Bott, T. R.,** Extraction of rape-seed oil using compressed CO_2: evaluation of mass transfer coefficients, in *Proc. Int. Symp. on Supercritical Fluids, Nice, France, 1988,* Perrut, M., Coordinator, Institut National Polytechnique de Lorraine, 1988, 651.

18. **King, M. B., Bott, T. R., Barr, M. J., Mahmud, R. S., and Sanders, N.,** Equilibrium and rate data for the extraction of lipids using compressed carbon dioxide, *Sep. Sci. Technol.,* 22, 1103, 1987.

19. **Brunner, G.,** Mass transfer from solid material in gas extraction, *Ber. Bunsenges. Phys. Chem.,* 88, 887, 1984.

20. **Rao, V. S. G. and Mukhopadhyay, M.,** Mass transfer studies for supercritical fluid extraction of spices, *Proc. Int. Symp. on Supercritical Fluids, Nice, France, 1988,* Perrut, M., Coordinator, Institut National Polytechnique de Lorraine, 1988, 643.

21. **Schaeffer, S. T., Zalkow, L. H., and Teja, A. S.,** Modelling of the supercritical extraction of monocrotaline from *Crotalaria spectabilis, J. Supercritical Fluids,* 2, 15,1989.

22. **Schaeffer, S. T., Zalkow, L. H., and Teja, A. S.,** Solubility of monocrotaline in supercritical carbon dioxide and carbon dioxide-ethanol mixtures, *Fluid Phase Equilibria,* 43, 45, 1988.

23. **Lim, G.-B., Holder, G. D., and Shah, Y. T.,** Solid-fluid mass transfer in a packed bed under supercritical conditions, in *Supercritical Fluid Science and Technology,* ACS Symp. Ser. 406, Johnston, K. P. and Penninger, J. M. L., Eds., American Chemical Society, Washington, D.C., 1989, chap. 24.

24. **Debenedetti, P. G. and Reid, R. C.,** Diffusion and mass transfer in supercritical fluids, *AIChE J.,* 32, 2034, 1986.

25. **Wakao, N. and Kaguei, S.,** *Heat and Mass Transfer in Packed Beds,* Gordon & Breach Science Publishers, New York, 1982.

26. **Martin, H.,** Low Peclet number particle-to-fluid heat and mass transfer in packed beds, *Chem. Eng. Sci.,* 33, 913, 1978.

27. **Ruthven, D. M.,** *Principles of Adsorption and Adsorption Processes,* John Wiley & Sons, New York, 1984.

28. **Ely, J. F. and Huber, M. L.,** NIST Standard Reference Database 4, Thermophysical Properties of Hydrocarbon Mixtures (SUPERTRAPP), National Institute of Standards and Technology, Boulder, CO, 1990.

29. **Reid, R. C., Prausnitz, J. M., and Poling, B. E.,** The Properties of *Gases and Liquids,* 4th Ed., McGraw-Hill, New York, 1987, 592.

30. **Tomida, T. and McCoy, B. J.,** Polynomial profile approximation for intraparticle diffusion, *AIChE J.,* 33, 1908, 1987.

31. **Do, D. D. and Rice, R. G.,** Validity of the parabolic profile assumption in adsorption studies, *AIChE J.,* 32, 149, 1986.

32. **Van der Laan, E. T.,** *Chem. Eng. Sci.,* 6, 227, 1958.
33. **Crank, J.,** *The Mathematics of Diffusion,* Oxford University Press, London, 1975, 96.
34. **Suzuki, M. and Smith, J. M.,** Kinetic studies by chromatography, *Chem. Eng. Sci.,* 26, 221, 1971.

Chapter 11

DESIGN AND CONTROL OF SUPERCRITICAL EXTRACTION PROCESSES—A REVIEW

Miriam L. Cygnarowicz and Warren D. Seider

TABLE OF CONTENTS

ABSTRACT

Design studies are important in assessing the feasibility of supercritical extraction (SCE) processes since they properly weigh factors such as the scale of the process, the value of the product, the need for a nontoxic solvent, etc. Once the most promising design is located, the decision to invest in the SCE process can be made with confidence. In this chapter, both experimental and simulation approaches to design are discussed. The experimental studies aim to determine the effect of the key process parameters, and where possible, to collect fundamental data to aid in the development of more rigorous models for simulation and design optimization. Examples discussed include the decaffeination of coffee beans, the extraction of edible oils, and the dehydration of alcohols. Applications of simulation and optimization methods have been less common, due primarily to the difficulty of modeling the complex systems. Good success, however, has been obtained when these techniques have been applied to the dehydration of alcohols and ketones. Several examples are discussed that indicate that SCE can be competitive for these separations on the basis of energy consumption. A more complete study, which includes the cost of the equipment, demonstrates that SCE is not competitive since the products are low-value, high-volume chemicals. Other work is described that attempts to model more competitive SCE processes, such as the extraction of lecithin from soya oil and the isolation of β-carotene from fermentation broths; however, the designs for the recovery of these high-value, low-volume chemicals are more uncertain since simplified models were used. Finally, the transient behavior and control of SCE processes is discussed. Although this challenging problem is only beginning to be addressed, initial studies indicate that maintenance of the proper phase distribution in the extractors and separators is difficult to achieve with conventional control schemes. As design models are extended to simulate the dynamics of SCE processes, new model predictive control algorithms can be expected to significantly improve the control of the phase distribution.

I. INTRODUCTION

SCE is an increasingly important technology in the food and pharmaceutical industries because it allows the substitution of nontoxic, environmentally safe solvents such as CO_2 for traditional liquid solvents such as methylene chloride and hexane. At the present time, this is most applicable in the food and pharmaceutical industries where the use of toxic solvents is regulated; however, more stringent regulations regarding toxic waste disposal may eventually broaden the significance of this technology to include many segments of the chemical process industry.

In Europe, and on a more limited scale in the U.S., SCE is used commercially to decaffeinate coffee and tea[1] and to extract hops[2] and spices.[3] These commercial successes indicate that SCE is a viable alternative for the preparation of some food products; however, as elucidated by Krukonis,[4] its feasibility is determined by the scale of the process, the value of the product, the need for a nontoxic solvent, etc. Design studies properly weigh these factors, and identify the most promising SCE processes for a particular application. They provide the best designs upon which to base the decision to invest in SCE processes.

II. EVOLUTION OF A PROCESS DESIGN

In designing a SCE process, consideration must be given to the properties of the feed material (i.e., solid or liquid), the solute concentration in the feed, the required production rate, and the thermodynamic and mass transfer relationships between the solute and solvent.[5] Marentis[6] suggests a design protocol in which these concerns are addressed through a series of experimental steps from which the process design evolves.

First, the feasibility of using SCE to affect a particular separation is assessed in "screening unit testing". Then, larger vessels and solvent recycle are incorporated in "process development unit testing". At this stage, the extractor and separator temperatures and pressures, solvent-to-feed ratios, and processing modes (i.e., batch, semi-batch, or continuous) are varied to establish favorable operating conditions. Experience is also gained in the control of the proposed process. Manipulated and controlled parameters are selected and a control scheme (e.g., PID) is proposed.

Finally, when a design is promising, pilot-plant testing is initiated to minimize scale-up uncertainties. Pilot plant tests verify vessel configurations and sizes, and identify mechanical design problems, such as clogging of pipes, valves, and heat exchangers due to precipitation of the solute. Materials handling and clean-up are also considered. The resiliency (i.e., controllability) of the process is studied further, as well as the sensitivity of the proposed design to changes in operation (i.e., flexibility).

At each stage, the economics are estimated at the reduced scale of the process, and unfavorable designs are modified or eliminated. When the pilot-plant studies for a particular design appear promising (i.e., have favorable economics), development of the process proceeds through the construction of the full-scale plant. A good discussion of the economic considerations in designing a SCE plant are given by Novak and Robey.[7] They utilized proprietary data to prepare a preliminary design for a multiproduct spice and herb extraction plant. The base case design used two 973 L extractors to process 0.8 million kg of spices or herbs per year. In the design, the extractors are operated batch-wise, in staggered cycles. The sensitivities of the capital and operating costs to the size of the plant were determined. The capital costs were shown to decline 40% as the plant size is increased to four times the base case. Although the operating costs per hour increase 87% in the larger capacity plant, the operating cost per unit feed decreases 54%. Process costs per kilogram of extract are given for 12 spices and 8 herbs. The costs range from $4.00/kg for nutmeg extract to $113/kg for Arnica extract. No comparisons were given to the costs of conventional extraction. It should be noted that since the process parameters (i.e., pressure, temperature, solvent-to-feed ratio, etc.) were not optimized, more economical operating regimes may exist.

Additional discussions of the construction and operation of full-scale SCE plants are given by Eggers,[8] Koerner,[9] Herderer and Heidemeyer,[10] Eggers and Sievers,[11] and Marentis and Vance.[12]

Because there are relatively few commercial installations of SCE, most of the design and control studies for these processes have been at the feasibility or process development stage. Consequently, this chapter concentrates on process development and design. These studies often involve experimentation to determine the effect of the key process parameters and, where possible, to develop more rigorous models for simulation and design optimization.

III. EXPERIMENTAL STUDIES FOR DESIGN

Because SCE processes typically involve complex mixtures for which fundamental phase equilibria and mass transfer data are not available, many workers study the effect of different process parameters for a particular system in small-scale pilot plant experiments.

For example, Brunner[13,14] used this approach to study the extraction of caffeine from coffee using supercritical CO_2. The effect of extractor temperature, pressure, and bed height was considered for the semi-batch extraction of caffeine from moist, roasted coffee beans. Figure 1 shows the caffeine extracted as a function of the solvent-to-feed ratio for two different extraction times. Note that the rate of extraction is nearly twice as high after 30 min than after 240 min. Also, the rate of extraction increases more rapidly with the solvent-to-feed ratio at shorter extraction times; i.e., at 30 min, the caffeine extracted increases sharply as the solvent-to-feed ratio is increased to 90. At longer extraction times, however,

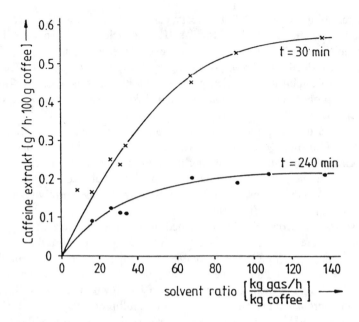

FIGURE 1.   Caffeine extract as a function of solvent-to-feed ratio. (From Brunner, G., *Proc. Int. Symp. Supercritical Fluids*, Perrut, M., Ed., Société Francaise de Chimie, Nice, 1988. With permission.)

only small changes were recorded for solvent-to-feed ratios above 30 because the extraction efficiency is limited by intraparticle diffusion. Once the caffeine in the outer edges of the bean is depleted, it must diffuse from the center to be solubilized by the CO_2, reducing the extraction rate. Brunner also studied the mass transfer effects for the caffeine-CO_2 system and estimated the mass transfer coefficients.

The extraction of edible oils from seed materials using supercritical CO_2 has been studied extensively by Friedrich and co-workers,[15-18] Stahl,[19] and others.[20,21] These investigators examined the impact of the extractor and separator temperatures and pressures, the condition of the feed material (crushed, flaked, etc.) and the design of the extractor (e.g., extruder-type for continuous processing), on the quality of the product (i.e., the taste, appearance, and color). Unfortunately, SCE is probably not economically feasible. Soybean oil, for example, is produced in large quantities (approximately 2000 tons/d) and sells for a relatively low price ($0.25/lb). Although detailed economic analyses have not been published, it seems clear that SCE is too expensive for this application, even though the SCE product is superior to that obtained using traditional extraction methods.

Supercritical solvents have also been considered for the dehydration of dilute alcohols and ketones. Moses et al.[22] studied the extraction of ethanol, isopropanol, and sec-butanol from water using CO_2 at 65 atm and 40°C in a 0.16 L sieve-tray extractor. An energy-efficient configuration was proposed in which the vapor from the distillation column is compressed and subsequently condensed as it vaporizes the bottoms liquid in the reboiler. Extractor tray efficiencies were determined, and the effect of changing the solvent-to-feed ratio was explored. The most favorable solvent-to-feed ratios (beyond which little appreciable increase in the solute recovery was recorded) were 10 for ethanol, 4 for isopropanol, and 1.5 for sec-butanol. Furthermore, the mass transfer efficiency was observed to increase with the molecular weight of the solute. Tray efficiencies measured for an 8 tray (7-in. tray spacing) extractor were 12 to 20% for ethanol, 35 to 40% for isopropanol, and approximately 50% for sec-butanol. It was concluded that extraction with CO_2 may be advantageous since it can save 10 to 40% of the energy required for distillation.

Seibert and Moosberg[23] and other researchers at the University of Texas[24,25] used the alcohol/water system to examine the rates of mass transfer in high-pressure towers. Their experimental apparatus is shown in Figure 2. The stainless-steel extractor had an inside diameter of 9.88 cm, was 215 cm high, and had 32 high-pressure sapphire windows to permit viewing of the contacting phases. The pilot plant included an extensive process control system to regulate temperatures, pressures, and liquid levels by adjusting the four valves shown. Analyses were performed by on-line gas chromatography (GC). Extraction conditions ranging from 82 atm and 297 K to 150 atm and 318 K were studied for the dehydration of isopropanol using supercritical CO_2. Assuming a linear equilibrium relationship, dilute solute concentration, a pure solvent, and a straight operating line, the number of transfer units (NTU_{oc}) was calculated from:

$$NTU_{oc} = \frac{\ln\left(\left(\frac{x_f}{x_r}\right)\left(1 - \frac{1}{\lambda}\right) + \frac{1}{\lambda}\right)}{1 - \frac{1}{\lambda}} \tag{1}$$

$$\lambda = \left(\frac{U_D}{U_C}\right)M_{dc} \tag{2}$$

and x_f is the mass fraction of isopropanol in the recycled solvent, x_r is the mass fraction of isopropanol in the raffinate, U_C is the superficial velocity of the continuous (aqueous) phase based on the column area, U_D is the superficial velocity of the dispersed (CO_2) phase, and M_{dc} is the equilibrium distribution coefficient. Since the contact height (Z) is known, the overall height of a transfer unit (based on the continuous phase, HTU_{oc}) was calculated from:

$$HTU_{oc} = \frac{Z}{NTU_{oc}} \tag{3}$$

Finally, the height equivalent of a theoretical stage (HETS) was determined from:

$$HETS = \left(\frac{\lambda \ln \lambda}{\lambda - 1}\right)HTU_{oc} \tag{4}$$

Note that the HETS varies inversely with the efficiency of the separator. Figure 3 shows the HETS for four contacting devices: a spray tower, a tower with sieve trays, and packed towers with Raschig rings (RR) and metal Intalox saddles (IMTP). Note that the sieve trays are most efficient, followed closely by the Raschig rings and Intalox saddles. The inefficiency of the spray tower is attributed to increased back-mixing in the continuous phase and a smaller residence time for the CO_2 droplets. In general, column efficiencies are higher for supercritical than liquid-liquid extractors due to higher mass transfer coefficients in the supercritical phase. These studies provide a more reliable basis for the design of full-scale extractors and separators. Although not competitive economically, alcohol/water systems continue to be studied, since they are well characterized and, unlike most food systems, do not exhibit variability from batch to batch.

Bohm et al.[26] discuss critical factors in the design and operation of multipurpose plants for the continuous extraction of liquids with supercritical fluids (SCFs). A design strategy was employed in which the height equivalent of a theoretical plate (HETP) was determined by stagewise computer simulation of pilot plant experiments. Care was taken to ensure flexibility in the design of the mechanical and electrical units in the plant. For example, the

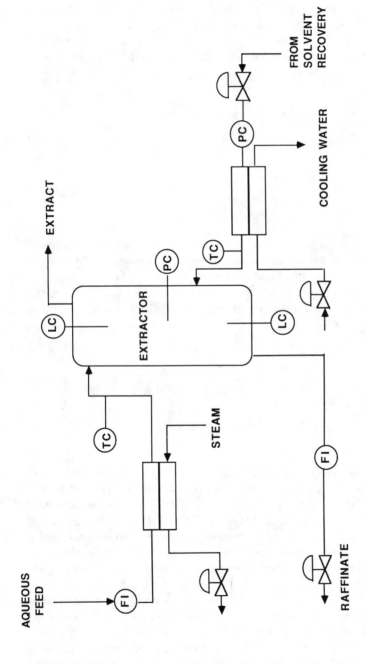

FIGURE 2.   Extractor section of Seibert and Moosberg apparatus.[23] LC, TC, and PC are level, temperature and pressure controllers, respectively. FI are flow indicators.

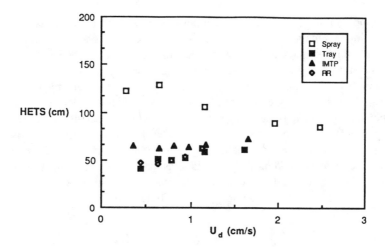

FIGURE 3. Efficiency of contacting devices for carbon dioxide/isopropanol/ water at 313K and 102 atm. Dispersed phase is carbon dioxide and $U_c = 0.048$ cm/s. (Reprinted from Ref. 23, p. 2049 by courtesy of Marcel Dekker, Inc.)

extractor consisted of six modules, each with length of 1 m, linked together with connecting elements. Each connecting element has six ports that can be used for feed input, temperature and pressure measurement, reflux, or sampling. The modular nature of this design allows for relatively easy changes in the column height or feed position.

In summary, experimentation has been very important in the development of SCE processes. This is because the solubilities, vapor pressures, rates of mass transfer, etc., are very difficult to estimate for large organic molecules in carbon dioxide with various cosolvents. Yet, thermodynamic models (using pseudocomponents to represent complex mixtures), when coupled with experimental data, can be adapted to apply over limited ranges of T and P. Similarly, mass and momentum balances for complex hydrodynamic systems can be approximated, for example, with equilibrium-stage models. The benefits of utilizing these models for design optimization are discussed in the next section.

IV. SIMULATION AND OPTIMIZATION FOR THE DESIGN OF SUPERCRITICAL EXTRACTION PROCESSES

Process simulation and optimization are widely used in the chemical industry to reduce the number of pilot plant studies needed to develop the final design. Process simulation allows a broad range of operating conditions and process configurations to be explored quickly and easily, and optimization systematically determines the most favorable operating conditions for various economic objectives. These techniques are advantageous for the design of SCE processes, since there are many process parameters and because the costs associated with pilot plant construction and operation are high. In addition, dynamic simulation allows the controllability of the process to be explored in the design stage, where designs with control problems can be rejected easily.

Reliable process simulations of complex SCE systems require more accurate models, and their development presents several challenges. First, the solutes and the natural substrates from which they are extracted are difficult to characterize. The solute is often a large, organic compound for which vapor pressures and critical properties are not known, and are difficult (or impossible) to measure, and the substrate is often a "natural" solid or liquid that may contain several hundred species. Furthermore, more reliable thermodynamic models are needed that can accurately predict the complex interactions between the small, nonpolar solvents and the large, organic, often-polar solutes. These models should not be overly complex, since phase equilibrium calculations are performed many thousands of times in an optimization or simulation. Finally, because SCE processes operate in the critical region, in the vicinity of mathematical singularities, the phase equilibrium calculations are inherently difficult to converge and require more sophisticated solution techniques.

Research is proceeding to address these difficulties. Thermodynamicists are working to develop simple, yet powerful equations of state (EOSs) that utilize density-dependent, local-composition mixing rules to bridge the gap between the EOS and activity coefficient models.[27,28] The group contribution equation of state (GC-EOS) of Skjold-Jorgensen[29,30] is particularly attractive since it predicts the phase behavior given only the chemical structure of the species. Another useful model is the Peng-Robinson EOS as modified by Panagiotopoulos and Reid,[31] who added a second interaction coefficient for each binary pair, k_{ji}, and a composition-dependent term. Their mixing rules for the a coefficient are

$$a = \sum_i \sum_j x_i x_j a_{ij} \tag{5}$$

$$a_{ij} = \sqrt{a_i a_j} \, [1 - k_{ij} + (k_{ij} - k_{ji})x_i] \tag{6}$$

Note that when $k_{ji} = k_{ij}$, Equations 5 and 6 are the mixing rules for the Peng-Robinson equation. This relatively minor modification significantly improves the predictions of the liquid phase compositions for mixtures such as acetone/CO_2 and water/CO_2.[32]

In the computation of high-pressure phase equilibria, Michelsen[33-36] has made substantial progress toward developing efficient and reliable algorithms. For example, his implementation of the tangent-plane method for stability analysis not only indicates whether a phase is stable, but also gives excellent initial guesses for the new phase. This is especially important for phase equilibrium calculations in the critical region where the phase distribution is uncertain. In addition, his continuation algorithm for tracing phase envelopes and computing critical points is useful in bounding the feasible regions for extraction and solute recovery.

Finally, researchers are continuing to measure solubilities in SCFs, and an increasing number of experimental phase equilibria studies are being published. Vapor-liquid, vapor-solid, and vapor-liquid-liquid equilibrium data are being collected for many solutes of interest in the food and pharmaceutical industries. A few studies are demonstrating that the solubility of large, organic species in SCFs is often increased significantly with the addition of small amounts of cosolvents.[37-39] For example, Wong and Johnston[38] measured the solubilities of cholesterol, stigmasterol, and ergosterol in supercritical CO_2 and computed an enhancement factor, E, given by:

$$E = \frac{Y_{ternary}}{Y_{binary}} \tag{7}$$

where $Y_{ternary}$ is the mole fraction of solute in the CO_2/cosolvent mixture, and Y_{binary} is the mole fraction of solute in CO_2. The enhancement factor for cholesterol was shown to range from 2.4 for a mixture containing 3.5% ethanol to 7.2 for a mixture with 3.5% methanol. Although much work is needed to describe the mechanism of solubility enhancement, the increased solubility may render SCE processes more competitive.

At the present time, the most successful design and optimization studies have been accomplished for the well-characterized alcohol/water and acetone/water systems. For example, Brignole et al.[40] used the GC-EOS and the SEPSIM[41] flow sheet program to investigate the dehydration of ethanol and 2-propanol. Several supercritical solvents were considered, with propane selected for ethanol recovery, and isobutane for 2-propanol recovery. Three process configurations were proposed, including cycles with vapor recompression and feed preconcentration. A flow sheet for the former cycle is shown in Figure 4. In this design,

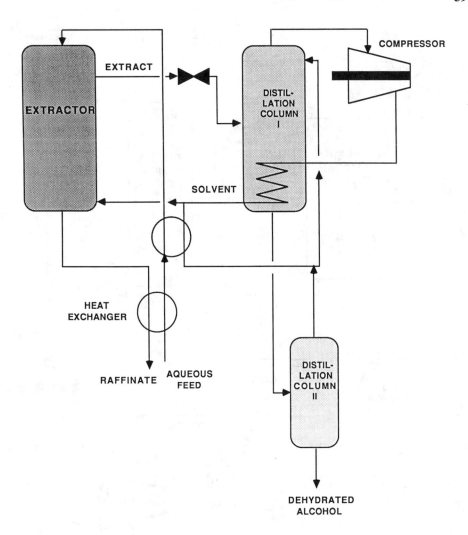

FIGURE 4. Vapor recompression cycle for alcohol dehydration. (Based upon Brignole et al.[40])

the aqueous feed is preheated by exchange with the raffinate and solvent recycle streams, and fed to the extractor. The extract is fed to a distillation column, where the overhead (nearly pure solvent) is compressed and condensed to satisfy the reboiler duty. The bottoms product is sent to a secondary distillation column for recovery of absolute alcohol as the liquid product. Brignole et al. concluded that this and similar SCE processes significantly reduce the energy consumption as compared to azeotropic distillation. It should be noted, however, that although attempts were made to optimize the important parameters such as the solvent-to-feed ratio and the extractor temperature and pressure, formal optimization strategies were not utilized, and thus the interdependence of these parameters was not properly considered. Furthermore, only the consumption of utilities was minimized; i.e., the installed cost of the equipment was not included.

Recently, Cygnarowicz and Seider[42] formulated a strategy for designing cost-efficient SCE processes. The approach involves the development of a model for the process flow sheet (using SEPSIM) and the creation of a nonlinear program to generate designs that minimize either the annualized or utility costs. The strategy was applied to design a process to dehydrate acetone using supercritical CO_2. A flow sheet for the dehydration of acetone, with operating conditions at the global minimum in utility cost, is shown in Figure 5. As

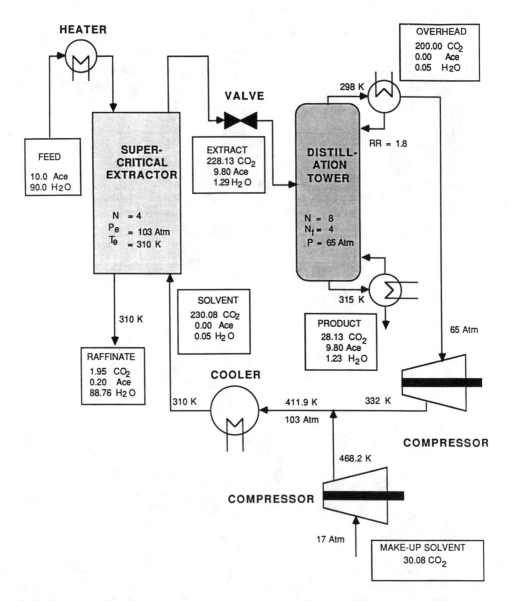

FIGURE 5. Dehydration of acetone with supercritical carbon dioxide. Stream flow rates are in mol/min. (Reprinted with permission from Cygnarowicz, M. L. and Seider, W. D., *Ind. Eng. Chem. Res.*, 28(10), 1497, 1989. © 1989 American Chemical Society.)

the supercritical CO_2 flows through the extractor, which is modeled using equilibrium stages, it preferentially dissolves the acetone, leaving the bulk of the water in the raffinate. The extract is expanded across a valve and fed to a distillation column where the acetone is concentrated at the bottom with nearly pure CO_2 in the distillate. The CO_2 stream is recompressed, mixed with make-up solvent, and recycled to the extractor.

The variables that have the greatest impact on the process cost and performance were determined by Cygnarowicz and Seider to be the extractor pressure (P_e) and temperature (T_e), the flow rate of the recirculated solvent ($F_{CO_2}^{rec}$), the distillation tower pressure (P_d), the reflux ratio in the distillation column, and the number of stages in the extractor and distillation column. In their approach, several of these variables are fixed to make the optimization more tractable. For example, the number of stages in the extractor and distillation column

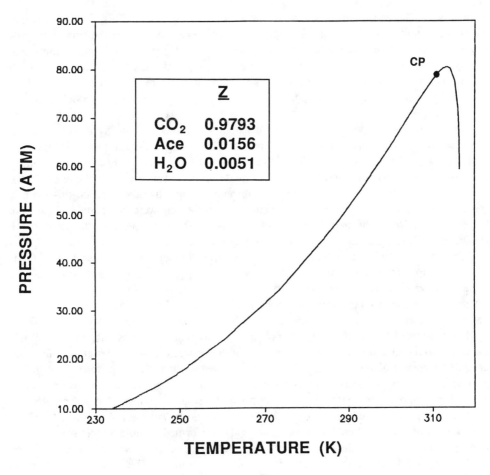

FIGURE 6. Phase envelope of typical feed to distillation column. (Reprinted with permission from Cygnarowicz, M. L. and Seider, W. D., *Ind. Eng. Chem. Res.*, 28(10), 1497, 1989. © 1989 American Chemical Society.)

are specified because their inclusion as design variables would require the solution of a mixed-integer, nonlinear program (with the number of stages as integer variables), which would significantly increase the computational complexity of the problem. Although the reflux ratio is a continuous design variable, in their experience, it is normally reduced to its lower bound and need not be varied continuously in the optimization. The distillation tower pressure is an important variable in two respects. First, P_d determines whether expensive refrigeration is needed in the condenser. In addition, the range of feasible values for P_d is tightly constrained. Figure 6 is a phase envelope calculated using the GC-EOS for a mixture with composition typical of the feed to the tower. Note that at such compositions, the two-phase region exists only at high pressures (close to the critical point) over a narrow range. The unavoidable proximity to the critical point makes the simulation of the distillation tower difficult. The Newton-based method that solves the equations is extremely sensitive to the initialization, and convergence failures are common. For this reason, P_d is fixed, and the optimal design is located at discrete values of this variable.

When minimizing the cost of utilities only, SCE was shown to be competitive with conventional separation techniques when the process was operated with low compression ratios and the distillation column pressure was close to the critical point in order to allow cooling water to be used in the condenser. This favorable result is in agreement with those of Brignole and Moses; however, at such pressures, the densities of the two contacting phases approach each other, resulting in small flooding velocities. This sharply increases the volume required to disengage the vapor and liquid phases and raises the equipment cost significantly. Thus, despite the low operating costs for this design, the minimum annualized cost is not competitive. Acetone, like ethanol and propanol, is a high-volume, low-cost chemical. These species are poor candidates for concentration by SCE, but are well suited for the development of design methodologies since their phase equilibria are modeled reliably by an EOS.

Some successes have been achieved in modeling more realistic systems. Moricet[43] simulated the extraction of raw lecithin from soya oil using a CO_2/propane mixture. The Redlich-Kwong EOS was used to model the phase equilibria, and a hybrid of the sum-rates and Tomich methods was used to solve the equations for the extractor. Although only the extractor was simulated, and no attempts were made to optimize the process, this work is important because simulated and experimental concentrations were compared. Reasonable agreement was found for the liquid product, but not for the vapor product, the errors being attributed to "inaccuracies" in modeling the phase equilibria.

Cygnarowicz and Seider[44] developed a simplified model for a process to isolate β-carotene from fermentation broths using supercritical CO_2. Pure component properties for β-carotene were estimated, and solubility data for β-carotene in CO_2[45] were fit to estimate the interaction coefficients for the modified Peng-Robinson EOS. Then, models were prepared to simulate the three-phase (solid, liquid, SCF) extraction and separation steps. These models were used to optimize the SCE process. A design strategy was utilized to minimize the annualized cost while producing β-carotene of the desired quality. A design for minimum annualized cost is given in Figure 7. The results demonstrate that despite high operating pressures, SCE can be competitive for the recovery of high-value products at low production rates. Furthermore, the increases in solubility with small amounts of cosolvent have a substantial impact on the economics of the proposed design. An example is given in which the addition of 1 wt% ethanol to the CO_2 solvent transformed an unfavorable design into a competitive design. Note, however, that these results were obtained with a highly simplified model (i.e., the liquid in the fermentation broth was assumed to be pure water, and mass transfer effects were ignored). Extractions from fermentation broth are needed to quantify the modeling errors. Yet, this analysis is useful because it shows that SCE can be competitive for the recovery of high-value products at low production rates.

In summary, models for simulation and design optimization are beginning to bear fruit. Advances in thermodynamic modeling and algorithms for computing high-pressure phase equilibrium, coupled with new phase equilibrium data, are leading to improved models and more reliable designs.

V. SPECIAL TOPICS IN THE DESIGN OF SUPERCRITICAL EXTRACTION PROCESSES

While most advances in the design of SCE processes have focused on process development and parameter optimization, a few other approaches and observations are worthy of consideration.

Chimowitz and co-workers[46,47] have based an approach to process synthesis on the "retrograde phenomenon" associated with SCFs. The retrograde phenomenon or "temperature inversion effect" is a manifestation of the complex relationship between the solubility and the solvent temperature and density. At a fixed pressure, the solubility decreases with increasing temperature close to the critical point due to a decrease in the density of the solvent, but increases with increasing temperature far from the critical point due to the increased vapor pressure of the solute.

For the separation of mixtures of isomeric solids into pure species, Chimowitz and co-workers show that when the solubilities of solids in supercritical solvents at several tem-

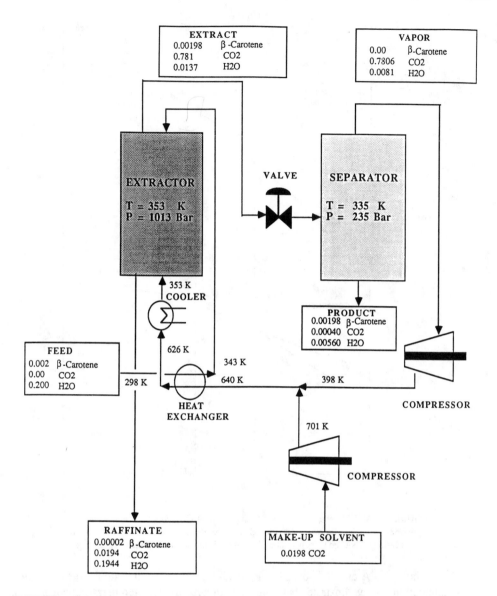

FIGURE 7. Dehydration of β-carotene with supercritical CO_2. Minimum annualized cost = \$19.48/kg. Flow rates are in kmol/h.

peratures are plotted as a function of the pressure, the isotherms intersect at "crossover points", as illustrated in Figures 8 and 9. Consequently, separation processes for solid mixtures can be designed in which the extractor and separator conditions lie between the crossover points of the two species. For example, consider a gas phase at a pressure P_o, intermediate to crossover points P_1^* and P_2^*, which is cooled from temperature T_H to T_L; i.e., the process from A to B in Figure 9. Note that the solubility of component 2 increases, while the solubility of component 1 decreases. Hence, a process can be synthesized in which a mixture of components 1 and 2 is extracted at P_o and T_H, and separated by a simple temperature decrease to T_L. This approach can be extended to multicomponent mixtures. Figure 10 shows a flow sheet for a process to isolate A, B, and C, and Figure 11 shows the accompanying isotherms. The process has two parallel isobaric cascades with temperature cycling in each one. In the first cascade, the temperature cycles between T_H and T_L to isolate component C at a fixed pressure, P_2, and in the second cascade, the same temperature cycle

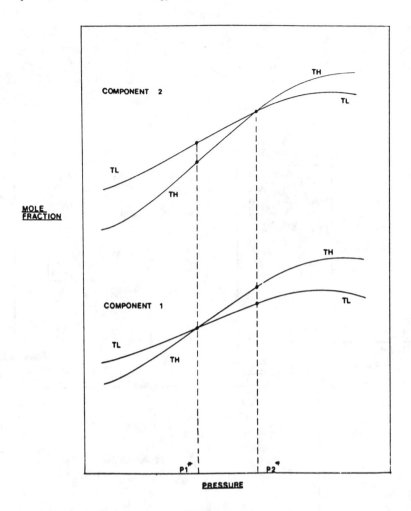

FIGURE 8.   Mole fractions of solutes in supercritical CO_2. Isotherm intersections are "crossover points". (From Chimowitz, E. H. and Pennisi, K. J., *AIChE J.*, 32(10), 1665, 1986. Reproduced by permission of the American Institute of Chemical Engineers.)

is used to isolate pure A and B at pressure P_1. Although these studies stop short of detailed design calculations, they illustrate that solubility inversions with temperature can provide potentially useful designs.

Cygnarowicz and Seider[48] have shown the impact of the retrograde effect on the optimization of a process to dehydrate acetone using supercritical CO_2. When minimizing the utility costs, a local minimum was found, which was related to the retrograde effect. Figure 12 shows how moles of acetone extracted vary with the temperature and pressure for a solvent-to-feed ratio of 8, and a feed flow rate of 100 mol/min (10% acetone, 90% water). Note that nearly 100% of the acetone is recovered near the critical point of CO_2 (approximately 307 K). The recovery decreases with increasing temperature until it reaches a minimum near 316 K, beyond which it increases, in a pattern consistent with the retrograde effect. Thus, when high acetone recoveries are specified, Newton-based methods converge to either the "high-temperature" or "low-temperature" solution, depending on the initialization.

As expected, since SCE processes operate in the critical region, they are particularly sensitive to uncertainties in phase equilibria and PVT data. Enthalpy and entropy data are needed to estimate heating, cooling, and refrigeration loads, but these are derivative prop-

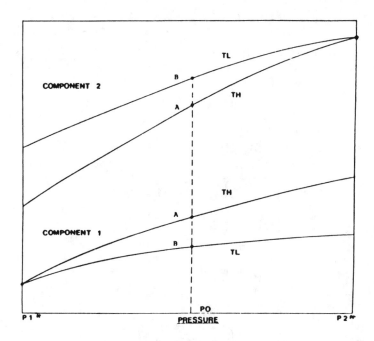

FIGURE 9. Crossover region. (From Chimowitz, E. H. and Pennisi, K. J., *AIChE J.*, 32(10), 1665, 1986. Reproduced by permission of the American Institute of Chemical Engineers.)

erties, and significant errors may be introduced when they are predicted using an EOS. To investigate this, Todd and Howat[49] considered the separation of butadiene from isoprene using supercritical trifluoromethane. This solvent has a "convenient" critical point, and volumetric data in the critical region and vapor pressure data are available. They compared two methods to estimate the heating, cooling, and compression loads for the process. The method based upon PVT data was considerably more accurate than that utilizing an EOS. It was found that the EOS predictions deviate significantly from the PVT predictions at high pressures, and these results suggest that economic analyses are sensitive to errors in the enthalpies and entropies.

VI. TRANSIENT BEHAVIOR AND CONTROL OF SUPERCRITICAL EXTRACTION PROCESSES

The previous sections have emphasized pilot plant studies, simulation, and optimizaton in the design of SCE processes that operate continuously and in the steady state. The design problem is complicated by modeling difficulties and the desire to operate in the critical region, where small changes in T and P can shift the extractor or separator from two to three or just one phase; however, solution of the tightly constrained design optimization does not account for the difficulties in maintaining steady operation in the face of disturbances and uncertainties. The disturbances must normally be dealt with in the short term as feed conditions, heat duties, etc., fluctuate. In addition, uncertainties can occur on a longer time frame, often over weeks and months, and arise in the fouling of heat exchangers, changes in the sources of feedstock, changes in product demand, etc. Hence, SCE processes are prime candidates for dynamic resiliency analysis,[50] to check the controllability of potential designs, and flexibility analysis,[50] to ascertain their ability to operate over a broad range of uncertain parameters.

Unfortunately, thus far little has been done to check the controllability of potential SCE

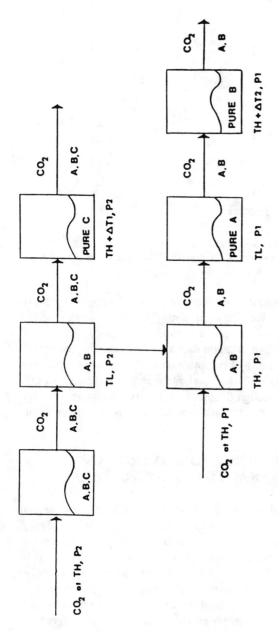

FIGURE 10. Process to isolate the species in a ternary solid mixture. (From Chimowitz, E. H. and Pennisi, K. J., *AIChE J.*, 32(10), 1665, 1986. Reproduced by permission of the American Institute of Chemical Engineers.)

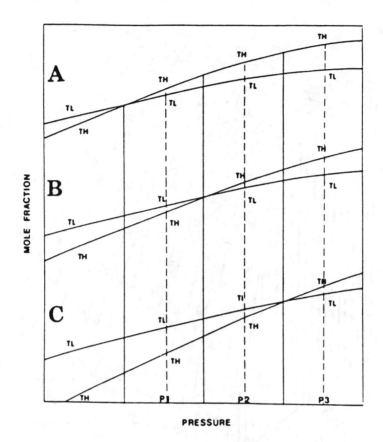

FIGURE 11.   Crossover regions for a ternary solid mixture. (From Chimowitz, E. H. and Pennisi, K. J., *AIChE J.*, 32(10), 1665, 1986. Reproduced by permission of the American Institute of Chemical Engineers.)

designs. One of the first control studies was recently reported by Cesari et al.[51] who studied the semi-batch extraction of ethanol from water and citral from lemon oil using supercritical CO_2. A dynamic model was prepared for the extractor, which included the mass and energy balances and the equilibrium equations, with proportional controllers used to maintain the extractor pressure and temperature by adjusting the product flow rate and the heat input to the extractor. Both the model and the physical system were sensitive to the controller actions with substantial changes causing the phases to coalesce. Such conventional controllers cannot predict the consequences of their actions, and therefore, cannot be expected to perform well in maintaining the proper phase distributions. Yet, the measurements by Cesari et al. of the vapor and liquid concentrations are in good agreement with the dynamic model, at least for the conditions considered (308 K and 17.2 MPa). Note that the extraction of citral from lemon oil was not studied experimentally.

This study represents an important first step in tackling the control problems associated with SCE processes. Unfortunately, it is a poor application of proportional control because it is applied to a semi-batch process. Such processes are better regulated with optimal control strategies, in which the model plays an important role.

Cygnarowicz and Seider[44] have also performed a dynamic analysis of a process to recover β-carotene using supercritical CO_2. The analysis focused on the product separator, and addressed the question of whether conventional (i.e., proportional (P) and proportional-integral (PI)) controllers can maintain acceptable product quality. The dynamic model included a control loop in which the vapor flow rate was manipulated to maintain the desired

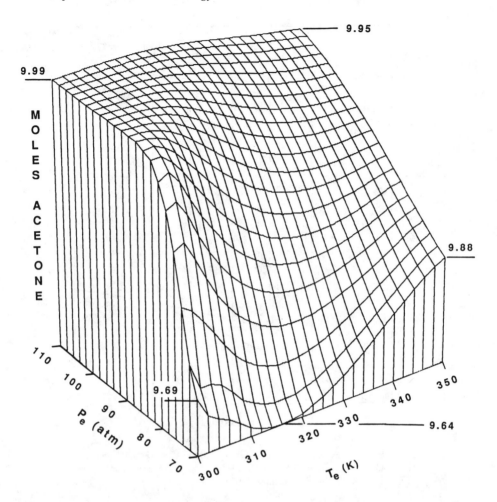

FIGURE 12. Moles of acetone extracted as a function of temperature and pressure. The solvent to feed ratio = 8. (Reprinted with permission from Cygnarowicz, M. L. and Seider, W. D., *Ind. Eng. Chem. Res.*, 28(10), 1497, 1989. © 1989 American Chemical Society.)

pressure. Although the servo response (i.e., response to a set-point change) was acceptable, the ability of the controller to reject disturbances was poor. A 10% decrease in the feed flow rate caused the product quality to drop below its design specification. The product specifications remained satisfied when the separator was overdesigned (i.e., when the separator pressure was increased), but the overdesign increased the annualized cost. A potentially better alternative to satisfy specifications while rejecting disturbances without overdesign is the implementation of advanced control strategies.

In this regard, for processes that are difficult to control, whether batch, semi-batch, or continuous, the process model is becoming an important element of more effective control strategies. New model predictive controllers (MPC),[52] implemented in high-speed computer workstations, are providing more effective disturbance rejection and set-point response in tightly constrained operating regions. As models for the design of SCE processes improve, it can be expected that they will be incorporated in the design of model predictive controllers. This should yield designs that can be operated reliably, closer to the economic optimum in tightly constrained regions.

In yet another example, Sunol et al.[53] developed a process that is a hybrid of adsorption and extraction, called "supercritical exsorption", in which the solute is removed from a liquid phase by simultaneous extraction with a supercritical solvent and adsorption on a porous solid bed. A dynamic model was developed for exsorption, consisting of four coupled, nonlinear partial differential equations with algebraic constraints, and was applied to simulate the dehydration of phenol. The model predicts that exsorption is superior to either extraction or adsorption in the efficiency and time for phenol removal, and experimental measurements are in qualitative agreement. Supercritical exsorption is more efficient since the adsorbant adds capacity for solute removal, and the high diffusivity of the supercritical solvent promotes rapid mass transfer from the solvent to the adsorbant. More experimental and theoretical work are needed to describe the mass transfer mechanism, but the preliminary results are promising.

VI. CONCLUSIONS

Most design studies have been oriented toward finding feasible SCE processes. Although experimentation has been indispensable in the development of new SCE processes, improvements in modeling and computational techniques are beginning to increase the role of simulation and optimization in SCE design studies. Furthermore, as advances in model predictive control permit more reliable operation, closer to the economic optimum, MPC can be expected to be applied and to increase the reliability and the profitability of SCE processes.

NOMENCLATURE

a	Mixture parameter in the Peng-Robinson EOS
a_i, a_j	Pure component parameters for Peng-Robinson EOS
D	Diameter
E	Enhancement factor
$F_{CO_2}^{rec}$	Flow rate of recirculated solvent
HTU_{oc}	Height of a transfer unit
k_{ij}, k_{ji}	Interaction coefficients for the modified Peng-Robinson EOS
M_{dc}	Equilibrium distribution coefficient
N	Number of equilibrium stages
NTU_{oc}	Number of transfer units
P	Pressure
P^*	Crossover pressure
P_d	Distillation tower pressure
P_e	Extractor pressure
T	Temperature
T_e	Extractor temperature
T_H	High temperature
T_L	Low temperature
U_C	Superficial velocity of the continuous phase
U_D	Superficial velocity of the dispersed phase
x_f	Mass fraction of alcohol in the solvent
x_i	Mole fraction of component i in the modified Peng-Robinson EOS
x_r	Mass fraction of alcohol in the raffinate
Y_{binary}	Mole fraction of solute in CO_2
$Y_{ternary}$	Mole fraction of solute in CO_2-cosolvent mixtures
Z	Contact height, feed mole fraction

REFERENCES

1. **Zosel, K.,** Process for the Decaffeination of Coffee, U.S. Patent 4,247,580, 1981.
2. **Vollbrecht, R.,** Extraction of hops with supercritical CO_2, *Chem. Ind.,* 19, 397, 1982.
3. **Hubert, P. and Vitzthum, O. G.,** Fluid extraction of hops, spices, and tobacco, with supercritical gases, *Agnew. Chem. Int. Ed. Eng.,* 17, 710, 1978.
4. **Krukonis, V. J.,** Processing with supercritical fluids: overview and applications in *Supercritical Fluid Extraction and Chromatography: Techniques and Applications,* Charpentier, B. A. and Sevenants, M. R., Eds., American Chemical Society, Washington, D.C., 1988, chap. 2.
5. **Lack, E. and Marr, R.,** Estimation of the process parameter for high-pressure (supercritical fluid) carbon dioxide extraction of natural products, *Sep. Sci. Technol.,* 23, 63, 1988.
6. **Marentis, R. T.,** Steps to developing a commercial supercritical carbon dioxide processing plant, in *Supercritical Fluid Extraction and Chromatography: Techniques and Applications,* Charpentier, B. A. and Sevenants, M. R., Eds., American Chemical Society, Washington, D.C., 1988, chap. 7.
7. **Novak, R. A. and Robey, R. J.,** Supercritical fluid extraction of flavoring material: design and economics, in *Supercritical Fluid Science and Technology,* Johnston, K. P. and Penninger, J. M. L., Eds., American Chemical Society, Washington, D.C., 1989, 511.
8. **Eggers, R.,** Large-scale industrial plant for extraction with supercritical gases, *Agnew. Chem. Int. Ed. Engl.,* 17, 751, 1978.
9. **Koerner, J. P.,** Design and construction of full-scale supercritical gas extraction plants, *Chem. Eng. Prog.,* 81(4), 63, 1985.
10. **Herderer, H. and Heidemeyer, H.,** Process design in high pressure extraction with supercritical gases in chemical plant construction, *Chem. Ing. Technol.,* 56, 418, 1984.
11. **Eggers, R. and Sievers, U.,** Current state of extraction of natural materials with supercritical fluids and developmental trends, in *Supercritical Fluid Science and Technology,* Johnston, K. P. and Penninger, J. M. L., Eds., American Chemical Society, Washington, D.C., 1989, 478.
12. **Marentis, R. T. and Vance, S. W.,** Selection of components for commercial supercritical fluid food processing plants, in *Supercritical Fluid Science and Technology,* Johnston, K. P. and Penninger, J. M. L., Eds., American Chemical Society, Washington, D.C., 1989, 525.
13. **Brunner, G.,** Extraction of caffeine from coffee with supercritical solvents, in *Proc. Int. Symp. Supercritical Fluids,* Perrut, M., Ed., Societe Francaise de Chimie, Nice, France, 1988, 691.
14. **Brunner, G.,** Mass Transfer in Gas Extraction, paper presented at AIChE Annual Meeting, San Francisco, 1984.
15. **Friedrich, J. P. and List, G. R.,** Characterization of soybean oil extracted by supercritical carbon dioxide and hexane, *J. Agric. Food Chem.,* 30, 192, 1982.
16. **Friedrich, J. P., List, G. R., and Heakin, A. J.,** Petroleum-free extraction of oil from soybeans with supercritical CO_2, *J. Am. Oil Chem. Soc.,* 59, 7, 288, 1982.
17. **Christianson, D. D., Friedrich, J. P., List, G. R., Warner, K., Bagley, E. E., Stringfellow, A. C., and Inglett, G. E.,** Supercritical fluid extraction of dry-milled corn germ with carbon dioxide, *J. Food Sci.,* 49, 229, 1984.
18. **King, J. W., Eissler, R. L., and Friedrich, J. P.,** Supercritical fluid-adsorbate-adsorbent systems, in *Supercritical Fluid Extraction and Chromatography: Techniques and Applications,* Charpentier, B. A. and Sevenants, M. R., Eds., American Chemical Society, Washington, D.C., 1988, chap. 4.
19. **Stahl, E., Schutz, E., and Mangold, H. K.,** Extraction of seed oils with liquid and supercritical carbon dioxide, *J. Agric. Food Chem.,* 28, 1153, 1980.
20. **Bulley, N. R., Fattori, M., Meisen, A., and Moyls, L.,** Supercritical fluid extraction of vegetable oil seeds., *J. Am. Oil Chem. Soc.,* 61, 8, 1362, 1984.
21. **Eggers, R.,** On the situation of continuous extraction of solids by means of supercritical gases, in *Proc. Int. Symp. Supercritical Fluids,* Perrut, M., Ed., Societe Francaise de Chimie, Nice, France, 1988, 595.
22. **Moses, J. M., Goklen, K. E., and de Filippi, R. P.,** Pilot Plant Critical-Fluid Extraction of Organics from Water, Paper 127c, AIChE Annual Meeting, Los Angeles, 1982.
23. **Seibert, A. F. and Moosberg, D. G.,** Performance of spray, sieve-tray and packed contractors for high pressure extraction, *Sep. Sci. Technol.,* 23, 2049, 1988.
24. **Seibert, A. F., Moosberg, D. G., Bravo, J. L., and Johnston, K. P.,** Spray, sieve-tray and packed high pressure columns-design and analysis, in *Proc. Int. Symp. Supercritical Fluids,* Perrut, M., Ed., Societe Francaise de Chimie, Nice, France, 1988, 561.
25. **Lahiere, R. J. and Fair, J. R.,** Mass-transfer efficiencies of column contactors in supercritical extraction service, *Ind. Eng. Chem. Res.,* 26, 2086, 1987.
26. **Bohm, F., Heinisch, R., Peter, S., and Weidner, E.,** Design, construction and operation of a multipurpose plant for commercial supercritical gas extraction, in *Supercritical Fluid Science and Technology,* Johnston, K. P. and Penninger, J. M. L., Eds., American Chemical Society, Washington, D.C., 1989, 499.

27. **Mathias, P. M. and Copeman, T. W.,** Extension of the Peng-Robinson equation of state to complex mixtures: evaluation of the various forms of the local composition concept, *Fluid Phase Equilibria,* 13, 91, 1983.

28. **Whiting, W. B. and Prausnitz, J. M.,** Equations of state for strongly nonideal fluid mixtures: application of local compositions toward density-dependent mixing rules, *Fluid Phase Equilibria,* 9, 119, 1982.

29. **Skjold-Jorgensen, S.,** Gas solubility calculations II. Application of a new group contribution equation of state, *Fluid Phase Equilibria,* 16, 317, 1984.

30. **Skjold-Jorgensen, S.,** Group contribution equation of state (GC-EOS): a predictive method for phase equilibrium computations over wide ranges of temperatures and pressures up to 30 MPa, *Ind. Eng. Chem. Res.,* 27, 110, 1988.

31. **Panagiotopoulos, A. Z. and Reid, R. C.,** New mixing rule for cubic equations of state for highly polar, asymmetric systems, in *Equations of State: Theories and Applications,* Robinson, R. L. and Chao, K. C., Eds., American Chemical Society, Washington, D.C., 1986, 571.

32. **Panagiotopoulos, A. Z. and Reid, R. C.,** High-pressure phase equilibria in ternary fluid mixtures with a supercritical component, in *Supercritical Fluids: Chemical and Engineering Applications,* Paulaitis, M. E. and Squires, T., Eds., American Chemical Society, Washington, D.C., 1987, chap. 10.

33. **Michelsen, M. L.,** Calculation of phase envelopes and critical points for multicomponent mixtures, *Fluid Phase Equilibria,* 4, 1, 1980.

34. **Michelsen, M. L.,** The isothermal flash problem. I. Stability analysis, *Fluid Phase Equilibria,* 9, 1, 1982.

35. **Michelsen, M. L.,** The isothermal flash problem. II. Phase split calculation, *Fluid Phase Equilibria,* 9, 21, 1982.

36. **Michelsen, M. L.,** Calculation of critical points and phase boundaries in the critical region, *Fluid Phase Equilibria,* 16, 57, 1984.

37. **Joshi, D. K. and Prausnitz, J. M.,** Supercritical fluid extraction with mixed solvents, *AIChE J.,* 30, 3, 522, 1984.

38. **Wong, J. M. and Johnston, K. P.,** Solubilization of biomolecules in carbon dioxide based supercritical fluids, *Biotechnol. Prog.,* 2 (1), 29, 1986.

39. **Dobbs, J. M., Wong, J. M., Lahiere, R. J., and Johnston, K. P.,** Modification of supercritical fluid phase behavior using polar cosolvents, *Ind. Eng. Chem. Res.,* 26, 56, 1987.

40. **Brignole, E. A., Anderson, P. M., and Fredenslund, Aa.,** Supercritical fluid extraction of alcohols from water, *Ind. Eng. Chem. Res.,* 26, 254, 1987.

41. **Andersen, P. M. and Fredenslund, Aa.,** Process simulation with advanced thermodynamic models, in *Proc. Chem. Eng. Fund., XVIII Cong.,* Sicily, 1987.

42. **Cygnarowicz, M. L. and Seider, W. D.,** Optimal design of supercritical extraction processes, in *Proc. Int. Symp. Supercritical Fluids,* Perrut, M., Ed., Societe Francaise de Chimie, Nice, France, 1988, 193.

43. **Moricet, M.,** Simulation of multistage separation by supercritical gases, in *Proc. Int. Symp. Supercritical Fluids,* Perrut, M. Ed., Societe Francaise de Chimie, Nice, France, 1988, 177.

44. **Cygnarowicz, M. L. and Seider, W. D.,** Design and control of a process to extract β-carotene with supercritical carbon dioxide, *Biotechnol. Prog.,* 6(1), 82, 1990.

45. **Cygnarowicz, M. L., Maxwell, R. J., and Seider, W. D.,** Equilibrium solubilities of beta carotene in supercritical carbon dioxide, *Fluid Phase Equilibria,* 59, 57, 1990.

46. **Chimowitz, E. H. and Pennisi, K. J.,** Process synthesis concepts for supercritical gas extraction in the crossover region, *AIChE J.,* 32(10), 1665, 1986.

47. **Chimowitz, E. H. and Kelley, D.,** Experimental separations in the crossover region of a supercritical fluid, in *Proc. Int. Symp. Supercritical Fluids,* Perrut, M., Ed., Societe Francaise de Chimie, Nice, France, 1988, 845.

48. **Cygnarowicz, M. L. and Seider, W. D.,** Effect of the retrograde phenomena on the design optimization of supercritical extraction processes, *Ind. Eng. Chem. Res.,* 28(10), 1497, 1989.

49. **Todd, J. N. and Howat, C. S.,** Sensitivity of supercritical fluid extraction process design to phase equilibria and PVT data, in *Proc. Int. Symp. Supercritical Fluids,* Perrut, M., Ed., Societe Francaise de Chimie, Nice, France, 1988, 185.

50. **Grossman, I. E. and Morari, M.,** Operability, resiliency, and flexibility—process design objectives for a changing world, *Proc. 2nd Int. Conf. on Found. Comp.-Aided Proc. Des.,* CACHE, 1984, 937.

51. **Cesari, G., Fermeglia, M., Kikic, I., and Policastro, M.,** A computer program for the dynamic simulation of a semi-batch supercritical fluid extraction process, *Comput. Chem. Eng.,* 13(10), 1175, 1989.

52. **Brengel, D. D. and Seider, W. D.,** A multi-step nonlinear model predictive controller, *Ind. Eng. Chem. Res.,* 28(12), 1812, 1989.

53. **Sunol, A. K., Shojaei, S., Bracey, W., and Akman, U.,** A new unit operation: supercritical exsorption, in *Proc. Int. Symp. Supercritical Fluids,* Perrut, M., Ed., Societe Francaise de Chimie, Nice, France, 1988, 603.

Chapter 12

MICROEMULSIONS IN NEAR-CRITICAL AND SUPERCRITICAL FLUIDS

Eric J. Beckman,* John L. Fulton, and Richard D. Smith

TABLE OF CONTENTS

---

* Present address: Chemical Engineering Department, University of Pittsburgh, Pittsburgh, Pennsylvania 15261.

I. INTRODUCTION

Although the discovery that micelles and microemulsion phases could be stabilized in supercritical fluid (SCF) solutions was made in the authors' laboratory less than 5 years ago, our knowledge of these systems has progressed rapidly. The combination of the unique properties of SCFs (e.g., viscosity, diffusion rates, solvent properties, etc.) with those of a dispersed microemulsion (or reverse micelle) phase creates a whole new class of solvents. The microemulsion phase adds to the properties of the SCF what amounts to a second solvent environment, which is highly polar and which may be manipulated using pressure. This second phase can itself manifest a wide range of solvent properties. Although SCFs are very attractive for separation and reaction processes owing to their density-dependent properties, they are limited at moderate temperatures and pressure by their inability to appreciably solvate most moderately polar solutes and nearly all ionic materials.[1] The addition of a dispersed droplet phase (forming a microemulsion) provides a convenient means of solubilizing highly polar or ionic species into the low polarity environment of the SCF phase. Hence, the combination of supercritical solvents with microemulsion structures provides a new type of solvent with some unusual and important properties of potential interest to a range of technologies.

A. NEAR-CRITICAL AND SUPERCRITICAL FLUIDS

In the SCF region, where the fluid temperature and pressure are above those of the critical point, the properties of the fluid are uniquely different from either the conventional gas or liquid states. The primary interest of this chapter is microemulsion systems in which a near-critical fluid or SCF constitutes the continuous phase and the dispersed "droplets" consist of surfactant and, generally, water. The near-critical region ($T/T_c \gtrsim 0.75$), where the temperature is just slightly below the critical point temperature, is also of considerable interest for microemulsion formation because these liquids are still quite compressible due to their proximity to the critical point. Table 1 gives the critical pressures, temperatures, and densities of several fluids whose moderate critical temperatures make them attractive candidates for microemulsion formation in the near-critical or supercritical states. Microemulsions having SCF continuous phases have been reported to exist in most of these fluids.[2-6] Such microemulsions would be expected to have unusual properties since the physical properties of near-critical fluids and SCFs are intermediate between those of a normal liquid and a gas.[7,8] Diffusion coefficients are 10 to 100 times higher in SFCs than in liquids, which is a significant advantage for applications (e.g., reactions or separations) that depend upon transport processes. Conventional liquid microemulsion systems often contain nanometer-sized droplets with very narrow size distributions and diffusion rates perhaps 10 to 100 times lower than molecularly dispersed species.[9] The high diffusion rates of these droplets in near-critical fluids and SCFs offset the transport limiting effect due to size for these relatively large droplets.

A second advantage to forming microemulsions in SCFs is that the properties of the continuous phase can be readily controlled by manipulation of system pressure. The density of near-critical fluids and SCFs, in particular, is strongly dependent on the pressure of the system. Because the properties of the fluid phase in large part dictate the thermodynamic stability of the microemulsion, the phase behavior of the microemulsions formed in SCFs is also strongly dependent on the pressure of the system.[10] Indeed, this is more true for SCFs than liquids because the magnitude of the pressure effect or the phase behavior is related to the fluid compressibility.

B. MICROEMULSIONS

Microemulsions are clear, thermodynamically stable solutions generally containing water, a surfactant, and an oil (i.e., a nonpolar or low polarity fluid). The surfactant molecules

TABLE 1
Critical Parameters of Some Near-Critical and SCFs that have Moderate Critical Points Suitable for Microemulsion Formation

Compound	Critical temp (°C)	Critical pressure (bar)	Critical density (g/cm³)
CO_2	31.3	73.8	0.448
Ethane	32.3	48.8	0.203
Ethylene	9.2	50.4	0.218
Propane	96.7	42.5	0.217
Propylene	91.8	46.2	0.232
n-Butane	152.0	37.5	0.228
n-Pentane	196.6	33.3	0.232

form organized molecular aggregates and preferentially occupy the interface between the oil and water phases. One should note that thermodynamic stability implies nothing concerning the kinetics of microemulsion formation, and sufficient energy input (i.e., mixing) may be necessary to speed equilibration for systems that do not form spontaneously. The oil and water microdomains have characteristic structural dimensions between 2 and 100 nm. Aggregates of this size are poor scatterers of visible light and hence these solutions are optically clear. Water-in-oil (w/o) microemulsions (predominantly oil) and oil-in-water (o/w) microemulsions (mostly water) can have a multitude of different microscopic structures; sphere, rod, or disc-shaped structures and layered or bicontinuous structures are commonly attributed to specific systems. Figure 1 shows idealized structures for a few of the commonly invoked microemulsion structures. The existence of such structures is supported experimentally by various particle (i.e., photon, neutron, or X-ray) scattering techniques. Micelles are the smallest aggregate entity and are generally defined as binary systems containing a surfactant that forms organized "clusters" in either water (normal micelles) or in oil (reverse or inverted micelles). When small amounts of water are added to a reverse micelle system, the water preferentially partitions into the hydrophilic core, causing them to "swell" and yielding (at some arbitrary water content) a w/o microemulsion. At low water concentrations such inverted microemulsions typically contain spherical, nanometer-sized droplets of water dispersed in the oil continuous phase.

The shape, size, and structure of these dispersed droplets depends upon a multitude of variables including the surfactant type, ionic strength, the presence of cosurfactant(s), and the amount of water. Figure 2 shows some commonly used surfactants of the four general categories; anionic, cationic, nonionic, and zwitterionic surfactants. Cosurfactants are small amounts of somewhat amphiphilic substances, typically low molecular weight alcohols, having solubility in both oil and water, which greatly enhance the water- or oil-solubilizing capacity of the microemulsion. The amount of water added to a w/o microemulsion system is generally defined by the molar water-to-surfactant ration, W_m. When the amount of the dispersed aqueous phase exceeds about 26% of the total microemulsion volume a transition occurs from an inverted structure to what is postulated as a bicontinuous structure and finally to normal structures at aqueous volume fractions greater than 74%.[11] This type of transition is illustrated in Figure by the sequence a→c→e.

A universal property of surfactant solutions is the existence of a critical micelle concentration (CMC) representing the minimum amount of surfactant required to form aggregates. The CMC also necessarily represents the solubility of surfactant monomer in the oil or water continuous phase solvent. At surfactant concentrations above the CMC, surfactant monomers exist in equilibrium with the micellar aggregates (for a second surfactant phase). For normal micelle systems the CMC occurs at a very sharp transition, approximately 10^{-4}

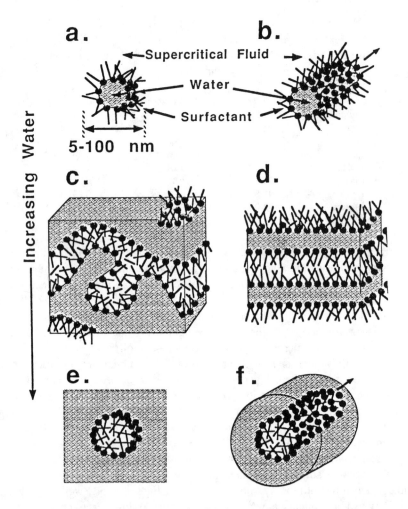

FIGURE 1. Microemulsion structure in oil and water solvents. At low water volume fractions a water-in-oil microemulsion exists in which possible structures such as (a) spherical or (b) rod-shaped aqueous regions are dispersed in the the oil-like solvent. At volume fractions of the aqueous phase between 26 to 74% a microemulsion consisting of (c) bicontinuous or (d) lamellar liquid crystalline phases is possible. For low oil volume fractions, an oil-in-water microemulsion can also contain either (e) spherical or (f) rod-shaped structures which in this case are filled with the oil-like solvent.

to 10^{-2} M, whereas for reverse micelle systems the CMC is less clearly defined because a more gradual transition is believed to occur, including an intermediate region containing premicellar aggregates.[12] The CMC in reverse-type systems is typically 10^{-4} to 10^{-3} M. For both normal and reverse micelle systems the CMC is dependent upon the type or density of the oil, the amount of water, the temperature, and pressure of the system. The CMC is related to the energy of formation[13] of the aggregate and therefore there is much interest in determining its value (and dependence upon pressure) for microemulsions formed in near-critical fluids and SCFs.

The study of microemulsions in SCFs is a relatively new area and there are still only relatively few publications that deal with this subject. These first studies have dealt with investigations of some simple properties of these systems such as phase behavior, surfactant solubility, and CMC determination. The sections that follow review what is currently known about microemulsions in fluids as has been determined from light scattering, spectroscopic, and phase behavior studies. Also discussed are the thermodynamic aspects of these systems,

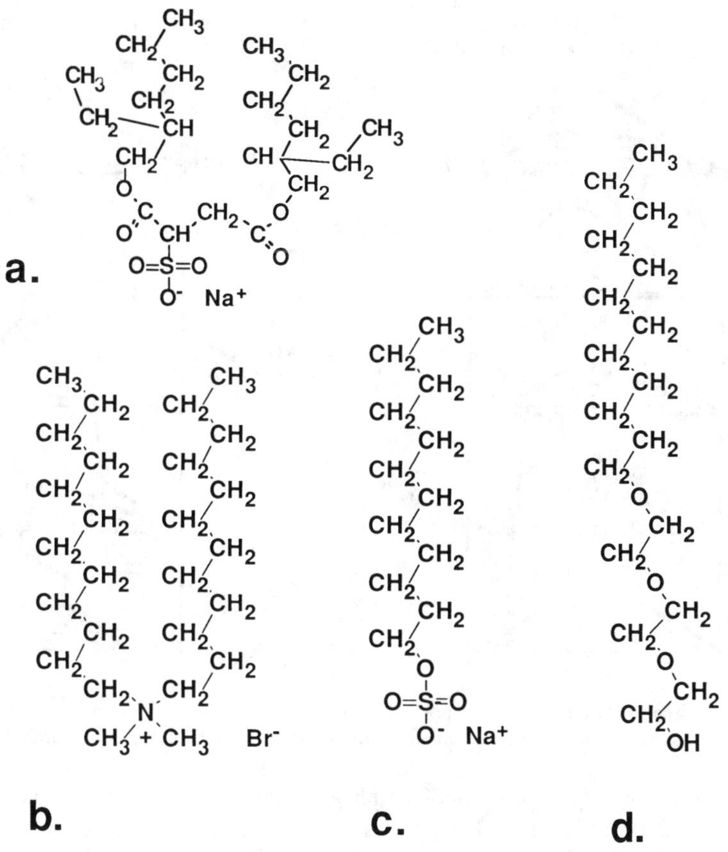

FIGURE 2. Structure of four commonly used surfactants: (a) AOT (sodium bis(2-ethylhexyl) sulfosuccinate), (b) DDAB (didodecyl dimethyl ammonium bromide), (c) SDS (sodium dodecyl sulfate), and (d) $C_{12}E_3$ (triethylene glycol dodecyl ether).

which set them apart from liquid systems, and the authors describe current modeling efforts to characterize the phase behavior of these complex systems.

II. PHENOMENOLOGICAL MODEL OF MICROEMULSION PHASE BEHAVIOR

Qualitative descriptions of the phase behavior of microemulsion systems can be constructed from a combination of phase rule principles and experimental results. General classes of phase behavior in these surfactant systems can be identified. For conventional liquid microemulsion systems, Kahlweit et al.[14-16] have described a phenomenological approach that relates the effects of composition and temperature to the phase behavior of oil/surfactant/water mixtures. In this approach, the phase behavior of the ternary system can be qualitatively predicted given knowledge of the properties of the surfactant/water and surfactant/oil binary phase diagrams. For example, the simplest system to describe is one that contains a single miscibility gap on the oil/water side of the ternary phase diagram. In this system, a one-phase region exists across the entire transition from the oil-rich to the water-rich side of the phase diagram. Such systems are often not realized in real ternary mixtures because of the existence of ''secondary'' phases, such as a second micelle phase or optically active liquid crystal phases. Even so, the general features of the phase behavior of many systems can often be successfully described using the phenomenological approach. Therefore, one can

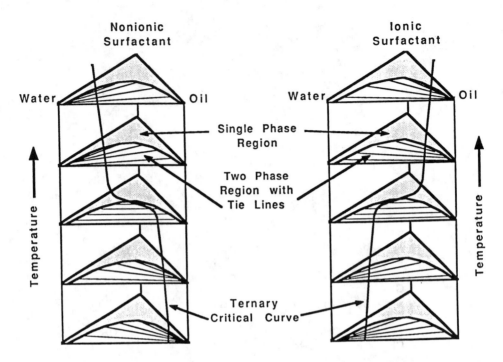

FIGURE 3.   Temperature-composition ''prisms'' for oil/surfactant/water systems in which a line connects the series of plait points of the liquid-liquid immiscibility regions of each ternary phase diagram.[15]

use this phenomenological description of phase behavior developed for liquid systems as a starting point to describe phase behavior of near-critical fluids and SCF systems.

For liquid ternary mixtures, a key to this description is the path of the critical line, which is the line connecting the series of liquid-liquid critical points illustrated in Figure 3. A series of triangular ternary phase diagrams, each at a different temperature but the same pressure, are stacked vertically to construct the prism shown in Figure 3. The most common ternary phase diagram contains a single region of immiscibility that touches the o/w axis. The tie lines within the miscibility gap represent the equilibrium concentrations on the two liquid phases. At the critical point, the two liquid phases become one. In Figure 4, the corresponding two-dimensional projections of the critical line onto the temperature/composition (oil and water) plane are shown. The characteristic shape of this critical curve reflects the phase behavior of the two classes of surfactant solutions: ionic and nonionic. Nonionic surfactants are generally miscible with water at lower temperatures, where they produce o/w microemulsions in equilibrium with an excess oil phase. Increasing temperature decreases the affinity of these surfactants for water (presumably due to a decrease in hydrogen bonding), which will eventually induce a phase inversion to form a w/o microemulsion in equilibrium with excess water.[17] This is shown graphically in Figure 4a where, at low temperature, the critical line appears on the oil-rich side of the phase diagram and curves across to the water-rich side as temperature is increased. As the temperature is decreased, the critical line will ultimately intersect the upper critical solution temperature (UCST) of the oil/surfactant binary. Likewise, by raising the temperature, the critical line will intersect the lower critical solution temperature (LCST) of the surfactant/water binary system. As can be seen in Figure 4c, the critical line of a generalized ionic surfactant mixture traces a path that is a mirror image of that for the nonionic surfactant system. At lower temperatures a w/o microemulsion in equilibrium with excess water is the preferred state. This inverts to an o/w microemulsion as temperature increases. It is important to note that the greatest solubilization of water (in the case of w/o microemulsions) or oil (for o/w microemulsions)

411

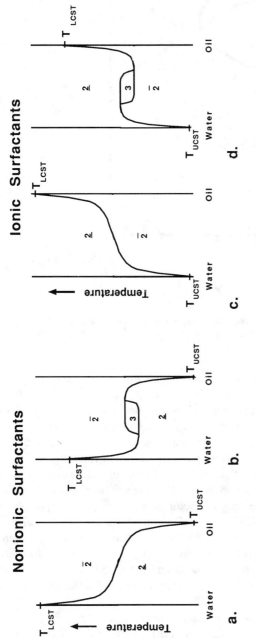

FIGURE 4. Projection of the critical curves given in Figure 3 onto the temperature/composition (water and oil) plane and the effect of changes in the LCST and UCST on the phase behavior of the ternary system.[15]

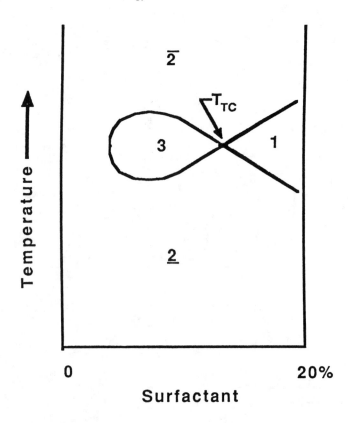

FIGURE 5. Typical "fish"-shaped phase diagram from a slice, at a water-to-oil ratio of 1:1, through a temperature/composition prism that contains a three-phase region.[15]

is observed to occur in the vicinity of the phase inversion temperature represented in Figures 4a and 4c by the temperature of the inflection point of the critical line.[14,18] The phase inversion temperature interval can be systematically shifted by changing the chain length of the oil, changing the surfactant hydrophobe structure, adding salt, or changing pressure.[14,15]

A complication to the phase behavior of ternary surfactant systems arises with the formation of three-phase regions. In Figure 4b, raising the UCST and/or lowering the LCST reduces the slope of the critical line at the inflection point. Ultimately, the slope drops to zero and the critical line breaks, indicating the formation of a three-phase region. The point of zero slope is a tricritical point, representing the point of emergence of a three-phase region. Changes in the UCST and the LCST are a reflection of changes to the oil and surfactant structures. For instance, if one increases the oil chain length, the oil/surfactant UCST will rise.[16] Furthermore if the number of ethylene oxide units in a C_iE_j surfactant (Figure 2) is reduced, the solubility in water is reduced and the LCST will drop. Using analogous methods, three-phase regions can also be constructed for ionic surfactant systems (see Figure 4d).

A phase diagram representing a vertical section through a temperature/composition prism, like the one shown in Figure 3, can be utilized describing the interrelationship of one-, two-, and three-phase regions. Variation of the surfactant concentration, while maintaining a constant o/w ratio at a value close to 1:1, will produce the characteristic "fish"-shaped diagram shown in Figure 5 for those systems that contain three-phase regions. Kahlweit[14-16] and others[19] have demonstrated that both ionic and nonionic surfactant systems will produce qualitatively similar phase diagrams. At the critical point (T_{TC}), the three liquid phases merge to one as shown in Figure 5.

A. EFFECT OF PRESSURE ON PHASE BEHAVIOR IN LIQUID SYSTEMS

The role of pressure in determining phase behavior in liquid systems is obviously far less dramatic than in SCF-based systems, but subtle and often significant effects can be observed. Within the framework of the above model, the effect of pressure on phase behavior of ternary surfactant-containing mixtures can be viewed as resulting from the pressure effects on the surfactant/oil and surfactant/water binary systems. This requires some assumptions regarding the nature of these binary phase diagrams. The effect of pressure on the phase behavior of aqueous solutions of both ionic and nonionic surfactants is relatively well documented,[15,16] whereas analogous data for oil-surfactant systems are relatively sparse. Liquid-liquid phase behavior for aqueous nonionic surfactants (of the C_iE_j type) usually takes the form of a closed-loop miscibility gap that shrinks (in temperature/composition space) as pressure increases.[16] The LCST thus shifts to higher temperature with increasing pressure. Although Schneider[20-22] has shown that the miscibility gap can reappear at higher pressure in liquid systems displaying the closed loop miscibility gap, insufficient high pressure data are available to make a reliable prediction for C_iE_j/water systems. The liquid-liquid phase behavior of the ionic surfactant/water system is observed to be of the UCST variety, or an upwardly pointing dome in temperature/composition space.[15] The critical point moves to lower temperatures as pressure is increased,[15] thereby increasing the size of the one-phase region in temperature/composition space. Based upon limited data for the effect of pressure on the phase behavior of the oil/surfactant binary system, some qualitative predictions can be made. For nonionic systems, the phase behavior of the oil/surfactant system is assumed to be of the UCST type, for which there is ample justification in liquids.[14,16] Increasing pressure is said to raise the critical point of a nonionic surfactant/oil mixture, increasing the size of the miscibility gap. For ionic surfactant/oil mixtures, a UCST exists at very low temperatures, and usually does not influence the phase behavior of the ternary; an LCST is observed at higher temperatures. There are too few data on the phase behavior of ionic surfactants in oils at high pressure to yet draw meaningful conclusions as to the effect of pressure upon such mixtures.

1. Nonionic Surfactant Systems

Clearly, in the phenomenological model, the net effect of pressure on the ternary oil/nonionic surfactant/water phase behavior will depend on both the direction of the pressure-induced shifts of the oil-surfactant and water-surfactant binary critical points and the relative magnitudes of these two shifts. This latter characteristic will in large part depend on the relative compressibilities of the two binary solutions. For example, in the C_4E_2-water-phenylalkane system, Kahlweit[14] observed that both the C_4E_2/water (an LCST) and the C_4E_2/phenylalkane (a UCST) binary critical points are shifted to higher temperatures as pressure is increased, the shift of the C_4E_2/oil UCST being the greater. For this system the predicted net result is that the phase inversion region, or the three-phase interval, will widen and be shifted to higher temperatures, which is indeed observed. In such a case, if at a temperature where an oil-in-water microemulsion and excess oil exist in equilibrium, an increase in pressure will shift the three-phase region away from the ambient conditions resulting in an increase in oil solubilization of the microemulsion phase. If, on the contrary, the ambient conditions produce a w/o microemulsion and excess water, increasing pressure will first decrease water solubilzation as the three-phase region shifts upward in temperature, but will eventually induce an inversion to an o/w microemulsion and excess oil. Curves in pressure-temperature space at constant solubilized oil or water content will be positively sloped for this type of system.

2. Ionic Surfactant Systems

Kahlweit[15] also measured the effect of pressure on the phase behavior of systems based on an ionic surfactant, that of the decane/AOT/water/NaCl quaternary. It was suggested that the UCST of the AOT/water binary moves to lower temperatures as pressure increases, whereas no results are reported for the AOT/decane mixture. Nevertheless, the net effect of pressure is to shift the three-phase region of the quaternary mixture to lower temperatures, implying that either the magnitude of the pressure effect on the water/AOT binary is the limiting factor or that the shift of the relevant conventional AOT/decane and AOT/water miscibility gaps exhibit the same sign. Therefore, for liquid ternary mixtures at a temperature where a w/o microemulsion is formed with an excess water phase, increases in pressure will decrease the water-solubilization capacity of the microemulsion phase (w_{max}), and will eventually cause phase inversion to an o/w microemulsion along with an excess oil phase. The initial increase in solubilization comes as the phase inversion point is shifted closer to the ambient temperature by the pressure increase. For the same reasons, under conditions where o/w water microemulsions are formed, more oil will be solubilized in the microemulsion as pressure increases. Curves in the pressure-temperature space at constant solubilization will be negatively sloped for this system.

In the simplest case, the effect of pressure is effectively to shift the characteristic phase diagrams for ionic and nonionic surfactants vertically along the temperature axis; hence, the effect of pressure is similar to the effect of added salt on the microemulsion phase behavior.[16] An effect due to salt concentration in a particular system can consequently be counteracted or magnified by increases in pressure, depending on the specifics of the pressure effect for the system in question.

Results by a number of researchers confirm the qualitative trends outlined above. Fotland[23] examined the phase behavior of a five-component [sodium dodecyl sulfate (SDS)/cyclohexane/ water/butanol/NaCl] microemulsion system at 20°C and at pressures up to 500 bar. When compositions were employed such that three phases formed at atmospheric conditions, a transition to a two-phase o/w microemulsion in equilibrium with excess oil was observed as pressure increased. Likewise, a one-phase system at atmospheric pressure was also transformed to an o/w microemulsion/excess oil two-phase system at higher pressure. Interestingly, at pressures above approximately 350 to 370 bar, a solid precipitate was formed, which was assumed by Fotland to be crystalline SDS with some associated salt.

Kim et al.[24] measured the location of the critical point as a function of temperature and pressure for a decane/AOT/water microemulsion. At a constant composition near the critical point, the critical temperature was found to decrease with increasing pressure, again consistent with a shift of the critical curve to lower temperatures with increasing pressure.[15] Kim et al.[25] also measured the phase behavior of five-component microemulsions (using the monoethanolamine salt of dodecyl-o-xylene sulfonate in the decane/surfactant/water/t-amyl alcohol/NaCl system) at pressures up to 250 bar and temperatures from 25 to 45°C, where the o/w ratio was approximately 1. Again, a three-phase microemulsion system at 1 bar is transformed to an o/w microemulsion and excess oil as pressure increases and, as suggested by the phenomenological model, lines of constant solubilization in temperature-pressure space exhibit a negative slope.

B. PHASE BEHAVIOR IN NEAR-CRITICAL AND SUPERCRITICAL FLUIDS

The first report of reverse micelles and microemulsions in SCF systems, based on the anionic surfactant AOT, was that of Gale et al. in 1987. Since that time several additional groups, in addition to the authors' laboratory, have begun studies of these or related systems. The present review calls upon all such work.

Within the framework of the phenomenological model described above, replacement of a liquid oil component by an SCF will affect the phase behavior of the ternary mixture insofar as it changes the relative positions of the oil/surfactant and water/surfactant critical points. For example, Kahlweit[15] predicts that decreasing the size of the oil molecule will raise the UCST of an oil/ionic surfactant binary mixture. Thus, in oil/ionic surfactant/water ternary mixtures, where the oil/surfactant UCST is usually so low that an electrolyte must be added in order to observe three-phase behavior, use of a low molecular weight SCF may allow study of three-phase behavior in simple ternary mixtures at room temperature.

The effect of pressure on the miscibility gaps may also be different for SCF/surfactant mixtures than for conventional liquid/surfactant mixtures. First, the SCF/surfactant binary solution is much more compressible than the water/surfactant binary. Unlike the liquid systems examined by Kahlweit,[14] the pressure-induced shift of the surfactant/oil binary could very well play the dominant role in determining the shift of the three-phase region within the temperature/composition prism. Second, increasing pressure could shift the UCST of a surfactant/SCF solution to lower temperatures, rather than the reverse situation observed by Kahlweit[14] for C_iE_j/liquid alkane mixtures. Although there are little data available on the phase behavior of C_iE_j-type materials in SCF, Zeman and Patterson[26] have shown that the UCST of propylene oxide oligomer/propane mixtures shifts to lower temperatures as pressure increases. Finally, the relevant liquid-liquid miscibility gap for a surfactant/SCF mixture may be of the LCST, rather than the UCST form.

The differences in the phase behavior of surfactant/SCF solutions as compared with analogous liquid mixtures may invalidate some of the generalizations of previous phenomenological models regarding pressure effects on the ternary oil/surfactant/water phase diagrams. For the case of nonionic surfactants, an SCF mixture displaying a UCST that drops as pressure increases will, in the ternary mixture, exhibit trends opposite to those described by Kahlweit[14] for the phenylalkane/C_4E_2/water system. Unfortunately, detailed studies of the phase behavior of SCF-surfactant mixtures are still quite few in number, making difficult *a priori* predictions (and generalizations) of ternary phase behavior using this model. Nevertheless, the phenomenological model can be used to explain many facets of SCF/surfactant/water phase behavior and provides a useful starting point in understanding these systems and their similarities and differences with conventional liquid systems.

1. Nonionic Surfactant Systems

Studies of nonionic surfactant-based microemulsions in SCF continuous phases are few, but a few reports of related systems have appeared. Ritter and Paulaitis[2] measured the phase compositions of two water/C_iE_j/CO_2 mixtures in the 3 and 4 phase regions at temperatures (25 to 50°C) and pressures (10 to 100 bar) where CO_2 exhibits large changes in compressibility. Multiple regions of three-phase behavior were observed. For example, at low temperature and pressure (<6 bar), a C_4E_1/water solution exists in equilibrium with essentially pure CO_2 vapor. Increasing the pressure induces a phase separation to a C_4E_1-rich, water-rich, and CO_2 vapor phase. Further increases in pressure lead to a four-phase system containing (1) a predominantly aqueous phase, (2) a CO_2 vapor phase, (3) a C_4E_1-rich phase containing 70% CO_2, and (4) a CO_2-rich liquid phase with 5% C_4E_1 and 4% H_2O. At pressures above the four-phase line, depending on the concentration, either three liquid phases or, at very low surfactant concentrations, dilute aqueous and CO_2 solutions plus a vapor phase will form. Eventually, increased pressure causes the surfactant and CO_2-rich phases to merge, leaving a CO_2/C_4E_1 solution in equilibrium with an excess water phase. Pressure and temperature limitations of the view cell apparatus prevented investigations at higher pressures where larger single-phase regions might exist. These observations suggest that as pressure increases, the behavior of the middle liquid phase involves a normal microemulsion that solvates an increasing amount of CO_2. Dye solubilization experiments on the four-phase regions revealed that the second liquid (middle) phase will absorb the bulk of either a nonpolar (Sudan III) or polar (Methyl Red) material. Binary mixtures of CO_2 and C_8E_3 exhibit three-phase liquid-liquid-gas behavior near the critical point of CO_2, unlike the C_4E_1 mixture, resulting in phase behavior for this ternary system that is quite different. This is not surprising, as the higher molecular weight and greater hydrophilicity of C_8E_3 vs.

C_4E_1 would tend to increase the extent of the miscibility gap in a CO_2 mixture. The effect of pressure on both water and Methyl Red solubilization by the middle phase was similar to that observed for the C_4E_1 system. These experiments leave many questions still unresolved regarding the possible existence of either normal or inverted microemulsion phases in these systems.

Beckman and Smith[27] examined the phase behavior of an alkane/C_iE_j/water/acrylamide system at pressures up to 500 bar in preparation for an investigation of the microemulsion polymerization of a water-soluble monomer within a supercritical continuous phase. The surfactant, an 80/20 blend of $C_{16}E_2$ and $C_{12}E_4$ materials, was chosen in order to conduct the polymerization in a one-phase region and over a potential temperature range of 25 to 70°C. Although the surfactant blend is soluble up to approximately 10% in either ethane or propane at high pressure, a simple binary mixture with ethane or propane solubilizes very little water at 25°C. Fortunately, the acrylamide monomer acts as a cosurfactant with the C_iE_j system in *n*-alkanes, allowing solubilization of up to 8 mol of water per mole of surfactant at 500 bar. Such cosurfactant behavior by acrylamide has also been observed by Candau and co-workers in C_iE_j/heptane[28] and AOT/toluene[29] mixtures. Unpublished observations from the authors' laboratories showed no increase in water solubilization due to added acrylamide in an AOT/propane mixture, indicating that the cosurfactant activity of acrylamide is dependent on the particular oil and surfactant under consideration. Cosurfactant behavior by acrylamide clearly suggests localization of the monomer in the interfacial region rather than in the aqueous core.

At the vapor pressure of propane (or ethane), the alkane/surfactant/water/acrylamide mixture displays two phases that merge at higher pressure to form one clear fluid phase. At the clearing point the microemulsion appears deep red in color (by transmitted light); further pressure increases change the color from red to orange to yellow while the microemulsion remains a single phase. These color changes are evidence for the existence of large structures that scatter visible light, and they perhaps represent the formation of large clusters of individual micelles or droplets.

The clearing pressure of the system is a function of the temperature, continuous phase composition, surfactant, and acrylamide concentrations. Figure 6 shows a series of cloud point curves in which the composition of the continuous phase is varied from pure propane to pure ethane. By replotting these data as continuous phase density-temperature curves (Figure 7), the effects of temperature and density can be separated. The stability of this microemulsion is likely governed by the interdroplet attractive interactions where the density (i.e., solvent strength of the fluid) should be a key parameter controlling the strength of the interactions, and hence density determines the phase behavior of the system.

As mentioned previously, Kahlweit[14] observed that an increase in pressure in the phenylalkane/C_4E_2/water system is effectively equivalent to changes observed for the phase diagram by going to higher temperatures. If it can be assumed that the acrylamide-based system behaves similarly, and that at room temperature this system is above the phase inversion zone, then indeed an increase in pressure would be predicted to stabilize the system by shifting the one-phase region (the "tail" area in Figure 5) toward higher temperature and moving away from the phase boundary.

The effect of surfactant composition can be seen in Figure 8, where the clearing pressure is plotted against the dispersed phase volume fraction (defined as the total volume of water, surfactant, and acrylamide divided by the view cell volume). The clearing pressure drops sharply as dispersed phase volume increases, approaching a constant value of approximately 305 bar. Returning to the generalized "fish"-shaped diagram in Figure 5, one can see that the results in Figure 8 resemble a portion of the one-phase "tail" region in Figure 7.

2. Ionic Surfactant Systems

The first report of the existence of microemulsions phases in SCFs[3] involved the use of

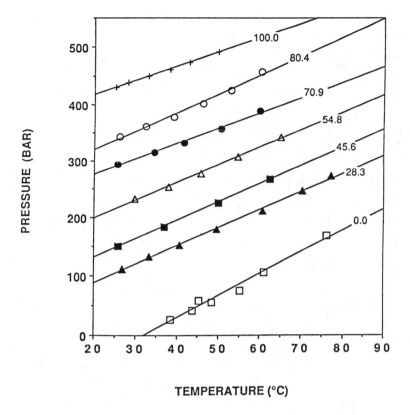

FIGURE 6.   Cloud point curves of Brij 52/Brij 30, 80/20 mixture with a water/surfactant ratio of 5.0, acrylamide/surfactant ratio of 1.0, total dispersed phase volume fraction of 0.136, and seven continuous phase ethane concentrations (wt%). (Reprinted with permission from Beckman, E. J. and Smith, R. D., *J. Phys. Chem.*, 94, 345, 1990. © 1990 American Chemical Society.)

the surfactant AOT in alkane/water systems. AOT is an attractive choice for near-critical fluids and SCF studies since the phase behavior of alkane/AOT/water has been extensively studied in the liquid alkanes and AOT readily forms w/o microemulsions as a simple ternary system. Such systems do not require the addition of a cosurfactant, which would increase the complexity of the systems.

Since the first report of microemulsion formation in SCF, subsequent studies have extensively examined the phase behavior of various near-critical fluids and SCF/AOT/water systems.[5,10] These studies revealed a strong dependence of the microemulsion phase behavior on the density or pressure of the near-critical fluid or SCF phase. For supercritical ethane at 37°C the amount of water that can be solubilized in a 150-mM AOT microemulsion varies from W_m = 3 at 250 bar up to W_m = 40 at 850 bar.[10] For a near-critical propane microemulsion, the effect of pressure is even more dramatic, since W_m = 5 at 10 bar and increases to W_m = 35 at 300 bar.[10] This is a somewhat surprising finding since liquid propane at 25°C is only moderately compressible, but it can be explained (as discussed later) by the attractive interactions between microemulsion droplets.

Other more recent studies of phase behavior include a work by Fulton et al.[5] and one by Steytler and co-workers[30] in which the effect of pressure on the water-solubilization capacity of AOT microemulsions in a variety of n-alkanes was determined. Steytler measured the pressure-temperature cloud point of AOT solutions at constant water content and found that for the lower alkanes (C_2, C_3, and C_4), the cloud point temperatures increased with increasing density or pressure (see Figure 9). As the oil carbon number increases above 4,

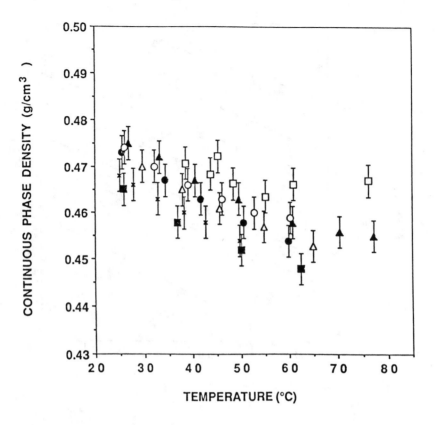

FIGURE 7. Data from Figure 6 replotted as density of continuous phase vs. cloud point temperature; symbols same as in Figure 6. (Densities for pure ethane and propane were taken from the literature; those for the 80.4/19.6 mixture were measured, and those for the other mixtures were calculated using the Starling variant of the Benedict-Webb-Rubin EOS with literature values for the ethane and propane parameters). (Reprinted with permission from Beckman, E. J. and Smith, R. D., *J. Phys. Chem.*, 94, 345, 1990. © American Chemical Society.)

the relationship is reversed; the cloud point temperature decreases with increasing density. In a related study, Fulton et al. determined the maximum water-to-surfactant ratio, W_o, for various alkanes from C_1 to C_8. In addition, they looked at the noble gases, krypton and xenon, at room temperature and various pressures. These data, shown in Figure 10, are plotted in terms of the dielectric constant of the continuous phase solvent. Fulton et al. observed a maximum in W_o occurring at dielectric constants equivalent to a C_4 or C_5 alkane, the same point at which Steytler observed a maximum in the cloud point curves. The results shown in Figure 10 correlate the behavior of not only alkanes (ranging down to methane!), but also that of krypton and xenon. Within the framework of the phenomenological model, these observations suggest a change in the nature of the AOT/alkane binary phase diagram in the C_4 to C_5 region which manifests itself as a change in the ternary system. Fulton et al. have determined that on a molecular level at least part of the observed phase behavior can be ascribed to strong interdroplet attractive interactions in a model that is discussed in more detail in the following section.

Figure 11a shows the temperature/composition prism for the ethane/AOT/water system; it is analogous to the general temperature/composition prisms given in Figure 3 except that the walls of the prism are unfolded so as to examine the temperature/composition phase behavior of the three binary systems: AOT/ethane, AOT/water, and ethane/water. Also shown is the effect of pressure on the position of the phase boundary for the AOT/ethane binary

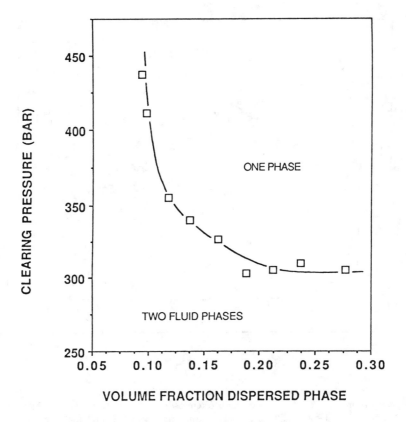

FIGURE 8. Clearing pressure (cloud point) of Brij 52/Brij 30 in a 80.4/19.6 w/w ethane-propane with a water/surfactant ratio of 5.0 and an acrylamide/surfactant ratio of 1.0 vs. dispersed phase volume fraction. (Reprinted with permission from Beckman, E. J. and Smith, R. D., *J. Phys. Chem.*, 94, 345, 1990. © 1990 American Chemical Society.)

system. (The binary AOT/ethane phase diagram was mapped using the experimental techniques given in Reference 10.) The AOT/water data are from Reference 31 and are for 1 bar; however, the effect of pressure on solubility of AOT in water over the range of pressure from 1 to 500 bar is expected to result in only small changes in the UCST of this solution. Likewise, the solubility of water in ethane or of ethane in water is less than 0.2% for the range of conditions given in Figure 11. As discussed at the beginning of this section, it is the path that the critical line takes from the UCST or LCST of the surfactant/water wall across the prism to the LCST or UCST of the surfactant/oil wall that is a key feature in the description of the phase behavior.

For the AOT/ethane binary, the LCST extends down to 25°C at 200 bar, and increasing the pressure greatly increases the LCST. Likewise, increasing the alkane chain length greatly increases the LCST, as shown in Figure 12a, where at 50 bar the AOT/propane LCST is at 85°C increasing to 120°C at 100 bar. For higher alkanes the LCST is far above 120°C. Surprisingly, a UCST was not found at temperatures down to −25°C in either the propane/AOT or the ethane/AOT binary solutions. Remarkably, even at −40°C a 1:1 volume fraction of AOT and propane is a single-phase, clear liquid solution.

The existence of the LCST at moderate temperatures, and the degree to which the LCST changes with either pressure or alkane carbon number, is the feature of this temperature/composition prism that dominates the phase behavior. As pressure is reduced the LCST moves to lower temperatures and eventually ''pulls'' the phase boundary, in the ternary phase diagram shown in Figure 11b, to the ethane/AOT axis at 200 bar. The existing phase

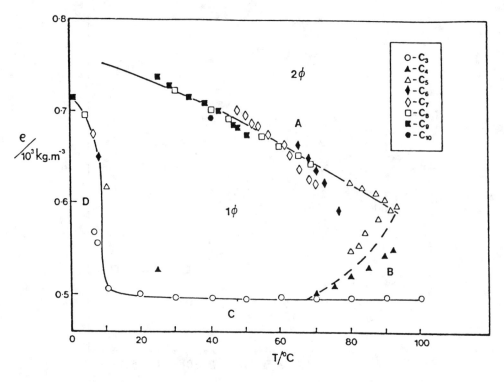

FIGURE 9.   The phase behavior of the *n*-alkane/AOT/water system for a fixed droplet size (W_m = 30) represented on a density-temperature surface. [AOT] = 50 mM. (From Steytler, D. C., Lovell, D. R., Moulson, P. S., Richmond, P., Eastoe, J., and Robinson, B. H., *Proc. Int. Symp. Supercritcal Fluids*, Société Francaise de Chimie, Nice, France, 1988, 67. With permission.)

behavior data allow the construction of a small portion of the temperature/composition prism, but there are insufficient data at this time to construct the path of the critical line through the prism. One must also keep in mind that the potential existence of various liquid crystalline phases (which may be stabilized by short, linear alkanes) can upset attempts to describe in detail the phase behavior on the basis of simple liquid-liquid-gas equilbria. Even with these limitations the general features of this phase behavior should be predictable using this approach. For instance, when the UCST of the AOT/water binary system and the LCST of the AOT/propane binary system occur near the same temperature, one would predict a three-phase region to exist near the center of the ternary triangle (see Figures 3, 4, and 5), and indeed a three-phase region exists around [AOT] = 140 mM, [H$_2$O] = 3.5 M, T = 103°C, and P = 300 bar. The extent of this three-phase region is not known.

III. STRUCTURAL MODELS FOR MICROEMULSION PHASE BEHAVIOR

The phenomenological model described in the previous section uses basic phase rule principles combined with experimental results to describe broad qualitative features of the phase behavior of surfactant-based microemulsion systems; however, in order to describe in detail the multiphase behavior of simple ternary systems in which aggregating surfactant molecules form a diverse range of microstructures, one needs to consider the underlying intermolecular forces. This is an area of intense, ongoing research in liquid surfactant systems, and the section that follows briefly reviews these approaches and then discusses them in terms of near-crictical fluid and SCF microemulsions.

Isotropic microemulsion solutions can be classified by three general categories of struc-

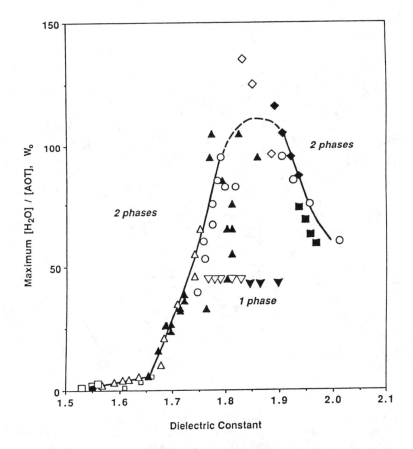

FIGURE 10. The maximum water-to-surfactant ratio, W_o, of AOT, water-in-oil microemulsions as a function of the dielectric of the continuous phase solvent for (□) krypton, (□) xenon, (●) methane, (△) ethane, (▲) propane, (▽) *n*-butane, (▼) *n*-pentane, (◇) 2-methylbutane, (○) isobutane, (◆) *n*-hexane, and (■) isooctane at 25°C and 87 m*M* AOT. (Reprinted with permission from Fulton, J. L, Tingey, J. M., and Smith, R. D., *J. Phys. Chem.*, 94, 1997, 1990. © 1990 American Chemical Society.)

tures spanning the range from low to high water content: (1) w/o, (2) bicontinuous, and (3) o/w microemulsions. Experimental studies in near-critical fluids and SCFs have thus far been limited to the high oil content, w/o type of microemulsions such as the alkane/AOT/water systems. For this reason, the discussion of the thermodynamic parameters of interest concerning microemulsions in near-critical fluids and SCFs centers around the discussion of w/o microemulsions. The two other isotropic microemulsion structures in near-critical fluids and SCFs, the o/w and the bicontinuous, have recently been demonstrated in the authors' laboratory and are discussed elsewhere.

The development of thermodynamic models that can be used to describe the complex phase behavior of liquid microemulsion systems is still in its infancy. For liquid microemulsion systems most of the relevant intermolecular interactions are understood; the difficulty in describing these systems, however, arise in assessing the relative importance of each type of interaction and, in particular, describing the thermodynamics of the interfacial region. A number of thermodynamic models have been proposed to describe the phase behavior of liquid microemulsion systems. Early models were generally limited to either a description of o/w or w/o microemulsions, but more recent models have included a description of the entire region from low to high water content. Each of the various approaches have their strengths and weaknesses. Common to most of the approaches[32-36] are free energy terms that

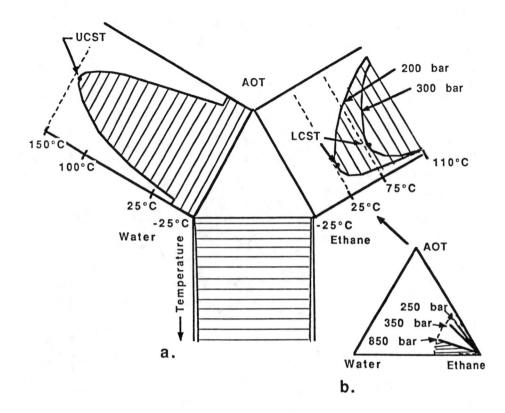

FIGURE 11.   (a) Binary temperature/composition diagrams for the AOT/water,[31] AOT/ethane (this work), and ethane/water binary systems. (b) Oil-rich corner of the ternary phase diagram for the ethane/AOT/water system showing the location of the phase boundary between a two-phase (to the left) and a single-phase (to the right) region at 25°C.

account for (1) bending energy associated with changes in the "natural" radius of interfacial curvature (derived from the geometric packing of molecules at the interface), (2) surface tension at the interface, (3) mixing of droplets in the continuous phase (including attractive and repulsive interactions), and (4) an electrostatic term to account for headgroup interactions of ionic surfactants.[37-41] Minimization of total system free energy thus provides the solubilization in each phase. Conventional molecular statistical thermodynamics, often in the form of a lattice model, have also been employed to describe microemulsion phase behavior.[42-48] An advantage of these models is that the mass action law,[49] which describes a smooth rather than a discontinuous CMC and a polydisperse rather than a monodisperse droplet size distribution, can readily be included.[42,44,45,50] Finally, some interesting goemetric models have appeared recently; they are designed to account for the phase behavior of bicontinuous microemulsions.[48,51-52]

w/o microemulsions containing uniformly sized, spherical droplets dispersed in an oil-continuous phase can be described by a model[35,53-55] in which the total free energy per unit volume, f, is composed of contributions from an interfacial bending term, f_b, and droplet-mixing terms, which account for both the entropy of mixing, f_s, and the attractive interactions between the droplets, f_a, such that:

$$f = f_b + (f_s + f_a) \qquad (1)$$

Various forms of the free energy due to the entropy of mixing have been described based on either the Carnahan-Starling approximation for hard spheres,[38] given in Equation 2, or upon a lattice model[35]

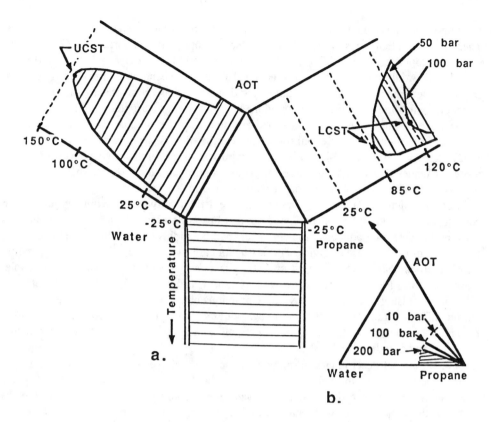

FIGURE 12. (a) Binary temperature/composition diagrams for the AOT/water,[31] AOT/propane (this work), and propane/water binary systems. (b) Oil-rich corner of the ternary phase diagram for the propane/AOT/water system showing the location of the phase boundary between a two-phase (to the left) and a single-phase (to the right) region at 25°C.

$$f_s = k \, T \, n \, (\ln \theta - 1 + \theta((4-3\theta)/(1-\theta)^2) + \ln (v_1/v_d) \qquad (2)$$

where n represents the number of droplets per unit volume, Θ is the volume fraction of hard spheres, v_1 is the molecular volume of the continuous phase solvent, and v_d is the volume of one droplet. The mixing term due to the attractive interactions between droplets can be written in a phenomenological form as

$$f_a = (n/6) \, A(T,r)g(r,\theta) \qquad (3)$$

where $A(T,r)$ is a material-dependent parameter that represents the strength of interactions between droplets and, in the absence of attractive interactions due to overlap and solvation of hydrocarbon tails, is equal to the Hamaker constant[5] for simple van der Waals-type attraction between the droplets. $g(r,\Theta)$ is a geometric term that accounts for the droplet radius and interdroplet spacing. The interfacial free energy is described by a curvature or bending term only, since for a microemulsion, the interfacial tension goes to zero. The bending term, f_b, is related to the radius of the droplet, r,[55]

$$f_b = 16 \, \pi \, n \, K \, (1 - (r/r_o))^2 \qquad (4)$$

where K is the "rigidity" constant of the interface and has dimensions of energy, and r_o is the natural bending radius, a parameter that is primarily related to geometric shape on packing of the surfactant molecules. Deviations from the natural radius of curvature result from a bending of the interface due to a gradient of interactions as one moves across the interface. These include interactions (1) of water with the hydrophile of the surfactant, (2) of neighboring hydrophiles, (3) of neighboring hydrophobes, and (4) of the hydrophobe of surfactant with the oil. Additions of salt or a cosurfactant also affect the bending energy of the interfacial region. When one adds a small amount of the dispersed phase component (water), droplets of radius r are formed; however, r_o provides an upper limit of droplet size so that as more of the disperse phase component is added and the droplet size exceeds r_o, the interfacial free energy is minimized by expulsion of the excess dispersed phase as a separate phase. In this case, w/o microemulsion is in equilibrium with a predominantly water lower phase. System changes that tend to increase the natural radius of curvature, r_o, should therefore increase the maximum solubilization of water by the microemulsion as well.

For w/o microemulsions, certain conditions exist where two oil-continuous phases can coexist. Several authors[53-59] have pointed out that in some systems the attractive interactions between the dispersed aqueous droplets rather than the radius of curvature is the overriding factor governing the phase behavior. This is true for cases where $r < r_o$. Rather than separating into a microemulsion and a pure oil or water phase, systems exhibiting strong interdroplet attractive forces (described by the free energy term f_a in Equation 3) will separate into two w/o microemulsion phases. These two oil-continuous phases can contain identical size droplets but different droplet concentrations, a behavior that is analogous to vapor-liquid equilibria for single-component systems. Such systems allow a simplified analysis of experimental phase behavior studies.[5]

For the model describing w/o microemulsion given above, the phase behavior is controlled by either the natural radius of curvature of the droplet or by the strength of attractive interactions between the droplets. For curvature-dominated phase separations, where the radius of the droplet exceeds the natural radius, $r > r_o$, a w/o microemulsion will exist in equilibrium with an excess water phase. If $r < r_o$, then the phase stability can be dominated by the attractive interactions of the droplets and such systems phase separate into two oil-continuous phases. Pressure or density of the continuous phase solvent will affect the microemulsion phase behavior insofar as it affects the natural radius of curvature and/or the attractive interactions between the droplets. These considerations are directly relevant to the phase behavior described in Figure 10.

A. EFFECT OF PRESSURE IN THE STRUCTURAL MODEL

Although not discussed explicitly here, this model can be used to account for the effect of pressure on microemulsion phase behavior by examining the expected pressure effects on the two governing variables, the spontaneous radius of curvature and the interdroplet attractive forces.

A number of groups have observed that the radius of curvature is a strong function of the extent to which the continuous phase molecules penetrate into the surfactant interfacial layer.[53,60,61] The greater the extent of penetration, the smaller the radius of curvature, and consequently the lower the maximum solubilization (if the radius of curvature is the limiting factor). In situations where the radius of curvature defines the phase behavior, the extent to which pressure influences continuous phase solvation of the surfactant layer will thus determine the dependence of solubilization on pressure. For a w/o microemulsion using a low molecular weight alkane continuous phase, if the mixing behavior of large alkanes can be used as a guide, increasing pressure should increase the affinity of the surfactant alkane tails for the continuous phase. An increase in pressure should therefore enlarge the penetration zone of the continuous phase molecules into the surfactant interfacial layer and thus decrease the spontaneous radius of curvature. In the region where the radius of curvature is the limiting factor for water content, increasing pressure should then decrease solubilization and lower W_o.

Solubilization can also be constrained by the interdroplet attractive forces. A number of studies have shown that the attractive interactions between droplets are a function of temperature, the droplet radius, and the oil chain length,[53,57] and have utilized the Hamaker constant concept to account for the attractions between droplets in liquid systems. Calje et al.[62] found that reasonable values of the Hamaker constant produced attractive forces that were too weak to adequately describe effects in a w/o microemulsion. Calje et al., and subsequently Lemaire et al.,[56,57] found it necessary to include the effect of droplet-droplet overlap (surfactant tail-tail interactions) on the attractive forces in their calculations of microemulsion phase behavior. The effect of the pressure or density of the continuous phase upon these two types of interdroplet attractive interactions will be determined by how well the longer- and shorter-range van der Waals-type attractive interactions between droplets are screened by the continuous phase solvent.

The authors use the following section as a forum to discuss the effect of pressure on microemulsion phase behavior in near-critical fluids and SCFs. Results on various systems are disclosed in the context of the structural model for microemulsion phase behavior. Contrasts are drawn between the behavior of liquid systems and mixtures containing much more compressible near-critical and supercritical components.

B. STRUCTURAL ASPECTS OF MICROEMULSIONS IN NEAR-CRITICAL AND SUPERCRITICAL FLUIDS

On a molecular level, three simple classes of interaction can be identified in oil/surfactant/water solutions. Listed in order of decreasing importance they are (1) ionic interactions of surfactant head groups, (2) hydrogen bonding within the aqueous core, and (3) London-van der Waals forces between the hydrocarbon tails of the surfactant and the continuous phase solvent and between hydrocarbon tails of adjacent droplets. The ionic interactions of the headgroups (for ionic surfactants only) determine the area occupied by the surfactant headgroup. The headgroup spacing controls the total interfacial area of the solution and together with curvature limitations is a major component controlling the microemulsion structure. Second, the aqueous core region affects the ionic headgroup interactions (only to a small degree) only at low water contents (W <5) where hydration of the headgroup and solvation of the counterion occurs. At higher water contents (W ≥5), the amount of water only affects the interfacial curvature and not the headgroup spacing. Finally, solvation of the hydrocarbon tails by the continuous phase solvent, a relatively weak short-range, van der Waals-type attraction, probably does not significantly affect the headgroup spacing, which is fixed by stronger ionic interactions. Thus, for ionic surfactant systems, the density or type of continuous phase solvent probably only affects the droplet structure insofar as it affects the natural curvature of the interfacial region by changes in the degree of solvation of the hydrocarbon tails by the continuous phase solvent. The fact that the headgroup spacing is unaffected by either the continuous phase solvent or the amount of water allows one to develop a simple relationship between the droplet size and the water-to-surfactant ratio[5] for single-phase systems. Additionally, whereas temperature will have an appreciable effect on headgroup spacing, the effect of pressure is expected to be negligible at pressures of general interest (<500 bar). In contrast to water content or continuous phase solvation, the addition of small amounts of salt greatly decreases the headgroup spacing by screening the repulsive interaction of the headgroups, resulting in much more dramatic effects on phase behavior.

Increased solvation or penetration of the interfacial layer by the near-critical fluid or SCF can only affect the microemulsion structure by decreasing the maximum allowable radius of curvature, r_o. In contrast, the degree of solvation of the microemulsion droplet is strongly affected by the properties of the continuous phase solvent. The density or solvent power of the continuous phase solvent will strongly affect how well the droplets are solvated and the degree to which the interdroplet attractive interactions are screened.

For nonionic surfactant systems, the headgroup spacing is determined by the degree of hydrogen bonding between the headgroups and between the headgroups and the aqueous core. In these systems, the degree of penetration of the hydrocarbon region by the "oil" phase may affect the amount of hydrogen bonding and hence the surfactant headgroup spacing. Thus, for nonionic systems, the continuous phase solvent could potentially affect the headgroup spacing, the natural curvature, and the degree of solvation of the dispersed droplet phase. Clearly, the effects of fluid pressure upon phase behavior for nonionic systems can be more complex than for ionic surfactant systems.

Thus, it appears that continuous phases consisting of near-critical fluids or SCFs will differ from liquids primarily in the degree of "solvation" of the droplet phase, which will be manifested as large changes in the attractive term, f_a, of Equation 1. In addition, the degree of solvation of the surfactant hydrocarbon region by the near-critical fluid or SCF may affect the bending of the interface (for both ionic and nonionic surfactants) and possibly even the surfactant headgroup spacing (for nonionic surfactants).

One measure of the solvent power of the fluid continuous phase is the dielectric constant of the solvent. It is well known that near-critical fluids and SCFs are relatively poor solvents when compared to their liquid analogs, in part due to the low dielectric constants of such fluids. Hence, the ability of a C_1 to C_3 alkane at a given pressure to solvate aqueous microemulsion droplets will be significantly lower than for liquid C_6 to C_{10} alkanes. This effect is demonstrated in experimental studies of near-critical fluids and SCFs where both the phase behavior and diffusion coefficients (as measured by dynamic light scattering[63]) are strongly dependent upon the density or dielectric constant of the continuous phase solvent. A phase separation from a clear, single-phase system to a two-phase fluid-liquid system can be induced in such dilute microemulsions (<1% dispersed phase volume) simply by lowering the density of the SCF through a reduction of pressure.[10] The droplet diffusion coefficient is observed to decrease and hydrodynamic diameter to increase exponentially as this phase boundary is approached from higher pressures.[63] This behavior suggests that strong inter-droplet attractive forces are important in these dense gas continuous phases, and that this droplet attraction is the first step leading to coalescence to form a second phase.

The magnitude of these longer-range, van der Waals attractive interactions has recently been calculated for near-critical fluids and SCFs.[5] In this work, the effect of the dielectric constant of the continuous phase was related to the magnitude of the interdroplet Hamaker constant. The microemulsions were modeled as droplets of water covered by surfactant shells, and uniformly dispersed in the continous phase fluid. The attractive force between droplets were assumed to be a sum of the shell-fluid-shell, core-shell-core, and core-shell-fluid interactions. The van der Waals-type interdroplet attraction was calculated using the method of Hamaker[64] and Deryaguin,[65] who derived an expression for the interparticle attractive potential based on droplet size and spacing. Fulton et al. utilized a substantially improved formulation of the Hamaker constant given by Lifshitz,[66] Ninham et al.,[67] and Gingell and Parsegian,[68] in which the dielectric response of each component in the system was utilized over the entire frequency range of the electromagnetic spectrum.

As shown in Figure 13, the Hamaker constant, which is proportional to the magnitude of the attractive potential, passes through a minimum value at a continuous phase dielectric constant of approximately 1.8. This minimum in the Hamaker constant occurs at approximately the same dielectric constant value for propane, ethane, isobutane, pentane, and xenon. The predicted minimum in the Hamaker constant shown in Figure 13 is due to the fact that this constant is based on the square of the difference between the dielectric constant of the droplet and the intervening fluid over the entire range of the electromagnetic spectrum. In the microwave region, water has a much larger dielectric constant ($\epsilon_o = 80$) than the intervening (continuous phase) fluids studied ($\epsilon_o = 1.8$ to 2.1); therefore, the microwave portion of the Hamaker constant is essentially identical for all the systems studied. In the

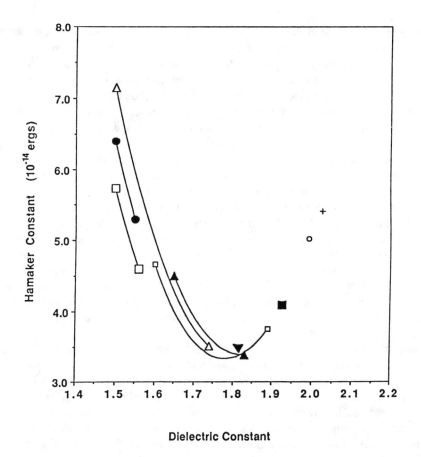

FIGURE 13.   Hamaker constant between two water droplets as a function of the dielectric of the intervening medium at 25°C for (☐) krypton, (▢) xenon, (●) methane, (△) ethane, (▲) propane, (▼) n-pentane, (■) octane, (○) decane, and (+) hexadecane. The symbols correspond to the dielectrics at 100 and 1500 bar for each of the supercritical and near-critical fluids and to 1 bar for the liquid alkanes. The lines represent the intermediate dielectric values between 100 and 1500 bar. (Reprinted with permission from Fulton, J. L., Tingey, J. M., and Smith, R. D., *J. Phys. Chem.*, 94, 1997, 1990. © 1990 American Chemical Society.)

UV-visible (UV-vis) region, on the other hand, the dielectric constants of the intervening fluids (1.5 to 2.0) and water (1.78) are similar. As the UV-vis dielectric constant of the intervening fluid approaches the UV-vis dielectric constant of water, the UV-vis portion of the Hamaker constant approaches zero and a minimum value in the total Hamaker constant is observed. At low dielectric constants, which are equivalent to lower molecular weight fluids or to lower fluid densities, the Hamaker constant rapidly increases, as shown in Figure 13. Hence, the effectiveness of low molecular weight, continuous phase solvents to screen the interdroplet interactions is greatly reduced at low pressures and densities.

In the work by Fulton et al., the interdroplet attractive forces were also found to have contributions arising from the hydrocarbon shells on the exterior of the droplets. (This is not attraction due to overlap and intersolvation of hydrocarbon tails of two or more micelles.) These forces become increasingly important when the droplets are small (W <5) and when the dielectric constant of the continuous phase solvent was much lower than the dielectric constant of the hydrocarbon tails. Fulton et al. found that in the low dielectric constant region (ϵ <1.65) the total interdroplet attractive potential must include contributions from both the water core and the hydrocarbon shell attractive interactions.

Fulton et al.[5] compared their predictions with experimental observations of phase behavior. The maximum water solubilization, in terms of W_o, was determined for AOT microemulsions in a variety of liquid and SCFs and it was observed that solubilization first increases with increasing pressure or density but then decreases (see Figures 9 and 10). In Figure 10, the gradual increase in W_o observed between a dielectric constant of 1.53 and about 1.66 is followed by a steeply rising envelope that reaches a maximum at a dielectric constant of about 1.81, in good agreement with predictions described earlier. Above a dielectric constant of 1.81, W_o values decrease with increasing dielectric constant (i.e., increasing fluid pressure). The propane and butane solutions at higher water contents were bounded by both an upper and lower cloud point pressure and thus displayed a maximum in their W_o vs. dielectric constant plot. Two exceptions to these general trends are the *n*-butane and *n*-pentane systems. It was believed that these small, linear alkanes stabilized a liquid crystalline phase and therefore deviated significantly from the phase behavior trends of the other alkanes. Near-critical fluids and SCFs have very low dielectric constants in comparison to liquid alkane solvents, and it is seen that if the fluid dielectric constant is below 1.65, the maximum amount of water that can be solubilized is not greater than $W_o = 5$. As the droplet size increases, the attractive potential between droplets increases, the low dielectric constant fluids cannot effectively screen these interactions, and the droplets coalesce to form a second phase. Thus, the relatively low W_o values for the noble gas solvents can be rationalized.

At higher fluid dielectric constants ($\epsilon \geq 1.65$), exceeding W_o results in the formation of a second phase that appears to also have a w/o microemulsion structure. From conductivity measurements, it was determined that in the two-phase regions both phases were oil-continuous, and it was concluded that both phases were of the dispersed droplet type. Such a system is dominated by the interdroplet attractive interactions.[55] Thus, Fulton et al. were able to establish a qualitative relationship between the magnitude of the calculated attractive interaction (in the form of a Hamaker constant) and the maximum size of a droplet that could be stably dispersed. This conclusion is illustrated in Figures 10 and 13; observe that the dielectric constant where the minimum in the calculated Hamaker constant occurs (1.81) corresponds to the dielectric constant where the maximum W_o is observed. At the minimum value of the Hamaker constant, the attractive interactions are minimized, allowing the largest droplets to be stably suspended. Interestingly, this approach did not consider the effect of the overlap and intersolvation of the hydrocarbon tails. Although this finding tends to discount an appreciable contribution from tail overlap, the relative contributions of the two different types of attractive interactions—the longer-range Hamaker type of interaction and the short-range interactions due to hydrocarbon tail overlap and intersolvation—were not determined; this issue remains an important question relevant to these near-critical fluid and SCF systems.

Experimental studies to date have shown that the for near-critical fluid and SCF microemulsions, strong interdroplet attractive interactions can dominate the phase behavior for systems in which two oil-continuous phases exist. The effect of the degree of solvation of the hydrocarbon tails by the fluid-continuous phase on the phase behavior of the system has yet to be determined. Such an investigation would probably require studies of two-phase systems in which a w/o microemulsion is in equilibrium with an excess aqueous phase. Under these conditions the radius of curvature would dominate the phase behavior so that the effect of the degree of solvation of the hydrocarbon tails by the continuous phase could be determined.

IV. EXPERIMENTAL STUDIES OF MICROEMULSIONS IN NEAR-CRITICAL AND SUPERCRITICAL FLUIDS

A. STRUCTURAL STUDIES OF MICROEMULSIONS

Pressure has a strong influence on the microstructure of microemulsion phases in compressible fluids as well as on the phase behavior. Microstructure includes the size, shape, and spatial distribution of the aqueous microdomains within the continuous phase. Furthermore, the microstructure of the microemulsion will in turn affect the nature of chemical environment experienced by the dispersed phase, and most importantly for applications, the solubilizing power of the microemulsion.

At the authors' laboratory, techniques have been developed allowing dynamic light scattering studies on microemulsions employing near-critical fluid and SCF components in which the time-dependent portion of the scattered intensity is utilized to directly measure droplet diffusion coefficients. Utilizing a specially designed, high-pressure sapphire light-scattering cell, capable of operation of pressures up to 600 bar and over 80°C, Blitz et al.[69] and Fulton et al.[63] measured the effect of pressure on the diffusion coefficient of dispersed AOT/water droplets at various water contents in both near-critical and supercritical ethane and xenon. The droplet apparent hydrodyamic diameter (or correlation length), as calculated using the Stokes-Einstein equation, appears to decrease as pressure increases up to 550 bar. For example, the apparent hydrodynamic diameter of a xenon microemulsion droplet decreased from 6.5 to 4.5 nm as pressure is increased from 350 to 550 bar (for [AOT] = 150 mM and W_m = 5). Rather than assuming deaggregation was occurring, Fulton et al. proposed that this apparent size decrease was in fact due to a decrease in the force of attraction between the dispersed droplets. This conclusion was later supported by the results of a solvatochromic probe study,[63] where, although the apparent droplet size decreased as pressure increased, the environment experienced by the hydrophilic probe in the droplet core did not change significantly over the same pressure range (as evidenced by the lack of shift in the UV peak maximum of the probe, thymol blue).

In another study, Smith et al.[70] measured the diffusion coefficients of AOT/water droplets (at W = 5) in alkanes ranging from propane to decane at 25°C and pressures up to 600 bar. As shown in Figure 14, the diffusion coefficients measured in liquids generally decrease by 10 to 15% as alkane length decreases (and viscosity decreases). Diffusion coefficients of the dispersed droplets in near-critical propane (25°C) and supercritical ethane are significantly higher than in the larger alkanes, which is to be expected given the significantly lower viscosities of the lower alkanes. In contrast to the larger alkanes, diffusion coefficients in ethane and propane show an initial increase as pressure increased. This is ascribed to the strong droplet-droplet interactions in these fluids at lower pressures. The interactions are relatively small for propane (but still much greater than for the larger alkanes), resulting in a maximum for the diffusion coefficient in propane at ~100 bar where attractions are significantly reduced so that the opposing effect of increased viscosity becomes apparent at pressures above >100 bar. The supercritical ethane microemulsion droplets shown in Figure 14 have diffusion coefficients that increase with pressure; due to the pressure limitations of the available instrumentation, however, the expected maximum was not observed. These results, again attributed to droplet-droplet interactions, suggest that significantly improved mass transport properties for these systems are often obtained at higher pressures, even though the viscosity of the fluid continuous phase is higher.

Figure 15 shows apparent hydrodynamic diameters for the dispersed droplet phase in the alkanes calculated from the diffusion coefficients given in Figure 14 using the Stokes-Einstein equation. The results show that apparent droplet diameter is generally in the 4 to 5 nm range for the larger alkanes, although slightly larger diameters were observed for decane. Generally, the larger alkanes show little or no change in the apparent hydrodynamic diameter with pressure. In contrast, propane and ethane show hydrodynamic diameters that decrease substantially as pressure is increased, which is attributed to decreased droplet-droplet interactions.

Beckman and Smith[71] investigated the structure of the propane/ethane/C_iE_j/water/acrylamide system (described earlier) at pressures up to 500 bar using dynamic light scattering.

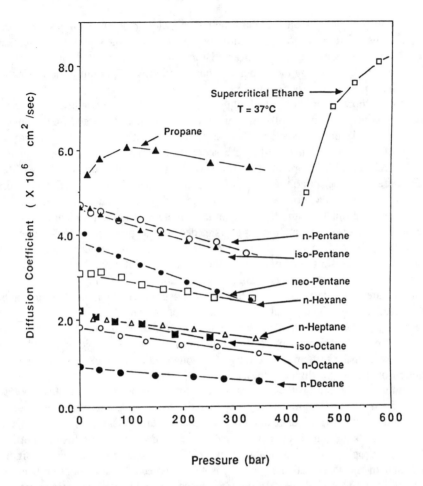

FIGURE 14.   Diffusion coefficients measured by dynamic light scattering of AOT/water droplets in various alkane/AOT/water microemulsions as a function of pressure. T = 25°C, W_m = 5, Y_{AOT} = 0.015 (mole fraction). (From Smith, R. D., Fulton, J. L., and Blitz, J. P., in *Supercritical Fluid Technology*, Johnston, K. P. and Penninger, J. L., Eds., American Chemical Society, Washington, D.C., ACS Symp. Ser. No. 406, 1989, 165. With permission.)

In the first 100 bar above the clearing pressure, the apparent diffusion coefficient increases by over 30%, as shown in Figure 16, even though the viscosity increases as pressure increases (by 15% over the 300 to 450 bar pressure range for the ethane/propane mixtures in this study). The apparent droplet diameter, as calculated from the Stokes-Einstein equation, decreases rapidly as pressure increases in the one-phase region above the clearing point (see Figure 17). As mentioned previously, a decrease in the apparent diameter at constant composition can be ascribed to a decrease in the actual hydrodynamic diameter or a decrease in interdroplet attractive interactions.

Along with micelle size, the strength of interactions between droplets depends on the degree to which the continuous phase molecules interact with, and penetrate into, the surfactant tail shell. As pressure is increased, it is expected that the ethane/propane continuous phase will exhibit an increasing affinity for the alkane tails of the surfactant, similar to behavior in liquid systems. In contrast to the AOT system in which the strong ionic interactions fix the headgroup spacing, solvation of the C_iE_j interface by the fluid continuous phase may increase the interfacial area, resulting in a reduction in the droplet size; however, considering the magnitude of the observed reductions in droplet sizes with pressure, Beckman

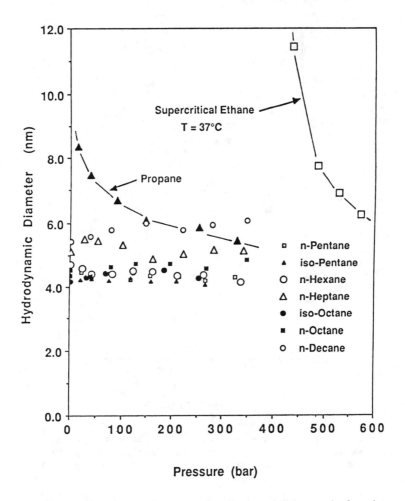

FIGURE 15. Hydrodynamic diameters measured by dynamic light scattering for various alkane/AOT/water microemulsions as a function of pressure. T = 25°C, W_m = 5, Y_{AOT} = 0.015 (mole fraction). (From Smith, R. D., Fulton, J. L., and Blitz, J. P., in *Supercritical Fluid Technology*, ACS Symp. Ser. No. 406, Johnston, K. P. and Penninger, J. L., Eds., American Chemical Society, Washington, D.C., 1989, 165. With permission.)

et al. assumed that the large decrease was primarily due to a reduction in the interdroplet attractive interactions. As the continuous phase-droplet interaction becomes more favorable, the droplet-droplet interaction will be increasingly screened as pressure increases, leading to a decrease in the apparent droplet diameter. Previous calculations by the authors for the AOT/water/alkane system have demonstrated that the attractive interactions between microemulsion droplets should decrease rapidly with increasing pressure in highly compressible SCFs.[5]

Beckman and Smith[71] also observed that the width of the distribution of apparent micelle sizes also decreases as the pressure increases above the clearing point. Both the apparent size and distribution width of micelles formed in fluids of various ethane/propane ratios tended to superimpose when plotted vs. fluid density. Similarly, cloud point curves of this system over the full range of ethane/propane ratios were also observed to correlate well with fluid density.[26]

The results of dynamic light scattering studies[72] given in Figure 18 provide insight into secondary structural changes that occur within the near-critical and supercritical microe-

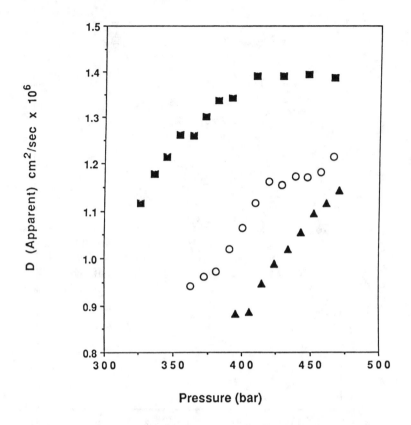

FIGURE 16.   Apparent diffusion coefficient vs. pressure in ethane/propane mixtures at three propane continuous phase wt% at 65°C: (■) 65%; (○) 56.7%; (▲) 49.9%. (Reprinted with permission from Beckman, E. J. and Smith, R. D., *J. Phys. Chem.*, 94, 3729, 1990. © 1990 American Chemical Society.)

mulsion very close to a phase boundary. On approaching a phase boundary, the apparent hydrodynamic size is seen to increase exponentially, an effect very much like behavior of a single component system near its critical point. This behavior is consistently observed upon approach to any such phase boundary in these systems. At lower pressures each system consists of two phases, where the lower phase may be either oil- or water-continuous dependng upon the total amount of water and the pressure. As pressure is increased in the two-phase region, the hydrodynamic diameter of the microemulsion droplets in the upper alkane continuous phase increases up to a maximum at which point the system becomes one phase. This behavior can also be observed visually in the single-phase region, where the transmitted portion of a white light source appears to be yellow, then orange, and finally deep red at a location very close to the phase boundary. These effects are evidence for the existence of large structures (>100 nm), and perhaps represent the formation of large clusters of individual micelles or droplets. As shown in Figure 18, the pressure defining the phase boundary, and hence the position of the observed dramatic increases in apparent size, depends upon the system water content, the surfactant concentration, the type of fluid and the temperature of the system.

The CMC is the minimum amount of surfactant required to form aggregates. For reverse micelle systems the CMC does not represent a distinct dividing line between monomer and micellar states. In oil-like solvents, micelle formation is a progressive aggregation process that begins with monomers, then dimers, then trimers, etc. until the initial micelle forms. The CMC region is a smooth, continuous transition region with some degree of aggregate

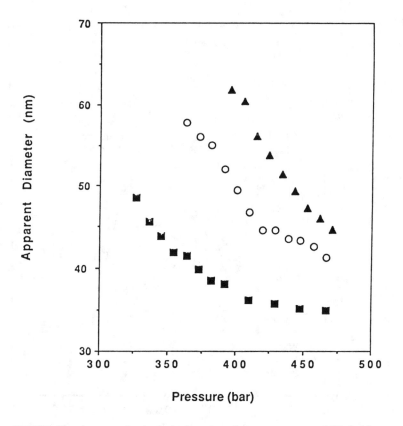

FIGURE 17.   Apparent droplet hydrodynamic radius vs. pressure at 56°C for three propane continuous phase wt%; symbols same as in Figure 16. (Reprinted with permission from Beckman, E. J. and Smith, R. D., *J. Phys. Chem.*, 94, 3729, 1990. © 1990 American Chemical Society.)

polydispersity. Both of these characteristics of the CMC can be described by the law of mass action.[49] Because the progressive aggregation occurs over a broad range of concentration and because the CMCs of reverse micelles are quite low, accurate measurements are difficult. For example, measurements of CMCs for AOT in benzene[11] span the range from 0.5 to 2.9 mM and depend strongly upon the measurement technique employed. Kotlarchyk et al.[11] report a CMC for AOT in insooctane of 0.2 mM, substantially lower than for the more polar benzene continuous phase solvent.

Determination of CMCs in near-critical and supercritical solvents is of interest since they will provide some fundamental thermodynamic information on micelle formation and since the microstructure of the microemulsion can depend upon the equilibrium concentration between monomer and micellar surfactant molecules. Changes in the near-critical fluid or SCF density at constant temperature should cause changes in the CMC since the CMC can be thought of as representing the solubility of the surfactant monomer in the fluid, and since the solvent power of the fluid strongly depends on the density.

In contrast to aqueous systems, CMC data at high pressure for w/o microemulsions, including those employing near-critical fluids or SCFs, are nearly nonexistent. Tamura and Suminaka[73] measured the spectroscopic absorbance of an alkyl ammonium propionate in carbon tetrachloride at pressures up to 981 bar and assigned the CMC to the discontinuity in the curve. The data for this w/o microemulsion appear to exhibit a decrease in the CMC as pressure is increased. The case for SCF systems is even less clear. Randolph et al.[74] have observed trends toward increased aggregation with increasing pressure in the cholesterol/

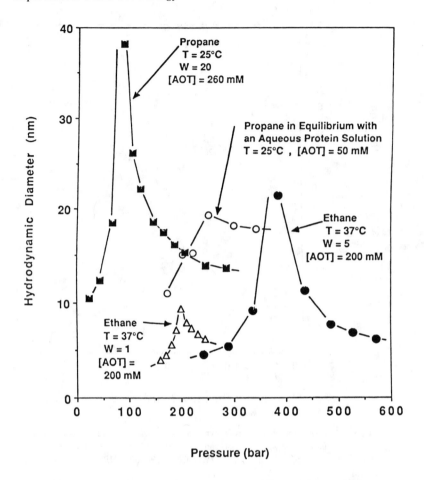

FIGURE 18.   Apparent hydrodynamic diameter as a function of pressure for the alkane/
AOT/water system in near-critical propane and ethane and in supercritical ethane. As
pressure is increased, the hydrodynamic diameters of the upper microemulsion phase
increase up to a maximum at which point the system becomes one phase. Also given are
results for the propane/AOT/water buffer system studied for the protein extraction which
is two phase under all conditions studied. (Reprinted with permission from Smith, R.
D., Fulton, J. L., Blitz, J. P., and Tingey, J. M., *J. Phys. Chem.*, 94, 781, 1990. ©
1990 American Chemical Society.)

CO_2 system at pressures between 80 and 85 bar. Johnston et al.,[6] by following the change
in the UV peak maximum of a pyridine N-oxide probe vs. surfactant concentration, reports
the CMC of AOT in ethane at 35°C and 345 bar to be approximately 1 mM. Olesik and
Miller,[75] using the change in a probe molecule, TCNQ, from neutral to radical anion, found
the CMC for AOT in ethane at 38.5°C and 272 bar to be 24 mM. Although these data
suggest the possibility that the CMC of AOT in a highly compressible fluid such as ethane,
decreases with pressure, clearly too few data exist to draw a meaningful conclusion. Such
a trend would also run counter to the expectation that the CMC would increase with pressure
due to the greater solubility of the monomer.[10] Measurement of the CMC in an SCF system
as a function of pressure remains a key experimental challenge.

B. CHEMICAL ENVIRONMENT WITHIN THE MICROEMULSION

The effect of pressure on the polarity of the solvent environment of the aqueous microdomains within a w/o microemulsion has been determined by measuring the pressure-induced
shift in the UV peak, or the fluorescence intensity maximum, for various probe molecules.

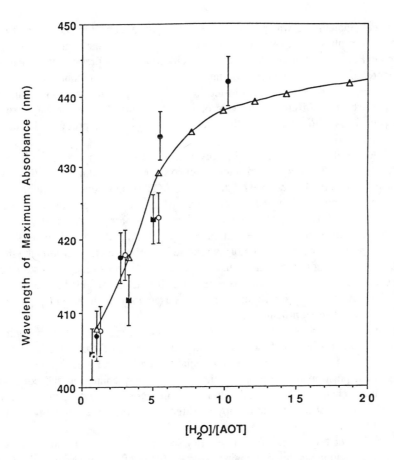

FIGURE 19. Solvatochromic shifts of thymol blue in reverse micelles as a function of water content for various continuous phases: ($\triangle$) isooctane at 1 bar and 25°C with best-fit line, ($\bigcirc$) supercritical xenon at 500 bar and 25°C, ($\bullet$) liquid ethane at 500 bar and 25°C, and ($\blacksquare$) supercritical ethane at 500 bar and 37°C. [AOT] = 50 mM. (Reprinted with permission from Fulton, J. L., Blitz, J. P., Tingey, J. M., and Smith, R. D., *J. Phys. Chem.*, 92, 4198, 1989. © 1989 American Chemical Society.)

Johnston et al.[6] and Fulton et al.[63] have used the shift in the UV peak maximum of hydrophilic probe molecules to evaluate the polarity of the core region of w/o microemulsions formed in a range of SCFs. It was observed that the solvatochromic shift of thymol blue in AOT/ water droplets dispersed in ethane and xenon at 500 bar exhibits a similar dependence on water (W_m) content to that in an AOT/isooctane system at atmospheric pressure (see Figure 19). At water to surfactant ratios up to 6, hydration of the surfactant headgroups was assumed to increase steadily, and this changed the polarity of the microemulsion environment substantially. At higher water contents, water pools formed in the micelle cores, whose environment was subsequently less sensitive to further increases in water concentration; however, selection of other spectroscopic probes for higher W_m could provide greater resolution. Increasing pressure of AOT microemulsions at constant W_o in either ethane or xenon produces a small blue shift, implying a slight decrease in micelle size. As mentioned previously, however, this reduction in size by itself is too small to account for the large decrease in apparent droplet hydrodynamic radius measured via dynamic light scattering. Xenon, although a slightly "better" solvent for AOT micelles than ethane, presents an added complication in that at high pressure and water content, thermodynamically stable clathrates (gas hydrates) form, extracting water from the droplet cores in the form of a precipitate.

Johnston et al. have measured the solvatochromic shift of a pyridine N-oxide probe in supercritical ethane/AOT/water as a function of pressure. Johnston showed that in the upper microemulsion phase of a two-phase ethane/AOT system without added water, the probe environment at low pressures was like that of chloroform. As pressure was increased to 200 bar, a more polar environment resembling ethanol was observed. Johnston interpreted these solvatochromic changes as occurring due to increased aggregate size in the upper microemulsion phase. The difference in the probe environment observed at high pressures may also be due in part to changes in the site of solvation of the probe molecule within the microemulsion structure.

Johnston et al.[6] also examined the solvatochromic shift of pyridine oxide in an ethane/ C_iE_j (C = 10 to 13; E = 5) w/o microemulsion, also in equilibrium with a lower liquid phase. Contrary to the behavior exhibited by the AOT system, the nonionic microemulsions display a polar environment at low pressures, which becomes progressively less polar as pressure increases. At a pressure of only 50 bar, Johnston et al. report the environment of the probe resembles that observed in bulk hexane. Added water increases the polarity somewhat, yet a cosurfactant (octanol) is required to produce an environment similar to that in bulk water. The polarity of the ethane/water/surfactant/cosurfactant system remains essentially constant as pressure increases up to 350 bar. These results are consistent with observations by Beckman and Smith[27] that a cosurfactant is required in C_iE_j-alkane systems in order to solubilize significant amounts of water.

Spectroscopic studies that utilize probe molecules to sense a microenvironment can be complicated by perturbation of the microemulsion by the probe itself. In addition, the precise location of the probe in the droplet or interfacial region is usually not known with certainty, and the radial position of the probe in the interface may change as the conditions of the experiment are changed. A favorable alternative in many instances is to use a nonintrusive technique such as infrared spectroscopy to directly examine the microenvironment of the water or surfactant domains.

Using Fourier transform-infrared (FT-IR) spectroscopy, Smith and co-workers[72] have shown that the aqueous microdomains in water/AOT/alkane microemulsions were little affected by the type (propane, pentane, and isooctane) or density of the continuous phase solvent. In one study, the microemulsion water core was monitored using the absorbance at ~3460 cm^{-1}, 5200 cm^{-1}, or 10,200 cm^{-1}. The ~3460 cm^{-1} IR band is a combination of the asymmetric stretch (~3500 cm^{-1}), symmetric stretch (~3450 cm^{-1}), and the first overtone of the bend (~3275 cm^{-1}). As shown in Figure 20, this band shifts to lower energy as W_m increases due to changes in the amount of hydrogen bonding. Therefore, the absorbance maximum and width of the ~3460 cm^{-1} band provide a convenient measure of W_m and the dispersed droplet size.

Figure 20 also gives the energy of maximum absorbance (v_{max}) of the ~3640 cm^{-1} IR band as a function of pressure for dispersed droplets in propane, n-pentane, and isooctane. The pressure studies were done at $W_m = 5$, where sensitivity to changes in the water core is large. The results show that large shifts in v_{max} (~60 cm^{-1} between $W_m = 2$ and 30) are observed as a function of W_m; however, a negligible change in v_{max} is observed as a function of the pressure or density of the continuous phase solvent. (The density of propane changes from 0.48 to 0.56) while that of isooctane changes from 0.69 to 0.73 as pressure is increased from 1 to 500 bar.) These results also show that the water core environment for AOT microemulsions is largely independent of the continuous phase properties.

Solvation of the hydrocarbon tails by propane, pentane, and isooctane, solvents that differ substantially in size and critical temperature (96.7, 196, and 288°C, respectively), has little effect on the properties of the aqueous core and hence the overall size of the AOT/water droplets for single phase solutions. Evidently, the electrostatic interactions of the ionic headgroups and the hydrogen bonding in the aqueous core largely determine droplet size and structure for single-phase solutions. Also, since the water band shows no shift as a

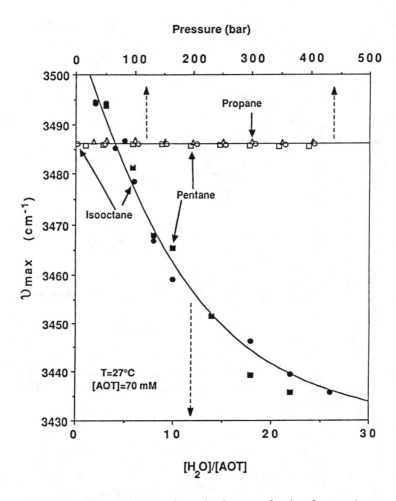

FIGURE 20. Wave number at maximum absorbance as a function of pressure (open symbols, top axis) and the water-to-surfactant ratio (closed symbols, bottom axis) for the ~3460 cm⁻¹ IR band of water in propane ($\triangle$), pentane ($\square$), and isooctane ($\bigcirc$) microemulsions at 25°C. [AOT] = 70 nM. (Reprinted with permission from Smith, R. D., Fulton, J. L., Blitz, J. P., and Tingey, J. M., *J. Phys. Chem.*, 94, 781, 1990. © 1900 American Chemical Society.)

function of continuous phase solvent or density, it appears that the core is shielded from the continuous phase solvent by the surfactant and that the solvent properties of the core are substantially dictated by W_m.

The above AOT microemulsion systems were also investigated using dynamic light scattering; the results are shown in Figure 21. The apparent hydrodynamic diameter shows a significant decrease with increasing pressure for a propane microemulsion at $W_m = 5$ and 27°C, whereas pentane and isooctane microemulsions show little or no change. Changes in droplet size of this magnitude for the propane microemulsion would clearly be inconsistent with the spectroscopic results. Thus, this is further evidence that the apparent increase in size can only be ascribed to strong interdroplet attractive interactions, as discussed earlier. In the dynamic light scattering technique, droplet "clustering" caused a measured hydrodynamic diameter (or correlation length) larger than an individual droplet. Such interactions would be expected to increase as the solvent power (or dielectric constant) of the fluid decreases. It is significant that this clustering takes place in a manner that presumably involves interactions of the surfactant shells of distinct droplets (i.e., the interfacial environment), while incurring a negligible effect on the solvent properties of the water core.

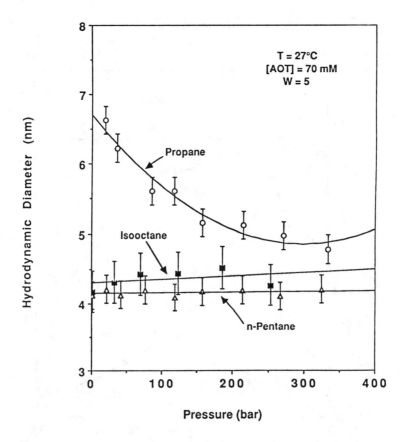

FIGURE 21.   Apparent hydrodynamic diameter in liquid propane, pentane, and isooctane at 27°C for [AOT] = 70 mM and W_m = 5, as measured by dynamic light scattering. (Reprinted with permission from Smith, R. D., Fulton, J. L., Blitz, J. P., and Tingey, J. M., *J. Phys. Chem.*, 94, 781, 1990. © 1990 American Chemical Society.)

V. POTENTIAL APPLICATIONS

A. SEPARATION PROCESSES

The study of microemulsion phases in SCFs has only recently been initiated; hence, the range of applications of these system is largely unknown. Preliminary studies have suggested several roles for these unique systems, however. Water-in-oil microemulsions formed in liquid alkanes such as isooctane have been recently identified as providing a possible shortcut to efficient separation of biomolecules from aqueous solutions, in particular, fermentation broths.[76-78] Many water-soluble biomolecules, including proteins, have been shown to be soluble in near-critical fluids and SCFs. Johnston and co-workers[6] have demonstrated that the solubility of tryptophan in an AOT/octanol/water/ethane microemulsion increases rapidly above 100 bar (see Figure 22), and approaches a constant value at approximately 300 bar.

The ability to extract a wide range of polar solutes, including proteins, from dilute aqueous solutions with reverse micelle phases is well established. In conventional liquids, changes in selectivity can be obtained by changes in pH or ionic strength of the aqueous phase. The studies reported here support the view that SCF reverse micelle phases have solvent powers similar to liquid systems at similar temperatures and W_m. It would, therefore, seem reasonable that variation of pressure might provide efficient extraction from an aqueous phase with significant selectivity without the need for variation of system composition. Such a process might have potential in the biotechnology industry where separations from dilute aqueous systems (e.g., fermentation broths) can be both difficult and expensive.

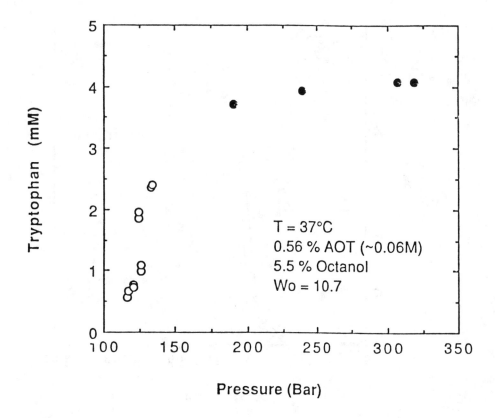

FIGURE 22. Solubility of tryptophan in an AOT/water/octanol/ethane microemulsion for a three-phase system (○) consisting of solid tryptophan, an aqueous-rich and an ethane-rich phase, and a two-phase system (●) consisting of solid tryptophan in equilibrium with the ethane microemulsion. Overall system concentrations are [AOT] = 60 mM. [H$_2$O] = 640 mM, and [octanol] = 0.6 M. (From Johnston, K. P., McFann, G. J., and Lemert, R. M., in *Supercritical Fluid Technology*, ACS Symp. Ser. No. 406, Johnston, K. P. and Penninger, J. L., Eds., American Chemical Society, Washington, D.C., 1989, 140. With permission.)

A fluid system for extraction processes should be functional over an acceptable range of temperature and pressure for the solutes of interest while attaining large W$_o$ values at moderate pressure. At 850 bar, W$_o$ of ~37 and ~35 can be achieved in near-critical (25°C) and supercritical (37°C) ethane, respectively; however, at pressures of 350 bar, W$_o$ is constrained to <5.[72] Supercritical propane (at 110°C) was used for extraction of highly polar solutes (e.g., Basic Red no. 5) from dilute aqueous systems. Both of these SCF systems, however, require compromises due to excessive temperatures or pressures. At high temperatures, many organic compounds are labile and the surfactant begins to degrade at significant rates; at high pressures some biological products tend to denature and the extraction system design is complicated.

In contrast, W$_o$ in the near-critical propane system (at room temperature and under 250 bar) can be varied from <7 to >40. Visual observation of the extraction process confirms that the solvent properties of microemulsions in the near-critical phase are pressure dependent. Additionally, the speed of extraction (compared to conventional liquids) is considerably faster due to higher diffusion coefficients. Plate 1 (p. 446a) shows a propane microemulsion extraction of cytochrome-c from a bulk aqueous solution containing a 0.1 N sodium phosphate buffer at pH 7 and 25°C. Cytochrome-c would be expected to require a microemulsion droplet size having W$_m$ ≳ 6 to 8 for substantial solubilization. At 100 bar, most of the AOT and cytochrome-c are in the turbid lower aqueous phase (which can split into two phases

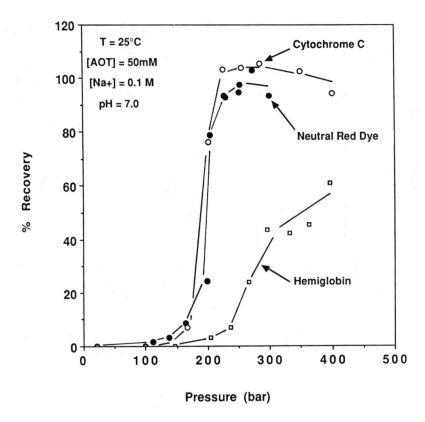

FIGURE 23. Extraction profiles as a function of pressure for cytochrome-*c* (10 μm), hemiglobin (15 μm), and neutral red dye (30 μm) from an aqueous sodium phosphate buffer (pH = 7) containing 50 m*M* AOT. The volume ratio of propane microemulsion to aqueous phase was 2:1. (Reprinted with permission from Smith, R. D., Fulton, J. L., Blitz, J. P., and Tingey, J. M., *J. Phys. Chem.*, 94, 781, 1990. © 1990 American Chemical Society.)

given sufficient time). When the pressure is increased the volume of lower phase decreases slightly as the AOT and water begin to partition into the propane phase and form reverse micelles. At pressures of ~140 bar the propane phase becomes slightly colored due to the extraction of cytochrome-*c*. The extraction efficiency increases dramatically above 190 bar and is virtually complete at 200 bar. Dynamic light scattering studies of this bulk aqueous extraction system are consistent with these results, showing a measurable micelle phase at ≳150 bar. Hydrodynamic diameter changes in a fashion qualitatively consistent with the extraction profiles. Since a two-phase system still exists, the observation of a maximum in the hydrodynamic diameter of ~18 nm at ~250 bar is somewhat surprising. The dynamic light scattering results suggest that for this system, micelle-micelle attractive interactions >250 bar are minimal since the hydrodynamic diameter of ~18 nm changes very little with pressure.

To determine whether this process could be exploited for selective extraction, a similar system with propane above a buffered aqueous AOT solution containing a low molecular weight dye (Neutral Red), cytochrome-*c* (mol wt 12,360), and hemiglobin (mol wt 64,500) was monitored using UV-vis spectroscopy with an estimated uncertainty of ±10%. These substances have no measurable solubility in the pure subcritical or supercritical alkane. The results in Figure 23 show substantial selectivity was obtained for the two proteins, although complete extraction was apparently not obtained for hemiglobin. Incomplete hemiglobin extraction might be attributed to the fact that the W_m for the upper microemulsion phase appears unchanged above ~250 bar. Both cytochrome-*c* and Neutral Red have similar extraction profiles and were not separated. Postextraction electrophoretic analysis of these proteins showed no change in molecular weight, consistent with reverse micelle extraction in conventional liquids (where little loss of enzymatic activity is generally observed if processing times are kept short).[78] These results suggest that opportunities exist for rapid, selective, and efficient extractions from dilute aqueous systems. As in liquids, pH, ionic strength, and temperature may also be useful variables for optimization. Such processes will depend on the selection of appropriate surfactants and the development of efficient recovery steps. Not only were high extraction efficiencies obtained but the propane microemulsion phase could be destroyed and recovered by a reduction of pressure from 250 to 150 bar.

A microemulsion of particular significance for potential applications is the ternary water/CO_2/surfactant system. Since water and CO_2 are two of the most abundant, inexpensive, and biologically compatible solvents, the discovery of such a system could have tremendous implications for the biochemical and chemical industries. A microemulsion and emulsion system based on CO_2 could also be an important solvent for enhanced oil recovery.[79] The probable existence of such a system was recently reported by Ritter and Paulaitis[2], in which a relatively simple nonionic surfactant (2-butoxyethanol) was used. It should be noted, however, that the surfactant-containing phase of this case was apparently water continuous and (therefore) certainly subcritical.

B. REACTION PROCESSES

The incorporation of a microemulsion phase into near-critical fluids and SCFs creates a new type of solvent, and one can speculate on the interesting potential advantages of these systems for reaction processes as well as separations. The high diffusivities and the low viscosities of near-critical fluids and SCFs are well established. These greatly improved mass transport properties are of particular interest for reactions that are diffusion limited. The improvement in mass transport properties also means that in separation processes the rate of uptake or recovery of the solubilizate will be greatly facilitated. Changes in the near-critical fluid or SCF density may be used to effect the partitioning of a reactant into different microdomains of the microemulsion. The hydrocarbon tail region provides the first part of a transition from the extremely low dielectric constant environment of the continuous, SCF phase to the high dielectric constant of the aqueous core. The interfacial region has intermediate properties. The surfactant alignment and strong interactions between the surfactant headgroup counterion and aqueous core can affect molecular orientation, a significant advantage for a catalyst contained in the aqueous core. Many potential reactants that have slightly polar functional groups, such as aromatic or alcohol moieties, will be preferentially solvated at this interface region. By decreasing the dielectric constant of the continuous phase solvent through a reduction of pressure (or utilizing a lower alkane), the slightly polar reactant may partition increasingly into the interface region from the continuous phase. The increased concentration of the reactant at the interface would result in reaction rate increases for those reactions in which the catalyst is contained with the droplet core. Finally, a potentially simple means of recovering the catalyst and/or products is available using the strongly pressure and density-dependent behavior of the near-critical fluid or SCF phase.

Several preliminary studies of reactions in the microemulsions formed in SCFs have been reported. In the work by Randolph et al.,[74] it was shown that a moderately amphiphilic biomolecule, cholesterol (mol wt 387), forms micellar-like aggregates in supercritical CO_2 at pressures near the critical point. These cholesterol aggregates were subsequently catalyzed by an enzyme immobilized on a solid surface. The higher the degree of cholesterol aggregation, the higher the rate of reaction, which is interpreted to result from the increased availability on the binding site of the hydrophobic region of cholesterol to the enzyme anchor site. Matson et al.[80] reported a reaction in a microemulsion formed in supercritical propane in which submicron particles of $Al(OH)_3$ were produced. Aqueous $Al(NO_3)_3$, which was present in the core of the AOT dispersed droplet phase, was reacted with NH_3 dissolved in the supercritical propane continuous phase and produced a particle in the micelle core. Beckman et al.[81] have studied the polymerization of the water-soluble monomer acrylamide in an ethoxylated alcohol/water/(ethane and propane) system both in the near-critical and supercritical regions. The molecular weight and yield of the polymer was apparently correlated to the degree of clustering of the micelles, itself a strong function of pressure. The wide range of reaction processes in liquid microemulsion systems suggests this will be a fertile area for future research.

C. CHROMATOGRAPHIC APPLICATIONS

The application of reverse micelle or microemulsion phases in SCF continuous phases in chromatography constitutes a potentially important area of application, as first demonstrated at this laboratory.[82] SCF chromatography provides somewhat different demands than other types of separations, although it may be argued that the same fundamental concerns are common to all.

For successful application of the SCF/reverse micelle chromatographic technique, the advantages and limitations of these systems must be understood. First, the phase behavior must be sufficiently well understood so that reverse micelle mobile phases (RMMPs) can be easily prepared, stored, and pumped as single-phase systems. Thus, the phase behavior of these systems must be determined for a useful range of temperature and pressure conditions. Second, the use of RMMPs must offer advantages over conventional SCF and liquid mobile phases. Specifically, their use offers a combination of solubilizing power and mass transport characteristics (i.e., diffusion coefficients) that allow compounds not amenable to gas chromatography (GC) to be addressed with higher chromatographic efficiencies than obtainable by liquid chromatography (LC). Finally, the use of RMMPs must be compatible with practical detection methods, the solvating power must be adjustable (ideally allowing gradient operation), and compatible stationary phases that offer reasonable sample loadings and capacity factors must be identified or developed.

Three independent variables are available in conventional SCF chromatography for manipulating the solvent power of the mobile phase and for changing retention: temperature, pressure, and mobile phase composition. Utilization of an RMMP adds a fourth variable related to the composition of the reverse micelle or microemulsion droplets and can strongly affect retention of compounds that partition significantly into either the micelle (surfactant) interface or the water core. It is reasonable to distinguish composition gradients of the continuous (supercritical) phase from those of the reverse micelle or microemulsion components.

Three new gradient methods can be based upon the use of RMMPs. These involve variation of (1) pH, ionic strength, or chemical composition of the water core; (2) water to surfactant ration (W_m); and (3) micelle or microemulsion droplet concentration (at constant W_m). The authors expect the first method to be of limited value in the near future; such a method leads to complex changes in the mobile phase, in CMCs, and in the thermodynamic stability of micelles. Both the second and third methods appear more practical and more readily related to the solvent properties of the RMMP. These methods may be considered analogous to the standard gradient methods in LC and GC, respectively.

W_m can be directly related to the solvent power of the microemulsion. As W_m increases the core evolves from a highly ordered state dictated by the surfactant headgroups and counterions (hydrogen bonding is minimal) to a situation at large W_m where solvent properties approach that of bulk water. Thus, the mixing of two micellar solutions can be used to program W_m in a manner strongly analogous to composition gradients in LC. Since the number of solute molecules solubilized in each micelle will be generally small (typically one for large molecules such as proteins), the micelle density (or concentration) is also an important consideration. If W_m is directly proportional to micelle core diameter, the number of surfactant molecules per micelle (N_s) is directly proportional to W_m^2. Similarly, the number of water molecules per micelle (N_w) is directly proportional to W_m^3. If one simply doubles the molar concentration of water (doubles W_m), the micelle core diameter doubles, and N_s and N_w increase by factors of 4 and 8, respectively. Because the number of surfactant molecules is unchanged, the density of micelles is reduced by a factor of 4. Thus, for the second method it may be useful to increase both N_s and N_w during a gradient, although practical limitations at high surfactant concentrations (e.g., restrictor plugging) may exist.

The third method explicitly involves manipulation of the micelle number density at constant W_m. In this situation, both surfactant and water concentrations (N_s and N_w) would be varied over (perhaps) several orders of magnitude such that while solvent environments remain unchanged, the capacity of the RMMP for hydrophilic solutes would increase dramatically. The micelle number density in the RMMP can be varied from low concentrations, where its effects will be negligible, to quite high concentrations in a readily manipulatable fashion. Such a gradient method can be considered analogous to temperature programming in GC, another method in which the solvent properties of the mobile phase are unchanged while the capacity for volatile substances changes dramatically. This gradient method has not yet been explored for RMMP in SCF chromatography.

A propane/AOT/water RMMP at W = 10 was used at 308 bar to study chromatographic capacity factors ($k' = (t_r - t_o)/t_o$) between 25 and 120°C for separation of two polar dyes using a column packed with 5 μm microporous silica particles.[83] The results in Figure 24 show the expected trend below 100°C; as temperature is increased the solvent power of the fluid is increased, and a decrease in k' is observed. Surprisingly, however, above 100°C retention again increases. It is tempting to ascribe this behavior to a change in solvent properties above the critical point of the fluid, but since no change in W_m takes place, it is difficult to rationalize how the solvent power of the fluid could decrease.

The use of fused silica capillary columns coated with porous silica has also been investigated for SFC with RMMPs. Propane was selected as the mobile phase to avoid the higher pressures required to obtain a large W_m (>10) with ethane. The high solubilizing power obtainable with RMMPs is illustrated by the chromatogram in Figure 25, which shows the results for an injection of an AOT isooctane reverse micelle solution containing the protein hemoglobin. These results were obtained using the silica column and a propane/AOT/water mobile phase with [AOT] = 50 mM at W = 7 and 130°C.

The use of reverse micelle and microemulsion mobile phases clearly has the potential for extension of SCF chromatography to nonvolatile hydrophilic compounds. The use of such systems offers clear attractions for extraction processes and other areas,[72] but application to SCF chromatography remains problematic. While most experimental difficulties associated with preparing, pumping, injection, and pressure restriction and detection (given certain limitations) appear tractable, major uncertainties related to the chromatographic process remain to be studied.

VI. CONCLUSIONS

Formation of a microemulsion within an SCF presents a means by which to overcome the difficulties in solubilizing polar substances in such fluids. The study of SCF-based microemulsion systems, although a recent phenomenon, has already provided some interesting details as to the effect of pressure on the phase behavior, micelle structure, and core environment. Sharp contrasts have been observed in several cases between the pressure-induced changes in liquid vs. SCF-based microemulsion systems.

Although the effect of pressure on liquid emulsion systems is qualitatively described by the phenomenological phase behavior model, too little is known about the phase behavior

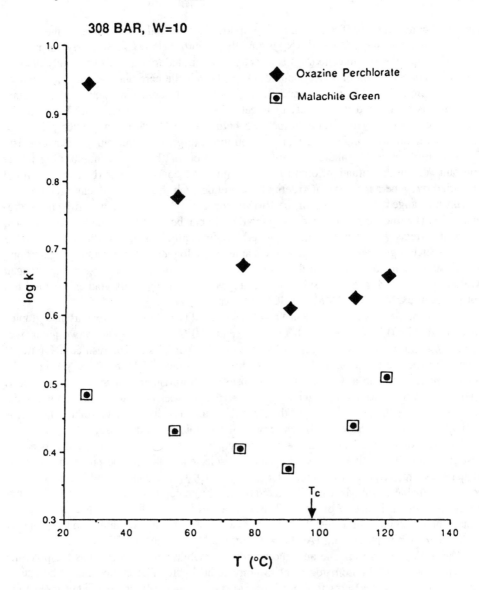

FIGURE 24.   Effect of temperature upon capacity factor (k′) for two hydrophilic dyes with a propane RMMP at 308 bar and $W_m = 10$. (From Smith, R. D., Fulton, J. L., Jones, H. K., Gale, R. W., and Wright, B. W., *J. Chromatogr. Sci.*, 27, 309, 1989. Reproduced by permission of Preston Publications, a division of Preston Industries, Inc.)

of surfactants in SCFs to make confident predictions regarding these systems. For example, in liquid systems employing ionic surfactants, the upper and lower miscibility gaps of the surfactant-oil binary generally occur at subzero and very high temperatures. Thus, the effect of pressure in these systems hinges upon its effect on the water-surfactant phase diagram. By contrast, the small size and lower density of an SCF can produce miscibility gaps well within the temperature range 0 to 100°C, and possibly one connected miscibility gap. These facts, coupled with the higher compressibility of a SCF solution as compared to an analogous liquid oil mixture can quite easily lead to a situation where the effect of pressure on the phase behavior of a microemulsion is due primarily to pressure-induced shifts in the oil-surfactant miscibility gaps. Furthermore, the shifts in the critical solution temperatures of liquid and SCF-surfactant mixtures can exhibit opposite signs.

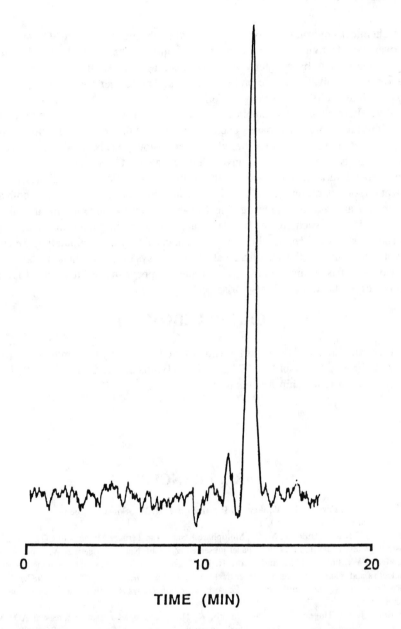

TIME (MIN)

FIGURE 25.   Supercritical propane RMMP (at W_m = 7, 247 bar, and 130°C) chromatogram obtained from the injection of a hemoglobin solution, with the silica capillary column. (From Smith, R. D., Fulton, J. L., Jones, H. K., Gale, R. W., and Wright, B. W., *J. Chromatogr. Sci.*, 27, 309, 1989. Reproduced by permission.)

The high compressibility of SCFs also produces stark contrasts between these and liquid systems when observing the effect of pressure on the structure and chemical environment of micellar systems. In aqueous systems, pressure up to approximately 1000 bar is observed to increase the CMC and decrease the apparent hydrodynamic radius of micelles, facts consistent with pressure-induced deaggregation. In SCF-based micellar systems, experimental evidence indicates that the observed drop in the apparent hydrodynamic radius is due to a decrease in micelle-micelle interactions, or effective declustering. Work in progress, including high-pressure neutron scattering studies and the measurement of the effect of pressure on the CMC of SCF-based systems, should help clarify the situation.

The chemical environment, or polarity, within the micelle in an SCF system can be varied over a wide range via pressure tuning, quite unlike an analogous liquid system. Results summarized in this chapter vividly demonstrated this feature of SCF microemulsions using a variety of solvatochromic probe molecules. The further advantages of nonintrusive methods, such as FT-IR, are clearly evident.

Finally, although potential applications for SCF microemulsion systems abound, the surface of this area of study has barely been scratched. The feasibility of performing reactions, both organic and inorganic, in SCF-based microemulsions, has been demonstrated and some interesting effects due to pressure have been revealed. It has also been shown that the pressure-induced variability of the polarity within the micelle core facilitates the selective solubilization, or separation, of biological molecules. Several research groups, both academic and industrial, are currently investigating CO_2-based microemulsions for use in enhanced oil recovery. Indeed, such studies may ultimately yield an improved understanding of microemulsion systems in general, due to the unique capability of manipulating the continuous phase without changing the chemical nature of the system. As fundamental data on the behavior of SCF-based microemulsions continue to appear in the literature, it is likely that the number of applications will also arise.

ACKNOWLEDGMENT

We thank the Chemical Sciences Division of the U. S. Army Research Office for support of this work through Contract P-24974-CH. E. J. B. thanks N. O. R. C. U. S. for support of his postdoctoral fellowship for this period.

REFERENCES

1. **Smith, R. D., Frye, S. L., Yonker, C. R., and Gale, R. W.,** Solvent properties of supercritical Xe and SF$_6$, *J. Phys. Chem.,* 91, 3059, 1987.
2. **Ritter, J. M. and Paulaitis, M. E.,** Multiphase Behavior in Ternary Mixtures of CO_2, H_2O and Nonionic Amphiphiles at Elevated Pressures, paper presented at AIChE National Meeting, Washington, D.C., 1988.
3. **Gale, R. W., Fulton, J. L., and Smith, R. D.,** Organized molecular assemblies in the gas phase: reverse micelles and microemulsions in supercritical fluids, *J. Am. Chem. Soc.,* 109, 920, 1987.
4. **Beckman, E. J. and Smith, R. D.,** Phase behavior of the system anionic amphiphile-water-propylene, submitted.
5. **Fulton, J. L., Tingey, J. M., and Smith, R. D.,** Inter-droplet attractive forces in AOT water-in-oil microemulsions formed in subcritical and supercritical solvents, *J. Phys. Chem.,* 94, 1997, 1990.
6. **Johnston, K. P., McFann, G. J., and Lemert, R. M.,** Pressure tuning of reverse micelles for adjustable solvation of hydrophiles in supercritical fluids, in *Supercritical Fluid Technology,* Johnston, K. P. and Penninger, J. L., Eds., ACS Symp. Ser. No. 406, American Chemical Society, Washington, D.C., 1989, 140.
7. **McHugh, M. A. and Krukonis, V. J.,** *Supercritical Critical Extraction,* Butterworths, Boston, 1986.
8. **Paulaitis, M. E., Penninger, J. M. L., Gary, R. D., and Davidson, R., Eds.,** *Chemical Engineering at Supercritical Fluid Conditions,* Ann Arbor Science Publishers, Ann Arbor, MI, 1984.
9. **Mazer, N. A.,** Laser light scattering in micellar systems, in *Dynamic Light Scattering,* Pecora, R., Ed., Plenum Press, New York, 1985, chap. 8.
10. **Fulton, J. L. and Smith, R. D.,** Reverse micelle and microemulsion phases in supercritical fluids, *J. Phys. Chem.,* 92, 2903, 1988.
11. **Kotlarchyk, M., Huang, J. S., and Chen, S. H.,** Structure of AOT reversed micelles determined by small-angle neutron scattering, *J. Phys. Chem.,* 89, 4382, 1985.
12. **Tanford, C.,** in *Micellization, Solubilization and Microemulsion,* Mittal, K. L., Ed., Plenum Press, New York, 1977, 119.

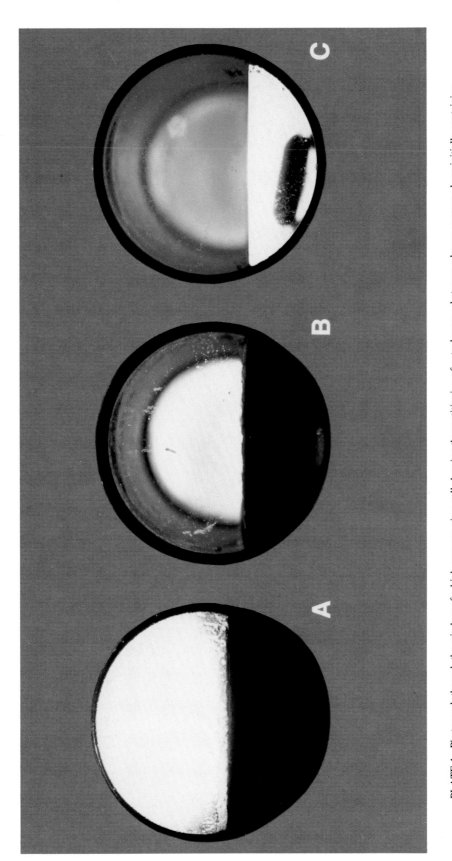

PLATE 1. Photograph through the window of a high-pressure view cell showing the partitioning of cytochrome-*c* between a lower aqueous phase initially containing 0.1 N sodium phosphate buffer (pH 7), 50 m*M* AOT, and 10 μ*M* cytochrome-*c* and an upper propane continuous phase at (A) 100 bar, (B) 200 bar, (C) 300 bar. (Reprinted with permission from Smith, R. D., Fulton, J. L., Blitz, J. P., and Tingey, J. M., *J. Phys. Chem.*, 94, 781, 1990. © American Chemical Society.)

13. **Kaler, E. W. and Chang, N. J.,** The structure of sodium dodecyl sulfate micelles in solutions of H_2O and D_2O, *J. Phys. Chem.*, 89, 2996, 1985.

14. **Kahlweit, M., Strey, R., Firman, P., Haase, D., Jen, J., and Schomacker, R.,** General patterns of the phase behavior of mixtures of H_2O, nonpolar solvents, amphiphiles, and electrolytes. *I., Langmuir,* 4, 499, 1988.

15. **Kahlweit, M., Strey, R., Schomacker, R., and Haase, D.,** General patterns of the phase behavior of mixtures of H_2O, nonpolar solvent, amphiphiles, and electrolytes. II. *Langmuir,* 5, 305, 1989.

16. **Kahlweit, M. and Strey, R.,** Phase behavior of ternary systems of the type H_2O-oil-nonionic amphiphile (microemulsions), Kahlweit, *Angew. Chem. Int. Ed. Engl.,* 24, 654, 1985.

17. **Sjoblom, J., Stenius, P., and Danielsson, I.,** Phase equilbria of nonionic surfactants and for the formation of microemulsions, in *Nonionic Surfactants Physical Chemistry,* Surfactant Science Ser. No. 23, Schick, M. J., Ed., Marcel Dekker, New York, 1987, 369.

18. **Mackay, R. A.,** Solubilization, in *Nonionic Surfactants Physical Chemistry,* Surfactant Science Ser. No. 23, Schick, M. J., Ed., Marcel Dekker, New York, 1987, 297.

19. **Wormuth, K. R. and Kaler, E. W.,** Microemulsifying polar oils, *J. Phys. Chem.,* 93, 4855, 1989.

20. **Schneider, G. M.** High pressure thermodynamics of mixtures, *Pure Appl. Chem.,* 47, 277, 1976.

21. **Schneider, G. M.,** Fluid mixtures at high pressures, *Pure Appl. Chem.,* 53, 479, 1983.

22. **Schneider, G. M.,** Physicochemical principles of extraction with supercritical gases *Angew. Chem. Int. Ed. Engl.,* 17, 716, 1978.

23. **Fotland, P.,** Some observations on the effect of pressure on a five-component anionic microemulsion, *J. Phys. Chem.,* 91, 6396, 1987.

24. **Kim, M. W., Bock, J., Huang, J. S., and Gallagher, W.,** Pressure induce phase transition of an oil external microemulsion, in *Surfactants in Solution,* Vol. 6, Mittal, K. L. and Bothorel, P., Eds., Plenum Press, New York, 1986, 1193.

25. **Kim, M. W., Gallagher, W., and Bock, J.,** Pressure dependence on multiphase microemulsions, *J. Phys. Chem.,* 92, 1226, 1988.

26. **Zeman, L. and Patterson, D.,** Pressure effects in polymer solution phase equilibria. II. Systems showing upper and lower critical solution temperatures, *J. Phys. Chem.,* 76, 1214, 1972.

27. **Beckman, E. J. and Smith, R. D.,** Phase behavior of inverse microemulsions for the polymerization of acrylamide in near-critical and supercritical continuous phases, *J. Phys. Chem.,* 94, 345, 1990.

28. **Candau, F. and Holtzscherer, C.,** Polymerization of Water-Soluble Monomers in Nonionic Bicontinuous Microemulsions, presented at American Chemical Society Fall Meeting, New Orleans, 1987.

29. **Candau, F., Leong, Y. S., Pouyet, G., and Candau, S.,** Inverse microemulsion polymerization of acrylamide: characterization of the water-in-oil microemulsions and the final microlatexes, *J. Coll. Interface Sci.,* 101, 167, 1984.

30. **Steytler, D. C., Lovell, D. R., Moulson, P. S., Richmond, P., Eastoe, J., and Robinson, B. H.,** Microemulsion stability in near-critical fluids, in *Proc. Int. Symp. Supercritical Fluids,* Societe Francaise de Chimie, Nice, France, 1988, 67.

31. **Rogers, J. and Winson, P. A.,** Change in the optic sign of the lamellar phase (G) in the aerosol OT/water system with composition or temperature, *J. Coll. Interface Sci.,* 30, 247, 1969.

32. **Robbins, M. L., Bock, J., and Huang, J. S.,** Model for microemulsions. III. Interfacial tension and droplet size correlation with phase behavior of mixed surfactants, *J. Coll. Interface Sci.,* 126, 114, 1988.

33. **Lam, A. C., Falk, N. A., and Schechter, R. A.,** The thermodynamics of microemulsion, *J. Coll. Interface Sci.,* 120, 30, 1987.

34. **Miller, C. A. and Neogi, P.,** Thermodynamics of microemulsions: combined effects of dispersion entropy of drops and bending energy of surfactant films, *AIChE J.,* 26, 212, 1980.

35. **Ruckenstein, E.,** The surface of tension, the natural radius, and the interfacial tension in the thermodynamics of microemulsions, *J. Coll. Interface Sci.,* 114, 173, 1986.

36. **Mukherjee, S., Miller, C. A., and Tomlinson, F., Jr.,** Theory of drop size and phase continuity in microemulsions. I. Bending effects with uncharged surfactants, *J. Coll. Interface Sci.,* 91, 223, 1983.

37. **Jansson, M. and Jonsson, B.,** Influences of counterion hydrophobicity on the formation of ionic micelles, *J. Phys. Chem.,* 93, 1451, 1989.

38. **Overbeek, J. Th. G., Verhoeckx, G. J., De Bruyn, P. L., and Lekkerkerker, H. N. W.,** On understanding microemulsions. II. Thermodynamics of droplet-type microemulsions, *J. Coll. Interface Sci.,* 119, 442, 1987.

39. **Overbeek, J. Th. G.,** The first rideal lecture, microemulsions, a field at the border between lyophobic and lyophillic colloids, *Faraday Disc. Chem. Soc.,* 65, 7, 1978.

40. **Evans, D. F. and Ninham, B. W.,** Ion binding and the hydrophobic effect, *J. Phys. Chem.,* 87, 5025, 1983.

41. **Jonsson, B. and Wennerstrom, H.,** Phase equilibria in a three-component water-soap-alcohol system. A thermodynamic model, *J. Phys. Chem.,* 91, 338, 1987.

42. **Blankschtein, D., Thurston, G. M., and Benedek, G. B.,** Phenomenological theory of equilibrium thermodynamic properties and phase separation of micellar solutions, *J. Chem. Phys.,* 12, 7268, 1986.
43. **Ciach, A., Hoye, J. S., and Stell, G.,** Microscopic model for microemulsions. I. Ground state properties, *J. Chem. Phys.,* 90, 1214, 1989.
44. **Hu, Y. and Prausnitz, J. M.,** Molecular thermodynamics of partially-ordered fluids: microemulsions, *AIChE J.,* 34, 814, 1988.
45. **Thurston, G. M., Blankschtein, D., Fisch, M. R., and Benedek, G. B.,** Theory of thermodynamic properties and phase separation of micellar solutions with lower consolute points, *J. Chem. Phys.,* 84, 4558, 1986.
46. **Widom, B.,** Lattice-gas model of amphiphiles and of their orientation at interfaces, *J. Phys. Chem.,* 88, 6508, 1984.
47. **Widom, B.,** Lattice model of microemulsions, *J. Chem. Phys.,* 84, 6943, 1986.
48. **Talmon, Y. and Prager, S.,** Statistical thermodynamics of phase equilibria in microemulsions, *J. Chem. Phys.,* 69, 2984, 1978.
49. **Wennerstrom, H. and Lindman, B.,** Micelles, physical chemistry of surfactant association, *Phys. Rep.,* 52, 1, 1979.
50. **Stecker, M. M. and Benedek, G. B.,** Theory of multicomponent micelles and microemulsions, *J. Phys. Chem.,* 88, 6519, 1984.
51. **Hyde, S. T., Ninham, B. W., and Zemb, T.,** Phase boundaries for ternary microemulsions. Predictions of a geometric model, *J. Phys. Chem.,* 93, 1464, 1989.
52. **Hyde, S. T.,** Microstructure of bicontinuous surfactant aggregates, *J. Phys. Chem.,* 93, 1458, 1989.
53. **Hou, M.-J. and Shah, D. O.,** Effects on the molecular structure of the interface and continuous phase on solubilization of water in water/oil microemulsions, *Langmuir,* 3, 1086, 1987.
54. **De Gennes, P. G. and Taupin, C.,** Microemulsions and the flexibility of oil/water interfaces, *J. Phys. Chem.,* 86, 2294, 1982.
55. **Safran, S. A. and Turkevich, L. A.,** Phase diagrams for microemulsions, *Phys. Rev. Lett.,* 50, 1930, 1983.
56. **Lemaire, B., Bothorel, P., and Roux, D.,** Micellar interactions in water-in-oil microemulsions. I. Calculated interaction potential, *J. Phys. Chem.,* 87, 1023, 1983.
57. **Lemaire, B., Bothorel, P., and Roux, D.,** Micellar interactions in water-in-oil microemulsions. II. Light scattering determination of the second virial coefficient, *J. Phys. Chem.,* 87, 1028, 1983.
58. **Huh, C.,** Equilibrium of a microemulsion that coexists with oil or brine, *Soc. Petrol. Eng. J.,* 23, 829, 1983.
59. **Hall, D. G.,** in *Nonionic Surfactants Physical Chemistry,* Schick, M. J., Ed., Marcel Dekker, New York, 1987, chap. 5.
60. **Fourche, G. and Bellocq, A.-M.,** Light scattering evidence for a critical behavior in microemulsions, *J. Coll. Interface Sci.,* 88, 302, 1982.
61. **Roux, D., Bellocq, A.-M., and Bothorel, P.,** Effect of the molecular structure of components on micellar interactions in microemulsions, in *Surfactants in Solution,* Mittal, K. L. and Lindman, B., Eds., Plenum Press, New York, 1984, 1843.
62. **Calje, A., Agterof, W., and Vrij, A.,** in *Micellization, Solubilization, and Microemulsions,* Mittal, K. L., Ed., Plenum Press, New York, 1977, 780.
63. **Fulton, J. L., Blitz, J. P., Tingey, J. M., and Smith, R. D.,** Reverse micelle and microemulsion phases in supercritical xenon and ethane: light scattering and spectroscopic probe studies, *J. Phys. Chem.,* 92, 4198, 1989.
64. **Hamaker, H. C.,** The London-Van der Waals attraction between spherical particles, *Physica IV,* 10, 1058, 1937.
65. **Deryaguin, B. V. and Kussakov, M.,** *Acta Physicochim, URSS,* 10, 25, 1939.
66. **Lifshitz, E. M.,** *Sov. Phys.-JETP,* 2, 73, 1956.
67. **Smith, E. R., Mitchell, D. J., and Ninham, B. W.,** Deviations of the van der Waals energy for two interacting spheres from the predictions of Hamaker theory, *J. Coll. Interface Sci.,* 45, 55, 1973.
68. **Gingell, D. and Parsegian, V. A.,** Computation of van der Waals interactions in aqueous systems using reflectivity data, *J. Theor. Biol.,* 36, 41, 1972.
69. **Blitz, J. P., Fulton, J. L., and Smith, R. D.,** Dynamic light scattering measurements of reverse micelle phases in liquid and supercritical ethane, *J. Phys. Chem.,* 92, 2707, 1988.
70. **Smith, R. D., Fulton, J. L., and Blitz, J. P.,** Structure of reverse micelle and microemulsion phases in near-critical and supercritical fluids as determined from dynamic light scattering studies, in *Supercritical Fluid Technology, ACS Symp. Ser. No. 406,* Johnston, K. P. and Penninger, J. L., Eds., American Chemical Society, Washington, D.C., 1989, 165.
71. **Beckman, E. J. and Smith, R. D.,** Microemulsions formed from nonionic surfactants in near-critical and supercritical alkanes: quasi-elastic light scattering investigations, *J. Phys. Chem.,* 94, 3729, 1990.

72. **Smith, R. D., Fulton, J. L., Blitz, J. P., and Tingey, J. M.,** Reverse micelles and microemulsions in near-critical and supercritical fluids, *J. Phys. Chem.,* 94, 781, 1990.

73. **Tamura, K. and Suminaka, M.,** Effects of pressure on reversed micellar systems; rates of the keto-enol transformation of pentane-2,4-dione, *J. Chem. Soc. Faraday Trans. I,* 81, 2287, 1985.

74. **Randolph, T. W., Clark, D. S., Blanch, H. W., and Prausnitz, J. M.,** Enzymatic oxidation of cholesterol aggregates in supercritical carbon dioxide, *Science,* 238, 387, 1988.

75. **Olesik, S. and Miller, C. J.,** Critical micelle concentration of AOT in supercritical alkanes, *Langmuir,* 6, 183, 1990.

76. **Goklen, K. E. and Hatton, T. A.,** Liquid-liquid extractions of low molecular-weight proteins by selective solubilization in reversed micelles, in *Separation Science and Technology,* Giddings, J. C., Ed., Marcel Dekker, New York, 1988, 22.

77. **Luisi, P. L.,** Enzymes hosted in reverse micelles in hydrocarbon solutions, *Angew. Chem. Int. Ed. Engl.,* 24, 439, 1985.

76. **Goklen, K. E. and Hatton, T. A.,** Liquid-liquid extractions of low molecular-weight proteins by selective solubilization in reversed micelles, in *Separation Science and Technology,* Giddings, J. C., Ed., Marcel Dekker, New York, 1987, 22.

77. **Luisi, P. L.,** Enzymes hosted in reverse micelles in hydrocarbon solutions, *Angew. Chem. Int. Ed. Engl.,* 24, 439, 1985.

78. **Goklen, K. E. and Hatton, T. A.,** Protein extraction using reverse micelles, *Biotech. Progr.,* 1, 69, 1985.

79. **Fulton, J. L. and Smith, R. D.,** Organized surfactant assemblies in supercritical fluids, in *Surfactant-Based Mobility Control,* Smith, D. H., Ed., ACS Symp. Ser., American Chemical Society, Washington, D.C., 1988, 373.

80. **Matson, D. W., Fulton, J. L., and Smith, R. D.,** Formation of fine particles in supercritical fluid micelle systems, *Mat. lett.,* 6, 31, 1987.

81. **Beckman, E. J., Fulton, J. L., Matson, D. W., and Smith, R. D.,** Inverse emulsion polymerization of acrylamide in near-critical and supercritical continuous phases, in *Supercritical Fluid Technology,* Johnston, K. P. and Penninger, J. L., Eds., ACS Symp. Ser. No. 406, American Chemical Society, Washington, D.C., 1989, 184.

82. **Gale, R. W., Fulton, J. L., and Smith, R. D.,** Reverse micelle supercritical fluid chromatography, *Anal. Chem.,* 59, 1977, 1987.

83. **Smith, R. D., Fulton, J. L., Jones, H. K., Gale, R. W., and Wright, B. W.,** The potential of reverse micelle mobile phases for supercritical fluid chromatography, *J. Chromatogr. Sci.,* 27, 309, 1989.

Chapter 13

SUPERCRITICAL FLUID EXTRACTION AND RETROGRADE CONDENSATION (SFE/RC) APPLICATIONS IN BIOTECHNOLOGY

Karl Schulz, Eloy E. Martinelli, and G. Ali Mansoori*

TABLE OF CONTENTS

* To whom correspondence concerning this article should be addressed.

ABSTRACT

The fundamentals and techniques of supercritical fluid extraction (SFE) and retrograde condensation (RC) processes are described. Supercritical (CO_2 and other biologically non-toxic supercritical fluids (SCFs) are found to be effective extracting agents of many compounds of interest in biotechnology. It is demonstrated that supercritical solubility calculation methods can be improved through the use of statistical mechanical theories of asymmetric mixtures. Such computation techniques could be successfully applied to improve engineering design. In addition, the effect of mixed supercritical solvents on the solubility of heavy solutes, at different pressures, temperatures, and solvent compositions is studied. The prospects for the commercial application of SFE/RC phenomena in biological processes are discussed and a review of important biotechnological applications of SFE/RC processes is presented.

I. INTRODUCTION

There has been a great deal of interest in recent years on separation techniques that could be of use in separating components of biological fluids in large industrial operations.[1-4] While there exist numerous separation techniques that have been used in the laboratory for bioseparation purposes, few such techniques are applicable for large-scale separations. Of the potential large-scale separation techniques bioseparation by partitioning[2,3] and supercritical fluid extraction/retrograde condensation[4] (SFE/RC) processes may be mentioned. There exist a number of biological separations in which either the partitioning or SFE/RC processes can be used. Partitioning is especially useful for separation of biologically active or live macromolecules, while the SFE/RC process is particularly applicable for separation of heat-sensitive biological fluids and compounds. In this chapter, the SFE/RC process as applied to biotechnical processes is presented and reviewed.

The SFE/RC phenomenon occurred in 1879, with Hannay and Hogarth's[5] initial observation. A typical SFE/RC separation system is depicted in Figure 1. A high-pressure solvent gas tank is connected to an extraction vessel. In the extractor, the solute is put into contact with a supercritical solvent (i.e., with a solvent at conditions above its critical pressure and temperature). The gaseous mixture, saturated with the extracted substance, leaves the extraction chamber through an expansion valve where the pressure is greatly reduced. Because of the sudden expansion, the mixture temperature undergoes a significant drop; hence, a heat exchanger is usually necessary to rewarm the fluid to at least its initial temperature. Since the solubility of the solute in the gas is appreciably lessened due to the pressure decrease, some of the extracted material will be separated at this stage. Retrograde condensation, or depressurizing to precipitate the extract, can be accomplished through one large, or several smaller, pressure reduction steps. After each step, a heat exchanger and collection tank may be necessary. Upon completion of the extraction cycle, the solvent, now almost totally devoid of extracted material and low in pressure, is returned to its initial pressure and temperature, thus preparing it for further solute extraction.

An alternative to the use of an extraction chamber in SFE/RC systems is the jet extraction method,[6] a technique developed primarily to facilitate extraction in the case of viscous mixtures of solids dispersed in liquids (which tend to clog the apparatus). In the jet extraction technique (see Figure 2), the starting viscous material is forced through a narrow capillary tube, often as fine as 0.2 mm i.d. The exit of this capillary tube is then inserted into a second capillary tube of slightly larger diameter. The second tube overlaps only a certain extent of the first tube. The extraction gas enters the larger capillary tube through the interstice between the two capillaries, thereby contacting the solute. Because the interstice is so small, the solvent achieves high velocity and a great deal of turbulence, thus effectively mixing

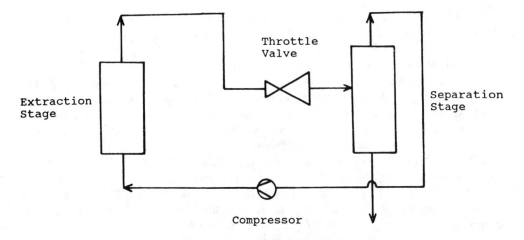

FIGURE 1. Schematic diagram of an SFE/RC cycle.

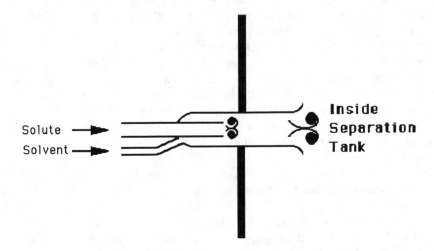

FIGURE 2. Jet mixing and extraction of viscous mixtures of solids dispersed in liquids.

itself with the solute in only a very short distance. The extraction is completed in this short distance, after which the undissolved material is collected in a separation tank while the gas phase passes through a second vessel where it undergoes retrograde condensation. Thereafter, the solvent gas is repressurized and recycled to the system.

An SCF can be an effective solvent for a condensed solute (liquid or solid). This requires a large molecular weight and size difference between the supercritical solvent and the condensed solute. To understand the behavior of SCFs, their physical properties must be compared to those of normal gases and liquids. The density of SCFs (about 0.3 to 0.7 g cm^{-3}) is much greater than gas densities at normal conditions, but nearly equal to liquid densities at the same conditions (about 10^{-3} and 1 g cm^{-3}, respectively). The viscosity of SCFs (10^{-6} to 10^{-3} g cm^{-1} s^{-1}) lies between the gas and liquid viscosities (10^{-6} and 10^{-2} g cm^{-1} s^{-1}, respectively). The diffusion coefficient of SCFs ranges between 10^{-6} and 10^{-3} cm^2 s^{-1}, which is below 10^{-1} (gas diffusivity) and similar to 10^{-5} (liquid diffusivity). Because the supercritical diffusivities are very small, molecular motion is relatively unimpeded, resulting in convective heat transfer as much as two orders of magnitude higher for SCFs than for liquids at normal conditions. The relatively low viscosities of SCFs make them easier to handle than liquids, and the loading capacity can be regulated through temperature and pressure control. A sudden temperature or pressure reduction often causes the loss of the special characteristics of the SCF. Therefore, a substance can be extracted at supercritical conditions of a solvent and precipitated by reducing the pressure or the temperature below the critical conditions of the solvent.

CO_2 has been found to be the most often used extraction solvent in the majority of SFE/RC systems in general and biosystems in particular. CO_2 is useful because it is not toxic when exposed to biological materials and is inexpensively produced, thus making it an ideal candidate for an extraction system. At room temperature and atmospheric pressure CO_2 is virtually useless as an organic solvent; however, at its supercritical conditions (T >304 K and P >74 bar) many molecules become quite soluble in CO_2. When CO_2 is used as a supercritical solvent, solubilities of solutes in it tend to increase markedly as the temperature rises above 323 to 333 K, especially when the pressure exceeds 400 bar.[7,8] A possible reason for this strong solubility increase at high pressures is that above 400 bar the CO_2 solvent is nearly incompressible. When the temperature is increased, the solvent density undergoes only minor reductions, thus tending to decrease solute solubility. Simultaneously, the temperature increases induces a major vapor pressure increase, thus tending to increase solute solubility. Because the vapor pressure effect dominates the density effect, the net result is a large solute solubility increase. At pressures below 400 bar, the density of CO_2 will tend to decrease appreciably as the temperature rises. When this effect is combined with the increase in solute vapor pressure the result is only small solute solubility increases.

The solubility of a solute in CO_2 is hindered by the presence of polar functional groups in the molecular structure of the solute, such as carbonyl and hydroxyl groups. The solute solubility is more strongly hindered by hydroxyl groups, so much so that few solutes containing more than three hydroxyl groups per molecule exhibit appreciable solubility in supercritical CO_2. The high molecular weight of the solutes tend to inhibit solubility in CO_2, to a degree that only a few solutes with a molecular weight above 400 g mol^{-1} show appreciable solubility in supercritical CO_2.

The use of cosolvents can sometimes increase solute solubility. Solutes containing polar groups show solubility enhancement when cosolvents such as ethanol or methanol are used. This solubility increase can be explained in terms of the hydrogen bonding between the solute and the cosolvent. On the other hand, there exist cosolvents that lower solute solubility due to the formation of solid complexes between them. Cosolvents that enhance solute solubility are called entrainers. Usually it is hard to find the optimum cosolvent for a certain solute because each possible system must be individually studied.

Having reviewed the properties of SCFs, the solubility trends of solutes in supercritical solvents, and the hardware of the extraction system, the theoretical basis of SFE/RC phenomena and some current biotechnological applications of SFE/RC are examined.

II. THERMODYNAMIC MODELING OF SFE/RC PHENOMENA

Figure 3 illustrates the phenomena of SFE and RC from the thermodynamic point of view. A simple pressure vs. temperature (PVT) diagram of a fluid mixture exhibiting retrograde condensation is presented in Figure 3A. In Figure 3B, pressure vs. solubility of a heavy compound in the gas phase, a three different temperatures, is shown. According to Figure 3, above the critical solution temperature and at low pressures, solubility in the supercritical gas will decrease with an increase of pressure, at a given temperature. As pressure rises, solubility passes through a minimum and then increases rapidly until it reaches a maximum at a pressure slightly above retrograde.

In general, to model the thermodynamic behavior of SFE/RC phenomena, two different statistical mechanical procedures may be utilized:[9] (1) the use of equations of state (EOSs) for pure fluids combined with appropriate conformal solution mixing rules, and (2) the use

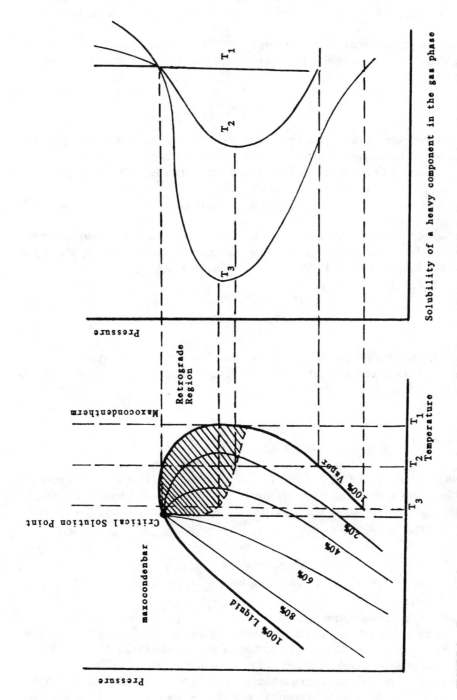

FIGURE 3.   Pressure against temperature diagram of a representative fluid mixture exhibiting retrograde condensation, and pressure against solubility of a heavy component in the gas phase at three different temperatures.[40-42]

of detailed expressions for the intermolecular potential energy functions and rigorous statistical mechanics of mixtures.[10,11] Conformal solution mixing rules and rigorous statistical mechanics of mixtures were developed substantially in the last decade;[12-17] however, there is little information about intermolecular interaction parameters for the asymmetric molecules found in the SFE/RC processes. Therefore, utility of the second procedure is presently limited.

The concept of conformal solutions refers to substances whose intermolecular potential energy functions, ϕ_{ij}, are related to those of a reference pure fluid, indicated by the subscript (oo), according to:[12,16]

$$\phi_{ij} = f_{ij} \, \phi_{oo}(r/h_{ij}^{1/3}) \tag{1}$$

where ϕ_{oo} is the potential energy function of the reference pure fluid, and f_{ij} and h_{ij} are the conformal interaction parameters of species i and j in the mixture.

For substances with an intermolecular potential energy function of the form

$$\phi_{ij} = E_{ij} \, [(L_{ij}/r)^n - (L_{ij}/r)^m] \tag{2}$$

and with exponents m and n equal to those for the reference substance, conformal parameters f_{ij} and h_{ij} will be related to the intermolecular potential energy parameter E_{ij} and to the intermolecular length parameter L_{ij} by the following equations:

$$f_{ij} = E_{ij}/E_{oo} \tag{3a}$$

$$h_{ij} = (L_{ij}/L_{oo})^3 \tag{3b}$$

The conformal solution theory for mixtures requires the definition of mixture conformal parameters f_{xx} and h_{xx}. These are related to the conformal parameters of the components of the mixture and to mixture composition:

$$f_{xx} = f_{xx}(f_{ij}, h_{ij}, x_i) \tag{4a}$$

$$h_{xx} = h_{xx}(f_{ij}, h_{ij}, x_i) \tag{4b}$$

Equations 4a and 4b are the conformal solution mixing rules. For different conformal solution theories of mixtures, different functional forms of the mixing rules will be obtained. The combining rules for unlike-interaction conformal parameters are also needed to formulate a mixture theory. The usual expressions are

$$f_{ij} = (1 - \tau_{ij}) \, (f_{ii}f_{jj})^{1/2} \tag{5a}$$

$$h_{ij} = (1 - \lambda_{ij}) \, [(h_{ii}^{1/3} + h_{jj}^{1/3})/2]^3 \tag{5b}$$

where τ_{ij} and λ_{ij} are fitting parameters.

In order to apply conformal solution mixing rules for the calculation of mixture thermodynamic properties, one needs to combine them with a pure-fluid EOS. Varieties of pure-fluid EOSs are available in the literature; however, for SFE/RC modeling, the use of semi-empirical EOSs is presently recommended because of the vast amount of database that is available for EOS parameters in the chemical and engineering industries. These equations are generally modifications of the van der Waals EOS:

$$Z = v/(v - b) - a/(vRT) \tag{6}$$

which was proposed by van der Waals[18] in 1873.

The Peng-Robinson conformal solution (PRCS) EOS is found to be quite capable of predicting properties of asymmetric mixtures of interest in SFE/RC processes. The PRCS EOS is in the following form:[9]

$$Z = v/(v - b) - \{A/(RT) + C - 2[AC/(RT)]^{1/2}\}/[(v + b) + (b/v)(v - b)] \tag{7}$$

$$A = a_C (1 + \kappa)^2, \quad C = a_C \kappa^2/(RT_C) \tag{8}$$

$$a_C = a(T_C) = 0.45724 (RT_C)^2/P_C \tag{9}$$

$$\kappa = 0.37464 + 1.54226 \, \omega - 0.26992 \, \omega^2 \tag{10}$$

$$b = 0.0778 \, RT_C/P_C \tag{11}$$

There exist three independent constants (A, b, and C) in this set of equations. Parameter A is proportional to (molecular volume)·(molecular energy), or (A ∝ f·h), while parameters b and C are proportional to the molecular volume, or b ∝ h, C ∝ h). The conformal solution van der Waals mixing rules for the constants of the PRCS EOS are[14]

$$A = \Sigma_i \Sigma_j \, x_i x_j A_{ij} \tag{12}$$

$$b = \Sigma_i \Sigma_j \, x_i x_j b_{ij} \tag{13}$$

$$C = \Sigma_i \Sigma_j \, x_i x_j C_{ij} \tag{14}$$

According to Equations 5a and 5b, the combining rules for the unlike interaction parameters of this EOS are as:

$$A_{ij} = (1 - \tau_{ij})(A_{ii} A_{jj}/b_{ii} b_{jj})^{1/2} b_{ij} \tag{15}$$

$$b_{ij} = (1 - \lambda_{ij}) [(b_{ii}^{1/3} + b_{jj}^{1/3})/2]^3 \tag{16}$$

$$C_{ij} = (1 - \zeta_{ij}) [(C_{ii}^{1/3} + C_{jj}^{1/3})/2]^3 \tag{17}$$

As demonstrated previously, similar procedures can be used to derive conformal solution mixing rules for other EOSs.

In this chapter, the PRCS EOS is used in order to calculate the solubility of solids and liquids in supercritical pure and mixed solvents. The PRCS EOS can be utilized to perform accurate solubility calculations for heavy solutes in supercritical solvents.

The solubility of a condensed phase in a vapor phase can be expressed as:[19]

$$y_2 = (P_2^{sat}/P) \, (1/\Phi_2) \Phi_2^{sat} \exp\left\{ \int_{P_2^{sat}}^{P} (v_2^{solid}/RT) \, dP \right\} \tag{18}$$

where Φ_2^{sat} is the fugacity coefficient of the condensed phase at the saturation pressure, P_2^{sat}. The vapor pressure of solids is small, and the fugacity coefficient of pure solids at saturation pressure is almost unity. If it is also assumed that the solute specific volume is independent of pressure, the above equation reduces to:

$$y_2 = (P_2^{sat}/P) \, (1/\Phi_2) \exp\left\{ v_2^{solid} (P - P_2^{sat})/RT \right\} \tag{19}$$

In order to calculate the fugacity coefficient, the following equation,[19] combined with an EOS and appropriate mixing rules, is generally used:

$$RT\ln\Phi_i = \int_V^\infty [(\partial P/\partial n_i)_{T,V,nj\neq i} - (RT/V)] \, dV - RT\ln Z \qquad (20)$$

With the PRCS EOS, Equation 7, and the correct version of the van der Waals mixing rules (Equations 12 to 14), the following expression for the fugacity coefficient is obtained:

$$\ln\Phi_i = [(2 \, \Sigma_j \, x_j b_{ij} - b)/b] \, (Z - 1) - \ln(Z - B) -$$

$$- [A^*/(8^{1/2} B)] \{[2 \, \Sigma_j \, x_j A_{ij} + 2 \, RT \, \Sigma_j \, x_j C_{ij} -$$

$$- 2 \, (A \, \Sigma_j \, x_j C_{ij} + C \, \Sigma_j \, x_j A_{ij}) \, (RT/AC)^{1/2}]/\beta^* -$$

$$- (2 \, \Sigma_j \, x_j b_{ij} - b)/b\} \ln[(Z + (1+2^{1/2}) \, B)/(Z + (1-2^{1/2}) \, B)] \qquad (21)$$

where $\beta^* = A + CRT - 2 \, (ACRT)^{1/2}$, $A^* = \beta^* P/(RT)^2$, and $B = bP/RT$.

III. MODEL SFE/RC SYSTEM CALCULATIONS

In the calculations reported here, solubilities of a number of model condensed solutes in supercritical solvents are predicted using the PRCS EOS for mixtures, in order to demonstrate the kind of general behavior that may be expected in SFE. PRCS EOS is shown to predict the phase behavior of model supercritical systems reported here accurately.[9] The solubilities of heavy solutes in supercritical solvents are calculated using Equations 19 and 21.

The solubility of 2,3-dimethylnaphthalene in supercritical CO_2 as a function of pressure at three different temperatures, compared with the calculations based on the PRCS EOS for mixtures, can be seen in Figure 4. According to this figure, the PRCS EOS is quite suitable for the model mixtures studied here. The same trends in Figure 4 are observed in Figure 5 for the solubility of 2,3-dimethylnaphthalene in supercritical ethylene. Interaction parameters τ_{ij}, λ_{ij}, and ζ_{ij} are constant, and they can be used as well for other temperatures consistent with the molecular theory of unlike interactions. Figures 6 to 8 are presented to study the effect of mixed supercritical solvents on the solubility of pure solutes. In these figures, the solvents consist of different mixtures of CO_2 and ethylene in contact with three different solutes. The solvent-cosolvent interaction parameters are taken into account to predict the solubility of solutes in mixed solvents. According to these figures, mixed solvents can either enhance or reduce the solubility of a solute, depending on the molecular interactions among them. This fact emphasizes the necessity of finding a proper solvent/cosolvent combination and their compositions for optimum design and operation of SFE/RC systems.

The phenomenon of the supercritical solubility peak, which has been observed experimentally for several condensed phase-dense gas systems, can be predicted using the PRCS EOS. Figures 9 to 12 demonstrate solubility peaks that are observed for 2,3-dimethylnaphthalene and 2,6-dimethylnaphthalene at temperatures above and below the critical temperature of the solvent. In addition, it is demonstrated that solubility decays are more pronounced at temperatures above the critical temperature of the solvent. The solubility peak was sharper when ethylene, instead of CO_2, was the supercritical solvent.

IV. CURRENT BIOTECHNOLOGICAL APPLICATIONS

The major requirement in the design of SFE/RC systems is the choice of a solvent that will cause a sharp change in solute solubility due to rather small changes in pressure or

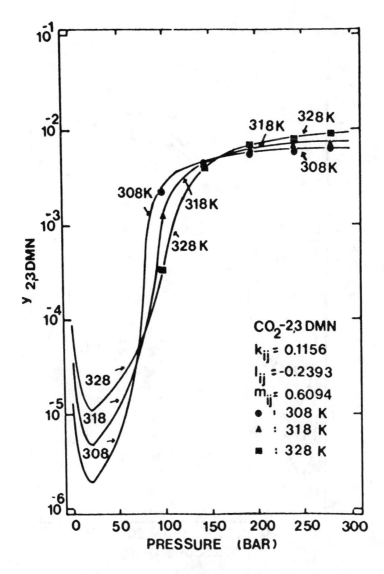

FIGURE 4.   Solubility of 2,3-dimethylnaphthalene in supercritical CO_2 as a function of pressure, at different temperatures, as calculated by the PRCS EOS (solid lines), and compared to the experimental data.[43]

temperature. Since an essentially infinite number of supercritical pure and mixed solvents can be chosen from the currently known compounds, there is little hope of ever generating a sufficient amount of experimental data to meet present, much less future, industrial needs. Nevertheless, there have been a number of key experimental measurements regarding biotechnological applications of SFE/RC phenomena which are reviewed in this section. Practical application of SFE/RC phenomena requires the knowledge of theoretical correlation and prediction techniques that could reduce the heavy burden of such experimental measurements.[20] This justifies the importance of the development of statistical mechanical techniques that can be used for predictive models of SFE/RC phenomena.

A. AMINO ACIDS

The influence of humid supercritical CO_2 and nitrogen on amino acids, namely L-glutamine, L-methionine, L-leucine, L-alanine, β-alanine, and L-lysine, was investigated.[25]

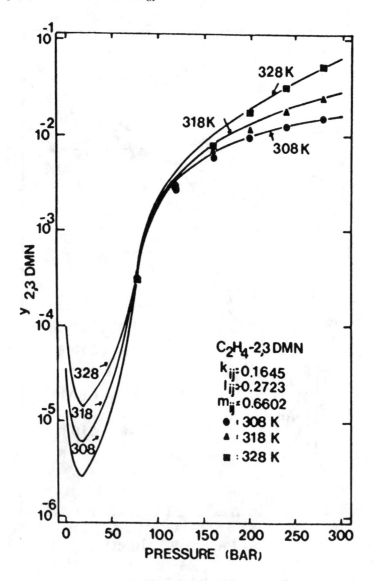

FIGURE 5.   Solubility of 2,3-dimethylnaphthalene in supercritical ethylene as a function of pressure, at different temperatures, as calculated by the PRCS EOS (solid lines), and compared to the experimental data.[43]

After exposure to humid CO_2 at 300 bar and 353 K for 6 h, only L-glutamine showed a significant composition decay (15 to 23%). The L-glutamine loss was 10% when treated with nitrogen under the same conditons. The decomposition product of L-glutamine is 2-pyrrolidinone 5-carboxylic acid (PCA), which is an intermediate in the γ-glutamyl cycle of mammalian metabolism. The other amino acids did not show a significant decay in composition after exposure to humid CO_2 at 300 bar and 353 K for 6 h. When L-glutamine was exposed to supercritical CO_2 and nitrogen at room temperature and 300 bar for 6 h, no PCA was detected, thus implying that L-glutamine decomposes due to heating. Pressure seems to have little effect on the structural integrity of L-glutamine, since Vickery et al.[22] have shown that an L-glutamine solution at atmospheric pressure will almost quantitatively convert to PCA and ammonia if exposed for 2 h at 373 K and pH 6.5. These results indicate that amino acids remain structurally sound after exposure to high pressure, and among tested amino acids, only L-glutamine is seriously damaged by heating.

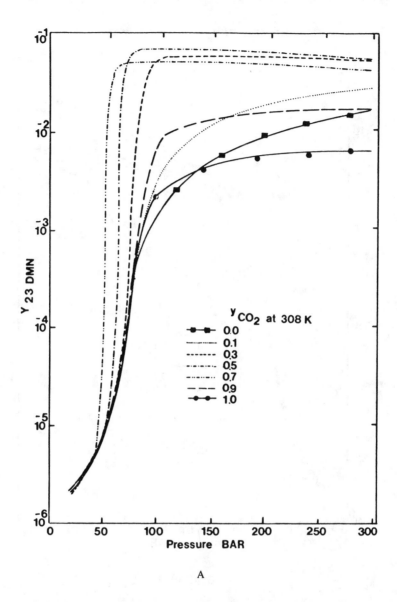

A

FIGURE 6. Solubility of 2,3-dimethylnaphthalene in different mixtures of CO_2 and ethylene according to the PRCS EOS (lines). In A, it is assumed that for CO_2 and ethylene $\tau_{12} = \lambda_{12} = \zeta_{12} = 0$. In B, the unlike interaction parameters of CO_2 and ethylene are assumed to be nonzero. They are calculated from CO_2-C_2H_4 vapor-liquid equilibrium data[44] and the results show that $\tau_{12} = 0.1829$, $\lambda_{12} = 0.6053$, $\zeta_{12} = 0.2153$.

The behavior of amino acids is very important for biological applications of supercritical systems because such molecules are the building blocks of life. The fact that the decomposition products of exposing amino acids to supercritical CO_2 are also naturally found in many food components makes SFE/RC application a potentially lucrative process for the food industry.

B. ESSENTIAL OILS FROM PLANTS

Limonene and eugenol, essential oils extracted from plants, are demonstrated to be soluble in supercritical CO_2, exhibiting solubilities according to their respective molecular

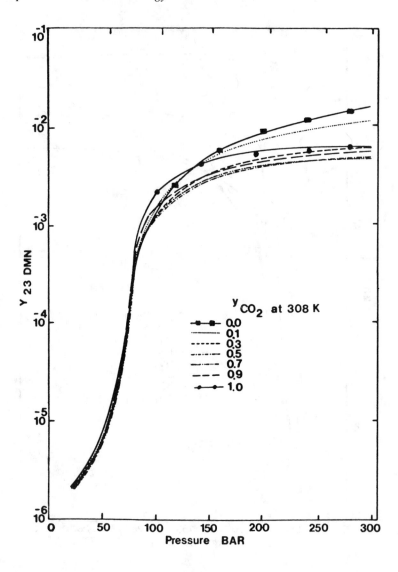

FIGURE 6B.

polarities. Limonene contains no functional groups that could drastically increase molecular polarity, but eugenol has an ether and a hydroxyl group. The experimental evidence[23] indicates that eugenol is much less soluble in CO_2 than limonene, as shown in Figures 13 and 14. The sharp rise in solubility of both limonene and eugenol at low temperatures can be explained by the fact that CO_2 becomes a liquid and, therefore, attains a strong solvent power for these essential oils. This is one of the few cases where an organic molecule is found to be soluble in subcritical CO_2. Solubility data of these compounds in CO_2 (Figure 14) indicate that both limonene and eugenol isothermal solubilities pass through a minimum. They also show strong solubility increases as the pressure rises above the point of minimum solubility in supercritical CO_2. A third essential oil, caryophyllene (Figure 14), also exhibits a minimum solubility, it being more soluble in CO_2 than eugenol. The rate of solubility increase of caryophyllene is less extreme than that of eugenol, and the solubilities of eugenol and caryophyllene in CO_2 become equal at 80 bar.

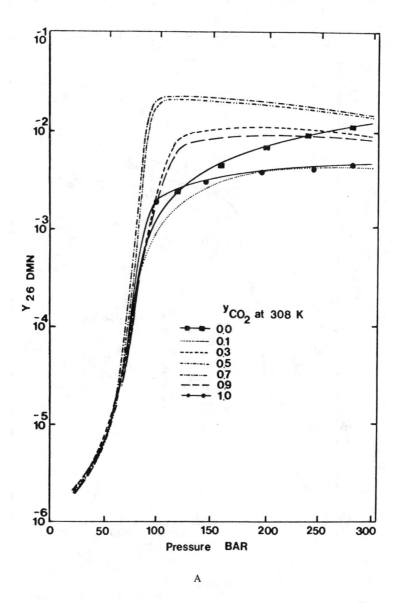

A

FIGURE 7. Solubility of 2,6-dimethylnaphthalene in different mixtures of CO_2 and ethylene according to the PRCS EOS (lines). In A, it is assumed that for CO_2 and ethylene $\tau_{12} = \lambda_{12} = \zeta_{12} = 0$. In B, the unlike interaction parameters of CO_2 and ethylene are assumed to be nonzero. These parameters are calculated from CO_2-C_2H_4 vapor-liquid equilibrium data[44] and they are the same as those used in Figure 6.

C. HOPS

The extraction of hops in beer production has proved to be an important application of the SFE/RC process.[24] The desirable components of the hops are soft resins, consisting of humulones and lupulones. In the brewing process, humulones are isomerized to give beer its bitter flavor, thus it is necessary to control the beer taste by extracting only the soft resins while minimizing the production of undesirable by-products. Here, subcritical CO_2 has proven to be a superior extractor because it does not extract undesirable hard resins like supercritical CO_2 does.[25,26] The SFE/RC of hops using liquid CO_2 requires more time than does SFE/RC with supercritical CO_2 because the liquid solvent extracts the hops components more selectively than does the supercritical solvent.

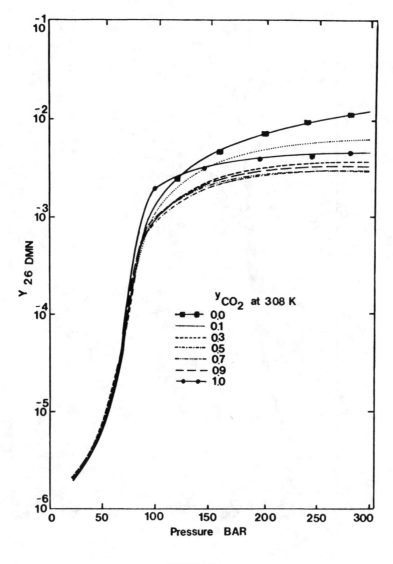

FIGURE 7B.

Subcritical CO_2 extracts only the humulones, lupulones, and essential hop oils, while supercritical CO_2 extraction includes the undesirable hard resins in addition to those components extracted by subcritical CO_2. Many patents cover this technique, so its exploitability is somewhat limited.

D. JOJOBA OIL

Jojoba oil is extracted by SFE/RC using supercritical CO_2 as the solvent. This oil is obtained from crushed seeds, and it is mostly a wax ester mixture. When the solubility of jojoba oil as a function of temperature at pressures varying between 100 and 2600 bar is measured,[27] one can see that at 100 bar the solubility of jojoba oil in CO_2 is near zero (Figure 15). A sharp increase in solubility does not occur until the pressure rises above 200 bar. As the pressure continues rising, the solubility increases more quickly until the solubility peak is attained at 700 bar for temperatures above 313 K. After the peak, the solubility decreases slowly as the pressure rises further. At temperatures below 313 K, the maximum

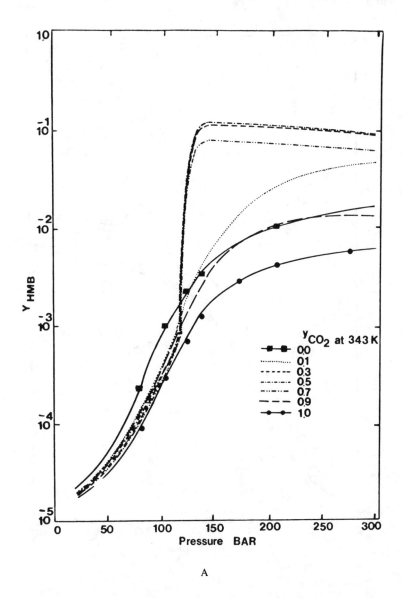

A

FIGURE 8.   Solubility of hexamethylbenzene in different mixtures of CO_2 and ethylene according to the PRCS EOS (lines). In A, it is assumed that for CO_2 and ethylene $\tau_{12} = \lambda_{12} = \zeta_{12} = 0$. In B, the unlike interaction parameters of CO_2 and ethylene are assumed to be nonzero. These parameters are calculated from CO_2-C_2H_4 vapor-liquid equilibrium data[44] and they are the same as those used in Figure 6.

solubility of jojoba oil in supercritical CO_2 can be observed at pressures slightly below 700 bar. Beyond 700 bar, the solubility of jojoba oil decreases at all temperatures. For example, at 2600 bar and 333 K, this solubility is less than one third of the solubility at the same temperature and 700 bar.

E. LYSOZYME

Lysozyme, a protein found in chicken egg white, was treated with humid supercritical CO_2 and nitrogen, and showed no alterations, as was observed in the case of amino acid analysis. Studies were made at 300 bar, varying both temperature and time of exposure from 353 K to room temperature and 6 to 2 h, respectively.[21,28]

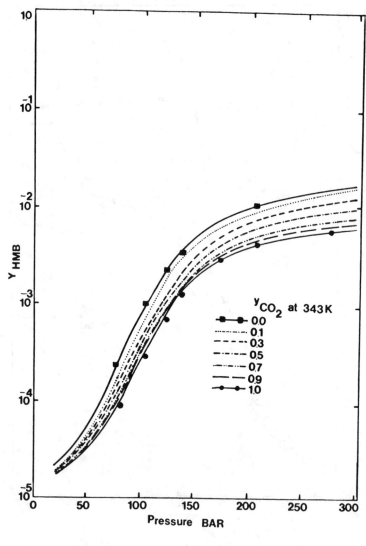

FIGURE 8B.

As indicated in Table 1, only minimal decay in protein content was observed even at high temperatures (the same as in the case of amino acids). The lowest total protein composition after treatment was 86.6%, indicating only a small loss of protein. Digestibility of lysozyme by trypsin, an indicator of the structural integrity of lysozyme, was fairly independent of the type of solvent gas utilized in the experiment. When the humid gas was brought to 300 bar and 353 K, the digestibility of lysozyme by trypsin was markedly increased, compared to the rate of digestibility at room temperature and 300 bar. The amount of digestion at 300 bar and 353 K was similar to the amount observed heating lysozyme with 15% water at 353 K and atmospheric pressure, during either 2 or 6 h. Heating dry lysozyme at 353 K and atmospheric pressure showed a much smaller increase in trypsin digestibility compared to the increase observed with humidity at 353 K and 1 bar. Thus, the increased digestibility was due mainly to the presence of water and not to the supercritical conditions. Because lysozyme (and the other proteins studied) can withstand supercritical conditions, SFE/RC may prove to be an effective technique for the extraction of food products.

CARBON DIOXIDE - 2,3DMN

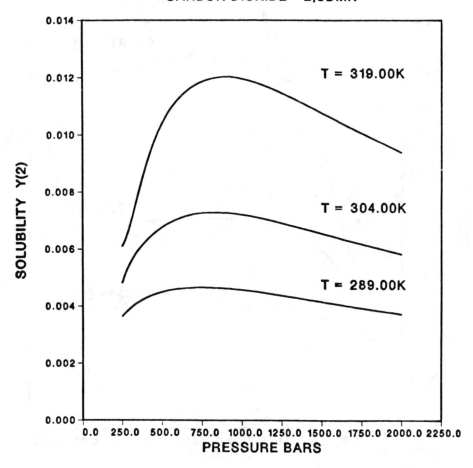

FIGURE 9.   Solubility of 2,3-dimethylnaphthalene in CO_2 at supercritical, critical, and subcritical temperatures of CO_2 according to the PRCS EOS.

F. NICOTINE AND CAFFEINE

Nicotine, a major toxic agent found in tobacco, is shown to be soluble in supercritical CO_2.[29] At 323 K, the solubility peak of nicotine in CO_2 is 80 mg/g of solvent, at a gas density of about 0.650 g cm^{-3}. At 343 K, the maximum solubility falls to about 65 mg nicotine per gram of CO_2, at a gas density of 0.480 g cm^{-3}. Nicotine, unlike most other molecules that are soluble in supercritical gases, does not tend to exhibit larger solubilities as the temperature rises. The high solubility of nicotine in supercritical CO_2 is due to the fact that nicotine is a relatively nonpolar molecule possessing no carbonyl or hydroxyl functional groups that inhibit solubility. By contrast, caffeine, which is more polar than nicotine due to its two carbonyl groups per molcule, has a maximum solubility in CO_2 of <5 mg caffeine per gram of CO_2 even at 353 K and 390 bar. One industrial option in applying the SFE/RC process for extraction of caffeine consists of treating green coffee beans, presoaked in water, with CO_2 in a pressure vessel operated between 160 and 220 bar.[26] After adequate mixing, the caffeine-rich solvent is allowed to leave the pressure vessel and enter into a washing tower. In the washing tower, water at 343 to 363 K is used as an entrainer for the caffeine, stripping the CO_2 of the caffeine. The solvent is then recycled, and caffeine is recovered by distillation. The caffeine content of coffee beans can be reduced from 0.7 to 3% to as low as 0.2% βΨ using this industrial option.

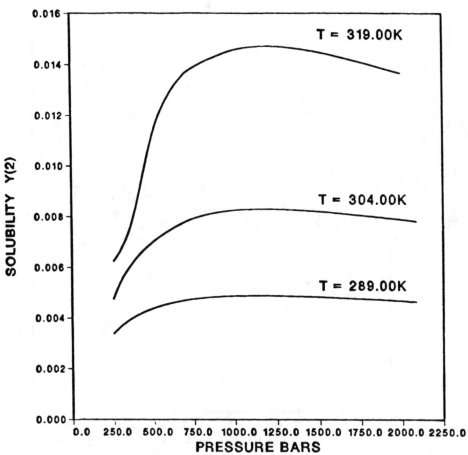

CARBON DIOXIDE - 2,6DMN

FIGURE 10.   Solubility of 2,6-dimethylnaphthalene in CO_2 at supercritical, critical, and subcritical temperatures of CO_2 according to the PRCS EOS.

G. STEROIDS

Steroids, a class of compounds with varius biological applications, have proven to be soluble in supercritical CO_2. There exist many different types of steroids, all of which have a cyclopentaneperhydrophenanthrene skeleton. The difference in the steroids' function lies in the variation of the groups attached to this skeleton. These compounds are quite stable and they survive after the death of the organism far longer than any other biological molecule. Steroids (cholesterol) cause heart disease. They contribute to the control of sexual development (progesterone), pain reduction (cortisone), and a number of other biological processes. Experiments have shown that sterols with the least number of hydroxy and carboxyl functional groups have the highest solubilities in CO_2.[30] Some examples are ergosterol and stigmasterol, each containing only one hydroxyl group. At 313 K, sitosterol, containing only one hydroxyl substituent, begins exhibiting solubility at 80 bar. Progesterone, containing two carbonyl groups, also becomes extractable at this pressure. Androsterone, with one hydroxyl and one carbonyl group, displays measurable solubility at 90 bar. Pregnanediol, containing two hydroxyl groups, requires a pressure of 120 bar to be extracted, while digitoxigenin, possessing two hydroxyl and one carbonyl group in an α,β-unsaturated lactone ring, does not become appreciably extractable until the pressure is raised to 150 bar.[7]

ETHYLENE - 2,3DMN

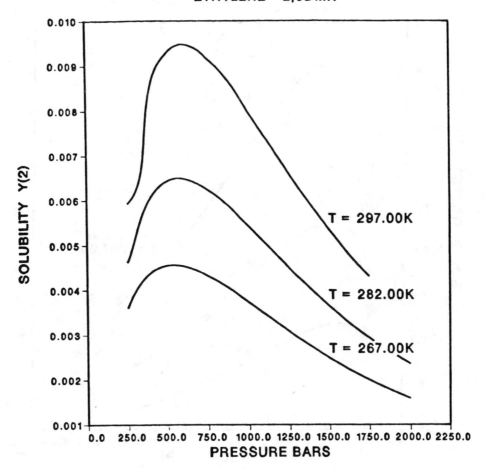

FIGURE 11.   Solubility of 2,3-dimethylnaphthalene in ethylene at supercritical, critical, and subcritical temperatures of ethylene according to the PRCS EOS.

H. THUJONE

Wormwood, which has a bitter taste, is toxic because of its thujone content. SFE/RC with supercritical CO_2 can remove thujone from wormwood, leaving most of the bitter flavor which is mainly due to absinthine.[31] Thujone has only one carbonyl functional group per molecule, quite low polarity, and thus it is rather easy to extract. Absinthine and artabsine, which are also present in wormwood, have lower solubilities in supercritical CO_2 compared to thujone. By examining the solubility curves of thujone and artabsine (Figure 16), it can be observed that thujone exhibits a solubility in CO_2 much greater than artabsine. At around 90 bar, thujone is ten times more soluble than artabsine, but for all other pressures, the solubility ratio increases rapidly. Artabsine has one hydroxyl and one carbonyl functional group per molecule, and is therefore more polar and less soluble in supercritical CO_2 than thujone. Absinthine is nearly unextractable by SFE/RC using CO_2 as the solvent because it contains two hydroxyl and two carbonyl functional groups per molecule. The high molecular weight of absinthine (497 g mol^{-1}) also hinders its solubility in supercritical CO_2. Experiments have shown that use of the SFE/RC process for extraction of wormwood can be quickened by increasing the mass ratio of CO_2 to wormwood in the process cycle. At mass ratios of six parts CO_2 to one part wormwood or greater, thujone is nearly 100% extracted. While at a mass ratio of two parts carbon dioxide to one part thujone, only about 80% of the thujone will be dissolved in the gas phase.

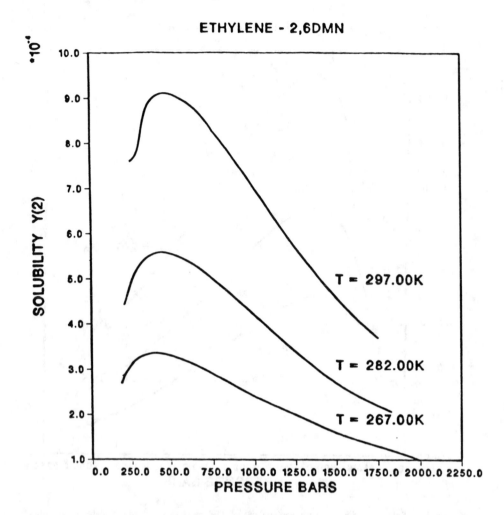

FIGURE 12. Solubility of 2,6-dimethylnaphthalene in ethylene at supercritical, critical, and subcritical temperatures of ethylene according to the PRCS EOS.

I. THYMOL

The essential oil of thyme (a plant of the mint family whose leaves are used for seasoning) is called thymol. Structurally, thymol consists of benzene ring with hydroxyl, isopropyl, and methyl substituents. The fact that thymol contains only one strongly polar functional group (OH) allows thymol to be easily extractable by SFE/RC with CO_2 as the solvent. The oil extract obtained by SFE/RC of thyme with CO_2 has been compared to that from steam distillation and to commercially available red thyme oil.[32] The results indicate that CO_2 at 40 bar and 323 K extracts as many as 50 different compounds from thyme. The extract contains 36.6% thymol and 19% stearic acid and other high molecular weight compounds, while steam distillation produces an extract containing 48.3% thymol and only 0.27% stearic acid and heavier molecular weight compounds. Analysis of red thyme oil indicated no heavy compounds and 30.9% thymol. Thus, the SFE/RC of thyme with supercritical CO_2 produces an extract containing a larger percentage of high molecular weight compounds than the red thyme oil or the oil from the steam distillate.

J. TRIGLYCERIDES

Triglycerides are composed of three organic acids attached to glycol. Triglycerides are considered neutral fats easily mobilized as an energy source in mammals. The extraction of

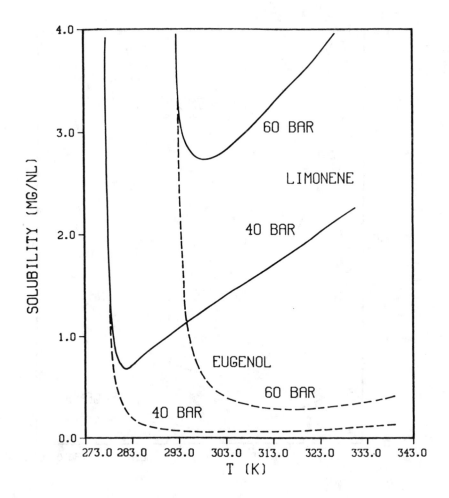

FIGURE 13.   Solubility of limonene and eugenol in CO_2 as a function of temperature at 40 and 60 bar.[23]

triglycerides with supercritical CO_2 is found to be a promising procedure for the modification of butter oils in order to lower their cholesterol level and improve their spreadability at refrigeration temperatures.[33] Triacylglycerols from plant tissue have been shown to be extractable by supercritical CO_2. Samples of dried coconut flesh, or Copra, containing triacylglycerols were extracted in pressurized batches weighing 700 g using supercritical CO_2 at various conditions. The results of this analysis are reported in Table 2.

Examination of the results in Table 2 indicate that at the conditions of lowest pressure and temperature (300 bar and 313 K), a CO_2/extract mass ratio of 64.9 was required for an exposure time of 8 h to achieve a triacylglycerol yield of around 60%. The experiment utilizing the highest pressure and temperature (900 bar and 333 K) required a CO_2/extract ratio of only 6.2 for an exposure time of 1 h in order to achieve the same yield. This case is 83 times more effective in its extraction of triacylglycerols than the 300 bar, 313 K example and helps prove the general rule that solubility increases with an increase in pressure and temperature.[34]

A major source of triglycerides produced on a world scale is soybean oil. Several important observations can be made when the solubility data of soybean oil in CO_2 (Figure 17) are studied. Examination of the data indicates that the solubility of soybean oil in CO_2 increases as the temperature rises and when the pressure is above 280 bar.[35] The solubility increase becomes more pronounced with rising temperature, reaching a peak of about 65

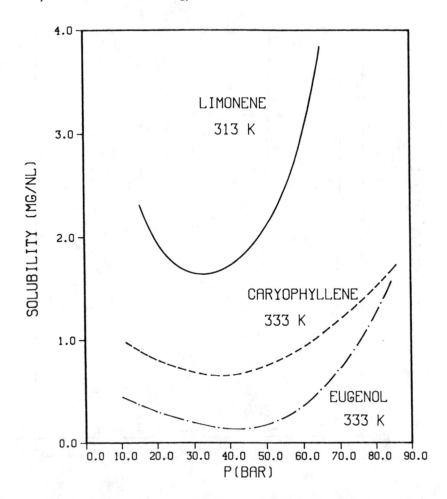

FIGURE 14.   Solubility of limonene at 313 K, eugenol at 333 K, and caryophyllene at 333 K, in CO_2 as a function of pressure.[23]

mg nl^{-1} at 333 K and more than 140 mg nl^{-1} at 353 K (Figure 17). The solubility of soybean oil in CO_2 exhibits the crossover effect at 280 bar. The existence of the crossovers was also observed in the theoretical and experimental results of the model systems reported in Figures 4 and 5. The existence of solubility maxima of soybean oil in CO_2 is also observed for each isotherm studied. According to Figure 17, the solubility maxima are reached at higher pressurees as the temperature is raised. The solubiliy maxima for soybean oil in CO_2 and for jojoba oil in the same solvent occur at a gas phase density between 1.04 and 1.08 g cm^{-3}.

Presently, industrial extraction of soybean oil is mainly carried out by using liquid hexane as the solvent. Soybean oil extracted by SFE/RC with Co_2 as the supercritical solvent tends to be lighter in color and exhibits lower oxidation stability than the oil extracted by hexane, although extraction yields are similar. The lack of oxidation inhibiting phosphatides in the supercritical CO_2 extract seems to be the cause of the instability of the extract.[36]

K. VIABILITY OF BACTERIA UNDER HIGH PRESSURE

A set of experiments have been performed to test the viability of certain bacteria in a supercritical CO_2 environment. These experiments were conducted at room temperature and 20 min of exposure time at pressures of 500, 1500, and 2500 bar.[37] The results showed that *Pseudomonas aeruginosa* and *Candida albicans* were completely destroyed after exposure

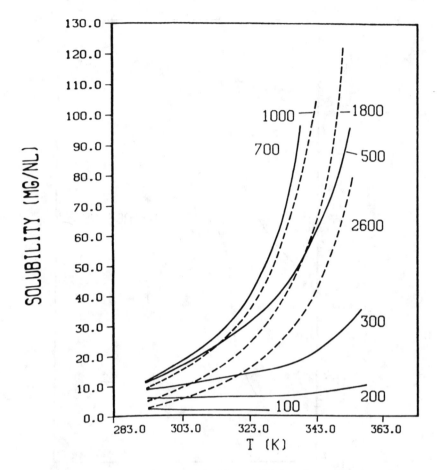

FIGURE 15. Solubility of jojoba oil in CO_2 as a function of temperature at various pressures.[27]

TABLE 1
Results from the Treatment of Lysozyme with CO_2

Protein content (%)

Solvent	Untreated	RmT[a], 6 h	353 K, 2 h	353 K, 6h
CO_2	90.3	89.7	93.0	92.9
N_2	90.3	93.1	90.4	86.6

[a] RmT: room temperature.

to CO_2 at 500 bar. *Bacillus subtillis* showed 30% logarithmic viability at all three pressures. *Escherichia coli,* which lives in human intestines, showed a 25% logarithmic viability at 500 bar, 10% at 1500 bar, and around 0% at 2500 bar *Coliphage* showed less sensitivity to pressure with 75% logarithmic viability at 2500 bar. Implications of these experiments could be far-reaching. For example, if it was found that certain bacteria in milk were pressure sensitive, a new and possibly more economical milk sterilization technique could be developed. Other applications would be the destruction of harmful bacteria in a wide variety of foodstuffs, such as seeds or goods earmarked for long-term packaging. These studies

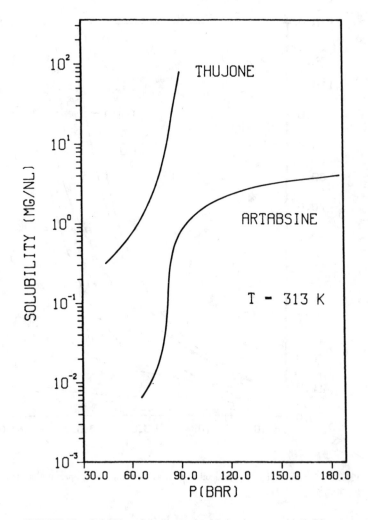

FIGURE 16.   Solubility of thujone and artabsine in supercritical CO_2 as a function of pressure at 313 K.[31]

TABLE 2
Results of the Extraction of Triacylglycerols from Copra

P (bar)	T (K)	g CO_2/g extract	t (h)	Yield (%)
300	313	64.9	8	59.9
600	313	34.1	5	58.6
600	333	12.8	2	61.4
900	333	6.2	1	60.8

could also indicate whether the SFE/RC process is applicable to certain products (mainly foods) without altering their useful microbiological contents. At the present time, the industrial potential of such a process is unknown. Numerous bacteriological experiments need to be performed on each specific bacteria strain in order to test the pressure sensitivity of these organisms.

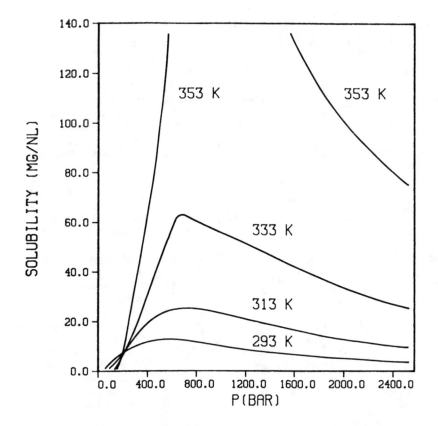

FIGURE 17.   Solubility of soybean oil in CO_2 as a function of pressure at various temperatures.[31]

V. CONCLUSIONS AND RECOMMENDATIONS

In Table 3 the SFE/RC biotechnological applications discussed above are summarized. According to the above discussion and the summary reported in Table 3, biological molecules exhibit different degrees of solubility in supercritical solvents, especially CO_2. These solubility properties can be exploited by extracting, in some cases selectively, the specific molecules by carefully controlling the pressure and temperature. The industrial and commercial implications of these investigations are still to be determined. One major possibility is the purification of substances vital to human health. Since in many SFE/RC cases efficient extraction can be carried out at normal to low temperatures (below 318 K), many thermoliable compounds found in natural systems could be, for the first time, extracted without a loss. Possible applications are the extraction of hormones, the regulation of nutrients in the blood, and the separation of toxic contaminants. Further research is necessary in this direction in order to determine the solubility and temperature sensitivity of fluids found in the human body. Another possible research route is the study of humid supercritical CO_2 solvents. It is possible that water can enhance solubilities in such systems as those extracting nicotine or oleic acid.

In order to tackle the problems of utilizing the SFE/RC phenomena in biotechnology and related disciplines, the following general tasks would have to be accomplished:

1.   Experimental studies for measurement of solubilities, compositions, and solute-solvent interaction characteristics by utilizing high pressure SFE/RC cycles, gas chromatography, high pressure liquid chromatography, gel permeation chromatography, etc.

TABLE 3
Biological SFE/RC Applications Summary

Solute	Solvent	T(K)	P(bar)	Notes
Amino acids	CO_2 N_2	298—353	300	SFE/RC conditions have no effect on their structural integrity
Caffeine	CO_2	313—353	150—400	Maximum solubility at highest temperature
Essential oils from plants (limonene, eugenol and caryophyllene)	CO_2	278—333	15—80	Soluble in subcritical CO_2
Hops	CO_2	<304	<74	Extraction with subcritical CO_2 is more selective than SFE
Jojoba oil	CO_2	293—353	100—2600	Maximum solubility occurs at around 700 bar for all temperatures
Lysozyme	CO_2 N_2	298—353	300	Small decay in protein content due to SFE conditions
Nicotine	CO_2	323—343	70—250	Higher solubilities at lower temperatures
Steroids	CO_2	313	80—250	Steroids have the lowest solubilities
Thujone (artabsin, absinthine)	CO_2	313	50—200	Thujone can be selectively extracted from wormwood due its much higher solubility
Thymol	CO_2	>304	>74	SFE process is more effective than steam distillation technique
Triglycerides (triacylglycerols, soybean oil)	CO_2	313—333	300—1000	Solubility greatly increases above 333 K and 400 bar

2.   Due to the high pressure conditions of SFE/RC processes and the exhaustive number of solvent and cosolvent compositions that need to be tried, the extensive experimental studies alone will not be practical from an economic point of view. As a result, development of computational schemes for the prediction of solubilities of solutes (pure or mixture) in pure or mixed supercritical solvents will be needed to reduce the burden of experimental measurements. Such schemes[13,38] would be quite useful to search for solvents to be used in a particular supercritical separation problem or feasibility study of separating new candidate biological compounds by supercritical fluids.

3.   Engineering design and economic comparative study of separation, purification, and stripping of heat-sensitive biological fluids and close-molecular weight compounds through SFE/RC and other techniques will establish the feasibility of each process for commercial applications.[39]

4.   Experimental studies and modeling of flow and diffusion in pipes and conduits of pastes and slurries in the presence of supercritical solvents.

Overall, the extraction and separation of biotechnological material using the SFE/RC processes is a growing field of research and development with unlimited potential. The future commercialization of such technology could prove to be quite rewarding.

ACKNOWLEDGMENT

This research was supported by the Gas Research Institute, contract no. 5086-260-1244.

NOMENCLATURE

A, a, b, C   EOS parameters
E   Intermolecular energy parameter
f   Conformal solution energy parameters
h   Hour
h   Conformal solution volume parameter
L   Intermolecular length parameter
mg   Milligram
K   Degree Kelvin
n   Number of moles
NL   Normal liter (at 1 atm and 25°C)
P   Pressure
PRCS   Peng-Robinson conformal solution EOS
r   Intermolecular distance
R   Universal gas constant
RC   Retrograde condensation
SFE   Supercritical fluid extraction
t   Time
T   Absolute temperature
V   Total volume
v   Specific volume
x   Mole fraction
y   Solubility (solute mole fraction in the gas phase)
Z   Compressibility factor ($Z = Pv/RT$)

Subscripts and Superscripts
2   Condensed phase (solute)
i, j   Component (or molecule) identifiers
c   Critical property
sat   Saturation

Greek Letters
τ, λ, ζ Binary unlike-interaction parameters
ϕ Intermolecular potential energy function
Φ Fugacity coefficient
ω Acentric factor
Σ Summation over the number of components in the mixture

GLOSSARY OF TERMS

Critical Point: The condition at which a gas and a liquid are at equilibrium, and have identical composition and densities.

Critical Temperature: Temperature at the critical point, above which the liquid and vapor phases are indistinguishable.

Critical Pressure: Pressure of a fluid at its critical point.

Entrainer: Fluid that enhances the solubility of a dense phase.

Supercritical Gas: A gas at conditions above its critical temperature and pressure.

REFERENCES

1. **Olien, N. A.** *Chem. Eng. Progr.*, p. 45, October 1987.
2. **Mansoori, G. A. and Ely, J. F.**, A Preliminary Research Report on Partitioning of Monodisperse/Poly-disperse Polymers and Biological Macromolecules in Aqueous Two-Phase Systems, National Bureau of Standards Technical Note, NBS, Washington, D.C., 1987.
3. **Hariri, H., Ely, J. F., and Mansoori, G. A.**, *Bio/Technol. J.*, 7, 686, 1989.
4. **Mansoori, G. A., Schulz, K., and Martinelli, E.**, *BIO/TECHNOLOGY*, 6, 393, 1988.
5. **Hannay, J. B. and Hogarth, J.**, *Proc. R. Soc. London Ser. A*, 29, 324, 1879.
6. **Stahl, E. and Quirin, K. W.**, *Fette Seifen Anstrichm.*, 87, 219, 1985.
7. **Stahl, E. and Quirin, K. W.**, *Naturwissenschaften*, 71, 181, 1984.
8. **Friedrich, J. P.**, Rep. No. P. C. 6817, Agricultural Research Service/USDA, Peoria, IL.
9. **Kwak, T. Y. and Mansoori, G. A.**, *Chem. Eng. Sci.*, 41, 1303, 1986.
10. **Leach, J. W., Chappelear, P. S., and Leland, T. W.**, *AIChE J.*, 14, 568 1968, *Proc. Am. Petrol. Inst., Sect. III*, 46, 223, 1966.
11. **Leland, T.W., Rowlinson, J. S., and Sather, G. A.**, *Trans. Faraday Soc.*, 64, 1447, 1968.
12. **Brown, W. B.**, *Proc. R. Soc. London Ser. A*, 240, 561, 1957, *Philos. Trans. R. Soc. London Ser. A*, 250, 221, 1957.
13. **Mansoori, G. A. and Ely, J. F.**, *J. Chem. Phys.*, 82, 406, 1985.
14. **Mansoori, G. A.**, in *Equations of State: Theories and Applications*, Chao, K. C. and Robinson, R. L., Jr., Eds., ACS Symp. Ser., American Chemical Society, Washington, D.C., 1986, 314.
15. **van Konynenburg, P. H. and Scott, R. L.**, *Philos. Trans. R. Soc. London*, 298, 495, 1980.
16. **Massih, A. R. and Mansoori, G. A.**, *Fluid Phase Equilibria*, 10, 57, 1983.
17. **Scott, R. L. and van Konynenburg, P. H.**, *Disc. Faraday Soc.*, 49, 87, 1970.
18. **van der Waals, J. D.**, Over de Continuiteit van den Gasen Vloeistoftoestand, Ph.D. dissertation, University of Leiden, Leiden, 1873.
19. **Prausnitz, J. M.**, *Molecular Thermodynamics of Fluid-Phase Equilibria*, Prentice-Hall, New York, 1969, chap. 5.
20. **Ely, J. F. and Baker, J. K.**, A Review of Supercritial Fluid Extraction, NBS Tech. Note 1070, National Bureau of Standards, Washington, D.C., 1983, 16.
21. **Weder, J. K. P.**, *Food Chem.*, 15, 175, 1984.
22. **Vickery, H. B., Pucher, G. W., Clark, H. E., Chibnall, A. C., and Westall, R. G.**, *Biochem. J.*, 29, 2710, 1935.
23. **Stahl, E. and Gerard, D.**, *Perfum. Flavor*, in press.
24. **Caragay, A. B.**, *Perfum. Flavor.*, 6, 43, 1981.
25. **Harold, F. V. and Clarke, B. J.**, *Brewer's Dig.*, September, 45, 1979.
26. **Hubert, P. and Vitzthum, O. G.**, *Angew. Chem. Int. Ed. Engl.*, 17, 710, 1978.
27. **Stahl, E., Quirin, K. W., and Gerard, D.**, *Fette Seifen Anstrichm.*, 85, 458, 1983.
28. **Weder, J. K. P.**, *Z. Lebensm. Untersch. Forsch.*, 171, 95, 1980.
29. **Wong, J. M. and Johnston, K. P.**, *Biotechnol. Progr.*, 2, 29, 1986.
30. **Stahl, E. and Glatz, A.**, *Fette Seifen Anstrichm.* 86, 346, 1984.
31. **Stahl, E. and Gerard, D.**, *Z. Lebensm. Untersch. Forsch.*, 176, 1, 1983.
32. **Bestmann, H. J., Erler, J., and Vostrowsky, O.**, *Z. Lebensm. Untersch. Forsch.*, 180, 491, 1985.
33. **Shishikura, A., Fujimoto, K., Kaneda, T., Arai, K., and Saito, S.**, *Agric. Biol. Chem.*, 50, 1209, 1986.
34. **Brannolte, H. D., Mangold, H. K., and Stahl, E.**, *Chem. Phys. Lipids*, 33, 297, 1983.
35. **Stahl, E., Quirin, K. W., Glatz, A., Gerard, D., and Gau, G.**, *Ber. Bunsenges. Phys. Chem.*, 88, 900, 1984.
36. **List, G. R. and Friedrich, J. P.**, *JAOCS*, 62, 82, 1985.
37. **Stahl, E., Rau, G., and Kaltwasser, H.**, *Naturwissenschaften*, 72, 144, 1985.
38. **Park, S. J., Kwak, T. Y., and Mansoori, G. A.**, *Int. J. Thermophys.*, 8, 449, 1987.
39. Proceedings of the Supercritical Fluid Extraction/Retrograde Condensation with Applications—A Tutorial Symposium, Mansoori, G. A., Chairman, Gas Research Institute, Chicago, 1985.
40. **Katz, D. L. and Kurata, F.**, *Ind. Eng. Chem.*, 32, 817, 1940.
41. **Katz, D. L., Cornell, D., Kobayashi, R., Poettmann, F. H., Vary, J. A., Elenbaas, J. R., and Weinaug, C. F.**, *Handbook of Natural Gas Engineering*, McGraw-Hill, New York, 1959.
42. **Stalkup, F. I.**, SPE Monograph, Society of Petroleum Engineers of the American Institute of Mechanical Engineers, Dallas, June 1983.
43. **Kurnik, R. T., Holla, S. J., and Reid, R. C.**, *J. Chem. Eng. Data*, 26, 47, 1981.
44. **Mollerup, J.**, *J. Chem. Soc. Faraday Trans. I*, 71, 2351, 1975.

Chapter 14

SUPERCRITICAL EXTRACTION IN ENVIRONMENTAL CONTROL

Aydin Akgerman, Robert K. Roop, Richard K. Hess, and Sang-Do Yeo

TABLE OF CONTENTS

I. INTRODUCTION

Over the past decade there has been a substantial interest in supercritical fluids (SCFs) from both the academic and industrial communities. An SCF is a fluid above its critical temperature and pressure. Three factors have contributed to the recent attention given to SCFs: (1) the environmental problems associated with common industrial solvents (mostly chlorinated hydrocarbons), (2) the increasing cost of energy-intensive separation techniques (for example distillation), and (3) the inability of traditional techniques to provide the necessary separations needed for emerging new industries (microelectronics, biotechnology, etc.). The availability of inexpensive, nontoxic SCF solvents such as CO_2 and their attractive properties has renewed interest in SCF extraction as a viable commercial separation technique. SCF extraction has been applied to a wide variety of areas, including most recently various aspects of environmental control.

SCF extraction has been proposed as both a clean-up technique and as an analytical analysis technique for liquid and solid environmental samples. Focus has been on three major areas: extraction of organic contaminants from water, extraction of organics from contaminated soil, and extraction of organics from adsorbents. The principal advantage of using SCF extraction as a clean-up process is the availability of nontoxic SCF solvents. The use of such solvents eliminates concern over residual solvent concentration in the processed environmental matrix. The principle advantages of using SCF extraction as an analytical technique are increased efficiencies and shorter analysis times as compared to standard techniques.This chapter reviews the use of SCF technology in environmental control. Section II is a brief background on SCFs.

II. BACKGROUND

The critical point of a pure substance is defined as the highest temperature and pressure at which the substance can exist in vapor-liquid equilibrium.[1] At temperatures and pressures above this point, a single homogeneous fluid is formed and is said to be supercritical. The relative properties of SCFs in this region compared to gases and liquids are shown in Table 1. There are several advantages to using a SCF as a solvent, including ease of solvent recovery, lower pressure drops, and lower mass transfer resistance than liquids.

The existence of the critical point has been known since the first half of the 19th century, when Cagniard de la Tour discovered the critical point in 1822. At that time, the general consensus was that materials above their critical points would be gaseous in nature and thus poor solvents. It was not until the latter half of that century that Hannay and Hogarth[2] reported on the ability of a SCF to dissolve low vapor pressure solid materials. In their study of several inorganic salts dissolved in supercritical ethanol, they observed that the solubilities were greater than those expected based on the vapor pressure of the salts, even after correcting the vapor pressures for the elevated pressures. In addition, solubilities were seen to increase with increasing pressure.

Since the initial work by Hannay and Hogarth,[2] many authors have investigated the solubility phenomena of SCFs. A wide variety of SCF solvents have been studied, including air, hydrogen, water, helium, ammonia, nitrogen, nitrous oxide, and most extensively CO_2 and a number of light hydrocarbons (e.g., ethane, ethylene, propane, etc.). Solutes investigated have ranged from all types of organic compounds to inorganic compounds to complex natural products. A thorough review (with references) is given by Paulaitis et al.[3] In addition, Francis[4] reported solubility data of 261 organic compounds in near-critical liquid CO_2. Many classes of organic compounds were included, such as heterocyclics, aliphatics, aromatics, and compounds with a variety of functional groups. Nearly half of the substances tested were miscible. He concluded that homologs differ only slightly in miscibility. Hydroxyl,

TABLE 1
Order-of-Magnitude Comparison of Physicochemical Properties of a Typical Gas, Liquid, and SCF

Property	Gas	SCF[a]	Liquid
Density (kg/m³)	1	700	1000
Viscosity (ns/m²)	10^{-5}	10^{-4}	10^{-3}
Diffusion coefficient (cm²/s)	10^{-1}	10^{-4}	10^{-5}

[a]   At $T_r \sim 1$ and $P_r \sim 2$.

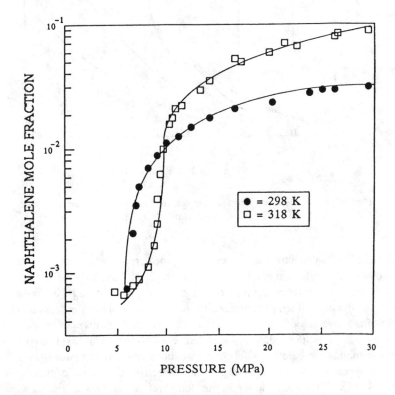

FIGURE 1.   Solubility of naphthalene in supercritical ethylene. (From Paulaitis, M. E., Krukonis, V. J., and Kurnik, R. T., *Rev. Chem. Eng.*, 1, 185, 1983. With permission.)

amino, and nitro groups diminished solubility, especially if two or more are present, while halogen atoms and carbonyl and ether groups had a slight effect.

Perhaps the most extensively studied solute has been naphthalene. Buchner[5] was the first to investigate the solubility behavior of naphthalene in CO_2. Since then, the solubility of naphthalene has been measured in ethylene,[6-8] ethane,[9] hydrogen, helium, argon,[10] methane,[8] xenon,[11] and CO_2.[7,8,12-14] The extensive investigation of this solute has revealed the often-complex solubility behavior in SCFs.

The solubility of naphthalene in supercritical ethylene is shown in Figure 1.[6,7] For a given isotherm, the solubility increases with increasing pressure. The effect of temperature at a given pressure, is more complex. At higher pressures (>100 atm), increasing the temperature increases the solubility; however, at lower pressures (<100 atm), increasing the temperature decreases the solubility. Since solubility is a function of solvent density for

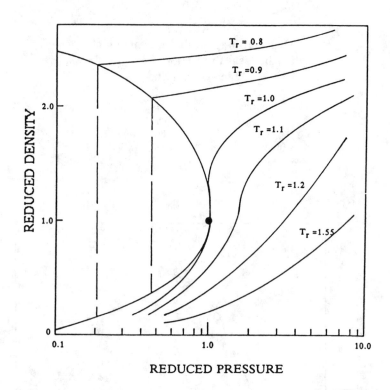

FIGURE 2.   Reduced density of a pure substance near the critical point. (From Paulaitis, M. E., Krukonis, V. J., and Kurnik, R. T., *Rev. Chem. Eng.*, 1, 184, 1983. With permission.)

SCFs, this behavior can be explained by examining the density behavior of a pure component in the vicinity of the critical point, shown in Figure 2. For the reduced temperature range between 1.0 and 1.1 and reduced pressure range between 1.0 and 2.0, the reduced density is a strong function of both temperature and pressure. This explains the pressure dependency of solubility for a given isotherm. Also, in this region a small increase in temperature causes a large reduction in reduced density, and hence solubility. This explains the solubility behavior of naphthalene below 100 atm. For reduced pressures greater than 2.0, a small change in temperature causes much less of an effect on the reduced density. Here, the increased volatility of the solute at the higher temperature can be significant, as is the case of naphthalene above 100 atm. Van Leer and Paulaitis[15] observed a similar behavior for phenol. The solvent characteristics of a SCF can therefore be adjusted using both temperature and pressure to provide the desired solvent power. These effects can be capitalized on most easily in the region $0.9 < T_r < 1.2$ and $P_r > 1.0$, which defines the SCF region for practical applications.[3]

The use of a compressed gas in a separation process was first proposed by Wilson et al.[16] for the deasphalting of oil. Although the process was not strictly supercritical, it did take advantage of the change in solubility associated with a pressure reduction. Later, Elgin and Weinstock[17] reported on a phase-splitting technique for recovering methyl ethyl ketone (MEK) from water using supercritical ethylene. The presence of compressed ethylene caused the mutual solubility of MEK and water to decrease, yielding a salting-out effect. Since then, SCFs have been used as solvents to extract a variety of materials such as coal,[18] caffeine from coffee,[19,20] tobacco,[21] fruit aromas,[22] alcohols from water,[23,24] and most recently toxic contaminants from environmental matrices.[25]

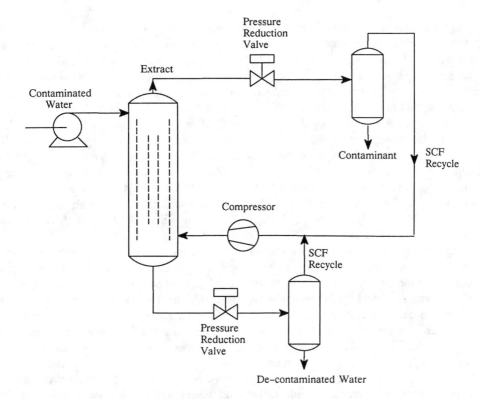

FIGURE 3. Schematic of SCE process.

III. AQUEOUS EXTRACTIONS

Increasing amounts of hazardous wastes are being produced in the U.S. each year. Much of this waste exists in the form of water contaminated with toxic organic compounds. Many methods currently available for large-scale clean-up of contaminated water are applicable over a limited range of contaminated levels. Distillation and incineration are best suited for highly concentrated solutions. Distillation, however, is energy intensive for dilute aqueous solutions. Incineration is both energy intensive and requires proper disposal of the solid residue (ash). Adsorption and biodegradation are usually limited to very low level contamination (parts per million levels) and air stripping is most effective for volatile compounds. Adsorption requires regeneration of the adsorbent. Biodegradation, although used widely, is slow and necessitates sludge disposal. Liquid extraction is a viable technique, but has limited use due to concern over residual solvent present in the processed water. SCF extraction offers an attractive alternative to these conventional techniques.

The broad range of organic solutes that can be extracted from aqueous systems by SCF extraction as well as the availability of nontoxic SCF solvents such as CO_2, has directed attention to this process as a viable method for removing toxic organic waste present in industrial waste streams, surface water reservoirs, and ground water reserves. The advantages of extracting contaminants from water using an SCF instead of a liquid solvent can be seen by considering a typical extraction process (Figure 3). Contaminated water enters the extraction column at the top and is contacted countercurrently with the solvent. The extract (solvent plus some contaminant) exits from the top of the column. In order for the process to be economical, the solvent must be regenerated. If a liquid solvent is used, distillation must be used that requires significant thermal input; however, if an SCF is used as the solvent, a simple pressure reduction or an increase in temperature (pressure or a temperature swing) will decrease the solubility of the contaminant in the SCF, allowing it to be collected. The SCF can then be recompressed and recycled. In addition, residual solvent in the raffinate is minimized upon expansion. The availability of nontoxic SCF solvents such as CO_2 makes this process particularly attractive for environmental applications. The upper bound for the amount of CO_2 lost with water would be the solubility of CO_2 in water at the process conditions and the lower bound would be the solubility at ambient conditions.

Most work to date dealing with the removal of organic compounds from water using SCFs has focussed on the extraction of single-component oxygenated organic compounds. Fueled by the energy shortages of the 1970s, a great deal of attention has been given to the extraction of ethanol from water using various SCF solvents.[24,26,27] Examination of ternary phase diagrams depicting equilibrium tie lines for ethanol/water/CO_2 mixtures indicate that CO_2 is a selective solvent for ethanol, but that the distribution coefficients are relatively small (~0.09 on a weight basis). Reduction in energy costs in the 1980s made this separation economically unattractive using SCF extraction.

Other alcohol-water mixtures investigated have included n-propanol,[24] isopropyl alcohol,[28,29] and n-butanol.[30] Results of the latter system indicate that the equilibrium phase behavior of the ternary system depends significantly on the pressure and temperature of the system. Paulaitis et al.[28] discussed the phase equilibrium behavior of alcohol/water/SCF systems in detail. DeFilippi and Vivian[27] described an alternative process for removing ethanol from water and presented a general correlation for the distribution coefficients of normal aliphatic alcohols as a function of carbon number. Distribution coefficients tend to increase with increasing carbon number.

Various other compounds than alcohols have been extracted from water using SCF solvents. Panagiotopoulos and Reid[31] reported on the extraction of acetone from water using supercritical CO_2. Schultz and Randall[22] measured the distribution coefficients of aroma constituents of fruit and other foods in supercritical CO_2. Compounds investigated included various aromatics, higher weight alcohols, and esters. Distribution coefficients ranged up to 10 or more. Ehntholt et al.[32] reported on the effectiveness of supercritical CO_2 for removing 23 organic compounds (typically found in aqueous systems) from water as an analysis technique. A flow system was used and no equilibrium data were reported.

Application of SCF extraction to aqueous solutions contaminated with hazardous wastes has been limited due to the lack of fundamental thermodynamic parameters, such as the distribution coefficient, needed for design and scale-up. Roop[33] and Roop and Akgerman and associates[34,35] have investigated the SCF extraction of three model toxic organic contaminants from water (phenol, a phenolic mixture, and petroleum creosote), and recently Yeo and Akgerman[36] completed studies on supercritical extraction of organic contaminants both as single components and as a mixture from water. Static equilibrium SCF extractions were conducted as a function of pressure and temperature using supercritical CO_2 as the solvent. CO_2 is the preferred solvent for environmental applications since it is nontoxic, inexpensive, and readily available. In addition, CO_2 has a conveniently low critical temperature (304 K) and moderate critical pressure (7.39 MPa).

A. THERMODYNAMICS OF AQUEOUS EXTRACTIONS

Thermodynamic models describing phase equilibrium associated with SCF extraction are based on standard thermodynamic principles which equate fugacities for each constituent in all phases. For a two-phase system consisting of a liquid phase (L) and a supercritical phase (SC):

$$\hat{f}_i^L = \hat{f}_i^{SC} \qquad (1)$$

For fluid-phase equilibria at low or moderate pressures, the conventional approach is to calculate gas phase fugacities using fugacity coefficients and liquid phase fugacities using activity coefficients. At elevated pressures, however, the clear distinction between gas or fluid phases and liquid phase is lost, and either can be used for each phase at equilibrium.

Application of activity coefficients to either the supercritical phase or the liquid can be difficult. If activity coefficients are used, then:

$$\hat{f}_i = x_i \gamma_i f_i^o \tag{2}$$

$$\hat{f}_i = x_i \gamma_i^o f_i^o exp \int_{P_o}^{P} \frac{\overline{V}_i}{RT} dP \tag{3}$$

where γ_i^o is defined at the reference pressure and the Poynting correction (exponential term) corrects it to the pressure of the mixture. Caution must be used when applying Equation 3 to the supercritical phase, since the partial molar volume (PMV) of compressible fluids near the critical region is sensitive to both pressure and mixture composition.[37] For the liquid phase, Equation 3 can be simplified for dilute solutions using Henry's law:

$$f_i^o = H_{im} \tag{4}$$

and

$$\gamma_i^o exp \int_{P_o}^{P} \frac{\overline{V}_i^L}{RT} dP \approx 1 \tag{5}$$

$$\hat{f}_i^L = x_i H_{im} \tag{6}$$

where H_{im} is the Henry's law constant for component i in the mixture. Even in this simplified form, it is clear that a substantial amount of data (Henry's law constants) are needed for calculating fugacities of multicomponent mixtures.

Alternatively, fugacity coefficients can be used to calculate fugacities for both phases:

$$\hat{f}_i = x_i \hat{\phi}_i P \tag{7}$$

Application of Equation 7 to Equation 1 yields the usual expression for the distribution coefficient:

$$K_i = \frac{x_i^{SC}}{x_i^L} = \frac{\hat{\phi}_i^L}{\hat{\phi}_i^{SC}} \tag{8}$$

Calculation of the fugacity coefficients requires the application of an equation of state (EOS) consistent with the rigorous thermodynamic relationship:

$$\ln\hat{\phi}_i = \int_v^{\infty} \left[\left(\frac{\partial P}{\partial n_i} \right)_{T,v,n_j} - \frac{RT}{v} \right] dv - \ln Z \tag{9}$$

There are many such EOSs available in the literature such as the Redlich-Kwong[38] and the Soave modification to the Redlich-Kwong.[39] The Peng-Robinson[40] EOS

$$P = \frac{RT}{V - b_m} - \frac{a_m}{V(V - b_m) + b_m(V - b_m)} \tag{10}$$

has been successfully applied to correlate solid solubilities in SCFs.[13] Calculation of the mixture attraction (a_m) and size (b_m) parameters requires the use of mixing rules to combine pure component parameters (a_i and b_i), which are calculated from corresponding-states correlations based on critical properties and acentric factors. The common van der Waals-1 fluid mixing rules are given by:

$$a_m = \sum_i \sum_j x_i x_j a_{ij} \qquad b_m = \sum_i x_i b_i \qquad (11)$$

with the usual combining rule:

$$a_{ij} = \sqrt{a_i a_j}(1 - k_{ij}) \qquad (12)$$

where k_{ij} are binary interaction parameters regressed from binary phase equilibria data. Equations 11 and 12 do not adequately describe the mixture behavior of systems containing highly nonideal polar compounds.[41,42] This is particularly true when compounds are present that hydrogen bond, such as water.

A great deal of work has been done on developing new mixing rules for these systems. Local composition density-dependent mixing rules that give the necessary quadratic dependence of the second virial coefficient at low density have received much attention, but are significantly more complex.[43-45] Recently, Panagiotopoulos and Reid[46] utilized the common van der Waals-1 fluid mixing rules (Equation 11) but with a new combining rule. They relaxed the assumption that the binary interaction parameter k_{ij} is equal to k_{ji}, yielding a second interaction parameter per binary:

$$a_{ij} = \sqrt{a_i a_j}(1 - k_{ij} + (k_{ij} - k_{ji})x_i) \qquad (13)$$

One feature of this combining rule is that it reduces to Equation 12 if $k_{ij} = k_{ji}$, allowing the interaction parameters available in the literature for many compounds to be used. It does not reproduce the correct functional dependence of the mixture second virial coefficient on composition but gives good results in the high density region where SCF extractions are conducted.

B. SUPERCRITICAL FLUID EXTRACTION OF SINGLE COMPONENT CONTAMINANTS FROM WATER

As a model system for the SCF extraction of a single-component contaminant from water, phenol (a phenolic component), benzene and toluene (single-ring aromatic hydrocarbon solvents), naphthalene (a double-ring aromatic hydrocarbon which may represent polycyclic aromatics), and parathion (a pesticide) were extracted from an aqueous solution using near-critical and supercritical CO_2.[33,36] Phenol is a common pollutant in the wastewater from coal-coking processes used by the steel industry and coal gasification plants and is a priority pollutant. Water containing 6.8% phenol was extracted using pure CO_2. The experimentally measured distribution coefficients for phenol are shown in Figure 4 as a function of pressure at 298 and 323 K. Values of the distribution coefficient ranged from 0.4 at 6.9 MPa to 1.2 at 27.6 MPa, representing a 12 to 32% reduction in the phenol concentration in the water. Reproduction of the data yielded a standard deviation of 1.5%. Compared to the distribution coefficients previously reported for aliphatic oxygenated organics such as ethanol, values of the distribution coefficients for phenol (a more hydrophobic compound) were considerably higher.

The distribution coefficients (K value) of benzene, toluene, naphthalene, and parathion between supercritical CO_2 and water were obtained as a function of pressure at 318 and 330 K. The operating pressure ranged from 7.8 to 11.0 MPa, which is slightly above the critical

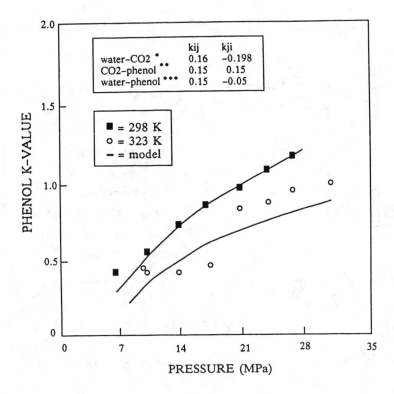

FIGURE 4.   Distribution coefficients for phenol at 298 and 323 K, dimensionless. (*From Ref. 46; **from Ref. 47; from ***k_{ij} Ref. 47 with k_{ji} adjusted to the data.)

pressure of CO_2. The results are shown in Figures 5 to 8. The distribution coefficients of the four compounds differed by several orders of magnitude and provided a comparison of removal of the four compounds from water. The K value of benzene and toluene, which are highly soluble in supercritical CO_2, are of the order 10^3, and indicate an up to 99% reduction in the concentration of aqueous phase in a single equilibrium stage. The maximum reduction rates of aqueous phase concentration of naphthalene and parathion at 318 K are 95% at 10.0 MPa and 54% at 10.8 MPa, respectively. These results show that supercritical CO_2 is a better extractant for relatively light compounds (benzene, toluene, and naphthalene) than heavy compounds (parathion). The distribution coefficients increased with pressure for each isotherm, and decreased overall at a higher temperature. Because of the nature of a dilute system, the influence of temperature on volatility of the compounds is negligible, and the extent of extraction depends mainly on the density of supercritical CO_2, which becomes lower at higher temperature.

Using the Peng-Robinson EOS (Equation 10), van der Waals 1-fluid mixing rules (Equation 11), and Panagiotopoulos and Reid's[46] combining rule (Equation 13), the phenol data were correlated by performing a flash calculation similar to that described by McHugh and Krukonis.[23] The results are also shown in Figure 4, with the binary interaction parameters used. For a ternary system, six interaction parameters are needed for Equation 13. The k_{ij} and k_{ji} for the two major components (water and CO_2) are the most critical to describe the phase equilibria and were obtained from Panagiotopoulos and Reid.[46] Tsonopoulos[47] provides general values for CO_2 and phenol, as well as the interaction parameter for water and phenol. The parameter for phenol and water (a single parameter out of six), which was not available in the literature, was adjusted to fit the data at 298 K and held constant throughout the rest of the study. Correlation of the data at 298 K gave quantitative results, except near the critical point. Without adjusting the interaction parameters, the model qualitatively predicted

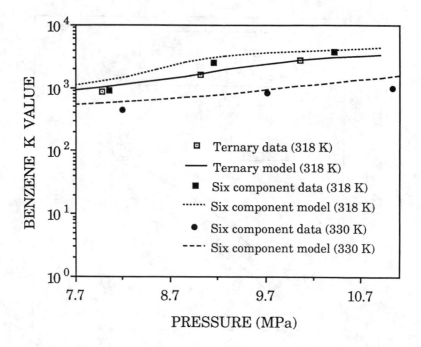

FIGURE 5.   Distribution coefficient of benzene at 318 and 330 K, dimensionless.

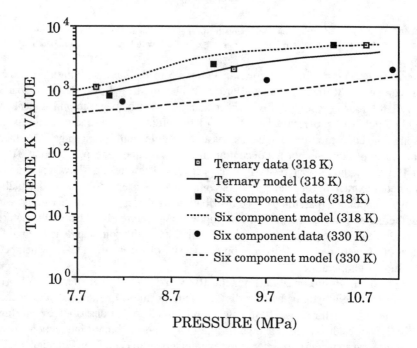

FIGURE 6.   Distribution coefficient of toluene at 318 and 330 K, dimensionless.

the reduction in the distribution coefficients when the temperature is increased to 323 K. These predictions were verified experimentally and the results also shown in Figure 4. Because temperature control was approximately ± 3 K, there was more scatter in the data for the 323 K experiments than at room temperature.

Using the same procedure as phenol data correlation, the data of benzene, toluene,

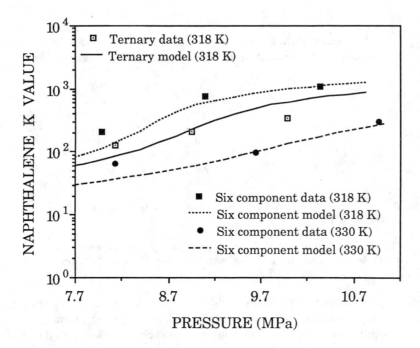

FIGURE 7.   Distribution coefficient of naphthalene at 318 and 330 K, dimensionless.

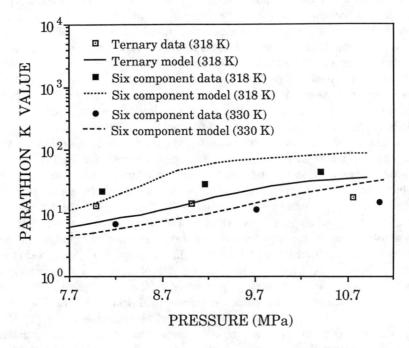

FIGURE 8.   Distribution coefficient of parathion at 318 and 330 K, dimensionless.

naphthalene, and parathion at 318 K were also correlated (see Figures 5 to 8). The interaction parameters, which were not available in the literature, were adjusted to fit the data, and were used for modeling the each ternary system (see Table 2).

C. SCF EXTRACTION OF ORGANIC MIXTURES FROM WATER

As an environmental organic mixture system, water was initially saturated with a mixture of benzene, toluene, naphthalene, and parathion, and then brought to equilibrium with

TABLE 2
Interaction Parameters (k_{ij} and k_{ji}) of Six Compounds for
Composition Dependent Mixing Combining Rule (Equation 11)*

Component	1	2	3	4	5	6
1	0.0	0.16	0.26	0.26	0.40	0.40
2	−0.198	0.0	0.075	0.081	0.115	−0.428
3	−0.024	0.075	0.0	0.0	0.0	0.0
4	−0.106	0.081	0.0	0.0	0.0	0.0
5	−0.110	0.115	0.0	0.0	0.0	0.0
6	−0.097	0.428	0.0	0.0	0.0	0.0

Component: 1 = water, 2 = CO_2, 3 = benzene, 4 = toluene, 5 = naphthalene, 6 = parathion.

*k_{12} and k_{21} from Reference 46, $k_{23} = k_{32}$ and $k_{24} = k_{42}$ from Reference 85, $k_{25} = k_{52}$ from Reference 86, k_{13} and k_{14} from Reference 87, k_{15} from Reference 47, and other values from fitting ternary data at 318 K.

supercritical CO_2, forming a six-component dilute system. The thermodynamic model, which was used for single-contaminant systems, was extended to the six-component system. The model successfully predicted the six-component system behavior using only the information based on ternary system data. The mixing rule parameters adjusted from one-temperature data were directly applied to the other temperatures and the predictions were acceptable.

In order to model the six-component system, a total of 36 interaction parameters are required; however, the number of parameters to be determined can be reduced by applying the following assumptions: (1) interaction parameters are zero for hydrocarbon pairs[48] and for very dilute component pairs, and (2) for nonpolar components pairs that include CO_2, benzene, toluene, and naphthalene, $k_{ij} = k_{ji}$, because the given combining rule was developed for polar compounds. Since the first parameter k_{ij} becomes a binary interaction parameter in the case of $k_{ij} = k_{ji}$, the authors used the binary interaction parameters, which were available in the literature, for k_{ij}. The second parameter, k_{ji}, which is usually not available in the literature, was determined by regressing the experimental data. Only the ternary system data were used to adjust the interaction parameters. Diandreth and Paulaitis[49] also used ternary instead of binary data to determine their interaction parameters, and provided better results. Interaction parameters determined from ternary data at 318 K were applied to predict the six-component system behavior at 318 and 330 K. Table 2 shows the determined interaction parameters k_{ij}, k_{ji} as well as the values obtained from the literature. The negative interaction parameters are common for component pairs if one of the components is polar.

The distribution coefficients of all compounds in six-component systems were increased compared to those of the ternary system and the rate of increase was dependent on the compounds. Figure 9 shows the average percentage increase in K value of the four components in six-component systems based on experimental data at 318 K. The percentage increase in K value of benzene was 37.3%; toluene, 11.5%; naphthalene, 201.5%; and parathion, 113.6%, respectively. These results show that the effect of benzene and toluene on the K value of other components was large, and that of naphthalene and parathion was relatively small.

As an extension of the single-component work with phenol, a phenolic mixture was chosen as a model multicomponent contaminant. The composition and estimated properties of the mixture are shown in Table 3. Water containing 0.87 wt% (8700 ppm) of the phenolic mixture was extracted using pure CO_2. Water samples were tested for total organic content (TOC) before and after extraction. The ratio of the TOC of an extracted aqueous sample to the TOC of the original solution is equivalent to the ratio of the moles of phenolic mixture

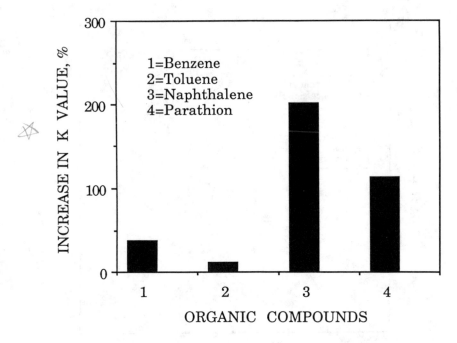

FIGURE 9.   Percentage increase in distribution coefficient of each compound in multicomponent system compared to ternary system at 318 K.

TABLE 3
Composition and Estimated Properties of Phenolic Mixture

Component	wt%	$T_c(k)$[a]	$P_c(bar)$[b]	Acentric
Phenol	0.76	692.1	61.3	0.461
Phenol, 2-methyl	10.1	692.4	50.3	0.480
Phenol, 2-methoxy	77.5	700.2	49.1	0.566
Phenol, 2,3-dimethyl	5.5	716.6	43.2	0.530
Phenol, methoxy, methyl	5.08	708.7	40.6	0.598
Unidentified	~1			
Average mixture properties		700.3	48.5	0.556

[a]   Joback's method.[48]
[b]   Lee-Kesler.[48]

in the sample to the total moles of phenolic mixture. With this ratio (r) the distribution coefficient lumped in terms of TOC content can be calculated using

$$K = \left(\frac{1-r}{r}\right)\left(\frac{\text{total mole water}}{\text{total mole } CO_2}\right) \qquad (14)$$

The experimentally measured distribution coefficients for the phenolic mixture distributed between water and CO_2 are shown in Figure 10 as a function of pressure at 298 K. Values of the distribution coefficients ranged from 2.4 at 6.9 MPa to 5.7 at 27.6 MPa. Reproduction of the data yielded an average standard deviation of 6%. As expected, the values of the distribution coefficients for the more hydrophobic phenolic mixture were considerably higher than for phenol.

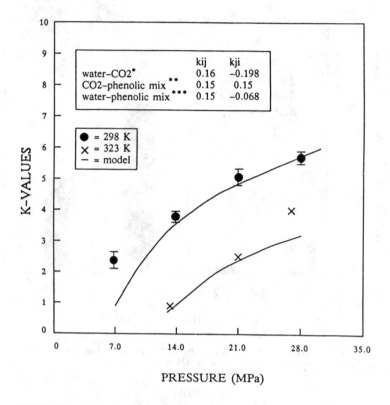

FIGURE 10.   Distribution coefficients for the phenolic mixture at 298 and 323 K, dimensionless. (*From Ref. 46; **from Ref. 47; ***k_{ij} from Ref. 47, with k_{ji} adjusted to the data.)

The data for the pseudo-ternary system (phenolic mixture/water/CO_2) were also correlated using the thermodynamic model. Binary interaction parametrs for the phenolic mixture were assumed to be the same as those obtained for phenol. The k_{ij} and k_{ji} for the two major components (water and CO_2) are the most critical to describe the phase equilibria, and were again obtained from Panagiotopoulos and Reid.[46] The parameter for the phenolic mixture and water (a single parameter out of six) was adjusted to fit the data at 298 K and held constant throughout the rest of the study. The results of the model are also shown in Figure 10, with the binary interaction parameters used. The model was then used to predict the effect of increasing the temperature of the system. As is shown in Figure 10, the predicted decrease in the distribution coefficients at an increased temperature of 323 K was verified experimentally.

The final contaminant mixture investigated was petroleum creosote. The composition and estimated properties of the petroleum creosote are shown in Table 4. This system was chosen as a model environmental contaminant of current concern. Petroleum creosote is much more hydrophobic than the other contaminants tested. In addition, contamination levels were again reduced and were an order of magnitude less than the phenolic mixture and two orders of magnitude less than the phenol. Most of the compounds that make up petroleum creosote are not readily soluble in water. In fact, TOC analysis shows that water saturated with petroleum creosote contained only 500 ppm organics. Extracted water samples contained as little as 380 ppm. At these concentration levels, trace levels of contamination in the extractor cause significant error in TOC analysis. In addition, sensitivity of the TOC analyzer (± 10 ppm) also begins to affect the measurements.

Water containing 500 ppm petroleum creosote was extracted using pure CO_2. The experimentally measured distribution coefficients of the contaminant lumped in terms of

TABLE 4
Composition and Estimated Properties of Petroleum Creosote

Component	wt%	$T_c(K)$[a]	$P_c(bar)$[b]	Acentric
Naphthalene	21.9	737.2	39.0	0.371
Phenanthrene	37.1	877.0	32.4	0.517
Acenaphthene	8.9	800.0	35.2	0.484
Pyrene	12.7	927.9	45.4	0.833
Fluorene	5.0	815.0	33.8	0.500
Fluoranthrene	9.6	902.8	30.7	0.646
Phenol	2.0	692.1	61.3	0.461
Other phenolics	~1	~704	~46	~0.54
Average mixture properties		824.2	35.8	0.524

[a] Joback's method.[48]
[b] Lee-Kesler.[48]

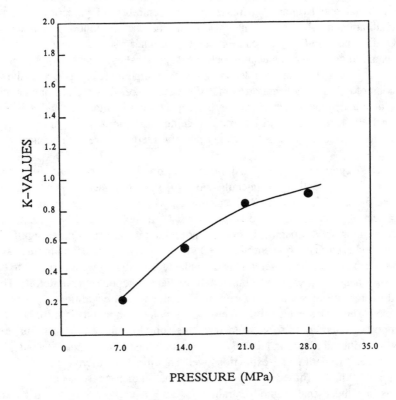

FIGURE 11.   Distribution coefficients of petroleum creosote at 298 K, dimension-less. (*From Ref. 47; **from Ref. 48; ***k_{ij} adjusted to the data.)

TOC content are shown in Figure 11 as a function of pressure. Values of the distribution coefficient ranged from 0.22 at 6.9 MPa to 0.91 at 27.6 MPa, indicating the same trends seen for the other contaminants; however, due to the sources of error discussed above, the results are only semi-quantitative.

A limited number of the extracted samples were tested using gas chromatgraphy/mass spectrometry (GC/MS). Analysis of the saturated water solution (500 ppm) revealed that the bulk of the organics dissolved in the water were phenolics, specifically phenol and a smaller quantity of methoxy phenol (which constitutes only a few percent of the creosote). The major components of petroleum creosote dissolved only sparingly in the water. Therefore, the resulting water solution was essentially phenol-contaminated water. This explains why the behavior of this system and the resulting distribution coefficients were similar to the single-component phenol extractions. Analysis of the extracted water samples indicated that the trace quantities of the more hydrophobic compounds such as naphthalene are virtually completely removed from the water in a single equilibrium extraction.

D. THE ENTRAINER EFFECT

If SCF extraction is going to be implemented on a commercial scale to remove contaminants from water, high costs associated with poor solvent power of certain SCFs and the use of high pressure equipment must be reduced. These costs can be substantially lowered with the use of an entrainer (or cosolvent) to increase the solubility of the solute in the supercritical phase, reducing the size of the required extractor and/or the pressure needed to effect the desired extraction. The literature contains only a few examples of liquid-fluid equilibrium data for liquid/entrainer/SCF systems. Most deal with the separation of two organic compounds with similar boiling points which are difficult to separate by conventional distillation. Peter and Brunner[50] increased the concentration of glycerides in propane by adding small amounts of acetone. They also found that the distribution coefficient for palm oil in CO_2 doubled with the addition of ethanol.[51] Brunner[52] studied the effect of cosolvents on the separation factor of hexadecanol, octadecane, and salicylic acid phenyl ester in several SCFs including CO_2. He found that depending on the temperature and pressure, the solubility as a function of entrainer concentration may decrease, increase, or run through a maximum. Hacker[53] obtained good selectivities for the separation of butadiene and 1-butene using supercritical ethylene, CO_2, and ethane containing small amounts of ammonia. Ethylene was found to provide the best selectivity, probably due to its greater chemical affinity for the olefins.

Roop and Akgerman[54] investigated the application of entrainers to aqueous extractions. After successfully modeling the phenol/water/CO_2 system using the Peng-Robinson EOS with a modified mixing rule, the flash calculation was extended to a quaternary system so that the effect of an entrainer could be predicted. Methanol, a commonly used entrainer in studies concerning solid organics, has been reported to increase solubilities significantly and was therefore chosen as the first entrainer to study. Interaction parameters for the base ternary system were not altered. The interaction parameters for methanol were assumed to be similar to those of ethanol, which were obtained from Panagiotopoulos and Reid.[46] The model predicted a slight decrease in the K values of phenol with the addition of a small amount of methanol. This is expected, since methanol (a polar compound) is miscible with water at all proportions and could serve to increase the solubility of phenol in water. Further, it is known that normal aliphatic alcohols preferentially distribute to the aqueous phase over the CO_2 phase.[23] Experimental extractions were then performed with methanol at concentrations up to 10 mol% (based on CO_2) to verify the predicted response.

It was concluded that a good entrainer for an aqueous system would be soluble in supercritical CO_2 but insoluble in water. Most nonpolar organic solvents fit this criteria. To qualitatively predict which compound would be the best entrainer, general values of the binary interaction parameters for a hydrocarbon were chosen and flash calculations performed for a variety of compounds at 298 K and 27.6 MPa with 5 mol% entrainer present and the relative magnitude of the phenol distribution coefficient calculated. Compounds with structures similar to phenol, such as benzene, were predicted to increase the phenol distribution coefficient. To confirm the model predictions, benzene was added as an entrainer (up to 6 mol%) to the phenol extractions at 17.3 and 27.6 MPa. The results are shown in Figure 12. As predicted, the distribution coefficients of phenol increased (as much as 50%) over the

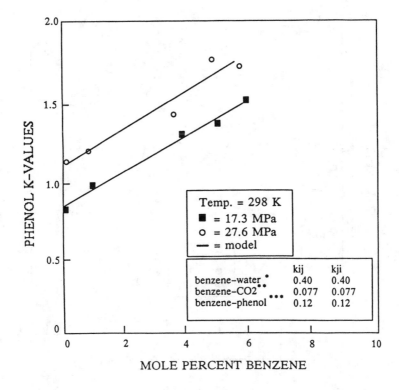

FIGURE 12.   Effect of benzene as an entrainer on the distribution coefficients of phenol.

pure CO_2 extractions, providing up to 41% removal of the phenol from the aqueous phase in a single equilibrium stage. The data were quantitatively correlated using the interaction parameters shown in the figure (the binary interaction parameter for benzene-phenol[47] was adjusted slightly from 0.15 to 0.12).

Using the same interaction parameters from the phenol study, methanol was predicted to have no effect on the distribution of the phenolic mixture and therefore was not tested experimentally. Figure 13 shows the predicted effect of benzene on the extraction of the phenolic mixture (again using the same interaction parameters used in the phenol study). Since a favorable effect was predicted, experiments were performed using up to 4 mol% benzene (based on CO_2). The experimental results are also shown in Figure 13. As predicted, benzene did increase the distribution of the phenolic mixture up to 50%. These results agree with the results of the single-component study, in which the distribution coefficient of phenol was increased with the addition of benzene.

The entrainer effect was also observed in the dilute organic mixture system mentioned in Section III.B. The distribution coefficient of each compound was increased in the six-component system compared to ternary system value. For example, the rate of increase in the distribution coefficient of parathion was greater than that of benzene, which could be regarded as an entrainer for extraction of parathion. This multicomponent effect can be explained by interactions among supercritical CO_2, water, and each mixture component. The variation of CO_2 density with temperature had a dominating effect on extractability in comparison to the change of volatility of dilute organic components with temperature. These results show that as far as a dilute system is concerned, ternary system data will be very useful for verifying the multicomponent system behavior in supercritical extraction.

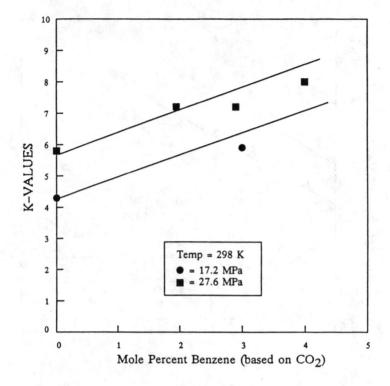

FIGURE 13.   Prediction and experimental verification of the effect of benzene on distribution coefficients of the phenolic mixture.

E. DISCUSSION

In aqueous extraction, most people report only the K value, i.e., the ratio of mole fraction of the contaminant in the two phases. Thus, a K value of 1 would indicate that the contaminant is roughly equally distributed between the two phases, where as a K value of 1000 would indicate almost all of it will go to the supercritical phase.

The authors performed some experiments using a Jerguson gauge and observed that the position of the interface does not change, i.e., not much water ends up in the supercritical phase. Thus, in all their calculations, the amount of water was taken as constant.

IV. SOIL EXTRACTIONS

SCF extraction of organic compounds from environmental solids such as soil and sediment is a promising new technique for analysis of solid samples and for solid waste clean-up.

A. ANALYSIS TECHNIQUE FOR SOLID SAMPLES USING SUPERCRITICAL FLUIDS

Extraction and recovery of organic species from environmental solids for analysis traditionally has been done with liquid solvent extraction or thermal desorption. Liquid solvent extraction techniques such as Soxhlet and sonication methods are time consuming and selection of solvents is currently a trial-and-error process; no universal solvent exists for all classes of chemicals commonly found in environmental solids. Soxhlet extraction protocols also suffer from the evaporation of volatile components during extraction, resulting in inaccurate analysis. Thermal desorption is limited by the thermal stability of the organic species and can result in degradation of some organics. Supercritical extraction (SCE) offers the advantages of moderate temperatures and fast recovery times.

Several investigators have used SCE to analyze organics in environmental solids. Various SCFs have been used to extract the natural organic matter from soil samples.[55,56] Schnitzer and Preston[57] used SCF and SCF mixtures of increasing polarity to extract organics from untreated soil which was predominately inorganic. In addition to soil, various other environmental solids have been analyzed using SCE. Hawthorne and Preston[58] extracted poly-aromatic hydrocarbons (PAHs) and other organics from diesel exhaust particulate using supercritical CO_2 and a CO_2/methanol mixture. Hawthorne and Miller[59] later extracted PAHs from three other environmental solids: urban dust, fly ash, and river sediment. They concluded that SCE is as good or better than Soxhlet extraction or sonication. Schantz and Chesler[60] also removed comparable or better amounts of PAHs from urban particulate using SCE as compared to Soxhlet extraction. In addition, they extracted polychlorinated biphenyls (PCBs) from sediment with comparable results with Soxhlet. Wright et al.[61] extracted PAHs from an air particulate sample with capillary GC for analysis. Analysis was done on-line, but after the expansion of the SCF.

B. SOIL CLEAN-UP APPLICATIONS

SCF extraction has been investigated as an alternative clean-up technique to remove organic compounds from soil. Brady et al.[62] demonstrated the ability of SCFs to extract PCBs and DDT from contaminated topsoil (12.6% natural organic matter) and subsoils (0.74% natural organic matter) using supercritical CO_2. Over 90% of the PCB was extracted from the subsoil in less than 1 min and 70% DDT was leached from the topsoil in less than 10 min. They also investigated the effect of 20% water content and concluded that longer extraction times were required, but the same amounts of the contaminants could still be recovered. Eckert et al.[63] studied the removal from soil of chlorinated aromatics such as trichlorophenol as model compounds for PCBs and dioxins. They were able to remove virtually all trichlorophenol from a soil sample using supercritical ethylene.

Bound residues have also been extracted from soil using SCFs. Capriel et al.[64] used supercritical methanol to extract bound ^{14}C pesticide residues from several types of soil. Compared to high-temperature distillation, supercritical methanol was able to remove a significantly larger amount of the bound compounds. They also concluded that the amount of the residual ^{14}C remaining after SCE depended on the chemical structure of the compound rather than the organic matter content of the soil.

The authors recently investigated the extraction of phenol from six different soils (soil:phenol = 100:1 weight ratio). The characteristics of the various soils are given in Table 5 and were determined by nitrogen BET adsorption and thermal gravimetric analysis (TGA). Surface area of the soils ranged from 1.1 to 67.3 m^2/g. The percent weight loss of a sample between 175 and 350°C was found to correlate with known organic carbon content for three of the soils determined by traditional wet chemistry techniques.[65] The percentage of organic matter present in the soils was therefore determined by TGA.[66] These values are also shown in Table 5.

Distribution coefficients for phenol distributed between the soils and supercritical CO_2 were measured at 13.8 MPa and 310 K. Values of the distribution coefficients ranged from 0.6 to 11.1 for the six soils and are also shown in Table 5. In addition, the distribution coefficients for phenol were also measured for sand and clay (montmorillonite). Extraction of sand contaminated with phenol resulted in nearly all of the phenol distributing to the CO_2 phase, indicating that sand has little adsorption capability for phenol. Extraction of phenol from clay yielded the lowest distribution coefficient (0.08) and took approximately 5 d to reach equilibrium. A time series study of the sand and soil systems indicated that equilibrium was reached for these systems within 3 h.

Examination of the resulting distribution coefficients of phenol with the physical characteristics of the different soils indicate that two parameters, surface area and natural organic

TABLE 5

Soil Characteristics and Experimental Distribution Coefficients

Material	Surface area[a] (m^2/g)	% Organic[b] (by weight)	% Organic[c] (by weight)	K value[d]
Sand	0.67	0.0	NA	~120[e]
Montmorillonite	256.0	0.0	NA	0.08 ± 0.01
Soil 1	6.4	1.0	1.0	2.22 ± 0.19
Soil 2	17.2 ± 0.1	1.4	1.4	1.52 ± 0.07
Soil 3	15.4 ± 0.1	NA	0.7	1.61 ± 0.11
Soil 4	67.3 ± 0.5	NA	2.1	0.78 ± 0.05
				0.73 ± 0.03
Soil 5	52.7 ± 0.3	NA	1.5	0.81 ± 0.04
Soil 6	1.1 ± 0.1	0.4	0.5	5.7 ± 0.6
				5.9 ± 1.8

NA: not available.

[a]   Surface areas of the six soils were determined by N_2 BET.
[b]   As determined by wet chemistry.
[c]   As determined by TGA.
[d]   System conditions were 13.8 MPa, 310 K, and 100:1 ratio of solid to phenol mass.
[e]   The ratio of phenol in the CO_2 phase to the total phenol in the system was 0.995 ± 0.022; total number of samples was 25.

content, have a significant effect. Figure 14 shows the effect of natural soil organic content on the distribution coefficient of phenol. In this case, larger percentages of natural organic material in the soil again resulted in lower distribution coefficients. This indicates that the natural organic material in soil can strongly adsorb the phenol, trapping it in the soil phase. Figure 15 shows the effect of the surface area of the soil on the distribution coefficient of phenol. Solids with higher surface areas (and hence more adsorption sites) had relatively lower distribution coefficients (i.e., more phenol stayed in the soil phase) as expected.

For the six soils used in this research, it was observed that the organic content and the surface areas were correlated: the higher the organic content, the larger the surface area. Whether this relationship is due to the surface area of the organic matter or the inorganic composition of the soil has yet to be determined. Surface area seems to correlate the K values better than organic content since sand and montmorillonite together with all types of soil fall on the same smooth curve.

For soil 4, the authors have determined how K varies with temperature and pressure by running experiments at a fixed pressure of 15 MPa and temperatures of 300 to 350 K, and at fixed temperatures of 310 and 328 K in the pressure range 8 to 30 MPa. The amount of soil was reduced from 100 to 50 g in both of these experimental series. The amount of phenol was also reduced to 0.5 g to keep the ratio of phenol: soil the same as with previous experiments.

The results of the temperature dependency are shown in Figure 16. The error bars plotted with the K values are for the worst-case errors. As the temperature was increased, the K value increased from 0.65 to 1.3, an increase by a factor of 2, yet the distribution of phenol between the two phases remained constant as indicated by the extraction percentage remaining at 75%. Thus, the increasing distribution coefficient indicates that the amount of CO_2 needed to remove the phenol is less at higher temperatures than at room temperature. The increase in the K value is due to changes in the ratio of the masses of the solid phase to the supercritical phase, which is a direct result of the density of CO_2 decreasing from 0.87 to 0.45 g/cm^3 as the temperature is increased from 298 to 359 K. Though the amount of CO_2 needed for the extraction has decreased, the distribution of phenol remained constant between the two

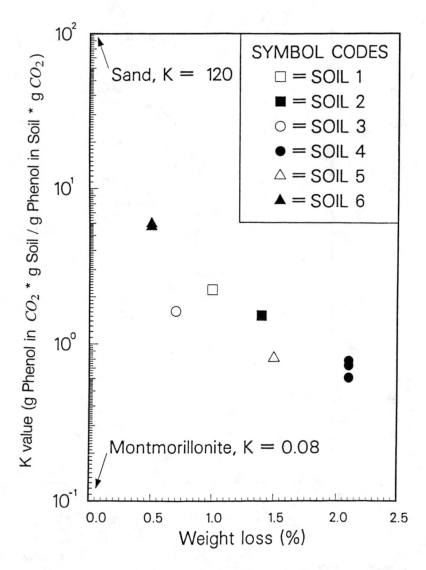

FIGURE 14. The effect of organic material in soil (as measured by weight loss between 175 and 350°C during TGA analysis) on the distribution coefficient of phenol for various soils at 13.8 MPa and 310 K, g soil/g CO_2 (soil:phenol = 100:1).

phases. These results indicate that the K value must be used with caution since the ratio of the masses of the two phases affects it. The K value defined in this work has the units weight soil per unit weight SCF. The thermodynamic distribution coefficient, on the other hand, is based on the ratio of mole fractions of the compound of interest between two phases. Without a definitive molecular weight for soil, the thermodynamic distribution coefficient for the extraction of soil will remain undefined.

The second series of experiments investigated how the extraction was influenced by pressure. An initial constant temperature of 310 K was employed. Figure 17 shows that the extraction percentage increased from 68 to 78% when the pressure was increased from 8 to 30 MPa. At 7.5 MPa, a broken line is drawn to represent the solubility limit of phenol in CO_2.[15] For this series of experiments, 0.5 g of phenol will saturate the CO_2 phase at 7 MPa. The 7 MPA phenol/CO_2 value shows an abrupt decrease from the previous trend of this ratio of the higher pressures (8 to 30 MPa). This same behavior is observed for a higher

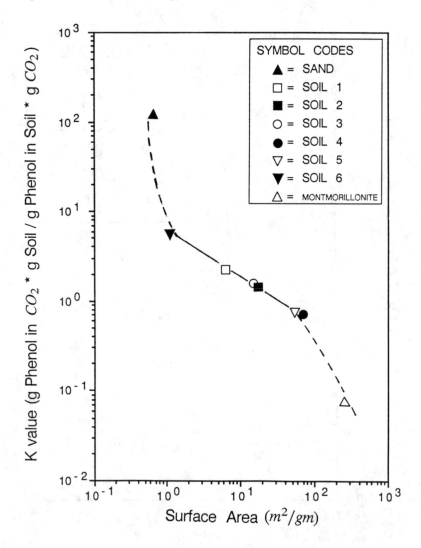

FIGURE 15.   The effect of surface area of soil, sand, and clay on the distribution coefficient of phenol at 13.8 MPa and 310 K, g soil/g CO_2, (soil:phenol = 100:1).

temperature of 328 K (see Figure 18). The experimental conditions were 100 g of soil and 1 g of phenol. Again, the extraction percentage decreases slightly as the pressure is decreased, and as observed in the 310 K experiments, the phenol/CO_2 ratio indicates that a solubility limit is present below 11 MPa.

C. ENTRAINERS

The majority of the information in the literature dealing with the "entrainer effect" focuses on the increased solubility of solids in SCF containing small amounts of cosolvents. Kurnik and Reid[67] measured the solubility of several binary mixtures of nonvolatile solids. They found that the solubility of a solid component of a multicomponent solute system increased as much as 300% compared to the pure component solubility at the same operating conditions. Gopal et al.[68] found that the selectivity of the naphthalene-phenanthrene system was considerably different than the pure component solubility ratio (~ 300%). Dobbs et al.[69,70] studied the effect of nonpolar and polar cosolvents on the solubility of solid organics in supercritical CO_2. They successfully modeled their system using a hard sphere van der

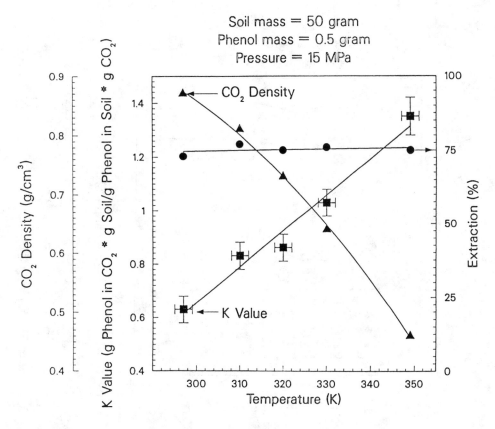

FIGURE 16.   The effect of temperature on the extraction of phenol from soil at a constant pressure of 15 MPa.

Waals EOS with van der Waals type 1 mixing rules. Joshi and Prausnitz[71] did a theoretical investigation altering the critical properties of an SCF by the addition of an entrainer to achieve higher solubilities for solid solutes. Walsh et al.[72] suggests the entrainer effect is caused by a chemical association between cosolvent and solute. They concluded that spectroscopic measurement in liquid solvents and published solvatochronic parameters and dissociation constants could be used as a qualitative prediction of the entrainer effect of solids in SCFs.

Dooley et al.[73] reported on the use of entrainers (or cosolvents) for the SCE of DDT from contaminated soil. They compared the results of extracting with pure CO_2 with CO_2/methanol and CO_2/toluene mixtures. The most effective solvent system, supercritical CO_2 with 5 wt% methanol, was able to leach approximately 95% of the DDT from soil in under 5 min, as compared to either pure CO_2 or CO_2 with 5 wt% toluene at the same extraction conditions. This system could only extract 70% in 10 min, thus demonstrating the advantages of using supercritical modifiers (or entrainers).

Roop et al.[74] have investigated the use of entrainers to enhance the extraction of phenol from a model soil (dry and wetted). Water was added to the system because natural soils will have water present. Methanol was tested as an entrainer. Finally, benzene was added to the system to see if a co-pollutant would affect the extraction. The temperature for this series of experiments was 310 K and the pressure was held at 15 MPa. The water was added to the soil in an end-over-end rotating vessel prior to loading the soil into the autoclave. (If water were added to the soil in the autoclave, the top portion of the soil would form a mud and the water would not be evenly distributed.) Dry soil, 50 g, was used in each of the 14 different experiments. The results of the composition experiments are shown in Figure 19.

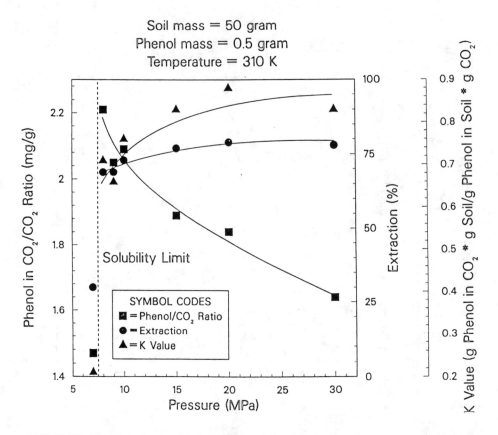

FIGURE 17.   The effect of pressure on the extraction of phenol from soil at a constant temperature of 310 K.

The figure shows which chemicals were present and their amount. The shading of the bar graph (in the extraction portion) indicates which chemical was being investigated.

1. Benzene Experiments

Runs 1, 2, and 3 determined the extraction of benzene in the presence of soil and soil-phenol. The results indicated that benzene does not interact with the soil.

2. Dry Soil Experiments with Phenol

Runs 4 and 5 are the reference conditions for the dry soil experiments with phenol. Comparing these results to runs 6 and 7, one observes that the presence of methanol increased the extraction of phenol from 78 to 87%; however when benzene is present (runs 8 and 9), a slight decrease in the extraction of phenol was observed. These results can be understood by examining the hydrophobic/hydrophilic natures of phenol and CO_2. Phenol is composed of an aromatic ring with one hydroxyl group. The aromatic ring is hydrophobic, as demonstrated by the low solubility of benzene in water. The hydroxyl group is hydrophilic, as demonstrated by the relatively high solubility of phenol in water (-8% b.w.). These structures influence the behavior of phenol in CO_2. The aromatic structure increases the solubility of phenol in CO_2, a hydrophobic solvent. Benzene, for example, is completely miscible with CO_2;[4] however, the hydroxyl group limits the solubility of phenol in CO_2 as demonstrated by the data of Van Leer and Paulaitis.[15] The hydroxyl group also makes phenol acidic, reactive, and capable of hydrogen bonding. These three features allow phenol to interact with the many different soil surface structures.

Methanol is completely soluble in CO_2[4] and makes the CO_2 phase more hydrophilic.

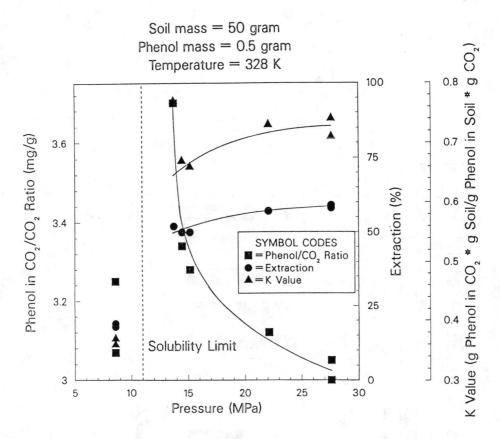

FIGURE 18. The effect of pressure on the extraction of phenol from soil at a constant temperature of 328 K.

Thus, when the results of runs 4 through 9 are examined (when methanol is added to the system), there is a distinct increase in the extraction of phenol. More phenol is extracted because the CO_2 phase is more hydrophilic. When benzene is added to the system, the benzene goes directly into the CO_2 phase (based on runs 1 through 3), causing the CO_2 to be more hydrophobic. This shift of the CO_2 phase to more hydrophobic state drives phenol out of the CO_2 phase and onto the soil.

3. Wet Soil Experiments with Phenol

Runs 10 and 11 are the reference conditions for the wet soil studies with phenol. When water is added to the system, the extraction of phenol is better than for dry soil. There are three possible processes that could account for this observed increase. First, water could selectively absorb to some of the sites where phenol would adsorb if no water were present. This free phenol is easier to extract, being no longer bound to the soil surface, and an increase in the extraction should be observed. Second, water is slightly soluble in CO_2 and its presence would shift the CO_2 to a more hydrophilic state, and more phenol would be extracted as previously explained in the dry soil section. The third is that the water on the soil dissolves some of the phenol on the soil surface. This "aqueous" phenol is extracted by the CO_2 phase, and again, more phenol is seen in the CO_2. Any or all three of these processes could be occurring to increase the extraction of phenol.

When methanol is present for the set soil experiments, runs 12 and 13, there is a slight decrease in the extraction compared to runs 10 and 11. Here, the most reasonable explanation is that some of the methanol is dissolving into the water on the soil surface. This methanol

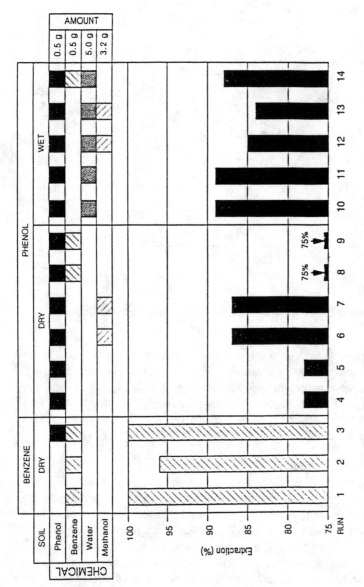

FIGURE 19. The extraction of benzene or phenol in the presence of other chemicals. The amount of soil was 50 g, temperature was 310 K, and pressure was 15 MPa for each run.

would increase the affinity of the phenol for the water on the soil surface. This idea is confirmed by Roop and Akgerman,[54] and it was shown that the presence of methanol did not improve the extraction of phenol from water. It appears that the presence of methanol may hinder the removal of hydrogen-bonding compounds from wet soil. For run 14, when benzene was present, again there is a slight decrease in the extraction with respect to runs 10 and 11. Assuming that the benzene is only in the CO_2 phase (runs 1 through 3), the presence of benzene makes the CO_2 phase more hydrophobic. This hydrophobic state results in a decrease in the phenol in the CO_2.

V. EXTRACTION OF ADSORBENTS

Adsorbents are another solid matrix to which SCE has been applied. Organic compounds can be extracted from various adsorbents using SCFs as a method of regenerating the adsorbents or as an analytical method.

Activated carbon is a common industrial adsorbent which, for economical reasons, must be regenerated and recycled. SCE has recently been proposed for this purpose.[3,23] Groves et al.[25] provides a detailed discussion of the regeneration of activated carbon using SCFs. Kander and Paulaitis[75] studied the desorption of activated carbon loaded with phenol by supercritical CO_2. They concluded that supercritical CO_2 offers no significant thermodynamic advantages for regenerating carbon loaded with phenol, a strongly adsorbing compound. DeFilippi et al.[76] studied the regeneration of activated carbon loaded with pesticides by supercritical CO_2 and concluded that the supercritical regeneration would be economical. Picht et al.[77] also studied the desorption of phenol, as well as acetic acid and alachlor, from activated carbon and four other adsorbents into supercritical CO_2. Recently, Tan and Liou[78] studied the regeneration of activated carbon loaded with ethyl acetate by supercritical CO_2 at different operating conditions. They found that the adsorptive capacity of the regenerated activated carbon was close to that of the virgin carbon and stable after several regeneration cycles. They also found that higher operating pressures were more favorable for regeneration due to the increased density and that viscosity may play an important role in determining the optimum temperature for extraction.

Adsorbents are commonly used to trap organic compounds in gas and liquid streams for analysis. Initially, a certain volume of the stream is passed over the adsorbent. Once the organics are trapped on the adsorbent, they must be released for analysis. Common techniques such as Soxhlet extraction and thermal desorption suffer from the previously mentioned problems. Desorption of these compounds using SCFs such as CO_2 is gentle, more efficient, and potentially more widely applicable to recovering compounds of lesser volatility and greater polarity than other methods.

One of the most widely used adsorbents is 2,6-diphenyl-p-phenylene oxide porous polymer, or Tenax-GC.[79] Raymer and Pellizzari[80] extracted a model chlorinated hydrocarbon (hexachlorocyclohexane), chlorinated aromatic compound (hexchlorobiphenyl), polynuclear aromatic compound (anthracene), and organophosphate (parathion) from Tenax-GC using supercritical CO_2. They were able to recover in excess of 90% of each compound (significantly higher than thermal desorption). Later, Raymer et al.[81] extracted the same compounds from more polar sorbents (polyimide based) which provide for the analysis of compounds with a broader range of polarities than possible using Tenax-GC. They found that both supercritical and thermal methods were more difficult from the polyimide sorbents than from the Tenax-GC. Supercritical recoveries ranged from 58 to 100%.

Wright et al.[82] extracted model PAHs (with a range of molecular weights) from various adsorbents, including XAD-2 resin, polyurethane foam, and Spherocarb. CO_2, isobutane, and methanol modified (20 mol%) CO_2 fluid systems were evaluated and compared to liquid Soxhlet extraction. Analysis was done on the extract after the SCF was expanded. The authors investigated various extract collection techniques and found that if the collection vessel is left open to the atmosphere, significant solute losses can occur, possibly due to aerosol formation.

Another approach to the extraction of adsorbents has been proposed by Akman et al.[83] Their approach involves the simultaneous SCE and adsorption of the contaminant in a countercurrent contacting system with an adsorbing bed. The major advantage of this approach is a prolonged bed activity.

VI. INDUSTRIAL APPLICATIONS

The CF Systems Corporation has developed an SCE process to separate and recover oils from refinery sludges and to extract hazardous organic compounds from wastewater, sludge, sediment, and soil.[84] The process uses supercritical propane on contaminated soil and sludge and CO_2 to treat wastewater. The solid feed materials are reduced in size and are slurried so that they can be pumped to the extractor. Wastewater is used directly. The process closely resembles the flow chart presented in Figure 1. Reportedly, up to 90% of the solvent is recycled in the system; the remaining 10% retains the extracted contaminants.[84]

The CF Systems process was demonstrated at pilot scale for the U.S. EPAs SITE program and shown to be capable of removing PCBs from sediments. Recently, a 300 barrel per day unit was installed at the Star Enterprise Refinery at Port Arthur, TX, to treat refinery sludge. Other applications include wood treatment processes waste, industrial waste dumps, and manufacturing processes wastewaters. The process can extract up to 99% of liquid hydrocarbons from waste.

VII. CONCLUSIONS

SCE has been shown to be capable of removing of organic contaminants from water, soil, and a variety of adsorbents. Though the data are still very limited, the behavior of SCFs and their solvent powers have only recently begun to be understood. As the ability to predict the behavior of contaminants in an SCF increases, the use of SCE will find new applications. CO_2, one of the most widely used fluids in the literature, has been demonstrated as a fluid capable of extracting organic contaminants from water, soil, and adsorbents. As the need continues to arise, the use of SCE will continue to find use in environmental control.

REFERENCES

1. **Smith, J. M. and Van Ness, H. C.,** *Introduction to Chemical Engineering Thermodynamics,* 4th ed., McGraw-Hill, New York, 1975, 54.
2. **Hannay, J. B and Hogarth, J.,** On the solubility of solids in gases, *Proc. R. Soc. London,* 30, 324, 1879.
3. **Paulaitis, M. E., Krukonis, V. J., and Kurnik, R. T.,** Supercritical fluid extraction, *Rev. Chem. Eng.,* 1, 179, 1983.
4. **Francis, A.,** Ternary systems of liquid carbon dioxide, *J. Phys. Chem.,* 58, 1099, 1954.
5. **Buchner, E. G.,** Die beschrankte Mischbarkeit von Flussigkeiten das System Diphenylamin und Kohlensaure, *Z. Phys. Chem.,* 56, 257, 1906.
6. **Diepen, G. A. and Scheffer, F. E. C.,** The solubility of naphthalene in supercritical ethylene, *J. Am. Chem. Soc.,* 70, 4085, 1948.
7. **Tsekhanskaya, Yu. V., Iomtex, M. B., and Mushkina, E. V.,** Solubility of naphthalene in ethylene and carbon dioxide under pressure, *Russ. J. Phys. Chem.,* 38, 1173, 1964.
8. **Najour, G. C. and King, A. D.,** Solubility of naphthalene in compressed methane, ethylene, and carbon dioxide: evidence for a gas-phase complex between naphthalene and carbon dioxide, *J. Chem. Phys.,* 45, 1915, 1966.

9. **Van Welie, G. S. A. and Diepen, G. A. M.,** The solubility of naphthalene in supercritical ethane, *J. Phys. Chem.,* 67, 755, 1963.

10. **King, A.D., Jr. and Robertson, W. W.,** Solubility of naphthalene in compressed gases, *J. Chem. Phys.,* 37, 1453, 1962.

11. **Krukonis, V. J., McHugh, M. A., and Seckner, A. J.,** Xenon as a supercritical solvent, *J. Phys. Chem.,* 88, 2687, 1984.

12. **McHugh, M. A. and Paulaitis, M. E.,** Solid solubilities of naphthalene and biphenyl in supercritical carbon dioxide, *J. Chem. Eng. Data,* 25, 326, 1980.

13. **Kurnik, R. T., Holla, S. J., and Reid, R. C.,** Solubility of solids in supercritical carbon dioxide and ethylene, *J. Chem. Eng. Data,* 26, 47, 1981.

14. **Sako, S., Ohgaki, K., and Katayama, T.,** Solubilities of naphthalene and indole in supercritical fluids. *J. Supercritical Fluids,* 1, 1, 1988.

15. **Van Leer, R. A. and Paulaitis, M. E.,** Solubilities of phenol and chlorinated phenols in supercritical carbon dioxide, *J. Chem. Eng. Data,* 25, 257, 1980.

16. **Wilson, R. E., Keith, P. C., and Haylett, R. E.,** Liquid propane—use in dewaxing, deasphalting, and refining heavy oils, *Ind. Eng. Chem.,* 28, 1065, 1936.

17. **Elgin, J. C. and Weinstock, J. J.,** Phase equilibria molecular transport thermodynamics, *J. Chem. Eng. Data,* 4, 3, 1959.

18. **Wilhelm, A. and Hedden, K.,** Nonisothermal extraction of coal with solvents in liquid and supercritical state, *Proc. Int. Conf. Coal Sci.,* Pittsburgh, 1983, 6.

19. **Zosel, K.,** Separation with supercritical gases: practical applications, *Angew. Chem. Int. Ed. Engl.,* 17, 702, 1978.

20. **Vitzthum, O. G. and Hubert, P.,** U.S. Patent No. 3,879,569, 1975.

21. **Hubert, P. and Vitzthum, O. G.,** Fluid extraction of hops, spices, and tobacco with supercritical gases, *Angew. Chem. Int. Ed. Engl.,* 17, 710, 1978.

22. **Schultz, W. G. and Randall, J. M.,** Liquid carbon dioxide for selective aroma extraction, *Food Technol.,* 24, 94, 1970.

23. **McHugh, M.A. and Krukonis, V. J.,** *Supercritical Fluid Processing: Principles and Practice,* Butterworths, Stoneham, MA, 1986, 1.

24. **Kuk, M. S. and Montagna, J. C.,** Solubility of oxygenated hydrocarbons in supercritical carbon dioxide, in *Chemical Engineering at Supercritical Conditions,* Paulaitis, M. E., Penninger, J. M. L., Gray, R. D., Jr., and Davidson, P., Eds., Ann Arbor Science, Ann Arbor, MI, 1983, 101.

25. **Groves, F. R., Brady, B., and Knopf, F. C.,** State-of-the-art on the supercritical extraction of organics from hazardous wastes, *Rev. Environ. Control,* 15, 237, 1985.

26. **McHugh, M. A., Mallet, M. W., and Kohn, J. P.,** High pressure fluid phase equilibria of alcohol-water-supercritical solvent mixtures, in *Chemical Engineering at Supercritical Conditions,* Paulaitis, M. E., Penninger, J. M. L., Gray, R. D., Jr., and Davidson, P., Eds., Ann Arbor Science, Ann Arbor, MI, 1983, 113.

27. **DeFilippi, R. P. and Vivian, J. E.,** U.S. patent 4,349,415, 1982.

28. **Paulaitis, M. E., Kander, R. G., and DiAndreth, J. R.,** Phase equilibria related to supercritical-fluid solvent extractions, *Ber. Bunsenges. Phys. Chem.,* 88, 869, 1984.

29. **Radosz, M.,** Variable-volume circulation apparatus for measuring high-pressure fluid-phase equilibria, *Ber. Bunsenges. Phys. Chem.,* 88, 859, 1984.

30. **Panagiotopoulos, A. Z. and Reid, R. C.,** Multiphase high pressure equilibria in ternary aqueous systems, *Fluid Phase Equilibria,* 29, 525, 1986.

31. **Panagiotopoulos, A. Z. and Reid, R. C.,** High-pressure phase equilibria in ternary fluid mixtures with a supercritical component, Prep. Paper, American Chemical Society, Div. Fuel Chem., Vol. 30, p. 46, 1985.

32. **Ehntholt, D. J., Thrun, K., and Eppig, C.,** The concentration of model organic compounds present in water at parts-per-billion levels using supercritical fluid carbon dioxide, *Int. J. Environ. Anal. Chem.,* 13, 219, 1983.

33. **Roop, R. K.,** Supercritical Extraction of Organic Compounds from Aqueous Solutions, Ph.D. thesis, Texas A&M University, College Station, 1989.

34. **Roop, R. K., Akgerman, A., Stevens, E. K., and Irvin, T. R.,** Supercritical extraction of creosote from water with toxicological validation, *J. Supercritical Fluids,* 1, 31, 1988.

35. **Roop, R. K., Akgerman, A., Dexter, B., and Irvin, T. R.,** Extraction of phenol from water with supercritical carbon dioxide, *J. Supercritical Fluids,* 2, 57, 1989.

36. **Yeo, S. and Akgerman, A.,** Supercritical extraction of organic mixtures from aqueous solutions, *AIChE J.,* 36, 1743, 1990.

37. **Paulaitis, M.E., Johnston, K. P., and Eckert, C. A.,** Measurement of partial molar volumes at infinite dilution in a supercritical-fluid solvent near its critical point, *J. Phys. Chem.,* 85, 1770, 1981.

38. **Redlich, O. and Kwong, J. N. S.,** On the thermodynamics of solutions, *Chem. Rev.,* 44, 233, 1949.

39. **Soave, G.,** Equilibrium constants from a modified Redlich-Kwong equation of state, *Chem. Eng. Sci.,* 27, 1197, 1972.

40. **Peng, D. Y. and Robinson, D. B.,** A new two-constant equation of state, *Ind. Eng. Chem. Fundam,* 15, 59, 1976.

41. **Huron, M. and Vidal, J.,** New mixing rules in simple equations of state for representing vapor-liquid equilibria of strongly non-ideal mixtures, *Fluid Phase Equilibria,* 3, 255, 1979.

42. **Won, K. W.,** Thermodynamic calculation of supercritical-fluid equilibria: new mixing rules for equations of state, *Fluid Phase Equilibria,* 10, 191, 1983.

43. **Luedecke, D. and Prausnitz, J. M.,** Phase equilibria for strongly nonideal mixtures from an equation of state with density dependent mixing rules, *Fluid Phase Equilibria,* 22, 1, 1985.

44. **Vidal J.,** Equations of state—reworking the old form, *Fluid Phase Equilibria,* 13, 15, 1983.

45. **Whiting, W. B. and Prausnitz, J. M.,** Equations of state for strongly nonideal fluid mixtures: application of local compositions towards density-dependent mixing rules, *Fluid Phase Equilibria,* 9, 119, 1982.

46. **Panagiotopoulos, A.Z. and Reid, R. C.,** *New Mixing Rule for Cubic Equations of State for Highly Polar, Asymmetric Systems, ACS Symp. Ser. No. 300, American Chemical Society,* Washington, D.C., 1986, 571.

47. **Tsonopoulos, C.,** An empirical correlation of second virial coefficients, *AIChE J.,* 20, 263, 1974.

48. **Reid, R. C., Prausnitz, J. M., and Poling, B. E.,** *The Properties of Gases & Liquids,* McGraw-Hill, New York, 1987.

49. **Diandreth, J. R. and Paulaitis, M. E.,** Multiphase behavior in ternary fluid mixtures: a case study of the isopropanol-water-CO_2 system at elevated pressure, *Chem. Eng. Sci.,* 44, 1061, 1989.

50. **Peter, S. and Brunner, G.,** The separation of nonvolatile substances by means of compressed gases in countercurrent processes, *Angew. Chem. Int. Ed. Engl.,* 17, 746, 1978.

51. **Brunner, G. and Peter,S.,** State of art of extraction with compressed gases (gas extraction), *Ger. Chem. Eng.,* 5, 181, 1982.

52. **Brunner, G.,** Selectivity of supercritical compounds and entrainers with respect to model substances, *Fluid Phase Equilibria,* 10, 298, 1983.

53. **Hacker, D. S.,** Separation of butadiene butene mixture with mixtures of ammonia and ethylene in near critical conditions, *Prep. Paper, American Chemical Society, Div. Fuel Chem.,* Vol. 30(3), p. 213, 1985.

54. **Roop, R. K. and Akgerman, A.,** The entrainer effect for supercritical extraction of aqueous solutions, *Ind. Eng. Chem. Res.,* 28, 1542, 1989.

55. **Spiteller, M. and Goettingen, F. R. G.,** Extraction of soil organic matter by supercritical fluids, *Org. Geochem.,* 8, 111, 1985.

56. **Schnitzer, M., Hindle, C. A., and Meglic, M.,** Supercritical gas extraction of alkanes and alkanoic acids from soils and humic materials, *Soil Sci. Soc. Am. J.,* 50, 913, 1986.

57. **Schnitzer, M. and Preston, C. M.,** Supercritical gas extraction of a soil with solvents of increasing polarities, *Soil Sci. Soc. Am. J.,* 51, 639, 1987.

58. **Hawthorne, S. B. and Preston, C. M.,** Extraction and recovery of organic pollutants from environmental solids and Tenax-GC using supercritical CO_2, *J. Chromatogr. Sci.,* 24, 258, 1986.

59. **Hawthorne, S. B. and Miller, D. J.,** Extraction and recovery of polycyclic aromatic hydrocarbons from environmental solids using supercritical fluids, *Anal. Chem.,* 29, 1705, 1987.

60. **Schantz, M. and Chesler, S. N.,** Supercritical fluid extraction procedure for the removal of trace organic species from solid samples, *J. Chromatogr.,* 363, 397, 1986.

61. **Wright, B.W., Frye, S. R., McMinn, D. G., and Smith, R. D.,** On-line supercritical fluid extraction-capillary gas chromatography, *Anal. Chem.,* 59, 640, 1987.

62. **Brady, B. O., Kao, C. C., Dooley, K. M., Knopf, F. C., and Gambrell, R. P.,** Supercritical extraction of toxic organic from soils, *Ind. Eng. Chem. Res.,* 26, 261, 1987.

63. **Eckert, C. A., Van Alsten, J.G., and Stoicos, T.,** Supercritical fluid processing, *Environ. Sci. Technol.,* 20, 319, 1986.

64. **Capriel, P., Haisch, A., and Khan, S. U.,** Supercritical methanol: an efficacious technique for the extraction of bound pesticide residues from soil and plant samples, *J. Agric. Food Chem.,* 34, 70, 1986.

65. **Kunze, G. W. and Dixon, J. B.,** Pretreatment for mineralogical analysis, in *Methods of Soil Analysis, Part 1, Physical and Mineralogical Methods,* 2nd ed., Klute, A., Ed., American Society of Agronomy/ Soil Science Society of America, Madison, WI, 1986, 91.

66. **Tan, K. H., Hajek, B. F., and Barshad, I.,** Thermal analysis techniques, in *Methods of Soil Analysis, Part 1, Physical and Mineralogical Methods,* 2nd ed., Klute, A., Ed., American Society of Agronomy/ Soil Science Society of America, Madison, WI, 1986, 151.

67. **Kurnik, R. T. and Reid, R. C.,** Solubility of solid mixtures in supercritical fluids, *Fluid Phase Equilibria,* 8, 93, 1982.

68. **Gopal, J. S., Holder, G. D., and Kosal, E.,** Solubility of solid and liquid mixtures in supercritical carbon dioxide, *Ind. Eng. Chem. Process Des. Dev.,* 24, 697, 1985.

69. **Dobbs, J. M., Wong, J. M., and Johnston, K. P.,** Nonpolar co-solvents for solubility enhancement in supercritical fluid carbon dioxide, *J. Chem. Eng. Data,* 31, 303, 1986.

70. **Dobbs, J. M., Wong, J. M., Lahiere, R. J., and Johnston, K. P.,** Modification of supercritical fluid phase behavior using polar cosolvents, *Ind. Eng. Chem. Res.,* 26, 56, 1987.

71. **Joshi, D. K. and Prausnitz, J. M.,** Supercritical fluid extraction with mixed solvents, *AIChE J.,* 30, 522, 1984.

72. **Walsh, J. M., Ikonomou, G. D., and Donohue, M. D.,** Supercritical phase behavior: the entrainer effect, *Fluid Phase Equilibria,* 33, 295, 1987.

73. **Dooley, K. M., Kao, C., Gambrell, R. P., and Knopf, F. C.,** The use of entrainers in the supercritical extraction of soils contaminated with hazardous organics, *Ind. Eng. Chem. Res.,* 26, 2058, 1987.

74. **Roop, R. K., Hess, R. K., and Akgerman, A.,** *Supercritical Extraction of Priority Pollutants from Water and Soil, ACS Symp. Ser.* No. 406, American Chemical Society, Washington, D.C., 1989, 468.

75. **Kander, R. G. and Paulaitis, M. E.,** The adsorption of phenol from dense carbon dioxide onto activated carbon, in *Chemical Engineering at Supercritical Fluid Conditions,* Paulaitis, M. E., Penninger, J. M. L., Gray, R. D., Jr., and Davidson, P., Eds., Ann Arbor Science, Ann Arbor, MI, 1983, 461.

76. **DeFilippi, R. P., Kurkonis, V. J., Robey, R. J., and Modell, M.,** Supercritical Fluid Regeneration of Activated Carbon for Adsorption of Pesticides, EPA-600/2-80-054, Environmental Protection Agency, Washington, D.C., March 1980.

77. **Picht, R. D., Dillman, T. R., and Burke, D. J.,** *AIChE Symp. Ser.,* 78, 136, 1982.

78. **Tan, C. and Liou, D.,** Desorption of ethyl acetate from activated carbon by supercritical carbon dioxide, *Ind. Eng. Chem. Res.,* 27, 988, 1988.

79. **Krost, K. J., Pellizzari, E. D., Walburn, S. G., and Hubbard, S. A.,** Collection and analysis of hazardous organic emissions, *Anal. Chem.,* 54, 810, 1982.

80. **Raymer, J. H. and Pellizzari, E. D.,** Toxic organic compound recoveries from 2,6-diphenyl-*p*-phenylene oxide porous polymer using supercritical carbon dioxide and thermal desorption methods, *Anal. Chem.,* 59, 1043, 1987.

81. **Raymer, J. H., Pellizzari, E. D., and Cooper, S. D.,** Desorption characteristics of four polyimide sorbent materials using supercritical carbon dioxide and thermal methods, *Anal. Chem.,* 59, 2069, 1987.

82. **Wright, B. W., Wright, C. W., Gale, R. W., and Smith, R. D.,** Analytical supercritical fluid extraction of adsorbent materials, *Anal. Chem.,* 59, 38, 1987.

83. **Akman, U., Bracey, W., and Sunol, A. K.,** Supercritical Exsorption of Waste Streams, presented at the 1988 Winter AIChE meeting, Washington, D.C., 1988, 1.

84. **Hall, D. W., Sandrin, J. A., and McBride, R. E.,** An overview of solvent extraction treatment technologies, *Environ. Progr.,* 9(2), 98, 1990.

85. **Nishiumi, H. and Arai, T.,** Generalization of the binary interaction parameter of the Peng-Robinson equation of state by component family, *Fluid Phase Equilibria,* 42, 43, 1988.

86. **Haselow, J.S., Han, S. J., Greenkorn, R. A., and Chao, K. C.,** Equation of State for Supercritical Extraction, ACS Symp. Ser. No. 300, American Chemical Society, Washington, D.C., 1986, 157.

87. **Tsonopoulos, C. and Wilson, G. M.,** High-temperature mutual solubilities of hydrocarbons and water, *AIChE J.,* 29(6), 990, 1983.

Chapter 15

REACTIONS IN AND WITH SUPERCRITICAL FLUIDS—A REVIEW

Benjamin C. Wu, Stephen C. Paspek, Michael T. Klein, and Concetta LaMarca

TABLE OF CONTENTS

ABSTRACT

The interest in supercritical fluid (SCF) solvents in chemical reaction processes derives from both their participation in reactions as well as their pressure-dependent solvent effects. Herein are described: (1) a set of hydrolysis reactions of aryl-alkyl ethers in supercritical water; (2) the influence of SCF solvent effects on rate constants; (3) a contrived example of the influence of SCF illustrating the compromise between (1) and (2) that can lead to the specification of an optimal SCF solvent; and (4) a real example of the enhancement of shale oil upgrading in the presence of supercritical H_2O and HCl.

I. INTRODUCTION

Many chemists and engineers have heard about SCF solvents because of their promise in separation processes.[5,8,16,26,36-38,48,57] Often viewed as "dense gases," these fluids possess physicochemical properties, such as density, viscosity, and dielectric constant, intermediate between those of liquids and gases. Moreover, these properties can be easily manipulated by pressure. The object of this chapter is to highlight the less obvious promise of SCF solvents as reaction media, and therefore view their solvent properties from the perspective of "light liquids".

Many examples of the use of SCF solvents as a medium for reaction have been reported.[4,13,21,25,39,51,53,62,66] These solvents may often participate in chemical reactions such as hydrolysis by supercritical water as well.[1-3,24,30,45,46,56,59,60,66] In both cases, a myriad of solvent effects can be expected. As outlined in Table 1, the chemistry of ionic, solvolytic, and free-radical reactions is discussed herein in relation to the solvent properties which can be dramatically altered in the supercritical region.

II. BACKGROUND

The three classical states of matter, solid, liquid, and vapor, are shown in Figure 1. The critical point is defined as the point where the two fluid phases, liquid and vapor, become indistinguishable. A fluid is supercritical above the critical temperature (T_c) and the critical pressure (P_c), shown in the shaded region of Figure 1. The unique property of an SCF is its pressure-dependent density. This is illustrated as an isotherm between points a and b in Figure 1. The density can be adjusted from that of a vapor to that of a liquid without discontinuity. It is common to focus on the region were the reduced temperature ($T_r = T/T_c$) and the reduced pressure ($P_r = P/P_c$) are approximately equal to 1, because in this region, large changes in fluid density and related properties, such as a material's solubility, are observed with small changes in pressure. These characteristics make the SCF phase attractive for solubility-based separation processes.

For example, Figure 2 shows the pressure dependence of the solvation power of supercritical ethylene for naphthalene.[12,61] At pressures far above P_c, the ethylene exhibits a liquid-like density and therefore a considerable amount of naphthalene can be dissolved. As the pressure is decreased below P_c, the density becomes gas-like and the solubility of naphthalene decreases significantly.

Other applications that exploit the pressure-controllable density of SCFs in separations include decaffeination of coffee and enhanced oil recovery with supercritical CO_2 and solid formation from supercritical solutions.[68] Examples of reactions in and with SCF solvents include applications of wet-air oxidation with supercritical water;[21,62,66] coal liquefaction in supercritical toluene;[5,36,37] and catalytic disproportionation of near-critical toluene.[9]

TABLE 1
Reactions In and With
Supercritical Fluids

Chemistry	Solvent effects
Ionic	Mechanical pressure
Solvolytic	Diffusional limitations
Radical	Electrostatic
	Phase equilibrium

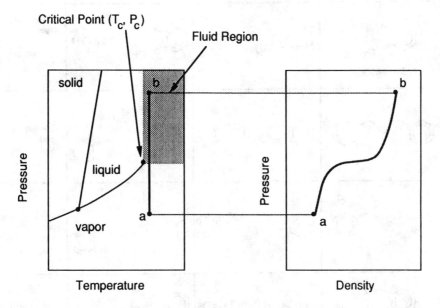

FIGURE 1. Pressure dependence of SCF properties.

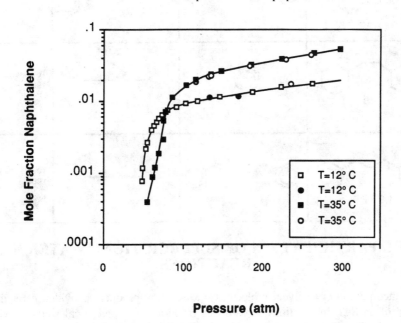

FIGURE 2. Solubility of naphthalene in supercritical ethylene.[12,61]

TABLE 2
Reaction of Coal Model Compounds in Supercritical Water

Reactant	Major Neat Pyrolysis Products	Major Solvolysis Products	Reference
Guaiacol			Lawson and Klein, 1985 Huppert, et al., 1988
DBE			Townsend and Klein, 1985
PPE			Townsend et al., 1988
BPE			Townsend et al., 1988
BB		———	Townsend et al., 1988
1,3-DPP		———	Townsend et al., 1988

III. PARTICIPATION OF SUPERCRITICAL WATER IN REACTIONS

The incorporation of solvent molecules into reaction products is evidence for the participation of an SCF in reactions. Hydrolysis reactions in supercritical water provide the examples shown in Table 2.

Pertinent observations from the literature that are of mechanistic significance are as follows:

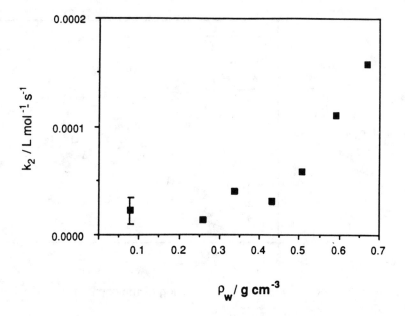

FIGURE 3. Dependence of apparent second-order guaiacol hydrolysis rate constant on water loading at 383°C.[24]

1. The reaction always involves a saturated carbon with a heteroatom-containing leaving group (LG).
2. The reaction is in parallel with a free-radical pyrolysis route.
3. The selectivity to hydrolysis increases with water density, with no discontinuity at the critical point.
4. The rate constant often increases with water density.
5. The addition of salts increases the hydrolysis rate constant.

As regards observation 1, neither 1,2-diphenylethane ($PhCH_2CH_2Ph$) nor 1,3-diphenylpropane ($Ph(CH_2)_3Ph$) underwent the hydrolysis seen from their heteroatom-containing analogs (Townsend, 1989). These fully hydrocarbon molecules decompose by free-radical pathways[19,23,52,55] at 400 ± 25°C. This confirms that the thermally unfavorably event of free-radical hydrogen abstraction from water does not occur. Therefore, while the neat experimental results with heteroatom-containing compounds are consistent with a free-radical pathway scheme, reaction with water must occur through a different mechanism.

The increased selectivity to this parallel hydrolysis route with water loading is illustrated in Figure 3 for the reaction of guaiacol at 383°C, where the guaiacol concentration was kept constant in single-phase experiments while increased pressure was generated with increased water loading.

The hydrolysis occurred at low as well as high ρ_w, and therefore did not require the solvent power required for the stabilizing reaction-derived ions. This rendered S_N1 reactions unlikely and focused attention on the S_N2 mechanism illustrated in Equation 1, where a pentavalent transition state is implied.

$$H_2O + R_3\overset{R_1}{\underset{R_2}{\ddot{C}}}LG \longrightarrow H_2\overset{\delta+}{O}\cdots\overset{R_1}{\underset{R_2\ R_3}{C}}\cdots\overset{\delta-}{LG} \longrightarrow R_2\overset{R_1}{\underset{R_3}{\ddot{C}}}OH + LG\text{-}H \tag{1}$$

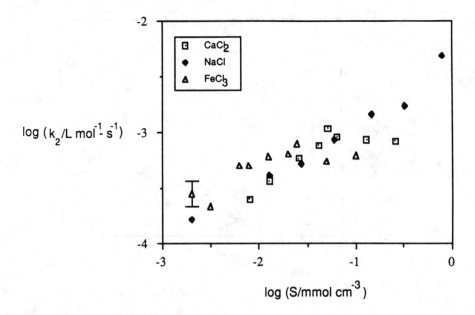

FIGURE 4. Variation of guaiacol hydrolysis rate constant with salt concentration at 383°C and $\rho_w = 0.5$ g/cm^{-3}.[24]

This required overall second-order kinetics, first order in each of water and reactant concentrations.

The second-order hydrolysis rate constant also increased with ρ_w. As detailed below, this was interpreted in terms of the increase in dielectric constant, ϵ[60] or solubility parameter, δ,[24] that occurs with increases in ρ_w.[18,69] Both would stabilize a "polar" transition state.[40]

This electrostatic explanation is not unique since changes in ρ_w are invariably accompanied by changes in pressure. Therefore, the increase in the hydrolysis rate constant can also be interpreted in terms of an overall activation volume, derived below from the transition state theory formalism. Experiments in liquid systems allow ϵ to be changed at essentially constant ρ_w and P by changing mole fractions;[14,15,63] this was not possible in the experiments reported above.

The addition of salts at constant ρ_w allowed changing the solution polarity without significantly altering the pressure. Hydrolysis data are illustrated in Figure 4 for the reaction of guaiacol in water at 383°C with varying concentrations of NaCl, CaCl$_2$, and FeCl$_3$. The hydrolysis rate constant increased by a factor of 25 with an NaCl addition of 0.71 mmol/cm^3, providing the most compelling indication of the polar nature of the hydrolysis transition state.

The preceding examples illustrate that SCF solvent effects contribute to both the rate constant and the concentration terms in the rate expression. A somewhat subtle feature in the above discussion is that the observed reaction is in part a result of the good mixing between supercritical water and the organic reactant with which water generally would be immiscible at room temperaure. Appreciable hydrolysis kinetics were observed in these examples because of significant concentrations of the reacting species. The discussion now turns to the effect of SCFs on the rate constant in this rate law.

IV. SCF SOLVENT EFFECTS

Transition state theory provides a convenient formalism for interpreting the effects of SCF solvents on the reaction rate constant. Within this essentially thermodynamic theory of rates, an activated transition state complex, $M^{\ddagger}$, is in virtual equilibrium with ground state reactants. The number or concentration of $M^{\ddagger}$ drives the rate. For bimolecular (elementary) reactions,

$$A + B \underset{}{\overset{K_a}{\rightleftharpoons}} M^{\ddagger} \longrightarrow products \tag{2}$$

The reaction rate is given by

$$r = \nu[M] = \frac{1}{C_T} \frac{\kappa T}{h} K_a \left(\frac{\gamma_A \gamma_B}{\gamma_M^{\ddagger}}\right) [C_A][C_B] = k_c[C_A][C_B] \tag{3}$$

where K_a is based on a standard state of 1 atm and the temperature of interest.

The concentration-based rate constant is therefore

$$k_c = k_o\left(\frac{\gamma_A \gamma_B}{\gamma_M^{\ddagger}}\right) \tag{4}$$

with k_o, the rate constant in an ideal solution, given by

$$k_o = \frac{1}{C_T} \frac{\kappa T}{h} K_a \tag{5}$$

The dependence of k_c on solvent properties is a useful quantitative measure of the SCF solvent effects. The overall pressure dependence of k_c defines V_{act}, akin to the temperature dependence measured as the activation energy, E_{act}. Differentiating Equation 4 with respect to pressure gives

$$V_{act} \equiv -RT\left(\frac{\partial \ln k_c}{\partial P}\right)_T = -\Delta V^{\ddagger} + (1-n)RT \kappa_T \tag{6}$$

where $\Delta V^{\ddagger}$ is difference in partial molar volumes (PMVs) of the transition state species M, and the reactants A and B.

$$\Delta V^{\ddagger} = \overline{V}_M - \overline{V}_A - \overline{V}_B \tag{7}$$

The activation volume has been frequently used to study the transition state of reactions in the condensed phase.[7,11,15,17,20,29,31-35,41-43,49] Because the liquid phase is essentially incompressible, pressure changes leave the solvent properties of the liquid virtually unchanged. In these cases, the activation volume can be construed as an intrinsic, or molecular, property. Typical values of an activation volume in a condensed phase generally range between -50 to 50 cm^3/mol for a truly elementary reaction step.[29] In these studies $\Delta V^{\ddagger}$ is essentially the "mechanical" effect of pressure on the rate constant.

The apparent activation volume for reactions in SCF solvents is a bit more complex because their solvent properties (η, D_{12}, δ, ϵ, ρ) are also strongly influenced by pressure. Thus, the "mechanical" pressure effects can be difficult to discriminate from others. Near the critical region, PMV often diverge which can cause excessively large activation volumes. In such cases, the activation volume begins to lose a qualitative connection with the actual molecular volume.

Along these lines, Simmons and Mason[50] studied the dimerization of chlorotrifluoroethylene near its critical point and found activation volumes in excess of 3000 cm^3/mol. Johnston and Haynes[25] observed activation volumes as low as -6000 cm^3/mol for the unimolecular decomposition of α-chlorobenzyl methyl ether in supercritical 1,1-difluoroethane. In both cases, the apparent activation volume is a manifestation of more than just the "mechanical" pressure effect.

It seems quite reasonable to partition the apparent activation volume into components. Johnston and Haynes[25] suggested that the transition state activation volume can be separated into a repulsive and an attractive term. The pressure effect would be treated in the repulsive term and the electrostatic interactions would be treated in the attractive term. For reactions in tubing bombs it is convenient to further subdivide the activation volume into intrinsic (int), compressibility (com), diffusion (diff), and electrostatic (el) parts, or

$$\Delta V_{app}^{\ddagger} = \Delta V_{int}^{\ddagger} + \Delta V_{com}^{\ddagger} + \Delta V_{diff}^{\ddagger} + \Delta V_{el}^{\ddagger *} \tag{8}$$

where $\Delta V_{int}^{\ddagger} = V_m^{\circ} - V_A^{\circ} - V_b^{\circ}$ (the pure molar, or molecular, property of order -50 to 50 cm^3/mol), and $\Delta V_{com}^{\ddagger} = (1-n)\,RT\,\kappa_T$.

The compressibility of an SCF solvent affects bimolecular reactions as shown in Equation 6. Alexander[4] investigated the bimolecular Diels-Alder cycloaddition reaction of isoprene and maleic anhydride in supercritical CO_2 and found that near the mixture critical temperature and pressure, the rate constant, based on mole fractions k_x**, was strongly influenced by the pressure. At conditions far above the critical point, the effect of pressure on k_x was comparable to that in a liquid phase ($\Delta V_{int}^{\ddagger}$); however, $\Delta V_{com}^{\ddagger}$ became extremely large near the critical point where $V_{act} = \Delta V_{act}^{\ddagger} + \Delta V_{com}^{\ddagger}$ became quite large.

Diffusional limitations would be expected at high SCF densities where the mobility of molecules becomes restricted. The notion that diffusional limitations could influence V_{act} has been advanced by Neuman.[43] The essential issue here seems to be whether the effect is on the transition state species $M^{\ddagger}$ or reaction intermediates.

Irrespective of the mechanistic detail, experimental evidence of the influence of transport on rates in SCF solvents continues to accumulate. Helling and Tester[21] found that the oxidation rate of carbon monoxide in supercritical water was much slower than predicted by existing intrinsic models. They suggested that water formed a cage about the carbon monoxide acting as a diffusional barrier. Abraham and Klein[2] also discussed a solvent cage in reactions of benzylphenylamine in supercritical water and methanol. Lawson and Klein[30] thermolyzed guaiacol, a lignin model compound, in the presence of supercritical water and found that the guaiacol conversion decreased with increasing supercritical water density up to reduced densities of approximately 0.4 g/cm^3. They also attributed the decrease to diffusional limitations, or "cage effects".

Electrostatic interactions ($\Delta V_{el}^{\ddagger}$) affect the reaction rate constant by shifting the equi-

---

* Another contribution, although not strictly a solvent effect, is $\Delta V_{phase}^{\ddagger}$. The partitioning of the reactant between phases can be manipulated by the pressure, thereby affecting both conversion and selectivities. The apparent activation volume can be as high as 2000 cm^3/mol (Wu et al., 1989).

** The reaction rate can be based on either concentration or mole fraction. The mole fraction reaction rate analog to Equation 3 is

$$r_x = v x_M = \frac{\kappa T}{h} K_a \left(\frac{\gamma_A \gamma_B}{\gamma_M} \right) x_A x_B$$

where the relationship between k_c and k_x is

$$k_x = k_c \cdot C_T$$

$$\Delta V_x^{\ddagger} = \Delta V_c^{\ddagger} - \Delta V_{com}^{\ddagger}$$

librium between the reactants and the transition state toward the latter. Johnston and Haynes[25] correlated the unimolecular decomposition of α-chlorobenzyl methyl ether to a function of the dielectric constant (ϵ) of supercritical 1,1-difluoroethane. Townsend et al.[60] followed the development of Kirkwood[28] and partitioned $\Delta G^{\ddagger}$ into electrostatic and nonelectrostatic parts and correlated the hydrolysis rate constant with ϵ. The literature also provides other examples of the correlation of rate data with δ[14,15,22,63] and E_t.[25,27] The essential issue is whether $\Delta G^{\ddagger}$ or $\Delta V^{\ddagger}$ should be partitioned.

Ionic mechanisms involving acid/base catalysis[6,47] and the Debye-Huckel formalism[46] have also been discussed.

The ostensibly diverse SCF solvent effects noted above are generally realized in terms of the variation of a reaction rate constant, k. These solvent effects provide valuable mechanistic information. It is also possible to increase or decrease k, and thus manipulate overall rates or selectivities, by adjusting P and, therefore ρ, ϵ, δ, or any other measure of solvent power.

This section focused on the effect of SCF solvents on the rate constant, while the preceding section focused on the concentration effects ($C_A C_B$), more subtly described as solubility, since kinetically significant hydrolysis rates required good mixing (solubility, in these cases) of the hydrocarbon and supercritical water. For mechanistic information, focusing on k is appropriate. For practical processing purposes the product $k C_A C_B$ is central, and it is interesting that strategies to increase k may decrease $C_A C_B$ and vice versa.

The authors now proceed to link these two contributions to the rate in a contrived example of combined phase equilibrium and reaction in and with an SCF. The ultimate aim is to show that, as in heterogeneous catalysis, overall activity is the product of activity (k) and availability ($C_A C_B$).

V. A CONTRIVED EXAMPLE

Consider a two-phase system composed of pure reactant (A) in equilibrium with a second phase of supercritical water and A. The hydrolysis of A occurs only in the second phase with rate proportional to its concentration of A in phase II, i.e.,

$$r = k \, C_A^{II} \qquad (9)$$

Solution polarity affects both the rate constant, k, and the availability of reactant A, C_A^{II}.

The hydrolysis rate constant increases with solution polarity (ϵ, δ, or the addition of salts), as demonstrated in experiments with guaiacol,[24] dibenzyl ether, and phenethyl phenyl ether.[60] This increase in k can be explained in light of Equations 2 and 3 as a shift in the transition state equilibrium toward $M^{\ddagger}$, where the transition state stability increases more with polarity than the stability of A and B.

The reactant availability, C_A^{II}, is also influenced by the solution polarity. The maximum solubility of a reactant is where its polarity is equal to that of a dilute solution. As the solution polarity exceeds that of the reactant, the availability of the reactant in the supercritical phase decreases, thereby offsetting the ever-increasing rate constant and decreasing the overall reaction rate. This suggests an optimum SCF solvent polarity exists for the maximum overall rate. Therefore, it is possible to observe simultaneous increases in k and decreases in C_A^{II} with increasing polarity, suggesting an optimum SCF solvent polarity exists for maximum overall rate. This is shown qualitatively in Figure 5 for thermodynamic calculations based on regular solution theory.

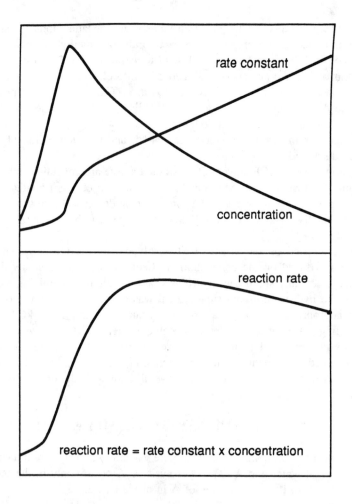

Pressure

FIGURE 5.   Effect of pressure on reaction rate, rate constant, and concentration.

VI. A REAL EXAMPLE

Experiments for upgrading Paraho shale oil in supercritical water/HCl systems provide a real example of the influence of solvent effects[44] on reactions in supercritical water. This solvent was used to enhance selectivity to the distillate fraction over that of resid and coke; HCl was used as a catalyst for nitrogen removal.

The reaction of shale oil with supercritical water at 425°C revealed an increased distillate formation and a decreased coke yield with increased water density. This is illustrated in Figure 6. In contrast, the resid yield was relatively insensitive to water density, and nitrogen removal was negligible without catalyst. Upon addition of HCl to the shale oil/supercritical water reaction system, increased distillate yields, decreased resid yields, and high levels of nitrogen removal were observed with increased HCl loading up to a HCl/SO ratio of approximately 0.1. Nitrogen removal is illustrated in Figure 7. Accompanying these upgrades, however, were increased yields of coke and gas. Possible solvent effects contributing to such behavior include "cage effects", dissolution of HCl, and HCl assisted N-removal.

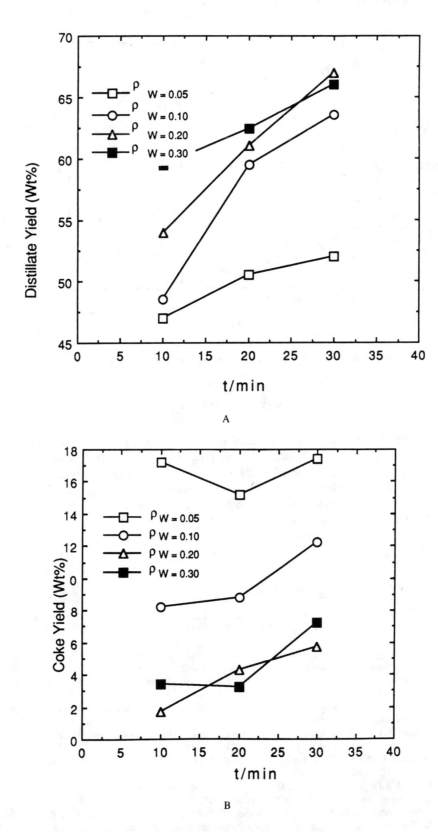

FIGURE 6. (A) The caging of the distillate lump by supercritical water. (B) The decrease in coke yield due to increased water density at 425°C.

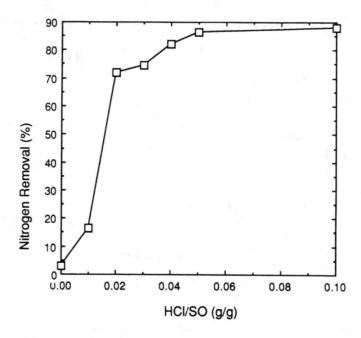

FIGURE 7. The increase in catalyst activity with increases in water density.

VII. SUMMARY

The interest in the use of SCFs in reaction processes derives from their participation in reactions and their pressure-adjustable solvent properties. The chapter has listed a battery of hydrolyses involving supercritical water and aryl-alkyl ethers and considered the effects of compressibility, "mechanical" pressure, electrostatic interactions, diffusional limitations, and phase behavior on reaction rates. The potential applications of SCF solvents in reaction processes is only beginning to emerge.

REFERENCES

1. **Abraham, M. A.,** Reactions with Supercritical Fluid Solvents of Heteroatom-containing Model Compounds of Heavy Feedstocks, Ph.D. thesis, University of Delaware, Newark, 1987.
2. **Abraham, M. A. and Klein, M. T.,** Pyrolysis of benzylphenylamine neat and with tetralin, methanol, and water solvents, *Ind. Eng. Chem. Res.,* 24, 300, 1985.
3. **Abraham, M. A., Wu, B. C., Paspek, S. C., and Klein, M. T.,** Reactions of dibenzylamine neat and in supercritical fluids solvents, *Fuel Sci. Technol. Int.,* 6, 557, 1988.
4. **Alexander, G.,** Thermodynamic Solvent Effects on a Diels-Alder Reaction in Supercritical Carbon Dioxide, Ph.D. thesis, University of Delaware, Newark, 1985.
5. **Amestica, L. A., and Wolf, E. E.,** Supercritical toluene and ethanol extraction of an Illinois no. 6 coal, *Fuel,* 63, 227, 1984.
6. **Antal, M. J., Jr., Brittain, A., DeAlmeida, C., Ramayya, S., and Jiben, C. R.,** Heterolysis and homolysis in supercritical water, *ACS Symp. Ser.,* 329, 77, 1987.
7. **Asano, T. and Okada, T.,** New simple functions to describe kinetic and thermodynamic effects of pressure. Application to Z-E isomerization of 4-(dimethylamino)-4'-nitroazobenzene and other reactions, *J. Phys. Chem.,* 88, 238, 1984.

8. **Barton, P.,** Supercritical separation in aqueous coal liquefaction with impregnated catalyst, *Ind. Eng. Chem. Process Des.Dev.,* 10, 315, 1983.

9. **Collins, N. A., Debenedetti, P. G., and Sundaresan, S.,** Disproportionation of toluene over ZSM-5 at Near-Critical Conditions, presented at the AIChE 1988 Annual Meeting, Washington, D.C., Nov. 27-Dec. 2.

10. **Debenedetti, P. G. and Reid, R. C.,** Diffusion and mass transfer in supercritical fluids, *AIChE J.,* 32, 2034, 1986.

11. **Dickson, S. J. and Hyne, J. B.,** The pseudo thermodynamics of solvolysis. A detailed study of the pressure and temperature dependence of benzyl chloride solvolysis in *t*-butyl alcohol-water mixtures, *Can. J. Chem.,* 49, 2394, 1971.

12. **Diepen, G. A. and Scheffer, F. E. C.,** *J. Phys. Chem.,* 57, 575, 1953.

13. **Dooley, K. M. and Knopf, F. C.,** Oxidation catalysis in a supercritical fluid medium, *Ind. Eng. Chem. Res.,* 26, 1910, 1987.

14. **Eckert, C. A.,** Molecular thermodynamics of chemical reactions, *Ind. Eng. Chem.,* 59, 20, 1967.

15. **Eckert, C. A., Hsieh, C. K., and McCabe, J. R.,** Molecular thermodynamics for chemical reaction design, *AIChE J.,* 20, 20, 1974.

16. **Eisenbach, W., Gottsh, P. J., Niemann, N., and Zosel, K.,** Extraction and supercritical gases: the first twenty years, *Fluid Phase Equilib.,* 10, 315, 1983.

17. **Fleischmann, F. K., and Kelm, H.,** Cycloaddition reaction between tetracyanoethylene and n-butyl vinyl ether: the solvent dependence of the volume of activation, *Tetrahedron Lett.,* 39, 3773, 1973.

18. **Giddings, J. C., Myers, M. N., McLaren, L., and Keller, R. A.,** High pressure gas chromatography of nonvolatile species, *Science,* 162, 67, 1968.

19. **Gilbert, K. E.,** *J. Org. Chem.,* 49, 6, 1984.

20. **Glugla, P. G., Byon, J. H., Eckert, C. A.,** High-pressure kinetics and solvent effects on halide exchange, *Ind. Eng. Chem. Fundam.,* 24, 379, 1985.

21. **Helling, R. H. and Tester, J. W.,** Oxidation kinetics of carbon monoxide in supercritical water, *Energy Fuels,* 1, 417, 1987.

22. **Herbrandson, H. F. and Neufeld, F. R.,** Organic reactions and the critical energy density of the solvent. The solubility parameter, δ, as a new solvent parameter, *J. Org. Chem.,* 31, 1140, 1966.

23. **Hung, M. and Stock, L. M.,** *Fuel,* 61, 1161, 1982.

24. **Huppert, G. L., Wu, B. C., Townsend, S. H., Klein, M. T., and Paspek, S. C.,** Hydrolysis in supercritical water identification and implications of a polar transition state, *Ind. Eng. Chem. Res.,* 27, 143, 1988.

25. **Johnston, K. P. and Haynes, C.,** Extreme solvent effects on reaction rate constants at supercritical fluid conditions, *AIChE J.,* 33, 2017, 1987.

26. **Johnston, K. P., Ziger, D. H., and Eckert, C. A.,** Solubilities of hydrocarbon solids in supercritical fluids. The augmented van der Waals treatment, *Ind. Eng. Chem. Fundam.,* 21, 191, 1982.

27. **Kim, S. and Johnston, K. P.,** Effects of supercritical solvents on the rates of homogeneous chemical reactions, *ACS Symp. Ser.,* 329, 42, 1987.

28. **Kirkwood, J. G.,** *J. Chem. Phys.,* 2, 351, 1952.

29. **Kohnstam, G.,** The kinetic effects of pressure, in *Progress in Reaction Kinetics,* Pergamon Press, Oxford, 1970, 335.

30. **Lawson, J. R. and Klein, M. T.,** Influence of water on guaiacol pyrolysis, *Ind. Eng. Chem. Res.,* 24, 203, 1985.

31. **le Noble, W. J.,** *Chem. Rev.,* 78(4), 409, 1978.

32. **le Noble, W. J.,** Kinetics of reactions in solution under pressure, *Progr. Phys. Org. Chem.,* 5, 207, 1967.

33. **le Noble, W. J., Guggisberg, H., Asano, T., Cho, L., and Grob, C. A.,** Pressure effects in solvolysis and solvolytic fragmentation. A correlation of activation volume with concertedness, *J. Am. Chem. Soc.,* 98(4), 920, 1976.

34. **MacDonald, D. D. and Hyne, J. B.,** The pressure dependency of benzyl chloride solvolysis in aqueous acetone dimethylsulfoxide, *Can. J. Chem.,* 487, 2494, 1970.

35. **MacKinnon, M. J., Lateef, A. B., and Hyne, J. B.,** Transition state volumes and solvolysis mechanisms, *Can. J. Chem.,* 48, 2025, 1970.

36. **Maddocks, R. R., Gibson, J., and Williams, D. F.,** Supercritical extraction of coal, *Chem. Eng. Prog.,* 73, 55, 1977.

37. **Maddocks, R. R., Gibson, J., and Williams, D. F.,** Supercritical extraction of coal, *Chem. Eng. Prog.,* 49, 1979.

38. **McHugh, M. A.,** An Experimental Investigation of the High Pressure Fluid Phase Equilibrium of Highly Asymmetric Binary Mixtures, Ph.D. thesis, University of Delaware, Newark, 1981.

39. **McHugh, M. A. and Occhiogrosso, R. N.,** Critical-mixture oxidation of cumene, *Chem. Eng. Sci.,* 42, 2478, 1987.

40. **Moore, J. W. and Pearson, R. G.,** *Kinetics and Mechanism,* John Wiley & Sons, New York, 1981.

41. **Neuman, R. C., Jr.,** Pressure effects as mechanistic probes of organic radical reactions, *Acc. Chem. Res.,* 5, 381, 1972.
42. **Neuman, R. C., Jr.,** High pressure studies. IX. Activation volumes and solvent internal pressure, *J. Org. Chem.,* 37, 495, 1971.
43. **Neuman, R. C., Jr. and Amrich, M. J.,** High pressure studies. X. Activation volumes for homolysis of single bonds, *J. Am. Chem. Soc.,* 94(8), 2730, 1972.
44. **Paspek, S. C. and Klein, M.T.,** Shale Oil Upgrading in Supercritical Water Solutions, 1989.
45. **Penninger, J. M. L.,** Reactions of di-n-butylphthalate in water at near-critical temperature and pressure, *Fuel,* 67, 490, 1988a.
46. **Penninger, J. M. L.,** Chemistry of Methyl Naphthylether in Near-critical Water, presented at AIChE 1988 Annual Meeting, Washington, D.C., Nov. 27-Dec. 2, 1988b.
47. **Ramayya, S., Brittain, A., DeAlmeida, C., Mok, W., and Antal, M. J., Jr.,** Acid-catalysed dehydration of alcohols in supercritical water, *Fuel,* 66, 1364, 1987.
48. **Schneider, G. M.,** Physicochemical aspects of fluid extraction, *Fluid Phase Equilib.,* 10, 141, 1983.
49. **Sera, A., Miyazawa, T., Matsuda, T.,Togawa, Y., and Matuyama, K.,** Effect of pressure on the rate of solvolysis. Hydrolysis of 1-aryl-a-methylethyl chlorides, *Bull. Chem. Soc. Jpn.,* 46, 3490, 1973.
50. **Simmons, G. M. and Mason, D. M.,** Pressure dependency of gas phase reaction rate coefficients, *Chem. Eng. Sci.,* 27, 89, 1972.
51. **Squires, T. G., Venier, C. G., and Aida, T.,** Supercritical fluid solvents in organic chemistry, *Fluid Phase Equilib.,* 10, 261, 1983.
52. **Stein, S. R.,** A fundamental chemical kinetics approach to coal conversion, *New Approaches to Coal Conversion,* ACS Symp. Ser. 169, Washington, D.C., 1981, 97-129.
53. **Subramaniam, B. and McHugh, M. A.,** Reactions in supercritical fluids—a review, *Ind. Eng. Chem. Process Des. Dev.,* 25, 1, 1986.
54. **Suppes, G. J. and McHugh, M. A.,** Analysis and Modelling of Partial Oxidation in Supercritical Media, presented at AIChE 1988 Annual Meeting, Washington, D.C., Nov. 27-Dec. 2, 1988.
55. **Sweeting, J. W. and Wilshire, J. F. K.,** *Aust. J. Chem.,* 15, 89, 1961.
56. **Takemura, Y., Itoh, H., and Ouchi, K.,** Hydrogenolysis of coal-related model compounds by carbon monoxide-water mixture, *Fuel,* 60, 379, 1981.
57. **Towne, S. E., Shah, Y. T., Holder, G. D., Deshpande, G. V., and Cronauer, D. C.,** Liquefaction of coal using supercritical fluid mixture, *Fuel,* 64, 883, 1985.
58. **Townsend, S. H.,** Ph.D. thesis, University of Delaware, Newark, 1989.
59. **Townsend, S. H. and Klein, M.T.,** Dibenzyl ether as a probe into the supercritical fluid solvent extraction of volatiles from coal with water, *Fuel,* 64, 635, 1985.
60. **Townsend, S. H., Abraham, M. A., Huppert, G. L., Klein, M. T., and Paspek, S. C.,** Solvent effects during reactions in supercritical water, *Ind. Eng. Chem. Res.,* 27, 143, 1988.
61. **Tsekhanskaya, Yu. V., Iomtev, M. B., and Mushinka, E. V.,** *J. Phys. Chem.,* 38, 1173, 1964.
62. **Webley, P. A. and Tester, J. W.,** Fundamental Kinetic Studies of Oxidation in Supercritical Water, presented at the AIChE 1988 Annual Meeting, Washington, D.C., Nov. 27-Dec. 2, 1988.
63. **Wong, K. F. and Eckert, C. A.,** Solvent design for chemical reactions, *I & EC Proc. Des. Dev.,* 8, 568, 1969.
64. **Wu, B. C., Klein, M. T., and Sandler, S. I.,** The Effect of Fluid Phase on BPE Pyrolysis, presented at AIChE 1988 Annual Meeting, Washington, D.C., Nov. 27-Dec. 2, 1988.
65. **Wu, B. C., Klein, M. T., and Sandler, S. I.,** Reactions in and with supercritical fluids: effect of phase behavior on dibenzyl ether pyrolysis kinetics, *Ind. Eng. Chem. Res.,* 28, 255, 1989.
66. **Yang, H. H. and Eckert, C. A.,** Homogeneous catalysis in the oxidation of p-chlorophenol in supercritical water, *Ind. Eng. Chem. Res.,* 27, 2009, 1988.
67. **Ziger, D. H. and Eckert, C. A. ,** Correlation and prediction of solid-supercritical fluid phase equilibria, *Ind. Eng. Chem. Process Des. Dev.,* 22, 582, 1983.
68. **Zosel, K.,** Separation Process, U.S. Patent 3969196, 1976
69. **Franck, E. W.,** Supercritical water as electrolytic solvent, *Ang. Chem.* (transl.), 10, 309, 1961.

A Summary of the Patent Literature of Supercritical Fluid Technology

(1982 - 1989)

Edited by: Thomas J. Bruno
Thermophysics Division
National Institute of Standards and Technology
Boulder, CO 80303

Introduction

The major patents issued in the field of supercritical fluid technology between 1982 and 1989 are summarized. In each case, the title, the names of the inventors, the critical dates and the assignee are provided. The abstract of each patent is also supplied.

Title: **METHOD FOR SEPARATING UNDESIRED COMPONENTS FROM COAL BY AN EXPLOSION TYPE COMMINUTION PROCESS**

Inventors: Abel, William A.   (US)
                          Brabets, Robert I. (US)
                          Massey, Lester G.   (US)

Assignee: Consolidated Natural Gas Service Company, Incorporated
Assignee Code: 19772
United States Class Code: 241001000

Application Number: US 127740   Patent Number: US 4313737
Application Date: 03/06/80   Issue Date: 02/02/82

Abstract:

A process for the fractionation of a porous or fluid-permeable hydrocarbonaceous solid, such as coal, containing an admixture of mineral matter and hydrocarbonaceous matter, into a separate mineral enriched fraction and a separate hydrocarbonaceous enriched fraction is disclosed. In this process, the hydrocarbonaceous solid is comminuted to convert the hydrocarbonaceous matter in the coal into discrete particles having a mean volumetric diameter of less than about 5 microns without substantially altering the size of the mineral matter originally present in the coal. As a result of this comminution, the hydrocarbonaceous particles can be fractionated from the mineral particles to provide a hydrocarbon fraction having a lesser concentration of minerals than in the original uncomminuted material and a mineral fraction having a higher concentration of minerals than in the original uncomminuted material. A preferred method for comminuting the porous or fluid-permeable hydrocarbonaceous solid, i.e. coal, is to first form a slurry of coal and a fluid such as water. This slurry is then heated and pressurized to temperatures and pressures in excess of the critical temperature and pressure of the fluid. The resultant supercritically heated and pressurized slurry is then passed to an expansion zone maintained at a lower pressure, preferably about ambient pressure, to effect comminution or shattering of the solid by the rapid expansion or explosion of the fluid forced into the coal during the heating and pressurization of the slurry. The supercritical conditions employed produce a shattered product comprising a mixture of discrete comminuted hydrocarbonaceous particles having a volumetric mean particle size equivalent to less than about 5 microns in diameter and discrete inorganic and mineral particles having a mean particle size substantially unchanged from that in the original solid. This material fraction, in turn, is then fractionated from the hydrocarbonaceous fraction.

Title: **DECAFFEINATION OF AQUEOUS ROASTED COFFEE EXTRACT**

Inventors: Gottesman, Martin   (US)
                          Prasad, Ravi   (US)
                          Scarella, Robert A. (US)

Assignee: General Foods Corporation
Assignee Code: 33888
United States Class Code: 426387000

Application Number: US 187223   Patent Number: US 4341804
Application Date: 09/15/80   Issue Date: 07/27/82

Abstract:

Aqueous extracts of roasted coffee are stripped of aroma, concentrated and thereafter decaffeinated by means of contact with a decaffeinating fluid such as liquid or supercritical carbon dioxide. Aroma loss is minimized by using water to remove caffeine and aroma from the CO_2 stream, recovering aromatics from this caffeine-containing aqueous stream and adding-back these aromatics to the decaffeinated extract. Preferably equipment cost is minimized by use of a single pressure vessel to transfer the caffeine from the extract stream to the CO_2 and from the CO_2 to the water stream.

Title: **SUPERCRITICAL TAR SAND EXTRACTION; BITUMENS, COUNTERCURRENT FLOW, HEAT EXCHANGING, FLASHING**

Inventor: Poska, Forrest L. (US)

Assignee: Phillips Petroleum Company
Assignee Code: 65688
United States Class Code: 208390000

Application Number: US 176749   Patent Number: US 4341619
Application Date: 08/11/80   Issue Date: 07/27/82

Abstract:

An integrated process for the recovery of carbonaceous material from tar sands by supercritical extraction involving countercurrent flow of the tar sand and the solvent is disclosed.

Title: **PROCESS FOR THE PRODUCTION OF PURE LECITHIN DIRECTLY USABLE FOR PHYSIOLOGICAL PURPOSES**

Inventors: Heigel, Walter (DE)
                          Hueschens, Rolf (DE)

Assignee: Kali-Chemie Pharma GMBH De
Assignee Code: 02143
United States Class Code: 260403000

Application Number: US 238704   Patent Number: US 4367178
Application Date: 02/27/81   Issue Date: 01/04/83

Abstract:

Disclosed is a process for the production from raw lecithin of pure lecithin
directly usable for physiological purposes, comprising the steps of contacting
raw lecithin with gas as the extraction medium under supercritical conditions
with respect to pressure and temperature in an extraction stage to produce a gas
containing an extract; passing the gas containing the extract from the extraction
stage into a separation stage; varying at least one of the pressure and the
temperature of the gas in the separation stage to separate the extract-containing
gas into the gas and the extract; recycling the gas after the step of varying the
pressure and/or temperature; and removing pure lecithin from the extraction.

Title: **PROCESS FOR THE TREATMENT OF ORGANIC AMINE COMPOSITIONS**

Inventors: Ahmed, Moinuddin   (US)
                          Gibson, Charles A.   (US)
                          Habenschuss, Michael (US)

Assignee: Union Carbide Corporation
Assignee Code: 87136
United States Class Code: 203091000

Application Number: US 307223   Patent Number: US 4381223
Application Date: 09/30/81   Patent Date: 04/26/83

Abstract:

This invention provides for the separation of an amine composition undergoing
processing at high temperatures and pressures which amine composition contains
volatile components and less volatile components where the volatile components
are desirable for further processing and/or utilization and the less volatile
components are less desirable for further processing and/or utilization. The
separation serves to divide the amine composition into two streams, one which is
enriched in the volatile components and the other which is enriched in the less
volatile components.

Title: **SUPERCRITICAL MULTICOMPONENT SOLVENT COAL EXTRACTION; EFFICIENT LIQUEFACTION OF COAL WITH MIXTURE OF SOLVENTS**

Inventors: Chan, Paul C.   (US)
                          Corcoran, William H.   (US)
                          Fong, William S.   (US)
                          Lawson, Daniel D.   (US)
                          Pichaichanarong, Puvin (US)

Assignee: Administrator, United States National Aeronautics and Space Administration
Assignee Codes: 86504
United States Class Code: 208435000

Application Number: US 315584   Patent Number: US 4388171
Application Date: 10/30/81   Issue Date: 06/14/83

Abstract:

The yield of organic extract from the supercritical extraction of coal with larger diameter organic solvents such as toluene is increased by use of a minor amount of from 0.1 to 10% by weight of a second solvent such as methanol having a molecular diameter significantly smaller than the average pore diameter of the coal.

Title: **RECOVERY OF HYDROCARBON VALUES FROM LOW ORGANIC CARBON CONTENT CARBONACEOUS MATERIALS VIA HYDROGENATION AND SUPERCRITICAL EXTRACTIONS**

Inventors: Hart, Peter J. (US)
                          Scinta, James (US)

Assignee: Phillips Petroleum Company
Assignee Codes: 65688
United States Class Code: 208390000

Application Number: US 250445   Patent Number: US 4390411
Application Date: 04/02/81   Issue Date: 06/28/83

Abstract:

Hydrocarbon values are recovered from low organic carbon content materials via treatment with hydrogen and extraction with supercritical solvents.

Title: **PROCESS FOR PREPARING ETHYLENE GLYCOL; EXTRACTION OF ETHYLENE OXIDE WITH SUPER-CRITICAL CARBON DIOXIDE CATALYTIC CARBONATION TO FORM ETHYLENE CARBONATE, THEN HYDROLYSIS**

Inventor: Bhise, Vijay S. (US)

Assignee: Halco SD Group, Incorporated
Assignee Code: 07714
United States Class Code: 568858000

Application Number: US 388395   Patent Number: US 4400559
Application Date: 06/14/82   Issue Date: 08/23/83

Abstract:

Ethylene glycol is prepared by a process in which ethylene oxide is extracted from an aqueous solution with near-critical or super- critical carbon dioxide. Thereafter an ethylene oxide-carbon dioxide-water mixture is contacted with a catalyst to form ethylene carbonate, which is then hydrolyzed to ethylene glycol in the presence of the same catalyst. The ethylene glycol is separated as product and the carbon dioxide and the catalyst are recycled.

Title: **METHOD OF OBTAINING AROMATICS AND DYESTUFFS FROM BELL PEPPERS; SOLVENT EXTRACTION UNDER CRITICAL TEMPERATURE AND PRESSURE CONDITIONS**

Inventors: Coenen, Hubert (DE)
                        Hagen, Rainer (DE)
                        Knuth, Manfred (DE)

Assignee: Krupp, Fried GMBH de
Assignee Code: 47065
United States Class Code: 426429000

Application Number: US 365473   Patent Number: US 4400398
Application Date: 04/05/82   Issue Date: 08/23/83

Abstract:

Method for obtaining aromatics and/or dyestuffs from bell peppers wherein red pepper is extracted with a solvent which is in a supercritical state and is gaseous under normal conditions. The extraction takes place at a pressure of > PK to 350 bar and a temperature of > TK to 70°C. The extracted aromatics and/or dyestuffs are separated from the separated supercritical gas phase by lowering the density of the gas phase.

Title: **SEPARATION OF NEUTRALS FROM TALL OIL SOAPS;**
                          USING ETHYLENE IN SUPERCRITICAL STATE

Inventor: Amer, Gamal I. (US)

Assignee: Union Camp Corporation
Assignee Codes: 87104
United States Class Code: 260097600

Application Number: US 476599   Patent Number: US 4422966
Application Date: 03/18/83   Issue Date: 12/27/83

Abstract:

The disclosure is of a process for separating neutrals from salts of fatty/resin
acids by extraction of tall oil soaps with supercritical fluid solvents.

Title: **METHOD FOR THE PRODUCTION OF FOOD ADDITIVES**
                          WITH IMPROVED TASTE; LIQUID CARBON DIOXIDE
                          EXTRACTION OF LOCUST BEAN POWDER OR GUM, AND
                          GUAR GUM POWDER

Inventors: Heine, Christian (DE)
                          Wust, Reinhold   (DE)

Assignee: Henkel KGAA De
Assignee Code: 01324
United States Class Code: 426312000

Application Number: US 361781   Patent Number: US 4427707
Application Date: 03/25/82   Issue Date: 01/24/84

Abstract:

Locust bean pod powder, locust bean gum powder and guar gum powder originally
having odor and taste characteristics making them unsuitable as food additives,
can be made suitable therefore by extraction with gases at supercritical
temperatures.

Title: **APPARATUS FOR REMOVING ORGANIC CONTAMINANTS FROM INORGANIC-RICH MINERAL SOLIDS; SOLVENT EXTRACTIONS**

Inventors: De Filippi, Richard P. (US)
                         Eppig, Christopher P. (US)
                         Putnam, Bruce M.   (US)

Assignee: Critical Fluid Systems, Incorporated
Assignee Code: 08492
United States Class Code: 196014520

Application Number: US 255037   Patent Number: US 4434028
Application Date: 04/17/81   Issue Date: 02/28/84

Abstract:

Method and apparatus for removing oil and other organic constituents from particulate inorganic-rich mineral solids. The method and apparatus are particularly suitable for removing oil from oil-contaminated drill cuttings. The solids to be treated are transferred into pressure vessel means wherein they are contacted with an extractant which is normally a gas but is under conditions of pressure and temperature to provide the extractant in a fluidic solvent state for the constituents to be removed, whereby the constituents are transferred to the extractant. The extractant containing the constituents is withdrawn from the pressure vessel and depressurized to render it a nonsolvent for the constituents and to form a two-phase system which can then be separated into extractant for repressurizing and recycling with proper handling of the constituents removed. In the case of removing oil from drill cuttings, the essentially oil-free cuttings can be disposed of in any suitable manner including dumping overboard from an offshore drilling rig.

Title: **PROCESS FOR RECOVERING ETHYLENE OXIDE FROM AQUEOUS SOLUTIONS; SOLVENT EXTRACTION WITH SUPERCRITICAL CARBON DIOXIDE**

Inventors: Bhise, Vijay S. (US)
                         Hoch, Robert   (US)

Assignee: Halcon SD Group, Incorporated
Assignee Code: 07714
United States Class Code: 203014000

Application Number: US 284153   Patent Number: US 4437938
Application Date: 07/17/81   Issue Date: 03/20/84

Abstract:

Ethylene oxide is recovered from aqueous solutions by extracting with carbon dioxide in the near-critical or super-critical state, thereby selectively removing the ethylene oxide from water, and thereafter recovering ethylene oxide from the carbon dioxide by distillation or other suitable means.

Title: **SUPERCRITICAL SOLVENT COAL EXTRACTION**

Inventor: Compton, Leslie E. (US)

Assignee: Administrator, United States National Aeronautics and
                              Space Administration
Assignee Codes: 86504
United States Class Code: 208435000

Application Number: US 322312   Patent Number: US 4443321
Application Date: 11/17/81   Issue Date: 04/17/84

Abstract:

Yields of soluble organic extract are increased up to about 50% by the
supercritical extraction of particulate coal at a temperature below the
polymerization temperature for coal extract fragments (450°C) and a pressure from
500 psig to 5,000 psig by the conjoint use of a solvent mixture containing a low
volatility, high critical temperature coal dissolution catalyst such as
phenanthrene and a high volatility, low critical temperature solvent such as
toluene.

Title: **PRODUCTION OF DISTILLATES BY THE INTEGRATION
                              OF SUPERCRITICAL EXTRACTION AND GASIFICATION
                              THROUGH METHANOL TO GASOLINE**

Inventors: Derbyshire, Francis J. (US)
                              Whitehurst, Darrell D. (US)

Assignee: Mobil Corporation
Assignee Codes: 56432
United States Class Code: 208415000

Application Number: US 391334   Patent Number: US 4447310
Application Date: 06/23/82   Issue Date: 05/08/84

Abstract:

A process for producing a wide slate of fuel products from coal is provided by
integrating a methanol-to-gasoline conversion process with coal liquefaction and
coal gasification. The coal liquefaction comprises contacting the coal with a
solvent under supercritical conditions whereby a dense-gas phase solvent extracts
from the coal a hydrogen-rich extract which can be upgraded to produce a
distillate stream. The remaining coal is gasified under oxidation conditions to
produce a synthesis gas which is converted to methanol. The methanol is
converted to gasoline by contact with a zeolite catalyst. Solvent for coal
extraction is process derived from the upgraded distillate fraction or gasoline
fraction of the methanol-to-gasoline conversion.

Title: **HYDROCARBON RECOVERY FROM DIATOMITE; SOLVENT EXTRACTION USING SUPERCRITICAL TEMPERATURE AND PRESSURE**

Inventor: Scinta, James (US)

Assignee: Phillips Petroleum Company
Assignee Codes: 65688
United States Class Code: 208415000

Application Number:	US 475991	Patent Number:	US 4448669
Application Date:	03/16/83	Issue Date:	05/15/84

Abstract:

Supercritical extraction of diatomaceous earth results in a much more significant improvement in hydrocarbon recovery over Fischer retorting than achievable with tar sands. Process and apparatus for supercritical extraction of diatomaceous earth are disclosed.

Title: **CRITICAL SOLVENT SEPARATIONS IN INORGANIC SYSTEMS; PREFERENTIAL EXTRACTION OF METAL HALIDE WITH NONMETAL HALIDE**

Inventor: Rado, Theodore A. (US)

Assignee: Kerr-McGee Chemical Corporation
Assignee Codes: 45702
United States Class Code: 204066000

Application Number:	US 514499	Patent Number:	US 4457812
Application Date:	07/18/83	Issue Date:	07/03/84

Abstract:

A process for separating inorganic substances involving their abstraction from a mixture with near-supercritical inorganic fluids. One or more inorganic substances are abstracted and then separatively recovered by retrograde condensations. The process particularly is applicable with mixtures obtained from the chlorination of metalliferous ores and may be conjoined to many ancillary metal abstraction processes such as volatizations, distillations or electrolyses.

Title: **OLIGOMERIZATION OF OLEFINS IN SUPERCRITICAL WATER; GASOLINE**

Inventor: Paspek, Stephen C., Jr. (US)

Assignee: Standard Oil Corporation
Assignee Code: 79776
United States Class Code: 585520000

Application Number: US 485094   Patent Number: US 4465888
Application Date: 04/14/83   Issue Date: 08/14/84

Abstract:

Fuel range liquid hydrocarbons are produced by the oligomerization of olefins in a process comprising contacting the olefins containing 5 or less carbon atoms with a water containing medium at temperature sufficient to cause oligomerization and at a pressure sufficient to maintain the density of the medium at about 0.5 to about 1.0 grams per milliliter.

Title: **SUPERCRITICAL CO_2 EXTRACTION OF LIPIDS FROM LIPID-CONTAINING MATERIALS**

Inventor: Friedrich, John P. (US)

Assignee: Secretary, United States Department of Agriculture
Assignee Code: 86512
United States Class Code: 260412400

Application Number: US 364290   Patent Number: US 4466923
Application Date: 04/01/82   Issue Date: 08/21/84

Abstract:

In the extraction of lipid-containing substances with supercritical CO_2, triglyceride solubilities of up to 20% or more are obtainable by the simultaneous application of temperatures in excess of about 60°C and pressures of at least 550 bar.

Title: **METHOD OF EXTRACTING THE FLAVORING SUBSTANCES FROM THE VANILLA CAPSULE; USING CARBON DIOXIDE UNDER CRITICAL CONDITIONS; ONE STEP**

Inventors: Muhlnickel, Peter   (DE)
                                Sand, Theodor   (DE)
                                Sandner, Klaus   (DE)
                                Schutz, Erwin   (DE)
                                Volbrecht, Heinz-Rudiger (DE)

Assignee: Haarmann and Reimer GMBH DE
                                SKW Trostberg AG DE
Assignee Code: 03643
                                36592
United States Class Code: 426425000

Application Number: US 416732   Patent Number: US 4470927
Application Date: 09/10/82   Issue Date: 09/11/84

Abstract:

A process for recovering flavoring substances from the vanilla capsule with carbon dioxide under fluid conditions is described, which consists in performing the extraction at temperatures between 10 and 30°C and pressures of 80 to 350 bar, and producing the separation of the extract at temperatures of 0 to 30°C and pressures of 30 to 60 bar. In this manner it is possible by a one-step process to extract all of the flavoring substances of the vanilla capsule with the recovery of a highly concentrated vanilla flavoring, and to obtain a material which is considerably superior to the products obtained by conventional extraction with organic solvents.

Title: **PURIFICATION OF VANILLIN; EXTRACTION WITH CARBON DIOXIDE**

Inventor: Makin, Earle C. (US)

Assignee: Monsanto Company
Assignee Codes: 56920
United States Class Code: 568436000

Application Number: US 417314   Patent Number: US 4474994
Application Date: 09/13/82   Issue Date: 10/02/84

Abstract:

Crude vanillin is purified by supercritical extraction of impurities. The process is especially useful in purifying crude vanillin obtained from paper will waste liquors. Preferred supercritical extraction fluid is CO_2.

Title: **FRACTIONATION PROCESS FOR MIXTURES BY ELUTION
                              CHROMATOGRAPHY WITH LIQUID IN SUPERCRITICAL
                              STATE AND INSTALLATION FOR ITS OPERATION;
                              INTERMITTENT INJECTION OF MATERIAL TO BE
                              DISTILLED INTO FLUID FLOW OF COLUMN, THEN
                              SEPARATION**

Inventor: Perrut, Michel (FR)

Assignee: Nationale Elf Aquitaine (Production) FR
Assignee Codes: 00627
United States Class Code: 210659000

Application Number: US 499597   Patent Number: US 4478720
Application Date: 05/31/83   Issue Date: 10/23/84

Abstract:

The present invention concerns a fractionation process of a mixture by elution
chromatography. It is characterized in that the collected eluent is purified,
restored to supercritical   state and recycled in the head of the column.   It
concerns a cracking process of mixtures by elution chromatography with liquid in
the supercritical state and an installation for its operation.

Title: **OPEN-TUBULAR SUPERCRITICAL CHROMATOGRAPHY**

Inventors: Fjeldsted, John C.   (US)
                                    Lee Milton L.   (US)
                                    Novotny, Milos   (US)
                                    Peaden, Paul A.   (US)
                                    Springston, Stephen R. (US)

Assignee: Brigham Young University
Assignee Codes: 11264
United States Class Code: 073061100C

Application Number: US 352890   Patent Number: US 4479380
Application Date: 02/26/82   Issue Date: 10/30/84

Abstract:

Apparatus for and method of open-tube supercritical fluid chromatography. The
apparatus comprises an elongated passageway having inlet and outlet ends such as
a capillary column, coated or not with a coating having affinity for solute
molecules to be analyzed, means such as a high-pressure pump operatively
connected to an electronic pressure controller/programmer for supplying fluid at
high pressure to said inlet end, means such a flow controller/restrictor for
removing fluid from said outlet end, means such as a constant temperature oven
for subjecting the fluid between said inlet and outlet ends to a temperature near
(below, at or above) the critical temperature of the fluid, means for introducing
a sample material into the fluid upstream from the inlet end for chromatographic
analysis, such means preferably being an injector of the split injection type or
optionally of the on-column type, and means for subjecting fluid to which a
sample material has been added to detection by means of an ultraviolet detector,
an on-column spectrofluorimeter, a mass spectrometer, and the like. The method
comprises carrying out the detection of material after subjecting it to
supercritical fluid chromatography in a long stream of fluid, preferably of
capillary diameter, having an inlet end and an outlet end and which is subjected
between said ends to near supercritical temperature and pressure for the fluid,
the detection being carried on the fluid out near the outlet end by any suitable
method including ultraviolet detection, spectrofluorimetric detection and mass
spectrometric detection.

Title: **SUPERCRITICAL EXTRACTION AND SIMULTANEOUS CATALYTIC HYDROGENATION OF COAL; COAL LIQUIDS**

Inventors: Coenen, Hubert (DE)
                          Hagen, Rainer   (DE)
                          Kriegel, Enrst (DE)

Assignee: Krupp, Fried GMBH DE
Assignee Codes: 47065
United States Class Code: 208417000

Application Number: US 402933   Patent Number: US 4485003
Application Date: 07/29/82   Issue Date: 11/27/84

Abstract:

A process for producing liquid hydrocarbons from coal comprises treating comminuted coal at 380° to 600°C and 260 to 450 bar with water in a high pressure reactor to form a charged supercritical gas phase and a coal residue. Simultaneously with the water treatment, hydrogenation with hydrogen takes place in the presence of a catalyst. The catalyst is selected from the group consisting of $NaOH$, KOH, NA_4SiO_4, $NaBO_4$, or KOB_2. Then, the gas phase is divided into several fractions by lowering its pressure and temperature. Energy and/or gas is generated from the coal residue.

Title: **SUPERCRITICAL EXTRACTION PROCESS**

Inventors: Coombs, Daniel M. (US)
                                    Willers, Gary P. (US)

Assignee: Phillips Petroleum Company
Assignee Codes: 65688
United States Class Code: 208309000

Application Number: US 408964   Patent Number: US 4482453
Application Date: 08/17/82   Issue Date: 11/13/84

Abstract:

The recovery of hydrocarbon values from high metals content feeds can be carried
out more efficiently via supercritical extraction with recycle of a portion of
the asphalt product and proper control of the use of countercurrent solvent flow
to said extraction.

Title: **UPGRADING   HEAVY   HYDROCARBONS   WITH
                                    SUPERCRITICAL WATER AND LIGHT OLEFINS**

Inventor: Paspek, Stephen C. Jr. (US)

Assignee: Standard Oil Corporation
Assignee Code: 79776
United States Class Code: 208106000

Application Number: US 510524   Patent Number: US 4483761
Application Date: 07/05/83   Issue Date: 11/20/84

Abstract:

Heavy hydrocarbons are upgraded and cracked in a process comprising contacting
the heavy hydrocarbons with olefins containing 5 or less carbon atoms and a
solvent, at a temperature both sufficient for cracking and greater than or equal
to the critical temperature of the solvent.

Title: **PROCESS FOR THE PREPARATION OF SPICE EXTRACTS; LIQUID AND GASEOUS EXTRACTIONS**

Inventors: Behr, Norbert   (DE)
                          Der Mei Henk, Van   (DE)
                          Schnelgelberger, Harald   (DE)
                          Sirtl, Wolfgang   (CH)
                          Von Ettingshausen, Othmar (DE)

Assignee: Henkel KGAA DE
Assignee Code: 01324
United States Class Code: 426312000

Application Number: US 337388   Patent Number: US 4490398
Application Date: 01/06/82   Issue Date: 12/25/84

Abstract:

This invention is directed to a process for the production of extracts from spices in two stages by extraction with a non-toxic gas as solvent, the improvement which comprises in a first stage extracting from the spices essential oils functioning as aroma components by contacting the spices with liquid solvent, at a pressure in the supercritical range and a temperature in the subcritical range, separating the solvent from the spice, evaporating the solvent and separating essential oils, and in a second stage extracting from the spices the portions acting as flavor carriers by contacting the spices from the first stage with gaseous solvent, at a pressure and a temperature both in the supercritical range, separating the gaseous solvent from the spice, reducing the pressure and temperature, and separating the flavor carriers from the gaseous solvent.

Title: **METHOD FOR SEPARATING ETHANOL FROM AN ETHANOL CONTAINING SOLUTION; SOLVENT EXTRACTION**

Inventors: Hagen, Rainer   (DE)
                       Hartwig, Jurgen (DE)

Assignee: Krupp, Fried GMBH de
Assignee Code: 47065
United States Class Code: 568916000

Application Number: US 600301   Patent Number: US 4492808
Application Date: 04/03/84   Issue Date: 01/08/85

Abstract:

A process is disclosed for the separation of ethanol from an ethanol containing water solution, wherein the ethanol containing solution is extracted by means of a solvent which is in the liquid or supercritical state, the ethanol containing solvent phase is separated into its components by being conducted over an adsorption medium without changing the pressure or temperature, and the ethanol is recovered by treating the ethanol containing adsorption medium with the solvent used for the extraction at a pressure from 1 to 30 bar and at a temperature from 150 to 300°C.

Title: **APPARATUS AND METHOD INVOLVING SUPERCRITICAL FLUID EXTRACTION; EXTRACTION OF LIGNIN FROM BLACK LIQUOR**

Inventor: Avedesian, Michael M. (CA)

Assignee: Unassigned or Assigned to Individual
Assignee Codes: 68000
United States Class Code: 530507000

Application Number: US 564500   Patent Number: US 4493797
Application Date: 12/22/83   Issue Date: 01/15/85

Abstract:

An autoclave extraction apparatus using supercritical fluid is used for supercritical fluid extraction of one or several compounds. The supercritical fluid containing the compound(s) may then be processed in a pressurized bed reactor under supercritical conditions. The bed being a fluidizable catalytic bed to carry out a catalytic reaction of the compound(s). The method is particularly applicable to recover valuable lignin and other extractable components from Kraft black liquor.

Title: **PRODUCTION OF DEFATTED SOYBEAN PRODUCTS BY
                     SUPERCRITICAL   FLUID   EXTRACTION;   LIPID
                     EXTRACTION WITH CARBON DIOXIDE**

Inventors: Eldridge, Arthur C. (US)
                     Friedrich, John P.   (US)

Assignee: Secretary, United States Department of Agricultural
Assignee Codes: 86512
United States Class Code: 426629000

Application Number: US 534015   Patent Number: US 4493854
Application Date: 09/20/83   Issue Date: 01/15/85

Abstract:

The raw grassy and bitter principles in soybeans are reduced to acceptable levels
for purposes of human consumption without significant degradation of the
nutritional properties. This result is achieved by a lipid extraction process
in which raw soybean material is treated with carbon dioxide, under carefully
controlled supercritical conditions. Of particular importance are the moisture
content of the bean material as well as the pressure, temperature, and contact
time of the carbon dioxide extractant.

Title: **PRODUCTION OF FOOD-GRADE CORN GERM PRODUCT BY
                     SUPERCRITICAL   FLUID   EXTRACTION;   CARBON
                     DIOXIDE   SOLVENT,   DECREASED   LIPID   AND
                     PEROXIDASE ACTIVITY**

Inventors: Christianson, Donald D. (US)
                     Friedrich, John P.   (US)

Assignee: Secretary, United States Department of Agriculture
Assignee Codes: 86512
United States Class Code: 426312000

Application Number: US 436541   Patent Number: US 4495207
Application Date: 10/25/82   Issue Date: 01/22/85

Abstract:

A high-protein, food-grade product is prepared by defatting drymilled corn germ
fractions with carbon dioxide under supercritical conditions. The residual lipid
and peroxidase activity responsible for development of off-flavors during storage
are reduced to a fraction of the levels obtainable by conventional hexane
extraction methods.

Title: **ACIDULATION AND RECOVERY OF CRUDE TALL OIL FROM TALL OIL SOAPS; SUPERCRITICAL CARBON DIOXIDE**

Inventors: Amer, Gamal I.   (US)
                       Lawson, Nelson E. (US)

Assignee: Union Camp Corporation
Assignee Codes: 87104
United States Class Code: 260097700

Application Number: US 481811   Patent Number: US 4495095
Application Date: 04/04/83   Issue Date: 01/22/85

Abstract:

In the present invention, tall oil acids are prepared by acidulating tall oil soap with supercritical fluid carbon dioxide. The method of preparation is carried out at a temperature of from about 31° to 400 °C and the supercritical carbon dioxide is under a pressure of from about 1075 to about 50,000 psi. The acidulate is extracted into the fluid phase of the supercritical carbon dioxide. The resultant tall oil acids are then recovered from the carbon dioxide fluid phase.

Title: **AUTOMATED SAMPLE CONCENTRATOR FOR TRACE COMPONENTS; ADSORPTION OF SOLUTES WITH SUPERCRITICAL FLUID**

Inventors: Bell, Raymond J.   (US)
                       Gere, Dennis R.   (US)
                       Nickerson, Mark A. (US)
                       Poole, John S.   (US)

Assignee: Hewlett-Packard Company
Assignee Codes: 38859
United States Class Code: 210659000

Application Number: US 538832   Patent Number: US 4500432
Application Date: 10/04/83   Issue Date: 02/19/85

Abstract:

Concentration of the solutes contained in a liquid solvent is effected by passing the solvent through a first trapping means that adsorbs them, flushing the first trapping means with supercritical fluid and reducing its solubility parameter in a second trapping means so that the latter can adsorb desired solutes.

Title: FRACTIONATION OF BUTTERFAT USING A LIQUEFIED GAS OR A GAS IN THE SUPERCRITICAL STATE

Inventors: Biernoth, Gerhard (DE); Merk, Werner (DE)

Assignee: Lever Brothers Company
Assignee Code: 49528
United States Class Code: 426312000

Application Number: US 413937
Application Date: 09/01/82
Patent Number: US 4504503
Issue Date: 03/12/85

Abstract:

A process for producing a mixture of triglycerides displaying butter-like properties by fractionating fats with a liquefied gas or gas under supercritical conditions. Said mixture of triglycerides predominantly consists of triglycerides with a carbon number ranging from 24 to 42 and can be used as one of the fat components of a margarine fat blend in order to improve its butter-like properties.

Title: METHOD OF REMOVING WATER FROM ALKALI METAL HYDROXIDE SOLUTIONS; EXTRACTION WITH ORGANIC LIQUID AT ELEVATED TEMPERATURE AND PRESSURE AND ADDITION OF HYDROXIDE

Inventors: Hoyer, Gale G. (US); Kozak, William G. (US); Sumner, William C. Jr. (US)

Assignee: Dow Chemical Company
Assignee Code: 24712
United States Class Code: 423592000

Application Number: US 505639
Application Date: 06/20/83
Patent Number: US 4505885
Issue Date: 03/19/85

Abstract:

A method removing water from aqueous alkali metal hydroxide solutions by contacting the solution with an organic liquid at elevated temperatures and pressures to form an organic liquid water phase and a hydroxide solution phase and thereafter separating the organic water phase from the hydroxide solution phase.

Title: **RECOVERY OF ORGANIC SOLVENTS FROM LIQUID MIXTURES; SEPARATION OF VAPOR AND LIQUID PHASE, MIXING LIQUID PHASE WITH STEAM**

Inventor: Roach, Jack W. (US)

Assignee: Kerr-McGee Refining Corporation
Assignee Code: 02887
United States Class Code: 203079000

Application Number: US 366040   Patent Number: US 4508597
Application Date: 04/05/82   Issue Date: 04/02/85

Abstract:

A method of recovering light organic solvent from a liquid mixture containing the solvent and a product material, such as asphaltenes or coal liquefaction products. The solvent-product material mixture is treated to separate a first vapor phase rich in solvent and a first liquid phase rich in product material. The first liquid phase is then intimately contacted with steam under shearing conditions, in a static or dynamic mixer. The steam liquid phase mixture is then treated to separate a second vapor phase, rich in steam and solvent, and a second liquid phase, rich in product material and substantially depleted of solvent. Solvent is recovered from the first and second vapor phases.

Title: **ETHANOL EXTRACTION PROCESS; CONTACTING WITH PROPENE SOLVENT**

Inventor: Victor, John G. (US)

Assignee: Institute of Gas Technology
Assignee Code: 42208
United States Class Code: 568916000

Application Number: US 374402   Patent Number: US 4508928
Application Date: 05/03/82   Issue Date: 04/02/85

Abstract:

A process for the separation of ethanol from water using solvent extraction at elevated pressures is disclosed. Separation is effected by contacting aqueous ethanol with either Propylene (Propene), Allene (Propadiene), Methyl Acetylene (Propyne), or Methyl Allene (1,2-Butadiene). This produces two liquid phases which separate because of the difference in their densities, and are easily drawn off as separate streams. The solvent is recovered by distillation and condensation using a heat pump to transfer heat. The ethanol and water remain in a liquid state and are substantially recovered.

Title: **METHOD FOR GROWTH OF CRYSTALS BY PRESSURE
                          REDUCTION OF SUPERCRITICAL OR SUBCRITICAL
                          SOLUTION; ISOTHERMAL OR ADIABATIC CONDITION
                          TO CONTROL HEAT EXCHANGING**

Inventor: Shlichta, Paul J. (US)

Assignee: Administrator, United States National Aeronautics
                          Space Administration
Assignee Code: 86504
United States Class Code: 1566230000

Application Number: US 342944   Patent Number: US 4512846
Application Date: 01/26/82   Issue Date: 04/23/85

Abstract:

Crystals (51) of high morphological quality are grown by dissolution of a
substance (28) to be grown into the crystal (51) in a suitable solvent (30) under
high pressure, and by subsequent slow, time-controlled reduction of the pressure
of the resulting solution (36). During the reduction of the pressure interchange
of heat between the solution (36) and the environment is minimized by performing
the pressure reduction either under isothermal or adiabatic conditions.

Title: **PROCESS FOR PRODUCING HIGH YIELD OF GAS
                          TURBINE FUEL FROM RESIDUAL OIL**

Inventor: Zarchy, Andrew S. (US)

Assignee: General Electric Company
Assignee Codes: 33808
United States Class Code: 210634000

Application Number: US 547275   Patent Number: US 4528100
Application Date: 10/31/83   Issue Date: 07/09/85

Abstract:

The acceptability of residual oil as a gas turbine fuel is greatly enhanced in
a two step process which significantly decreases the vanadium content of the
residual fuel. In the process, and the residual oil is first broken down into
an oil phase and asphaltene phase by either conventional or supercritical
extraction. In this step, the majority vanadium remains in the asphaltene phase.
The vanadium is then removed from the asphaltenes by a supercritical solvent
extraction process in which the vanadium free asphaltene phase is then re-
dissolved in the oil for use as a gas turbine fuel. This fuel possesses
significantly lower vanadium content, and thus permits gas turbine operation for
greater periods of time without maintenance.

Title: **EXTRACTION OF DEPOLYMERIZED CARBONACEOUS MATERIAL USING SUPERCRITICAL AMMONIA; WOOD, LIGNITE, COAL**

Inventors: Miller, Robert N.   (US)
                           Sunder, Swaminathan (US)

Assignee: Air Products and Chemicals, Incorporated
Assignee Codes: 01184
United States Class Code: 208429000

Application Number: US 601890   Patent Number: US 4539094
Application Date: 04/19/84   Issue Date: 09/03/85

Abstract:

Improved extraction and recovery of carbonaceous products are achieved by the aqueous alkali depolymerization of the carbonaceous materials, low rank coals, followed by the supercritical extraction with ammonia when the depolymerized material retains a significant aqueous presence.

Title: **SUPERCRITICAL FLUID EXTRACTION AND ENHANCEMENT FOR LIQUID-LIQUID EXTRACTION PROCESSES; SOLVENT EXTRACTION OF FEED SOLUTION; SUBSEQUENT EXTRACTION WITH SUPERCRITICAL SOLVENT**

Inventor: Zachary, Andrew S. (US)

Assignee: General Electric Company
Assignee Codes: 33808
United States Class Code: 210634000

Application Number: US 574274   Patent Number: US 4547292
Application Date: 10/31/83   Issue Date: 10/15/85

Abstract:

Liquid-liquid extraction processes employing a liquid solvent are enhanced through the utilization of a subsequent supercritical fluid extraction process performed on the liquid extract from the liquid-liquid process. The second, supercritical solvent is used in the supercritical fluid extraction process step. The extract from the supercritical fluid extraction process is further processed to separate the supercritical solvent and the liquid solvent, both of which are returned in a closed cycle flow to the supercritical fluid extraction vessel and the liquid-liquid reaction vessel respectively.

Title: **FRACTIONATION OF POLYMERIZED FATTY ACIDS;
                              SOLVENT EXTRACTION WITH SUPERCRITICAL FLUIDS**

Inventor: Frihart, Charles R. (US)

Assignee: Union Camp Corporation
Assignee Codes: 87104
United States Class Code: 260428500

Application Number: US 495096   Patent Number: US 4568495
Application Date: 05/16/83   Issue Date: 02/04/86

Abstract:

Polymeric fatty acids are fractionated by extraction with supercritical fluid
solvents.

Title: **ELECTROCHEMICAL   CELL   OPERATING   NEAR   THE
                              CRITICAL   POINT   OF   WATER;   FUELS   CELLS,
                              ELECTROSYNTHESIS CELLS**

Inventors: Chao, Mou S.   (US)
                              Hoyer, Gale G.   (US)
                              Paulaitis, Michael E. (US)
                              Varjian, Richard D.   (US)

Assignee: Dow Chemical Company
Assignee Codes: 24712
United States Class Code: 204001000R

Application Number: US 622803   Patent Number: US 4581105
Application Date: 06/20/84   Issue Date: 04/08/86

Abstract:

The invention is a method for operating an electrochemical cell of the type
having an anode and a cathode which are in contact with aqueous electrolyte in
the presence of at least one electroactive species, said method comprising:
electrochemically converting at least a portion of at least one electroactive
species to at least one electrochemical product while maintaining said aqueous
electrolyte at supercritical fluid conditions.

Title: **PURIFICATION OF TEREPHTHALIC ACID BY SUPERCRITICAL FLUID EXTRACTION; DISSOLVING, DEPRESSURIZATION TO VAPORIZE FLUID AND PRECIPITATE TEREPHTHALIC ACID**

Inventor: Myerson, Allan S. (US)

Assignee: Georgia Institute of Technology
Assignee Codes: 12423
United States Class Code: 562486000

Application Number: US 439329   Patent Number: US 4550198
Application Date: 11/04/82   Issue Date: 10/29/85

Abstract:

A process for the purification of terephthalic acid comprising (1) contacting an impure terephthalic acid with an amount of supercritical fluid, at a temperature and pressure, and for a period of time, sufficient to dissolve the terephthalic acid and its impurities in said supercritical fluid and (2) allowing said solution of terephthalic acid and impurities in supercritical fluid to expand into a reaction chamber, whereby the supercritical fluid becomes gaseous and the terephthalic acid, freed of impurities, precipitates out as a solid.

Title: **METHOD FOR DENSIFICATION OF CERAMIC MATERIALS; DISSOLVING PRECURSOR IN SUPERCRITICAL FLUID, IMPREGNATION OF POROUS CERAMIC**

Inventors: Berneburg, Philip L.   (US)
                 Krukonis, Val J.   (US)

Assignee: Babcock and Wilcox Company
Assignee Codes: 06920
United States Class Code: 427249000

Application Number: US 659139   Patent Number: US 4552786
Application Date: 10/09/84   Issue Date: 11/12/85

Abstract:

Supercritical fluids may be used to carry ceramic precursor materials into the pores of a ceramic host. Reducing the solubility of the ceramic precursor in the supercritical fluid will cause deposition of the ceramic precursor in the pores of void spaces of the ceramic host and accomplish densification of the ceramic host material.

Title: **SUPERCRITICAL FLUID MOLECULAR SPRAY FILM DEPOSITION AND POWDER FORMATION; EXPANDING THROUGH ORIFICE AND DIRECTING AGAINST SURFACE**

Inventor: Smith, Richard D. (US)

Assignee: Battelle Memorial Institute
Assignee Codes: 08120
United States Class Code: 427421000

Application Number: US 528723
Application Date: 09/01/83

Patent Number: US 4582731
Issue Date: 04/15/86

Abstract:

Solid films are deposited, or fine powders formed, by dissolving a solid material into a supercritical fluid solution at an elevated pressure and then rapidly expanding the solution through a short orifice into a region of relatively low pressure. This produces a molecular spray which is directed against a substrate to deposit a solid thin film thereon, or discharged into a collection chamber to collect a fine powder. Upon expansion and supersonic interaction with background gases in the low pressure region, any clusters of solvent are broken up and the solvent is vaporized and pumped away. Solute concentration in the solution is varied primarily by varying solution pressure to determine, together with flow rate, the rate of deposition and to control in part whether a film or powder is produced and the granularity of each. Solvent clustering and solute nucleation are controlled by manipulating the rate of expansion of the solution and the pressure of the lower pressure region. Solution and low pressure region temperatures are also controlled.

Title: **PROCESS FOR SEPARATING FATTY MATERIALS FROM SUPPORTED NICKEL CATALYSTS**

Inventors: Blewett, Charles W. (US)
Turner, Stephan W. (US)

Assignee: National Distillers and Chemical Corporation
Assignee Codes: 58320
United States Class Code: 260412800

Application Number: US 682404
Application Date: 12/17/84

Patent Number: US 4584140
Issue Date: 04/22/86

Abstract:

A process for recovering fatty materials, such as fatty acids and lower alkyl esters thereof, from supported nickel catalyst compositions is provided. For the process, the supported nickel catalyst composition, which may also contain non-nickel containing clays/earths, is extracted with a supercritical fluid, preferably supercritical carbon dioxide, to separate the fatty materials from the supported nickel catalyst composition.

Title: **LIGHT-PIPE FLOW CELL FOR SUPERCRITICAL FLUID CHROMATOGRAPHY**

Inventors: Allhands, Daniel R.   (US)
                          Vidrine, D. Warren   (US)

Assignee: Nicolet Instrument Corporation
Assignee Codes: 12918
United States Class Code: 250428000

Application Number: US 704739   Patent Number: US 4588893
Application Date: 02/25/85   Issue Date: 05/13/86

Abstract:

A light-pipe flow cell (10) for high pressure fluids is disclosed which has a main support body (15) with a gold lightpipe element (22) mounted therein. Infrared transmissive windows (26,27) are mounted to the main support body against sealing rings (31, 32) to seal off the polished central bore (23) of the light-pipe element from ambient atmosphere while allowing an infrared beam to be passed therethrough. Pressure plates (14, 18) are mounted to the main support body (15) to apply pressure to the windows over the sealing rings to tightly seal the windows without exerting undue stress thereon. Flow of liquid or supercritical fluid from a chromatography column is directed through inlet channels (35, 38) in the support body and lightpipe element to one end of the light-pipe bore (23) and out of the opposite end of the bore through a channel (41) in the lightpipe element and a communicating channel (36) in the main support body. The flow cell (10) may be connected in a supercritical fluid chromatography system wherein effluent from the chromatography column (64) is passed through the flow cell (10) and is subjected to an infrared beam to allow infrared spectrometric analysis.

Title: **APPARATUS FOR ANALYZING SOLID SAMPLE WITH SUPERCRITICAL FLUID; CHROMATOGRAPHY, SOLVENT EXTRACTION, LIQUID GAS**

Inventors: Saito, Muneo   (JP)
                  Sugiyama, Kenkichi (JP)
                  Wada, Akio   (JP)

Assignee: Japan Spectroscopic Company, Ltd JP
                  Morinaga & Company, Ltd JP
Assignee Codes: 01426
                  07521
United States Class Code: 422070000

Application Number: US 676200   Patent Number: US 4597943
Application Date: 11/29/84   Issue Date: 07/01/86

Abstract:

An apparatus for analyzing a sample with a supercritical fluid includes a fluid container containing as an extraction solvent a fluid obtained by compressing and liquefying a substance which is a gas at ambient temperature and atmospheric pressure. A pump is provided for drawing the fluid from its container through a suction line and delivering it through a delivery line, while the heads of the pump are cooled by a cooling device. An extraction mechanism is provided for bringing the fluid in a supercritical state into contact with the sample to be analyzed and extracting a specific component or components from the sample. A trapping mechanism is provided downstream of the extraction mechanism for collecting the extracted component or components from the fluid. An analyzing mechanism can be connected to the trapping mechanism by changeover valves for analyzing the collected component or components.

Title: **EXTRACTION PROCESS; OIL FROM SEEDS**

Inventor: Shindler, Brian D. (GB)

Assignee: Imperial Chemical Industries, LTD GB
Assignee Codes: 41248
United States Class Code: 424195100

Application Number: US 659527   Patent Number: US 4601906
Application Date: 10/10/84   Issue Date: 07/22/86

Abstract:

An extraction vessel A is charged with a particulate solid, e.g. oilseed, and a
transfer liquid, e.g. water, in an amount sufficient to pressurize vessel A to
an elevated pressure. Extraction solvent, e.g. carbon dioxide, is introduced,
preferably from a storage vessel B, to displace transfer liquid from vessel A.
After extraction by the solvent, the latter may be displaced into vessel B by
transfer liquid which is preferably stored under pressure in vessel B during the
extraction. The extraction solvent is gaseous at ambient conditions but
preferably is liquid or a supercritical fluid at the displacement conditions.

Title: **COAL LIQUEFACTION WITH PREASPHALTENE RECYCLE;
                          MINIMIZATION OF HYDROGEN REQUIREMENT**

Inventors: Miller, Robert N. (US)
                          Weimer, Robert F. (US)

Assignee: International Coal Refining Company
Assignee Codes: 06646
United States Class Code: 208412000

Application Number: US 703858   Patent Number: US 4609455
Application Date: 02/21/85   Issue Date: 09/02/86

Abstract:

A coal liquefaction system is disclosed with a novel preasphaltene recycle from
a supercritical extraction unit to the slurry mix tank wherein the recycle stream
contains at least 90% preasphaltenes (benzene insoluble, pyridine soluble
organics) with other residual materials such an unconverted coal and ash. This
subject process results in the production of asphaltene materials which can be
subjected to hydrotreating to acquire a substitute for No. 6 fuel oil. The
preasphaline-predominant recycle reduces the hydrogen consumption for a process
where asphaltene material is being sought.

Title: **SUPERCRITICAL   AMMONIA   TREATMENT   OF
                          LIGNOCELLULOSIC   MATERIALS;   CELLULASE
                          HYDROLYSIS TO MONOSACCHARIDES; LIVESTOCK FEED
                          FROM AGRICULTURAL WASTES**

Inventor: Chou, Yu-Chia T. (US)

Assignee: E.I. Du Pont De Nemours, & Company
Assignee Codes: 25048
United States Class Code: 536030000

Application Number: US 736386   Patent Number: US 4644060
Application Date: 05/21/85   Issue Date: 02/17/87

Abstract:

The bioavailability of polysaccharide components of lignocellulosic materials can be increased substantially by treatment with ammonia in a supercritical or near-supercritical fluid state.

Title: **PROCESS AND DEVICE FOR PURIFYING BENZOIC
                          ACID; EXTRACTION**

Inventors: Goorden, Josephus J.   (NL)
                          Kleintjens, Ludovicus A. (NL)
                          Simons, Antonius J.   (NL)

Assignee: Stamicarbon B V NL
Assignee Codes: 79552
United States Class Code: 562494000

Application Number: US 795641   Patent Number: US 4652675
Application Date: 11/06/85   Issue Date: 03/24/87

Abstract:

Process for purifying benzoic acid obtained by oxidation of toluene with a gas containing molecular oxygen, which purification is carried out at a pressure of at least 3 MPa by means of a supercritical extraction using a gas or gas mixture the critical temperature of which is lower than 435 K. The process being characterized in the benzoic acid to be purified is supplied in a liquid state to a crystallizer in which the prevailing temperatures is below the solidification temperature of that the benzoic acid and in which the benzoic acid is treated during and possibly after the crystallization process with the said gas or gas mixture.

Title: **PROCESS OF RECOVERING PURIFIED BENZOIC ACID;**
                                  SUPERCRITICAL EXTRACTION, MELTING

Inventors: Goorden, Josephus J.   (NL)
                                  Kleintjens, Ludovicus A. (NL)
                                  Simons, Antonius J.   (NL)

Assignee: Stamicarbon B V NL
Assignee Codes: 79552
United States Class Code: 562494000

Application Number: US 795642   Patent Number: US 4654437
Application Date: 11/06/85   Issue Date: 03/31/87

Abstract:

Process for recovering benzoic acid purified by means of a supercritical
extraction of solid benzoic acid, characterized in that after the extraction, the
purified benzoic acid is remelted at least in part and is carried off from the
melting device in at least partly liquid form.

Title: **REMOVAL OF TEXTURED VEGETABLE PRODUCT OFF-**
                                  FLAVOR BY SUPERCRITICAL FLUID OR LIQUID
                                  EXTRACTION

Inventor: Sevenants, Michael R. (US)

Assignee: Proctor and Gamble Company
Assignee Codes: 68128
United States Class Code: 426425000

Application Number: US 687612   Patent Number: US 4675198
Application Date: 12/31/84   Issue Date: 06/23/87

Abstract:

The invention is a method for the removal of off-flavor from a textured vegetable
product by extraction with a gas in the supercritical fluid or liquid state. The
method for supercritical fluid extraction of off-flavors comprises the steps of
(A) extracting off-flavors from the textured vegetable product by contacting it
with a supercritical fluid gas in a pressurized container; and (B) removing the
gas and off-flavors from the textured vegetable product. Liquid extraction
comprises the steps of: (A) extracting the off-flavors from the textured
vegetable product by contacting it with a liquid gas in a pressurized container;
and (B) removing the gas and off-flavors from the textured vegetable product.
The temperature and pressure of the gas may be varied to maintain it in the
supercritical fluid or liquid state. Carbon dioxide is a preferred gas.

Title: **STEP-WISE   SUPERCRITICAL   EXTRACTION   OF CARBONACEOUS RESIDUE; DEPRESSURIZATION**

Inventor: Warzinski, Robert P. (US)

Assignee: United States Department of Energy
Assignee Codes: 01715
United States Class Code: 107118188

Application Number: US 863494   Patent Number: US 4675101
Application Date: 05/15/86   Issue Date: 06/23/87

Abstract:

A method of fractionating a mixture containing high boiling carbonaceous material and normally solid mineral matter includes processing with a plurality of different supercritical solvents. The mixture is treated with a first solvent of high critical temperature and solvent capacity to extract a large fraction as solute. The solute is released as liquid from solvent and successively treated with other supercritical solvents of different critical values to extract fractions of differing properties. Fractionation can be supplemented by solute reflux over a temperature gradient, pressure let down in steps and extractions at varying temperature and pressure values.

Title: **SAMPLE DILUTION SYSTEM FOR SUPERCRITICAL FLUID CHROMATOGRAPHY; FOR ANALYZING A PROCESS STREAM**

Inventors: Leaseburge, Emory J. (US)
                                Melda, Kenneth J.   (US)

Assignee: Combustion Engineering, Incorporated
Assignee Codes: 19080
United States Class Code: 210101000

Application Number: US 917628   Patent Number: US 4681678
Application Date: 10/10/86   Issue Date: 07/21/87

Abstract:

An apparatus for analyzing a process stream (1) via supercritical fluid chromatography including container means (10), extraction means (50) for drawing a known sample volume from the process stream and injecting the sample volume into a known volume of supercritical fluid to form a dilute mix (5), analyzer means (70) for analyzing the dilute mix via supercritical fluid chromatography, sample inlet valve means (60) for receiving the dilute mix from the extraction means (50) and supplying a portion of the dilute mix to the analyzer means (70). Pneumatically driven piston-type pump means (30) for drawing the substance from the container means (10) and delivering same under pressure either to the extraction means (50) or to the analyzer means (70) as desired, and control means (40) for supplying low pressure air to the pump means (30) in a controlled manner so as to draw the substance from the container means (10) and maintain the substance at a desired pressure above the supercritical pressure for delivery to the analyzer means (70) and the extraction means (50) as desired.

Title: **SUPERCRITICAL FLUID CHROMATOGRAPH WITH PNEUMATICALLY CONTROLLED PUMP; FOR ANALYZING A PROCESS STREAM**

Inventors: Leaseburge, Emory J. (US)
Thomas, Thomas J. (US)

Assignee: Combustion Engineering, Incorporated
Assignee Codes: 19080
United States Class Code: 210198200

Application Number: US 917466   Patent Number: US 4684465
Application Date: 10/10/86   Issue Date: 08/04/87

Abstract:

An apparatus for analyzing a process stream (1) via supercritical fluid chromatography including container means (10), extraction means (50) for drawing a sample volume from the process stream and injecting the sample volume into the supercritical fluid to form a dilute mix (5), analyzer means (70) for analyzing the dilute mix via supercritical fluid chromatography, sample inlet valve means (60) for receiving the dilute mix from the extraction means (50) for supplying a portion of the dilute mix to the analyzer means (70), pneumatically driven piston-type pump means (30) for drawing the substance from the container means (10) and delivering same under pressure either to the extraction means (50) or to the analyzer means (70) as desired, and control means (40) for supplying low pressure air to the pump means (30) in a controlled manner so as to draw the substance from the container means (10) and maintain the substance at a desired pressure above the supercritical pressure for delivery to the analyzer means (70) and the extraction means (50) as desired.

Title: **CONDITIONING OF CARBONACEOUS MATERIAL PRIOR TO PHYSICAL BENEFICIATION; SOLVENT EXTRACTION WITH METHANOL OR CYCLOHEXANE**

Inventors: Ruether, John A.   (US)
                             Warzinski, Robert P. (US)

Assignee: United States Department of Energy
Assignee Codes: 01715
United States Class Code: 208311000

Application Number: US 863650   Patent Number: US 4695372
Application Date: 05/15/86   Issue Date: 09/22/87

Abstract:

A carbonaceous material such as coal is conditioned by contact with a supercritical fluid prior to physical beneficiation. The solid feed material is contacted with an organic supercritical fluid such as cyclohexane or methanol at temperatures slightly above the critical temperature and pressures of 1 to 4 times the critical pressures. A minor solute fraction is extracted into critical phase and separated from the solid residuum. The residuum is then processed by physical separation such as by froth flotation or specific gravity separation to recover a substantial fraction thereof with reduced ash content. The solute in supercritical phase can be released by pressure reduction and recombined with the low-ash, carbonaceous material.

Title: **APPARATUS AND METHOD INVOLVING SUPERCRITICAL FLUID EXTRACTION; RECOVERING LIGNIN TO BE CONVERTED TO USEFUL PRODUCTS**

Inventor: Avedesian, Michael M. (CA)

Assignee: Domtar, Incorporated
Assignee Codes: 24496
United States Class Code: 422140000

Application Number: US 871752   Patent Number: US 4714591
Application Date: 06/09/86   Issue Date: 12/22/87

Abstract:

An autoclave extraction apparatus where a supercritical fluid is used for supercritical fluid extraction of one or several compounds. The supercritical fluid containing the compound(s) may then be processed in a pressurized, fluidized, bed reactor under supercritical conditions. The fluidized bed reactor is used to carry out a catalytic reaction of the compound(s). The method is particularly applicable to recover valuable lignin and other extractable components from Kraft black liquor.

Title: **SUPERCRITICAL FLUID EXTRACTION METHOD FOR MULTI-COMPONENT SYSTEMS**

Inventors: Chimowitz, Eldred (US)
Pennisi, Kenneth (US)

Assignee: University of Rochester
Assignee Codes: 72104
United States Class Code: 203049000

Application Number: US 26603   Patent Number: US 4714526
Application Date: 03/17/87   Issue Date: 12/22/87

Abstract:

A process for extracting pure components from a multi-component system, said system comprising a mixture of at least two solids or liquids in a supercritical fluid at constant pressure, which involves making use of the cross-over pressure points of the components which comprise the system.

Title: **PROCESS FOR RECOVERING PRIMARY NORMAL ALIPHATIC HIGHER ALCOHOLS; EXTRACTION OF SUGAR CANE WITH SUPERCRITICAL FLUID**

Inventors: Furukawa, Kurune   (JP)
Honda, Keijiro   (JP)
Inada, Shoshichiro (JP)
Masui, Takachika   (JP)
Ogasawara, Joji   (JP)
Tsubakimoto, Giichi(JP)

Assignee: Seitetsu Kagaku Company, Ltd, JP;
Shinko Seito Company, Ltd JP;
Shinko Sugar Production Company, Ltd JP
Assignee Codes: 17489
17490
75540
United States Class Code: 568913000

Application Number: US 864246   Patent Number: US 4714791
Application Date: 05/19/86   Issue Date: 12/22/87

Abstract:

Primary normal aliphatic higher alcohols are selectively recovered with a high efficiency by contacting sugarcanes, or products obtained from the sugarcanes, or processed products from production of sugar as an extraction raw material with a fluid in a subcritical or supercritical state as an extractant, thereby extracting a trace amount of primary normal aliphatic higher alcohols contained in the extraction raw materials as an extract and separating the extracted alcohols from the extract.

Title: **UPGRADING PETROLEUM ASPHALTENES; DESULFURIZATION, SUPERCRITICAL TEMPERATURE AND PRESSURE**

Inventor: Beckberger, Lavern H. (US)

Assignee: Atlantic Richfield Company
Assignee Codes: 06096
United States Class Code: 208044000

Application Number: US 596173   Patent Number: US 4719000
Application Date: 04/02/84   Issue Date: 01/12/88

Abstract:

A process for reducing the sulfur content of petroleum asphaltenes containing sulfur comprising: (1) forming a mixture of petroleum asphaltenes containing sulfur and a liquid selected from the group consisting of water, methanol, carbon dioxide and mixtures thereof; (2) raising the temperature and pressure of the mixture to a temperature and pressure above the critical temperature and pressure of the liquid to convert the liquid to a supercritical fluid; (3) maintaining the mixture above the critical temperature and pressure of the liquid for a time sufficient to effect sulfur reduction; (4) reducing the pressure to a second

Title: **DEVICE FOR CARRYING OUT EXTRACTION-SEPARATION-CRACKING PROCESSES BY SUPERCRITICAL FLUIDS; DENSITY REDUCTION USING NEEDLE VALVE, CYCLONE SEPARATOR**

Inventor: Perrut, Michel (FR)

Assignee: Nationale Elf Aquitaine (Production) FR
Assignee Codes: 00627
United States Class Code: 210788000

Application Number: US 881106   Patent Number: US 4724087
Application Date: 07/02/86   Issue Date: 02/09/88

Abstract:

Separation/extraction/cracking method using a supercritical fluid in which the density of the fluid is reduced in a needle valve and directly brought into an assembly formed by a cyclonic chamber and a recovery pot and of which the walls are heated.

Title: **METHOD OF REMOVING BINDER MATERIALS FROM A SHAPED CERAMIC PERFORM BY EXTRACTING WITH SUPERCRITICAL FLUID; DISSOLVING**

Inventors: Ishira, Mamoru   (JP)
Nakajima, Nobuaki (JP)
Yasuhara, Seiji   (JP)

Assignee: Sumitomo Heavy Industries, Ltd. (JP)
Assignee Codes: 81572
United States Class Code: 264037000

Application Number: US 812986   Patent Number: US 4731208
Application Date: 12/24/85   Issue Date: 03/15/88

Abstract:

The invention relates to a method of removing binder materials from a shaped perform (green body) in the preparation procedure for the manufacture of ceramics from particulate materials. The new method comprises exposing the green body to a supercritical fluid to dissolve the binder materials in the supercritical fluid without deforming the shape of the article. In the method, the binder materials can be removed without swelling of the article because the green body is not exposed to a rapid temperature increase and the binder does not volatilize in the body. In addition, the binder material can be removed in a drastically shorter period of time from the entire body.

Title: **METHOD   OF   MAKING   SUPERCRITICAL   FLUID
                              MOLECULAR SPRAY FILMS, POWDER AND FIBERS;
                              EXPANSION, PRECIPITATION**

Inventor: Smith, Richard D. (US)

Assignee: Battelle Memorial Institute
Assignee Code: 08120
United States Class Code: 264013000

Application Number: US 838932   Patent Number: US 4734227
Application Date: 03/12/86   Issue Date: 03/29/88

Abstract:

Solid films are deposited, or fine powders formed, by dissolving a solid material
into a supercritical fluid solution at an elevated pressure and then rapidly
expanding the solution through a heated nozzle having a short orifice into a
region of relatively low pressure. This produces a molecular spray which is
directed against a substrate to deposit a solid thin film thereon, or discharged
into a collection chamber to collect a fine powder. In another embodiment, the
temperature of the solution and nozzle is elevated above the melting point of the
solute, which is preferably a polymer, and the solution is maintained at a
pressure such that, during expansion, the solute precipitates out of solution
within the nozzle in a liquid state. Alternatively, a secondary solvent mutually
soluble with the solute and primary solvent and having a higher critical
temperature than that of primary solvent is used in a low concentration (< 20%)
to maintain the solute in a transient liquid state. The solute is discharged in
the form of long, thin fibers. The fibers are collected at sufficient distance
from the orifice to allow them to solidify in the low pressure/temperature
region.

Title: **SUPERCRITICAL FLUID MOLECULAR SPRAY THIN FILMS AND FINE POWDERS**

Inventor: Smith, Richard D. (US)

Assignee: Battelle Memorial Institute
Assignee Codes: 08120
United States Class Code: 524493000

Application Number: US 839079   Patent Number: US 4734451
Application Date: 03/12/86   Issue Date: 03/29/88

Abstract:

Solid films are deposited, or fine powders formed, by dissolving a solid material in a supercritical fluid solution at an elevated pressure and then rapidly expanding the solution through a short orifice into a region of relatively low pressure. This produces a molecular spray which is directed against a substrate to deposit a solid thin film thereon, or discharged into a collection chamber to collect a fine powder. The solvent is vaporized and pumped away. Solution pressure is varied to determine, together with flow rate, the rate of deposition and to control in part whether a film or powder is produced and the granularity of each. Solution temperature is varied in relation to formation of a two-phase system during expansion to control porosity of the film or powder. A wide variety of film textures and powder shapes are produced of both organic and inorganic compounds. Films are produced with regular textural feature dimensions of 1.0 to 2.0 mu m down to a range of 0.01 to 0.1 mu m. Powders are formed in very narrow size distributions, with average sizes in the range of 0.02 to 5 mu m.

Title: **DEPOSITION OF THIN FILMS USING SUPERCRITICAL FLUIDS**

Inventors: Bekker, Alex Y.   (US)
                                Murthy, Andiappan K. (US)
                                Patel, Kundanbhai M. (US)

Assignee: Allied-Signal, Incorporated
Assignee Codes: 01960
United States Class Code: 427369000

Application Number: US 793935   Patent Number: US 4737384
Application Date: 11/01/85   Issue Date: 04/12/88

Abstract:

A process of depositing thin film onto a substrate using supercritical fluids.

Title: **SOLVENT   EXTRACTOR;   HYDROCARBONS   FROM DIATOMACEOUS EARTH**

Inventor: Scinta, James (US)

Assignee: Phillips Petroleum Company
Assignee Codes: 65688
United States Class Code: 196014520

Application Number: US 591757   Patent Number: US 4741806
Application Date: 03/21/84   Issue Date: 05/03/88

Abstract:

Supercritical extraction of diatomaceous earth results in a much more significant improvement in hydrocarbon recovery over Fischer retorting than achievable with tar sands. Process and apparatus for supercritical extraction of diatomaceous earth are disclosed.

Title: **SUPERCRITICAL   FLUID   EXTRACTION   OF   ANIMAL DERIVED   MATERIAL: LIPIDS,   PROTEIN, POLYSACCHARIDES, AND NUCLEOTIDES FROM TISSUE; PURIFICATION OF SEPARATING WITH GASES**

Inventor: Kamarei, Ahmad R.   (US)

Assignee: Angio Medical Corporation
Assignee Code: 17350
United States Class Code: 260412800

Application Number: US 793622   Patent Number: US 4749522
Application Date: 10/31/85   Issue Date: 06/07/88

Abstract:

Supercritical fluids (SCF) are found to be useful in extracting desired materials from animal tissues, cells, and organs. By varying the choice of SCF, experimental conditions, and animal source material, one may obtain lipids, proteins, nucleotides, saccharides, and other desirable components or remove undesirable components.

Title: **SULFUR REMOVAL AND COMMINUTION OF CARBONACEOUS MATERIAL; SLURRYING, HEATING, PRESSURIZATION, SUDDEN DEPRESSURIZATION TO CAUSE EXPLOSIVE FRACTURING**

Inventors: Narian, Nand K.   (US)
                        Ruether, John A. (US)
                        Smith, Dennis N. (US)

Assignee: United States Department of Energy
Assignee Codes: 01715
United States Class Code: 044624000

Application Number: US 105166   Patent Number: US 4775387
Application Date: 10/07/87   Issue Date: 10/04/88

Abstract:

Finely divided, clean coal or other carbonaceous material is provided by forming a slurry or coarse coal in aqueous alkali solution and heating the slurry under pressure to above the critical conditions of steam. The supercritical fluid penetrates and is trapped in the porosity of the coal as it swells in a thermoplastic condition at elevated temperature. By a sudden explosive release of pressure the coal is fractured into finely divided particles with release of sulfur-containing gases and minerals. The finely divided coal is recovered from the minerals for use as a clean coal product.

Title: **METHOD FOR EXTRACTING A SUBSTANCE FROM ANIMAL DERIVED MATERIAL; CRYOGRINDING AT OR BELOW BRITTLENESS TEMPERATURE; PRETREATMENT**

Inventors: Kamarei, Ahmad R. (US)
                        Sinn, Robert   (US)

Assignee: Angio Medical Corporation
Assignee Codes: 17350
United States Class Code: 062063000

Application Number: US 148127   Patent Number: US 4776173
Application Date: 01/26/88   Issue Date: 10/11/88

Abstract:

Animal or plant derived materials are prepared for extraction of desired substances there from by grinding said materials at or below their brittleness temperature. This treatment allows fracture of the materials into small particles with high surface area to volume, as well as high volume to mass ratios, and disrupts membranes of tissues, organs, cells or organelles which would otherwise prevent or limit separation of desired biomolecules.

Title: **METHOD FOR DETERMINING THE AMOUNT OF OIL IN**
                          A SPONGE CORE; EXTRACTION WITH CYCLOALKANES,
                          ETHERS, OR FREONS

Inventors: Dangayach, Kailash C. (US)
                          Difoggio, Rocco   (US)
                          Ellington, William E. (US)

Assignee: Shell Oil Corporation
Assignee Codes: 76232
United States Class Code: 210656000

Application Number: US 122622   Patent Number: US 4787983
Application Date: 11/17/87   Issue Date: 11/29/88

Abstract:

The oil lost by the core sample and captured by the sponge during sponge coring
is extracted from the sponge using a solvent selected from the group consisting
of cycloalkanes, ethers, and freons.

Title: **CHROMATOGRAPHY COLUMNS WITH CAST POROUS PLUGS**
                          AND METHODS OF FABRICATING SAME; FUSING
                          SILICATE IN SITU

Inventors: Cortes, Hernan   (US)
                          Pfeiffer, Curtis D. (US)
                          Richter, Bruce E.   (US)
                          Stevens, Timothy S. (US)

Assignee: Lee Scientific, Incorporated
Assignee Codes: 19646
United States Class Code: 210198200

Application Number: US 111147   Patent Number: US 4793920
Application Date: 10/20/87   Issue Date: 12/27/88

Abstract:

Porous ceramic plugs are cast in place in chromatographic columns in order to
provide supports for chromatographic beds in liquid chromatography devices and
restrictors in supercritical fluid chromatographic devices. The supports are
cast in place by fusing a silicate containing solution, such as one containing
potassium silicate, which has been drawn into the outlet end of the column.

Title: **PROCESS FOR THE WORKING UP OF SALVAGE OIL;
           SUPERCRITICAL SOLVENT EXTRACTION, FILTRATION,
           DISTILLATION,   CATALYTIC   HYDROGENATION,
           GASIFICATION**

Inventors: Coenen, Hubert   (DE)
                    Kreuch, Winfried   (DE)
                    Wetzel, Rolf   (DE)

Assignee: Krupp-Koppers GMBH DE
Assignee Codes: 47063
United States Class Code: 208181000

Application Number: US 7358   Patent Number: US 4797198
Application Date: 01/27/87   Issue Date: 01/10/89

Abstract:

A process is disclosed for the working up of salvage oil, in which the salvage
oil is subjected to an extraction under supercritical conditions. The halogen
compounds contained in the produced extract are removed by catalytic
hydrogenation. The extraction residue is eliminated by deposition or thermal
treatment (gasification). In the case of a thermal treatment of the extraction
residue, other residues can be simultaneously converted, so that the process is
performed without yield of environmentally burdensome residues or by-products.
Ethane in particular and/or propane is employed as solvent for the supercritical
extraction.

Title: **CHROMATOGRAPHIC SEPARATION METHOD AND ASSOCIATED APPARATUS SUPERCRITICAL FLUID**

Inventor: Kumar, M. Lalith (US)

Assignee: Suprex Corporation
Assignee Codes: 20171
United States Class Code: 210659000

Application Number: US 157020
Application Date: 02/17/88

Patent Number: US 4814089
Issue Date: 03/21/89

Abstract:

In one embodiment of the invention supercritical fluid chromatographic separation may be accomplished by apparatus which includes a column, an oven for maintaining the temperature of a sample in the oven, an injector for delivering a sample containing fluid to the column and a pump for delivering carrier fluid to the injector. A discharge outlet for receiving processed fluid from the column contains one or two restrictors and a nozzle for discharge of processed fluid. The pump also delivers fluid which is the carrier fluid not containing the sample to a position in the pressure control inlet so as to alter the linear velocity of the sample containing fluid through the column. A controller controls operation of the oven and pump. A valve and a transducer may be positioned in the line between the pump and the discharge outlet in order to permit adjustment of the pressure of the unprocessed carrier fluid being introduced into the discharge outlet by the controller. The method of this embodiment may employ equipment of this type to effect chromatographic separation at the desired flow rate.

Title: **SUPERCRITICAL FLUID CHROMATOGRAPHY PACKING
                          MATERIAL; METAL OXIDES, HYDROXIDES, ALUMINA**

Inventors: Burr, Richard R.   (US)
                          Fraser-Milla, Karen R.   (US)
                          Khosah, Robinson P.   (US)
                          Novak, John W.   (US)
                          Weaver, Douglas G.   (US)

Assignee: Aluminum Corporation of America
Assignee Codes: 02456
United States Class Code: 210635000

Application Number: US 90880   Patent Number: US 4816159
Application Date: 08/31/87   Issue Date: 03/28/89

Abstract:

Disclosed is a method of separating organic or organometallic materials under
supercritical fluid conditions. The method comprising the steps of providing a
bed of packing material selected from a metal oxide/hydroxide support material
having phosphorous-containing organic molecules bonded to reactive sites on said
support material, alumina and alumina-containing mixtures. The materials are
introduced to the bed and a fluid is added to the bed under supercritical fluid
conditions. The fluid removes one of the materials from the bed.

Title: **METHOD FOR DECAFFEINATING COFFEE WITH A
                          SUPERCRITICAL FLUID CARBON DIOXIDE, FROM
                          GREEN COFFEE BEANS**

Inventor: Katz, Saul N. (US)

Assignee: General Foods Corporation
Assignee Codes: 33888
United States Class Code: 426481000

Application Number: US 166748   Patent Number: US 4820573
Application Date: 03/08/88   Issue Date: 04/11/89

Abstract:

A method of extracting caffeine from green coffee beans whereby an essentially
caffeine-free supercritical fluid is continuously fed to one end of an extraction
vessel containing green coffee beans and caffeine-laden supercritical fluid is
continuously withdrawn from the opposite end. A portion of decaffeinated beans
is periodically discharged while a fresh portion of undecaffeinated beans is
essentially simultaneously charged to the extraction vessel. The caffeine-laden
supercritical fluid is fed to a countercurrent water absorber. Supercritical
carbon dioxide is the preferred supercritical fluid. The method of the present
invention is more efficient than batch processes and produces an improved
decaffeinated coffee.

Title: **APPARATUS FOR THE EXTRACTION OF CONSTITUENTS BY A SUPERCRITICAL FLUID OR PRESSURIZED LIQUID**

Inventors: Bethul, Louis   (FR)
              Carles, Maurice (FR)
              Neige, Roger   (FR)

Assignee: Commissariat A L Energie Atomique FR
Assignee Codes: 19200
United States Class Code: 210511000

Application Number: US 46915   Patent Number: US 4824570
Application Date: 04/02/87   Issue Date: 04/25/89

Abstract:

An apparatus for the extraction of constituents present in a substance by means of an extraction fluid constituted by a supercritical fluid or a pressurized liquid. According to this apparatus, in an exterior (3) contacting takes place between the substance and the extraction fluid in order to dissolve the constituents in fluid. The fluid leaving the extractor is then treated to separate the extracted constituents. The fluid is expanded at a pressure p_1. Firstly, the less volatile constituents are separated in a liquid-gas separator (41). The separated gas is then liquefied in gas separator-liquefier (13) and the thus liquefied gas is rectified in column (19) to concentrate the extracted constituents in the liquid phase. The extraction fluid can be carbon dioxide gas.

Title: **SUPERCRITICAL FLUID CHROMATOGRAPHY**

Inventor: Berger, Terry A. (US)

Assignee: Hewlett-Packard Company
Assignee Codes: 38859
United States Class Code: 073023000

Application Number: US 157344   Patent Number: US 4845985
Application Date: 02/17/88   Issue Date: 07/11/89

Abstract:

In a system for analyzing a chemical sample by chromatographic separation of said sample into components a method and apparatus are provided for maximizing detector resolution and minimizing duration by maintaining the linear velocity of the mobile phase at a desired level by affecting the viscosity and density characteristics of the mobile phase as it passes through a restrictor by programming the restrictor temperature.

Title: **SUPERCRITICAL FLUID METAL HALIDE SEPARATION PROCESS**

Inventors: Tolley, William K. (US)
                   Whitehead, Alton B. (US)

Assignee: Secretary, Department of Interior
Assignee Codes: 86576
United States Class Code: 423472000

Application Number: US 25252   Patent Number: US 4853205
Application Date: 03/12/87   Issue Date: 08/01/89

Abstract:

Process of using supercritical fluid to selectively separate, purify and recover metal halides.

Title: **PROCESS FOR PRODUCING ALCOHOL-REDUCED OR ALCOHOL-FREE BEVERAGES MADE BY NATURAL FERMENTATION**

Inventors: Kolb, Erich   (DE)
                   Marr, Rolf   (AT)
                   Schildmann, Jens A. (DE)
                   Weisrock, Reinhard (DE)
                   Wiesenberger, Alfred (DE)

Assignee: Peter de Eckes
Assignee Codes: 05591
United States Class Code: 426387000

Application Number: US 940138   Patent Number: US 4867997
Application Date: 12/10/86   Issue Date: 09/19/89

Abstract:

A process is described for preparing in particular alcohol-free wine in which by extraction with supercritical CO_2 firstly a specific aroma fraction with limited ethanol content is separated from the starting wine. The residual wine is then subjected to a vacuum distillation in which apart from the complete separation of the ethanol from the residual wine a fraction of more difficulty extractable aromatic substances not affected by the extraction is recovered which in the end together with the extract is added to the dealcoholized residual wine again to obtain an alcohol-free wine. The method permits an almost complete removal of the alcohol content without detrimentally affecting the basic wine and with retention of all the aromatic and flavoring substances characteristic of the wine.

Title: **CHROMATOGRAPHIC   SEPARATION   METHOD   AND ASSOCIATED APPARATUS**

Inventor: Kumar, M. Lalith (US)

Assignee: Suprex Corporation
Assignee Codes: 20171
United States Class Code: 21098200

Application Number: US 274753   Patent Number: US 4871453
Application Date: 11/22/88   Issue Date: 10/03/89

Abstract:

In one embodiment of the invention supercritical fluid chromatographic separation may be accomplished by apparatus which includes a column, an oven for maintaining the temperature of a sample in the oven, an injector for delivering a sample containing fluid to the column and a pump for delivering carrier fluid to the injector. A discharge outlet for receiving processed fluid from the column contains one or two restrictors and a nozzle for discharge of processed fluid. The pump also delivers fluid which is the carrier fluid not containing the sample to a position in the pressure control inlet so as to alter the linear velocity of the sample containing fluid through the column. A controller controls operation of the oven and pump. A valve and a transducer may be positioned in the line between the pump and the discharge outlet in order to permit adjustment of the pressure of the unprocessed carrier fluid being introduced into the discharge outlet by the controller. The method of this embodiment may employ equipment of this type to effect chromatographic separation at the desired flow rate.

Index

INDEX

A

Absinthine, unextractability of, 469
Absorption spectra, 202
Acetone
  dehydration of, flow sheet for, 391—392
  extraction of, 394, 400
Acetone/water systems, design and optimization
  studies in, 390—391
Acid/base catalysis, 519
Acridine, solubility of, 174
Acrylamide
  monomer of, 416
  polymerization of, 442
Activation volumes, 517—518
Activity coefficients, 485
Adiabatic compressibility, 26
Adjustable parameters
  changes with variations in, 115
  proliferation of, 106—107
  structure of, 101—107
Adsorbents, extraction of, 505—506
Adsorption-desorption kinetics, 375
Adsorption equilibrium data, 368—369
Adsorption mass-transfer model, 375—377
Adsorption techniques, 483
Adsorptive equilibrium, fluid phase in, 376
Adsorptive systems, 368—369
Alachlor, adsorption of, 368
Alcohol dehydration, 386
  vapor recompression cycle for, 391
Alcohol-reduced beverages, production of, 573
Alcohol/water systems, 387, 484
  design and optimization studies in, 390—391
Alder functions, second-order, 175
Aliphatic higher alcohols, recovering of, 561
Alkali metal hydroxide solution, water removal
  from, 545
Alumina, in chromatography, 571
Amine composition, 528
Amino acids
  importance of behavior of, 461
  supercritical carbon dixoide and nitrogen effects
    on, 459—461
Amplitudes, of pure fluids, 4—6
Androsterone, solubility of, 468
Anemometer, 312
Animal derived material extraction, 566
  methods for, 567
AOT/ethane binary system, temperature/composition
  prism for, 418—419
AOT microemulsions, water-solubilization capacity
  of, 417
AOT/propane binary system, LCST of, 420
AOT/water droplets
  diffusion coefficient of, 429

hydrodynamic diameters of, 429
Apparent molar heat capacity, 44—45
  infinite-dilution, 45
Apparent molar properties, 38—39
Apparent molar volume, infinite-dilution, 42
Aqueous extractions, 483—484
  entrainer effect in, 494—495
  K value in, 496
  of organic mixtures, 489—494
  of single component contaminants, 486—489
  thermodynamics of, 484—486
Argon mixtures, 87
Aris-Taylor method, 320
  extension of, 321—322
Aromatics, from bell peppers, 530
Artabsine, solubility of, 469, 474
Associated perturbed anisotropic chain theory
    (APACT), 178
Association models, 178
Attractive behavior, 213—214
Attractive near-critical systems, correlation functions
    in, 215
Azeotropic mixtures
  asymptotic expansions of dew-buble curves of,
    132—135
  consolute point of, 138—139
  Leung-Griffiths correlations for, 129—135
    current status of, 135—139
  phase diagram structures of, 131
  in P-T space, 130—131
Azeotropy
  maximum-boiling, 130
  relation of to minimum mixture critical tempera-
    ture, 129

B

Bacillus subtillis, viability of under high pressure,
  473
Bacteria, viability of under high pressure, 472—474
Beer-Lambert law, 315—316
Bell peppers, extractions from, 530
Benedict-Webb-Rubin equation, modified, 299, 301
Benzene
  distribution coefficients of, 486—488
  effect of as entrainer on phenol distribution coef-
    ficients, 495
  effect of on phenolic mixture distribution coeffi-
    cients, 496
  as entrainer in soil extraction, 501—502
  extraction of from soil, 504
Benzoic acid
  purification of, 555
  recovering of purified, 556
Beta-carotene
  dehydration of with supercritical carbon dioxide,
    395

P

Pade approximation, 175
Pair correlation function, 13
  in attractive systems, 215
  in repulsive systems, 215—217
  for solute-solute, solute-solvent, and solvent-solvent interactions, 205, 209—211
  solute-solvent, 205—206, 208, 212, 216
  sources of, 204
Pair distribution function, 196
Parallel plates method
  groups of corrections in, 348—349
  for measuring thermal conductivity, 348—349
Parathion, distribution coefficients of, 486—487, 489
Partial molar enthalpy, 43—44
  divergence of, 52
  and excess enthalpy of mixing, 180
Partial molar heat capacity, 44—45
  nonclassical path dependence of, 45—46
Partial molar properties, 38—39, 52
  definition of, 38
  divergence of in solute, 39—42
  enthalpy, 43—44
  heat capacity, 44—45
  path dependence of critical anomalies of, 45—46
  of supercritical infinitely dilute Lennard-Jones mixtures, 229—232
  volume, 39—43
Partial molar volumes
  classical path dependence of, 42—43
  cluster size calculations based on, 207
  difference in, 517
  experimental conditions in measurements of, 197
  experiments of, 197—199
  implications of measurement of, 202
  infinite-dilution, 52, 197—199
  large, positive, 230
  large negative value of, 181—182
  method of measuring, 181
  nonclassical path dependence of, 45—46
  of solutes in SCFs, 180—181
  and supercritical solubility, 50—52
  in terms of KB factors, 196
Partitioning, bioseparation by, 452
Patents, supercritical fluid technology, 526—574
Path dependence, 6—7
PCBs
  extraction of from sediments, 506
  extraction of from soil, 497
Peak broadening method, 319
Peng-Robinson correlations, 143
Peng-Robinson equation of state, 59, 118, 174, 306, 390, 485—486
  conformal solution, 457—458
  modified, 394
Percus-Yevick closure, 204
Perturbation equations of state, 174—175
Perturbed anisotropic chain theory (PACT), 178
Perturbed hard-chain theory, 175

Petroleum asphaltenes, upgrading of, 562
Petroleum creosote
  composition and estimated properties of, 493
  distribution coefficients of, 493
  extraction of from water, 492—493
Phase behavior
  effect of pressure on in liquid systems, 413—414
  in near-critical and supercritical fluids, 414—420
Phase diagrams
  classification of, 165
  for supercritical fluids, 164—166
    type 1, 166—167
    type 4, 167—168
  for supercritical fluid-solid equilibria, 168—172
Phase equilibria
  computer simulations of, 179—180
  description of surface of, 99
  molecular analysis of, 164—187, 228
  simulation techniques of, 228, 243
  simulations of in supercritical extraction, 233—242
Phenol
  dehydration of, 401
  distribution coefficients of, 487, 494—495, 497—498
    effect of soil surface area on, 500
  distribution of in soil, 498—499
  dry soil experiments with, 502—503
  effect of pressure on extraction of from soil, 502—503
  effects of entrainer on, 495
  extraction of from soil, 497, 504
  extraction of from water, 486, 487—489
  interaction of with water, 487—488
  wet soil experiments with, 503—505
Phenolic mixture
  benzene effect on distribution coefficients of, 496
  binary interaction parameters for, 492
  distribution coefficients of, 491—492
  effects of entrainer on, 495
  extraction of from water, 490—491
  testing of extracts of, 493—494
Photon correlation spectroscopy, 269
Plant essential oils, supercritical carbon dioxide effect on, 461—462
Platinum resistance thermometer (PRT), 299
Polar gases, correlation of excess viscosity to reduced density in, 263—264
Polarity
  effects of, 110, 519
  miscibility and, 90
  salts and, 516
Polar-nonpolar fluid mixtures, 110
Polar transition state, 516
Polyaromatic hydrocarbons, extraction of, 497
Polyatomic fluids, thermal conductivity expressions for, 279—280
Polyatomic gas
  estimating thermal conductivity of, 280—282
  viscosity collision integral for, 256
Polymerized fatty acids, fractionation of, 550